Regional to Wellbore Analog for Fluvial-Deltaic Reservoir Modeling: The Ferron Sandstone of Utah

Edited by

Thomas C. Chidsey, Jr.
Roy D. Adams
Thomas H. Morris

AAPG Studies in Geology 50
Published by
The American Association of Petroleum Geologists
Tulsa, Oklahoma, U.S.A.

Printed in the U.S.A.

ISBN: 0-89181-057-9

AAPG Editor: Neil F. Hurley, 1997–2001; John C. Lorenz, 2001–2004; and Ernest A. Mancini, 2004–2007.
Geoscience Director: J. B. "Jack" Thomas
Production Manager: Gerald Buckley
Special Publications Coordinator: Beverly Molyneux
Copy Editor: Thomas C. Chidsey, Jr., Utah Geological Survey
Design and Production: Vicky Clarke, Utah Geological Survey
Printing: The Covington Group, Kansas City, MO

On front cover: View to the southwest of cliffs formed by the Ferron Sandstone along the southern part of Molen Reef. Most of the tan, cliff- and ledge-forming units are progradational shoreline sandstones, several of which are capped by beds of coal. The dark slope beneath the Ferron cliffs is formed on the Tununk Shale; minor dark, slope-forming units separating the sandstone cliffs are beds of marine shale contained in the Ferron. Photograph by Thomas A. Ryer.

On back cover: Anisotropic permeability models (left) and computed oil saturation after water injection of selected Ferron Sandstone permeability models, Ivie Creek area. See Forster et al., this volume.

This and other AAPG publications are available from:

The AAPG Bookstore
P.O. Box 979
Tulsa, OK 74101-0979
U.S.A.
Telephone: 1-918-584-2555 or 1-800-364-AAPG (U.S.A./Canada)
Fax: 1-918-560-2652 or 1-800-898-2274 (U.S.A./Canada)
www.aapg.org

Utah Department of Natural Resources Map & Bookstore
1594 W. North Temple
Salt Lake City, UT 84116
U.S.A.
Telephone: 1-801-537-3320 or 1-888-UTAH MAP
Fax: 1-801-537-3395
E-mail: geostore@utah.gov
http://mapstore.utah.gov

About the Editors

Thomas C. Chidsey, Jr. is currently Petroleum Section Chief for the Utah Geological Survey. His responsibilities include managing the petroleum program and conducting research in Utah's petroleum geology. This work involves studying carbonate oil reservoirs in the Paradox Basin of southeastern Utah, reservoir outcrop analogs of the Green River Formation of the Uinta Basin, and hydrocarbon plays in the Utah thrust belt. Tom was the Principal Investigator of a major Ferron Sandstone project titled *Geological & Petrophysical Characterization of the Ferron Sandstone for 3-D Simulation of a Fluvial-Deltaic Reservoir*, funded by the U.S. Department of Energy.

A native of the Washington, D.C. metropolitan area, Tom received his B.S. degree in 1974 and M.S. degree in 1977, both in geology from Brigham Young University. During his 26-year career, Tom has worked as a production geologist for Exxon Company U.S.A. in the Kingsville District of the South Texas Division and an exploration geologist in the Rocky Mountains for Wexpro/Celsius Energy Company before joining the Utah Geological Survey in 1989. He has served as Rocky Mountain Section President of the American Association of Petroleum Geologists (AAPG), President of the Utah Geological Association (UGA), and General Chairman for the 2003 AAPG Annual National Convention. Tom has led numerous field trips to the Ferron Sandstone outcrop and other areas of Utah for UGA, GSA, and AAPG.

Tom has wide interest in geology and has published on petroleum geology, carbon dioxide resources and sequestration, economic mineral potential, hydrogeology, and the general geology of Utah's parks. He has served as an editor or co-editor of three other geological guidebooks.

Roy D. Adams entered the petroleum industry in 1979 and has worked in the oil patch ever since, with the exception of two sojourns: one to earn his Ph.D. in geology at Massachusetts Institute of Technology and the other to teach for a year as a Visiting Assistant Professor at Tennessee Technological University. Roy is a petroleum geologist, sequence stratigrapher, and sedimentologist with a broad professional background that includes exploration, exploitation, consulting, research, and training experience in the petroleum industry; university research and teaching; and state and federal government positions. His domestic work experience includes the Western Interior of the U.S., the Gulf Coast, and west Texas, and his international work experience includes Mexico, Saudi Arabia, West Africa and Brazil, the Caspian Sea, Pakistan, the Canadian Arctic, and Venezuela.

Roy has worked for Exxon, the Utah Geological Survey, the Energy & Geoscience Institute at the University of Utah, and has run his own consulting firm, A & R Resources LLC. He is presently employed by Saudi Aramco, working in Saudi Arabia, building detailed reservoir models based on carbonate core and other subsurface data.

In addition to a Ph.D. in geology, Roy earned a B.S. from the University of California at Riverside (1974) and a M.A. from Rice University (1979), both in geology. He has also been adjunct faculty at the University of Utah and at Westminster College in Salt Lake City. His publications include papers and presentations on carbonate and siliciclastic sequence stratigraphy, depositional facies, and reservoirs and reservoir analogs. Roy is active in AAPG, the Utah Geological Association, and the Association for Women Geoscientists and has held volunteer positions in each.

Thomas H. Morris is a professor of geology and currently the Associate Chair for the Department of Geology at Brigham Young University (BYU). As a member of the Petroleum Research Group at BYU, his research interests have primarily focused on siliciclastic sedimentology and its importance to reservoir characterization. Tom's graduate and undergraduate students have studied rocks throughout the Colorado Plateau of Utah. He has received funding from the U.S. Department of Energy for research involving the Ferron Sandstone in east-central Utah and the Green River Formation of the Uinta Basin. Tom has published more than 20 articles on the geology of Utah.

Born in St. Paul, Minnesota, Tom's first career was in minor league baseball. It was through his travels with his college baseball team that he became interested in geomorphology and how natural things came to be. After taking a geology class, he was hooked. Tom received his B.S. degree in geology from BYU in 1981 before heading to the University of Wisconsin-Madison for his M.S. (1983) and Ph.D. (1986). Upon completing his formal education he spent nearly four years drilling wells on-shore in Louisiana for the Eastern Production Division, Exxon Company U.S.A. He joined the faculty at

BYU in 1990. Currently, Tom continues to work on reservoir characterization studies looking at thin sand-shale, basin floor fan successions on the coast of New Zealand. He also has students mapping in Capitol Reef National Park, where he has been involved in educating the public for a number of years.

Tom is an active member of AAPG, SEPM, GSA, and the Utah Geological Association. He has served as the technical program chair for the Rocky Mountain Section of AAPG in 1993, as SEPM Field Trip Chair at the 1998 AAPG/SEPM Annual Meeting, and most recently as the 2003 Annual Meeting AAPG Poster Program co-chair.

Acknowledgments

In 1995, Dr. M. Lee Allison, former Director of the Utah Geological Survey, proposed a concept for an AAPG volume devoted entirely to the Cretaceous Ferron Sandstone of east-central Utah. At that time, the Utah Geological Survey was conducting a major study of the Ferron Sandstone, funded by the U.S. Department of Energy, as a three-dimensional outcrop analog for fluvial-deltaic reservoirs. This study involved three universities, two industry partners, geologic consultants, as well as geoscientists from the Survey. The results of that study are included in this volume. However, many other workers from industry and academia were conducting studies of the Ferron or leading field trips to its many excellent outcrops. In addition, the Ferron itself was being developed as a major coalbed methane resource. We envisioned an AAPG volume that not only included papers from the Utah Geological Survey study, but from all Ferron workers, designed to serve as a reference for comparative studies of deltaic oil and gas reservoirs in the Gulf Coast, Rocky Mountain basins, Alaskan North Slope, North Sea, and many other areas worldwide.

With this concept in mind, our Ferron Sandstone volume represents the work of a diverse group of authors, who volunteered time and talents to share their unique knowledge of these fascinating rocks. We sincerely appreciate their contributions, support of their employers, and all funding and grants, that made this publication possible.

We also thank our colleagues (listed individually with each paper) who provided careful technical reviews and constructive criticism of the manuscripts. Their prompt efforts improved the value of each paper and ensured that the scientific contents were sound.

We are grateful to Bradshaw B. Lupton Jr. who kindly provided photographs from the family collection of U.S. Geological Survey geologist Charles T. Lupton.

We sincerely appreciate the help and support of the Utah Geological Survey, managers Kimm Harty and Rick Allis, and Vicky Clarke, Jim Parker, and Jim Stringfellow of the Survey's editorial staff, who designed, formatted, and assembled the volume. The CD-ROM was prepared and designed by Sharon Wakefield, Kevin McClure, and Cheryl Gustin of the Survey.

Finally, we thank the AAPG, particularly Rick Fritz, Executive Director, and the Executive Committee for endorsing this volume. We most sincerely appreciate the help, enduring patience and encouragement of the AAPG editors and production staff, both past and present, which ultimately ensured that this volume was published. These individuals include Neal Hurley, John Lorenz, Ken Wolgemuth, Jack Thomas, Gerald Buckley, and Beverly Molyneux.

Thomas C. Chidsey, Jr.
Roy D. Adams
Thomas H. Morris

AAPG

wishes to thank the

Utah Geological Survey

for proposing, producing, and cost sharing

Regional to Wellbore Analog for Fluvial-Deltaic Reservoir Modeling: The Ferron Sandstone of Utah

The Utah Geological Survey's cost share was applied against the production costs of publication, thus directly reducing the book's purchase price and making the volume available to a greater audience.

Contents

About the Editors iii
  Thomas C. Chidsey, Jr., Roy D. Adams, and Thomas H. Morris

Acknowledgments v

Dedication: Charles Thomas Lupton xi
  Thomas C. Chidsey, Jr.

Preface xiii

The Ferron Sandstone — Overview and Reservoir Analog

Previous Studies of the Ferron Sandstone 3
  Thomas A. Ryer

Searching for Modern Ferron Analogs and Application to Subsurface Interpretation 39
  Janok P. Bhattacharya and Robert S. Tye

Facies of the Ferron Sandstone, East-Central Utah 59
  Thomas A. Ryer and Paul B. Anderson

Integrated Analysis of the Upper Ferron Deltaic Complex, Southern Castle Valley, Utah 79
  Richard J. Moiola, Joann E. Welton, John B. Wagner, Larry B. Fearn, Mike E. Farrell,
  Roy J. Enrico, and Ron J. Echols

Regional Sequence Stratigraphic Interpretations

Stacking Patterns, Sediment Volume Partitioning, and Facies Differentiation in
Shallow-Marine and Coastal-Plain Strata of the Cretaceous Ferron Sandstone, Utah 95
  Michael H. Gardner, Timothy A. Cross, and Mark Levorsen

High-Resolution Depositional Sequence Stratigraphy of the Upper Ferron
Sandstone Last Chance Delta: An Application of Coal-Zone Stratigraphy 125
  James R. Garrison, Jr. and T. C. V. van den Bergh

Stratigraphic Architecture of Fluvial-Deltaic Sandstones
from the Ferron Sandstone Outcrop, East-Central Utah 193
  Mark D. Barton, Edward S. Angle, and Noel Tyler

Regional Stratigraphy of the Ferron Sandstone 211
Paul B. Anderson and Thomas A. Ryer

General Geology of the Ferron Sandstone

Petrophysics of the Cretaceous Ferron Sandstone, Central Utah 227
Richard D. Jarrard , Carl H. Sondergeld, Marjorie A. Chan,
and Stephanie N. Erickson

Facies and Permeability Relationships for Wave-Modified and
Fluvial-Dominated Deposits of the Cretaceous Ferron Sandstone, Central Utah 251
Ann Mattson and Marjorie A. Chan

Outcrop Case Studies

Sedimentology and Structure of Growth Faults at the Base of
the Ferron Sandstone Member Along Muddy Creek, Utah 279
Janok P. Bhattacharya and Russell K. Davies

Geologic Framework of the Lower Portion of the Ferron Sandstone
in the Willow Springs Wash Area, Utah: Facies, Reservoir Continuity,
and the Importance of Recognizing Allocyclic and Autocyclic Processes 305
John A. Dewey, Jr. and Thomas H. Morris

Geologic Framework, Facies, Paleogeography, and Reservoir Analogs
of the Ferron Sandstone in the Ivie Creek Area, East-Central Utah 331
Paul B. Anderson, Thomas C. Chidsey, Jr., Thomas A. Ryer, Roy D. Adams,
and Kevin McClure

Reservoir Permeability, Modeling, and Simulation Studies

Modeling Permeability Structure and Simulating Fluid Flow
in a Reservoir Analog: Ferron Sandstone, Ivie Creek Area, East-Central Utah 359
Craig B. Forster, Stephen H. Snelgrove, and Joseph V. Koebbe

Facies Architecture and Permeability Structure of
Valley-Fill Sandstone Bodies, Cretaceous Ferron Sandstone, Utah 383
Mark D. Barton, Noel Tyler, and Edward S. Angle

3-D Fluid-Flow Simulation in a Clastic Reservoir Analog: Based on
3-D Ground-Penetrating Radar and Outcrop Data from the Ferron Sandstone, Utah 405
Craig B. Forster, Stephen H. Snelgrove, Siang Joo Lim, Rucsandra M. Corbeanu,
George A. McMechan, Kristian Soegaard, Robert B. Szerbiak, Laura Crossey, and Karen Roche

Three-Dimensional Architecture of Ancient Lower Delta-Plain Point Bars
Using Ground Penetrating-Radar, Cretaceous Ferron Sandstone, Utah 427
    Rucsandra M. Corbeanu, Michael C. Wizevich, Janok P. Bhattacharya, Xiaoxiang Zeng, and
    George A. McMechan

The Geometry, Architecture, and Sedimentology of Fluvial and Deltaic Sandstones
Within the Upper Ferron Sandstone Last Chance Delta: Implications for Reservoir Modeling 451
    T. C. V. van den Bergh and James R. Garrison, Jr.

The Ferron Coalbed Methane Play

Coalbed Gas in the Ferron Sandstone Member of the
Mancos Shale: A Major Upper Cretaceous Play in Central Utah 501
    Scott L. Montgomery, David E. Tabet, and Charles E. Barker

Hydrodynamic and Stratigraphic Controls for a Large
Coalbed Methane Accumulation in Ferron Coals of East-Central Utah 529
    Robert A. Lamarre

Helper Field: An Integrated Approach to Coalbed Methane Development, Uinta Basin, Utah 541
    Andre Klein, Keith Buck, and Steve Ruhl

Coalbed Methane Production from Ferron Coals at Buzzard Bench Field, Utah 551
    Robert A. Lamarre

Index 559

Charles T. Lupton (from Hares, 1936).

Dedication

Charles Thomas Lupton
1878–1935

by Thomas C. Chidsey, Jr., Senior Editor
Utah Geological Survey

During 1911 and 1912, Charles Thomas Lupton led a team from the U.S. Geological Survey to investigate the quality and quantity of coal, and collect geologic information in Castle Valley of east-central Utah. This included studying the age, character, thickness, and stratigraphic relations of the Ferron Sandstone, which he defined as a member of the Upper Cretaceous Mancos Shale. Lupton and his field assistants, as other geologists of their day, did not have sport-utility vehicles to transport themselves and equipment. Their field equipment did not consist of laser alidades or global positioning systems, nor were topographic maps available. They did not have drill-hole core, geophysical well logs, aerial photos, and photo mosaics to aid in correlating and describing the rocks. The few roads were not paved and they did not return to motels at the end of long days or to home very often. Instead, as seen in the various photographs throughout this book, kindly provided by the Lupton family, they rode on horseback and used horse-drawn wagons for transportation. Their field equipment included plane tables and telescopic alidades used to map and determine location by stadia and triangulation methods, and earlier survey markers. They lived in rustic field camps. Yet, Lupton and other early geologists produced excellent scientific contributions.

From this field work, that included correlating numerous measured stratigraphic sections, Lupton devised a system to designate and describe the various coals and sand units in the Ferron Sandstone. Lupton's work on the Ferron Sandstone, Geology and Coal Resources of Castle Valley in Carbon, Emery, and Sevier Counties, Utah, was published by the U. S. Geological Survey in 1916. It represents the first definitive work on the Ferron, and provided a firm basis for subsequent researchers as the study of these fascinating rocks continued. Lupton's coal designations, although many correlations of coal units have been modified and reinterpreted, are still used by Ferron workers nearly one hundred years later. Many of the papers in this book refer to Lupton's original work. We, therefore, have decided to dedicate this latest compilation of Ferron research to Charles Thomas Lupton.

Charles Thomas Lupton was born February 28, 1878, on a farm near Mount Pleasant, Ohio, and died of a heart attack May 8, 1935, at home, in Denver, Colorado, at the untimely age of 57. He was known as a notable and able geologist to both the scientific community and the oil and gas industry. Besides his Ferron work, Lupton published 23 papers on coal, oil and gas, and gypsum resources in the Western U.S. Lupton was an active member of the American Association of Petroleum Geologists, Geological Society of America, American Institute of Mining Engineers, American Mining Congress, American Association for Advancement of Science, Geological Society of Washington, Colorado Academy of Sciences, and Colorado Historical Society. As a man he was said to be kind, quiet, upright, and humorous, and a thoughtful father and husband.

Addie White Bradshaw Lupton, wife of Charles T. Lupton (photograph courtesy of the family of C. T. Lupton).

Lupton received his B.S. Degree in geology, a perfect fit for his love of the outdoors and inquiring nature, from Oberlin College in 1907. He joined the U. S. Geological Survey that same year and resigned in 1916 to join Cosden Oil Company of Tulsa, Oklahoma and the rapidly expanding oil industry. While working full time for the U. S. Geological Survey in Washington D.C., Lupton received degrees in law from the National University of Law in 1916, and although admitted to the bar, never practiced law. During this period, he married Addie White Bradshaw of Sedley, Virginia in 1914. They had four children: Martha, Charles T., Jr., Franz Russell, and Bradshaw Babb.

Lupton's early work with the U. S. Geological Survey was as a field assistant in Wyoming, mapping the Glenrock-Douglas coal field and the anticline that would later become Big Muddy oil field — his first introduction to petroleum geology. He also assisted with coal resource evaluations in Montana's Bull Mountain and Wyoming's coal fields. Lupton's ability, eagerness, friendliness, and charm were quickly recognized. After two years, he was made field party chief studying coal resources in Wyoming, Montana, Idaho, and of course, Utah. As the nation's attention turned to oil resources on public lands, Lupton led field studies into oil possibilities in these same States. However, he was always particularly fascinated with the Colorado Plateau geology and ancient Indians of the San Rafael Swell in east-central Utah and surrounding areas. Here, Lupton and his colleagues experienced everything from the hardships of long waterless pack trips to being marooned by snow, while still managing to complete the geologic work at hand.

After leaving the U. S. Geological Survey in 1916, Lupton became a leading geologist in the oil and gas industry where he stayed for the remainder of his career. While at Cosden Oil Company, Lupton was in charge of the Rocky Mountain region. In 1919, Lupton joined Franz Oil Corporation, organized by ex-Governor Franz of Oklahoma, where a year later he was directly responsible for drilling the Cat Creek anticline that became the first major oil discovery in Montana. Lupton was involved in the discovery of several other oil and gas fields in Montana and Wyoming. He also evaluated the oil potential of western Kansas and even the state of Washington. Lupton shared his geologic knowledge of these areas through various publications. In generating and drilling oil and gas prospects, he optimistically relied on facts rather than theories. Although Lupton was a frugal and thrifty person with acute business acumen, he was also charitable and helped many fellow geologists who were struggling in the early years of the Great Depression up until the time of his death.

We the editors are sure Charles T. Lupton would be extremely excited to see how his early observations on the Ferron Sandstone in Utah have evolved, and that these rocks, which he found fascinating would become a favorite of many other geologists today. The AAPG dedicates this book to his significant contribution and his memory.

SOURCES

Ball, M. W., 1936, Memorial of Charles Thomas Lupton: Proceedings of the Geological Society of America for 1935, p. 273–282.

Lupton, C. T., 1916, Geology and coal resources of Castle Valley in Carbon, Emery, and Sevier Counties, Utah: U. S. Geological Survey Bulletin 628, 88 p., 12 pls.

Hares, C. J., 1936, Memorial — Charles T. Lupton: AAPG Bulletin, v. 20, no. 4, p. 513–515.

"Jack and his friends." Charles T. Lupton in center (circa 1910; photograph courtesy of the family of C. T. Lupton).

Preface

As Earth scientists, we espouse the "scientific method" as a valid approach to the discovery of truth or fact. "Truths" may initially come in the form of hypotheses. We require rigorous testing of hypotheses by different individuals, in a variety of settings, before they are generally accepted as truth. The history of the study of the Upper Cretaceous Ferron Sandstone in east-central Utah directly follows this approach, and recent advances in our understanding of this deltaic complex have been driven by one such hypothesis — sequence stratigraphy.

The compilation of papers within this volume, and photomosaics, cross sections, and other plates on the CD-ROM, illustrate the rigor involved in the scientific testing. From regional stratigraphic packaging to detailed fluid-flow modeling, this volume extends our knowledge of facies and facies fabrics, the fluctuation of depositional processes in both time and space, and the variety and degree of forcing mechanisms involved in creating the Ferron deltaic complex and many of its sequence stratigraphic entities. The volume also attests to the variety of interpretations involved in resolving the truth and, in so doing, it serves as a platform for further discussion and research.

To those who have toiled in the Ferron Sandstone over the decades, its true treasure lies in its three-dimensional exposures. With outcrops located in the semiarid climate of Utah and within the rain shadow of the Wasatch Plateau, study of the Ferron is hindered little by vegetation. Differential erosion of strata on the west flank of the San Rafael Swell created the Coal Cliffs, which contain the Ferron Sandstone. These cliffs trend southwest-northeast, which is largely parallel to the depositional dip of the Ferron deltaic complex. A labyrinth of cross-cutting canyons dissects the outcrop belt and exposes the details of the developing delta through time. Three-dimensional exposures act as windows through which we may "see" many of the gems in the Ferron's treasure chest. Undoubtedly, there are many more gems yet to be discovered. The Ferron has served and will continue to serve as a place for discovery of new ideas, as well as a "testing ground" for ideas developed elsewhere.

The often-quoted observation that "There is nothing more humbling than the outcrop" is certainly true of the Ferron Sandstone. The sedimentology of the Ferron, as well as its placement within a sequence stratigraphic framework, is at times subtle, complex, and even problematic. This volume explores some of this complexity and illustrates the variety of interpretations that can result.

Finally, this volume shouts out resoundingly for the need and usefulness of outcrop reservoir analogs. For example, the three-dimensional architecture of sequence- and sedimentary-bounding surfaces could be recognized only broadly in our most detailed log suites or remote sensing surveys. Details, which often dictate the style and degree of reservoir partitioning, can be most accurately observed on the outcrop. Lessons learned from study of the Ferron outcrops can raise flags of caution to simplistic interpretations, as well as provide answers to actual reservoir problems in other deltaic settings.

As editors, our profound appreciation goes to the individuals whose efforts produced the papers in this volume. Without their "drive to know," this volume would not exist. We also express our gratitude to our reviewers and to those who have worked diligently in getting the volume in publishable format. We greatly appreciate the support and insight of the American Association of Petroleum Geologists and the Utah Geological Survey in allowing this volume to be published. Our hope for the reader is that this volume will provide ideas and practical applications to a variety of problems dealing with deltaic systems. We anticipate that as you read and contemplate the relationships and concepts expressed in this volume, you will desire to one day see for yourself this world-class outcrop analog known as the Ferron Sandstone.

Thomas C. Chidsey, Jr,
Roy D. Adams
Thomas H. Morris

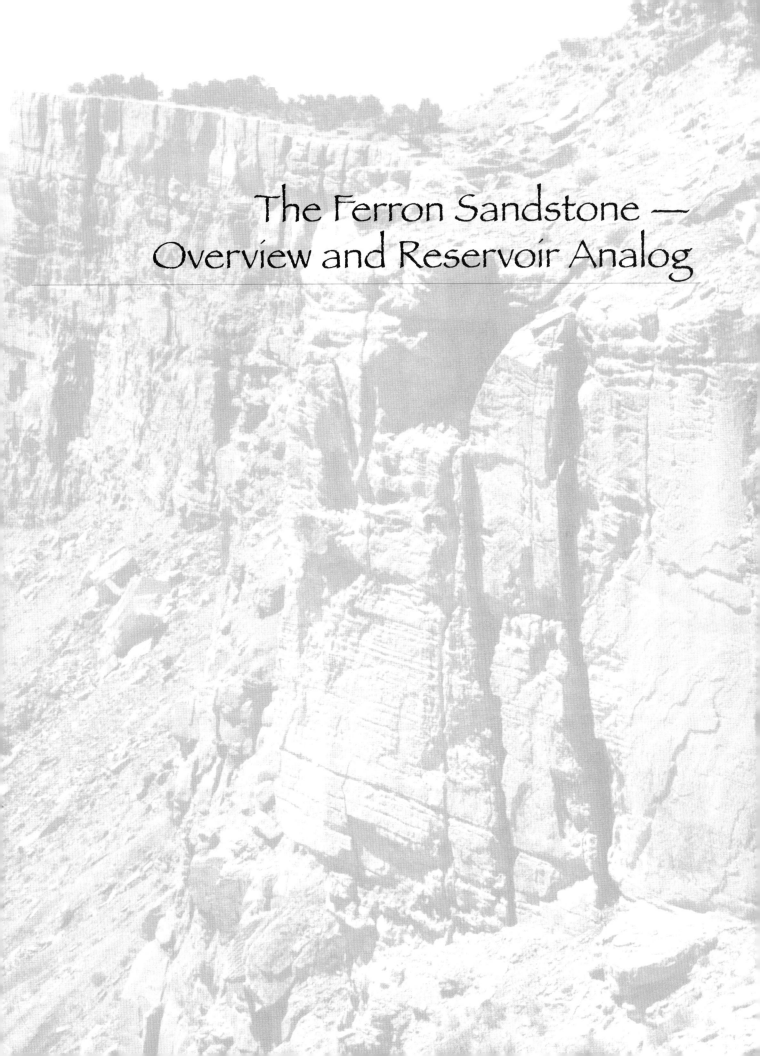

The Ferron Sandstone —
Overview and Reservoir Analog

"The bunch," Lupton's field team near Cleveland, Utah, circa 1910. Photograph courtesy of the family of C. T. Lupton.

Previous Studies of the Ferron Sandstone

Thomas A. Ryer[1]

ABSTRACT

The Ferron Sandstone has, at least by North American standards, a long and rich history of study. Early interest in the Ferron was based on the fact that it contains substantial amounts of mineable coal, the presence of which was noted in the geological literature as early as 1874. A comprehensive documentation of the many coal seams of the Ferron was published in 1916.

Several studies published in the late 1800s and early 1900s established basic correlations between the Ferron outcrops in Castle Valley and equivalent strata in other areas. Initial correlations were done entirely on the basis of similarity of sections — by means of what is commonly called "jump correlation." This approach led to some errors that were subsequently corrected as the role of biostratigraphy in stratigraphic correlation became increasingly important.

Early attempts to interpret the paleogeographic setting of the Ferron led to interesting but erroneous results. A model of Ferron deposition published in the 1920s evoked southward transport of clastics to form a large, elongated sand-rich body or "plume" centered on Castle Valley. The interpretation was incorrect, stemming from a miscorrelation between Castle Valley and scattered outcrops on the western side of the Wasatch Plateau, but it heralded a theme that would be recurrent in the interpretation of the Ferron.

Discovery of a significant accumulation of natural gas in the Ferron Sandstone at Clear Creek field on the Wasatch Plateau in 1951 accelerated the pace of Ferron study. The Ferron was described in much greater detail, but a correct interpretation of Ferron paleogeography remained elusive. The concept that Ferron strata were deposited by prograding deltas and on adjacent strand plains became firmly established in the 1970s and was substantiated in the literature by well documented sedimentological data. Cyclicity was recognized and the issue of whether the cycles resulted from auto- or allocylic processes was raised.

The basic stratigraphic framework of the Ferron, as we know it today, was constructed in the late 1970s and early 1980s. Recent studies have addressed the details and the numerous Ferron studies completed during the last 20 years, taken in their entirety, represent one of the most detailed stratigraphic and architectural frameworks available for any clastic unit in the world.

Many recent studies make use of the detailed stratigraphic framework to focus on the petroleum reservoir properties of the various facies recognized in the Ferron, which provides a wealth of well exposed and well understood analogs for petroleum reservoirs elsewhere. At the same time, the Ferron has become an important producer of coalbed methane from the northern part of Castle Valley.

[1]The ARIES Group, Inc., Katy, Texas

STRATIGRAPHY

Early Studies of the Ferron

The Ferron Sandstone was described as a member of the Mancos Shale by Charles T. Lupton (1916) in his comprehensive study of the coal resources of Castle Valley. Lupton mentioned the Ferron in an earlier publication (1914), making this the first formal use of the name, but the first meaningful description of the Ferron occurs in his 1916 work. Lupton did not designate a type locality or type section for the Ferron, stating simply that the member "is well developed in the vicinity of Ferron and Emery" (Lupton, 1916, p. 31.) In his description of the Ferron, he included a 475-foot- (145-m-) thick stratigraphic section that he had measured through the Ferron in Ivie Creek Canyon, about 8 mi (13 km) south of the town of Emery (Figure 1). It is clear that he considered the Ivie Creek section representative of the Ferron and its coalbeds. Had the designation of a type section been required in the naming of a stratigraphic unit then, as it is today, it is likely that Lupton would have chosen the Ivie Creek section as his type section for the Ferron.

At least two geologists, Hoxie (1874) and Taft (1906), had examined the Ferron outcrops in Castle Valley prior to Lupton's study. Both had misinterpreted its stratigraphic position, but in different ways. Hoxie (1874) interpreted the coal-bearing Ferron outcrops in the vicinity of Emery to be a down-faulted block of Montana Group strata (the coals, therefore, belonging to what is now known as the Blackhawk Formation — see account by Lupton, 1916, p. 9). Taft (1906) believed that the Ferron in Castle Valley belonged to the Jurassic section (what Lupton [1916] and other early workers referred to as the McElmo Formation, which apparently spanned the Carmel through Morrison Formations of today's terminology).

Taft's work was very much reconnaissance in nature, but he did descriptively name some of the stratigraphic units. The Tununk Shale was designated the "Shale of the San Rafael Swell", the Ferron the "Sandstone of the Red Plateau," and the Blue Gate Shale the "Shale of Castle Valley." According to Lupton (1916), Red Plateau is an antiquated name for Cedar Mountain, at the northwestern edge of the San Rafael uplift. The Ferron is not present on Cedar Mountain. Taft had miscorrelated the Ferron with the Jurassic section. These names were not used by any subsequent authors. Taft produced a generalized map of Castle Valley (Figure 2). On it is traced the contact between the Blue Gate Shale and the Ferron Sandstone. The contact is more or less correct between Ivie Creek and Ferron Creek, but farther north is misplaced toward the east.

The transgressions responsible for deposition of the Tununk Shale, the regression that led to deposition of the

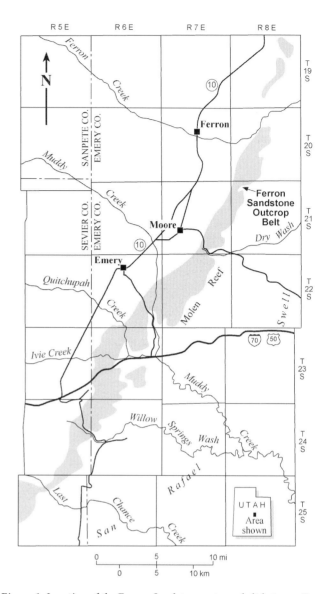

Figure 1. Location of the Ferron Sandstone outcrop belt between Ferron Creek and Last Chance Creek, east-central Utah.

Ferron, and the transgression that marks the change from Ferron to Blue Gate deposition were widespread events in the Western Interior. Sandy strata equivalent to the Ferron had been recognized in the Henry Mountains basin by Gilbert (1877). He named this unit the Tununk Sandstone and the underlying marine shale the Tununk Shale. The marine shale overlying his Tununk Sandstone he named the Blue Gate Shale and the ridge-forming sandstone above that the Blue Gate Sandstone (Gilbert's Blue Gate Sandstone, which is equivalent to the Star Point Sandstone of the Wasatch Plateau, has been renamed the Muley Canyon Sandstone by Eaton, 1990). Lupton (1916) miscorrelated the Castle Valley and Henry Mountains basin sections, equating the shale beneath the Ferron in Castle Valley to the Blue Gate Shale, and possibly the Tununk Sandstone and Tununk Shale as well, and the Ferron Sandstone of Castle Valley to the Blue Gate Sandstone of the Henry Mountains basin. The error

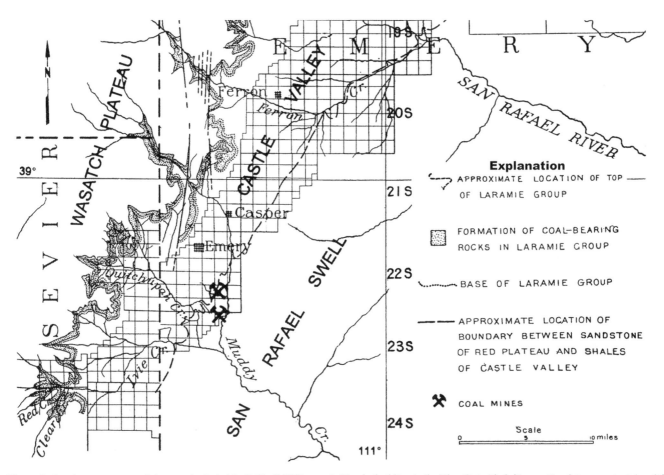

Figure 2. Southwestern part of the map included in Taft's (1906) report. The dashed line is the Blue Gate Shale/Emery Sandstone contact (modified from Taft, 1906). The northern of the two coal mines south-southeast of Emery is the Casper mine; the southern is the Bear Gulch mine.

probably stems from the very similar appearances of the ridges formed by the Ferron along the Molen Reef and Coal Cliffs outcrops on the west flank of the San Rafael uplift and those formed by Gilbert's Blue Gate Sandstone around Caineville, in the northern part of the Henry Mountains basin.

Lupton's (1916) study of the Ferron was very much focused on coal: his description of the rocks of the Ferron is rudimentary and lacks any discussion of facies. Lupton noted the pronounced northeastward thinning and loss of sandstone in the Ferron, but did not offer any interpretation he may have had of how and why such a change takes place. He did comment on the presence of "local unconformities" in the Ferron, one of which he documented in a photograph. The photo (his Plate IIIB) illustrates a fluvial channel scoured into the basal shallow-marine sandstone unit (Kf-1 of Anderson and Ryer, this volume) on exposures along the west side of Blue Trail Canyon, south of I-70. Discordance between the inclined strata of a point-bar deposit within the channel fill and the horizontally bedded sandstone of the underlying shoreface constitutes the evidence for an unconformity. Although Lupton did not discuss the origin of this angular contact in the Ferron, his discussion of a similar

feature that he recognized in the Dakota Sandstone makes it clear he understood that the angular "local unconformities" were products of channels cut into pre-existing sediments.

Lupton portrayed Ferron stratigraphy using a series of measured sections (Figure 3) ranging from the Paradise Lake area of the Fish Lake Plateau on the south to the Castle Valley outcrops south of Wellington on the north. Five of the eight sections are hung on his A coal bed. To the northeast, beyond the limit of deposition of the A coal, carbonaceous shale is identified and used as a datum in sections at Short Canyon and Horn Silver Gulch, east of Ferron. This carbonaceous shale, which forms a conspicuous black band on fresh outcrops, is of marine origin and occupies a position between the top of the Washboard unit and the base of Kf-2, gradually changing facies to normal, silty marine mudstone southward along the Molen Reef. Lupton's northern three sections are hung too high. Although he noted some of the principal lithologic differences between the Ferron sandstones in the northern part of Castle Valley and the coal-bearing Ferron strata in the southern part, he failed to recognized that they occupied slightly different stratigraphic positions.

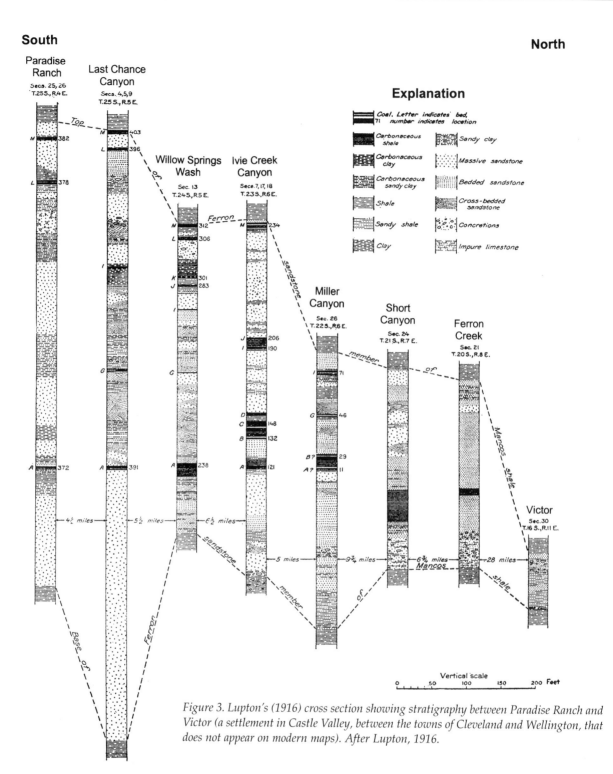

Figure 3. Lupton's (1916) cross section showing stratigraphy between Paradise Ranch and Victor (a settlement in Castle Valley, between the towns of Cleveland and Wellington, that does not appear on modern maps). After Lupton, 1916.

Despite the above-mentioned correlation problem, plus what are now known to be minor miscorrelations of the coals, Lupton's simple cross section identifies the fact that most of the northeastward thinning of the Ferron occurs in its upper part. It is fairly clear that Lupton did not recognize the lateral equivalence of the upper part of the Ferron and the lower part of the Blue Gate Shale to the northeast. His was a simple, "layer-cake" stratigraphy with the complication that one of the layers varied in thickness.

Because Lupton (1916) designated the Ferron as a member of the Mancos Shale and because this has become standard U.S. Geological Survey usage, subdivisions of the Ferron must be sub-member-level units. Assignment of the Ferron to member status came about as a result of the westward extension of stratigraphic terminology as geologists moved westward during the course of their studies. The Mancos Shale was originally named for outcrops in southwestern Colorado (Cross and Spencer, 1899), where the Mancos lacks sandy members. Upon encountering sandstones in the Mancos in Castle Valley, Lupton (1916) chose to distinguish members, rather than to elevate the Mancos to group status and recognize formations within it. In hindsight, this was an unfortunate choice. On the basis of its thickness and wide areal extent, the Ferron Sandstone, the Emery Sandstone, and the marine shale intervals that separate them certainly deserve formation status. Some authors (most notably Hale, 1972) have, in fact, described the Ferron as a formation, but without going through the exercise of formally proposing a change in status.

Although it lacked detailed stratigraphic description, Lupton's (1916) study of the Ferron provided a sound foundation on which subsequent investigators could build.

Paleoenvironments of Deposition

The first publications that attempted to place the Ferron Sandstone in a paleoenvironmental setting were authored by Spieker and Reeside (1925, 1926). The earlier publication focused on the stratigraphy of Castle Valley and the eastern side of the Wasatch Plateau; the second attempted to correlate the Castle Valley section westward to isolated Cretaceous outcrops on the western margin of the Wasatch Plateau near the towns of Manti and Salina. The papers are interesting for their descriptions of Cretaceous strata, but more importantly, represent a significant conceptual advance, portraying lateral as well as vertical relationships between the units.

Regional Stratigraphy

Spieker and Reeside's 1925 paper portrays northeastward interfingering of the upper part of the Ferron and the lower part of the Blue Gate Shale, the base of the Ferron being drawn as a horizontal surface in Castle Valley. Citing the presence of coal in drill holes near the town of Price, they noted (p. 438) the "regional tendency of the Cretaceous beds to change westward from marine to littoral, to continental and coal-bearing sediments." This is the situation pictured, albeit in a very general way, along the Castle Valley outcrops for the upper part of the Ferron.

Spieker and Reeside's correlations between the Henry Mountains basin, Castle Valley, and the western part of the Book Cliffs are shown in Figure 4. Clearly portrayed is northeastward interfingering of the upper part of the Ferron Sandstone and the lower part of the Blue Gate Shale. Section 1, at the left, is based on Gilbert's (1877) study of the Henry Mountains basin; sections 2 and 3 represent the southern and northern parts of Castle Valley; section 4, at the right, is at Horse Canyon, in the Book Cliffs. Lupton's (1916) miscorrelation of the Ferron Sandstone of Castle Valley with the Blue Gate Sandstone of the Henry Mountains Basin is corrected. The Ferron is properly shown as equivalent to

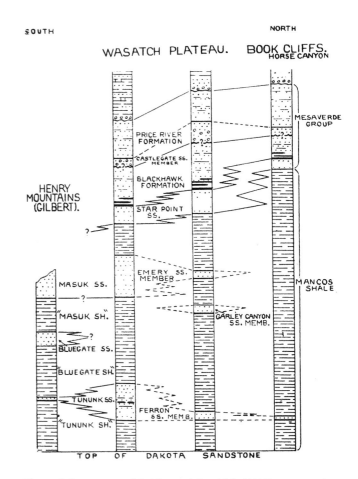

Figure 4. Lower part of Spieker and Reeside's (1925) cross section. The Masuk Sandstone of the Henry Mountains basin is miscorrelated with the Emery Sandstone Member of Castle Valley. Modified from Spieker and Reeside (1925).

Gilbert's Tununk Sandstone (the name Tununk Sandstone was discarded and replaced with the name Ferron Sandstone Member by Hunt [1946]).

Paleogeography

Spieker and Reeside's 1926 paper offers a broader paleoenvironmental interpretation of the Ferron than any of the earlier studies. They interpreted the Ferron Sandstone in Castle Valley to represent a sandy peninsula or shoal area in the sea (Figure 5) that extended southward from the area of the present-day Wasatch Mountains and Uinta Basin. A unit of marine shale recognized on outcrops along the western side of the Wasatch Plateau, and subsequently named the Allen Valley Shale by Spieker (1946), was correlated with the Ferron, as were shales of the Mancos east of Castle Valley. The correlation across the breadth of the Wasatch Plateau was incorrect, although the error was not recognized until the 1950s (it remains in Spieker's landmark 1946 paper detailing Mesozoic and Cenozoic stratigraphy and paleogeography in central Utah). Spieker and Reeside's (1926) interpretation of Ferron paleogeography is the first of several interpretations in which all or part of the

7

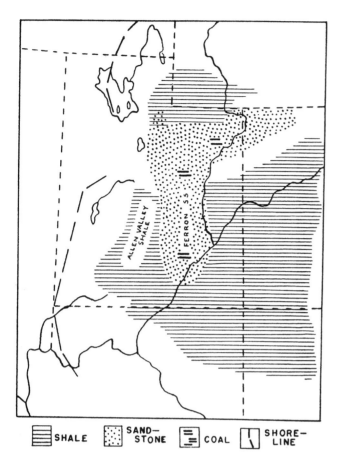

SHALE   SAND-STONE   COAL   SHORE-LINE

Figure 5. Paleogeography of the Ferron Sandstone as reconstructed by Spieker and Reeside (1926), as redrawn by Katich (1953). The Ferron Sandstone and the Allen Valley Shale were interpreted to be age-equivalent. From Katich, 1953; modified from Spieker and Reeside, 1926.

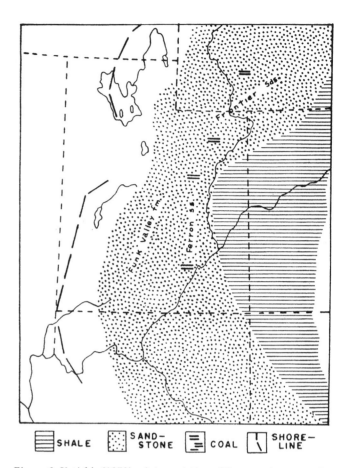

SHALE   SAND-STONE   COAL   SHORE-LINE

Figure 6. Katich's (1953) reinterpretation of Ferron paleogeography. The Ferron is correctly correlated with the basal part of the Funk Valley Formation, eliminating the peninsula and the bay that lay west of it that had been interpreted by Spieker and Reeside (1926). From Katich, 1953; modified from Spieker and Reeside, 1926.

Ferron in Castle Valley was reconstructed as a southward-extending body with a northern provenance surrounded on three sides by shaley strata.

Study of the Ferron Sandstone Accelerates

Gas was discovered in the Ferron Sandstone at Clear Creek field in the northeastern part of the Wasatch Plateau in 1951 (see Petroleum in the Ferron Sandstone section later in this paper). The discovery prompted the publication of several studies in the 1950s that led to a great improvement in the understanding of Ferron stratigraphy and depositional history.

Katich (1953, 1954) studied the Ferron in the vicinity of the Wasatch Plateau and came to a very different conclusion about Ferron paleogeography than had Spieker and Reeside (1926). The presence of the ammonite *Collignoniceras woollgari* in both the Allen Valley and Tununk Shales demonstrated their equivalence. The Ferron was shown (Katich, 1953) to be younger than the Allen Valley Shale, being equivalent, instead, to the lower part of the Funk Valley Formation (Spieker, 1946)

of the western Wasatch Plateau. Katich's (1953) paleogeographic map (Figure 6) represents Ferron paleogeography in a much simpler form and resembles most paleogeographic maps produced by later authors. His analysis of cross stratification in fluvial sandstone of the Ferron convincingly demonstrated that fluvial sands of the Ferron on the Castle Valley outcrop belt had a western and southwestern, not a northern, source.

Davis (1954) studied the Ferron and reached many of the same conclusions that Katich had. Davis confirmed a concept that had earlier been reached by Katich (1951) in his dissertation, but which had been retracted in his publication (Katich, 1954): that the Ferron consists of two distinct parts. Davis designated these the lower and upper units of the Ferron.

The lower Ferron consists of gray, fine-grained, calcareous, marine sandstone and siltstone and contains two distinctive horizons bearing large, "cannon-ball" concretions. It extends from "Molen Amphitheater" and the southern part of Molen Reef northeastward to the vicinity of Price, generally varying between 60 and 70 ft (18 and 21 m) in thickness. Davis believed that he could recognize the lower Ferron on wireline logs run in the gas wells of Clear Creek field. He correlated the lower Ferron to the uppermost, silty and sandy part of the Tununk Shale in the Henry Mountains basin. Southward thinning and loss of sandstone suggests a northern or northwestern source, although no paleocurrent or other definitive data was cited in support of this interpretation.

The upper Ferron comprises a much wider variety of rocks than does the lower Ferron: shallow-marine sandstone, coal, carbonaceous shale, and fluvial sandstone in lenticular bodies are common. The upper Ferron is about 500 ft (150 m) in thickness south of Emery, feathering out into marine shale northeastward along the Castle Valley outcrops. The paleocurrent data indicating a southwestern source described by Katich (1953) and confirmed by Davis (1954) are all from strata of the upper Ferron.

Davis' (1954) studies led him to the conclusion that "a decided change in sedimentation took place in the passage from the lower Ferron to upper Ferron time. Deposition from the northwest and western sources became practically insignificant as compared with a new strong supply that began to come from the southwest." This conclusion has been borne out by all later studies, with one notable exception discussed later.

Edson et al. (1954) described the productive Ferron section in Clear Creek field. Seven sands were recognized in the discovery well, leading to a "1st" (shallowest) to "7th" (deepest) sandstone terminology that was applied to the field. Detailed correlation of the well logs indicates that only the 7th sandstone is continuous throughout the field; all of the other sandstones are lenticular and their number and positions vary between well bores. Although Edson et al. (1954) did not interpret facies and paleoenvironments, it is now clear that the basal sandstone was deposited on the prograding shoreline and that the remaining, shallower sandstones are of fluvial origin.

Knight (1954) studied the heavy mineral content of samples of Ferron Sandstone collected from outcrops and from wells drilled on the Wasatch Plateau. Correlating sandstones between wells at Clear Creek field had proven to be very difficult. Knight's study was undertaken with the purpose of determining whether or not heavy minerals were of value in establishing correlations between individual sandstone beds of the Ferron in the subsurface. His results are negative, with the exception of one general conclusion that lends support to the paleoenvironmental interpretations of Katich (1953) and Davis (1954): that "confinement of ... garnet to the southern area [southern part of Castle Valley] seems to indicate a sedimentary realm separated by one means or another from that in which the rest of the Ferron was deposited and also a separate, local source area."

Important papers describing the petroleum geology of latest Paleozoic through Early Tertiary strata beneath the Wasatch Plateau and on the western flank of the San Rafael uplift were published by Walton (1954, 1955). The bearing of Walton's papers on hydrocarbons in the Ferron is discussed later, but one critical point must be mentioned here because it influenced the paleoenvironmental interpretation of some later investigators. Walton (1955, p. 399), in a brief description of the Ferron, stated that the member "has little or no coal [in the northeastern part of the Wasatch Plateau] and appears mostly marine throughout." Subsequent work has shown that only the basal sandstone section of the Ferron is of shallow-marine origin beneath the northeastern part of the Wasatch Plateau; the remainder is composed of continental strata. Coal is best developed in a position farther to the east. Much the same relationship can be documented on the Ferron outcrops in the southern part of Castle Valley and on the east flank of the Fish Lake Plateau.

Ferron Deltas

Cretaceous paleogeography was reconstructed for a broad area including northeastern Utah, northwestern Colorado and southwestern Wyoming by Hale and Van De Graaff (1964). Castle Valley lies at the southwestern edge of their maps. Hale and Van De Graaff (1964, p. 129) mapped a large "eastward bulging delta-like feature" centering on the Uinta Basin and Uinta Mountains and extending eastward just beyond the Utah-Colorado state line. They referred to it as the "Vernal delta." Their map shows a narrow embayment on the southern flank of the delta, stretching southwestward from the area north of Price to the vicinity of Clear Creek field. Although Hale and Van De Graaff offered no evidence to support the existence of this embayment, there can be little doubt that it's origin lies in Walton's (1954 and 1955) assertions that the Ferron at Clear Creek field is mostly or entirely of marine origin.

Hale (1972) studied subsurface data from wells drilled during the 1950s and early 1960s in the search for gas beneath the Wasatch Plateau and combined them with original outcrop studies to arrive at a new interpretation of Ferron paleogeography and depositional history. He elaborated on Davis' (1954) concept of two distinct members of the Ferron, although he defined them in a different way. He revived Spieker and Reeside's old idea of a southward-extending peninsula, though in a different location. Two paleogeographic maps (Figures 7 and 8) summarize Hale's concepts.

Hale (1972) recognized two geomorphic features on which predominantly sandy sediments accumulated during early Ferron time: the "Castle Valley bar" in the north, a west-southwestward extending peninsula supplied with sediment from the Vernal delta and fed by long-shore drift, and the "Last Chance delta," a northeastward extending deltaic complex fed from the southwest and named for exposures along Last Chance Creek

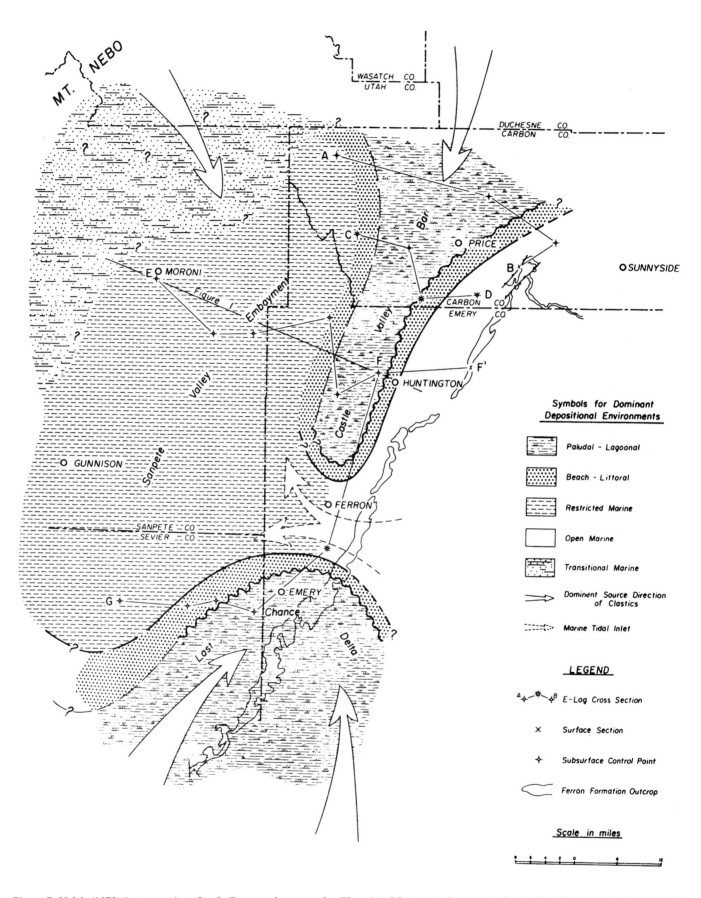

Figure 7. Hale's (1972) interpretation of early Ferron paleogeography. The critical feature in this map is the Castle Valley bar, which separated the large Sanpete Valley embayment from the Cretaceous seaway. From Hale, 1972.

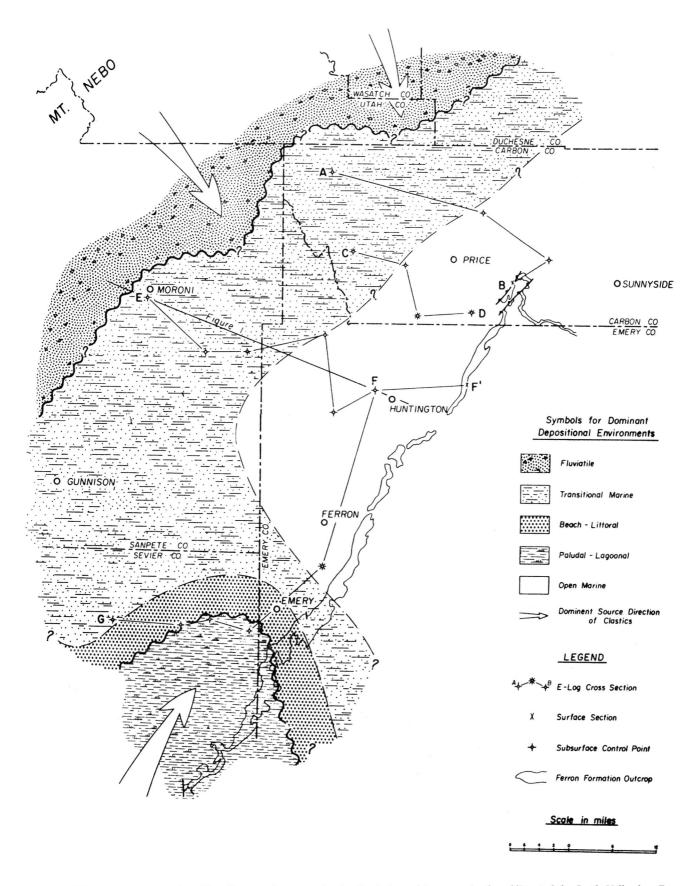

Figure 8. Hale's (1972) interpretation of late Ferron paleogeography. Sea-level rise and transgression has obliterated the Castle Valley bar. From Hale, 1972.

11

Southwest   **Northeast**

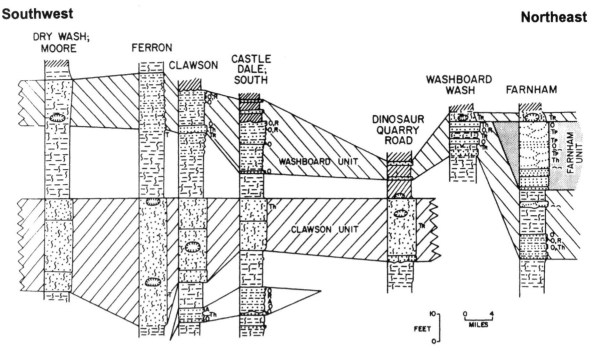

Figure 9. Part of Cotter's (1975a) cross section showing the stratigraphy of the lower Ferron in Castle Valley. The Clawson and Washboard units are the most widespread of the four units he defined. After Cotter, 1975a.

on the east side of the Fish Lake Plateau. A large inlet situated at the town of Ferron separated the two and allowed exchange of marine water with the "Sanpete Valley embayment," which lay to the west of both features. The Sanpete Valley embayment is the most significant feature in this interpretation, defining the northwestern margin of the Last Chance delta and the western margin of the Castle Valley bar. The only evidence Hale offered for the existence of the Sanpete Valley embayment is Walton's (1954, 1955) statement that the Ferron at Clear Creek field is entirely of marine origin. Aware that the Ferron includes coal in many of the wells that penetrated the embayment section, Hale (1972) rationalized that the "few thin carbonaceous shales and coaly streaks are ... carbonaceous matter derived from [an] adjacent paludal environment."

Hale defined the "Lower Ferron," a member-level unit, to include the deposits of the Castle Valley bar and the Sanpete Valley embayment, plus time-equivalent deposits in the lower part of the Last Chance delta. Rise of relative sea level subsequently submerged the Castle Valley bar and the area of the Sanpete Valley embayment opened to become part of the Mancos seaway. The "Upper Ferron" includes the younger part of the Last Chance delta. Davis (1954; and Katich, 1951, before him) had interpreted the lower Ferron to be an older, northern depositional system and the upper Ferron a younger, southern system. Hale's interpretation of contemporaneity of the northern Ferron system and the lower part of the southern deltaic system is quite different. Subsequent work would show the earlier interpretations of

Katich and Davis to be correct. Although Hale's (1972) interpretation of paleogeography was rejected, his Vernal and Last Chance deltas have been recognized, at least as generalized features, by most students of the Ferron who followed.

Cotter conducted a study of the entire Ferron outcrop belt during the early 1970s. His work culminated in of a series publications (Cotter, 1971, 1975a, 1975b, 1976) that offered interpretations of Ferron stratigraphy and depositional history that were considerably more refined than the preceding interpretations. Cotter's interpretations incorporated important concepts that evolved from numerous studies of depositional processes in modern alluvial plain, coastal, and shallow-marine settings that had been completed during the 1960s.

Cotter (1975a) defined four mappable "units" (Figure 9) on the Ferron outcrops of northern Castle Valley. These strata correspond to the lower Ferron of Davis (1954). Most important are the Clawson unit and overlying Washboard unit, which are laterally continuous over most of the length of the Castle Valley outcrops. They constitute two upward-coarsening depositional units composed predominantly of very fine to fine-grained, silty sandstone that has been extensively burrowed in most areas. They bear the distinctive "cannon ball" concretions described by previous authors. Cotter (1975a, 1975b, 1976) documented the fact that these two units disappear southward by facies change into normal marine shale in the uppermost part of the Tununk Shale. He clearly demonstrated that they occupy a position stratigraphically beneath the coal-bearing strata of the

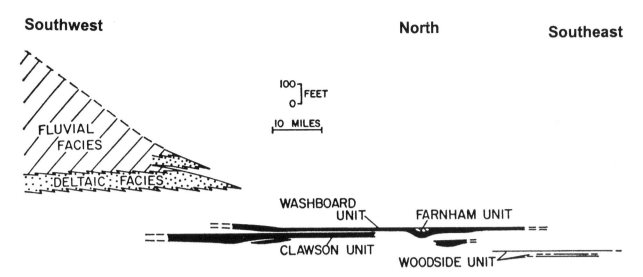

Figure 10. *Relationship between lower and upper Ferron strata (from Cotter, 1975a). The lower Ferron, composed primarily of the Clawson and Washboard units, is clearly distinguished as older than the upper Ferron deltaic strata. After Cotter, 1975a.*

upper Ferron near Emery (Figure 10). Cotter interpreted the Clawson and Washboard units as representing a low-energy coast where sand, transported southward by longshore drift from the Vernal delta (Figure 11), accumulated on barrier islands similar to Sapelo Island, Georgia.

Cotter (1975b, 1976), like Katich, Davis, and Hale before him, interpreted the coal-bearing strata of the upper Ferron to represent a generally eastward to northeastward prograding delta, to which he applied Hale's (1972) name "Last Chance delta." Cotter offered a very specific interpretation of the delta: it "formed as a broad, fan-shaped complex comprising numerous coalescing and overlapping subdelta lobes" (Cotter, 1976, p. 481) representing a high-constructive lobate delta. Cotter recognized the cyclic character of progradation of the Ferron shoreline (Figure 12) and found in his deltaic interpretation an explanation for the cyclicity: the process of deltaic progradation followed by subsequent abandonment and transgression. His interpretation (Figure 13) shows two hypothetical delta lobes superimposed on a map of the Ferron outcrops in southern Castle Valley. Although the figure is generalized, it clearly indicates that progradation and abandonment of delta lobes might explain shoreline regressions and transgressions of 5 mi (8 km) or more, enough to explain the youngest "deltaic facies" sandstone pictured in Figure 10. Cotter's work contains the first clear conceptualization of cyclicity in the deltaic deposits of the Ferron and represents the first attempt to explain the origin of the cyclicity.

Another important advance in Cotter's work was the recognition that the shoreline of the Last Chance delta did not advance uniformly eastward or northeastward. Some of the shorelines prograded northwestward or even westward, and these shoreline deposits share an important characteristic: whereas the shorelines that

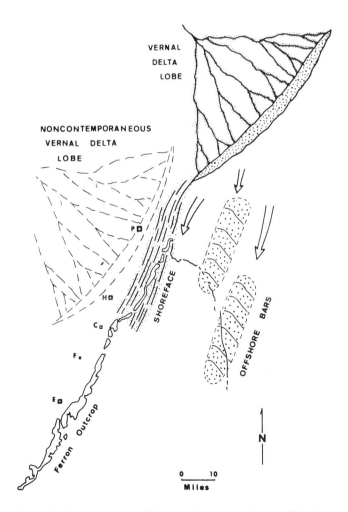

Figure 11. *Paleogeography of the lower Ferron as interpreted by Cotter (1976). The Clawson and Washboard units accumulated along a strand-plain/barrier-island coast and were nourished by sand transported southward from the Vernal delta. From Cotter, 1976.*

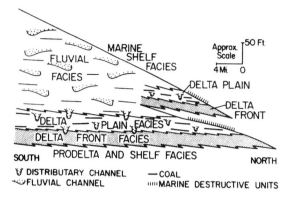

Figure 12. Diagrammatic cross section showing basic facies and style of cyclicity in the Ferron, as recognized by Cotter, 1975b. From Cotter, 1975b.

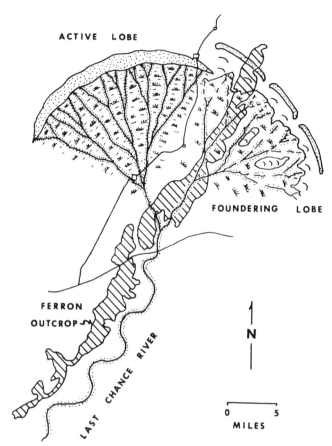

Figure 13. Cotter (1976) interpreted the Last Chance delta as a high-constructive, lobate delta system. Lobe progradation and abandonment, pictured diagrammatically here, was offered as an explanation for the transgressive-regressive cyclicity evident in the Ferron. From Cotter, 1976.

prograded generally eastward display very gentle seaward dips, those that prograded northwestward to westward commonly display much greater dips, in some instances reaching 15 degrees. Cotter (1975b) cited the outcrops along the north side of Ivie Creek as an example of this phenomenon, documenting the steep westward dips with several photographs. He interpreted the steep delta fronts as forming where fresh, sediment-laden water from the Last Chance distributary channels flowed into embayments in which the salinities were reduced.

Cotter used the term "Gilbert delta" to refer to steep delta fronts he observed locally in the Ferron. The term derives from the morphologies of Pleistocene deltas deposited in glacial Lake Bonneville and described by Gilbert (1885). Gilbert deltas are characterized by steep delta-front strata bounded below and above by low-angle to flat "bottom-set" and "top-set" beds. The "type" Gilbert deltas deposited on the shoreline of Lake Bonneville formed where streams characterized by high bed loads reached and rapidly mixed with the fresh water of the lake. Along marine coasts, fresh water from river mouths generally overflows the denser, saline water, resulting in salinity stratification and development of a fresh-water plume. Cotter speculated that a shallow, restricted embayment, perhaps like the Sanpete Valley embayment of Hale (1972), but more likely much smaller, contained (at least periodically) low-salinity water, facilitating rapid mixing and, therefore, steep slopes. This question is addressed further in the description of the Ivie Creek area by Anderson et al., this volume.

Details Emerge

Ryer conducted a comprehensive study of the Ferron outcrops in the southern part of Castle Valley and available drill-hole data from Castle Valley and the Wasatch Plateau during the late 1970s. The results of this study were presented in a series of papers published by Ryer and various collaborators during the early 1980s

(Ryer, 1980; Ryer et al., 1980; Ryer, 1981a, 1981b, 1982a, 1982b, 1984; Ryer and McPhillips, 1983). The emphasis of Ryer's work was the recognition and mapping of packages of strata representing "major cycles of sedimentation" in the Ferron. These cycles were designated "4th order" cycles (Ryer, 1984) building on the terminology of Vail et al. (1977). Five delta-front units representing 4th order cycles (Kf-1 through Kf-5 of Anderson and Ryer, this volume) were initially recognized (Figure 14). Each was distinguished on the basis of an overall upward-coarsening shoreline sandstone unit plus equivalent delta-plain strata. The deposits of each cycle contain a widespread bed of coal, for which the letter designations of Lupton (1916) were retained.

Two additional shoreline sandstone bodies were identified by Ryer in the vicinity of Emery (Kf-6 and Kf-7 of Anderson and Ryer, this volume), but were not associated with 4th-order cycles because of their very limited dimensions in the depositional dip direction. Kf-6 and Kf-7 were elevated to the same status as Kf-1 through Kf-5 in Ryer's later publications (Figure 15). Like Cotter, Ryer interpreted the Ferron to be the product of deltaic sedimentation, specifically river-dominat-

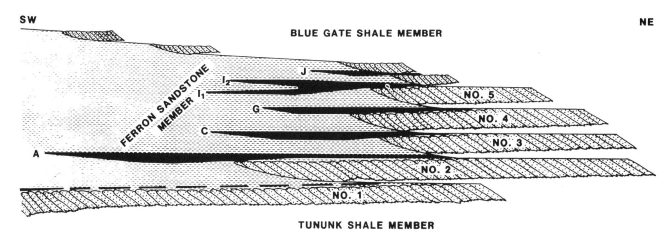

Figure 14. Generalized southwest to northeast cross section of the Ferron defining Ryer's (1981a) delta-front units. Two additional delta-front units lying near the landward edge of delta-front unit no. 5 were identified, but were not assigned numbers owing to their small size. The final two units, near the left edge of the diagram, do not exist. From Ryer, 1981a.

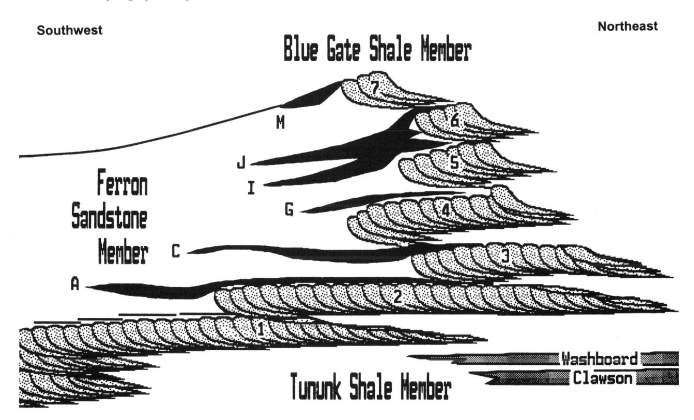

Figure 15. Ryer (1991) updated his generalized cross section of the Ferron, recognizing delta-front units no. 6 and no. 7 as of equal importance to nos. 1 through 5. Compare to Figure 14. After Ryer, 1991.

ed, lobate deltas. Wave-modified deposits of the shorefaces that lay between active delta lobes were also distinguished.

Ryer (1981a) analyzed the relationships between the coalbeds of the Ferron and the equivalent delta-front sandstone units. The concept of "stratigraphic rise" played a key part in his analysis, as it has in more recent studies. Progradation of a shoreline during a period of relative rise of sea level results in what can be termed stratigraphic rise of a shoreline sandstone unit. During progradation, the position of recognizable facies boundaries move upward relative to some horizontal datum. Recognizable horizontal data in the rock record are rarities (an example would be a layer of volcanic ash preserved in a peat deposit). In practice, stratigraphic rise is most easily recognized on the basis of seaward thickening of the unit of marine shale beneath and seaward of the shoreline deposit. Greater stratigraphic rise is reflected by more rapid basinward thickening of a marine shale unit. In Figure 16, stratigraphic rise is reflected by

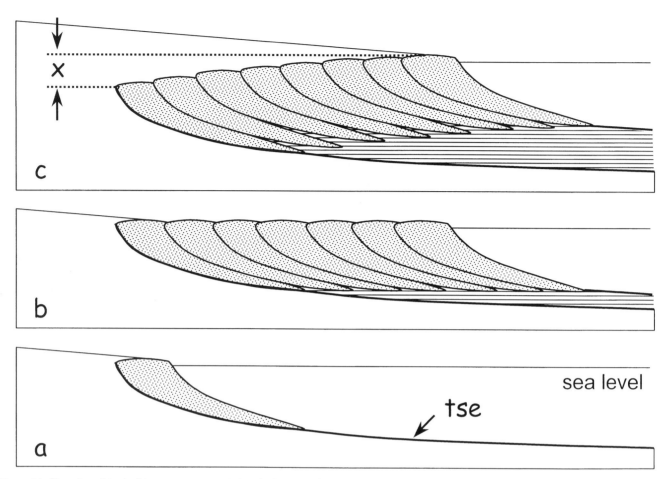

Figure 16. "Stratigraphic rise" is a consequence of rise of relative sea level during progradation. (A) Transgression of the sea onto a coastal plain has ended and progradation has just begun. A transgressive surface of erosion (tse), a shallow-water portion of the marine flooding surface, is marked be the heavy line. The stippled body represents a prograding shoreface. (B) Progradation has occurred during a period of relative sea level stability. No stratigraphic rise is discernable. (C) Progradation has occurred during a period of rising sea level. The amount of stratigraphic rise, "x," corresponds to the upward rise of a facies boundary, here the boundary between the shoreface sand and overlying coastal-plain sediments, relative to a datum parallel to the original sea level.

the upward shift of the boundary between upper shoreface and middle shoreface strata. The thicknesses of the delta-front sandstone units is closely related to stratigraphic rise: thicker sandstone units record higher rates of relative sea-level rise during progradation; thinner units record lower rates. The older delta-front sandstones of the Ferron (Kf-1 and Kf-2 of Anderson and Ryer, this volume) are thinner and extend farther basinward than the younger delta-front sandstone units (Kf-3 through Kf-7; Figure 15), suggesting that the rate of relative sea-level rise in central Utah increased through Ferron deposition. Ultimately, the rate of relative sea-level rise accelerated, resulting in a final, rapid transgression of the sea southwestward across the Last Chance delta. Several subsequent studies by other authors developed these concepts further.

Utilizing outcrop relationships as a guide, Ryer and McPhillips (1983) correlated the Ferron in the subsurface of the Wasatch Plateau. They rejected Hale's (1972) concepts of a Castle Valley bar and a Sanpete Valley embay-

ment, tying the lower Ferron sandstone westward and southwestward to the basal sandstone of the Ferron in the subsurface. They interpreted the Clawson and Washboard units of Cotter (1975a) as distal lower Ferron sands deposited on the shelf during a lowstand of sea level. Most importantly, they argued that the Vernal and Last Chance deltas were not true deltas, but larger-scale geomorphic features related to tectonics. The Ferron shoreline, they argued (Ryer and McPhillips, 1983) "received sediment from numerous rivers that built deltas on a much smaller scale than the Last Chance and Vernal deltas as [defined by Hale, 1972]. These small deltas coalesced laterally to produce continuous sheets of delta-front sandstone as the shoreline prograded."

The origin of the Vernal delta was addressed in greater detail by Ryer and Lovekin (1986). They argued that the conspicuous seaward bulge of the shoreline in northeastern Utah and southwestern Wyoming that had been named the Vernal delta by Hale (1972) owed its origin to movement on the ancestral Uinta Mountain uplift.

Uplift of this feature during earliest Late Cretaceous time had earlier been documented by Weimer (1962). As the Ferron-Frontier shoreline prograded eastward to what is now the Uinta Basin and Uinta Mountains, subsidence occurred, but at a slower rate than in areas to the north and south. This differential subsidence, Ryer and Lovekin argued, is the cause of the bulge in the shoreline. A negative line of evidence was also presented. They cited the northern Andes in South America as a modern analog for the Sevier Orogenic Belt in Utah and Wyoming and inferred the spacing of drainages that would have existed in Utah during Ferron-Frontier deposition. Features as large as the Vernal and Last Chance "deltas," they concluded, could not have been deposited by individual rivers (Figure 17).

A series of seven drill holes cored through the Ferron just west of the outcrop belt by ARCO in 1982 were the subject of a study by Thompson (1985). The principal findings are described in a paper by Thompson et al. (1986). The study focused primarily on the interpretation of facies and depositional environments represented in the core. Six outcrop sections located close to the cores were measured in order to augment the core interpretations. Thompson et al. (1986), like Cotter (1975b) and Ryer (1981a), interpreted the Ferron to represent lobate, river-dominated deltas (Figure 18). Although their description and interpretation of sedimentary features in cores are excellent, their study presents only a very general picture of Ferron stratigraphy. A cross section depicting the stratigraphy in and between five of the ARCO drill holes (Figure 19) distinguishes facies very broadly. No attempt was made to distinguish and correlate the coal seams mapped and named by Lupton (1916) or the individual delta-front sandstone bodies defined by Ryer (1981a).

Of particular interest is Thompson et al. (1986) identification of sandy shelf "plume" deposits in the Ferron. Their model, shown in Figure 20, recognizes bodies of sand deposited on the shelf by a south- or southwest-flowing current and nourished by sediment delivered from a delta front situated up-drift. The model is based on an interpretation by Coleman et al. (1981) of a sand body located in the Mediterranean Sea off and down-drift of the mouth of the Damietta Branch of the Nile Delta. The sand body was interpreted to be a Holocene deposit formed by transport of sand from the river mouth onto the adjacent shelf by an eastward flowing current. Palmer and Scott (1984) elaborated on the model (Figure 21) and applied the concept of a sandy shelf "plume" to explain sandstones in the Upper Cretaceous La Ventana Tongue of the Cliffhouse Sandstone in the San Juan Basin. Later, Scheihing and Gaynor (1991) re-examined data for the Damietta shelf "plume" and argued that it is not a Holocene deposit formed on the shelf, but rather a Pleistocene deposit that accumulated along the shoreline, brought by longshore drift from the

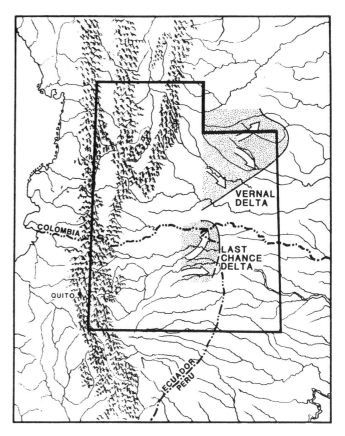

Figure 17. Comparison of Ferron paleogeography in Utah with the present-day drainage pattern on the eastern flank of the Andes in northern South America. The Andes are made to coincide with the Cretaceous Sevier orogenic belt. On the basis of this comparison, Ryer and Lovekin (1986) concluded it was unlikely that the Vernal and Last Chance deltas could have been the deltas of individual rivers. From Ryer and Lovekin, 1986.

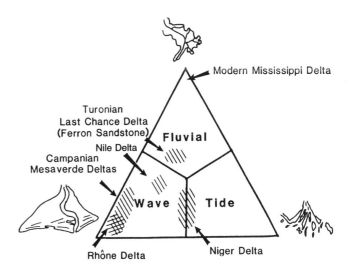

Figure 18. The Last Chance delta is interpreted by Thompson et al. (1986) to be a river-dominated delta with moderate wave influence (from Thompson et al., 1986). The ternary diagram they used was modified from Galloway, 1975.

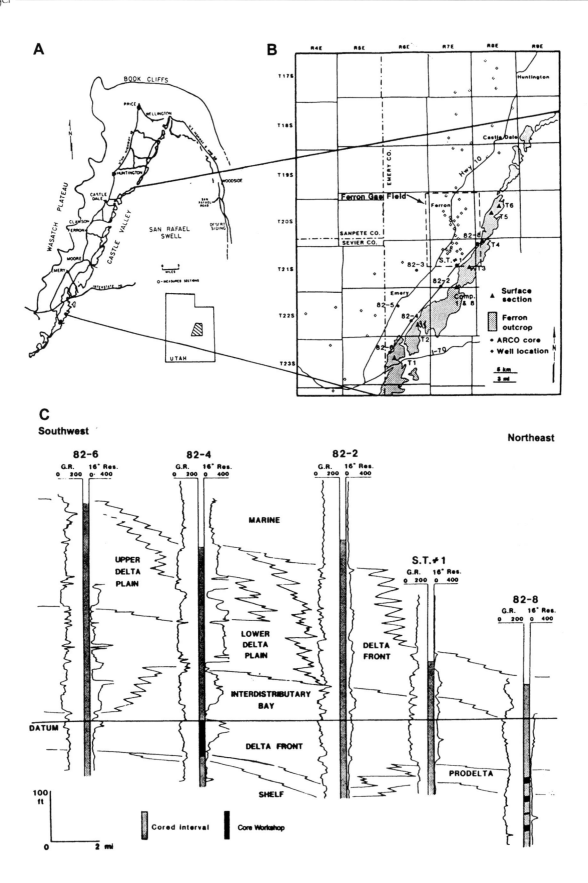

Figure 19. (A) Ferron Sandstone outcrop belt in Castle Valley (from Cotter, 1975b). (B) Location of Ferron drill holes, surface sections, and cross section. (C) Southwest to northeast cross section (shown on Figure 19B) showing correlations between five core holes drilled through the Ferron by ARCO. Drill hole 82-6 is situated near Ivie Creek; 82-8 is southeast of Ferron. Interfingering of facies is shown, but is highly generalized and, in some areas, simply incorrect. Modified from Thompson et al., 1986.

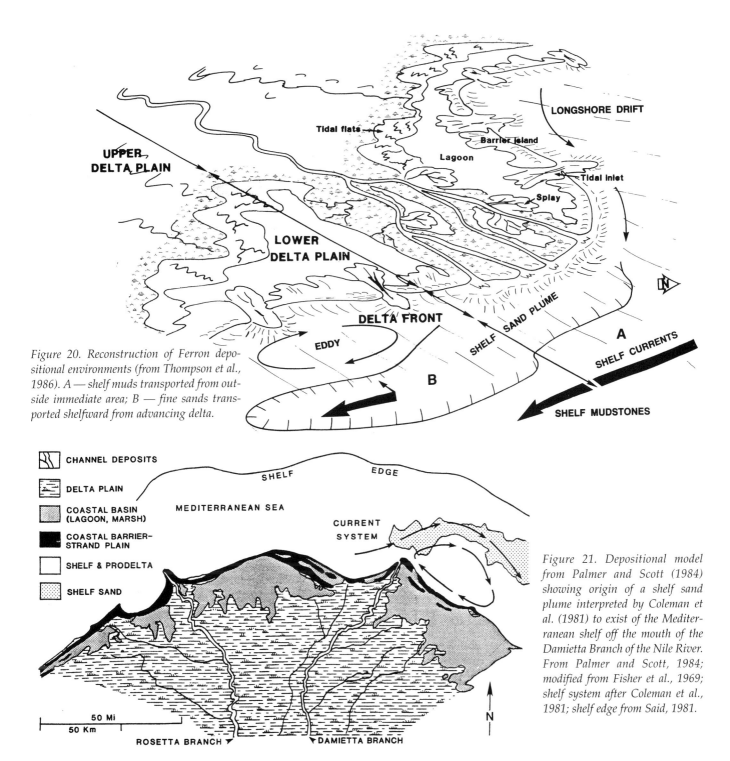

Figure 20. Reconstruction of Ferron depositional environments (from Thompson et al., 1986). A — shelf muds transported from outside immediate area; B — fine sands transported shelfward from advancing delta.

Figure 21. Depositional model from Palmer and Scott (1984) showing origin of a shelf sand plume interpreted by Coleman et al. (1981) to exist of the Mediterranean shelf off the mouth of the Damietta Branch of the Nile River. From Palmer and Scott, 1984; modified from Fisher et al., 1969; shelf system after Coleman et al., 1981; shelf edge from Said, 1981.

mouth of the Damietta Branch when sea level was lower. Regardless of whether or not the sand body associated with the Damietta Branch accumulated as a shelf plume, the shelf plume model itself has merit. For the Ferron, the sand plume model was invoked to explain the origin of the lower Ferron, which Thompson et al. (1986) interpreted as a large plume derived from the Vernal delta (Figure 22), and also to explain some more localized sandy strata in the basal part of the upper Ferron (their Last Chance delta; Figure 23). In some respects, the model of Thompson et al. (1986) is strikingly similar to

the Castle Valley bar interpretation of Hale (1972) and also resembles the peninsula interpretation of Spieker and Reeside (1925); Figure 5. The main difference is that the shelf plume is interpreted to have existed mainly to the east of the present-day outcrop belt, on what is now the San Rafael uplift, whereas Hale's Castle Valley bar was located mainly to the west of the outcrop, in the subsurface. Cotter (1976) (Figure 11) invoked southward transport on the shelf to explain some lenses of sandstone that he assigned to the lower Ferron (his Woodside unit; Figure 10) on the northeast flank of the San Rafael

19

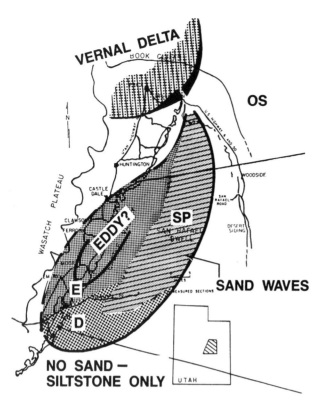

Figure 22. *Vernal delta and its shelf sand plume from Thompson et al. (1986). OS — open shelf; SP — sand plume; E — eddy; D — distal silty plume. Sandy plume deposits are shown as limited to the San Rafael uplift, from which they were eroded when the structure was uplifted in Tertiary time. For data-point locations see Figure 19.*

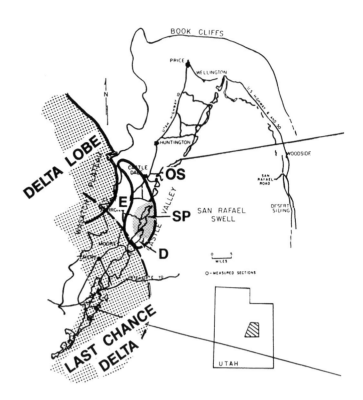

Figure 23. *Possible smaller-scale sand plumes deposited in front of prograding delta lobes from Thompson et al. (1986). OS — open shelf; SP — sand plume; E — eddy; D — distal silty plume. Note the large size of the hypothetical delta lobe pictured north of the Last Chance delta. It is comparable in size to the lobes pictured by Cotter (1976; see Figure 13). For data-point locations see Figure 19.*

uplift, but his shelf-sand interpretation was applied at a much smaller scale than was the Vernal delta plume model of Thompson et al. (1986).

Ryer (1987) proposed a general model that distinguished different origins for the lower Ferron and upper Ferron (the paper was finally published in 1993a; a revised version appeared in 1994.) Ryer argued that the effects of both sea-level change and variations in the flux of clastics from the Sevier orogenic belt can be distinguished in the Ferron. The lower Ferron was deposited primarily as the result of a lowering of eustatic sea level. Rise of sea level led to deepening of water and accumulation of marine mudstone above the Washboard unit. Slowing of subsidence and possibly structural rebound of the foredeep that lay immediately to the east of the Sevier orogenic belt led to an increase in the flux of sediment across the foredeep to the shoreline in central Utah during late Turonian time. The coarse-grained Calico Bed of the Kaiparowits Plateau records this event in southwestern Utah; the Coalville Conglomerate of the Frontier Formation records it in northern Utah. The increased flux of sediment caused progradation of the upper Ferron shoreline in Castle Valley, despite continued rise of sea level. The distinctive, river-dominated deltas of the upper Ferron, which are unusual in the Upper Cretaceous section of the Western Interior, are the

product of this rapid sediment influx. In this interpretation, relative sea level and the flux of sediment from the orogenic belt to the shoreline are decoupled, the latter being largely a function of regional tectonics instead of rejuvenation of streams during a period of relative sea level fall. Failure to take this phenomenon into account may explain the very disparate results of recent sequence stratigraphic studies of the Ferron, as described later.

Molenaar and Cobban (1991a, 1991b) collaborated to synthesize their knowledge, gained through many years of study, of Cretaceous strata on the south and east sides of the Uinta Basin. The Ferron outcrops lie at the southwestern margin of the area they considered in their report and there are no findings that materially altered previous interpretations of the Ferron in Castle Valley. Some of their conclusions regarding areas to the north and east, however, are of interest.

Molenaar and Cobban (1991a, 1991b) named a bed of sandstone that is present near the middle of the Tununk Shale on the northeast flank of the San Rafael uplift and into eastern Utah the Coon Spring Sandstone Bed. Included in it is the Woodside Unit of the Ferron defined by Cotter (1975a) (Figure 10). The Coon Spring Sandstone Bed contains the ammonite *Collignoniceras woollgari*, dating it as early middle Turonian in age.

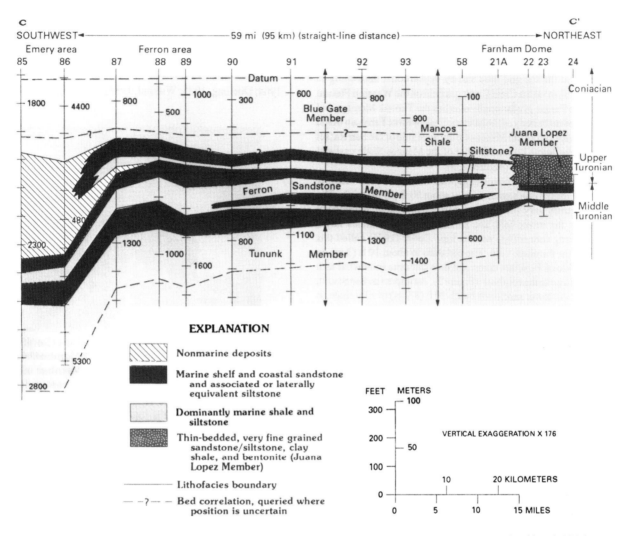

Figure 24. Stratigraphy of the Ferron Sandstone along the Castle Valley outcrops from Molenaar and Cobban (1991a).

A cuesta-forming unit of black marine shale containing varying amounts of very fine grained, platy, thin-bedded sandstone in eastern Utah had previously been assigned to the Ferron by many workers (for example Hintze and Stokes, 1964). Molenaar and Cobban reassigned it to the Juana Lopez Member of the Mancos, arguing that it is much more like the Juana Lopez of areas to the east and southeast than it is to the Ferron in Castle Valley. Their cross sections, one of which incorporates subsurface data between Emery and the northern plunge of the San Rafael uplift (Figure 24) show the Juana Lopez Member as being essentially equivalent to the upper Ferron.

Ferron Reservoir Analogs and Sequence Stratigraphy

Numerous studies of the Ferron undertaken in the late 1980s and early 1990s filled in many of the details that were lacking in earlier studies. Analysis was done at a considerably finer scale and the focus shifted to determining the distributions of properties relevant to pro-

duction of fluids from petroleum reservoirs. This was the result of two primary factors: (1) previously completed studies provided a sound stratigraphic framework within which such detailed studies could be efficiently conducted; and (2) analysis of petroleum reservoir types and the amounts of mobile but unrecovered petroleum that they contain pointed to fluvial-dominated sandstones as an important, underdeveloped resource (Tyler, 1988; Ray et al., 1991).

The landward pinchouts of two of the delta-front units of the Ferron, Kf-2 and Kf-6 of Anderson and Ryer (this volume), were described by Anderson (1991a, 1991b). This study addressed the character of stratigraphic traps that might be formed by these features and compared them with landward pinchouts of highly wave-dominated shoreline deposits of the Blackhawk Formation in the Book Cliffs. Anderson's description of the Kf-2 landward pinchout at the mouth of Willow Springs Wash (Figure 25) is notable in that it discriminates sub-facies within the landwardmost part of the Kf-2 delta-front sandstone body and characterizes them, at least in general terms, as to their relative reservoir quality.

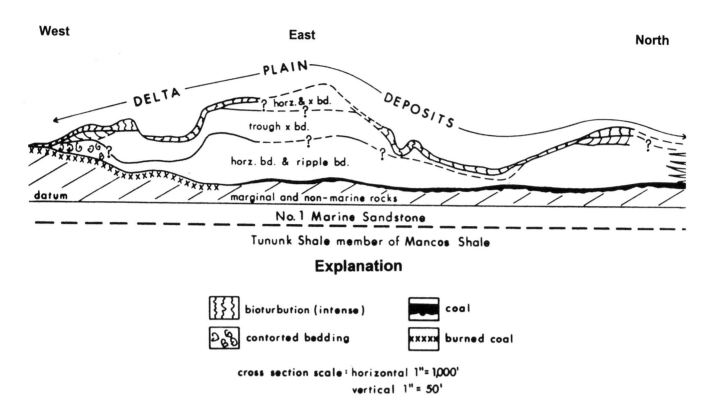

Figure 25. Facies that occur within the landward pinchout of Kf-2 near the mouth of Willow Springs Wash (modified from Anderson, 1991b).

Five drill holes were cored southeast of Emery as part of a study of Ferron unit Kf-2 in the Muddy Creek Canyon area by Ryer in 1992. The initial results of the study were presented in a field trip guidebook (Ryer, 1993b) and core workshop notes (Gustason et al., 1993). The study addressed the origin of several parasequences in Kf-2. Permeability data from the cores and adjacent outcrops in Muddy Creek Canyon were compiled and analyzed by Mobil Exploration, but these results remain proprietary.

Ryer studied outcrops in Indian Canyon, south of Willow Springs Wash, with the purpose of distinguishing autocyclic from allocyclic origins of several parasequence-level units defined there. Preliminary results were reported by Ryer (1993a, 1993b, 1993c). Exposures in Indian Canyon record an abrupt change from wave-modified shoreline to fluvial-dominated deltaic deposits. The bedding surface across which this change occurs has many of the characteristics of a marine-flooding surface, yet lacks any evidence of transgression of the shoreline or of relative sea level rise. Ryer interpreted the change to represent autocyclic shifting of a river system into the Indian Canyon area and subsequent progradation of a fluvial-dominated delta. This concept figures prominently in the approach discussed by Anderson and Ryer (this volume) to more regional analysis of the Ferron.

Comprehensive studies of the Ferron were conducted by Gardner as part of a Ph.D. dissertation (Gardner, 1993) at the Colorado School of Mines. Gardner contributed a large amount of new data and many innovative interpretations, as well as a complete review and, in some cases, rethinking of previous interpretations. Gardner's studies of the Ferron are summarized in two papers published in 1995 (Gardner, 1995a, 1995b). Both deal with tectonic and eustatic controls on sedimentation of mid-Cretaceous strata: the first addresses most of the Rocky Mountain region (exclusive of Montana); the second focuses on central Utah. Many of Gardner's detailed stratigraphic descriptions and interpretations of the Ferron remain unpublished and can be found only in his dissertation (Gardner, 1993) and in guidebooks reproduced for various field trips (Gardner, 1992; Gardner et al., 1994).

Gardner (1995b) defined a hierarchy of cycles related to base-level rise and fall (Figure 26). The Ferron represents the base-level fall-to-rise turnaround of a "long-term cycle" (the 2nd order cycle of Vail et al., 1977). Four "intermediate-term cycles" (3rd order) were distinguished and named for characteristic fossils (Gardner, 1995a). The lower Ferron of Castle Valley plus a thick, basal sandstone unit included in the Ferron on its southernmost outcrops near Last Chance Creek are assigned to the Hyatti sequence (named for the ammonite *Pyionocyclus hyatti*). The upper Ferron constitutes the Ferronensis sequence (named for the ammonite *Scaphites ferronensis*).

Gardner designated the level of cyclicity represented by the numbered delta-front sandstones of Ryer (1981a)

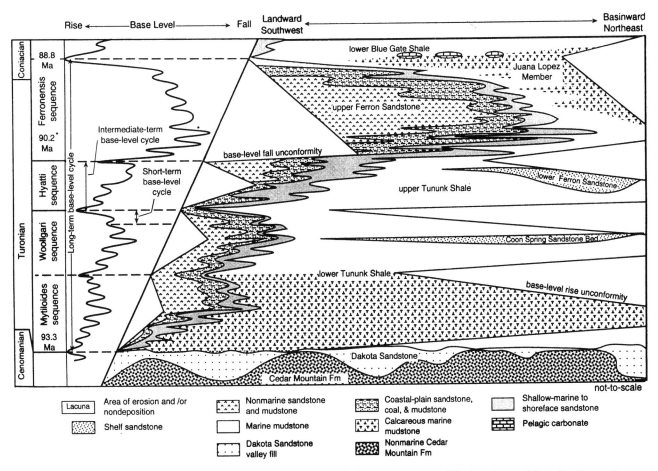

Figure 26. Gardner (1995a) distinguished four sequences in Turonian-Coniacian strata in central Utah and named them for diagnostic inverte-brate faunas. The Hyatti sequence includes strata of the lower Ferron; the Ferronensis sequences includes the deltaics of the upper Ferron (from Gardner, 1995a).

"short-term cycles" (4th order). Gardner designated these units as genetic sequences (GS) — GS 1 through GS 7 of the upper Ferron (Figure 27) following the numbering system of Ryer (1981a) (the deposits of genetic sequences are referred to in some of Gardner's work as "stratigraphic cycles" [SC], for example SC 1, SC 2, and so forth; in the remainder of this discussion, the Kf- terminology is substituted for the GS or SC designation). Kf-1 through Kf-3 were recognized as displaying a forward-stepping pattern that represents base-level fall; Kf-4 displays a vertically-stacked pattern representing the turnaround and beginning of base-level rise; and Kf-5 through Kf-7 display a backward-stepping pattern representing a sustained period of base-level rise. A smaller level of cyclicity, that is essentially equivalent in scale to the "parasequence" of Van Wagoner (1995), was also distinguished. Gardner's thesis and guidebooks include numerous, very well documented examples of this level of cyclicity (Gardner, 1992, 1993; Gardner et al., 1994).

Gardner, like previous students of the Ferron, recognized that the Ferron includes deposits of fluvial-dominated deltas. He noted a pattern, however, that previous workers had not described: the shoreline sandstones of Kf-1 through Kf-3 consist largely of fluvial-dominated deltaic deposits, whereas those of Kf-4 through Kf-7 consist predominantly or even entirely of wave-dominated deltaic deposits. One of Gardner's innovative contributions to understanding the Ferron is his analysis of this phenomenon in terms of volumetric partitioning of sediment between various paleoenvironments. Figure 28 summarizes the concept of volumetric partitioning. Volumetric partitioning offers an explanation for the distribution of fluvial- and wave-dominated delta-front deposits in the Ferron. Stated in the simplest possible terms, Gardner's explanation for the distribution of fluvial- and wave-dominated deltaic deposits is as follows:

During the lowering of base level associated with deposition of Kf-1 through Kf-3, most of the sediment was conveyed through the fluvial systems to the shoreline. Because of the high rate of delivery of sediments at the river mouths, wave energy was insufficient to cause major reworking of the deltaic deposits which, as a result, took on classic lobate to bird-foot, fluvial-dominated forms. Slow aggradation of sediment in the coastal plain behind the shoreline led to accumulation of relatively thin, widespread, largely single-storied channelbelt deposits.

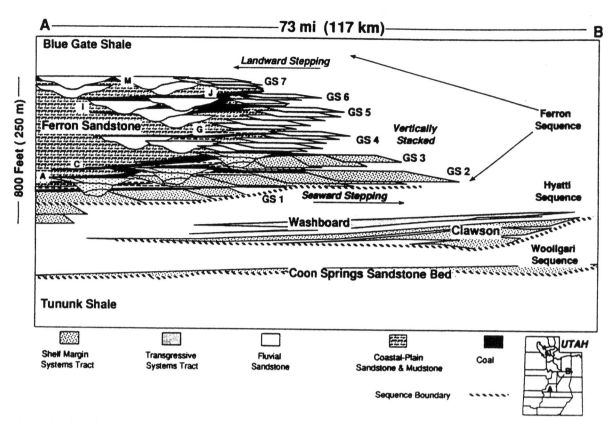

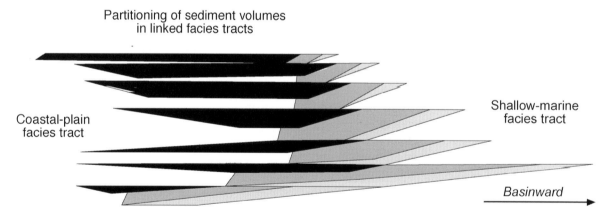

Figure 27. Gardner (1993) interpreted the shoreline sandstones of Kf-1 through Kf-4 (GS in this figure stands for "genetic sequence") as deposits of fluvial-dominated deltas and those of Kf-5 through Kf-7 as deposits of wave-dominated deltas (from Gardner et al., 1994).

Figure 28. Diagram showing the partitioning of sediment in various facies tracts of the Ferron. Rate of rise of relative sea level increased through Ferron deposition, resulting in increasing volumes of sediment deposition in coastal-plain facies tracts and decreasing volumes of sediment deposition in shallow-marine facies tracts (from Gardner et al., 1994).

During the rising of base level associated with deposition of Kf-4 through Kf-7, rivers aggraded the coastal plain more rapidly. More sediment was tied up in channelbelt deposits, which are thick and multistoried, and in the fine-grained deposits of the adjacent flood basins. The amount of sediment conveyed throughout the fluvial systems to the shoreline was reduced. Because of the lower rates of accumulation of sediment at the river mouths, wave energy was sufficient to rework the deltas, leading to cuspate, wave-dominated morphologies.

Gardner subsequently became an employee of the Bureau of Economic Geologists (BEG), The University of Texas at Austin. Based largely on Gardner's findings, the BEG identified the Ferron as a good analog for a number of important Neogene reservoirs in the Texas Gulf Coast productive province. The BEG, with funding from the Gas Research Institute (GRI) and DOE, and later with direct financial support from oil and gas companies, focused on the stratigraphy, sedimentology, and permeability structure of a number of stratigraphic units representing significant reservoir facies. Numerous geoscientists from the BEG contributed to the project. Barton, however, was the principal contributor of most of the new stratigraphic concepts. The initial object of study by the BEG was a sandy channelbelt deposit plus underlying delta-front strata of Kf-5 exposed in Muddy Creek Canyon. Study of Kf-5 was later extended northward to include all of the Kf-5 outcrops east and north of Muddy Creek Canyon. Additional case studies were initiated in deposits of Kf-2 at Ivie Creek and at Dry Wash. The final phase of the BEG study of the Ferron involved the detailed stratigraphy of the entire Ferron section between I-70 and Dry Wash.

Many results and conclusions of the BEG work can be found in various interim reports (Barton and Tyler, 1991), in reports published by the GRI (Fisher et al. 1993a, 1993b), in the Ph.D. dissertation of Barton (1994), and in unpublished field trip guidebooks (Barton and Tyler, 1995).

Barton, initially working jointly with Gardner and then independently, compiled very detailed stratigraphic information and large permeability data sets for a number of sites in the Ferron. Working at a more regional scale, Barton (1994) recognized two errors that had initially been made by Ryer and propagated by Gardner: (1) the seaward limit of Kf-1 occurs in the southern part of the Coal Cliffs north of I-70, not in the southern part of Molen Reef as previously mapped; and (2) the seaward limit of Kf-6 occurs just north of Dry Wash, close to the seaward limit of underlying Kf-5, farther northeastward than previously thought. Barton came to the realization that unit Kf-2, as previously defined, constitutes

more than one "4th order" or "short-term" cycle of sedimentation. The details of one of Barton's (1994) subdivisions of Kf-2 is summarized in Figure 29.

The report by Barton and Tyler (1995) includes a new and fundamentally different interpretation of the origins of parasequence-level units in the Ferron. Like many other workers, they interpreted parasequences as products of fluctuations in relative sea level. Ryer and Gardner had done the same, but had concluded that rates of lowering of sea level had always been exceeded by the overall rate of basin subsidence. Interpreted this way, eustatic fluctuations are represented by varying rates of relative sea-level rise: falling eustatic sea level results in a slower rate of relative sea-level rise and progradation; rising eustatic sea level results in a more rapid rate of relative sea-level rise and transgression. Barton and Tyler (1995) argued that the Ferron preserves evidence of relative sea-level lowering based on relationships within the lower, forward-stepping part of the Ferron deltaic complex. Following a period of progradation under conditions of rising relative sea level, in their interpretation, sea level fell, leading to incision of channel systems and seaward shift of the shoreline to a more distal and lower position. Subsequent rise of sea level led to upward and then forward building of the shoreline, formation and filling of a bay/lagoon complex, and filling of valleys. One of the most important implications of this interpretation is that rivers incised into pre-existing sediments during periods of relative sea-level fall to form valleys: many of the deposits, previously interpreted as simple channelbelts laid down on an aggrading alluvial plain or delta plain were reinterpreted as valley fills.

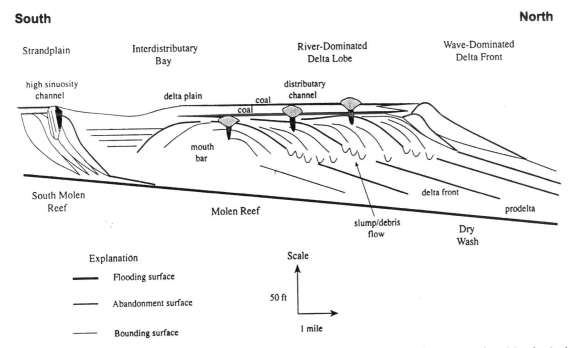

Figure 29. Detailed outcrop work and analysis of photomosaics allowed Barton (1994) to distinguish numerous depositional episodes, which he equated to parasequences, within Kf-2 between Miller Canyon and Dry Wash (after Barton, 1994).

Beginning in 1993, Utah Geological Survey (UGS) began an extensive study of the Ferron Sandstone from Last Chance Creek to Ferron Creek. The objective of the UGS project was to develop a comprehensive, quantitative characterization of the Ferron as a fluvial-deltaic reservoir analog. The study included reservoir-scale models and reservoir simulations (see Forster et al., Jarrard et al., and Mattson and Chan, in this volume). This project also involved regional stratigraphic and facies analysis and case-studies (see Anderson and Ryer, Anderson et al., Dewey and Morris, Ryer and Anderson, in this volume). Other results and conclusions of the UGS project can be found in various annual and open-file reports published by the DOE and UGS (Chidsey and Allison, 1996; Chidsey, 1997; Anderson et al., 1997a; Anderson et al., 1997b; Chidsey et al., 1998; Chidsey, 2001; Anderson et al., 2002).

An interpretation of Ferron stratigraphy published by Schwans (1995) is radically different than anything that preceded it. Schwans distinguished a major unconformity — a sequence boundary — on outcrops of the Ferron at I-70. His unconformity lies near the base of Kf-1, such that the Ferron exposed there lies with an erosional, unconformable contact on the Tununk Shale. Carried into the subsurface using well control, Schwans' unconformity defines a large, east-trending paleovalley. In general terms, Schwans' model calls for a pronounced drop of relative sea level in central Utah during middle Turonian time, an eastward shift in the position of the shoreline to the present-day San Rafael uplift, and fluvial erosion of a broad valley. During a subsequent rise of sea level, the valley flooded to form a large estuary within which the bulk of the Ferron, as exposed in Castle Valley, accumulated.

There is much disagreement with Schwans' interpretation and to fully critique it would require a lengthy discussion. In short, it can be argued that Schwans force fitted a sequence stratigraphic model that, despite its merits, cannot be properly applied to the Ferron. Particularly revealing in understanding the origin of this interpretation is the fact that the list of references cited in the paper (Schwans, 1995) does not include a single one of the studies of the Ferron that have been described here. It includes, instead, references to all of the widely recognized papers defining the principles of sequence stratigraphy, particularly the numerous papers published on the Blackhawk Formation in the nearby Book Cliffs. The implications of this are clear: only the standard sequence stratigraphic model can be correct; any previous studies of the Ferron are meaningless. Schwans transported the Blackhawk model to the Ferron without adequately studying the Ferron strata. This work epitomizes what the author regards as a period of chaos in the study of the Ferron, as discussed below.

Initial results of an on-going Ferron study conducted by Garrison and van den Bergh have been published in extended abstract form (Garrison and van den Bergh, 1996; van den Bergh and Garrison, 1996) and in a field trip guidebook (Garrison et al., 1997). Their project shared one of the same goals as the BEG and UGS studies: to divide the Ferron into genetic units at the level of the parasequence.

Figure 30 shows Garrison and van den Bergh's stratigraphic framework. It distinguishes four sequences within the Ferron consisting of 33 parasequences organized into 11 parasequence sets. There are some subtle, but important, differences between this and previous correlations schemes. Three unconformities (sequence boundaries) identified within the Ferron are defined primarily on the basis of erosional relief that is present at the bases of channelbelts. The relief on the contacts, apparently, is attributed to subaerial erosion and development of topography, rather than to simple local variation in the amount of scouring that occurred during emplacement of the channelbelts.

Another new aspect of Garrison and van den Bergh's interpretation is the arrangement of parasequences in Kf-7 and Kf-8 (Figure 30). In each, younger parasequences are positioned landward of older ones such that the parasequence sets are back-stepping or retrogradational. Previous workers who studied these units did not recognize the internal organization described by Garrison and van den Bergh.

Sequence Boundaries and Chaos in Ferron Stratigraphy

The last decade has seen a fundamental change in how geologists look at the Ferron: not just in how they go about looking at it, but also in what kinds of things they see when they look. This change is a direct result of the rise and widespread acceptance of sequence stratigraphy.

From the early work of Lupton (1916) through the fleshing-out of the details of Ferron stratigraphy provided by Gardner (1993) and Barton (1994), each student of the Ferron essentially built upon the work of his or her predecessors. Each publication provided new details and more refined interpretations. To be sure, there were misinterpretations that were subsequently set right, but overall, there was a smooth progression forward. During the 1990s, an increasing number of geologists approached study of the Ferron with a different question in mind: How does what we see in the Ferron fit the tenets of sequence stratigraphy? Many geologists have looked at the Ferron in this new way, although only a few have fully documented their conclusions in publications. The new way of looking at things has led to a very different result: each worker sees something different than the other; no two are in agreement.

Gardner (1993, 1995a) made the first well-documented effort to define the sequence stratigraphy of the Ferron. An essential conclusion of his analysis is that no

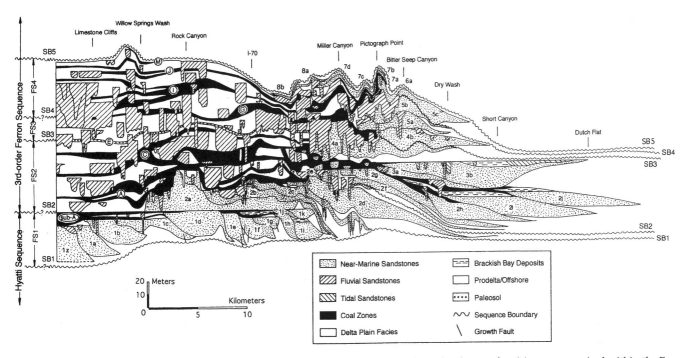

Figure 30. Ferron stratigraphic framework from Garrison et al. (1997). Three subaerial, erosional unconformities are recognized within the Ferron. They, plus two more undefined unconformities in the underlying Tununk Shale and overlying Blue Gate Shale Members, are interpreted to be sequence boundaries that define four sequences within the Ferron. Numerous parasequences are distinguished in the figure. They are grouped into 11 parasequence sets.

unconformities exist in the Ferron. The three sequences he distinguished in the Turonian-Coniacian section of central Utah are, instead, based on surfaces that he believes are correlative to unconformities defined in other parts of the Western Interior (e.g. "correlative conformities"). All of the deltaic, upper Ferron strata that have been the subject of recent outcrop studies belong to the youngest of Gardner's sequences (his Ferronensis sequence). The correlative conformity that serves as its base coincides with the top of Kf-Washboard (the top of the lower Ferron). The upper boundary of the sequence remains undefined.

Seven sequence boundaries that correspond to unconformities, listed here in descending stratigraphic order, have been identified recently in the deltaic strata of the upper Ferron:

1. Garrison and van den Bergh (1996) distinguished an unconformity in Kf-5 (their SB 4).
2. Garrison and van den Bergh (1996) distinguished an unconformity in Kf-4 (their SB 3).
3. Garrison and van den Bergh (1996) distinguished an unconformity in the seaward part of Kf-2 (their SB 2).
4. van den Bergh and Sprague (1995) recognized an unconformity in the more landward part of Kf-2 in the area south of Coyote Basin (north of Willow Springs Wash).
5. Barton and Tyler (1995) distinguished several minor erosional surfaces that they believe are characterized

by subaerial erosion in Kf-2. Though minor, they must still be considered sequence boundaries.
6. Shanley and McCabe (1991, 1995) extend an unconformity they recognize in the Straight Cliffs Formation of the Kaiparowits Plateau to a position somewhere low in the Ferron in the Castle Valley outcrops. Their initial interpretation (verbal communication, 1998) is that it corresponds to the top of Kf-1 at Ivie Creek.
7. Schwans (1995) defined a sequence boundary near the base of Kf-1 at Ivie Creek and subsequently correlated it over a broad area using wireline logs.

Thus, seven studies identified seven unconformities, no two of which are the same. There has been a total lack of agreement among those seeking to distinguish unconformity bounded sequences within the Ferron. Does this mean that no unconformities exist in the Ferron? Certainly not. It does, however, indicate that if unconformities are present, they are quite subtle. Study of the Ferron will certainly continue and perhaps a consensus will be reached.

COAL IN THE FERRON SANDSTONE

Coal has long played an important role in the economy of central Utah. The coalbeds of the Ferron have always been of minor importance compared to the coals of the Blackhawk Formation in the nearby Wasatch Plateau, but nonetheless have been important as a source of coal for local use. Only one mine, Consolidation Coal Company's Emery Mine, which was active from the mid-1970s to the mid-1990s, has produced coal on a large, commercial scale from the Ferron (8.5 million tons). The following is a brief review of the history of study of coal in the Ferron. Not included are studies addressing geochemical aspects of Ferron coal.

The first written documentation of coal in Castle Valley occurs in a report issued by the Wheeler Survey. R.L. Hoxie of the Army Corps of Engineers recorded in his journal of October 11, 1853, that "specimens of coal were brought in from the hills near the camp." Lupton interpreted the location of the camp as being near the head of the canyon of Muddy Creek. The J coalbed (Lupton correlated it to his I seam, defined farther south) is conspicuously exposed at many localities along the canyon walls in this area. It is almost certain that Hoxie collected the samples he described from this seam.

Forrester (1892) collected samples from coalbeds in Ivie Creek and Quitchupah canyons and published proximate analysis. It is likely that his samples came from the I coalbed and from either the A or C coal beds of Lupton's (1916) terminology. Forrester interpreted the coal-bearing strata as belonging to the Montana Group (that is, the Mesaverde Group), dropped into this position several miles east of the main Mesaverde outcrop belt by faulting. The Joe's Valley-Paradise graben lies between the Ferron and Mesaverde outcrop belts in the area south of Emery, although it is not apparent that the structure was recognized by Forrester. If it was, it would have lent credibility to his interpretation. It is likely, however, that a more important factor was the presence of coal. The presence of coal was known in Mesaverde Group strata throughout much of the Colorado Plateau, whereas strata between the base of the Mesaverde and the top of the Dakota are generally shaley and lack coal. That the coal-bearing strata at Ivie Creek and Quitchupah Canyons do not belong to the Dakota is obvious to anyone visiting the southern Castle Valley outcrops. It is likely that Forrester simply assumed any coal younger than the Dakota must belong to the Mesaverde and devised a structural interpretation to fit his preconception.

Taft (1906) briefly examined some of the Ferron outcrops near Emery. He reached the conclusion that "... coal, which is of little economic importance, is known to occur only to the east and southeast of Emery." His statement that coal is unimportant in the Emery area is somewhat surprising inasmuch as his map shows the locations of the Casper and Bear Gulch mines. Taft presumably visited these mines, both of which exploited sections of the C coal bed that exceed 6 ft (2 m) in thickness. Furthermore, he collected a sample of coal from what is now known as the Cowboy mine, south of Emery, where the I coalbed is 5–6 ft (1.5–2 m) thick.

Lupton's (1916) study of the coal geology and coal resources of Castle Valley stands as the first of two primary reference on coal in the Ferron. Lupton conducted his field work during two periods: July 17 through October 7, 1911, and September 9 through November 2, 1912. This represents a total of 138 days. His field party consisted of himself plus five other men, several of whom apparently were locals. One cannot help but be impressed by the amount of work completed, particularly given the poor condition of roads in the area at that time and the fact that the party lacked adequate topographic maps and had to survey most of their locations.

Lupton (1916) named the significant coal beds of the Ferron, assigning them letters of the alphabet, in ascending stratigraphic order. The thickest and, therefore, economically most important coal beds are the A, C, and I. The C and I coal beds were being mined at the time of Lupton's visits and he had encountered a prospect dug into the A coal along Quitchupah Creek. Based on his many outcrop measurements of coal thicknesses, Lupton (1916) estimated a total coal resource of 1.43 trillion tons $(1.3 \times 10^{12}$ Mg$)$.

Lupton's report includes some minor miscorrelations of coal beds. The most significant miscorrelation involves the A coal, stratigraphically the lowest seam designated. Another coal zone, older than Lupton's A coal, exists in the middle and southern parts of the coal field, but this fact was unrecognized (this lowest coal zone now has the awkward designation "Sub-A"). Coal of this zone in the Last Chance Creek and Paradise Lake areas were miscorrelated with the A coal, the miscorrelation occurring in the rugged outcrops of the Limestone Cliffs south of Willow Springs Wash.

Lupton documented the fact that coal is a minor component of the Ferron on outcrops north of a line extending east from Emery and becomes absent entirely a short distance north of Dry Wash as the Ferron becomes increasingly dominated by marine facies. Although coal is absent on many of the Ferron outcrops in the middle and northern parts of Castle Valley, it does occur in the Ferron to the west, in the subsurface. The presence of Ferron coal in the Price area was noted by Spieker and Reeside (1925). They reported that drilling along the Rio Grande Railroad about 10 mi (16 km) west of the Ferron outcrops indicated about 200 ft (61 m) of predominantly sandy Ferron section containing several coal beds, one of them 7.5 ft (2.3 m) thick. Additional data about the thicknesses and stratigraphic positions of the coals near Price were published by Gray et al. (1966). Situated at depths too great for economic mining, these coal beds would later become the focus of an active program of coalbed methane development.

Doelling (1972) published an analysis of the Emery coal field as part of his comprehensive treatise on coal in Utah. His is the second principal work on Ferron coal. Although it contributes little data on the thicknesses and areal extents of coal seams beyond what can be found in Lupton (1916), it does present a much broader picture of Ferron coal, including considerable information about coal quality. Doelling (1972) confirmed Lupton's estimate of coal resources, and in addition distinguished measured, indicated, inferred, and potential reserves. Measured and indicated reserves total 758 million tons (688 Mg). Color maps corresponding to U.S. Geological Survey 7.5 minute quadrangles (although with different names) portray the surface geology at a scale of approximately 1: 42,000.

Ryer (1981a) placed the principal coalbeds of the Ferron into their depositional setting and recognized an important genetic relationship between coals and their equivalent shoreline sandstone units: the greatest thicknesses of coal are present at and landward of the landward pinchouts of the shoreline sandstones (Figure 14). Thick coal tends to occur in pods whose long axes are perpendicular to the shoreline trend. These pods represent flood basin and swamp areas bounded along depositional strike by sand-rich channelbelt deposits. Ryer et al. (1980) used laterally continuous beds of altered volcanic ash to document these relationships in the C coal bed of the Ferron (Figures 31 and 32). The details of the physical and chemical properties of the altered volcanic ash layers are addressed by Triplehorn and Bohor (1981) and Bohor and Triplehorn (1993).

Coal in the Ferron in the subsurface near Price was analyzed by Bunnell and Hollberg (1991). Their work predated the development of coalbed methane in this area and their interest in the coal was from the perspective of a mineable resource, rather than in its contained methane. They utilized the model proposed by Ryer (1981a) to explain the thickness distributions of the four to five coal zones recognized there. The Ferron in the area they studied is about 200 ft (60 m) thick. The shoreline orientation was north-northeast to northeast. Of particular interest is the strongly "back-stepping" or "retrogradational" pattern that the shoreline sandstone bodies of the Ferron display (Figure 33). As is the case on outcrops of the upper Ferron farther south in Castle Valley, maximum thicknesses of coal occur in the vicinities of landward pinchouts (Figure 34).

PETROLEUM IN THE FERRON SANDSTONE

The Ferron Sandstone has been a target of oil and gas exploration since 1921, when the Phillips No. 1 Huntington well was completed as a dry hole on the Huntington anticline, located in Castle Valley near the town of Huntington (Kuehnert, 1954). The first significant shows of oil in the Cretaceous section were reported in the Dakota Sandstone in a well drilled at Gordon Creek in 1922. The first commercial success occurred on

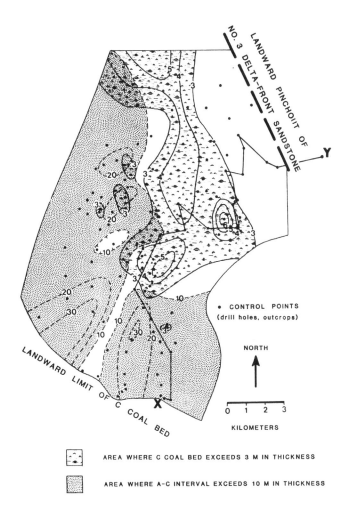

Figure 31. Comparison of thickness of the C coalbed and of laterally equivalent fluvial deposits. There is an inverse relationship, peat deposition having been suppressed in areas of fluvial aggradation (from Ryer et al., 1980).

the Clear Creek anticline, where Byrd-Front, Inc. discovered gas in sandstone reservoirs of the Ferron in 1951 (Edson et al., 1954; Tripp, 1991a, 1993a) (Figure 35). Gas was subsequently discovered in sandstones of the Ferron at Flat Canyon anticline (Flat Canyon field) in 1953 (Tripp, 1993b), at Joes Valley in 1953 (Tripp, 1993c), at Ferron anticline in 1957 (Tripp, 1990a, 1990b, 1991b, 1993a) and Indian Creek/East Mountain field, now part of Flat Canyon field, in 1981 (Laine and Staley, 1991) (Figure 35). The Ferron was found to contain gas and extremely light oil (56° API) on the southern part of the Flat Canyon field, this constituting the only significant occurrence of producible liquids from the unit (Laine and Staley, 1991; Sprinkel, 1993).

Of these discoveries, Clear Creek field is by far the most important. Clear Creek field has produced over 114 bcf (3.2 billion m³) of gas, far surpassing the 11.7 bcf (331 million m³) of gas produced from Ferron field, the second in importance (Utah Division of Oil, Gas and Mining, 2003). Flat Canyon and Joes Valley fields have pro-

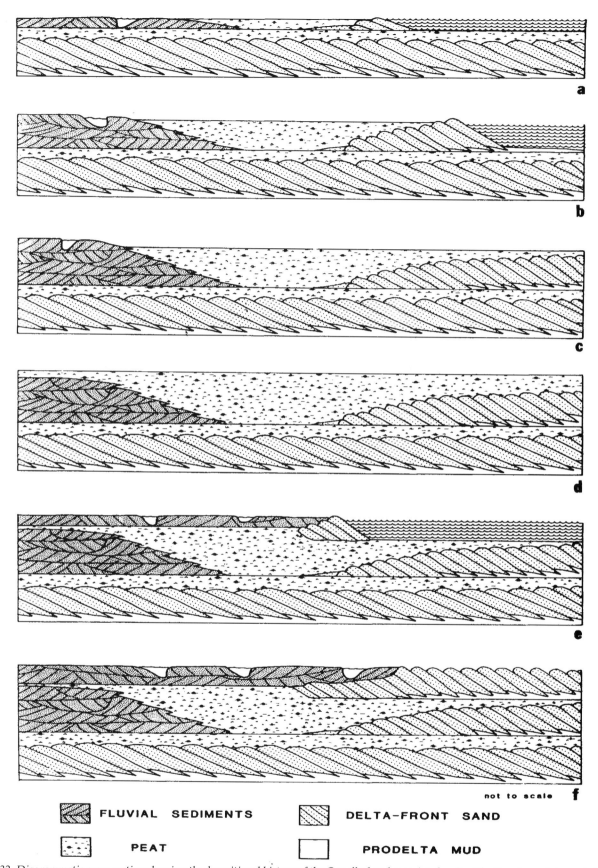

FLUVIAL SEDIMENTS   DELTA-FRONT SAND

PEAT   PRODELTA MUD

Figure 32. Diagrammatic cross section showing the depositional history of the C coalbed and associated units. The C coal was interpreted to have formed entirely during progradation of Kf-3, an interpretation that must be revised following discovery of the fact that the Kf-3 pinches out landward into the C coal, such that its lower part formed during the latest part of deposition of Kf-2 and perhaps partly during the transgression that ended Kf-2 (from Ryer et al., 1980.)

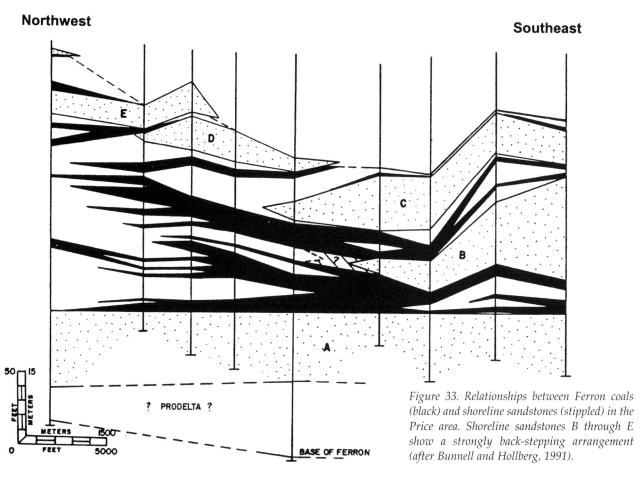

Northwest

Southeast

? PRODELTA ?

BASE OF FERRON

Figure 33. Relationships between Ferron coals (black) and shoreline sandstones (stippled) in the Price area. Shoreline sandstones B through E show a strongly back-stepping arrangement (after Bunnell and Hollberg, 1991).

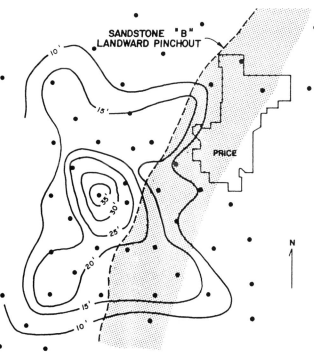

SANDSTONE "B" LANDWARD PINCHOUT

PRICE

Figure 34. Bunnell and Hollberg (1991) demonstrated that thick pods of coal are associated with the landward pinchouts of shoreline sandstone units in the subsurface near Price, as is the case on the Ferron outcrops in Castle Valley (from Bunnell and Hollberg, 1991).

duced 9.7 bcf (275 million m³) and 3 bcf (85 million m³), respectively, from commingled Ferron and Dakota reservoirs (Utah Division of Oil, Gas and Mining, 2003). Cumulative oil production from Flat Canyon field is 16,686 bbl (2653 m³) (Utah Division of Oil, Gas and Mining, 2003).

The discovery of Clear Creek field played a significant role in stimulating study of the Ferron. As discussed earlier, the Ferron at Clear Creek was found to include a basal sandstone unit characterized by good lateral continuity, overlain by a thicker section containing discontinuous sandstones. The difference in these two sandstone types was of considerable interest. The dissertation by Katich (1951), completed in the same year the field was discovered, provided a picture of the Ferron stratigraphy on outcrop. The discovery provided an impetus for studies integrating the outcrop and basic subsurface observations by Katich (1954) and Davis (1954).

Walton, who was instrumental in the discovery of Clear Creek (Edson et al., 1954) summarized the petroleum geology of the Wasatch Plateau and Castle Valley in papers published in 1954 and 1955. Numerous papers published by Tripp (1990a, 1990b, 1991a, 1991b, 1993a, 1993b, and 1993c), Laine and Staley (1991), and Sprinkel (1993) provide a more recent review of Ferron oil and gas exploration and production.

31

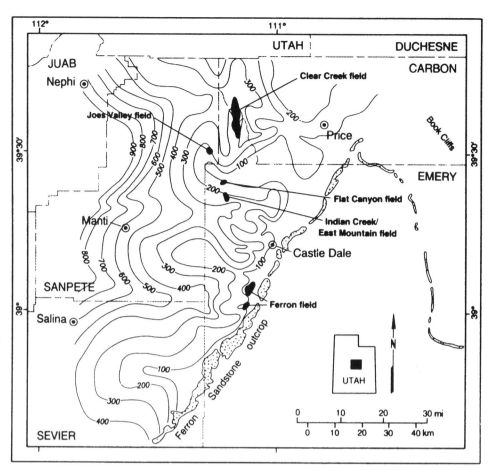

Figure 35. Location of gas fields (solid black) productive from the Ferron Sandstone and net sandstone-thickness map, Wasatch Plateau/Castle Valley. Contour interval is 100 ft. Thick lobes of sandstone indicate areas of deltaic deposition (from Sprinkel, 1993; modified from Tripp, 1989).

COALBED METHANE DEVELOPMENT IN THE FERRON SANDSTONE

Coal miners long knew that methane in coalbeds posed a serious risk of explosion, and the U.S. Bureau of Mines began research on this problem when it was created in 1910 (Irani et al., 1977). Early research by the U.S. Bureau of Mines focused on quantifying the amount of methane released in active coal mines to alleviate the health and safety problems created by the gas (Diamond and Levine, 1981). In the late 1970s, following two interruptions in petroleum supply from the Middle East, the DOE and the GRI began researching new and unconventional sources of natural gas. Based on the U.S. Bureau of Mines measurements of large volumes of methane in coal beds, the DOE and the GRI sponsored studies to quantify and develop techniques to economically produce gas from coal beds. Preliminary coalbed gas resource estimates by the DOE were completed for all major coal basins in the United States by 1982. These early studies focused subsequent research and development efforts on five basins: Black Warrior, central and northern Appalachians, Piceance, and San Juan (ICF, 1990). Utah's coalbed gas potential was discounted because little was known of the deep coal resources.

Early coalbed gas work in Utah, during the late 1970s and early 1980s, consisted of UGS studies to mea-

sure the gas content of coals in the major Utah coal fields (Davis and Doelling, 1976, 1977; Doelling et al., 1979; Smith, 1981; Smith, 1986; Keith et al., 1990a, 1990b, 1990c, 1990d; and Keith et al., 1991), and a few individual production test wells drilled in the Book Cliffs coal field by Mountain Fuel (Allred and Coates, 1980, 1982). Studies in the 1990s by the UGS helped define the structure, thickness, depth, extent, and maturity of the coal deposits, and the coalbed gas resources of the Ferron Sandstone (Figure 36) (Tripp, 1989; Gloyn and Sommer, 1993; Sommer et al., 1993; Tabet and Burns, 1996; Tabet et al., 1996).

The Ferron coalbed methane play (traps, reservoir characteristics, fields, production, and reserves) is described by Lamarre, Montgomery et al., and Klein et al., in this volume.

ACKNOWLEDGMENTS

This paper was supported by the Utah Geological Survey, contract number 94-2488, as part of a project entitled *Geological and Petrophysical Characterization of the Ferron Sandstone for 3-D Simulation of a Fluvial-Deltaic Reservoir*, M. Lee Allison and Thomas C. Chidsey, Jr., Principal Investigators. The project was funded by the U.S. Department of Energy (DOE) under the Geoscience/Engineering Reservoir Characterization Pro-

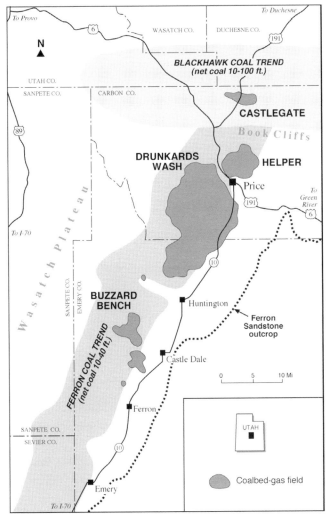

Figure 36. Location of fields producing coalbed methane in east-central Utah and the Ferron Sandstone coalbed gas "fairway" Carbon, Emery, Sanpete, and Sevier Counties, Utah (modified from Tabet et al., 1996).

gram of the DOE National Petroleum Technology Office, Tulsa, Oklahoma, contract number DE-AC22-93BC 14896. The Contracting Office Representative was Robert Lemmon.

I thank David E. Tabet and Craig D. Morgan, Utah Geological Survey, and Mary L. McPherson, McPherson Geologic Consulting, Grand Junction, Colorado, for their careful reviews and constructive criticisms of the manuscript. Special thanks are expressed to Thomas C. Chidsey, Jr. for his help and the help of his staff at the UGS in compiling the final version of the manuscript.

REFERENCES CITED

Allred, L. D., and R. L. Coates, 1980, Methane recovery from deep unminable coal seams, *in* Proceedings of the first annual symposium on unconventional gas recovery: Society of Petroleum Engineers Reprint 8962, p. 307–312.

—1982, Methane recovery from deep coalbeds in the Book Cliffs, *in* Proceedings of the 1982 SPE/DOE unconventional gas recovery symposium: Society of Petroleum Engineers Reprint 10819, p. 245–252.

Anderson, P. B., 1991a, Landward pinch-out of Cretaceous marine nearshore clastics in the Ferron Sandstone Member of the Mancos Shale and Blackhawk Formation, east-central Utah; potential stratigraphic traps: Utah Geological Survey Contract Report 91–12, 110 p.

—1991b, Comparison of Cretaceous landward pinch-outs of nearshore sandstones: wave- versus river-dominated deltas, east-central Utah, *in* T. C. Chidsey, Jr., ed., Geology of east-central Utah: Utah Geological Association Publication 19, p. 283–300.

Anderson, P. B., and T. A. Ryer, 1995, Proposed revisions to parasequence-set nomenclature of the Upper Cretaceous Ferron Sandstone Member of the Mancos Shale, central Utah (abs.): AAPG Bulletin, v. 79, no. 6, p. 914–915.

Anderson, P. B., T. C. Chidsey, Jr., and Kevin McClure, 1997a, Ferron Sandstone reservoir type logs and lithofacies, Emery County, Utah: Utah Geological Survey Open-File Report 358, 2 pl.

Anderson, P. B., T. C. Chidsey, Jr., and T. A. Ryer, 1997b, Fluvial-deltaic sedimentation and stratigraphy of the Ferron Sandstone, *in* P. K. Link and B. J. Kowallis, eds., Mesozoic to Recent geology of Utah: Brigham Young University Geology Studies, v. 42, pt. 2, p. 135–154.

Anderson, P. B., Kevin, McClure, T. C. Chidsey, Jr., T. A. Ryer, T. H. Morris, J. A. Dewey, Jr., and R. D. Adams, 2002, Photomosaics and measured sections from the Cretaceous Ferron Sandstone Member of the Mancos Shale, Emery and Sevier Counties, east-central Utah: Utah Geological Survey Open-File Report (in press), compact disk.

Barton, M. D., 1994, Outcrop characterization of architecture and permeability structure in fluvial-deltaic sandstones, Cretaceous Ferron Sandstone, Utah: Ph.D. dissertation, University of Texas, Austin, 255 p.

Barton, M. D., and Noel Tyler, 1991, Quantification of permeability structure in distributary channel deposits Ferron Sandstone, Utah, *in* T. C. Chidsey, Jr., ed., Geology of east-central Utah: Utah Geological Association Publication 19, p. 273–282.

—1995, Sequence stratigraphy, facies architecture, and permeability structure of fluvial-deltaic reservoir analogs — Cretaceous Ferron Sandstone, central Utah: University of Texas, Bureau of Economic Geology, Field Trip Guidebook, August 21–23, 1995, 80 p.

Bohor, B. F., and D. M. Triplehorn, 1993, Tonsteins — altered volcanic-ash layers in coal-bearing sequences: Geological Society of America Special Paper 285, 44 p.

Bunnell, M. D., and R. J. Hollberg, 1991, Coal beds of the Ferron Sandstone Member in northern Castle Valley, east-central Utah, *in* T. C. Chidsey, Jr., ed., Geology of east-central Utah: Utah Geological Association Publication 19, p. 157–172.

Chidsey, T. C., Jr., 1997, Geological and petrophysical characterization of the Ferron Sandstone for 3-D simulation of a fluvial-deltaic reservoir — annual report for the period October 1, 1995, to September 30, 1996: U.S. Department of Energy, DOE/BC/14896-15, 57 p.

Chidsey, T. C., Jr., editor and compiler, 2001, Geological and petrophysical characterization of the Ferron Sandstone for 3-D simulation of a fluvial-deltaic reservoir — final report: U.S. Department of Energy, DOE/BC/14896-24, 471 p.

Chidsey, T. C., Jr., and Allison, M. L., 1996, Geological and petrophysical characterization of the Ferron Sandstone for 3-D simulation of a fluvial-deltaic reservoir — annual report for the period October 1, 1994 to September 30, 1995: U.S. Department of Energy, DOE/BC/14896-13, 104 p.

Chidsey, T. C., Jr., P. B. Anderson, T. H. Morris, J. A. Dewey, Jr., Ann Mattson, C. B. Forster, S. H. Snelgrove, and T. A. Ryer, 1998, Geological and petrophysical characterization of the Ferron Sandstone for 3-D simulation of a fluvial-deltaic reservoir — annual report for the period October 1, 1996, to September 30, 1997: U.S. Department of Energy, DOE/BC/14896-22, 69 p.

Coleman, J. M., H. H. Roberts, S. P. Murray, and M. Salama, 1981, Morphology and dynamic sedimentology of the eastern Nile delta shelf: Marine Geology, v. 42, p. 301–326.

Cotter, Edward, 1971, Paleoflow characteristics of a Late Cretaceous river in Utah from analysis of sedimentary structures in the Ferron Sandstone: Journal of Sedimentary Petrology, v. 41, no. 1, p. 129–138.

—1975a, Deltaic deposits in the Upper Cretaceous Ferron Sandstone, Utah, *in* M. L. S. Broussard, ed., Deltas, models for exploration: Houston Geological Society, p. 471–484.

—1975b, Late Cretaceous sedimentation in a low-energy coastal zone: the Ferron Sandstone of Utah: Journal of Sedimentary Petrology, v. 45, p. 669–685.

—1976, The role of deltas in the evolution of the Ferron Sandstone and its coals: Brigham Young University Geology Studies, v. 22, pt. 3, p. 15–41.

Cross, C. W., and A. C. Spencer, 1899, Description of the La Plata quadrangle [Colorado]: U.S. Geological Survey Geologic Atlas of the United States, Folio 60.

Davis, L. J., 1954, Stratigraphy of the Ferron Sandstone, *in* A. W. Grier, ed., Geology of portions of the high plateaus and adjacent canyon lands, central and south-central Utah: Intermountain Association of Petroleum Geologists, Fifth Annual Field Conference Guidebook, p. 55–58.

Davis, F. D., and H. H. Doelling, 1976, Methane content of Utah coals — first annual report to the U.S. Bureau of Mines: Utah Geological and Mineral Survey Open-File Report 23, 34 p.

—1977, Methane content of Utah coals: second annual report to the U.S. Bureau of Mines (Carbon County): Utah Geological and Mineral Survey Open-File Report 24, 26 p.

Diamond, W. P., and J. R. Levine, 1981, Direct method determination of the gas content of coal: procedures and results: U.S. Bureau of Mines Report of Investigations 8515, 36 p.

Doelling, H. H., 1972, Central Utah coal fields: Sevier-Sanpete, Wasatch Plateau, Book Cliffs and Emery: Utah Geological and Mineral Survey Monograph Series, no. 3, 570 p.

Doelling, H. H., A. D. Smith, and F. D. Davis, 1979, Methane content of Utah coals, *in* Martha Smith, ed., Coal studies: Utah Geological and Mineral Survey Special Study 49, p. 1–43.

Eaton, J. G., 1990, Stratigraphic revision of Campanian (Upper Cretaceous) rocks in the Henry Basin, Utah: Mountain Geologist, v. 27, no. 1, p. 27–38.

Edson, D. J., Jr., M. R. Scholl, Jr., and W. E. Zabriskie, 1954, Clear Creek gas field, central Utah: Intermountain Association of Petroleum Geologists Guidebook, 5th Annual Field Conference, p. 89–93.

Fisher, W. L., L. F. Brown, Jr., A. J. Scott, and J. H. McGowen, 1969, Delta systems in the exploration for oil and gas: a research colloquium: University of Texas Bureau of Economic Geology, 210 p.

Fisher, R. S., M. D. Barton, and Noel Tyler, 1993a, Quantifying reservoir heterogeneity through outcrop characterization — architecture, lithology, and permeability distribution of a seaward-stepping fluvial-deltaic sequence, Ferron Sandstone (Cretaceous), central Utah: Gas Research Institute Report 93-0022, 83 p.

—1993b, Quantifying reservoir heterogeneity through outcrop characterization — architecture, lithology, and permeability distribution of a landward-stepping fluvial-deltaic sequence, Ferron Sandstone (Cretaceous), central Utah: Gas Research Institute Report 93-0023, 132 p.

Forrester, Robert, 1892, The coal fields of Utah: U.S. Geological Survey Report on Mineral Resources, p. 511–520.

Galloway, W. E., 1975, Process framework for describing the morphologic and stratigraphic evolution of deltaic depositional systems, *in* M. L. S. Broussard, ed., Deltas, models for exploration: Houston Geological Society, p. 87–98.

Gardner, M. H., 1992, Volumetric partitioning and facies differentiation in Turonian strata of central Utah — field illustrations from the Ferron Sandstone of base level and sediment accommodation concepts: unpublished guidebook, 215 p.

—1993, Sequence stratigraphy and facies architecture of the Upper Cretaceous Ferron Sandstone Member of the Mancos Shale, east-central Utah: Ph.D. dissertation T-3975, Colorado School of Mines, Golden, 528 p.

—1995a, Tectonic and eustatic controls on the stratal architecture of mid-Cretaceous stratigraphic sequences, central Western Interior foreland basin of North America, *in* Stratigraphic evolution of foreland basins: Society for Sedimentary Geology (SEPM) Special Publication No. 52, p. 243–281.

—1995b, The stratigraphic hierarchy and tectonic history of the mid-Cretaceous foreland basin of central Utah, *in* Stratigraphic evolution of foreland basins: Society for Sedimentary Geology (SEPM) Special Publication No. 52, p. 284–303.

Gardner, M. H., M. D. Barton, and R. S. Fisher, 1994, Sequence stratigraphy, facies architecture, and permeability structure of fluvial-deltaic reservoir analogs — a field guide to selected outcrops of the Upper Cretaceous Ferron Sandstone, central Utah: unpublished guidebook prepared for 1994 AAPG Annual Meeting, Denver, 257 p.

Garrison, J. R., Jr., and T. C. V. van den Bergh, 1996, Coal zone stratigraphy — a new tool for high-resolution depositional sequence stratigraphy in near-marine fluvial-deltaic facies — a case study from the Ferron Sandstone, east-central Utah: AAPG Rocky Mountain Section Meeting, expanded abstracts volume, Montana Geological Society, p. 31–36.

Garrison, J. R., Jr., T. C. V. van den Bergh, C. E. Barker, and D. E. Tabet, 1997, Depositional sequence stratigraphy and architecture of the Cretaceous Ferron Sandstone: implications for coal and coal bed methane resources — a field excursion, *in* P. K. Link and B. J. Kowallis, eds., Mesozoic to Recent geology of Utah: Provo, Brigham Young University Geology Studies, v. 42, pt. 2, p. 155–202.

Gilbert, G. K., 1877, Report on the geology of the Henry Mountains: U.S. Geographical and Geological Survey Rocky Mountain Region Report, 160 p.

—1885, The topographic features of lake shores: U.S. Geological Survey, 5th Annual Report, p. 69–123.

Gloyn, R. W., and S. N. Sommer, 1993, Exploration for coalbed methane gains momentum in Uinta Basin: Oil and Gas Journal, May 1993, p. 73–76.

Gray, R. J., R. M. Patalski, and N. Schapiro, 1966, Correlation of coal deposits from central Utah, in Central Utah coals: Utah Geological and Mineralogical Survey Bulletin 80, p. 81–86.

Gustason, E. R., T. A. Ryer, and P. B. Anderson, 1993, Integration of outcrop and subsurface information for geological model building in fluvial-deltaic reservoirs: unpublished Utah Geological Survey/AAPG Rocky Mountain Section Core Workshop Manual, September 16, 1993, 24 p.

Hale, L. A., 1972, Depositional history of the Ferron Formation, central Utah, *in* J. L. Baer and Eugene Callaghan, eds., Plateau-basin and range transition zone: Utah Geological Association Publication No. 2, p. 115–138.

Hale, L. A., and R. F. Van DeGraff, 1964, Cretaceous stratigraphy and facies patterns — northeastern Utah and adjacent areas: Intermountain Association of Petroleum Geologists, 13th Annual Field Conference Guidebook, p. 115–138.

Hintze, L. F., and W. L. Stokes, 1964, Geologic map of southeastern Utah: Salt Lake City, University of Utah, College of Mines and Mineral Industries, scale 1:250,000.

Hoxie, R. L., 1874, Geologic map included in G. M. Wheeler, Annual Report for 1874: U.S. Geological Surveys West of the 100th Meridian, p. 5.

Hunt, C. B., 1946, Guidebook to the geology and geography of the Henry Mountains region, *in* Guidebook to the geology of Utah: Utah Geological Society Publication No. 1, 51 p.

ICF Resources, Incorporated, 1990, The United States coalbed methane resource: Gas Research Institute Quarterly Review of Methane from Coal Seams Technology, v. 7, no. 3, p. 10–28.

Irani, M. C., J. H. Jansky, P. W. Jeran, and G. L. Hassett, 1977, Methane emission from U.S. coal mines in 1975, a survey: U.S. Bureau of Mines Information Circular 8733, 55 p.

Katich, P. J., Jr., 1951, The stratigraphy and paleontology of the pre-Niobrara Upper Cretaceous rocks of Castle Valley, Utah: Ph.D. dissertation, Ohio State University, Columbus, 208 p.

—1953, Source direction of Ferron Sandstone in Utah: AAPG Bulletin, v. 37, no. 4, p. 858–862.

—1954, Cretaceous and Early Tertiary stratigraphy of central and south-central Utah with emphasis on the Wasatch Plateau area: Intermountain Association of Petroleum Geologists, 5th Annual Field Conference Guidebook, p. 42–54.

Keith, A. C., J. S. Hand, and A. D. Smith, 1990a, Coalbed methane resource map, Castlegate A bed, Book Cliffs coal field, Utah: Utah Geological and Mineral Survey Open-File Report 176 A, 1 plate, 1:100,000.

—1990b, Coalbed methane resource map, Castlegate B bed, Book Cliffs coal field, Utah: Utah Geological and Mineral Survey Open-File Report 176 B, 1 plate, 1:100,000.

—1990c, Coalbed methane resource map, Castlegate C bed, Book Cliffs coal field, Utah: Utah Geological and Mineral Survey Open-File Report 176 C, 1 plate, 1:100,000.

—1990d, Coalbed methane resource map, Gilson bed, Book Cliffs coal field, Utah: Utah Geological and Mineral Survey Open-File Report 176 D, 1 plate, 1:100,000.

—1991, Coalbed methane resource map, Castlegate D bed, Book Cliffs coal field, Utah: Utah Geological and Mineral Survey Open-File Report 176 E, 1 plate, 1:100,000.

Knight, L. L., 1954, A preliminary heavy mineral study of the Ferron Sandstone, Utah: Brigham Young University Research Studies, Geology Series, v. 1, no. 4, 31 p.

Kuehnert, H. A., 1954, Huntington anticline, Emery County, Utah: Intermountain Association of Petroleum Geologists, 5th Annual Field Conference Guidebook, p. 94–95.

Laine, M. D., and Don Staley, 1991, Summary of oil and gas exploration and production in Carbon, Emery, and Sanpete Counties, east-central Utah, in T. C. Chidsey, Jr., ed., Geology of east-central Utah: Utah Geological Association Publication 19, p. 227–235.

Lupton, C. T., 1914, Oil and gas near Green River, Grand County, Utah: U.S. Geological Survey Bulletin 541, p. 115–133.

—1916, Geology and coal resources of Castle Valley in Carbon, Emery, and Sevier Counties, Utah: U.S. Geological Survey Bulletin 628, 88 p.

Molenaar, C. M., and W. A. Cobban, 1991a, Middle Cretaceous stratigraphy on the south side of the Uinta Basin, east-central Utah, in T. C. Chidsey, Jr., ed., Geology of east-central Utah: Utah Geological Association Publication 19, p. 29–43.

—1991b, Middle Cretaceous stratigraphy on south and east sides of the Uinta Basin, northeastern Utah and northwestern Colorado: U.S. Geological Survey Bulletin 1787-P, 34 p.

Palmer, J. J., and A. J. Scott, 1984, Stacked shoreline and shelf sandstone of La Ventana Tongue (Campanian), northeastern New Mexico: AAPG Bulletin, v. 68, no. 1, pt. 1, p. 74–91.

Ray, R. M., J. P. Brashear, and K. Biglarbgi, 1991, Classification system targets unrecovered U.S. oil reserves: Oil and Gas Journal, Sept. 30, 1991, p. 89–93.

Ryer, T. A., 1980, Deltaic coals of the Ferron Sandstone Member of the Mancos Shale — predictive model for Cretaceous coals of the Western Interior, in L. M. Carter, ed., Proceedings of the Fourth Symposium on the geology of Rocky Mountain coal: Colorado Geological Survey Resource Series no. 10, p. 4–5.

—1981a, Deltaic coals of Ferron Sandstone Member of Mancos Shale — predictive model for Cretaceous coal-bearing strata of Western Interior: AAPG Bulletin, v. 65, no. 11, p. 2323–2340.

—1981b, The Muddy and Quitchupah projects — a project report with descriptions of cores of the I, J, and C coal beds from the Emery coal field, central Utah: U.S. Geological Survey Open-File Report 81–460, 34 p.

—1982a, Possible eustatic control on the location of Utah Cretaceous coal fields, in K. D. Gurgel, ed., Proceedings — 5th ROMOCO symposium on the geology of Rocky Mountain coal: Utah Geological and Mineralogical Survey Bulletin 118, p. 89–93.

—1982b, Cross section of the Ferron Sandstone Member of the Mancos Shale in the Emery coal field, Emery and Sevier Counties, central Utah: U.S. Geological Survey Map MF-1357, vertical scale: 1 inch = 30 meters, horizontal scale: 1 inch = 1.25 kilometers.

—1984, Transgressive-regressive cycles and the occurrence of coal in some Upper Cretaceous strata of Utah, U.S.A.: International Association Sedimentologists Special Publication No. 7, p. 217–227.

—1987, Distinguishing eustatic and tectonic contributions to major clastic wedges in Cretaceous foreland basin, Wyoming and Utah (abs.): Geological Society of Canada Annual Meeting, Program with Abstracts, v. 12, p. 86.

—1991, Stratigraphy, facies, and depositional history of the Ferron Sandstone in the canyon of Muddy Creek, east-central Utah, in T. C. Chidsey, Jr., ed., Geology of east-central Utah: Utah Geological Association Publication 19, p. 45–54.

—1993a, Speculations on the origins of mid-Cretaceous clastic wedges, central Rocky Mountain region, United States, in W. G. E. Caldwell, and E. G. Kauffman, eds., Evolution of the Western Interior Basin: Geological Association of Canada Special Paper 39, p. 189–198.

—1993b, The Ferron Sandstone of central Utah — an overview, in T. A. Ryer, E. R. Gustason, and P. B. Anderson, P. B., eds., Stratigraphy and facies architecture of the Ferron Sandstone Member of the Mancos Shale: Utah Geological Survey/AAPG Rocky Mountain Section unpublished field trip guidebook, 9 p.

—1993c, The Upper Cretaceous Ferron Sandstone of central Utah: an overview (abs.): AAPG Bulletin, v. 77, no. 8, p. 1459.

—1994, Interplay of tectonics, eustasy, and sedimentation in the formation of mid-Cretaceous clastic wedges, central and northern Rocky Mountain regions, *in* J. C. Dolson, M. L. Hendricks, and S. A. Wescott, eds., Unconformity-related hydrocarbons in sedimentary sequences: Denver, Rocky Mountain Association of Geologists, p. 35–44.

Ryer, T. A., and P. B. Anderson, 1995, Parasequence sets, parasequences, facies distributions, and depositional history of the Upper Cretaceous Ferron deltaic clastic wedge, Utah (abs.): AAPG Bulletin, v. 79, no. 6, p. 924.

Ryer, T. A., and J. R. Lovekin, 1986, The Upper Cretaceous Vernal delta of Utah — depositional or paleotectonic feature? *in* J. A. Peterson, ed., Paleotectonics and sedimentation in the Rocky Mountain region, United States: AAPG Memoir 41, p. 497–510.

Ryer, T. A., and Maureen McPhillips, 1983, Early Late Cretaceous paleogeography of east-central Utah, *in* M. W. Reynolds and E. D. Dolly, eds., Mesozoic paleogeography of the west-central United States: Society of Economic Paleontologists and Mineralogists, Rocky Mountain Paleography Symposium 2, p. 253–272.

Ryer, T. A., R. E Phillips, B. F. Bohor, and R. M. Pollastro, 1980, Use of altered volcanic ash falls in stratigraphic studies of coal-bearing sequences — an example from the Upper Cretaceous Ferron Sandstone Member of the Mancos Shale in central Utah: Geological Society of America Bulletin, v. 91, no. 10., pt 1, p. 579–586.

Said, Rushdi, 1981, The geological evolution of the River Nile: New York, Springer-Verlag, 151 p.

Scheihing, M. H., and G. C. Gaynor, 1991, The shelf sand-plume model — a critique: Sedimentology, v. 38, p. 433–444.

Schwans, Peter, 1995, Controls on sequence stacking and fluvial to shallow-marine architecture in a foreland basin, *in* J. C. Van Wagner and G. T Bertram, eds., Sequence stratigraphy of foreland basin deposits: AAPG Memoir 64, p. 55–102.

Shanley, K. W., and P.J. McCabe, 1991, Predicting facies architecture through sequence stratigraphy — an example from the Kaiparowits Plateau, Utah: Geology, v. 19, no. 7, p. 742–745.

—1995, Sequence stratigraphy of Turonian-Santonian strata, Kaiparowits Plateau, southern Utah, U.S.A. — implications for regional correlation and foreland basin evolution, *in* J. C. Van Wagoner and G. T. Bertram, eds., Sequence stratigraphy of foreland basin deposits: AAPG Memoir 64, p. 103–136.

Smith, A. D., 1981, Methane content of Utah coals — progress report 1979-1980: Utah Geological and Mineral Survey Open-File Report 28, 9 p.

—1986, Utah core desorption project , final report: Utah Geological and Mineral Survey Open-File Report 88, 59 p.

Sommer, S. N., H. H. Doelling, and R. W. Gloyn, 1993, Coalbed methane in Utah, *in* Atlas of major Rocky Mountain gas reservoirs: New Mexico Bureau of Mines and Mineral Resources, p. 167.

Spieker, E. M., 1946, Late Mesozoic and Early Cenozoic history of central Utah: U.S. Geological Survey Professional Paper 205-D, p. 117–160.

Spieker, E. M., and J. B. Reeside, Jr., 1925, Cretaceous and Tertiary formations of the Wasatch Plateau, Utah: Geological Society of America Bulletin, v. 36, no. 3, p. 435–454.

—1926, The Upper Cretaceous shoreline in Utah: Geological Society of America Bulletin, v. 37, no. 3, p. 429–438.

Sprinkel, D. A., 1993, Wasatch Plateau [WP] play — overview, *in* Atlas of major Rocky Mountain gas reservoirs: New Mexico Bureau of Mines and Mineral Resources, p. 89.

Tabet, D. E., and Terry Burns, 1996, Drunkards Wash, *in* B. G. Hill and S. R. Bereskin, eds., Oil and gas fields of Utah: Utah Geological Association Publication 22 (Addendum): non-paginated.

Taft, J. A., 1906, Book Cliffs coal field, Utah, west of Green River: U.S. Geological Survey Bulletin 285, p. 289–302.

Thompson, S. L., 1985, Ferron Sandstone Member of the Mancos Shale; a Turonian mixed-energy deltaic system: M.S. thesis, University of Texas, Austin, 165 p.

Thompson, S. L., C. R. Ossian, and A. J. Scott, 1986, Lithofacies, inferred processes, and log response characteristics of shelf and shoreface sandstones, Ferron Sandstone, central Utah, *in* T. F. Moslow and E. G. Rhodes, eds., Modern and ancient shelf clastics — a core workshop: Society for Sedimentary Geology (SEPM) Core Workshop No. 9, p. 325–361.

Triplehorn, D. M., and B. F. Bohor, 1981, Altered volcanic ash partings in the C coal bed, Ferron Sandstone Member of the Mancos Shale, Emery County, Utah: U.S. Geological Survey Open-File Report, 45 p.

Tripp, C. N., 1989, A hydrocarbon exploration model for the Cretaceous Ferron Sandstone Member of the Mancos Shale, and the Dakota Group in the Wasatch Plateau and Castle Valley of east-central Utah, with emphasis on post-1980 subsurface data: Utah Geological and Mineral Survey Open-File Report 160, 82 p, scale 1:250,000.

—1990a, Ferron oil and gas field, T. 20–21 S., R. 7 E., Emery County, Utah: Utah Geological and Mineral Survey Open-File Report 191, 23 p., scale 1:24,000.

—1990b, Wasatch Plateau oil and gas field, T. 13-17 S., R. 6-7 E., Carbon, Emery, and Sanpete Counties, Utah: Utah Geological and Mineral Survey Open-File Report 192, 33 p., scale 1:100,000.

—1991a, Wasatch Plateau oil and gas fields, Carbon, Emery and Sanpete Counties, Utah, *in* T. C. Chidsey, Jr., ed., Geology of east-central Utah: Utah Geological Association Publication 19, p. 255–264.

—1991b, Ferron oil and gas field, Emery County, Utah, *in* T. C. Chidsey, Jr., ed., Geology of east-central Utah: Utah Geological Association Publication 19, p. 265–272.

—1993a, Ferron, *in* B. G. Hill and S. R. Bereskin, eds., Oil and gas fields of Utah: Utah Geological Association Publication 22, non-paginated.

—1993b, Flat Canyon, Indian Creek, East Mountain, *in* B. G. Hill and S. R. Bereskin, eds., Oil and gas fields of Utah: Utah Geological Association Publication 22, non-paginated.

—1993c, Clear Creek, *in* B. G. Hill and S. R. Bereskin, eds., Oil and gas fields of Utah: Utah Geological Association Publication 22, non-paginated.

Tyler, Noel, 1988, New oil from old fields: Geotimes, July, p. 8–10.

Utah Division of Oil, Gas and Mining, 2003, Monthly production report (August 2003), non-paginated.

Vail, P. R., R. M. Mitchum, and S. Thompson, III, 1977, Global cycles of relative changes of sea level, *in* C. E. Payton, ed., Seismic stratigraphy — applications to hydrocarbon exploration: AAPG Memoir 26, p. 83–98.

van den Bergh, T. C. V., and J. R. Garrison, Jr., 1996, Channel belt architecture and geometry — a function of depositional parasequence set stacking pattern, Ferron Sandstone, east-central Utah: AAPG Rocky Mountain Section Meeting, expanded abstracts volume, Montana Geological Society, p. 37–42.

van den Bergh, T. C. V., and A. R. Sprague, 1995, The control of high-frequency sequences on the distribution of shoreline fluvial reservoir facies: an example from the Ferron Sandstone, Utah (abs.): AAPG Annual Convention Official Program, v. 4, p. 99A.

Van Wagoner, J. C., 1995, Overview of sequence stratigraphy of foreland basin deposits — terminology, summary of papers, and glossary of sequence stratigraphy, *in* J. C. Van Wagoner and G. T Bertram, eds., Sequence stratigraphy of foreland basin deposits: AAPG Memoir 64, p. ix-xxi.

Walton, P. T., 1954, Wasatch Plateau gas fields, Utah: Intermountain Association of Petroleum Geologists, 5th Annual Field Conference Guidebook, p. 79–85.

—1955, Wasatch Plateau gas fields, Utah: AAPG Bulletin, v. 39, no. 4, p. 385–420.

Weimer, R. J., 1962, Late Jurassic and Early Cretaceous correlations, south-central Wyoming and northwestern Colorado: Wyoming Geological Association, 17th Annual Field Conference Guidebook, p. 124–130.

Analog for Fluvial-Deltaic Reservoir Modeling: Ferron Sandstone of Utah
AAPG Studies in Geology 50
T. C. Chidsey, Jr., R. D. Adams, and T. H. Morris, editors

Searching for Modern Ferron Analogs and Application to Subsurface Interpretation

Janok P. Bhattacharya[1] and Robert S. Tye[2]

ABSTRACT

A quantitative approach to selecting modern-depositional settings analogous to those of the Cretaceous Ferron Sandstone is presented as well as an approach to using these analogs to improve subsurface interpretations. Paleotectonic, paleogeographic, and climatic setting of the U.S. Western Interior are integrated to estimate the size of the Ferron drainage network to be 50,000 km^2 (19,000 mi^2). Estimates of flow depths, flow velocities, and channel cross-sectional areas suggest maximum trunk river paleo-discharge was on the order of 50 x 10^9 m^3/year (250 x 10^9 ft^3/year).

Analogous modern examples include moderate-sized rivers that drain active mountain belts like the Po (Italy), Rhône, (France), and Ebro (Spain). Continental-scale systems, such as the Mississippi (U.S.A.), Niger (West Africa), Amazon (South America), and Nile (North Africa) deltas, are not appropriate analogs. Incised-valley systems within the Ferron are comparable in depth (about 30 m [100 ft]) to distributary channels in more continental-scale systems. A key difference is that Ferron valleys are filled with multiple channel deposits, with individual channel fills less than about 9 m (30 ft) deep.

The relatively uniform size of distributary channels suggests that Ferron rivers experienced only a few orders of bifurcation as they flowed across the delta plain. Shoreline and delta-front deposits are wave-influenced. Locally, the basal Ferron deltas were fluvial dominated, although these fluvial-dominated lobes may lie on the downdrift side of asymmetric wave-influenced deltas, similar in plan to the Brazos, Ebro, and Rhône deltas, and to the southern St. George lobe of the Danube delta.

Sizes and geometries of various depositional bodies in modern deltas better constrain estimates of inter-well heterogeneity in subsurface correlations. Correlations of Ferron core and wireline-log datasets are compared with the more complete stratigraphy documented in outcrop. Although the broad clinoform geometry of strata in these deposits can be recreated from subsurface correlations, specific parasequences could not be reliably correlated using only the subsurface data. The subsurface-to-outcrop comparison demonstrates a risk of over-correlating reservoir compartments (i.e. non-connected bodies) separated by minor flooding shales within complex, offlapping fluvial-deltaic reservoir deposits. The Ferron outcrop data provides a measure of uncertainties in correlation of subsurface analogs.

[1]*Geosciences Department, University of Texas at Dallas, Richardson, Texas*
[2]*PetroTel Inc., Plano, Texas*

INTRODUCTION

The Ferron Sandstone Member of the Cretaceous Mancos Shale, a superbly exposed fluvio-deltaic clastic wedge in central Utah, U.S.A., has long been used as an outcrop analog for subsurface fluvial-deltaic reservoirs (e.g. Barton, 1994; Gardner, 1995; Knox and Barton, 1999; and numerous papers in this volume). There are numerous reservoir-characterization studies of Ferron outcrops (Lowry and Jacobsen, 1993; Barton, 1997; Knox, 1997; McMechan et al., 1997; Corbeanu et al., 2001; Novakovik et al., 2002; Forster et al., this volume).

Core and wireline-log data give lithofacies type and thickness (Z dimension), but distributing depositional facies, such as the deposits of channels, bars, splays, and other elements, between wells remains one of the most difficult challenges in subsurface interpretation. Studies of modern depositional settings can be helpful in reservoir analog studies because they provide plan-view geomorphic data about specific depositional elements, such as length and widths of channels and bars, that ideally can be linked to a specific vertical facies association or facies architectural element in an ancient example (e.g. Bridge, 1993). Sizes and shapes of architectural elements in modern systems are highly variable, although they commonly follow specific trends, such as increasing channel size with increasing drainage basin area. Such trends help constrain correlation lengths of analog subsurface facies elements.

A traditional practice in interpreting stratigraphic successions is to compare them with modern depositional settings. However, one of the challenges in using data from modern depositional systems is identifying an appropriate analog. Most comparisons between modern and ancient systems tend to be qualitative and anecdotal. Specifically, the Holocene delta lobes and distributary channels of the Mississippi delta have been cited as analogs to the Ferron (e.g. Cotter, 1975a; Moiola et al., this volume), but comparisons have not been rigorous.

We present a quantitative approach to determining the size of the Ferron rivers and associated deltas for comparison to modern systems. We specifically wish to address the question of whether the Ferron compares to the continental-scale Mississippi River and delta system, or whether a smaller system might be more appropriate.

Several recent papers provide a methodology for estimating the paleohydraulics and dimensions of ancient river deposits (e.g. Collinson, 1978; Lorenz et al., 1985; Williams, 1986; Bridge and Tye, 2000; Le Clair and Bridge, 2001). These can be used to estimate the scale of Ferron rivers. With knowledge of the size and discharge of the drainage network, one can constrain the size of the associated deltas, and choose modern-depositional systems most resembling the Ferron, recognizing that there may be several.

Following identification of modern analogs, dimensions of features in the appropriate modern systems are used to enhance subsurface correlations. The second aspect of this paper is a subsurface interpretation of the Ferron Sandstone, using wireline logs and core data, integrating facies-based concepts of stratigraphic correlation, and geomorphic insights from modern analogs in order to demonstrate the stratigraphic uncertainties that are inherent in subsurface studies.

Regional stratigraphic studies of the Ferron Sandstone document several regionally mappable clastic wedges (e.g. Cotter, 1975a, 1975b; Ryer and McPhillips, 1983; Thompson et al., 1986; Garrison and van den Bergh, this volume), each of which contains numerous smaller-scale stratigraphic units. Terminology is variable and includes sequences, parasequence sets, parasequences, and bedsets (Garrison and van den Bergh, this volume; Anderson and Ryer, this volume), and the long, intermediate, and short-term stratigraphic cycles of Gardner et al. (this volume).

As depositional facies vary greatly within the Ferron Sandstone (as described in this volume by Anderson et al.; Mattson and Chan; Ryer and Anderson), it is clear that more than one delta type is represented. Therefore, no single modern example will be entirely analogous to the Ferron.

METHOD FOR FINDING A MODERN ANALOG

Finding an appropriate modern analog for an ancient fluvio-deltaic system requires an assessment of temporal and spatial scales in the ancient system. Certain physical processes, such as the formation and migration of bedforms that produce distinctive stratification, operate over time scales of minutes to hours. For example, ripple and dune-scale cross-stratification can be readily compared between modern deposits and ancient rocks. Various bedform-phase diagrams provide a theoretical and experimental framework that can be used for interpreting paleohydraulic conditions. In particular, Rubin and McCulloch (1980) showed that specific bedforms (e.g. ripples and dunes) are stable within specific ranges of flow velocity, flow depth, and grain size (Figure 1). The occurrence of dune-scale bedforms, for example, is only weakly dependent on flow depth but is strongly dependent on grain size and average flow velocity. The size (height) of a dune, in contrast, is strongly dependent on flow depth (LeClair and Bridge, 2001). Dune-scale cross-stratification is therefore particularly useful in determining both velocity and water depth.

Other types of stratification, such as hummocky cross-stratification, may be related to longer-term seasonal processes, such as storms. However, determining storm or flood frequency in an ancient example can be difficult, because of an inability to date the rock succes-

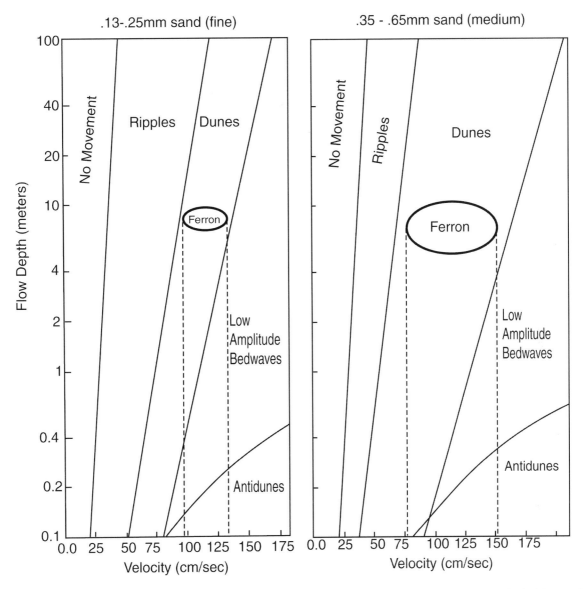

Figure 1. Bedforms, grain size, water depth, and velocity plots modified after Rubin and McCulloch (1980). Range of velocities matching Ferron observations are shown with the circles and suggest a flow velocity of between 75–150 cm/sec (29–59 in./sec). Data show that the presence of dunes is weakly dependent on flow depth, but strongly dependent on grain size and velocity. The size of the dunes, however, is strongly dependent on flow depth (see LeClair and Bridge, 2001).

sion at time scales at which the events occur. For example, large storms are thought to occur at decadal, centennial, or millennial frequencies, but it is very difficult to quantify this frequency in most ancient settings. Long-term processes controlling the distribution and style of sedimentation include tectonic setting, subsidence, and eustasy. Many of these parameters can be estimated. Environmental parameters such as climate, bedrock geology, and topographic relief control sediment flux and may correlate to tectonic and paleolatitudinal setting. Prevailing paleoclimatic and tectonic conditions within a drainage basin can be estimated using plate tectonic and paleo-oceanographic data to reconstruct the position, size, and elevation of land masses relative to the oceans. Additionally, paleoclimate can be interpreted from examination of floodplain paleosols (Wright, 1992; Mack et al., 1993; Mack and James, 1994).

CHARACTERIZING THE FERRON

Estimating Size of the Ferron Drainage Basin

The Ferron clastic wedge was built by rivers eroding and draining highlands uplifted during the later phase of the Sevier orogeny (Fouch et al., 1983). To the east of the Pavant thrust front, three Ferron depocenters, termed the Vernal, Last Chance, and Notom delta systems (Gardner, 1995; Garrison, this volume) formed in a

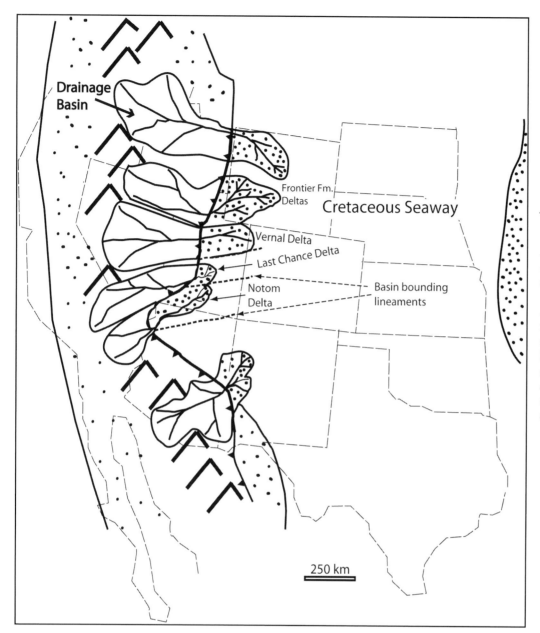

Figure 2. Paleogeographic reconstruction of mid-Cretaceous clastic wedges. Ferron delta complexes lie primarily in Utah. The drainage basin lengths are estimated to be about 1/2 the width of the western lands bordering the Cretaceous Seaway. Drainage basin widths are estimated based on the localization of wedges, assuming each is fed by a different major trunk river, and that drainage divides coincide with major basin lineaments. Aggradational alluvial and delta plains (close stipple) are assumed to begin seaward of the major thrust front. The Last Chance drainage is estimated to be about 500 km (300 mi) in length and about 100 km (60 mi) wide (50,000 km² [19,000 mi²]). Reconstructions are primarily based on Gardner (1995) and Williams and Stelck (1975).

foreland basin (Figure 2). Gardner (1995) suggests that owing to the segregations caused by major fault lineaments, three drainage basins developed. Gardner (1995) also shows that during Ferron deposition, the distance to the eroding Pavant thrust front was on the order of 100 km (60 mi). Integration of the tectono-stratigraphic interpretations of Gardner (1995) with the general paleogeographic reconstructions of the Cretaceous (Williams and Stelck, 1975) is shown in Figure 2. The drainage basin is interpreted to have been about 500 km (300 mi) long by about 100 km (60 mi) wide, suggesting that the Last Chance delta was fed by a river draining an area of about 50,000 km² (19,000 mi²). Comparing this with the other Ferron drainage basins suggests that they were all on the order of about 10^4 km² in area (Figure 2).

Ferron drainage networks appear to have been broadly similar in scale to modern intrabasinal rivers draining active mountain belts, like the Po (Italy — 70,000 km² [27,000 mi²]), the Rhône (France — 96,000 km² [37,000 mi²]), the Ebro (Spain — 83,000 km² [32,000 mi²]), and the Red River (Vietnam — 120,000 km² [46,000 mi²]). Ferron drainages were clearly orders-of-magnitude smaller than continental-scale drainages (10^4 km² versus 10^6 km² [1500 mi² versus 57,000 mi²]) such as is associated with the Mississippi (U.S.A.), Amazon (Brazil), Yellow (China), and Ganges-Brahmaputra (India and Bangladesh) Rivers and their associated deltas (Smith, 1966).

Estimating Climate and Nature of the Receiving Basin

The Ferron Sandstone succession developed at a time generally thought to represent global "greenhouse" conditions that enhanced a major tectono-eustatic sea-level highstand (Dean and Arthur, 1998). This area of Utah lay at a paleolatitude of about 40° N (Ryer and McPhillips, 1983). Abundant coals, gleysols, and a lack of aridisols, caliche, and evaporites in Ferron floodplain deposits indicate extensive lower coastal plain wetlands with a generally high water table (Corbeanu et al., this volume; Garrison and van den Bergh, this volume). Given the global greenhouse setting, the Ferron climate was thus likely humid and sub-tropical.

The intracratonic seafloor over which Ferron depocenters prograded was unlikely to be flat, but rather, consisted of a series of lows and highs that partly controlled local wave and tidal regimes. In the Last Chance delta complex, rivers predominantly flowed northeastward, parallel to the foreland axis (Ryer and McPhillips, 1983; Gardner, 1995; Ryer this volume). Locally, deltas prograded to the northwest, at nearly 90° to the more general northeast progradation (Anderson et al., this volume). The northwest-building deltas are the most fluvial-dominated, probably because they prograded into an embayed area, protected from waves (Anderson et al., this volume; Bhattacharya and Davies, this volume).

Searching for Ferron Trunk Rivers

Major trunk rivers supply sediment to the delta plain and delta front, and are commonly contained within incised valleys, especially where an incised tributary drainage network formed. Delta-plain streams and distributary channels tend to be confined only by aggradational levees, particularly when they build over flat wetlands. Once a trunk stream leaves the degradational region of its valley and becomes unconfined, the river is able to avulse. In many modern delta plains, upper delta-plain channels tend to be few (e.g. Bhattacharya and Giosan, 2003), whereas, multiple distributary channels of various scales occur on the marine-influenced lower delta plain (Figure 3).

Distributary channel bifurcation occurs at a point where the water can no longer flow over the distributary-mouth bar, therefore it splits into two small channels circumventing the bar crest. Channel-bifurcation frequency and branching patterns are strongly dependent on delta type. Multiple bifurcations are favored in low-gradient, fluvial-dominated deltas, where friction exerts a strong control on sediment dispersal and deposition (Welder, 1959; Wright, 1977). This is true whether or not the sediment load is sandy or muddy. Trunk-stream avulsion and distributary crevassing are common processes in fluvial-dominated deltas because hydraulic gradients decrease as rivers and distributaries extend their course.

In wave-modified deltas, much of the sediment delivered to the distributary-mouth bar is carried away by longshore transport. Thus, compared to fluvial-dominated deltas, the progradation rate of wave-influenced deltas is retarded. This allows rivers to maintain a greater slope that inhibits avulsion. As a consequence, wave-influenced deltas typically have only a few active distributary channels, whereas fluvial-dominated deltas have numerous active distributary channels (Figure 3).

Ferron Delta Types

Most Ferron researchers agree that the Ferron deltas were broadly wave-influenced, although the lower seaward-stepping parasequences show greater fluvial influence than the landward-stepping parasequences (e.g. Anderson et al., this volume). Tidal facies have also been recognized, particularly within the upper Ferron parasequences. Anderson and Ryer (this volume, Figure 4) suggest that the lower fluvial-dominated deltaic lobes within parasequence sets Kf-1 and Kf-2 of the Ferron possibly formed within embayments positioned on the downdrift margin of an asymmetric wave-influenced delta. A general model for wave-influenced deltas, presented by Bhattacharya and Giosan (2003; Figure 4), conforms to paleogeographic reconstructions of delta lobes associated with the lower Ferron parasequences, and in particular parasequence Kf-2-Mi (e.g. Figure 4 in Anderson and Ryer, this volume). Paleogeographic maps of Ferron delta-plain channels depict major trunk streams splitting into a few terminal distributary channels. At most, two orders of branching are interpreted (e.g. Anderson et al., this volume, Figures 31 and 33). These paleogeographic reconstructions are quite consistent with the plan-view geometries of modern wave-influenced deltas (Bhattacharya and Giosan, 2003) and contrast with the large, low slope, modern fluvial-dominated deltas, such as the muddy Mississippi and Atchafalaya deltas in the Gulf Coast and the more sandy Colville, Saganavirtok, and Lena Rivers in the Arctic that show up to 10 orders of branching. Truly fluvial-dominated delta lobes are largely formed as crevasse deltas (e.g. Welder, 1959; Wells et al., 1984; Bhattacharya and Davies, this volume) and generally form in bays protected from waves by a wave-formed barrier island (Bhattacharya and Giosan, 2003; Figure 4).

Gardner et al., Garrison and van den Bergh, and Barton (this volume), note a remarkable uniformity on the scale of preserved channel and bar deposits, which also suggests a low order of distributary branching. Cross sections through deltas with many orders of branching should show large variability in the size and distribution of channels than is generally observed in the Ferron Sandstone. This suggests that the largest of the documented Ferron channels are trunk streams.

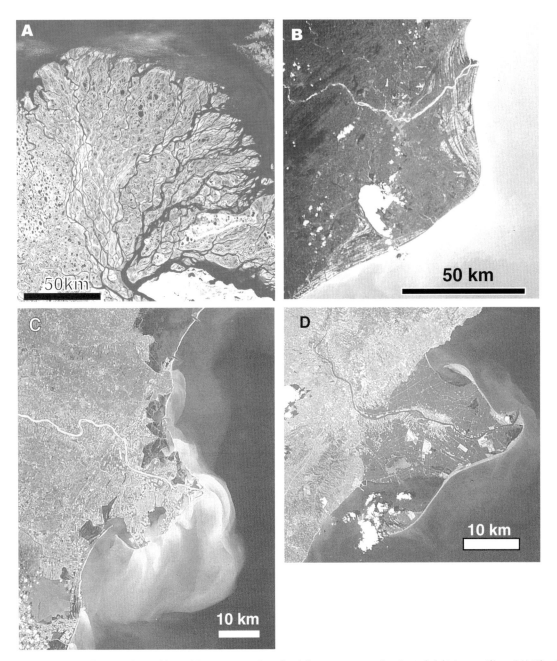

Figure 3. Comparison of distributary channel branching patterns in a fluvial- versus wave-dominated deltaic coastline. (A) Fluvial-dominated Lena River delta (Russian Arctic) shows numerous orders of branching with many tens of terminal distributary channels (photo from NASA Earth Observatory website — http://earthobservatory.nasa.gov/Newsroom/NewImages/images.php3?img_id=10291). (B) Wave-dominated coastline associated with the Paraibo do Sul, Brazilian coast. (C) Po delta, Italy. (D) Ebro delta, Spain. Bifurcation is inhibited in wave-dominated deltas because the river is unable to prograde into the basin as rapidly. This effectively allows the river to maintain its grade, which in turn inhibits avulsion. For the most part, Ferron deltas are interpreted as primarily wave-influenced. The Po and Ebro are considered to be likely modern analogs (photos of the Po, Ebro, and Paraibo do Sul deltas courtesy of W. R. Muehlberger).

Valleys or Channels in the Ferron?

Several Ferron studies (Barton et al., this volume, Figures 5 and 17; Garrison and van den Bergh, this volume) interpret multistory channel belts to lie within incised valleys. Barton et al. (this volume) show 22–24 m (70–80 ft) deep incisions filled by multiple, stacked channel stories each less than 6 m (20 ft) thick (Figure 5).

These observations satisfy the definition of incised valleys as "elongate erosional features larger than a single channel" (Dalrymple et al., 1994; Willis, 1997). The multistory fills clearly show that streams were underfit relative to the larger incised valley (Figure 5).

Rivers confined to valleys are candidates for the largest-scale trunk channels. Therefore, their multistory channel-belt deposits provide insight into the depth and

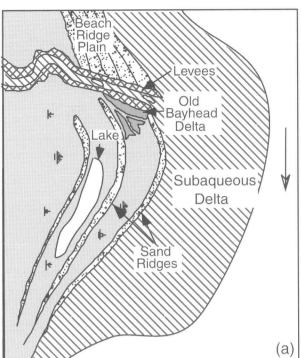

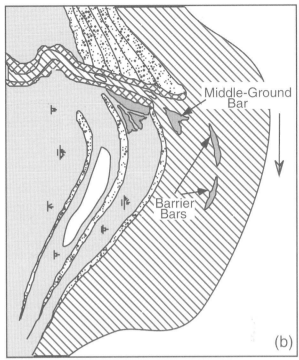

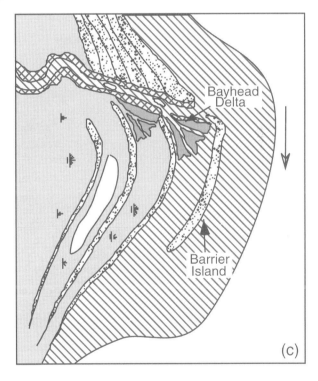

Figure 4. Conceptual model for evolution of a modern asymmetric delta lobe, such as the St. George of the Danube delta: (A) Subaquaeous Delta Phase — sediment deposition is primarily on the subaqueous part of the delta; the beach-ridge plain on the updrift flank is also advancing; (B) Middle-Ground Bar Phase — a middle-ground bar forms at the mouth, forcing the distributary to bifurcate; linear barrier bars form on the subaqueous delta; (C) Barrier Island Phase — the linear barrier bars coalesce and become emergent to form a barrier island that rolls over to attach to the mainland; a secondary, fluvial-dominated, bay-head delta may develop in the sheltered lagoon behind the barrier island. Longshore drift (represented by the open arrow) is southward. Modified after Bhattacharya and Giosan (2003).

width of the formative channel, although the amalgamated character means that channels may be incompletely preserved. Interpretations based on incomplete channel thicknesses may result in underestimation of bankfull channel depth (Bridge and Tye, 2000), although more complete channel fills are found in the higher storeys (Figure 5).

Although the Mississippi Delta has been used as a modern analog for the Ferron Sandstone (e.g. Cotter, 1975a; Moiola et al., this volume) comparisons have not

been rigorous. In fact, a significant Ferron stratigraphic debate is centered on the interpretation of channelized deposits on the order of 20 m (65 ft) deep and a few hundred meters wide. Moiola et al. (this volume) suggest that because these incised features are comparable in size to distributary channels in the modern Mississippi Delta, they are distributary channels. In contrast, Barton et al., and Garrison and van den Bergh (this volume) interpret these deposits as incised valleys, that contain the deposits of numerous, vertically stacked, small-scale

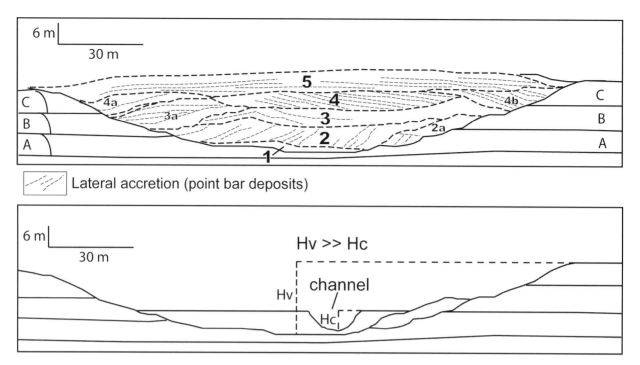

Figure 5. Facies architecture of an interpreted Ferron valley fill in cliffs along Interstate 70. Base of valley erodes into several upward-coarsening parasequences (A, B, C). Valley depth (Hv) is about 21 m (70 ft). In contrast, associated channel depths (Hc) are only about 6 m (20 ft). Valley is filled with five channel storeys (1–5). Lowest channel belt deposit (1) is largely eroded by migration of younger channels. Predominance of laterally accreting bars, define the internal facies architecture of each channel-belt deposit. The bedding geometry shows that the rivers were single thread, meandering streams that gradually filled the larger valley. Calculation of water depth from dune-scale cross-strata within the bar deposits suggests maximum bankfull depth of about 9 m (30 ft). Figure modified after Barton et al. (this volume).

channels. Thus, these differing opinions are manifested in numerous sequence stratigraphic interpretations of the Ferron. Ryer (this volume) points out that seven sequence boundaries identified in the Ferron by several research groups do not match!

In the previous section we concluded that the Ferron drainage area was far too small for the development of a Mississippi-scale river. Additionally, the ratio of bankfull-channel depth to the maximum-erosion depth defines a valley, not the absolute depth of the incision. In the Ferron Sandstone, the ratio of valley depth to channel depth is about 4.

Estimating Paleodischarge of Ferron Rivers

To compare interpreted Ferron channel belts with modern counterparts it is helpful to calculate a possible discharge for the Ferron rivers. Matthai (1990 [quoted in Mulder and Syvitsky, 1995]) demonstrates that peak-flood fluvial discharge is related to drainage basin area by the equation:

$$\text{Log } Q_{flood} = -0.070 \, (\log A^2) + 0.865 \log A + 2.084 \quad (1)$$

Given drainage area A = 50,000 km^2 (19,000 mi^2), we calculate log Q_{flood} = 5.49. This, in turn, yields a calculated peak flood (Q_{flood}) of 309,029 m^3/sec (10,914,854 ft^3/sec), although there is an order of magnitude uncertainty in this calculation. It is emphasized that this value

would not apply year round, would primarily reflect only times of high seasonal discharge, and should not be used to calculate average yearly discharge. Data compiled by Mulder and Syvitsky (1995) suggest that $Q_{flood}/Q_{Average}$ ratios can range over 4 orders of magnitude, from 10 to 10,000. This ratio suggests that average discharge for Ferron rivers could be as high as 30,000 m^3/sec (1,060,000 ft^3/sec) but as low as 30 m^3/sec (1060 ft^3/sec). Clearly the uncertainty in these estimates is large, and results must be interpreted with caution.

A more robust method of estimating discharge is based on the paleohydraulics of the river deposits themselves. We start with the equation for channel discharge:

$$Q = A \times U \quad (2)$$

where Q = discharge, A = cross-sectional area of the channel (width x depth), and U = average velocity. For this method, estimates of water depth, channel width, and flow velocity (U) are required. Channel velocity could be calculated from the slope, but ancient fluvial slopes are notoriously difficult to estimate, primarily because they are low and range over several orders of magnitude (typically 0.001 to 0.00001). Water depths of the interpreted trunk streams can be readily estimated from the mean preserved thickness of dune-scale cross-strata (Bridge and Tye, 2000; Le Clair and Bridge, 2001) using the equation:

$$H_m = 5.3\text{ß} + 0.001\text{ß}^2 \qquad (3)$$

where $\text{ß} = s_m/1.8$, s_m = mean cross-set thickness, and H_m = mean dune height.

To use this technique, it is essential that data on the measured thicknesses of cross-sets be accurately presented. Bridge and Tye (2000) stress that abnormally thick, isolated cross-sets formed by unit bars should be excluded from the data. Based on measured sections presented by Gardner et al. (this volume) and Garrison and van den Bergh (this volume), thickness of dune-scale cross-stratal sets are shown to average 20–30 cm (8–12 in.). Setting s_m = 20–30 cm, the above equation therefore yields a mean dune height of about 59–88 cm (23–35 in.). LeClair and Bridge (2001) further show that flow depth is typically 8 to 10 times mean dune height. This suggests flow depths of between 4.7–8.8 m (15.4–28.9 ft). We thus determine maximum (i.e. bankfull) flow depths to be about 9 m (30 ft).

Channel depth can also be independently estimated from thicknesses of fully preserved bar deposits (i.e. macroform) and channel-fill deposits, typically represented in the uppermost storey within the interpreted valley fills. Thickness of bar-scale macroforms reaches a maximum of about 8 m (26 ft) (Gardner et al., Figure 15, this volume). With the assumption that bar deposit thicknesses are roughly 90% of channel depth (c.f. Bridge and Mackey, 1993) we again estimate a flow depth of about 9 m (30 ft), similar to that calculated above. Maximum thicknesses of channel-fill deposits, based on measured sections presented by Gardner (this volume) and Garrison and van den Bergh (this volume) are also on the order of 9 m (30 ft). These independent estimates of maximum bankfull-channel depth (d_m) all suggest a value of about 9 m (30 ft). Taking the mean channel depth to be approximately 0.57 the maximum-bankfull depth (5.1 m [16.7 ft]) (Bridge and Mackey, 1993) and these equations from Bridge and Tye (2000):

$$w_c = 8.88(d_m)^{1.82} \qquad (4)$$

$$w = 59.86(d_m)^{1.8} \qquad (5)$$

$$w = 192.01(d_m)^{1.37} \qquad (6)$$

where d_m = mean channel depth, gives a channel width (w_c) of 174 m (571 ft) and a range of channel-belt widths from 1135–1800 m (3724–5900 ft). Several of the valley-scale channels in outcrop (9 m [30 ft] deep, 250 m [820 ft] wide) show laterally accreting point-bar deposits (Barton et al., this volume), suggesting single-channel, meandering streams. Equation 6 is most applicable for a high-sinuosity channel. The equations of Fielding and Crane (1987) (equation 7) and Collinson (1978) (equation 8):

$$w = 64.6(d_m)^{1.54} \qquad (7)$$

$$w = 65.6(d)^{1.57} \qquad (8)$$

where d = maximum channel depth, give a range of channel-belt widths from 650–2100 m (2130–6900 ft). The value in these calculations is that they give a range of possible channel and channel-belt sizes that compare favorably with outcrop observations where the Ferron channels are confined, or where channel margins are observable (channel not confined within a valley).

Garrison and van den Bergh (this volume) measured channel-belt widths, and thicknesses. They show that maximum thicknesses are 30 m (100 ft) and widths do not exceed 2 km (1.2 mi). Channel widths within the belts average 250 m (820 ft). There is no outcrop evidence that Ferron river channels were wider than a few hundred meters. Numerous unconfined distributary channels occur in the Ferron delta plain (e.g. Corbeanu et al., this volume), but it is impossible to estimate overall stream discharge from these types of channels because it is not known how many distributary channels were active simultaneously.

Sedimentological descriptions of the fluvial deposits show a predominance of dune-scale cross-stratified, pebbly, coarse- to medium-grained sandstone, fining upward into fine-grained sandstone and mudstone (Ryer and Anderson, this volume). Ryer and Anderson (this volume) present strong evidence for sinuous channels that migrated laterally (e.g. upward-fining abandoned-channel fills, lateral accretion surfaces). The abundant mud deposited in both the floodplain, prodelta, and off-shore areas demonstrate that the Ferron rivers carried a significant muddy load in suspension, as well as a sandy bedload moving primarily as dune-scale bedforms. Channel deposits become increasingly mud rich, down depositional dip and in higher stratigraphic positions (Barton et al., this volume; Gardner et al., this volume).

Flow velocities in the Ferron rivers can be estimated using the method of Rubin and McCulloch (1980). Their method is based on bedform stability as a function of grain size, water depth, and velocity (Figure 1). A flow velocity of between 50–110 cm/sec (20–43 in./sec) is estimated based on grain size (fine sand), water depths (9 m [30 ft]), and the fact that three-dimensional dunes were the primary stable bedforms. Assuming that grain size and sedimentary structures represent high-discharge or early, post-flood depositional conditions, we use 100 cm/sec (1 m/sec [3 ft/sec]) as a reasonable estimate of flow velocity.

A maximum channel depth of 10 m (33 ft) and a width of 250 m (820 ft) gives an average cross-sectional area of on the order of 1250 m^2 (13,460 ft^2). From this it is easy to calculate a discharge (Q_w) of 1250 m^3/sec (50 x 10^3 ft^3/sec). Of course the channel-forming discharge was likely related to times of major floods, so the average discharge is probably lower than 1250 m^3/sec. Assuming that discharge was this high year round, allows us to place an upper limit of maximum possible discharge of about 50 x 10^9 m^3/year or 50 km^3/year (about 250 x 10^9 ft^3/year). This contrasts with the discharge calculated using an estimated drainage area, which suggests maximum possible peak-flood discharge of 309,000 m^3/sec (10,900,000 ft^3/sec) compared to 1250 m^3/sec. The 1250 m^3/sec estimate is, however, within the range of possible average discharge values of 30–30,000 m^3/sec (1060–1,060,000 ft^3/sec) using the alternate approach of Matthai (1990).

COMPARISON OF FERRON WITH MODERN DELTAS

We now have a relatively complete characterization of the Ferron system (although focused primarily on the lower parasequences within the Last Chance delta). Width and depth estimates for Ferron rivers are an order-of-magnitude less than for those of the modern Mississippi, whereas the overall sediment caliber in the Ferron channel and bar deposits is considerably coarser grained. The Mississippi, therefore, should not be used as an analog in comparing the scale of the delta and its associated elements, particularly the major Mississippi distributary channels.

A narrow range in sizes of distributary channels suggests that there were only a few orders of bifurcation. This is consistent with predominantly wave-influenced shorelines and delta fronts in which only fluvial dominance is localized (Figure 3). The Ferron deltas were likely wave-modified and probably asymmetric, similar in plan to the Brazos delta, the Ebro, Rhône, or the southern St. George lobe of the Danube (see summary in Bhattacharya and Giosan, 2003; Figures 3 and 4). Fluvial-dominated lobes may have prograded into bays, protected by wave-formed barrier systems on the downdrift

margins of these systems. The low number of bifurcations is characteristic of highly wave-modified delta systems (Figure 3).

The Ferron river systems are more similar to moderate-sized, sandy-bedload rivers such as the Brazos (Texas), or the Po (Italy), Rhône, (France), and Ebro (Spain) that drain off the active alpine systems in central Europe (Figure 3). The Ferron rivers did not drain a craton, and therefore mud-dominated continental-scale systems, such as the Mississippi (U.S.A.), Niger (West Africa), Amazon (Brazil), and Nile (North Africa), are not appropriate analogs.

Our calculated estimates suggest that the Ferron rivers were an order-of-magnitude smaller in discharge than the smallest of the world's 25 largest rivers (Meade, 1996). The two smallest of the top 25 rivers (Niger River in Africa and Fly River in Papua New Guinea) have respective water discharges of 190 and 150 x 10^9 m^3/year (950 and 750 x 10^9 ft^3/year) (Meade, 1996) three times larger than the 50 x 10^9 m^3/year (250 x 10^9 ft^3/year) calculated for the Ferron rivers, which as discussed is probably still too high an estimate. Estimates of the Ferron drainage area are also two orders-of-magnitude lower than for continental-scale river systems like the Niger, Amazon, and Mississippi. Additionally, the estimated size of the mapped subaerial delta plains formed by each of the Ferron deltas (10^3 km^2 [240 mi^2]) is an order-of-magnitude less than that for major continental deltas (e.g. Gardner, 1995; Anderson et al., this volume).

A collection of data on modern deltas (Table 1) that may span the sizes and morphologic types of Ferron deltas include the Ebro (Spain), Colville and Sagavanirktok (Alaska), Rhône (France), Danube (Romania), Magdalena (Columbia), Mississippi River subdeltas (West

Table 1. Summary data for Holocene deltas possibly analogous to Ferron deltas. Data summarized from Russell (1942), Coleman and Wright (1975), Wells et al. (1984), Ecological Research Associates (1983), and van Heerden and Roberts (1988).

Climate		Basin Area (x 10^3 km^2)	Aluv. Channel Length (km)	Delta Area (km^2)	Discharge (m^3/sec)	No. of Mouths	Tidal Range (m)
Ebro (Spain)	Dry Trop.	89.8	67	624	552	2	0
Colville (Alaska)	Arctic	59.5	481	1687	491.7	19	0.21
Sagavanirktok (Alaska)	Arctic	11.8	55	1178	600	11	0.21
Cubits Gap (Louisiana)	Temp.	3344.6	-	200	2262	8	0.5
Atchafalaya (Louisiana)	Temp.	3344.6	-	125	937	10	0.5
Wax Lake (Louisiana)	Temp.	3344.6	-	107	1537	8	
West Bay (Louisiana)	Temp.	3344.6	-	300	750	6	0.5
Danube (Romania)	Cool Temp.	712.6	774	2740	6250.1	14	0
Magdalena (Columbia)	Humid Trop.	251.7	136	1689	7500	1	1.1
Rhône (France)	Dry Trop.	-	177	1813	13,000 (max.) 360 (min)	6	0
São Francisco (Brazil)	Humid Trop.	602.3	150	734	3420	1	1.86
Last Chance delta	Humid Sub-Trop.	50	100	1000	1250		<2.0

Delta, Cubits Gap, Atchafalaya, and Wax Lake), and São Francisco (Brazil).

USE OF MODERN ANALOGS FOR SUBSURFACE INTERPRETATION OF FERRON WELL DATA

Having restricted the population of possible modern river and delta analogs for the Ferron Sandstone, the validity and usefulness of using their geomorphic traits (Table 2) were tested on a dataset of 16 wells to guide a subsurface interpretation of the Ferron (Figure 6). Core and wireline-log data include research wells drilled at Ivie Creek (Utah Geological Survey), Muddy Creek (British Petroleum), and a regionally spaced set of wells drilled by ARCO. Well spacing is highly variable. At Ivie Creek and Muddy Creek, closely spaced wells (0.5 km [0.3 mi]) provide control on correlation of facies associations, whereas the distance between the remaining wells leaves much room for interpretation.

Each well has gamma-ray and density logs; most of the ARCO wells also have neutron-porosity logs. Existing core descriptions (Thompson et al., 1986; Utah Geological Survey; and Moiola et al., this volume) for the two lowermost Ferron parasequence sets (Kf-1 and Kf-2; Anderson et al. this volume) provide the basis for our stratigraphic interpretation.

Given the clustered wells at Ivie Creek and Muddy Creek, plus the widely spaced ARCO wells, our assumption is that the well data are a reasonably unbiased sample of the lower Ferron stratigraphy. Our hypothesis is that subsurface interpretations integrating core data with modern analog observations are superior to those made in which geomorphic guidance from modern

Table 2. Summary statistics for length and width of channels and distributary-mouth bars in various deltaic settings.

	Length (km)			Width (km)			
Colville	Min.	Mode	Max.	Min.	Mode	Max.	N
Channels	N/A	N/A	N/A	0.27	0.5	0.78	40
Ab. Channels	0.09	0.39	2.74	0.03	0.05	1.07	140
Distributary Channels	N/A	N/A	N/A	0.05	0.42	2.43	281
Dist.-Mouth Bars	0.14	0.43	6.73	0.09	0.16	2.93	109
Sagavanirktok	Min.	Mode	Max.	Min.	Mode	Max.	N
Channels	N/A	N/A	N/A	0.09	0.53	1.25	65
Ab. Channels	0.22	0.53	2.95	0.06	0.09	0.93	75
Distributary Channels	N/A	N/A	N/A	0.05	0.16	0.68	75
Dist.-Mouth Bars							164
Atchafalaya	Min.	Mode	Max.	Min.	Mode	Max.	N
Distributary Channels	N/A	N/A	N/A	0.07	0.09	1.78	51
Dist.-Mouth Bars	0.44	2.05	3.43	0.26	1.05	2.07	14
Wax Lake	Min.	Mode	Max.	Min.	Mode	Max.	N
Distributary Channels	N/A	N/A	N/A	0.03	0.12	0.78	102
Dist.-Mouth Bars	0.16	0.57	4.94	0.09	0.41	1.33	64
West Delta	Min.	Mode	Max.	Min.	Mode	Max.	N
Distributary Channels	N/A	N/A	N/A	0.06	0.22	1.22	73
Cubits Gap	Min.	Mode	Max.	Min.	Mode	Max.	N
Dist.-Mouth Bars	0.15	0.17	2.61	0.08	0.16	2.14	46
Rhône	Min.	Mode	Max.	Min.	Mode	Max.	N
Distributary Channels	N/A	N/A	N/A	0.5	0.6	1.0	11

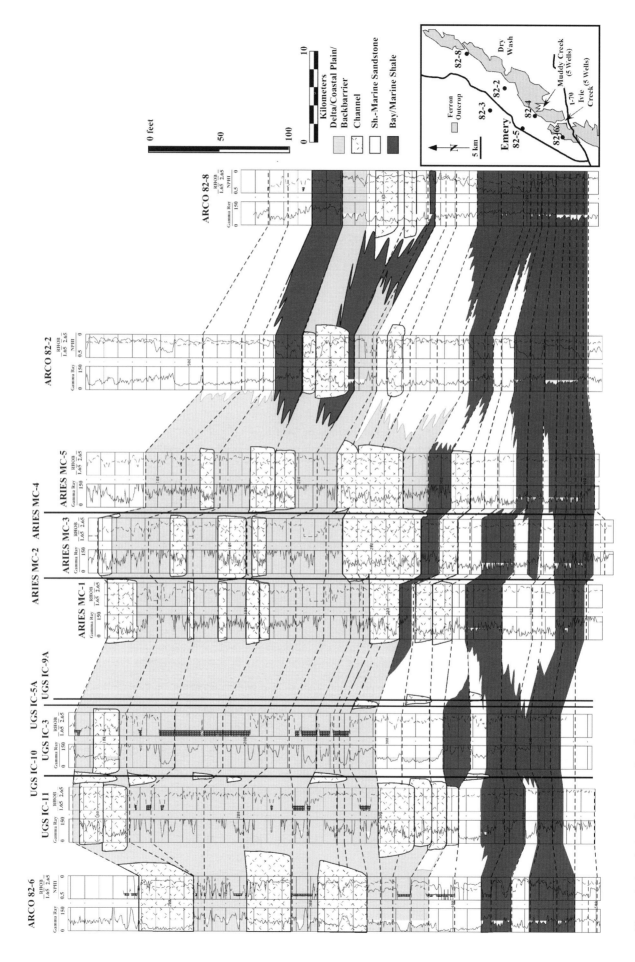

Figure 6. Correlation of Ferron subsurface data not incorporating dimensions based on modern analogs. Color version on CD-ROM.

Table 3. Fluvial channel, distributary channel, distributary-mouth bar, and shoreface dimensions summarized from 671 paralic sandstone bodies. From Reynolds (1999).

Sandstone Body Type	Length (km)			Width (km)		
	Min.	Mean	Max.	Min.	Mean	Max.
Fluvial Channels	N/A	N/A	N/A	0.06	0.76	1.4
Distributary Channels	N/A	N/A	N/A	0.02	0.52	5.9
Dist.-Mouth Bars	2.4	6.48	9.6	1.1	2.87	14
All Shoreface/Shelf Sandstones	47	93.2	190	1.6	25.4	106

analogs are lacking. The exercise also allows us to estimate the degree of stratigraphic uncertainty inherent in subsurface data sets compared to continuously exposed outcrop data sets.

To evaluate the hypothesis, our stratigraphic interpretation was compared to the cross sections based on nearly continuously exposed outcrops presented by Garrison and van den Bergh (this volume; their Figure 3). Parasequences within parasequence sets Kf-1 and Kf-2 were correlated in all the wells using genetic-sequence and allostratigraphic concepts (Figure 6; c.f. Van Wagoner et al., 1990; Cant, 1992; Bhattacharya, 1993; Posamentier and Allen, 1999). We deliberately did not refer to the detailed outcrop stratigraphies presented in this volume to constrain our correlations. The intent was to illustrate the degree to which a more-or-less random series of wells represents the actual stratigraphy.

A stratigraphic cross section paralleling the Ferron outcrop belt (Figure 6) shows the parasequences, bounding surfaces, and facies associations bundled into transgressive-regressive cycles, similar to those described by Anderson et al. (this volume) and Ryer (this volume). Moiola et al. (this volume) attribute many of the bounding surfaces to be localized flooding surfaces created through delta-lobe abandonment.

In making subsurface interpretations given only wireline logs and cores, one can measure facies associations, and bedset and bed thicknesses in addition to interpreting unconformities. Lorenz et al. (1985), Fielding and Crane (1987), Bridge and Tye (2000), and Le Clair and Bridge (2001), provide insights into how thickness data can facilitate estimating channel dimensions. Additionally, Reynolds (1999) gives good thickness-to-width relationships for paralic facies associations (Table 3).

In a revised Ferron cross section (Figure 7), the size, connectivity, position, and frequency of occurrence of specific facies elements (e.g. fluvial, distributary, and tidal channel; distributary-mouth bar; bay/lagoon; shoreface) were governed by geomorphic constraints based on these published thickness versus width datasets. Channel dimensions were varied according to whether the channel's interpreted environmental setting was fluvial, distributary, or tidal in nature. Fluvial and

distributary-channel dimensions were estimated using the methods previously described and those of Bridge and Tye (2000). Also, size ranges for distributary channels and distributary-mouth bars in various river-influenced deltas were incorporated (Table 2). The number of interwell channels and their connectivity were estimated using the net-to-gross observed in each stratigraphic interval (Bridge and Mackey, 1993). Correlations at Ivie Creek and Muddy Creek, where well spacing is as close as 0.5 km (0.3 mi), demonstrate the localized extent of the distributary channels (Figure 7). Had we assumed that distributary channels were at the scale of the Mississippi delta, we would have correlated them over longer distances. The proportion and dimensions for tidal-inlet channels are based on thickness-to-width relationships of Holocene examples (Tye and Moslow, 1993).

A third correlation (Figure 8) assumes that the multistory channel sandstones are incised valleys. A major difference between it and Figure 7 is the over-correlation of channel facies, especially in the uppermost part of the Ferron Sandstone.

The subsurface and outcrop cross sections (Figure 7, and Figure 3 of Garrison and van den Bergh, this volume) capture the progradational and aggradational stacking pattern of multiple Ferron parasequences and the flooding surfaces separating them. Three sequences (FS1, FS2, and FS3) identified by Garrison and van den Bergh (this volume) are projected onto Figure 7; although, their placement on the wireline logs is approximate.

This comparison illuminates interpretive differences and difficulties between outcrop and subsurface interpretations. Given the greater exposure and more complete sampling, the outcrop interpretation contains a greater diversity in interpreted facies associations, an increased number of identified parasequences and bedsets, and consequently greater stratigraphic fidelity. However, considering that in the absence of the outcrop data, Figures 7 and 8 could represent equi-probable stratigraphic interpretations, its similarity to the work of Garrison and van den Bergh (this volume) is reassuring. For instance, at the base of FS1 (Figure 7), both interpretations show bay/lagoonal deposits separating parasequence set 1 from parasequence set 2. In outcrop, the

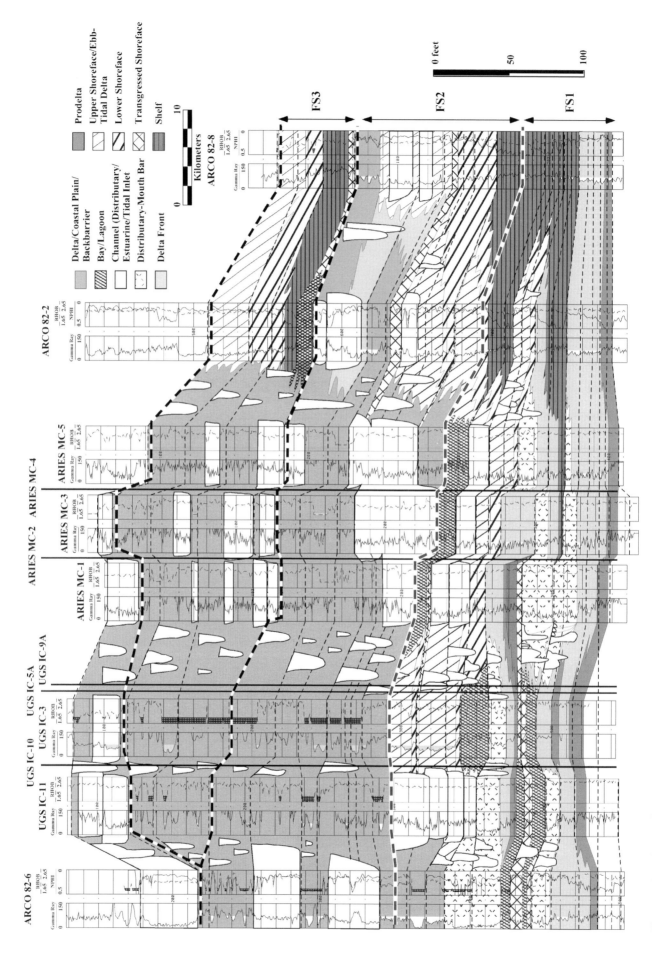

Figure 7. Correlation of same Ferron cores, interpolating detail from modern analogs. Note additional channel elements interpolated between wells. Color version on CD-ROM.

Table 4. *Distributary channel and distributary-mouth bar dimensions in river-dominated and wave-reworked progradational parasequences in the Last Chance delta. From Garrison and van den Bergh (this volume).*

Last Chance Delta	Length (km)	Width (km)	Thickness (m)	Length/Thickness Aspect Ratio	Width/Thickness Aspect Ratio	Length/Width Aspect Ratio
Distributary Channels	N/A	0.364	9.7	N/A	38.0	N/A
Dist.-Mouth Bars	3.67	2.69	8.9	512	169.4	2.5

bay/lagoonal deposits grade into prodelta/offshore sediments between Ivie Creek and Muddy Creek. In the subsurface section, the bay/lagoonal deposits are correlated continuously between Ivie Creek and Muddy Creek as if they are landward facies equivalents of distributary channel and distributary-mouth bar deposits in well MC-1. This illustrates the greatest difference between the two interpretations and a drawback to subsurface interpretations — the inability to recognize and correlate all parasequences and to potentially over-correlate potential reservoir compartments. This is a particularly common problem where subsurface data undersample complex stratigraphic features, such as offlapping clinoform strata. Data from modern environments can help make predictions about the maximum likely extent of a clinoform feature, such as a delta lobe, although to date, there are far more data on channelized facies elements. Additionally, although Ferron channel widths average less than 0.5 km [0.3 mi]) (Table 4), the lateral continuity of amalgamated channel deposits (channel belts) demonstrated by Garrison and van den Bergh (this volume) is generally greater than that shown in Figure 7, but not as much as shown in Figure 8. Thus, the subsurface interpretation in Figure 7 is optimistic in estimating the continuity of shallow-marine facies associations (i.e. delta front, shoreface) in parasequences but pessimistic in estimating the dimensions and number of channel deposits. The over-thick channel-belt deposits could also represent candidates for incised valley fills, as illustrated in Figure 8, which would result in interpretation of several additional sequence boundaries. The valley systems would correlate over larger distances than smaller distributary channels. However, even with such excellent datasets, correlation of sequence boundaries is far more uncertain than correlation of the flooding surfaces (Ryer, this volume).

In evaluating the probability that Figure 7 accurately represent the Ferron geology, and using Garrison and van den Bergh's work as the ground truth, this subsurface Ferron interpretation falls on the low side of a most-likely case scenario, with a greater level of compartmentalization of sandstone bodies than is seen in the outcrop. Figure 8 tends to over-correlate the channel sandstones.

Making multiple interpretations of the same subsurface data set can be a valuable method in determining the range of stratigraphic uncertainty. If the degree of uncertainty can be incorporated in building two- or three-dimensional reservoir models, then the risk associated with a development program can be better evaluated. In a sense, the question for a reservoir geologist should not be, "how big is a distributary channel (or any other facies architectural element)," but "what is the range of possible or most likely sizes." This is the scale of reservoir description at which data from modern depositional analogs can be most useful.

CONCLUSIONS

Integration of the tectono-stratigraphic interpretations of Gardner (1995) with the general paleogeographic reconstructions of the Cretaceous (Williams and Stelck, 1975) suggests that the Last Chance delta was fed by a trunk river draining an area of about 50,000 km^2 (19,000 mi^2). This drainage basin area can in turn be used to estimate maximum flood discharge (Matthai, 1990) which suggests that peak flood discharges could have been as high as 300,000 m^3/sec (10,600,000 ft^3/sec), although estimates using this method vary over several orders of magnitude. More robust calculations of river discharge, based on paleohydraulic estimates of channel-flow depths, cross-sectional channel areas, and flow velocity, suggest maximum discharge of $Q_w = 50 \times 10^9$ m^3/year (about 250×10^9 ft^3/year).

Discharge estimates for Ferron rivers are orders-of-magnitude less than for those of the modern Mississippi River, whereas the overall sediment caliber in the Ferron channel and bar deposits is considerably coarser grained. Some of the larger-scale Ferron trunk rivers lie within 20 m (60 ft) deep incised valleys. Although these valleys are comparable in scale to distributary channels in continental-scale deltas, like the Mississippi, such comparisons can be misleading. Anecdotal comparisons of ancient delta systems with well-studied modern continental-scale systems, such as the Mississippi, may result in erroneous interpretations of valleys as distributary channels (Bhattacharya et al., 2001). Ferron channels were on the order of 5–9 m (16-30 ft) deep, compared to the greater depths of continental-scale rivers. Recognition of valleys within the Ferron Sandstone suggests significant erosion and has important sequence stratigraphic implications.

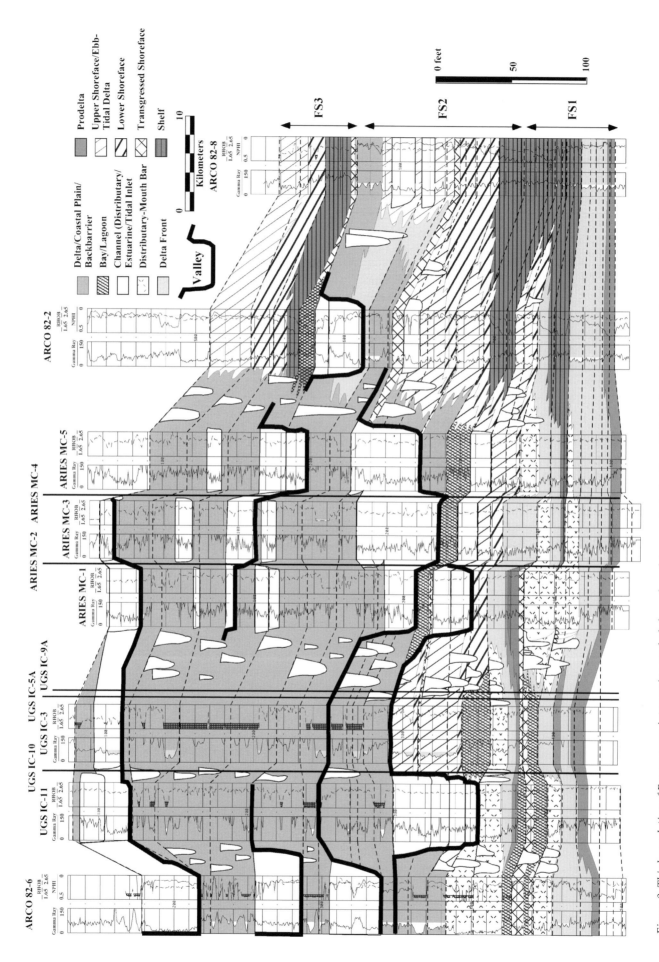

Figure 8. Third correlation of Ferron cores, interpreting stacked channel sandstones as incised valleys. Note tendency to over-correlate upper valley fill as a continuous sandstone. Color version on CD-ROM.

A lack of size variability of distributary channels suggests that there were only a few orders of bifurcation. The low number of bifurcations is characteristic of highly wave-modified delta systems. Shoreline and delta-front facies are predominantly wave-influenced, with only local fluvial dominance. The Ferron deltas were likely wave-modified and probably asymmetric, similar in plan view to the Brazos, Ebro, and Rhône deltas, or the southern St. George lobe of the Danube (see summary in Bhattacharya and Giosan, 2003; Figure 4). Fluvial-dominated lobes may have prograded into bays, protected by wave-formed barrier systems on the downdrift margins of deltaic depocenters.

Lastly, dimensional compilations from modern analog data can be used to better constrain interwell heterogeneity in subsurface correlations. Comparison of subsurface-type correlations of Ferron core and wireline log datasets with the detailed outcrop stratigraphy shows that despite capturing the clinoform geometry, several parasequences could not be reliably distinguished in the subsurface data. Over-correlating shallow-marine reservoir compartments in subsurface interpretations is a danger. In the Ferron subsurface example, the lateral continuity of shallow-marine facies associations was overestimated, whereas the sizes and number of channel facies associations interpreted were underestimated. The Ferron outcrops are an excellent guide to evaluating correlation uncertainties in subsurface settings.

ACKNOWLEDGMENTS

This paper has benefited from discussions with numerous experts, both in fluvial sedimentology and in Ferron stratigraphy. Mark Barton, John Bridge, Gus Gustason, Tom Ryer, Mike Gardner, John Holbrook, John Milliman, and James Syvitsky provided us with many valuable ideas. W. R. Muehlberger at the University of Texas at Austin kindly provided access to the NASA shuttle photographs of the Ebro, Po, and Paraibo do Sul deltas included in Figure 4. Brian Willis and James MacEachern provided helpful reviews. We also thank memoir editor, Tom Chidsey for inviting this contribution. Gus Gustason and Tom Ryer are also thanked for introducing us to these fabulous outcrops, back in 1991. Despite all of the input, we take full responsibility for any shortfalls, pitfalls, or errors in our interpretations. Funding for this research was provided to by BP, ChevronTexaco, and the American Chemical Society, ACS-PRF Grant # 35855-AC8 to J. P. Bhattacharya. This is contribution number 1003 of the Geosciences Department, University of Texas at Dallas.

REFERENCES CITED

Barton, M. D., 1994, Outcrop characterization of architecture and permeability structure in fluvial-deltaic sandstones, Cretaceous Ferron Sandstone, Utah: Ph.D. dissertation, University of Texas at Austin, Austin, Texas, 259 p.

—1997, Application of Cretaceous Interior Seaway outcrop investigations to fluvial-deltaic reservoir characterization — pt. 1, predicting reservoir heterogeneity in delta front sandstones, Ferron gas field, central Utah, in K. W. Shanley and R. F. Perkins, eds., Shallow marine and nonmarine reservoirs, sequence stratigraphy, reservoir architecture and production characteristics: Society for Sedimentary Geology (SEPM) Gulf Coast Section Foundation 18th Annual Research Conference, p. 3–40.

Bhattacharya, J. P., 1993, The expression and interpretation of marine flooding surfaces and erosional surfaces in core; examples from the Upper Cretaceous Dunvegan Formation in the Alberta foreland basin, in C. P. Summerhayes and H. W. Posamentier, eds., Sequence stratigraphy and facies associations: International Association of Sedimentologists Special Publication No. 18, p. 125–160.

Bhattacharya, J. P., and L. Giosan, 2003, Wave-influenced deltas — geomorphological implications for facies reconstruction: Sedimentology, v. 50, p. 1–24.

Bridge, J. S., 1993, The interaction between channel geometry, water flow, sediment transport and deposition in braided rivers, in J. L. Best and C. S. Bristow, eds., Braided rivers, Geological Society Special Publication No. 75, p. 13–71.

Bridge, J. S, and S. D. Mackey, 1993, A theoretical study of fluvial sandstone body dimensions, in S. S. Flint and I. D. Bryant, eds., The geological modeling of hydrocarbon reservoirs and outcrop analogues: International Association of Sedimentologists Special Publication No. 15, p. 3–20.

Bridge, J. S., and R. S. Tye, 2000, Interpreting the dimensions of ancient channel bars, channels, and channel belts from wireline-logs and cores: AAPG Bulletin, v. 84, no. 8, p. 1205–1228.

Cant, D. J., 1992, Subsurface facies analysis, in R. G. Walker and N. P. James, eds., Facies models — response to sea level change: Geological Association of Canada, St. John's, Newfoundland, p. 27–45.

Coleman, J. M., and L. D. Wright, 1975, Modern river deltas — variations of processes and sand bodies, in M. L. Broussard, ed., Deltas — models for exploration: Houston Geological Society, p. 99–149.

Collinson, J. D., 1978, Vertical sequence and sand body shape in alluvial sequences, in A. D. Miall, ed., Fluvial sedimentology: Canadian Society of Petroleum Geologists Memoir 5, p. 577–586.

Corbeanu, R. M., K. Soegaard, R. B. Szerbiak, J. B. Thurmond, G. A. McMechan, D. Wang, S. H. Snelgrove,

C. B. Forster, and A. Menitove, 2001, Detailed internal architecture of a fluvial channel sandstone determined from outcrop, cores and 3-D ground-penetrating radar — example from the Mid-Cretaceous Ferron Sandstone, east-central Utah: AAPG Bulletin, v. 85, p. 1583–1608.

Cotter, E., 1975a, Deltaic deposits in the Upper Cretaceous Ferron Sandstone, Utah, *in* M. L., Broussard, ed., Deltas, models for exploration: Houston Geological Society, p. 471–484.

—1975b, Late Cretaceous sedimentation in a low-energy coastal zone — the Ferron Sandstone of Utah: Journal of Sedimentary Petrology, v. 45, no. 3, p. 669–685.

Dalrymple, R. W., R. Boyd, and B. A. Zaitlin, 1994, History of research, valley types and internal organization of incised-valley systems, *in* R. W. Dalrymple, R. Boyd, and B. A. Zaitlin, eds., Incised-valley systems — origin and sedimentary sequences: Society for Sedimentary Geology (SEPM) Special Publication No. 51, p. 3–10.

Dean, W.E., and M. A. Arthur, 1998, Cretaceous Western Interior Seaway drilling project — an overview, *in* W. E. Dean and M. A. Arthur, eds., Stratigraphy and paleoenvironments of the Cretaceous Western Interior Seaway, U.S.A.: Society for Sedimentary Geology (SEPM) Concepts in Sedimentology and Paleontology No. 6, p. 1–10.

Ecological Research Associates, 1983, Sagavanirktok River — hydrology and hydraulics: Environmental Summer Studies (1982) for the Endicott Development, v. II : Physical Processes, p. 31–43.

Fielding, C. R., and R. C. Crane, 1987, An application of statistical modeling to the prediction of hydrocarbon recovery factors in fluvial reservoir sequences, *in* F. G. Ethridge, R. M. Flores, and M. D. Harvey, eds., Recent developments in fluvial sedimentology: Society for Sedimentary Geology (SEPM) Special Publication 39, p. 321–327.

Fouch, T. D., T. F. Lawton, D. J. Nichols, W. B. Cashion, and W. A. Cobban, 1983, Patterns and timing of synorogenic sedimentation in Upper Cretaceous rocks of central and northeast Utah, *in* M. W. Reynolds and E. D. Dolly, eds., Mesozoic paleogeography of west-central United States: Society for Sedimentary Geology (SEPM) Rocky Mountain Section, Rocky Mountain Paleogeography Symposium 2, p. 305–336.

Gardner, M. H., 1995, Tectonic and eustatic controls on the stratal architecture of mid-Cretaceous stratigraphic sequences, central western interior foreland basin of North America, *in* S. L. Dorobek and G. M. Ross, eds., Stratigraphic evolution of foreland basins: Society for Sedimentary Geology (SEPM) Special Publication No. 52, p. 243–281.

Knox, P. R., 1997, Application of Cretaceous Interior Seaway outcrop investigations to fluvial-deltaic reservoirs II — example from Gulf of Mexico reservoirs, Frio Formation, South Texas, *in* K. W. Shanley and R. F. Perkins, eds., Shallow marine and nonmarine reservoirs, sequence stratigraphy, reservoir architecture and production characteristics: Society for Sedimentary Geology (SEPM) Gulf Coast Section Foundation 18th Annual Research Conference, p. 127–138.

Knox, P. R., and M. D. Barton, 1999, Predicting interwell heterogeneity in fluvial-deltaic reservoirs — effects of progressive architecture variation through a depositional cycle from outcrop and subsurface observations, *in* R. Schatzinger and J. Jordan, eds., Reservoir characterization-recent advances: AAPG Memoir 71, p. 57–72.

Le Clair, S. F., and J. S. Bridge, 2001, Quantitative interpretation of sedimentary structures formed by river dunes: Journal of Sedimentary Research, v. 71, p. 713–716.

Lorenz, J. C., D. M. Heinze, J. A. Clark, and C. A. Searls, 1985, Determination of widths of meander-belt sandstone reservoirs from vertical downhole data, Mesaverde Group, Piceance Creek Basin, Colorado: AAPG Bulletin, v. 69, p. 710–721.

Lowry, P., and T. Jacobsen, 1993, Sedimentology and reservoir characterization of a fluvial-dominated delta-front sequence — Ferron Sandstone Member (Turonian), east-central Utah, U.S.A., *in* M. Ashton, ed., Advances in reservoir geology: Geological Society of America Special Publication No. 69, p. 81–103.

Mack, G. H., W. C. James, and H. C. Monger, 1993, Classification of paleosols: Geological Society of America Bulletin, v. 105, p. 129–136.

Mack, G. H., and W. C. James, 1994, Paleoclimate and the global distribution of paleosols: Journal of Geology, v. 102, p. 1360–1366.

Matthai, H. F., 1990, Floods, *in* M. G. Wolman and H. C. Riggs, eds., The geology of North America: Geological Society of America, Surface Water Hydrology, v. 0–1, p. 97–120.

McMechan, G. A., G. C. Gaynor, R. B. Szerbiak, 1997, Use of ground-penetrating radar for 3-D sedimentological characterization of clastic reservoir analogs: Geophysics, v. 62, p. 786–796.

Meade, R. H., 1996, River-inputs to major deltas, *in* J. D. Milliman and B. U. Haq, eds., Sea-level rise and coastal subsidence: Kluwer Academic, p. 63–85.

Mulder, T., and J. P. M. Syvistky, 1995, Turbidity currents generated at river mouths during exceptional discharges to the world's oceans: Journal of Geology, v. 103, p. 285–299.

Novakovic, D., C. D. White, R. M. Corbeanu, W. S. Hammon III, J. P. Bhattacharya, and G. A. McMechan, 2002, Hydraulic effects of shales in fluvial-deltaic deposits — ground-penetrating radar, outcrop observations, geostatistics and three-dimensional flow modeling for the Ferron Sandstone, Utah: Mathematical Geology, v. 34, p. 857–893.

Posamentier, H. W., and G. P. Allen, 1999, Siliciclastic sequence stratigraphy — concepts and applications: Society for Sedimentary Geology (SEPM) Concepts in Sedimentology and Paleontology, no. 7, 210 p.

Reynolds, A. D., 1999, Dimensions of paralic sandstone bodies: AAPG Bulletin, v. 83, no. 2, p. 211–229.

Rubin, D. M., and D. S. McCulloch, 1980, Single and superimposed bedforms — a synthesis of San Francisco Bay and flume observations: Sedimentary Geology, v. 26, p. 207–231.

Russell, R. J., 1942, Geomorphology of the Rhône Delta: ANNALS, Association of American Geographers, v. 32, no. 2, p. 49–254.

Ryer, T. A., and M. McPhillips, 1983, Early Late Cretaceous paleogeography of east-central Utah, *in* M. W. Reynolds, and E. D. Dolly, eds., Mesozoic paleogeography of west-central United States: Society for Sedimentary Geology (SEPM) Rocky Mountain Section, Rocky Mountain Paleogeography Symposium 2, p. 253–272.

Smith, A. R., 1966, Modern delta comparison maps, *in* M. L. Shirley and J. A. Ragsdale, eds., Deltas in their geologic framework: Houston Geological Society, p. 233–251.

Thompson, S. L., C. R. Ossian, and A. J. Scott, 1986, Lithofacies, inferred processes, and log response characteristics of shelf and shoreface sandstones, Ferron Sandstone, central Utah, *in* T. F. Moslow and E. G. Rhodes, eds., Modern and ancient shelf clastics — a core workshop: Society for Sedimentary Geology (SEPM) Core Workshop No. 9, p. 325–361.

Tye, R. S., and T. F. Moslow, 1993, Tidal inlet reservoirs — insights from modern examples, *in* E. G. Rhodes and T. F. Moslow, eds., Marine clastic reservoirs: New York, Springer-Verlag, p. 77–99.

van Heerden, I. L., and H. H. Roberts, 1988, Facies development of Atchafalaya delta, Louisiana — a modern bayhead delta: AAPG Bulletin, v. 72, no. 4, p. 439–453.

Van Wagoner, J. C., R. M. Mitchum, K. M. Campion, and V. D. Rahmanian, 1990, Siliciclastic sequence stratigraphy in well logs, cores and outcrops: AAPG Methods in Exploration Series No. 7, 55 p.

Welder, F. A., 1959, Processes of deltaic sedimentation in the lower Mississippi River: Defense Technical Information Center, Alexandria, Virginia, 90 p.

Wells, J. T., S. J. Chinburg, and J. M. Coleman, 1984, The Atchafalaya River delta — generic analysis of delta development: U.S. Army Corps of Engineers, New Orleans District, Technical Report HL-82–15, Contract No. DACW39-80-C-0082, 89 p.

Williams, G. D., and C. R. Stelck, 1975, Speculations on the Cretaceous paleogeography of North America, *in* W. G. E. Caldwell, ed., The Cretaceous System in the Western Interior of North America: The Geological Association of Canada Special Paper No. 13, p. 1–20.

Williams, G. P., 1986, River meanders and channel size: Journal of Hydrology, v. 88, p. 147–164.

Willis, B. J., 1997, Architecture of fluvial-dominated valley-fill deposits in the Cretaceous Fall River Formation: Sedimentology, v. 44, p. 735–757.

Wright, L. D., 1977, Sediment transport and deposition at river mouths — a synthesis: Geological Society of America Bulletin, v. 88, p. 857–868.

Wright, V. P., 1992, Paleopedology, stratigraphic relationships and empirical models, *in* I. P. Martini and W. Chesworth, eds., Weathering, soil and paleosols: Amsterdam, Elsevier, Developments in Earth Surface Processes, no. 2, p. 475–499.

U. S. Geological Survey field camp near Emery, Utah, circa 1910. Photograph courtesy of the family of C. T. Lupton.

Analog for Fluvial-Deltaic Reservoir Modeling: Ferron Sandstone of Utah
AAPG Studies in Geology 50
T. C. Chidsey, Jr., R. D. Adams, and T. H. Morris, editors

Facies of the Ferron Sandstone, East-Central Utah

Thomas A. Ryer[1] and Paul B. Anderson[2]

ABSTRACT

The Upper Cretaceous Ferron Sandstone represents a spectrum of depositional environments and facies spanning offshore marine to alluvial plain. Because they are very well exposed, are readily accessible, and have been extensively studied, these deposits serve as excellent analogs for many oil and gas reservoirs. Sediment was delivered to the Ferron depositional system by eastward- to northward-flowing rivers represented by sandy channelbelts. The rivers were generally meandering, although some were lower-sinuosity streams. Flood basins adjacent to the channelbelts accumulated predominantly muddy sediment, sandy crevasse-splay deposits, and, locally, peat. Peat accumulated in belts that roughly correspond to the lower part of the coastal plain and generally paralleled the shoreline. The geometries of individual, thick bodies of coal vary greatly. The dynamics of peat accumulation was controlled primarily by the rate of relative sea level rise.

The balance between sediment supply and wave energy was such that most Ferron shorelines were characterized by wave-dominated, cuspate deltas that graded laterally into strandplains. During the early part of Ferron deposition, however, the supply of sediment was great enough that fluvial processes predominated locally. Lobate deltas with numerous distributaries and interdistributary bays were well represented. Shoreface and delta-front deposits graded seaward into offshore marine mud. Sandy Ferron shoreline strata interfinger extensively with marine shale. Elongated sand bodies or sand plumes accumulated on a shallow-shelf area that lay east and northeast of the Ferron shoreline during early Ferron deposition.

Barrier islands developed during transgressions and are preserved, along with the lagoons that lay landward of them, at the "turnaround" points where episodes of transgression ended and shoreface progradation began. Landward pinchouts of the shoreface sandstone bodies are a key element for deciphering the transgressive-regressive history of the Ferron Sandstone. Tidal inlets and associated flood-tidal deltas are present locally in the vicinities of the landward pinchouts.

[1]The ARIES Group, Inc., Katy, Texas
[2]Consulting Geologist, Salt Lake City, Utah

INTRODUCTION

The Upper Cretaceous Ferron Sandstone consists predominantly of fluvial-deltaic deposits. The majority of Ferron sand bodies that could be reservoirs for substantial volumes of oil and gas and, therefore, may constitute useful analogs, were deposited on prograding shorelines and by the fluvial channels that supplied the sediment to the shorelines. As a whole, the Ferron Sandstone exposures in the southern part of Castle Valley (Figure 1) have roughly equal amounts of these two principal sand-body types. There is great variation, however, in the abundance of facies from place to place along the Ferron outcrop belt: fluvial strata and their associated overbank deposits predominate in the southwest, grading seaward to predominantly shoreline strata in the northeast.

Gardner (1993) included very detailed discussions of facies, their rock types, sedimentology, and ichnology: his dissertation stands as the most thorough treatment of Ferron facies. Here, a very generalized discussion of Ferron facies is offered with the goal of "setting the stage" for the papers that follow. Facies are discussed generally from seaward to landward, as they would occur in a stratigraphic succession recording progradation of the shoreline.

SYNOPSIS OF FERRON STRATIGRAPHY AND DEPOSITIONAL HISTORY

Stratigraphy

The Ferron Sandstone Member of the Mancos Shale is an eastward-thinning clastic wedge deposited during Turonian-Coniacian (Upper Cretaceous) time. Lower and upper parts of the Ferron are distinguished. The Lower Ferron consists, on outcrop, of shelf sandstone deposits that were transported generally from north to south; the Upper Ferron consists of fluvio-deltaic deposits that prograded generally from southwest to northeast. Nine correlatable and mappable units, designated Kf-Last Chance through Kf-8 (Figure 2), are recognized in the Upper Ferron. Each unit records a transgressive-regressive cycle of sedimentation and is defined on the basis of marine-flooding surfaces, most of which are widespread. The shoreline sandstone units that define the cycles are characterized by an initial forward-stepping arrangement, followed by vertical-stacking, and then back-stepping arrangements. This architecture indicates an initial strong supply of sediment relative to the accommodation space within which sediment could accumulate, followed by near-balance, and then a relative decrease in sediment supply. The older shoreline units, specifically Kf-1 and Kf-2, have much greater lengths along the Ferron outcrop belt, which roughly parallels deposition dip, than do the younger units.

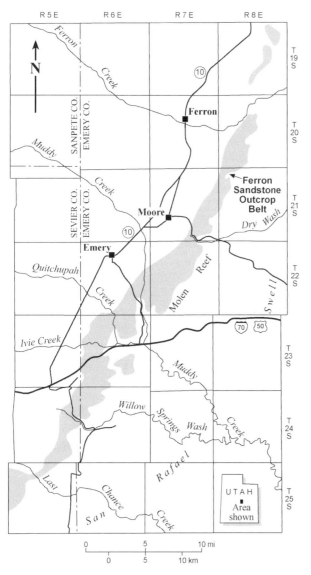

Figure 1. Location of the Ferron Sandstone study area in southern part of Castle Valley, between Ferron Creek and Last Chance Creek, east-central Utah. The Ferron outcrop belt is shown by shading.

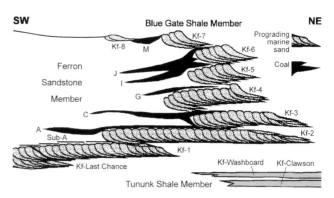

Figure 2. Diagrammatic cross section of the Ferron in the southern part of Castle Valley showing relative positions of Ferron depositional units and distinguishing their associated coal zones (black). The diagram has no scale.

Paleogeography

The Ferron Sandstone Member of the Mancos Shale and equivalent parts of the Frontier Formation in northern Utah and Wyoming record a widespread regression of the Western Interior Cretaceous seaway (Figure 3) during middle and late Turonian time. Rapid subsidence of a foredeep immediately east of the Sevier orogenic belt began in late Early Cretaceous time and continued through early Late Cretaceous time. Subsidence, however, was not uniform: during the Cenomanian (Figure 4A), the foredeep was strongly elongate and extended throughout Utah; during Turonian (Figure 4B) and Coniacian through Santonian time (Figure 4C), subsidence became concentrated in northern Utah and southwestern Wyoming.

The relatively straight, north-northeast-trending western shoreline of the seaway had transgressed to parts of western Utah during early Turonian time (Figure 5A). The shoreline encroached to within tens of miles of the Sevier orogenic belt in some areas, the closest that it was to come during the Cretaceous. The Tununk Shale accumulated at sub-wave-base depths in central Utah at this time. As the shoreline prograded eastward during middle Turonian time (Figure 5B), the shoreline configuration changed. A pronounced curve of the shoreline developed, reflecting interaction between sediment supply and subsidence. This regression is represented by the Ferron Sandstone in central Utah and by the Frontier

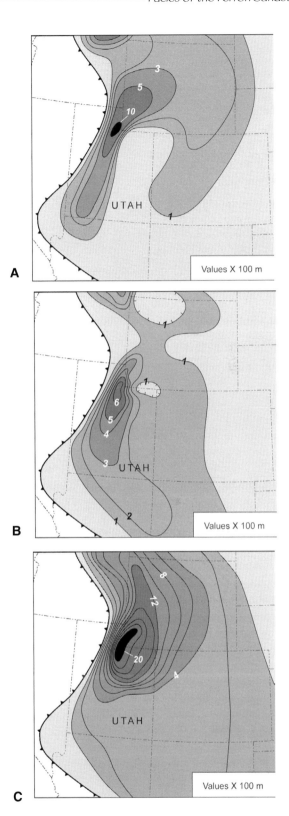

Figure 4. (A) Isopach map of Cenomanian strata. A pronounced fore-deep lay east of and parallel to the Sevier orogenic belt, whose eastern edge is approximated by the thrust fault symbol. (B) Isopach map of Turonian strata. The foredeep remained, but subsidence became concentrated in northeastern Utah and southwestern most Wyoming. (C) Isopach map of Coniacian and Santonian strata. Subsidence became strongly focused in northeastern Utah and southwestern Wyoming. Modified from Roberts and Kirschbaum, 1995.

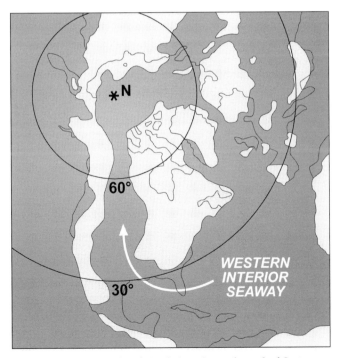

Figure 3. Paleogeography of North America at the peak of Cretaceous transgression during early Turonian time (modified from Hay et al., 1993). Light areas represent landmasses; dark areas represent water bodies.

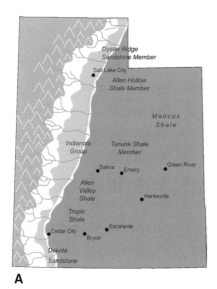

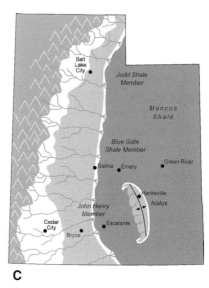

A   **B**   **C**

Figure 5. (A) Paleogeography of Utah during early Turonian time. Labels identify principal stratigraphic units deposited at this time. (B) Paleogeography during middle Turonian time. (C) Paleogeography during Coniacian time. Coniacian strata are missing in the Henry Mountains basin of southeastern Utah (Peterson et al., 1980). This is interpreted as evidence of erosion on an island or shoal located along the crest of a peripheral bulge.

Formation in northeastern Utah. In northern Utah and southwestern Wyoming, subsidence was rapid, but sediment supply was more than enough to fill the depositional space created by subsidence. In central and southern Utah, subsidence and sediment supply were more nearly balanced. The embayment that developed in central Utah during Turonian time probably reflects the northward shift of subsidence during the Turonian.

Hale (1972) and Cotter (1975a, 1975b) interpreted the bulges of the shoreline that developed in Utah during middle Turonian time to represent deltas, the Vernal delta to the north and the Last Chance delta to the south. Ryer and Lovekin (1986) argued that the embayment in central Utah was the result of a supply of sediment in central Utah that was inadequate to fill the space created by subsidence and owes its origin more to tectonics than to the distribution of rivers. The question of just what constitutes an ancient delta aside, all previous investigators have agreed that the Ferron in Castle Valley accumulated on deltaic to wave-dominated coastal environments in a rapidly subsiding portion of the Cretaceous foreland basin.

Transgression occurred during Coniacian time (Figure 5C). The sea transgressed very rapidly across the Ferron coastal plain to a position near the western edge of the Wasatch Plateau. Coniacian strata are not known in the Henry Mountains basin, to the southeast, indicating that an unconformity exists there between continental strata of the Ferron and overlying marine shales of the Blue Gate Shale. This unconformity may represent uplift of and erosion on a peripheral bulge. The erosion may have occurred subaerially, as pictured in Figure 5C, or entirely in marine environments.

Ferron Architecture

Ferron stratigraphy and architecture is shown diagrammatically in Figure 2. The Clawson and Washboard units, which constitute the Lower Ferron, are shelf sandstone deposits that drifted southward along a shoal that likely corresponded to the hinge of the foreland basin, sourced from the shoreline in the vicinity of the present-day Book Cliffs. The Upper Ferron consists of fluviodeltaics that were generally transported from southwest to northeast.

Ryer (1981) recognized that the Upper Ferron consists of a series of stacked, deltaic units that can be defined in outcrop on the basis of cliff-forming shoreline sandstone bodies. These record transgressive-regressive cycles of sedimentation. Five major, correlatable, sandy delta-front units, Kf-1 through 5 in ascending order, plus three less widespread delta-front sandstones, later designated Kf-6, Kf-7, and Kf-8, were mapped. Each delta-front unit, with the exception of Kf-8, was found to have an associated coal bed. The coals carry letter designations originally assigned to them by Lupton (1916). Gardner (1993, 1995a, 1995b), Barton (1994), Garrison et al. (1997) and Anderson et al. (this volume) have completed recent studies of the stratigraphy of the Ferron. All of these authors place the Ferron in sequence stratigraphic frameworks, although their interpretations, particularly with respect to the number and placement of sequence boundaries, differ significantly.

OFFSHORE-MARINE DEPOSITS

The Tununk and Blue Gate Shale Members of the Mancos Shale, which under- and overlie, respectively, the

Figure 6. Section exposed on the cliffs of Molen Reef, south of Dry Wash. Dark gray units are marine shales. The Lower Ferron consists of the Clawson and Washboard shelf-sand units: only the upper, Washboard Unit is visible is this photo. A unit of marine shale separates the Lower and Upper Ferron. Units Kf-2 through Kf-5 of the Upper Ferron are labeled.

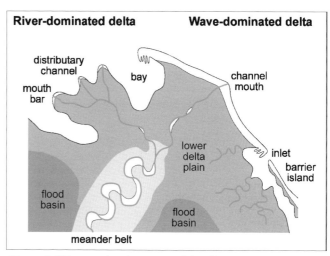

Figure 7. Diagram showing range of shoreline types present in the Ferron Sandstone.

Ferron Sandstone Member, were deposited in offshore-marine paleoenvironments at depths below storm-wave base. Offshore-marine shale is a relatively minor component of the Ferron in the study area, found in increasing amounts toward the northeast. Units of marine shale that separate the shoreline sandstone units of the Ferron in the northeastern part of the study area (Figure 2) may be considered as representing southwestward-thinning tongues of the main body of the Mancos Shale.

SHELF SAND DEPOSITS

The Clawson and Washboard units of the Ferron Sandstone (Kf-Clawson and Kf-Washboard, respectively, of our terminology; Figures 2 and 6) are commonly referred to together as the Lower Ferron. They were deposited on the shelf ranging from about storm-wave base to somewhat shallower than fair-weather wave base. They represent an elongate deposit (or "plume") of sand that was transported south-southwestward on the shelf (Thompson et al., 1986) from the bulge of the shoreline commonly referred to as the "Vernal delta."

The Clawson and Washboard units consist predominantly of very fine and fine-grained sandstone. Most of the sandstone has been bioturbated. Planar lamination and oscillation ripple lamination are the most common sedimentary structures in these sandstones where physical structures have not been obliterated by burrowing. Both units generally coarsen upward from a gradational base on marine shale; both the Clawson and Washboard have abrupt tops that correspond to marine-flooding surfaces.

In the southern part of Castle Valley, the Clawson and Washboard units form minor ledges beneath the

larger cliffs formed by the shoreline sandstone units of the Upper Ferron (Figure 6). As the shoreline sandstones of the Upper Ferron grade northeastward into marine shale, the ledges formed by the Clawson and Washboard units become relatively more significant until, in the northernmost part of the study area, the Clawson becomes the most conspicuous feature of the Ferron outcrop belt. The sandstone content of the units varies: in the north, the Clawson unit generally forms a more prominent ledge than does the Washboard unit, while in the south the situation is reversed.

SHORELINE DEPOSITS

Shoreline types preserved in the Ferron Sandstone range from wave-dominated to strongly fluvial-dominated. Figure 7 shows the range in shoreline types interpreted to have existed in the Ferron. Although highly diagrammatic, this figure shows that the shoreline orientation had a large influence on shoreline type. The western shoreline of the Interior Cretaceous Seaway generally trended northward (Figures 3 and 5A) but, because of the aforementioned embayment in central Utah (Figure 5B), the shoreline in the study area generally trended northwestward. Shoreline orientations varied locally, however, particularly on the deltas. Ferron shorelines that faced eastward or northeastward were reworked by waves; much of the sand from east-facing, wave-dominated deltas was carried along shore to accumulate in adjacent strandplains. Ferron deltas that built northward or northwestward, toward the subsiding area (Figure 4B and 4C), faced the embayment and were sheltered from waves. This explains why the most strongly fluvial-dominated deltaic deposits of the Ferron accumulated on deltas that prograded toward the north or northwest, an observation that was first made by Cotter (1975a, 1975b).

There is clear evidence of tidal deposits preserved in

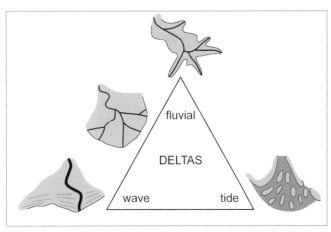

Figure 8. Ternary diagram showing the principal forces that shape delta morphology: fluvial input of sediment, wave reworking of sediment, and reworking of sediment by tidal currents (after Galloway, 1975).

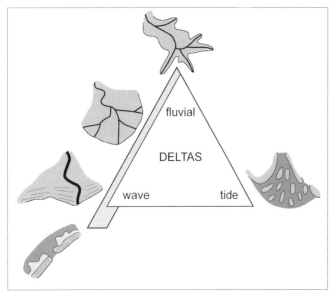

Figure 9. Shoreline types recognizable in the Ferron Sandstone include wave-dominated, non-deltaic shorelines deposited as prograding strand plains and on barrier islands (after Galloway, 1975).

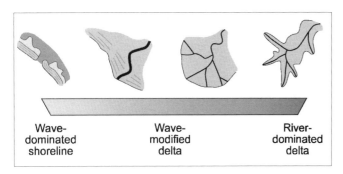

Figure 10. Diagram showing the three basic categories of shoreline types distinguished in the Ferron Sandstone. There is a gradation of types over the recognized range and the three types are arbitrarily defined.

Ferron strata (primarily in deposits of bays, lagoons, and tidal inlets, all of which are discussed later), but no evidence to indicate that the tidal range exceeded microtidal. Delta types in the Ferron Sandstone range from wave-dominated to fluvial-dominated (Figure 8). Shoreline strata of the Ferron as a whole, therefore, can be considered to lie along a linear axis extending from strongly wave-dominated deposits of the shoreface through wave-dominated deltas to strongly fluvial-dominated deltas (the later two being included as the upper apex and lower left apex in Figures 9 and 10).

Three basic shoreline types are distinguished in the Ferron Sandstone: wave-dominated (i.e. shoreface), wave-modified deltaic, and fluvial-dominated deltaic. The boundaries between these categories are arbitrary: examples representing the entire range between the end members can be recognized on the Ferron outcrops. The end members are described first, then the intermediate, wave-modified deltaic category.

WAVE-DOMINATED SHORELINES

Shoreface

Deposits of wave-dominated shorelines (Figure 11) predominate in the Ferron Sandstone. A typical outcrop of a wave-dominated, shoreface deposit is shown in Figure 12. The thickness of Ferron shoreface successions is generally from 30–60 ft (9–18 m). Stratigraphic successions of the Ferron deposited on wave-dominated shorelines (both wave-dominated deltas and shorefaces) have one or more shoreface subfacies: transition zone; lower, middle, and upper shoreface; and foreshore.

Most of the shoreline sandstone bodies that contain these subfacies accumulated on prograding shorefaces that were not directly influenced by rivers, although they were undoubtedly nourished largely with sediment transported by longshore drift from river mouths. Some

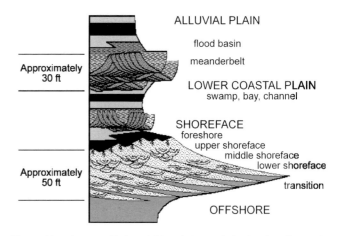

Figure 11. Diagram distinguishing facies and, in the shoreface, subfacies within a depositional sequence formed by progradation of a wave-dominated coast.

of the shoreline deposits may represent the margins of strongly wave-modified deltas, areas that morphologically would be included in the delta but which lack active distributary channels and that receive most of their sediment by longshore drift.

Transition Zone and Lower Shoreface

The transition zone and lower shoreface include interbedded sandstone and mudstone deposited between storm and fair-weather wave bases. Strata assigned to the transition zone consist of mudstone with thin beds of sandstone. Sandstone constitutes less than 50% of the rock. The sandstone typically occurs in thin beds that have abrupt bases, are predominantly planar laminated in their lower parts, and commonly grade upward to oscillation-ripple laminated tops. Thicker beds may include hummocky/swaley cross-stratification. The beds generally fine upward to mudstone. Burrows of invertebrate organisms are rare to common. Burrowing is most common in the uppermost parts of the sandstone beds, affecting the oscillation-rippled tops. The sandstone beds represent storm layers deposited below fair-weather wave base.

The term "lower shoreface," as used here, designates the section of interbedded sandstone and mudstone overlying the transition zone and underlying the middle shoreface. It is characterized by the same style of bedding as the transition zone, but the amount of sandstone exceeds 50% and the sandstone beds tend to be thicker and generally contain more hummocky/swaley cross-stratification (Figures 13 and 14). The combined transition zone and lower shoreface part of the succession is thin, commonly 10 ft (3 m) or less.

Middle Shoreface

The middle shoreface (Figure 15) consists of planar laminated and hummocky/swaley cross-stratified sand-

Figure 13. Lower shoreface strata of Kf-5 in Muddy Creek Canyon. The abrupt-based beds of fine-grained sandstone, averaging about 6 in. (14 cm) in thickness, were deposited during storms. The thinner layers of mudstone that separate them represent fair-weather sedimentation. A transgressive surface of erosion (tse) separates the basal storm layer from an underlying interval of floodbasin mudstone assigned to Kf-4.

Figure 12. Thick shoreface unit (parasequence Kf-2-Miller Canyon-b as defined by Anderson and Ryer, this volume) on the east side of Muddy Creek Canyon. The sandstone has a well-developed "white cap." The abrupt, flat contact at its top is a marine-flooding surface.

Figure 14. Hummocky and swaley cross-stratification are common in lower and middle shoreface strata (parasequence Kf-2-Muddy Canyon-b as defined by Anderson and Ryer, this volume). In this example, the strata are only sparsely burrowed.

Figure 15. Middle shoreface strata (parasequence Kf-1-Indian Canyon-a) in Indian Canyon. These strata are intensely burrowed. Remnant flat bedding is apparent. The top of the sandstone unit is a transgressive surface of erosion overlain by transition zone strata of the next parasequence.

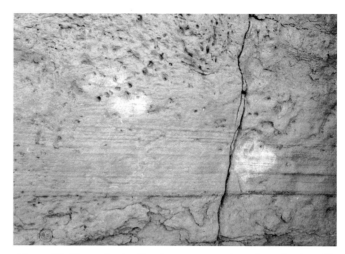

Figure 17. Planar lamination in a remnant storm layer in the middle shoreface (parasequence Kf-2-Muddy Canyon-b) in Muddy Creek Canyon. Burrowing has extended downward from the top of the storm bed, obliterating the original physical structures in its upper part. The burrows did not extend downward to the base of the storm layer, which abruptly places laminated sandstone on burrowed sandstone of the preceding storm layer.

Figure 16. Oscillation ripples in the middle shoreface of Kf-2-Miller Canyon-b in Muddy Creek Canyon. Gastropod grazing traces are visible in some of the ripple troughs. U.S. quarter for scale.

stone. These beds represent storm layers deposited above fair-weather wave base. Mudstone layers are generally absent in the middle shoreface. In the Ferron, middle-shoreface sandstone typically is very fine to fine-grained. Oscillation ripples are commonly preserved in the upper parts of individual storm beds (Figure 16). Most of the sandstone occurs in sharp-based beds whose tops are burrowed to bioturbated (Figure 17).

The middle shoreface is commonly thick, 15–30 ft (5–9 m), representing 50% or more of the total thickness of the shoreface succession. Middle-shoreface deposits may include abundant burrows. In any given vertical shoreface section, the degree of burrowing in the middle shoreface is usually greater than in the underlying, storm-layered sandstones of the lower shoreface. The diversity of burrow types is great, with *Ophiomorpha* and *Thalassi-noides* (Figure 18) being the most conspicuous forms.

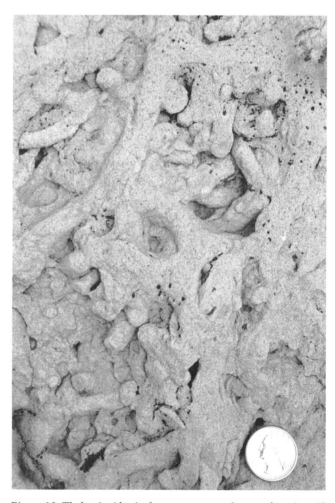

Figure 18. *Thalassinoides* is the most common burrow form in middle-shoreface deposits of the Ferron. This example of a bioturbated middle-shoreface deposit outcrops (parasequence Kf-2-Muddy Canyon-b) in Muddy Creek Canyon.

Figure 19. Upper-shoreface strata (parasequence Kf-2-Miller Canyon-b) in Muddy Creek Canyon. The strata are predominantly trough cross-stratified and display high variability in transport direction. Burrows at the top of the photo extend downward from a transgressive surface of erosion: these <u>Ophiomorpha</u> and <u>Thalassinoides</u> burrows are superimposed from the lower-shoreface to offshore environments that existed during and following the transgression. Strata in the lower, center part of the photo are shown in Figure 20.

Figure 20. Trough cross-stratification of the upper shoreface. The bed in the middle part of the photo includes burrows (although burrows are generally rare upper shoreface deposits of the Ferron). The burrows are truncated at the top of the bed, indicating that they existed in the upper-shoreface environment and were not superimposed, as were the burrows at the top of parasequence Kf-2-Miller Canyon-b, as shown in Figure 19.

Upper Shoreface

The upper shoreface is characterized by a distinctly different set of sedimentary structures than the underlying middle shoreface and the overlying foreshore. It consists of trough cross-stratified sandstone with only minor amounts of planar lamination (Figures 19 and 20). The upper shoreface is usually coarser than the middle shoreface. In the Ferron Sandstone, upper-shoreface strata are typically composed of medium-grained sandstone. The transport directions recorded by the trough cross-stratification are generally highly variable, but oblique onshore directions tend to predominate. The upper/middle shoreface contact is generally abrupt and slightly erosional, possibly as a result of scouring by rip-current channels.

Where the upper shoreface is well developed, it

commonly has a thickness of 5–10 ft (1.5–3 m). The thickness may reach 20 ft (6 m) or more near landward pinchouts of parasequence-level shoreline units, particularly where the underlying transgressive surface of erosion overlies carbonaceous shale or coal and syndepositional compaction of the underlying, organic-rich material occurred.

Foreshore

A foreshore can commonly be recognized at the top of shoreline successions that include upper-shoreface strata. The foreshore consists of planar-laminated sandstone (Figure 21) in sets that dip seaward at angles of a few degrees to as much as about 10–15 degrees. Foreshore deposits in the Ferron Sandstone generally consist of fine-grained sandstone. The lamination is formed under upper-flow-regime conditions in the swash zone. Foreshore deposits in the Ferron rarely exceed about 5 ft

Figure 21. Upper-shoreface (sf-u) and foreshore (sf-fs) deposits of Kf-6 in Muddy Creek Canyon. The boundary between planar, gently seaward inclined foreshore strata and trough cross-stratified upper-shoreface strata is usually abrupt, as is the case here.

Figure 22. Cliff formed by a shoreface unit (parasequence Kf-1-Indian Canyon-c) in Indian Canyon. It overlies marine shale of the Tununk Shale Member. The sandstone is extensively rooted at its top and is overlain directly by carbonaceous shale of the Sub-A coal zone. The very flat top is typical of wave-dominated shoreline units. The uppermost part of the sandstone cliff weathers to a very light color and constitutes a "white-cap."

(1.5 m) in thickness. A "white cap" is commonly present, affecting foreshore and upper-shoreface strata. The top of the succession, except where affected by channels, is abrupt. It represents the berm of the beach and generally forms a flat bench that is easily traced on outcrops. Such a surface is shown in Figure 22. Foreshore deposits are also commonly root-penetrated and burrowed at their tops. Interestingly, coastal dune fields, so common along modern coasts, have not been recognized in the

Ferron. The berms of the prograding Ferron beaches apparently were quickly vegetated and thus stabilized.

Tidal Inlet

Tidal-inlet deposits have been recognized in several units of the Ferron. They replace the wave-dominated shoreline deposits locally, removing mostly middle shoreface strata (Figure 23). Normal upper-shoreface and foreshore strata occur at the tops of most of the tidal-inlet deposits. Tidal-inlet deposits are characterized by conspicuous, sloping bedding surfaces that record migration of the inlets along the coast (Figures 23, 24, and 25). They include fine- to medium-grained, predom-

Figure 23. Tidal-inlet deposits exposed in the west wall of Muddy Creek Canyon (parasequence Kf-2-Muddy Canyon-b). Note the left to right inclination of large planar tabular beds. These beds record northward migration of the tidal inlet. Although not continuously exposed here, the upper, light-colored sandstone bed represents upper-shoreface (sf-u) and foreshore environments, which migrated across the inlet as it filled. Shoreface sandstone of parasequence Kf-2-Miller Canyon-b forms the vertical cliff at the bottom of the photo.

Figure 24. Steeply dipping, planar-tabular beds of fine- to medium-grained sandstone characterize Ferron tidal inlets. In this photo, which shows the same inlet deposit shown in Figure 23, the strata dip away from the viewer, toward the north.

Figure 25. Closer view of planar-tabular beds shown in Figures 23 and 24. Cross-stratification (X), ranging from planar-tabular to large-scale trough, is preserved locally among the large, inclined, planar-tabular beds.

inantly laminated sandstone. Trough cross-stratification is also present (Figures 25 and 26). Burrows are rare.

Distribution of Wave-Dominated Deposits

Wave-dominated deposits are present in all nine of the Upper Ferron shoreline units. It is the predominant type of shoreline deposit in Kf-4 through Kf-8, but is also an important component in Kf-Last Chance through Kf-3. An important observation is that the oldest of the shoreline strata deposited in each of the parasequences distinguished in the Ferron Sandstone are wave-dominated. Barrier islands were common features on transgressive Ferron shorelines. Following cessation of transgression, the barrier islands prograded and the lagoons behind them filled. Even after the lagoons had filled, progradation typically continued in a wave-dominated mode. Only following a mile to several miles of progradation did conditions suitable for construction of wave-modified or river-dominated deltas come into existence. Tidal inlets existed concurrently with lagoons and so are limited to the landward-most parts of the parasequence-level shoreface sandstone bodies.

FLUVIAL-DOMINATED DELTA

General Characteristics

In the Ferron, fluvial-dominated deltaic deposits are generally distinguished by the following characteristics (Figure 27).

1. Thickness of the delta-front succession is relatively small, typically about 20–30 ft (6–9 m), indicating a shallow "mud line" and implying low wave energy (Figure 28).

Figure 26. Closer view of cross-stratification in tidal-inlet deposit. U.S. quarter in lower right for scale.

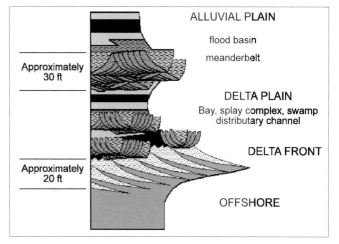

Figure 27. Diagram distinguishing facies within the depositional sequence formed by progradation of a fluvial-dominated delta.

Figure 28. Typical fluvial-dominated deltaic deposits (parasequence Kf-1-Ivie Creek-a as defined by Anderson et al., this volume) are exposed along the south escarpment of Mesa Butte, south of Interstate 70. The strongly bedded appearance of the fluvial-dominated strata contrasts with the massive appearance of the wave-dominated strata of overlying Kf-2.

Kf-2

Kf-1-Iv-a

Kf-Washboard

2. Mudstone interbeds are common in all but the uppermost part of the succession (Figure 29).

3. Inclination of beds is usually very gentle, but locally may become steep enough to be quite apparent on outcrops (Figure 30). Where this happens, sandstone beds and bedsets pinch rapidly seaward as they interfinger with mudstone.

4. The succession of sedimentary structures in sandstone beds varies greatly, but the most common pattern is upward gradation from storm layers characterized by planar lamination and hummocky/swaley cross-stratification to trough cross-stratification. Where present, trough cross-stratification indicates generally unidirectional, seaward flow.

5. Contemporaneous distributary channels and small channels that may represent tidal creeks or distributary channels (see Figure 30) are common, replacing the tops of the delta-front successions in many or most outcrops. As a result, the top of a fluvial-dominated delta-front succession generally lacks the abrupt, flat top characteristic of wave-dominated shoreline deposits.

6. Burrowing is sparse. *Planolites* and *Arenicolites* are the most common burrow forms. This suggests rapid sedimentation, reduction of salinities in the vicinity of the river mouth, or both.

The morphologies of fluvial-dominated deltas in the Ferron Sandstone cannot be defined precisely. The Kf-1-Ivie Creek-a parasequence at Ivie Creek and the Kf-1-Indian Canyon-c parasequence at Willow Springs Wash, as defined and described respectively by Anderson et al.

Figure 29. Typical fluvial-dominated deltaic-front facies represented by interbeds of fine-grained sandstone and mudstone (Kf-1-Ivie Creek-a in the Ivie Creek area), north side of Interstate 70, near the confluence of Quitchupah and Ivie Creeks.

(this volume) and Dewey and Morris (this volume) include well-documented examples of fluvial-dominated deltaic facies. In both cases, the deltas built to the northwest. Kf-1-Ivie Creek-a has been mapped as having a general lobate form. The delta or deltas included in Kf-1-Indian Canyon-c are less well defined, but may have had an elongate, possibly even "bird-foot" form.

Perhaps the most distinctive characteristic of fluvial-dominated deltaic deposits in the Ferron Sandstone is the presence of numerous beds of mudstone throughout much of the thicknesses of the delta-front depositional successions. This is the characteristic that is most notice-

Figure 30. *Steeply inclined beds of a fluvial-dominated delta-front deposit of Kf-1-Ivie Creek-a in the Ivie Creek area. Inclination is toward the west. Note the rapid thinning of individual sandstone beds as they descend westward and interfinger with marine mudstone. A small channel of the overlying unit replaces the uppermost delta-front strata near the center of the photo.*

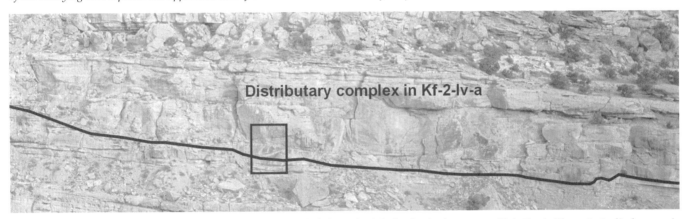

Figure 31. *Facies typical of the wave-modified delta front are included in Kf-2-Ivie Creek-a in the canyon of Ivie Creek. These strata display a variety of bedforms related to distributary channels and their mouth-bars. The rectangle indicates the position of the photo shown in Figure 32.*

able when viewing outcrops at a distance or when analyzing photomosaics. The presence of mudstone and the fact that beds of mudstone commonly occur within 10 ft (3 m) of the tops of the shoreline successions indicates that fair-weather periods were characterized by low wave energies on these deltas (Figure 31). By contrast, storm events, represented by the sharp-based, planar laminated to hummocky cross-stratified sandstone beds, impacted the deltas strongly, moving sand to water depths of about 30 ft (9 m).

Distribution of Fluvial-Dominated Deltaic Deposits

Fluvial-dominated deltaic deposits are an important component of Kf-1 and Kf-2. Their areal distributions have been documented by means of photomosaic mapping of the outcrop of these units within the study area (Anderson et al., 2003). They are not recognized in the other units.

WAVE-MODIFIED DELTA — INTERMEDIATE SHORELINE TYPES

General Characteristics

Many of the shoreline sandstone units of the Ferron Sandstone correspond to neither the wave-dominated nor the fluvial-dominated categories described above. They are commonly thick, like the shoreface sandstone sequences, but include more mudstone interbeds and have, as a result, more gradational bases. In the latter respect, they resemble the fluvial-dominated deltaic deposits. They commonly include distributary channels and associated mouth-bar deposits, collectively referred to as distributary complexes (Figures 31), indicating that river mouths were present during their deposition. This facies is dominated by trough cross-stratification, particularly in the upper portions of the delta-front successions. Stratigraphically lower in the successions are

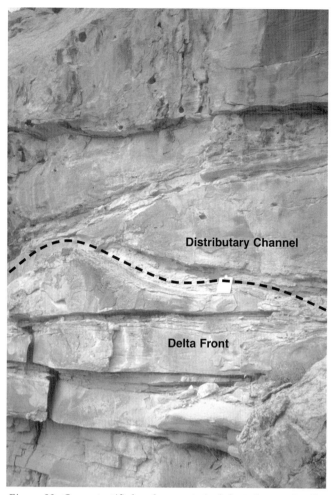

Figure 32. Cross-stratified and current-rippled sandstone of a distributary channel, part of the larger distributary complex shown in Figure 31, erosionally overlies hummocky/swaley storm deposits of the delta front in parasequence Kf-2-Ivie Creek-a.

Figure 33. Sandstone ledge representing filling of a lagoon formed by the final stages of transgression associated with Kf-6. Location is on the western wall of Muddy Creek Canyon, about 1 mi (1.6 km) south of the landward pinchout of the shoreface sandstone of Kf-6. The ledge consists of about 6 ft (2 m) of very fine and fine-grained, intensely burrowed sandstone. It is underlain by coal of the I coal zone and overlain, with a rooted contact, by coal of the J coal zone. The basal contact is shown in Figure 34.

Distribution of Wave-Modified Deltaic Deposits

Wave-modified deltaic deposits are common in Kf-1 through Kf-4 and are documented on photomosaics of the outcrop of these units within the study area (Anderson et al., 2003). They are not recognized in Kf-Last Chance or in Kf-5 through Kf-8.

BAY AND LAGOON

Mudstone, muddy sandstone, and sandstone were deposited in bays and lagoons associated with Ferron shorelines (Figures 33 and 34). Depositional successions representing bays and lagoons are usually no more than about 10 ft (3 m) thick. Sandstones are generally sparsely burrowed to bioturbated and contain horizontal bedding, commonly with oscillation ripples. In some cases, the oscillation ripples have flat tops, indicating intertidal to very shallow subtidal settings. Shells, particularly those of oysters and corbulid bivalve, are common locally.

It is generally difficult to distinguish deposits of bays and lagoons from each other on Ferron outcrops. Invertebrate faunas are sometimes of value in distinguishing whether a particular body of water contained brackish water or water of full marine salinity, but whether a particular deposit is interpreted to represent a bay or a lagoon is most often based on interpretations of contemporaneous shoreline deposits: bodies of marine water that lay landward of wave-dominated shorelines,

troughs complexly mixed with parallel and current ripple lamination (Figure 32). Commonly, these facies grade laterally into more typical wave-dominated shoreface deposits. Trace fauna are less abundant than in the shoreface deposits, but constitute the typical shoreface assemblage dominated by *Thalassinoides* and *Ophiomorpha*. We classify these as intermediate deposits representing wave-modified deltas.

Wave-modified deltaic deposits accumulated on delta-fronts that experienced more wave reworking than did the fluvial-dominated delta fronts. We have no examples in the Ferron Sandstone where the form of a wave-modified delta can be reconstructed with certainty. It is likely that these deltas were broad, gently lobate forms several miles in width. Some may have been arcuate or cuspate in form. Where the orientations of inclined strata deposited on the delta fronts can be determined accurately, they indicate that these deltas prograded toward the northeast or east. Oriented in this way, they were strongly affected by waves rolling in from the Interior Cretaceous Seaway.

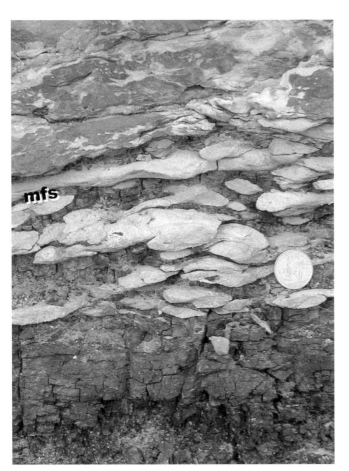

Figure 35. Kf-3 at Dry Wash consists of lower- to middle-shoreface sandstone erosionally overlain by a tidal inlet/tidal channel complex. The channel, which was probably closely associated with an inlet, migrated toward the north and displays prominent lateral accretion surfaces.

Figure 36. Lenticular, "ribbon" sands were deposited by minor, distributary channels that flowed into bays and lagoons on the lower coastal plain. The example shown here belongs to Kf-4 and crops out in a tributary of Quitchupah Canyon. The sandstone body, lens-shaped in cross section, occurs in a section of mudstone and carbonaceous shale.

Figure 34. Sand-filled burrows extend downward into the I coal from the marine-flooding surface (mfs) at the base of the lagoonal deposit of Kf-6. Unlike the transgressive surface of erosion (a portion of the marine-flooding surface) produced by wave erosion on a transgressing shoreface, the landward extension of the mfs into the lagoon represents a low-wave-energy, passive encroachment of bay waters over peat of the I swamp. The amount of erosion was minimal.

and specifically behind landward pinchouts of these bodies, are interpreted to be lagoons; bodies of brackish to marine water that are associated with fluvial-dominated or wave-modified deltaic deposits are interpreted to be bays.

Lagoonal deposits constitute only a very minor part of the Ferron Sandstone. The best-documented examples are located in Indian Canyon (landward of the landward pinchout of Kf-1-Indian Canyon-c), in the "Bear Gulch" and Miller Canyon exposures (Kf-2-Muddy Canyon-b), at Dry Wash (Kf-2-Dry Wash and Kf-2-Rochester), and in Muddy Creek Canyon (Kf-6; Figure 33).

Tidal channels are associated with bay and lagoon deposits and, like them, are relatively rare in the Ferron. A well-exposed example occurs in Dry Wash in strata of Kf-3 (Figure 35). The tidal-channel deposits share many of the features of tidal-inlet sandstones, the principal difference being the presence of numerous beds of mudstone. These range from a few feet to 20 ft (6 m) or more

in thickness. Figure 35 shows an example of a thick, slope-forming bed lying below the label marking the transgressive surface of erosion (tse) that bounds Kf-3 and Kf-4. This body consists mostly of inclined, tabular sets of medium-grained, well-sorted sandstone. Trough cross stratification, mostly ebb directed, is common. *Ophiomorpha* burrows are locally abundant in the sandstone beds.

Small, "ribbon" sandstone bodies were deposited by minor channels (Figure 36). Many of these are probably distributary channels associated with larger channels systems. They consist predominantly of fine-grained, current-rippled sandstone and commonly display lateral accretion surfaces. They were, presumably, the principal source of the sediments that ultimately filled the lagoons and bays.

MARSH AND SWAMP

General Characteristics

Carbonaceous strata ranging from mudstone to low-ash coal are common in the Ferron Sandstone. Coalbeds have been of considerable economic importance in the past; carbonaceous mudstone is presently being mined at several locations south of the town of Emery for use as a soil conditioner. Coalbeds and associated carbonaceous strata serve as extremely useful stratigraphic markers that allow correlation of shallow-marine and continental strata.

Thick beds of coal in the Ferron Sandstone (Figure 37; Doelling, 1972; Ryer, 1981) are the products of coalification of beds of sediment-free peat that accumulated in delta-plain settings and in swamps that lay between the principal meanderbelts in the alluvial plain. Carbonaceous shale, carbonaceous mudstone, and carbonaceous sandstone formed from peats that included substantial clastic components. These impure peats were deposited in settings where clastics were repeatedly introduced to the swamp, as occurs along the margins of channels.

Dynamics of Peat Accumulation

The dynamics of peat accumulation and preservation constitute a complicated subject that is addressed here only generally. Thick bodies of clean peat accumulate in settings where suitable vegetation flourishes; the swamp environments exists, uninterrupted, for a long period of time; clastic sediments are excluded; and the resulting sediment-free peat subsides below the water table.

The fourth term is probably the most critical. Subsidence of the peat, or rising of the water table, which facilitates the preservation of the peat, must occur at about the same rate as the generation of preservable organic material by the plants: if the water table rises faster than peat can be produced, the peat swamp is flooded to form a bay or shallow pond and peat deposition ceases; if the water table rises more slowly than peat accumulates, part of the peat oxidizes prior to burial, yielding a peat that is higher in impurities. In delta plains and in alluvial plains located along the margins of the sea, relative sea level is the primary factor that controls the long-term rise or fall of water level in swamps. It can be concluded that all of the conditions required for peat formation and preservation were optimal over long spans of time during Ferron deposition.

Coalbeds and Coal Zones

The Ferron Sandstone includes many beds of coal that can be correlated over considerable distances. Ferron coalbeds were originally named by Lupton (1916)

Figure 37. Coal beds are a conspicuous component of the Ferron. Several of them, including the C coal bed, shown here, are of economic importance. In the area of this outcrop, which occurs in a tributary of Quitchupah Canyon, the C coal attains a thickness of 18 ft (5.5 m). It is overlain by bay deposits, which in turn are capped by a "ribbon" sandstone representing a small channel.

and all subsequent workers have followed his terminology. The thickest and most widespread coals constitute the A, C, G, I, and J coalbeds. Coalbeds commonly split into two or more beds separated by clastic sediments. Traced laterally, coalbeds commonly thin and grade to carbonaceous mudstone. Coals, associated carbonaceous mudstones, plus non-carbonaceous strata included between the various splits of coal and beds of carbonaceous mudstone that are laterally equivalent to a principal coal bed are grouped into a "coal zone" (Doelling, 1972; Ryer, 1981; Garrison et al., 1997) that can be correlated over a wider area than the principal coalbed itself.

ALLUVIAL PLAIN

Alluvial-plain deposits laid down in river channels and in the adjacent flood plains constitute a major part of the Ferron Sandstone. They make up the bulk of the Ferron south of Interstate 70 (Figure 1). Channelbelt and overbank facies assemblages are distinguished.

Channelbelt

Predominantly sandy sediments deposited within active river channels constitute the channelbelt deposits of the Ferron Sandstone. Sand size generally ranges up to medium, but some coarse- and very coarse-grained sandstone is present, being particularly abundant in the channelbelt deposits of Kf-4. Trough cross-stratification

Figure 39. Meanderbelt deposits (parasequence Kf-2-Muddy Creek-a) in Muddy Creek Canyon. Vertical face is formed by trough cross-stratified sandstone. Person is sitting in upper part of deposit, where current rippled sandstone is interbedded with mudstone.

Figure 38. Unidirectional trough cross-stratified channel sandstone cutting into the shoreface of Kf-2 (parasequence Kf-2-Dry Wash at Dry Wash as defined by Anderson and Ryer, this volume). The base of the channel sandstone is at the person's feet. The height of cross-stratified lamina sets decreases upward.

Figure 40. Lateral accretion surfaces are conspicuous is a small channel deposit of Kf-2 exposed near the mouth of Rock Canyon. It existed during deposition of the A coal zone in adjacent areas, having eroded downward through carbonaceous sediments and into the underlying shoreline sand.

is the most common sedimentary structure preserved in the channelbelt sandstone (Figure 38). In a typical vertical profile through a channel deposit, the preserved height of the cross-stratification sets decreases upward. Ripple-drift cross-lamination is common in the highest part, where the sand fines upward, eventually interbedding with and grading to mudstone (Figure 39).

"Channelbelt" is a general term that includes the deposits of a variety of channel forms. Many, and probably the majority, of the Ferron channelbelts were meanderbelts laid down by highly sinuous rivers. Direct evidence of this can be found in the many lateral accretion surfaces (Figure 40), representing lateral migration of point bars. Lateral accretion surfaces and, locally, abandoned channel fills provide the best evidence for determining the size of the rivers that deposited the Ferron. They indicate that the rivers were typically up to about 30 ft (9 m) in depth and up to a few hundred feet in width where they curved around their point bars. Variations in size, however, are considerable. This stems from the fact that the Ferron was not deposited by a single river and probably not by a single river system: major rivers draining northeastward from the Sevier orogenic belt as well as smaller rivers that had their headwaters within the alluvial plain east of the orogenic belt all left their marks on the deposits of the Ferron.

Ferron channelbelts range up to about 80 ft (24 m) in thickness. Thick channelbelts are the result of stacking of individual channel depositional sequences, that is they are "multistoried" deposits. Successive channel bodies rarely stack one body directly above an immediately preceding one. Instead, they tend to be positioned one next to another, adding a "multilateral" component to the channelbelt. Continued deposition during rise of

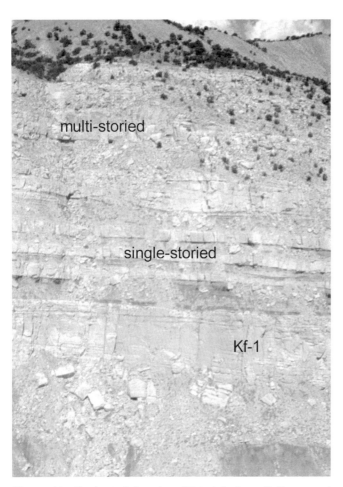

Figure 41. Single-storied and multistoried channelbelts exposed along the Limestone Cliffs. The massive cliff at the bottom of the photograph is formed by shoreline sandstone of Kf-1, beneath which the Tununk Shale is exposed. Overbank mudstone with three single-storied channelbelt sandstones of about equal thickness overlie the shoreline sandstone. Higher in the section, the style changes to thicker, multistoried channelbelt sandstones. The Blue Gate Shale forms the slopes at the top of the photo.

base level leads to aggradation and multiple stories of channel bodies. Channelbelts that are strongly multilateral are inferred to have been deposited during periods of more gradually rising base level; those that are more strongly multistoried are inferred to characterize periods of more rapidly rising base level (Figure 41).

The amount of mudstone included within Ferron channelbelts varies considerably. On outcrops, lateral accretion is most clearly displayed in "inclined heterolithic" channel deposits simply because of the contrast in weathering styles of the sandstone and mudstone interbeds. Lateral accretion surfaces are more difficult to recognize in point bar deposits that lack shale interbeds.

Gardner (1993) presented a detailed analysis and comparison of the arrangement of depositional "macroforms" (architectural elements representing recognizable bar forms or individual channel forms) within the

seaward-stepping, "low-accommodation" units Kf-1 through Kf-3 and the landward-stepping, "high-accommodation" units Kf-4 through Kf-8. He concluded that channelbelts associated with seaward-stepping shoreline units contain thinner and more highly interconnected macroforms and include a lower diversity of fluvial facies. The channels themselves were relatively deep and narrow. By contrast, channelbelts associated with landward-stepping shoreline units contain a higher diversity of fluvial facies and thicker macroforms that tend to be separated from one another by fine-grained sediments and are, therefore, less well interconnected. They include a lower percentage of sandstone. The individual channels were shallower and wider. Channelbelts associated with back-stepping shoreline units are flanked by much larger amounts of crevasse-splay deposits than are channelbelts associated with forward-stepping shoreline units.

Garrison et al. (1997) presented an analysis of the width-to-thickness ratios of Ferron channelbelts. They utilized data they collected in Willow Springs Wash (van den Bergh, 1995; and van den Bergh and Garrison, 1996) and data gathered from other Ferron outcrops by Lowry and Jacobsen (1993) and Barton (1994). Although there is a great deal of overlap of populations, they were able to demonstrate that width-to-thickness ratios for channelbelts generally increase upward in the Ferron, passing from seaward-stepping to aggradational to landward-stepping portions.

Overbank Deposits

Overbank deposits in the Ferron Sandstone consist predominantly of mudstone and sandy mudstone deposited in flood basins as fines that settled from suspension during and in the aftermath of floods. Lenticular bodies of sandstone, generally ripple-drift cross-laminated but also including trough cross-stratification, represent channels that fed crevasse-splay complexes. The crevasse-splay channels generally were much smaller than the principal channels from which they issued. Flood-basin clastics interfinger extensively with carbonaceous shales representing swamps formed in areas of lesser clastic input. On outcrop, overbank deposits generally are slope forming, although the more resistant, sandy crevasse-splay channel deposits locally form ledges.

ACKNOWLEDGMENTS

This paper was supported by the Utah Geological Survey, contract numbers 94-2488 and 94-2341, as part of a project entitled *Geological and Petrophysical Characterization of the Ferron Sandstone for 3-D Simulation of a Fluvial-Deltaic Reservoir*, M. Lee Allison and Thomas C. Chidsey, Jr., Principal Investigators. The project was funded by the U.S. Department of Energy (DOE) under the Geoscience/Engineering Reservoir Characterization Program of the DOE National Petroleum Technology Office, Tulsa, Oklahoma, contract number DE-AC22-93BC 14896. The Contracting Office Representative was Robert Lemmon.

We thank David E. Tabet and Craig D. Morgan, Utah Geological Survey, Salt Lake City, Utah, and Mary L. McPherson, McPherson Geologic Consulting, Grand Junction, Colorado, for their careful reviews and constructive criticisms of the manuscript.

REFERENCES CITED

Anderson, P. B., K. McClure, T. C. Chidsey, Jr., T. A. Ryer, T. H. Morris, J. A. Dewey, Jr., and R. D. Adams, 2003, Interpreted regional photomosaics and cross section, Cretaceous Ferron Sandstone, east-central Utah: Utah Geological Survey Open-File Report 412, compact disk.

Barton, M. D., 1994, Outcrop characterization of architecture and permeability structure in fluvial-deltaic sandstones, Cretaceous Ferron Sandstone, Utah: Ph.D. dissertation, University of Texas, Austin, 255 p.

Cotter, Edward, 1975a, Deltaic deposits in the Upper Cretaceous Ferron Sandstone, Utah, *in* M. L. S. Broussard, ed., Deltas, models for exploration: Houston Geological Society, p. 471–484.

—1975b, Late Cretaceous sedimentation in a low-energy coastal zone: the Ferron Sandstone of Utah: Journal of Sedimentary Petrology, v. 45, p. 669–685.

Doelling, H. H., 1972, Central Utah coal fields: Sevier-Sanpete, Wasatch Plateau, Book Cliffs and Emery: Utah Geological and Mineral Survey Monograph Series, no. 3, 570 p.

Galloway, W. E., 1975, Process framework for describing the morphologic and stratigraphic evolution of deltaic depositional systems, *in* M. L. S. Broussard, ed., Deltas, models for exploration: Houston Geological Society, p. 87–98.

Gardner, M. H., 1993, Sequence stratigraphy and facies architecture of the Upper Cretaceous Ferron Sandstone Member of the Mancos Shale, east-central Utah: Ph.D. dissertation T-3975, Colorado School of Mines, Golden, 528 p.

—1995a, Tectonic and eustatic controls on the stratal architecture of mid-Cretaceous stratigraphic sequences, central Western Interior foreland basin of North America, in Stratigraphic evolution of foreland basins: Society for Sedimentary Geology (SEPM) Special Publication No. 52, p. 243–281.

—1995b, The stratigraphic hierarchy and tectonic history of the mid-Cretaceous foreland basin of central Utah, in Stratigraphic evolution of foreland basins: Society for Sedimentary Geology (SEPM) Special Publication No. 52, p. 284–303.

Garrison, J. R., Jr., T. C. V. van den Bergh, C. E. Barker, and D. E. Tabet, 1997, Depositional sequence stratigraphy and architecture of the Cretaceous Ferron Sandstone: implications for coal and coal bed methane resources — a field excursion, *in* P. K Link and B. J. Kowallis, eds., Mesozoic to Recent geology of Utah: Provo, Brigham Young University Geology Studies, v. 42, pt. 2, p. 155–202.

Hale, L. A., 1972, Depositional history of the Ferron Formation, central Utah, *in* J. L. Baer and Eugene Callaghan, eds., Plateau-basin and range transition zone: Utah Geological Association Publication No. 2, p. 115–138.

Hay, W. W., D. L. Eicher, and R. Diner, 1993, Physical oceanography and water masses in the Cretaceous Western Interior Seaway, *in* W. G. E. Caldwell and E. G. Kauffman, eds., Evolution of the Western Interior Basin: Geological Association of Canada Special Paper 39, p. 297–318.

Lowry, P., and T. Jacobsen, 1993, Sedimentology and reservoir characteristics of a fluvial-dominated delta-front sequence — Ferron Sandstone Member (Turonian), east-central Utah, U.S.A., *in* M. Ashton, ed., Advances in reservoir geology: Geological Society (of London) Special Publication No. 69, p. 81–103.

Lupton, C. T., 1916, Geology and coal resources of Castle Valley in Carbon, Emery, and Sevier Counties, Utah: U.S. Geological Survey Bulletin 628, 88 p.

Peterson, Fred, R. T. Ryder, and B. E. Law, 1980, Stratigraphy sedimentology and regional relationships of the Cretaceous System in the Henry Mountains region, Utah, *in* M. D. Picard, ed., Henry Mountains symposium: Utah Geological Association Publication 8, p. 151–170.

Roberts, L. N. R., and M. A. Kirschbaum, 1995, Paleogeography of the Late Cretaceous of the Western Interior of Middle North America — coal distribution and sediment accumulation: U.S. Geological Survey Professional Paper 1561, 115 p.

Ryer, T. A., 1981, Deltaic coals of Ferron Sandstone Member of Mancos Shale — predictive model for Cretaceous coal-bearing strata of Western Interior: AAPG Bulletin, v. 65, no. 11, p. 2323–2340.

Ryer, T. A., and J. R. Lovekin, 1986, The Upper Cretaceous Vernal delta of Utah — depositional or paleotectonic feature? *in* J. A. Peterson, ed., Paleotectonics and sedimentation in the Rocky Mountain region, United States: AAPG Memoir 41, p. 497–510.

Thompson, S. L., C. R. Ossian, and A. J. Scott, 1986, Lithofacies, inferred processes, and log response characteristics of shelf and shoreface sandstones, Ferron Sandstone, central Utah, *in* T. F. Moslow and E. G. Rhodes, eds., Modern and ancient shelf clastics

— a core workshop: Society for Sedimentary Geology (SEPM) Core Workshop No. 9, p. 325–361.

van den Bergh, T. C. V., 1995, Facies architecture and sedimentology of the Ferron Sandstone Member of the Mancos Shale, Willow Springs Wash, east-central Utah: M.S. thesis, University of Wisconsin, Madison, 255 p.

van den Bergh, T. C. V., and J. R. Garrison, Jr., 1996, Channel belt architecture and geometry — a function of depositional parasequence set stacking pattern, Ferron Sandstone, east-central Utah: AAPG Rocky Mountain Section Meeting, expanded abstracts volume, Montana Geological Society, p. 37–42.

Analog for Fluvial-Deltaic Reservoir Modeling: Ferron Sandstone of Utah
AAPG Studies in Geology 50
T. C. Chidsey, Jr., R. D. Adams, and T. H. Morris, editors

Integrated Analysis of the Upper Ferron Deltaic Complex, Southern Castle Valley, Utah

Richard J. Moiola[1], Joann E. Welton[2], John B. Wagner[3], Larry B. Fearn[1], Mike E. Farrell[2], Roy J. Enrico[1], and Ron J. Echols[1]

ABSTRACT

The Upper Ferron Sandstone in southern Castle Valley, Utah, is a river-dominated deltaic complex made up of seaward-stepping, vertically stacked, and landward-stepping cycles. These cycles, which consist of delta plain, delta front, and prodelta/offshore facies associations, are partitioned by flooding surfaces that detailed biostratigraphic analysis indicates lack an open marine signature. They are thought to be abandonment flooding surfaces associated with delta lobe switching. The abundance of distributary channel belts associated with all cycles suggests that riverine processes controlled the evolution of their associated delta fronts and that the entire Upper Ferron is a river-dominated system in which marine processes (predominantly waves and storms) played a subordinate role. Autocyclic processes, channel avulsion and lobe switching, controlled the internal architecture and partitioning of the cycles. The stacking pattern was controlled by allocyclic processes, primarily decreasing sediment supply combined with increasing accommodation. No compelling evidence was found to confirm the presence of incised valleys in the Upper Ferron.

Distributary channel-belt sandstones and delta-front sandstones are the principal reservoir facies in the Upper Ferron deltaic complex. Channel-belt sandstones have the best reservoir quality, with an average porosity of 15.6% and average permeability of 116 md. By comparison, delta-front sandstones have an average porosity of 11.9% and an average permeability of 47 md. Rock physics and seismic modeling results indicate that the seismic response of the Upper Ferron is dominated by large acoustic impedance contrasts associated with coals and carbonaceous shales even though they are generally thin and well below typical tuning thicknesses. Distributary channel belts show up as amplitude dimouts of the strong response generated by associated coals and carbonaceous shales. They appear to be the only reservoir facies that can be imaged directly.

[1]*Mobil Technology Company — Retired, Dallas, Texas;* [2]*ExxonMobil Upstream Research, Houston, Texas;* [3]*Nexen Petroleum U.S.A., Dallas, Texas*

INTRODUCTION

The Cretaceous (Turonian) Ferron Sandstone Member of the Mancos Shale in Castle Valley, Utah, can be subdivided into two distinct units, the Upper and Lower Ferron (Ryer and McPhillips, 1983; Riemersma and Chan, 1995). This study focuses on the Upper Ferron that is well exposed in a 100 km (60 mi), dip-oriented belt along the western flank of the San Rafael Swell in southern Castle Valley. It comprises a southerly derived deltaic complex (170+ m [500+ ft] thick) of stacked delta-plain and delta-front deposits that prograded to the north-northeast into the foreland basin of the Western Interior Seaway (Davis, 1954; Katich, 1954; Hale, 1972; Cotter, 1975a, b; Ryer, 1981). The Upper Ferron has been subdivided into various numbers of stratigraphic cycles by different workers (e.g., Ryer, 1981; Gardner, 1993; Barton, 1995; Anderson et al., 1997; Garrison et al., 1997). Building on the original stratigraphy of Ryer (1981), Gardner (1993, 1995a, b; Gardner and Cross, 1994) developed a high-resolution stratigraphic framework for interpreting the depositional architecture of the Upper Ferron.

Gardner (1993, 1995a, b) subdivided the Upper Ferron into seven short-term stratigraphic cycles (parasequence sets in the terminology of Van Wagoner et al., 1990) partitioned by marine flooding surfaces or their updip correlative equivalents. These cycles are arranged in a seaward-stepping, vertically stacked, and landward-stepping pattern (Figure 1) that is interpreted to reflect the interplay between sediment supply and accommodation space. Each cycle contains a complete spectrum of marine shelf, shallow marine, and coastal plain facies tracts and their component facies associations (Gardner, 1993). Gardner bracketed the Upper Ferron (his Ferronensis sequence) between a sequence boundary at the base of the first seaward-stepping cycle and a maximum flooding surface within the overlying Blue Gate Shale Member.

According to Gardner (1993), the delta lobes of the Upper Ferron lack incised valleys or regionally extensive surfaces indicative of negative accumulation during times of relative sea-level fall. He also noted an evolution from river-dominated deltas in the seaward-stepping cycles to wave-dominated deltas in the landward-stepping cycles. Cotter (1975a, b), who did not partition the Upper Ferron, interpreted it as a highly constructive deltaic system made up of numerous coalescing and overlapping subdelta lobes. Recently, Barton (1995, 1997) and Knox and Barton (1999) concluded that the cycles in the Upper Ferron are not conformable successions as was advocated by Gardner (1993) and other workers (e.g., Ryer, 1981), but are divided by unconformities associated with fluvial incision and the development of incised valleys.

The purpose of our study was to analyze the Upper Ferron deltaic complex utilizing an integrated, multidisciplinary approach. Specifically, we were concerned with evaluating: (1) what type of deltaic system the Upper Ferron represents; (2) which facies has the best reservoir quality; (3) the nature of the "marine flooding surfaces" separating the cycles; (4) the signature of autocyclic versus allocyclic processes on its internal architecture/stacking pattern; (5) the presence or absence of incised valleys; and (6) its seismic response. Disciplines and approaches contributing to this study included sedimentology, sequence stratigraphy, petrography, biostratigraphy, and rock physics/seismic modeling. In addition to outcrop data, the study included core and log data from 13 wells (ARCO Stat. Test-1, 82-2, 4, 5, 6, 8; BP Muddy Creek 1 through 5; Unocal DW-1; UURI-1) whose location is shown in Figure 2. The stratigraphic framework of Gardner (1993) was adopted for the study.

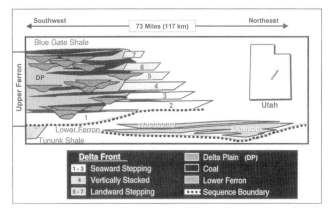

Figure 1. Schematic cross section showing stacking pattern of cycles in the Upper Ferron deltaic complex and stratigraphic relationships in Castle Valley, Utah. Cross section modified from Ryer (1981) and Gardner (1993).

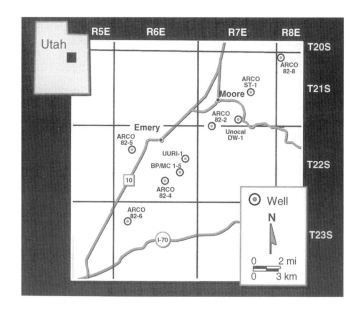

Figure 2. Map showing the location of 13 wells in southern Castle Valley whose cores and logs comprise the subsurface database for this study.

SEDIMENTOLOGY AND SEQUENCE STRATIGRAPHY

Gardner (1993, 1995a, b) has described the sedimentology and sequence stratigraphy of the Upper Ferron in considerable detail. In the following section, we briefly summarize our interpretations from the examination of outcrops from Willow Springs Wash, south of Interstate 70 (I-70), to as far north as Clawson and from the detailed description of cores from all of the wells shown in Figure 2. Our observations and interpretations are complemented by the observations and interpretations of geoscientists who attended Mobil's Sandstone Seminar from 1995 to 1999.

Cycles and Facies Associations

All cycles in the Upper Ferron are progradational even though the upper ones are landward-stepping. Each consists of delta-plain, delta-front, and prodelta/offshore facies associations. The delta-plain association comprises distributary channels, interdistributary bays, splays, and marshes. Distributary mouth bars and distal bars make up the delta-front association (following Coleman and Gagliano, 1965), and fine-grained shallow marine sediments deposited below storm wave base make up the prodelta/offshore association. Previous workers (cited above) have proposed varying scenarios for the nature of the cycles and their bounding surfaces, but there is a general sense of agreement that the delta fronts of the lower cycles are predominantly river-dominated, whereas those in the upper cycles are wave-dominated.

Delta Plain Association

Distributary Channels: Distributary channel-sand bodies are abundant in all of the seven cycles. They are erosionally based and consist predominantly of trough cross-stratified sandstone, often containing rip-up clasts. As documented by Gardner (1993), these channel-sand bodies are arranged into belts as much as 24 m (80 ft) thick and 610+ m (2,000+ ft) wide. Macroforms within these belts are arranged in multistory and multilateral geometries. Channel belts in the lower, seaward-stepping cycles have high sand to mud ratios and low width to depth ratios. In contrast, those in the landward-stepping cycles have lower sand to mud ratios but higher width to depth ratios. Low sinuosity and cut and fill macroforms are the dominant types observed in the seaward-stepping belts, whereas the landward-stepping belts are characterized by a higher diversity of macroforms (e.g., high and low sinuosity, abandonment fill).

A great number of channel belts that tie into the delta fronts of cycles above Cycle 1 can be seen in the outcrops at Willow Springs Wash and in cores from the ARCO 82-

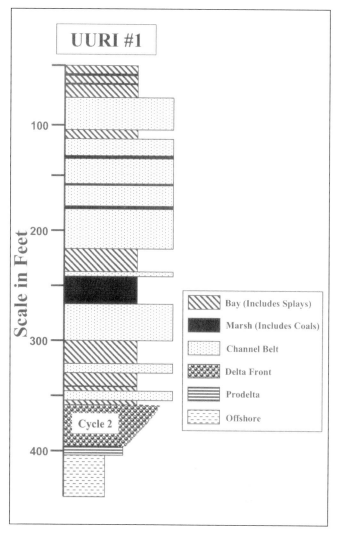

Figure 3. Interpreted depositional facies in cores of the Upper Ferron from the UURI-1 well. The delta-front succession of Cycle 2 is overlain by a delta-plain association that contains nine distributary channel belts. Downdip these channel belts have laterally equivalent delta-front successions. The abundance of channel belts in the Upper Ferron suggests that riverine processes exerted the primary control on the development of the delta fronts of all its cycles.

6 well, approximately 13 km (8 mi) downdip from Willow Springs Wash. Likewise, numerous channel belts (nine) that feed the delta fronts of cycles above Cycle 2 are present in the cores of the UURI-1 well (Figure 3) and in the cliffs above Muddy Creek, approximately 11 km (7 mi) downdip from the ARCO 82-6 well. The multitude of channel belts in all cycles of the Upper Ferron suggests to us that riverine processes dominated the development of the Upper Ferron delta front successions and that marine processes played a subordinate role.

Many of the distributary channel belts incise into and occasionally cut through underlying delta-front successions. Good examples can be seen in Cycle 1 at Willow Springs Wash, in Cycle 2 along I-70, and in Cycle 5 at

South Muddy Creek, Picture Flats, and Cedar Ridge. The channel belts referred to in Cycles 2 and 5, as well as channel belts in Cycles 3 and 4 were recently interpreted by Barton (1995, 1997) and Knox and Barton (1999) as incised valleys. As an example, the I-70 channel belt in Cycle 2, which is 24 m (80 ft) thick, 198 m (650 ft) wide, and incises approximately 14–15 m (45–50 ft) into underlying delta-front deposits, was interpreted as an incised valley by Barton (1995, 1997) and Knox and Barton (1999) on the basis that (1) the depth of incision is several times greater than estimated channel depth based on the thickness of channel bars and channel fills, and (2) deposits overlying the surface of incision display an abrupt basinward shift in facies associated with an inferred relative drop in sea level. Barton's and Knox and Barton's interpreted incised valleys in Cycle 5 are thinner, wider, and more mud-rich, and their depth of incision is less than those in Cycles 2 and 3. As a consequence of interpreting these channel belts as incised valleys, Barton (1995, 1997) and Knox and Barton (1999) concluded that the Upper Ferron cycles are not conformable successions but are partitioned by unconformities resulting from relative drops in sea level. Garrison and van den Bergh (1999) also reached similar conclusions.

Our observations, however, lead us to the conclusion that these distributary channel belts are not incised valleys. We believe they are too small to be true incised valleys, which are commonly several kilometers wide to many tens of kilometers wide (Van Wagoner et al., 1990). Their thickness, up to 24 m (80 ft) and depth of incision, as much as 15 m (50 ft), we suggest were controlled by autocyclic processes typical of deltaic systems. Specifically, we attribute the expanded thickness and depth of incision of these belts to channel avulsion and subsequent repeated reoccupation of the same site and to loading/compaction of prodelta muds. Autocyclic incision associated with avulsion and repeated reoccupation of abandoned channels is a common characteristic of many modern and ancient channel systems (Mohrig et al., 2000). Also, distributary channels in modern and ancient prograding deltaic complexes incise into delta-front deposits, often deeply, without a relative drop in sea level taking place (Coleman and Gagliano, 1964; Van Wagoner et al., 1990). Placing an unconformity/sequence boundary at their base decouples them from genetically related facies, and we see no compelling evidence to do so in the Upper Ferron.

We do not consider the occurrence of distributary channel deposits above delta-front deposits a valid reason for concluding that an abrupt basinward shift in facies has taken place. Also, the channel belts are encased in associated deltaic facies and not in deeper offshore facies; the "sequence boundaries" at the bases of the inferred incised valleys lack evidence of subaerial exposure; the presence of well-developed soil horizons in what would be the correlative subaerial exposure sur-

faces in the "interfluve" areas of the incised valleys is problematic; and a regionally significant extent for the inferred sequence boundaries has not been established. Lastly, the Upper Ferron distributary channel belts do not extend beyond the seaward limit of the deltas they feed; in contrast, valleys often extend kilometers beyond the seaward depositional limit of the deltas they incise (Gardner, 2001).

Interdistributary Bays, Marshes, Splays: A heterolithic assemblage of shales, mudstones, siltstones, sandstones, and coals of bay, splay, and marsh origin is intimately associated with the channel-belt deposits. The bay deposits typically comprise burrowed and bioturbated mudstones and siltstones with syneresis cracks. They contain a suite of trace fossils (*Skolithus, Lockeia, Palaeophycus, Planolites, Teichichnus, Thalassinoides, Cylindricus*) and body fossils (*Crassostrea, Corbicula, Varia*) suggestive of brackish water conditions. Interbedded with the bay deposits are thin siltstone and sandstone beds of overbank origin; upward-coarsening crevasse splay deposits; and marsh deposits consisting of rooted mudstones, carbonaceous shales/mudstones and coal. Coaly/carbonaceous deposits are common throughout the Upper Ferron, and a major coal bed/coal zone is associated with each of the seven cycles (Lupton, 1916; Ryer, 1981). The impact that these coaly marsh deposits have on the seismic response of the Upper Ferron will be discussed later.

Delta Front Association

Shallowing- and coarsening-upward successions of distal bar and distributary mouth bar deposits make up the delta fronts of the Upper Ferron cycles. They overlie fine-grained prodelta deposits and are capped by flooding surfaces. These successions, which range from less than 7.5 m (25 ft) to more than 18 m (60 ft) in thickness, are analogous to the "parasequence" of Van Wagoner et al. (1990), and stack in a systematic fashion to form the seaward- to landward-stepping cycles of the Upper Ferron. The distal-bar deposits consist of thin-bedded, very fine to fine-grained, parallel to ripple-laminated sandstones with interbeds of bioturbated mudstone. The distributary mouth bar deposits comprise thicker-bedded, predominantly fine- to medium-grained sandstones with low-angle parallel to horizontal stratification and tabular to trough cross-stratification. Burrowing and bioturbation is common throughout the delta-front succession and a variety of trace fossils is present (i.e., *Ophiomorpha, Thalassinoides, Teichichnus, Tebellina, Arenecolites, Skolithus, Monocraterion, Rosselia*).

The delta fronts of the seaward-stepping cycles are more mud-rich than those of the landward-stepping cycles and current ripples, graded bedding, soft-sediment deformation, and growth faults are prevalent features. By comparison, the delta fronts of the vertically stacked and landward-stepping cycles are more sand-rich and commonly exhibit wave ripples, hummocky stratification, and swaley stratification. As a consequence of these and other observations, Gardner (1993) and other workers (e.g., Barton, 1995) have concluded that the delta fronts of seaward-stepping cycles are predominantly river-dominated, whereas those of the landward-stepping cycles are predominantly wave-dominated.

We agree that the delta fronts of landward-stepping cycles display significantly more evidence of marine influence/modification that those of the seaward-stepping cycles. However, the delta fronts of all cycles are fed by numerous distributary channels, and according to Reading and Collinson (1996), wave-dominated deltas preserve few distributary channels. This suggests to us that riverine processes controlled the development of the delta fronts in all cycles and that the Upper Ferron is a river-dominated deltaic complex. As documented by Gardner (1993), sediment supply and total sediment volumes progressively decreased during the deposition of the Upper Ferron cycles and the position of maximum accommodation progressively shifted landward. This resulted in thinner delta-front successions in landward-stepping Cycles 5 through 7. As a consequence, we believe marine processes, predominantly wave and storm processes, were able to play a more pronounced role in modifying/influencing the delta fronts of these cycles. Also, as can be observed by looking at modern wave-dominated deltas (e.g., Nile, Tiber, Sao Francisco), they are fed by only one or two distributaries, whereas multiple distributaries are characteristic of river-dominated systems (James Coleman, personal communication, 1999). The older lobes of the modern Mississippi River delta complex, such as the Teche, Lafourche, and St. Bernard, are multichannel, river-dominated systems (Frazier, 1967) that serve as good analogs for the Upper Ferron deltas.

Prodelta/Offshore Association

The prodelta/offshore association consists of shales, mudstones, and siltstones with thin, very fine-grained sandstone interbeds. These sediments were deposited predominantly below storm wave base and represent the shallow marine environment into which the delta fronts of the Upper Ferron cycles prograded. Bioturbation is often intense and a diverse assemblage of trace fossils is present (i.e., *Chondrites, Palaeophycus, Teichichnus, Anconichnus, Terebellina, Thalassinoides, Scolicia, Schaubecylindrichnus, Helminthopsis, Planolites, Rosselia, Zoophycus*).

Flooding Surfaces

Flooding surfaces that partition the Upper Ferron cycles and their associated delta-front successions vary from non-erosive to erosive and are generally manifest

by the preservation of mud-dominated, presumably deeper marine strata overlying delta-front successions. Most commonly, the deposits above flooding surfaces are lithologically similar to the interdistributary bay deposits discussed above and most commonly consist of burrowed, bioturbated mudstones, sandy mudstones, and sandstones. Ravinement surfaces, where present, are characterized by a thin lag overlain by pervasively bioturbated sandstone. The sediments associated with flooding surfaces contain an assemblage of trace fossils (*Thasassinoides, Planolites, Palaeophycus, Teichichnus, Arenicolites, Cylindricus*) that is much more similar to the one found in the interdistributary bay deposits than to the one in the prodelta/offshore deposits. In total, the flooding-surface deposits thus appear to have a brackish, bay-like as opposed to an open marine aspect and they do not appear to mark a major dislocation in depositional environment or pronounced shift in facies.

Although flooding surfaces are commonly thought to result from allocyclic processes, they can also arise from autocyclic processes as straight forward as water-depth deepening occurring as a delta lobe subsides by the compaction of prodelta mudstones (Kamola and Van Wagoner, 1995). We interpret the Upper Ferron flooding surfaces as abandonment flooding surfaces generated by autocyclic lobe switching and not as open marine flooding surfaces produced by allocyclically induced transgressions. They appear to be incursions similar to those associated with the flooding and minor reworking of the upper veneer of the older, abandoned lobes of the modern Mississippi River deltaic complex (Frazier, 1967; Coleman, 1988; Penland et al., 1988). From their time of initiation, each of the Mississippi River delta lobes prograded and were abandoned/flooded in a period of only 1000 to 2000 years (Roberts, 1997), illustrating how dynamic river-dominated deltaic systems are and how the autocyclic processes that control them operate at a very high frequency. We believe the Upper Ferron deltaic system operated in a similar fashion.

Stacking Pattern

We believe, as did Gardner (1993), that the stacking pattern of the Upper Ferron cycles (Figure 1) exhibits a progression that was allocyclically controlled, primarily by decreasing sediment supply in a setting where accommodation, created by subsidence and relative sea-level rise, was progressively increasing. Initially, a high rate of sediment supply relative to available accommodation space resulted in rapid progradation (seaward-stepping cycles). This was followed by a period of aggradation and vertical stacking when sediment supply and accommodation were essentially in balance. When accommodation exceeded sediment supply, but sediment supply was still adequate to allow progradation, the system retrograded (landward-stepping cycles). Ultimately, sediment supply ceased and the Upper Ferron complex was transgressed by the advancing Blue Gate sea. Ryer (1991) also proposed a similar scenario for the evolution of the Upper Ferron stacking pattern.

Reservoir Quality

Distributary channel-belt deposits volumetrically make up the principal reservoir facies in landward-stepping cycles, whereas delta-front deposits volumetrically comprise the dominant reservoir facies in seaward-stepping cycles (Gardner, 1993; Barton and Angle, 1995). Barton and Angle (1995) investigated the petrography and reservoir quality of channel-belt and delta-front sandstones in seaward-stepping Cycle 2 and landward-stepping Cycle 5. Regarding their composition, they found that: (1) delta-front sandstones are more quartz rich and contain less feldspar and rock fragments than distributary channel sandstones; (2) cement volumes vary from a mean of 7.1% in distributary channel sandstones to a mean of 10.5% in delta-front sandstones; and (3) distributary channel sandstones have approximately equal amounts of authigenic quartz and kaolinite, whereas quartz and carbonates are the predominant authigenic phases in delta-front sandstones.

Barton and Angle (1995) noted that increases in grain size, grain sorting, and detrital quartz content corresponded to marked increases in permeability in sandstones of both Cycles 2 and 5. Overall, they found that well-sorted, quartz-rich wave-dominated upper delta-front sandstones (i.e., distributary mouth bar sandstones in our terminology) display the highest mean permeabilities (605 md) and that wave-dominated delta-front sandstones as a whole display the best reservoir quality (mean permeability of 240 md) in comparison to distributary channel and river-dominated delta-front sandstones (mean permeabilities of 212 md and 80 md, respectively) that are more poorly sorted and contain abundant feldspar and rock fragments. Barton and Angle (1995) attributed the quartz-rich nature and superior reservoir quality of the wave-dominated delta fronts and sandstones to physical reworking in the depositional environment, even though increases or decreases in quartz content could also be related to provenance.

Our analysis of the reservoir quality of channel-belt and delta-front sandstones in Cycles 1 through 7 comprised data derived from petrography, scanning electron microscopy, X-ray diffraction, core analysis, and minipermeametry. Among the principal factors controlling the reservoir quality of the Upper Ferron are its burial and thermal histories. The Upper Ferron was subjected to a burial depth of approximately 3000–4250 m (10,000–14,000 ft) and maximum burial temperatures estimated from coal composition and vitrinite reflectance data of 65 to 80° C (Fisher et al., 1993); consequently, the diagenetic overprint is mild.

Quartz, feldspar, and rock fragments are the basic detrital components of the Upper Ferron sandstones; in general, they can be classified as feldspathic lithic sandstones. As determined by counting 400 points per thin section, distributary channel-belt sandstones have an average Q:F:R composition of 79:12:9. By comparison, the average composition of delta-front sandstones is 84:11:5. Delta-front sandstones are mostly fine grained; in contrast, channel-belt sandstones are typically medium to coarse grained (Figure 4). Delta-front sandstones in general contain a higher percentage of authigenic cement. Carbonate cement, primarily calcite, is more abundant in the delta-front sandstones, whereas pore-filling kaolinite is more abundant in the channel sandstones; both contain quartz cement. Illite and mixed illite/smectite clay coatings that help preserve primary porosity by preventing the formation of quartz overgrowths occur on the detrital grains of both facies.

Our data (70 conventional permeability measurements from samples of the ARCO and UURI-1 wells and several thousand minipermeameter measurements on outcrops at Willow Springs Wash and Muddy Creek, and on cores from the BP wells) indicate that distributary channel-belt sandstones have the best overall reservoir quality. They have an average porosity of 15.6% (primary porosity is dominant) and an average permeability of 116 md as compared to 11.9% (secondary and microporosity are dominant) and 47 md for the delta-front sandstones. We found, as did Barton and Angle (1995), that both environments exhibit a broad range of variability and that the coarser grained, more quartz-rich members of each have the higher permeabilities and better reservoir quality. For example, in the ARCO 82-4 well a quartz-rich channel-belt sandstone (Q:F:R = 90:7:3) at 94 m (309 ft) in Cycle 4 has a permeability of 241 md compared to a feldspar/lithic-rich channel-belt sandstone (Q:F:R = 65:18:17) at 83 m (271 ft) in Cycle 5 that has a permeability of only 5 md.

Some distributary mouth bar sandstones, especially in the landward-stepping cycles, exhibit very good reservoir quality, but overall they are more calcite-cemented and display lower permeabilities than distributary channel sandstones. Also, the permeabilities of the distal-bar deposits are invariably very low, thus downgrading the reservoir quality of the delta front as a whole. In contrast, distributary channel-belt sandstones in landward- as well as seaward-stepping cycles exhibit better reservoir quality on a more consistent basis. We believe they are the primary reservoir facies in the Upper Ferron, even though volumetrically (as documented by Gardner, 1993) they are not the dominant reservoir facies in all cycles.

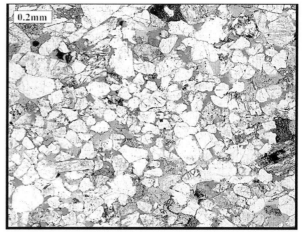

ARCO 82-6 (134 ft)   ARCO 82-2 (497 ft)

Figure 4. Thin-section photographs illustrating the relative reservoir quality of distributary channel-belt sandstones and delta-front sandstones. The channel sample (ARCO 82-6 well; Cycle 5, 41 m [134 ft]) is coarser grained, less cemented, and has a much higher permeability than the delta-front (distributary mouth bar) sample (ARCO 82-2 well; Cycle 4, 151 m [497 ft]). Distributary channel-belt sandstones are the primary reservoir facies in the Upper Ferron deltaic complex.

BIOSTRATIGRAPHY

Palynofacies and paleontological (foraminifera, dinocysts, pollen, and spores) analyses were conducted on a total of 231 core and outcrop samples from the Lower Ferron/Tununk Shale (48 samples), Upper Ferron (175 samples), and Blue Gate Shale (eight samples). The Upper Ferron samples comprise the following environments as determined from core and outcrop: offshore (10), prodelta (41), delta front (24), bay (84), and marsh (16). Included in the 84 bay samples are 45 flooding-surface samples. These include samples of deposits above flooding surfaces bounding delta-front successions within cycles as well as deposits above flooding surfaces partitioning cycles. Outcrop samples were collected from sections at I-70, Miller Canyon, Dry Wash, Dutch Flats, and Clawson, and subsurface samples were collected from most of the wells shown in Figure 2.

Our principal objective was to establish the environmental signature of the shales and fine-grained sediments associated with the flooding surfaces that partition Cycles 1 through 7. Comparison of their faunal and floral content relative to local marine benchmarks, the Lower Ferron/Tununk Shale and the Blue Gate Shale, was key to establishing their origin and determining if open-marine incursions occur in the Upper Ferron. As discussed above, our core and outcrop observations suggested to us that these sediments lack an open-marine signature (i.e., they contain a suite of trace fossils suggestive of brackish water conditions that is similar to that of the interdistributary bay deposits). This led us to the conclusion that the flooding surfaces in the Upper

Ferron were probably generated as a result of autocyclic lobe switching.

Foraminiferal Analysis

Foraminiferal analysis indicates that both the Lower Ferron/Tununk and the Upper Ferron samples were deposited in a marginal marine setting. Samples containing foraminifera (most of which contain fewer than 1000 total foraminifera) are dominated by a low-diversity assemblage of arenaceous benthic foraminifera. Calcareous benthic foraminifera are extremely rare and only present in a very few samples. The lack of calcareous foraminifera is not due to their tests having been dissolved because many of the samples contain abundant calcareous shell debris.

A few genera (*Guadryina ssp.*, *Trochaminina spp.*, *Haplophagmoides spp.*, *Saccammina spp.*, and *Reophax spp.*) make up the dominant or exclusive component in these low-diversity arenaceous faunas. The presence of other genera is spotty and in very low numbers. The dominance of these genera and lack of calcareous forms are suggestive of a hyposaline, dysoxic, marginal-marine rather than open-marine environment (Murray, 1973). These few genera can be grouped loosely into three faunas: a *Gaudryina spp.* fauna, a *Trochaminina spp./Haplophagmoides spp.* fauna, and a *Saccammina spp./ Reophlax spp.* fauna. The *Trochaminina spp./Haplophag-moides spp.* fauna dominates the Upper Ferron samples (including the flooding surface samples), except for the offshore samples which predominantly contain a *Gaudryina spp.* fauna. The Lower Ferron/Tununk samples are dominated by the *Gaudryina spp.* fauna. The *Saccammina spp./ Reophlax spp.* fauna occurs most commonly with the *Gaudryina spp.* fauna rather than with the *Trochaminina spp./Haplophagmoides spp.* fauna or by itself.

In contrast to the Lower Ferron/Tununk and the Upper Ferron, the Blue Gate samples contain good open marine microfossils. They are dominated by the planktic foraminifera *Hedbergella spp.* and *Heterohelix spp.* and also contain the arenaceous foraminifera *Bathysiphon spp.* These genera are indicative of outer neritic water depths.

Palynological Analysis

The principal aim of the palynological analysis was to achieve a higher level of discrimination of the terrestrial versus marine signature in the sample set, with an emphasis on monitoring dinocyst-acritarch genera. A 200-specimen count was attempted for each sample from the oxidized portion of the palynological residue, and taxa were identified to the generic level. The main phase of the study involved an objective assessment of the count data using principal component analysis (PCA). Normalized values derived from the raw data counts of 19 palynomorph taxa, 18 dinocyst-acritarch genera, and the entity "terrestrial palynomorphs" were used in the analysis. The PCA runs were completed for each groups of samples (i.e., Blue Gate — offshore; Upper Ferron — delta plain/front; Upper Ferron — prodelta; Lower Ferron/Tununk — offshore) with the palynomorph taxa being assigned eigenvalues, a measure of the amount each taxon contributes towards defining the overall character of the assemblage.

Results (Figure 5) show that terrestrial palynomorphs collectively define 95% of the total assemblage character in Upper Ferron delta plain/front samples and 53% of the total assemblage character in the Upper Ferron prodelta samples. In contrast, marine palynomorphs collectively define 60% of the total assemblage character in the Lower Ferron/Tununk and Blue Gate offshore samples. Samples of the sediments associated with flooding surfaces are part of the Upper Ferron delta plain/front population in which terrestrial palynomorphs define 95% of the total assemblage character.

In addition to PCA, the palynological evaluation also involved the quantitative and qualitative assessment of dinocysts using the Gonyaulacoid to Peridinioid ratio (GP) of Harland (1973) as modified by Mao and Mohr (1992). Using this environmental classification method, the higher the GP ratio, the more marine influence is attributed to samples. The GP ratio was calculated for all samples in which at least 20 dinocysts were counted, and the results are consistent with those using PCA. The GP ratios are highest in the Blue Gate samples (mean = 21.3), followed by the Lower Ferron/Tununk (mean = 14.8), the Upper Ferron prodelta (mean = 5.1), and the Upper Ferron delta plain/front (mean = 4.8).

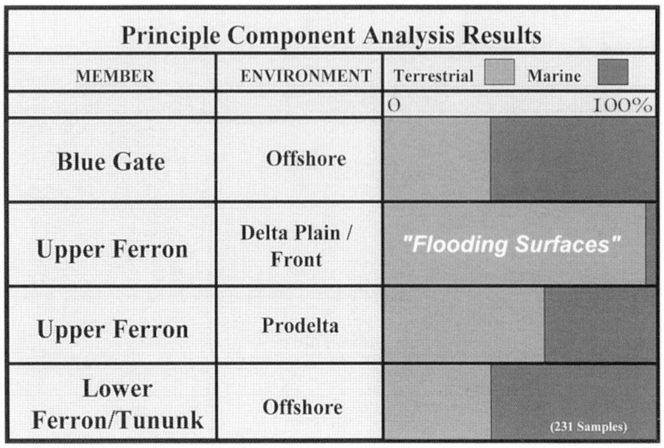

Principle Component Analysis Results			
MEMBER	ENVIRONMENT	Terrestrial ▢	Marine ▇
		0	100%
Blue Gate	Offshore		
Upper Ferron	Delta Plain / Front	*"Flooding Surfaces"*	
Upper Ferron	Prodelta		
Lower Ferron/Tununk	Offshore		(231 Samples)

Figure 5. Assessment of the terrestrial versus marine signature of palynomorph assemblages in samples from the Blue Gate Shale, Upper Ferron, and Lower Ferron/Tununk Shale using Principal Component Analysis. Refer to text for details.

These results are compatible with the PCA results and they collectively suggest that the flooding-surface samples have a terrestrial signature.

Palynofacies Analysis

The inorganic fraction of the 231 samples studied was eliminated by acid digestion and the assemblage of organic particles identified was classified as preserved, degraded, highly degraded phytoclasts; cuticle; resin; amorphous organic matter; pollen/spore; and dinocyst. An average of 300–400 particles was determined and counted per sample and their relative percentages calculated for palynofacies analysis. Also calculated were the concentrations of pollen/spore and dinocyct.

A total of nine palynofacies (1, 2, 2B, 3, 3B, 4, 4B, 5, 6), progressing from completely terrestrial-dominated (100% organic matter of terrestrial origin) to highly marine-dominated (as much as 90% organic matter of marine origin), were recognized (refer to Figure 6) and subsequently confirmed by cluster analysis. Five palynofacies (1, 2, 2B, 3, 3B) contain less than 5% amorphous organic matter and are interpreted as terrestrial-dominated. These five facies were subsequently differentiated by a progressive proportion of preserved and degraded plant remains (3, 3B) relative to the proportion of degraded and highly degraded phytoclasts (1, 2, 2B). Facies 4 and 4B contain 5–30% marine amorphous organic matter. Facies 4 is interpreted as a mixed terrestrial/marine facies because it also contains 60–80% plant debris, whereas facies 4B is considered marine-dominated because it also contains a surprisingly high percentage (10%) of dinocysts. Facies 5 and 6 are interpreted to be marine-dominated because of their high proportion of amorphous organic matter (30–60% and 70–90%, respectively).

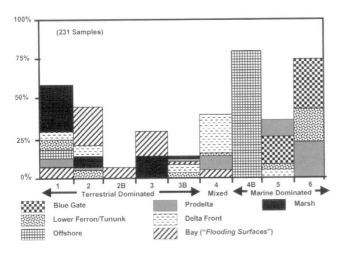

Figure 6. Diagram showing palynofacies types (1, 2, 2B, 3, 3B, 4, 4B, 5, 6) identified from sample sets of the Blue Gate Shale, Lower Ferron/Tununk Shale, and offshore, prodelta, delta-front, bay, and marsh facies of the Upper Ferron; palynofacies types present in each sample set; and percentage of samples in each set occurring in them. Refer to text for details.

Results of the palynofacies analysis are summarized in Figure 6, a diagram showing the nine palynofacies types identified from sample sets of the Blue Gate Shale, the Lower Ferron/Tununk Shale, and the offshore, prodelta, delta front, bay, and marsh facies of the Upper Ferron; the palynofacies types present in each sample set; and the percentage of samples in each set occurring in them. The Blue Gate set of samples, which is made up of Facies 5 (25% of the samples) and Facies 6 (75% of the samples), has a strong marine-dominated signature. By comparison, 50% of the Lower Ferron/Tununk samples have a marine-dominated signature (Facies 5 and 6), 33% have a terrestrial-dominated signature (Facies 1, 2, 3B), and 17% have a mixed signature (Facies 4).

Regarding the Upper Ferron, 80% of the offshore facies samples have a marine-dominated signature (Facies 4B) and 20% have a terrestrial-dominated signature (Facies 1). These samples were collected near Clawson from Gardner's (1993) marine-shelf facies tracts of Cycles 2 through 5 and represent offshore marine deposits that are the distal equivalents of the delta-front successions of these cycles. Fifty-five percent of the prodelta samples have a marine-dominated signature (Facies 5 and 6), 30% are terrestrial-dominated (Facies 1, 2, 3B), and 15% are mixed (Facies 4). In contrast to the prodelta samples, 55% of the delta-front samples have a terrestrial-dominated signature (Facies 1, 2, 3B); 40% a mixed signature (Facies 4); and 5% a marine-dominated signature (Facies 5). All of the marsh samples have terrestrial-dominated signatures (59% Facies 1, 15% Facies 2, 13% Facies 3, and 13% Facies 3B). And interestingly, 95% of the bay samples have a terrestrial-dominated signature (44% Facies 1, 5% Facies 2, 6% Facies 2B, 30% Facies 3, 10% Facies 3B), with

the remaining 5% having a mixed signature (Facies 4). As discussed previously, the bay suite of 84 samples includes 45 flooding-surface samples.

Overall, the palynofacies results indicate that the Upper Ferron offshore samples and especially the Blue Gate samples clearly have a marine-dominated signature. The Lower Ferron/Tununk samples and prodelta samples of the Upper Ferron are slightly marine-dominated. The Upper Ferron delta-front samples have a terrestrial-dominated to mixed signature and the Upper Ferron bay and marsh samples have a strong terrestrial-dominated signature. These results and those derived from the foraminiferal and palynological analysis complement each other and collectively indicate that the Upper Ferron flooding surfaces lack a definitive marine signature. Instead, they have a strong terrestrial-dominated, marginal marine/brackish signature that suggests they are abandonment flooding surfaces rather than open-marine flooding surfaces.

ROCK PHYSICS AND SEISMIC MODELING

Introduction

The objective of the rock physics study and seismic modeling was to develop a simple model that would characterize the acoustic response of the different depositional environments that make up the Upper Ferron deltaic complex. Data from 11 of the 13 wells was used for the rock physics and no attempt was made to remove any of the minor diagenetic effects associated with near surface processes. In the ARCO 82-4 well a pair of vertical core plugs representative of different depositional facies was taken. One plug was used to measure compressional and shear velocities, density, and vertical permeability and porosity; the other plug was used for petrographic, X-ray diffraction, and laser particle size analysis to determine rock mineralogy. These data were used to compare wireline log density values with those measured in the core plugs and to compare the wireline sonic with the measured sonic. In addition, the relationship between clay volume and gamma-log value was determined so estimates of clay volume could be determined for all of the wells lacking sonic curves. These sonic curves were estimated using a value for clay volume from the gamma log, a value for porosity from the density log, and an appropriate velocity-porosity-clay relationship (Han et al., 1986; Eberhardt-Phillips et al., 1989). Synthetic seismograms were then generated using in-situ values (ie., no correction for water table) and wavelets that ranged from low (20 Hz) to higher frequencies (125 Hz). One of our aims was to see if potential "reservoir" facies could be detected.

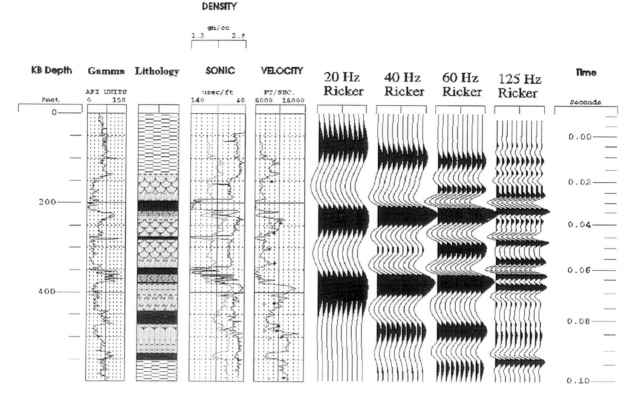

Figure 7. Synthetic seismogram generated for the ARCO 82-4 well. Synthetic responses were derived using wavelets ranging from 20 to 125 Hz. Note that coal beds dominate the seismic response independent of the wavelet used.

Results

The synthetic seismogram for the ARCO 82-4 well using in-situ measured values (no correction for water table effects) is shown in Figure 7. The seismic response is dominated by the coal beds regardless of wavelet frequency. Although the coal beds are thin and well below tuning thickness/seismic resolution, they are highly detectable by virtue of their strong impedance contrast with adjacent lithologies. The velocity of the sandstones is much slower in this in-situ case than it would under fully saturated conditions. This means that under fully saturated conditions the impedance contrast between the sandstones and coals would be even higher. This has a significant impact on our ability to image "reservoir" facies (i.e., delta-front and channel-belt sandstones). In most cases "reservoir" sandstones could not be imaged directly. An exception was the case of distributary channel-belt sandstones encased in coaly marsh deposits.

Figure 8 shows a seismic model for a channel-belt sand body encased in coaly deposits. This particular channel-belt sand body, which is in Cycle 5, is exposed along I-70 and was chosen to define the geometry of a typical distributary channel sand body in the Upper Ferron. It is approximately 6 m (20 ft) thick, exhibits well-developed lateral accretion surfaces, and forms a split in the I coal seam. This channel sand body is well below seismic resolution but appears in the modeled seismic profile as an amplitude dimout of the strong response generated by the coaly deposits. This suggests that one might interpret amplitude dimouts as indicating the presence of distributary channel sand bodies on seismic lines through the Upper Ferron deltaic complex. They appear to be the only reservoir facies that can be imaged directly.

CONCLUSIONS

This study used an integrated, multidisciplinary approach to better understand the Upper Ferron deltaic complex. Our principal conclusions are summarized below.

1. The Upper Ferron is a river-dominated deltaic complex that has been subordinately influenced and modified by marine processes (predominantly wave and storm). The delta fronts of all cycles are fed by numerous distributary channel belts, indicating that riverine processes dominated their evolution. The delta fronts of landward-stepping cycles, however, exhibit increasing marine modification as a result of a progressive decrease in sediment supply.

2. Foraminiferal, palynological, and palynofacies data indicate that the flooding surfaces partitioning the Upper Ferron cycles lack an open-marine signature. They have a bay-like, brackish aspect and appear to represent abandonment surfaces that were generated as a result of autocyclic lobe switching.

88

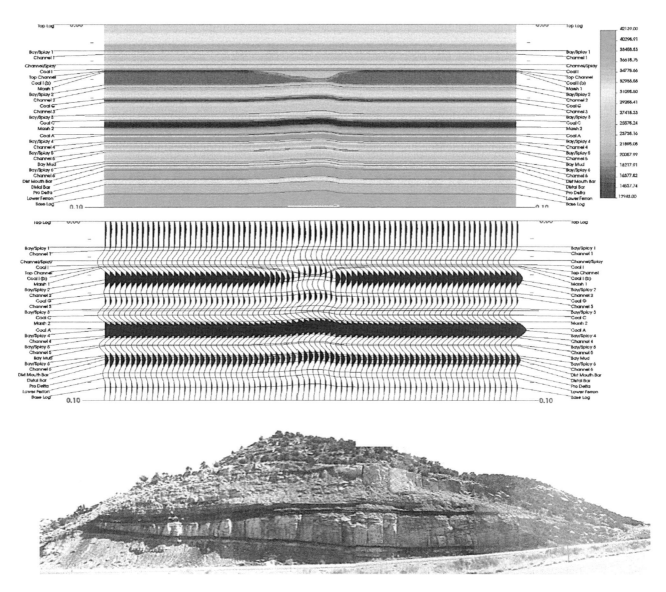

Figure 8. Acoustic impedance model (top), synthetic response (middle), and photograph of outcrop along I-70 used to define the geometry of a distributary channel-sand body encased in coaly deposits (bottom); the rest of the model was taken from the ARCO 82-4 well. Note the channel appears as an amplitude dimout on the modeled seismic.

3. Autocyclic processes, channel avulsion and lobe switching, controlled the internal architecture and partitioning of the Upper Ferron cycles. The stacking pattern (seaward-stepping, vertically stacked, landward-stepping) evolved in response to decreasing sediment supply combined with increasing accommodation.

4. Compelling evidence to support the presence of incised valleys in the Upper Ferron is lacking. Valleys interpreted by other workers are channel belts whose aspects (e.g., size, thickness, depth of incision) and relationships with associated facies are more characteristic of deltaic distributary channels than incised valleys.

5. Distributary channel-belt sandstones are the primary reservoir facies in the Upper Ferron deltaic complex, even though volumetrically they are not the dominant reservoir facies in the seaward-stepping cycles. They have an average permeability of 116 md compared to the 47 md of the delta-front sandstones.

6. Rock physics and seismic modeling demonstrate that the seismic response of the Upper Ferron deltaic complex is strongly dominated by the large acoustic impedance contrasts between sandstones and thin coals or carbonaceous shales that are well below seismic resolution. Distributary channel-sand bodies should be identifiable as amplitude dimouts of the strong response generated by associated coals and carbonaceous shales.

ACKNOWLEDGMENTS

This study was conducted while we were members of Mobil's Structure and Stratigraphic Process Group in Dallas. We are especially grateful to Mike Gardner for freely sharing his knowledge of the Ferron Sandstone with us on numerous enjoyable visits to the outcrop. We are indebted to Harvey Cohen, Martine Hardy, Christy Harrison, and Karen Roche whose studies during summer internships with Mobil contributed significantly to our understanding of the Upper Ferron deltaic complex. Frank Roof assisted us with all phases of the study, and the manuscript was improved by the thoughtful comments of reviewers David Mohrig and Paul Link. We thank ExxonMobil for permission to publish our study.

REFERENCES CITED

Anderson, P. B., T. C. Chidsey, Jr., and T. A. Ryer, 1997, Fluvial-deltaic sedimentation and stratigraphy of the Ferron Sandstone, in P. K. Link and B. J. Kowallis, eds., Mesozoic to Recent geology of Utah: Brigham Young Geology Studies, v. 42, part 2, p. 135–142.

Barton, M. D., 1995, Sequence stratigraphy, facies, architecture, and permeability structure of fluvial-deltaic reservoir analogs — Cretaceous Ferron Sandstone, central Utah: Gas Research Institute Fieldtrip Guidebook, Bureau of Economic Geology, The University of Texas at Austin, 139 p.

—1997, Application of Cretaceous Interior Seaway outcrop investigations to fluvial-deltaic reservoir characterization — Part I, predicting reservoir heterogeneity in delta front sandstones, Ferron gas field, central Utah, in K. W. Shanley and B. F. Perkins, eds., Shallow marine and nonmarine reservoirs — sequence stratigraphy, reservoir architecture and production characteristics: Gulf Coast Section SEPM, Eighteenth Annual Research Conference, p. 33–40.

Barton, M. D., and E. S. Angle, 1995, Comparative anatomy and petrophysical property structure of seaward- and landward-stepping deltaic reservoir analogs, Ferron Sandstone, Utah: Gas Research Institute Topical Report, Bureau of Economic Geology, The University of Texas at Austin, 59 p.

Coleman, J. M., 1988, Dynamic changes and processes in the Mississippi River delta: Geological Society of America Bulletin, v. 100, p. 999–1015.

Coleman, J. M., and S. M. Gagliano, 1964, Cyclic sedimentation in the Mississippi River deltaic plain: Gulf Coast Association of Geological Societies Transactions, v.14, p. 67–80.

—1965, Sedimentary structures: Mississippi River deltaic plain, in G. V. Middleton, ed., Primary sedimentary structures and their hydrodynamic interpretation: SEPM Special Publication 11, p. 133–148.

Cotter, E., 1975a, Deltaic deposits in the Upper Cretaceous Ferron Sandstone, Utah, in M. L. S. Broussard, ed., Deltas, models for exploration: Houston Geological Society, p. 471–484.

—1975b, The role of deltas in the evolution of the Ferron Sandstone and its coals, Castle Valley, Utah: Brigham Young University Geology Studies, v. 22, p. 15–41.

Davis, L. J., 1954, Stratigraphy of the Ferron Sandstone: Intermountain Association of Petroleum Geologists, Fifth Annual Field Conference Guidebook, p. 55–58.

Eberhardt-Phillips, D. M., D. Han, and M. D. Zoback, 1989, Empirical relationships among seismic velocity, effective pressure, porosity and clay content in sandstones: Geophysics, v. 54, p. 82–89.

Fisher, R. S., M. D. Barton, and N. Tyler, 1993, Quantifying reservoir heterogeneity through outcrop characterization — architecture, lithology, and permeability distribution of a seaward-stepping fluvial-deltaic sequence, Ferron Sandstone (Cretaceous) central Utah: Gas Research Institute Topical Report GRI-93-0023, 83 p.

Frazier, D. E., 1967, Recent deltaic deposits of the Mississippi River; their development and chronology: Gulf Coast Association of Geological Societies Transactions, v. 17, p. 287–315.

Gardner, M. D., 1993, Sequence stratigraphy of the Ferron Sandstone (Turonian) of east-central Utah: Ph.D. dissertation, Colorado School of Mines, Golden, 528 p.

—1995a, Tectonic and eustatic controls on the stratal architecture of mid-Cretaceous stratigraphic sequences, central Western Interior foreland basin of North America, in S. L. Dorobek, ed., Stratigraphic evolution of foreland basins: SEPM Special Publication 52, p. 243–282.

—1995b, The stratigraphic hierarchy and tectonic history of the mid-Cretaceous foreland basin of central Utah, in S. L. Dorobek, ed., Stratigraphic evolution of foreland basins: SEPM Special Publication 52, p. 283–303.

—2001, The distinction between ancient fluvial and valley-fill sandstone reservoirs — semantics or economics (abs.): AAPG Annual Convention Official Program, v. 10, p. A69.

Gardner, M. D., and T. A. Cross, 1994, Middle Cretaceous paleogeography of Utah, in M. V. Caputo, J. A. Peterson, and K. J. Franczyk, eds., Mesozoic systems of the Rocky Mountain region, U.S.A.: Rocky Mountain Section SEPM, p. 471–503.

Garrison, J. R., and T. C. V. van den Bergh, 1999, The high-resolution depositional sequence stratigraphy of the Turonian-Coniacian Upper Ferron Sandstone Last Chance delta, east-central Utah (abs): Geological Society of America Annual Meeting, Abstracts with Programs, v. 31.

Garrison, J. R., T. C. V. van den Bergh, C. E. Baker, and D. E. Tabet, 1997, Depositional sequence stratigraphy and architecture of the Cretaceous Ferron Sandstone: implications for coal and coalbed methane resources — a field excursion, in P. K. Link and B. J. Kowallis, eds., Mesozoic to Recent geology of Utah: Brigham Young University Geology Studies, v. 42, part 2, p. 155–202.

Hale, L. A., 1972, Depositional history of the Ferron Formation, central Utah, in J. L. Baer and E. Callaghan, eds., Plateau — Basin and Range transition zone: Utah Geological Association Guidebook 2, p. 115–138.

Han, D. H., A. Nur, and D. Morgan, 1986, Effects of porosity and clay content on wave velocities in sandstones: Geophysics, v. 51, p. 2093–2107.

Harland, R., 1973, Dinoflagellate cysts and acritarchs from the Bearpaw Formation (Upper Campanian) of southern Alberta, Canada: Paleontology, v. 16, p. 665–706.

Kamola, D. L., and J. C. Van Wagoner, 1995, Stratigraphy and facies architecture of parasequences with examples from the Spring Canyon Member, Blackhawk Formation, Utah, in J. C. Van Wagoner and G. T. Bertram, eds., Sequence stratigraphy of foreland basin deposits: AAPG Memoir 64, p. 27–54.

Katich, P. J., 1954, Cretaceous and Early Tertiary stratigraphy of central and south-central Utah with emphasis on the Wasatch Plateau area: Intermountain Association of Petroleum Geologists Fifth Annual Field Conference Guidebook, p. 42–54.

Knox, P. R., and M. D. Barton, 1999, Predicting interwell heterogeneity in fluvial-deltaic reservoirs: effects of progressive architecture variation through a depositional cycle from outcrop and subsurface observations, in R. Schatzinger and J. Jordan, eds., Reservoir characterization — recent advances: AAPG Memoir 71, p. 57–72.

Lupton, C. T., 1916, Geology and coal resources of Castle Valley in Carbon, Emery, and Sevier Counties, Utah: U.S. Geological Survey Bulletin 628, 88 p.

Mao, S., and B. Mohr, 1992, Late Cretaceous dinoflagellate cysts (? Santonian-Maestrichtian) from the southern Indian Ocean (Hole 748C), in S. W. Wise Jr.,

R. Schlich, and others: Ocean Drilling Program Proceedings, Scientific Results, v. 120 (1), p. 307–341.

Mohrig, D., P. L. Heller, C. Paola, and W. J. Lyons, 2000, Interpreting avulsion processes from ancient alluvial sequences — Guadalope-Matarranya system (northern Spain) and Wasatch Formation (western Colorado): Geological Society of America Bulletin, v. 112, p. 1787–1803.

Murray, J. W., 1973, Distribution and ecology of living benthic foraminiferids: New York, Crane, Russak and Company, Inc., 274 p.

Penland, S., R. Boyd, and J. R. Suter, 1988, Transgressive depositional systems of the Mississippi delta plain — a model for barrier shoreline and shelf sand development: Journal of Sedimentary Petrology, v. 58, p. 932–949.

Reading, H. G., and J. D. Collinson, 1996, Clastic coasts in H. G. Reading, ed., Sedimentary environments, third edition: Oxford, Blackwell Publishing, p. 154–231.

Reimersma, P., and M. Chan, 1995, Facies of the lower Ferron Sandstone and Blue Gate Shale Members of the Mancos Shale: lowstand and early transgressive facies architecture, in D. J. P. Swift, G. F. Oertel, R. W. Tillman, and J. A. Thorne, eds., Shelf sand and sandstone bodies: International Association of Sedimentologists Special Publication 14, p. 489–510.

Roberts, H. H., 1997, Dynamic changes of the Holocene Mississippi River delta plain — the delta cycle: Journal of Coastal Research, v. 13, p. 605–627.

Ryer, T. A., 1981, Deltaic coals of the Ferron Sandstone Member of Mancos Shale — predictive model for Cretaceous coal-bearing strata of Western Interior: AAPG Bulletin, v. 65, p. 2323–2340.

—1991, Stratigraphy, facies, and depositional history of the Ferron Sandstone in the canyon of Muddy Creek, east-central Utah, in T. C. Chidsey, Jr., ed., Geology of east-central Utah: Utah Geological Association Publication 19, p. 45–54.

Ryer, T. A., and M. McPhillips, 1983, Early Cretaceous paleogeography of east-central Utah, in M. W. Reynolds and E. D. Dolly, eds., Mesozoic paleogeography of the west-central United States: Rocky Mountain Section SEPM Paleogeography Symposium 2, p. 253–271.

Van Wagoner, J. C., R. M. Mitchum, K. M. Campion, and V. D. Rahmanian, 1990, Siliciclastic sequence stratigraphy in well logs, cores, and outcrops — concepts for high-resolution correlation of time and facies: AAPG Methods in Exploration Series 7, 55 p.

Charles T. Lupton and "Red" near Moore, Castle Valley, Utah, circa 1910. Photograph courtesy of the family of C. T. Lupton.

Regional Sequence Stratigraphic Interpretations

"Marooned" in Salina Canyon, Wasatch Plateau, Utah, circa 1910. Photograph courtesy of the family of C. T. Lupton.

Analog for Fluvial-Deltaic Reservoir Modeling: Ferron Sandstone of Utah
AAPG Studies in Geology 50
T. C. Chidsey, Jr., R. D. Adams, and T. H. Morris, editors

Stacking Patterns, Sediment Volume Partitioning, and Facies Differentiation in Shallow-Marine and Coastal-Plain Strata of the Cretaceous Ferron Sandstone, Utah

Michael H. Gardner, Timothy A. Cross, and Mark Levorsen[1]

ABSTRACT

Fluvial-deltaic strata of the Upper Cretaceous Ferron Sandstone, Western Interior Seaway, form a clastic wedge consisting of eight short-term stratigraphic cycles. The cycles are arranged consecutively in a seaward-stepping, vertically stacked, and landward-stepping stacking pattern. The stacking pattern is a product of fluctuations in accommodation-to-sediment supply (A/S) regimes described by intermediate-term, base-level cycles.

Each short-term stratigraphic cycle is a progradational/aggradational unit comprising a spectrum of coastal-plain, bay/lagoon/estuary, shoreface, and shelf facies tracts. Sediment volumes and sandstone:mudstone ratios were measured separately in coastal-plain and shoreface facies tracts in four of the cycles. Total sediment and total sandstone volumes are partitioned differentially into the two facies tracts in a systematic manner that follows the stacking pattern. The total sediment volume and total sandstone in the shoreface facies tract decreases regularly from seaward- to landward-stepping stacking patterns. The proportion of marine-to-nonmarine sandstone also decreases. This demonstrates increasing sediment storage in continental environments during the transition from seaward- to landward-stepping stacking patterns.

Sediment volume partitioning is accompanied by systematic changes in numerous other stratigraphic and sedimentologic attributes which illustrate the two types of facies differentiation. The first type — stratigraphic control on the types of geomorphic elements that occupy a geomorphic environment — is manifest by the transition from fluvial- to wave-dominated deltas in the progression from seaward- to landward-stepping cycles. The second type — a change in degree of preservation of original geomorphic elements — is illustrated by conspicuous differences in the facies that compose the shoreface and coastal-plain facies tracts. Shorefaces of high-accommodation, landward-stepping cycles comprise homogeneous, cannibalized and amalgamated sandstones, whereas shorefaces of low-accommodation, seaward-stepping cycles are lithologically heterogeneous containing diverse facies and well-preserved, original geomorphic elements. Distributary channelbelt sandstones of

[1]Department of Geology and Geological Engineering,
Colorado School of Mines, Golden, Colorado

landward-stepping cycles are composed of high diversity, well-preserved macroforms and bedforms, whereas those of seaward-stepping cycles are composed of strongly cannibalized, amalgamated, low-diversity macroforms and bedforms.

Sediment volume partitioning and facies differentiation are attributed to changing A/S conditions that accompany short- and intermediate-term base-level cycles. The A/S conditions control or influence the position and volume of sediment accumulation, the types of geomorphic elements in an environment, and the proportions and completeness of original geomorphic elements that enter the stratigraphic record.

INTRODUCTION

This study illustrates the well-organized behavior of the stratigraphic process-response system, and suggests underlying causes for this behavior. We observe systematic variations in numerous stratigraphic and sedimentologic attributes of siliciclastic coastal-plain to shelf strata that are coincident with stacking patterns of stratigraphic cycles. These organized, systematic variations of stratigraphic and sedimentologic attributes are analyzed from the perspectives of conservation laws and stratigraphic base level. Stratigraphic base level describes the balance between the energy required to change accommodation space and the energy used by surficial processes to erode, transport, and deposit sediment. Base-level changes are manifest by changes in the ratio of accommodation-to-sediment supply (A/S). Changes in A/S conditions and mass conservation determine the volumes and types of sediment which accumulate in different environments.

Sediment is partitioned differentially into coastal-plain and shoreface facies tracts through time and changing A/S conditions. Changes in the ratio of total sediment volumes within these two facies tracts are accompanied by changes in ratio of nonmarine-to-marine sandstone volumes. Sediment volume partitioning reflects the balance among rates of sediment delivery, rates of reworking and cannibalization of sediment, and rates of net sediment accumulation. Sediment volume partitioning can be explained by variations in the A/S conditions that accompany stratigraphic base-level cycles.

Sediment volume partitioning controls or influences sedimentologic and stratigraphic attributes of all scales including the constituents, associations and successions of facies, the degree of preservation of original geomorphic elements, petrophysical attributes, stratigraphic architecture, and frequency of occurrence of hiatal surfaces of different origins. Two types of facies differentiation are recognized and illustrated. One type is the change in original geomorphic elements that occupy the same environment under variable A/S conditions. The other is the variable degree of preservation of original geomorphic elements and their proportions that enter the stratigraphic record. This produces changes in facies diversity and lithologic heterogeneity in strata that accumulated in the same environment.

Heterogeneous shoreface strata of seaward-stepping cycles are responses to lower A/S conditions compared with homogeneous sandy shoreface strata of higher A/S, landward-stepping cycles. Similarly, heterogeneous distributary channelbelt deposits of landward-stepping cycles are responses to higher A/S conditions compared with homogeneous channelbelt sandstones of lower A/S, seaward-stepping cycles. In homogeneous strata relatively more time is represented by stratigraphic surfaces of discontinuity than by rock, whereas in heterogeneous strata relatively more time is represented by rock than by hiatal surfaces.

Stratigraphic and sedimentologic attributes of all scales and of many types show consistent, systematic patterns of change when viewed from the perspectives of conservation laws and stratigraphic base level. This organization produces transitional facies constituents, associations, and successions within a continuum of the preserved products of the same environment. The sedimentologic attributes of facies tracts commonly described in "facies models" are mixtures of the products of geomorphic elements that existed separately during base-level cycles. The next generation of facies models should be constructed from a stratigraphic perspective in which there is a continuum of transitional forms of facies associations and successions that are different products of the same parent.

GEOMORPHIC AND STRATIGRAPHIC BASE LEVEL CONCEPTS

We recognize three temporal and spatial scales of allogenic cyclicity in the upper Ferron Sandstone (Figure 1; Obradovich, 1991; Gardner, 1993; Gardner and Cross, 1994; Gardner, 1995a, 1995b, 1995c). Cycles of each scale record a complete stratigraphic base-level cycle *sensu* Wheeler (1959, 1964, 1966). Because our usage of stratigraphic base level follows Wheeler rather than the more common geomorphic usage, and because we incorporate some sequence stratigraphy concepts into Wheeler's original definition, this section presents our understanding of stratigraphic base level.

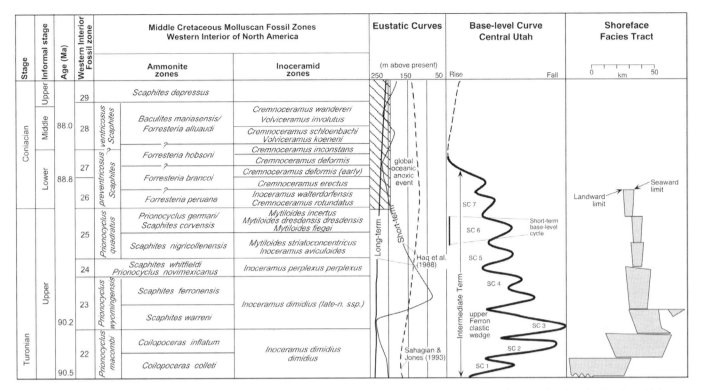

Figure 1. Chart showing biostratigraphy, radiometric dates, sea level curves, and hierarchy of stratigraphic base-level cycles for the mid-Cretaceous strata of central Utah (modified from Gardner, 1995a). Biostratigraphic chart showing ammonite and inoceramid fossil zones for late Cenomanian through middle Coniacian Stages from the Western Interior of North America. Because of potential discrepancies associated with different stage boundary dates, eustatic curves from Haq et al. (1988), Sahagian and Jones (1993), and Schlanger et al. (1986) are calibrated to biozones. Base-level curves based on stratigraphic relations in central Utah. Sources of data: Western Interior index fossils from Molenaar and Cobban (1991); upper Cenomanian and lower Turonian Substage biozones from Elder (1985); middle and upper Turonian Substage biozones from Kauffman et al. (1976), and Kauffman and Collom (unpublished data, 1990); lower and middle Coniacian zones from Collom (1991); argon-argon isotopic ages from Obradovich (unpublished data, 1991).

Originally, Powell (1875) defined base level in a geomorphic context as the lower limit to which the land may be degraded and equivalent to sea level. He also allowed for local and temporary base levels (i.e., multiple base levels) in addition to the grand base level or sea level. Most subsequent usage has maintained this geomorphic context, even though definitions and usages have multiplied and varied through time. Today, the term "geomorphic" base level includes one or more of the following concepts:

• base level is either a horizontal planar surface or multiple surfaces (Hayes, 1899; Barrell, 1917) or it is an inclined planar surface or multiple surfaces (Powell, 1875);

• this surface is either imaginary (e.g., Powell, 1875; Davis, 1902) or it is physical (Willis, 1895; Cowles, 1901);

• base level is either sea level (e.g., Powell, 1875; Davis, 1902; Schumm, 1993) or it is linked to sea level in some way;

• base level is either a process (Powell, 1875), a controlling surface (Barrell, 1917; Mackin, 1948; Posamentier and Vail, 1988), or a descriptor lacking any control; and

• base level is associated with or controls the graded profile of equilibrium (Davis, 1902; Mackin, 1948; Posamentier and Vail, 1988).

Despite the variability in usage of the term geomorphic base level, specific elements of its definitions clearly separate it from the term and concepts of stratigraphic base level. First, geomorphic base level does not consider an equilibrium or balance between erosion and deposition, as does stratigraphic base level; it is concerned only with the degradational part of the equation. Second, geomorphic base level does not consider or invoke conservation laws (specifically mass, space, and time), as does stratigraphic base level. Third, geomorphic base level in most contemporary usage invokes a fixed reference, sea level, as controlling fluvial geomorphology along a "graded" longitudinal profile. The balance of sediment flux across a topographic profile can't be accurately measured with a meter stick fixed to a point that controls or determines the outcome.

An alternative usage of base level in a stratigraphic context has an equally old history. In revising and expanding the fifth edition of Dana's Text-Book of Geology, Rice (1897) wrote that base level was "the condition of balance between erosion and deposition..." and there-

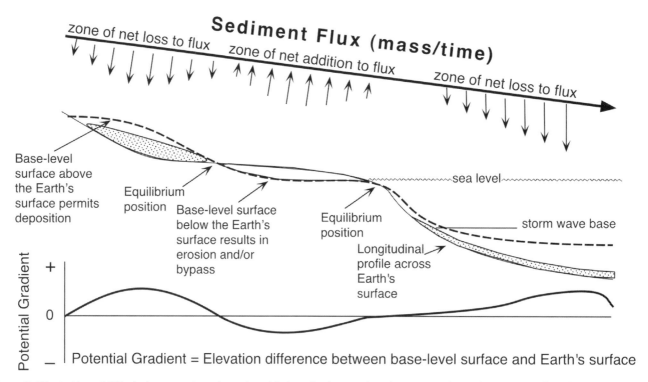

Figure 2. *Illustration of Wheeler's conception of stratigraphic base level as an imaginary potentiometric energy surface that undulates with respect to the Earth's surface. It relates the energy required to change accommodation to the energy required to erode, transport, and deposit sediment. Where stratigraphic base level is above the Earth's surface, sediment will accumulate, if available, building topography and bringing the Earth's surface closer to base level. Where stratigraphic base level is below the Earth's surface, erosion or bypass occurs and the removal of material reduces the topography bringing the Earth's surface closer to base level. Where stratigraphic base level is coincident with the Earth's surface, there is a state of equilibrium.*

by introduced the notion of considering not only degradation but sediment transport and accumulation in the definition of base level. (It is ironic that he incorrectly attributed this stratigraphic sense of base level to Powell [1875]; he either misread or misunderstood Powell, or simply chose to add this new dimension to Powell's degradational concept of base level.) This notion of an equilibrium or balance between erosion and sedimentation was supported through the years by others, but perhaps most influentially by Barrell (1917), Krumbein and Sloss (1951), and Sloss (1962).

Wheeler (1964, 1966) made significant additions and modifications to stratigraphic base-level concepts. He argued that base level was a single, continuous, nonhorizontal, undulatory, imaginary surface that rises and falls with respect to the Earth's surface. Where base level is above the Earth's surface, sediment will accumulate if it is available. Where base level is below the Earth's surface, sediment is eroded and transferred downhill to the next site where base level is above the Earth's surface. His stratigraphic base level is not a control, but a descriptor for measuring the energy budget between forces and processes that change sediment storage capacity (accommodation space) and those that erode, transfer, and deposit sediment across the surface of the Earth. In effect, but not explicitly, Wheeler defined stratigraphic

base level as a potentiometric energy surface that describes the energy required to move the Earth's surface up or down to a position where gradients, sediment supply, and accommodation are in equilibrium. If these forces are in balance neither deposition nor erosion occurs, and the Earth's surface is at equilibrium and coincident with base level (Figure 2). Base level describes the increase in topographic relief that results from deposition and its reduction by erosion. Base level accounts for deposition and erosion that occur at the same time in different parts of a basin. This is a necessary condition of a physical system where energy and mass are conserved. Stratigraphic base level may be expressed as the A/S ratio — a ratio of the energy required to change the accommodation at the Earth's surface and the energy required to erode, transport, and deposit sediment. The A/S ratio is expressed as the dimensionless term Nm/Nm. Or, if sediment volume is considered instead of energy, the A/S ratio is expressed as the dimensionless term m^3/m^3. Wheeler (1959, p. 701–702) states:

In contrast with the popular concept, baselevel is neither a "horizontal plane" nor can it be defined solely in terms of sea level or relationships on the sea floor... Moreover, baselevel should not be conceived solely in its relation-

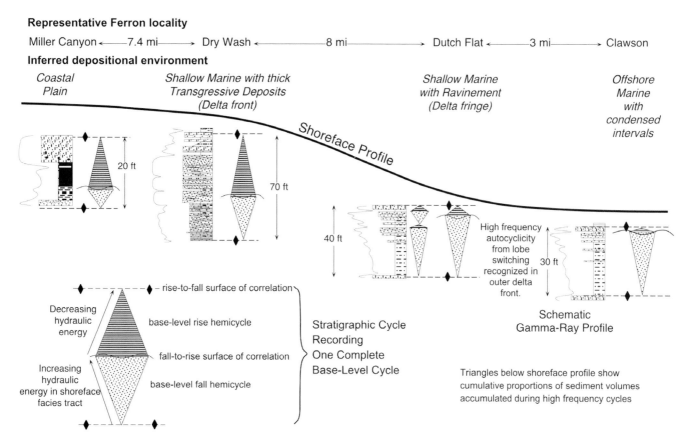

Figure 3. Illustration of changes in cycle symmetry caused by volumetric partitioning and correlation of base-level cycles across shelf, shoreface, and coastal-plain facies tracts. Note how surfaces correlate to rocks at various positions of the depositional profile during a base-level cycle.

ships to either erosion or aggradation alone, for its significance is best appreciated in stratigraphy, sedimentation, or geomorphology if it is conceived as a "surface at which neither erosion nor sedimentation (can take) place."

Stratigraphic base level describes the erosional decay and construction of topography through deposition that drives the partitioning or selective storage of sediment volumes across a topographic profile. This is the normal condition of sedimentation in a depositional system linked by a common dispersal mechanism. During a base-level cycle, the A/S ratio decreases unidirectionally to a limit (base-level fall minimum) that equates to a sequence boundary in geographic positions where there is no accommodation. It then increases unidirectionally to another limit (base-level rise maximum) that some place may be equivalent to the maximum flooding surface of sequence stratigraphic terminology.

Sedimentologic and stratigraphic attributes of all scales and numerous types respond consistently and coherently to these changes in A/S conditions, including small-scale attributes such as texture and petrophysical properties, meso-scale attributes such as sedimentary structures, facies diversity, and macroform types (sensu, Crowley, 1983) and mixtures, and large-scale attributes of stratigraphic architecture (Gardner, 1993; Ramon and

Cross, 1997; Cross, 2000). The limits, or "turnaround" points, of these unidirectional trends in A/S are correlated throughout the spatial extent of each stratigraphic cycle. Stratigraphic cycles of each scale are the time-bounded rock units that comprise all strata and hiatus produced during a base-level cycle (Figure 3). The initiation points for stratigraphic cycles of all scales are picked consistently at the same turnaround position. In this study, the initiation point was picked at the base-level rise-to-fall turnaround because it is the most practical; it is the position most easily recognized, frequently documented, consistently picked, and physically traceable.

Stratigraphic base-level concepts emphasize the correlation of all rocks and surfaces developed during a base-level cycle divided into base-level rise and fall time domains (Grabau, 1924; Busch, 1959; Wheeler, 1966; Gardner, 1993; Cross and Lessenger, 1998; Muto and Steel, 2000; Figure 3). This provides a more complete accounting for how time is represented in a stratigraphic cycle as either rock or surface of stratigraphic discontinuity. By contrast, a parasequence, defined as an asymmetric upward-shoaling succession bounded by a marine flooding surface (Van Wagoner et al., 1990), only accounts for the time of deposition during base-level fall in shallow-marine strata and does not recognize continental and marginal-marine tidal strata that accumulated during base-level rise; also see discussion by Arnott (1995). This more limited usage precludes consideration of sediment volume partitioning and excludes the possibility of a significant proportion of time represented by shallow-marine strata formed during base-level rise.

In theory, the relative proportions of sediment volumes in coastal-plain and shoreface facies tracts vary with position of the short-term stratigraphic cycle in the stacking pattern, as first drawn and explained in a subsidence context by Barrell (1912, Figure 4, p. 399). This is one of several important responses to changes in A/S conditions recorded by base-level cycles. As stratigraphic base level rises and intersects the Earth's surface progressively higher on the topographic profile, the A/S ratio increases and the sediment storage capacity in uphill positions increases. Since more sediment is stored uphill in continental environments, less sediment is available (conservation of mass) for downhill transport and accumulation in shoreface and shelf environments. Conversely, a decrease in the A/S ratio theoretically partitions more sediment volume into shoreface and shelf environments and less is stored uphill in continental environments. The degree of sediment volume partitioning in short-term stratigraphic cycles, as measured by the proportion of sediment volume stored in different facies tracts, produces systematic changes in cycle stacking patterns. These changes have been simulated and matched with field data using forward and inverse stratigraphic models (Cross and Lessenger, 1998, 1999, 2001).

STRATIGRAPHIC SETTING AND STACKING PATTERNS

The Upper Cretaceous (Turonian-Coniacian) Ferron Sandstone is a regressive-transgressive clastic wedge that accumulated at the site of a large delta in the foreland basin, east-central Utah (Speiker, 1949; Armstrong, 1968; Cotter, 1975; Ryer, 1981; Gardner, 1993; Gardner and Cross, 1994; Gardner, 1995a, 1995b, 1995c). During that period, sea level was near a maximum highstand, and both accommodation and sediment supply were high (Kauffman, 1977; Hancock and Kauffman, 1979; Schlanger et al., 1986; Haq et al., 1988; McDonough and Cross, 1991; Sahagian and Jones, 1993; Figure 1). Consequently, time in the Ferron is represented much more by rock than by stratigraphic surfaces of erosion and nondeposition. Regional unconformities are absent and distributary channel deposits do not extend beyond the delta they sourced, indicating an absence of incised valleys.

The upper Ferron Sandstone consists of eight progradational/aggradational stratigraphic units — short-term stratigraphic cycles — arranged consecutively in seaward-stepping, vertically stacked, and landward-stepping geometric patterns (Ryer, 1981; Gardner, 1995c; Figure 4). Seven of these cycles are recognized in outcrop, and the eighth and youngest cycle is recognized

from subsurface data. Each progradational/aggradational unit contains a spectrum of coastal-plain, estuary/bay/lagoon, shoreface, and shelf facies tracts. Exceptionally continuous, three-dimensional exposures plus a subsurface data base of geophysical well logs and coal core logs enabled physical correlation of short-term cycles across all facies tracts. These stratigraphic cycles are equivalent to a depositional episode (Frazier, 1974), a genetic increment of strata or genetic sequence (Busch, 1959, 1971), a fourth-order regressive-transgressive cycle (Ryer, 1983), and are comparable to parasequence sets of a fourth-order, high-frequency sequence (Van Wagoner et al., 1990). Subdelta lobes that compose the shoreface facies tract correspond to parasequences of other Ferron workers (e.g., Barton, 1994; Ryer and Anderson, 1995). They are considered autogenic because they are restricted to the shoreface facies tract.

The general motif of volumetric partitioning is shown along one example topographic profile from the Ferron Sandstone (Figure 3). The stacking pattern of progradational/aggradational stratigraphic units observed and mapped in this study and by Ryer (1981) is the product of changing A/S conditions of the intermediate-term, base-level cycle (1–2 m.y. duration). The short-term (0.3 m.y.), seaward-stepping units (SC1-3; Figure 4) accumulated during lowest A/S. The vertically stacked unit (SC4; Figure 4) accumulated during early intermediate-term, base-level rise. The landward-stepping units (SC5-8; Figure 4) accumulated during late intermediate-term base-level rise and highest A/S. The architecture of these short-term cycles is described below.

Seaward-Stepping SC2

The seaward-stepping shoreface facies tracts of SC1 and SC2 record offlap in excess of 70 km (43 mi). The SC2 shoreface extends over 60 km (37 mi) parallel to depositional dip and forms the thickest and widest shoreface facies tract capped by the most laterally continuous coal horizon (a coal bed) in the upper Ferron Sandstone (Figure 4). Discrete subdelta lobes are the dominant stratigraphic bodies and comprise upward-coarsening sandstone successions with gently inclined, dip-oriented clinothems 10–30 m (33–98 ft) thick and 1–3 km (0.6–2 mi) long. Amalgamated distributary mouthbar sandstones contain numerous inclined bedding surfaces that extend hundreds of meters along depositional dip and strike. Amalgamated sandstones are replaced laterally, on a kilometer-scale, by mudstone-dominated interdistributary bay strata.

Coastal-plain strata that overlie shoreface strata are thin but contain thick, isolated and disconnected channelform sandstone bodies. They consist of amalgamated, erosive-based compound macroforms arranged in single and amalgamated, multistory complexes with biconvex

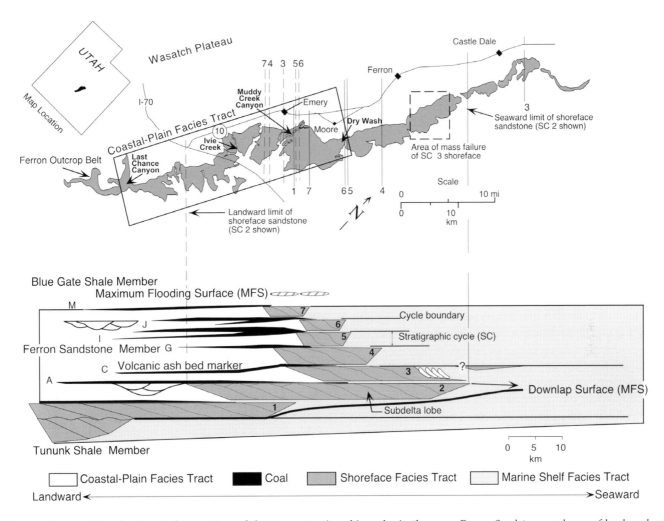

Figure 4. Cross section showing stacking pattern of short-term stratigraphic cycles in the upper Ferron Sandstone, and map of landward and seaward depositional limits of the shoreface facies tracts in the eight cycles.

to flat-based, convex-upward geometries. Bedform and macroform diversity is low with the latter dominated by highly amalgamated, cut-and-fill and low-sinuosity types. The principal bounding surfaces are thin, sandstone-rich, basal channel lags tens of meters wide, and numerous, shorter length reactivation surfaces. Thin mudstone-rich, lower delta-plain strata separate channelforms, and consist of laterally discontinuous coal seams and crevasse-channel and crevasse-splay sandstones. Multiple isolated distributary channel complexes produce a low distributary-channelbelt to delta-plain facies ratio in seaward-stepping stratigraphic cycles.

Seaward-Stepping SC3

The SC3 shoreface facies tract is in the most basinward position, and is the turning point between seaward- and landward-stepping stratigraphic cycles. The shoreface facies tract has a dip-elongate geometry centered over SC2 with landward and seaward limits positioned, respectively, 20 km (12 mi) and 15 km (9 mi) seaward from those of SC2. Discrete subdelta lobes are

sandier than older upper Ferron shoreface strata. Because sedimentary structures are highly amalgamated, resolution of internal sedimentary bodies within shoreface sandstones is greatly reduced. Mudstone-rich interdistributary bay strata are volumetrically subordinate to encasing SC2 and SC4 strata.

At Dry Wash (Figure 4), near the seaward limit of the coastal-plain facies tract, the shoreface facies tract is overlain by a thick heterolithic succession of tidally influenced strata that have approximately equal proportions by volume of base-level fall and rise strata. Tidal strata containing recurved spits, washover fans, barrier shorelines, bay strata, and other features indicative of tidal influence overlie a conspicuous transgressive surface of erosion that incises into progradational delta-front strata. These record in-phase, short- and intermediate-term, base-level rise (tidal and transgressive strata in this area were first described by Stalkup and Ebanks, 1986).

Large-scale deformation in the SC3 shoreface facies tract encompasses a tens of km^2 area near the seaward limit of underlying SC2 shoreface strata near the town of

Ferron, Utah (Figure 4). The most impressive features of this deformation are several isolated, 25-m (82-ft) thick, 100-m (328-ft) long, rotated sandstone lenses interpreted as slumped and rotated distributary bar-finger sandstones. These foundered bar-finger sandstones demonstrate that distributary channels did not extend beyond their delta during in-phase, short- and intermediate-term, base-level fall (the lowest A/S ratio) when unconformity development is most likely.

Channelform sandstone bodies resemble those in older seaward-stepping cycles and consist of erosive-based, cut-and-fill and low-sinuosity macroforms. The C coal bed (Figure 4) is laterally continuous from the top of the shoreface facies tract and across the coastal-plain facies tract. The C coal bed contains a regionally extensive volcanic ash bed (tonstein) that permits correlation between marine-shelf and coastal-plain facies tracts. Landward of the landward depositional limit of shoreface sandstones are the thickest and most extensive bay-fill strata in the upper Ferron Sandstone near the town of Emery, Utah (Figure 4). Heterolithic, mudstone-dominated, marginal-marine strata separate coeval coastal-plain and shoreface sandstones. This several kilometer-wide bay is inferred to have formed during intermediate-term, base-level rise, and filled prior to progradation of the shoreface facies tract of vertically stacked SC4.

Vertically Stacked to Landward-Stepping SC4

Relative to SC3, the shoreface facies tract of SC4 is narrower and offset landward; its landward and seaward limits are 5 and 30 km (3 and 19 mi), respectively, landward of those of the SC2 shoreface facies tract. Shoreface strata contain the thickest, coarsest and most fluvial-dominated facies of all landward-stepping, shoreface facies tracts. The shoreface facies tract geometry resembles landward-stepping SC5-SC7, but facies and geomorphic constituents resemble those in seaward-stepping SC1-SC2. Near the landward depositional limit of the shoreface facies tract, SC4 contains two vertically stacked subdelta lobe successions that thicken abruptly seaward to the northeast along the Molen Reef escarpment. Other than these subdelta lobes, a distinct stratigraphic break within the vertically stacked shoreface facies tract of SC4 was not identified, but two stratigraphic cycles may be represented. Vertically stacked facies successions are present in equivalent marine-shelf facies but occur only locally in the coastal-plain facies tract.

The shoreface facies tract of SC4 contains amalgamated sandstones separated by volumetrically subordinate interdistributary bay strata. Subdelta lobes up to 30 m (98 ft) thick are the thickest of all cycles other than SC2. South of Dry Wash, near the seaward depositional

limit of the shoreface facies tract, thick, mudstone-rich lower delta-front deposits are overlain by lenticular, 3-m-thick (10-ft) sandwaves inferred to represent small-scale mouthbar sandstones that prograded basinward. Comparison of these facies with SC2 shoreface facies at a similar distance from its seaward limit demonstrates increased partitioning of sediment landward in SC4. This may explain why coastal-plain strata that overlie the shoreface facies tract are the most organic-poor and coarsest grained in the upper Ferron Sandstone.

Landward-Stepping SC5

The SC5 shoreface facies tract contracted bidirectionally; its landward depositional limit is almost coincident with that of SC3 and its seaward limit is 7 km (4 mi) landward from that of SC4. Near the landward limit of the shoreface facies tract at Muddy Creek Canyon, distributary channelbelt strata extensively incise wave-dominated shoreface strata (Figure 4). Steeply inclined, seaward-dipping clinoforms segregate shoreface sandstones into 20–30 m (66–98 ft) thick and less than 1-km-wide (0.6-mi) subdelta lobes measured parallel to depositional dip. The paucity of distributary mouthbars and interdistributary bays strata reflects dominance of wave and storm processes along a nonembayed coastline, a significant contrast to shorefaces of seaward-stepping cycles. Coeval distributary channelbelt sandstones contain a diverse assemblage of moderately interconnected, low-sinuosity, high-sinuosity, and abandonment-fill macroforms, typically 3–10 m (10–32 ft) thick and hundreds of meters wide.

Local amalgamation of multiple thin coal seams produces irregular pod-shaped coals up to 8 m (26 ft) thick. Coals commonly overlie, but are replaced locally by, laterally extensive distributary and crevasse-channel sandstones. The ratio between distributary channelbelt and vertical-accretion flood-plain strata is high compared to older seaward-stepping cycles. Distributary channelbelts contain a diverse assemblage of moderately interconnected, low-sinuosity, high-sinuosity, and abandonment-fill macroforms, typically 3–10 m (10–32 ft) thick and hundreds of meters wide. Channel macroforms form moderately interconnected kilometer-wide channelbelts interdigitated along their margins with crevasse-splay and crevasse-channel strata. Macroforms have high bedform diversity, with basal channel lags hundreds of meters wide, and subordinate bounding surfaces of equal length represented by lateral and downstream accretion surfaces.

Shoreface strata of SC6 through SC8 record continued contraction of the shoreface facies tract. Shoreface contraction is associated with increased thickness of coeval coastal-plain strata. Discontinuous coal seams produce lower confidence shoreface to coastal-plain correlations, but the increased thickness of coastal-plain strata in

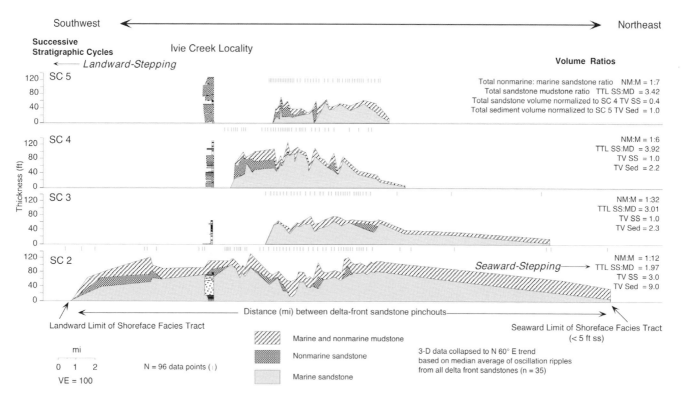

Figure 5. Thickness-distance plot from the Ferron Sandstone showing the areal distribution of various lithologies in facies tracts between the landward and seaward pinch-out of the shallow-marine facies tract of stratigraphic cycles 2–5. The orientation of this plot is parallel to the progradation direction of all delta lobes and to depositional dip. Three-dimensionally distributed data points are collapsed to this depositional dip line, and data points plotted are correlated by linear interpolation. The vertical axis is thickness (m) and the horizontal axis is distance (km), with the origin of the plot set a constant distance from the landward pinch-out of the shoreface facies tract.

the upper part of the upper Ferron Sandstone demonstrates increased landward partitioning of sediment volumes. Coastal-plain strata overlain by marine shelf mudstones of the Blue Gate Shale Member in the most proximal outcrops at Last Chance Canyon record accelerated transgression at the top of the upper Ferron in response to intermediate-term base-level rise (Figure 4).

SEDIMENT VOLUME PARTITIONING

To test whether sediment volumes change systematically with the stacking pattern of short-term stratigraphic cycles, sediment volumes were measured in four progradational/aggradational units in the upper Ferron Sandstone. These units were mapped over a 110 km^2 (68 mi^2) area using 97 data points from outcrop measured sections, geophysical well logs, core-derived lithology logs, and cliff-tape calibrated photomosaics (Figure 5). The three-dimensional distribution of these sediment volume measurements were normalized to account for paleogeographic variations in position and widths of facies tracts, and incomplete preservation of the depositional system from the study area westward to the thrust front. The most consistent and objective normalization procedure was to measure sediment volumes

in continental and marine facies tracts between the landward and seaward depositional limits of the shoreface (delta front) facies tract of each short-term stratigraphic cycle (Figure 4).

Between these paleogeographic limits, the total sediment volume, and the total sandstone and mudstone volumes were measured within shallow-marine and coastal-plain strata. This allows comparison of total sediment volumes and lithology ratios of shoreface and coastal-plain strata in seaward-stepping, vertically stacked, and landward-stepping progradational/aggradational units.

The three-dimensional distribution of these data is collapsed into a two-dimensional thickness-distance plot showing lithology distributions in coastal-plain and shallow-marine strata between the landward and seaward depositional limits of the shoreface facies tract (Figures 5 and 6). Plot orientation is parallel to depositional dip, as determined from mapped orientations of facies tract boundaries and paleoflow analysis. The plot origin is set a constant distance from the landward depositional limit of the shoreface facies tract. Plots show marine and nonmarine sandstone volumes, total sandstone and mudstone volumes, and total sediment volumes for each progradational/aggradational unit. In

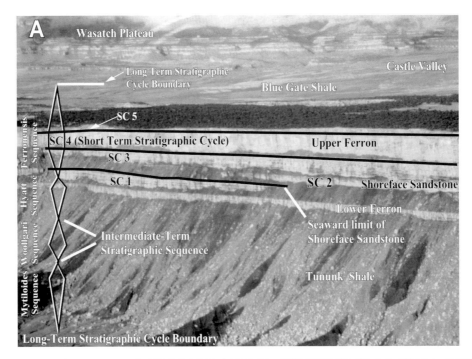

Figure 6. Examples of landward and seaward depositional limits of the shoreface facies tract of upper Ferron short-term cycles. (A) Seaward depositional limit of SC1 along the Molen Reef escarpment. View looking west toward the town of Emery, Utah, of the Tununk Shale and Ferron Sandstone Members along the Molen Reef in southern Castle Valley. Note the seaward depositional limit of shoreface sandstone in SC1 of the Ferronensis sequence near center of photo. From valley floor to cuesta top is approximately 330 m (1080 ft). (B) Landward depositional limit of SC6 shoreface at Muddy Creek Canyon. These limits constrain sediment volume calculations summarized in Figure 5.

this case total sediment volume only refers to the sediment volume between the landward and seaward depositional limits of the shoreface facies tract and not the total sediment volume of the entire depositional system. Although these sediment volumes may be related to changing sediment supply, it is important to emphasize that these volumes can't be determined by this method.

Sediment volumes are expressed as: (1) sandstone volume ratios of shallow-marine and coastal-plain strata; (2) stratigraphic cycle total sandstone/mudstone vol-

ume ratio; and (3) total sediment volume of each stratigraphic cycle. Sediment volume and lithology ratios were calculated for each locality to allow comparison with averaged values on thickness-distance plots. The total sediment and total sandstone volumes calculated in this manner decrease from seaward- to landward-stepping stratigraphic cycles (Figure 5). The nonmarine: marine sandstone volume ratios of seaward-stepping units 2 and 3 are 1:12 and 1:32, respectively; in vertically stacked unit 4 it is 1:6, and in landward-stepping unit 5 it is 1:7. Even though data were collected within a strike-oriented swath about 8 km (5 mi) broad, the sampling of channelbelt sandstones in the coastal-plain facies tract, which is the primary residence of sandstone, is subject to biased position of channelbelts; channelbelts could be in one position in one cycle and in another position in another cycle. Otherwise, measurements of sandstone in the shoreface facies tract and total sediment volume in both facies tracts are not biased. Seaward-stepping cycles record increased sediment volumes in shoreface strata, reduced sediment volumes in coastal-plain strata, and a basinward shift in accommodation; shoreface progradation dominates over coastal-plain aggradation. Vertically stacked cycles have little or no offset of facies tracts across cycle boundaries, little shift in the depositional tracts limits of successive stratigraphic cycles, and subequal sediment volumes in the two facies tracts. Conversely, landward-stepping stratigraphic cycles record increased coastal-plain sediment volumes, reduced shoreface sediment volumes, and a landward shift in accommodation; coastal-plain aggradation dominates over shoreface progradation. These results do not presume a constant sediment supply but rather reflect changes in the A/S ratio that allows for variable sediment flux and storage capacity across environments limited by accommodation.

Decreases in total sediment and total sandstone volumes reflect decreased shoreface facies tract widths in landward-stepping cycles. Compared with seaward-stepping cycles, progressively increased accommodation and storage capacity in the coastal plain of vertically stacked and landward-stepping cycles reduces the total sediment volume delivered to shallow-marine environments. The shift to increased sandstone storage in the coastal-plain facies tract of vertically stacked and landward-stepping cycles also reflects increased A/S conditions in these cycles.

Because accommodation measures the potential space available for sediment accumulation, the direction of sediment transport does not affect a thickness-distance plot relating sediment volume to accommodation. Sediment delivered from out of the plane (e.g., alongshore transport) of the thickness-distance plot may affect the aspect ratio but not the distribution of sediment volumes in linked facies tracts.

FACIES DIFFERENTIATION IN THE SHOREFACE FACIES TRACT

Accompanying sediment volume partitioning are differences in stratal architecture, facies associations and successions, lithologic diversity, stratification types, connectivity and continuity of lithosomes, and petrophysical attributes of strata which are preserved within identical facies tracts but in different portions of base-level

cycles. The term "facies differentiation" (Cross et al., 1993; Cross and Homewood, 1998) refers to these changes in sedimentological and stratigraphic attributes during base-level cycles as first noted by Van Siclen (1958). Facies differentiation reflects both the degree of preservation of original geomorphic elements, and the variations in types of geomorphic elements that existed within a depositional environment at different times and in different A/S regimes. The relative balance among rates of sediment addition, removal (cannibalization and winnowing), and net accumulation controls the degree of preservation. Rates of these processes are strongly influenced by sediment volume partitioning which accompanies changing A/S conditions during base-level cycles.

Strata in the shoreface facies tract of seaward- and landward-stepping progradational/aggradational units are very different in lithologic heterogeneity; facies associations and successions; angles, geometry, and aspect ratio of clinoforms; and relative dominance of current-formed versus wave-formed geomorphic elements. Landward-stepping shoreface sandstones deposited in higher A/S regimes are homogeneous, coarser, dominated by wave-generated and wave-reworked facies associations, and have narrower facies tract widths (Figure 7). Seaward-stepping shorefaces deposited in lower A/S regimes are heterolithic, characterized by much higher facies diversity of mixed wave and current origins, and contain many fully preserved bedforms and other paleogeomorphic elements (Figure 8). Observations of such differences in stratigraphic and sedimentologic attributes of a facies tract with respect to stacking patterns were made previously, but not explained, by Curtis (1970) for deltas in the Gulf Coast and by MacKenzie (1972) for shorefaces in the Western Interior Seaway.

Lower delta-front facies in landward- and seaward-stepping stratigraphic cycles record the same water-depth transition from storm to fair-weather wave base, but their stratigraphic and sedimentologic attributes are quite different. Lower delta-front successions of landward-stepping cycles are thinner (< 1–4 m [< 3–13 ft] thick), and sharp based because they prograde across the flat, shallow-water platform formed by the underlying progradational/aggradational unit. They consist of low-diversity, erosive-based co-sets of amalgamated hummocky cross-stratified sandstone capped by symmetrical ripples, and/or combined-flow asymmetrical ripples (Figure 9). This facies association records dominance of sediment reworking over sediment accumulation and burrowing, with limited preservation of individual bedforms and other paleogeomorphic elements on the seafloor.

By contrast, thicker (< 1–10 m [< 3–33 ft] thick), lower delta-front facies in seaward-stepping cycles consist of a mixture of shallow-water sediment gravity

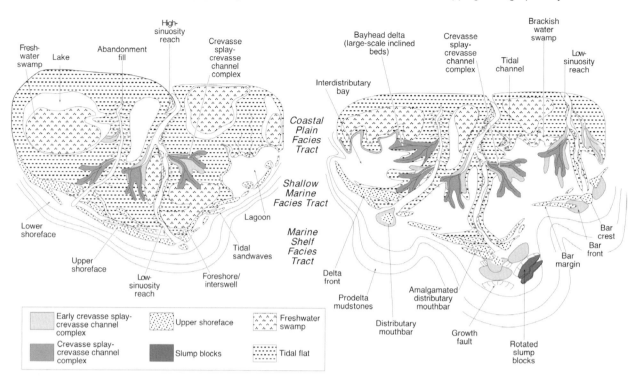

A. Facies Associations in Wave-Modified Deltas of Landward-Stepping Stratigraphic Cycles

B. Facies Associations in Fluvial Dominated Deltas of Seaward-Stepping Stratigraphic Cycles

Figure 7. Schematic diagram showing the variation in geomorphology of the shoreface depositional system of seaward- and landward-stepping cycles.

flows, wavy laminated to hummocky cross-stratified sandstone, amalgamated co-sets of symmetrical and asymmetrical ripple-laminated sandstone, and numerous mudstone drapes, partings, and beds (Figure 10). Sandstones and mudstones are approximately equal in proportion, contain more carbonaceous plant debris and soft-sediment deformation, and generally exhibit less burrowing. Bed geometry ranges from amalgamated to tabular and are broadly lenticular. Preservation of original geomorphic elements is high. These record numerous waning-flow river-flood and waning-flow storm events.

Upper delta-front facies in landward-stepping cycles consist of 1–20 m [3–66 ft] thick, upward-coarsening successions of well-sorted, fine to medium, amalgamated hummocky to swaley (or amalgamated trough cross-stratified) sandstones (Figure 8). The transition from lower to upper delta-front facies is sharp, reflecting a change in sandstone to mudstone ratio from about 5:1 to ≥ 10:1. Seaward-dipping clinoforms are more steeply inclined but cryptic because of lithologic homogeneity. Trace-fossil diversity is high and includes, in order of decreasing abundance, *Skolithos, Ophiomorpha, Thalassinoides, Planolites, Diplocraterion, Arenicolites, Rosselia,* and *Chondrites.*

Upward-coarsening successions of heterolithic mudstones and sandstones of upper delta-front facies in seaward-stepping cycles record increased fluvial influ-

ence. Facies and geomorphic constituents are diverse and include stacked distributary mouthbar sandstones, growth-faults and rotated-slump blocks, and interdistributary bay strata, as well as storm-generated hummocks and swales of the upper shoreface. Long, continuous mudstone drapes and beds separate sandstones deposited by waning-flow river-flood and storm events. Large- and small-scale bedforms are amalgamated to fully preserved. Burrows in sandstone tend to be restricted to the upper portions of beds. Seaward-dipping clinoforms are conspicuous due to the thick mudstone drapes and less steep than in shorefaces of landward-stepping units. However, steeper clinoforms are observed recording the lateral infilling of interdistributary bays.

Facies differentiation in the shoreface facies tract is explained by sediment volume partitioning in response to base-level cycles. In seaward-stepping stratigraphic cycles, less sediment is stored in the coastal plain. A proportionally greater volume of sediment is delivered by rivers to paralic, delta-front, shoreface, and shelf environments. Consequently, the shoreline is more irregular with lobate and elongate deltaic promontories and embayments. Interdistributary bay deposits are common and show enhanced tidal influence, and a complex interbedding of coal and carbonaceous muds, bay muds, crevasse splay/crevasse channel complexes, and washover sands. The delta front is fluvial dominated,

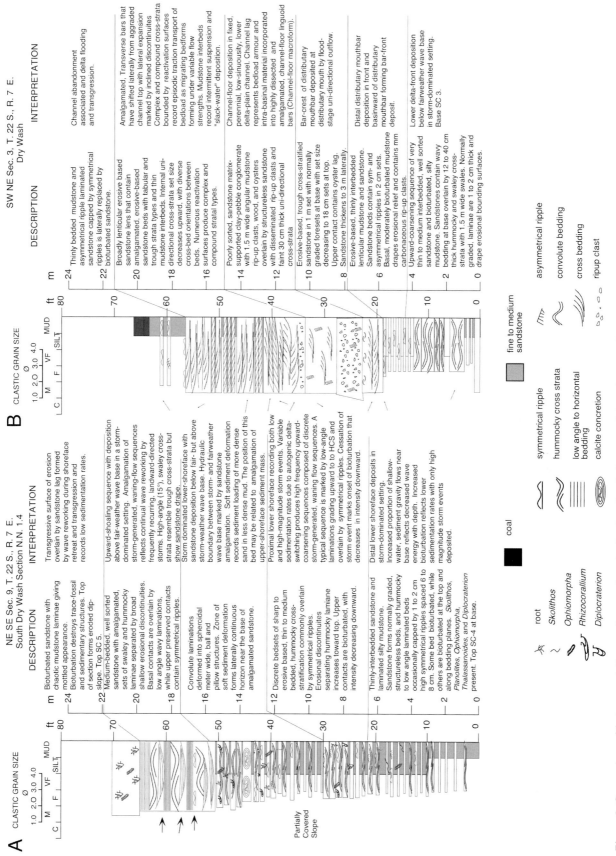

Figure 8. Comparison of facies associations and successions in the shoreface facies tract of river-dominated seaward- and wave-dominated landward-stepping short-term cycles.

Figure 9. Examples of facies and stratigraphic architecture from the shoreface facies-tract of landward-stepping, high-A/S, short-term cycles. (A) Outcrop photo of large scale, low-angle, seaward-dipping clinoforms in SC6 near the head of Muddy Creek Canyon. A prominent, laterally continuous clinoform in the middle of the cliff partitions multidirectional trough cross-stratified sandstone from overlying thinly bedded sandstones. The upper sandstone beds dip seaward at a slightly higher angle than the clinoform that separates the two facies types. (B) Highly amalgamated trough cross-stratified sandstone (lower 2/3) and horizontal planar laminated sandstone of the shoreface in SC6 near the head of Muddy Creek Canyon. (C) Hummocky cross-stratified sandstone from lower delta front of SC5, 1.6 km (1 mi) south of Dry Wash. In lower half of photo, lower- to upper-delta front facies contact is shown by vertical change from interbedded to amalgamated sandstone. (D) 30-cm (12-in) thick, hummocky cross-stratified sandstone bed with burrowed top. Sandstone bed consists of storm-generated, waning-flow succession of sedimentary structures. (E) Swaley cross-stratified sandstone consisting of shallow swales, several-meters wide, with gently dipping, low-angle, subparallel, concordant laminations decreasing in dip upward. Upper shoreface sandstone of SC6 from Picture Flats.

progrades rapidly, and rate of sediment accumulation is high. A broad, low-angle deltaic platform is constructed that increases the frictional drag of incoming waves, thus dissipating wave energy. Higher rates of sediment accumulation and dissipated wave energy provide proportionally less time for waves and currents to rework and cannibalize sediment delivered to the shoreface and delta front. The resultant strata comprise a well-preserved, diverse, heterolithic, mudstone-rich assemblage of river-flood and storm events of multiple shallow marine environments.

By contrast, landward-stepping stratigraphic cycles are characterized by more total sediment storage and increased proportion of mud to sand in the coastal plain. Consequently, a reduced sediment volume that is initially sandier is delivered to the delta front. Progradation and sediment accumulation rates are reduced in the sandier delta front, and waves and currents have proportionally more time to cannibalize and winnow shoreface sediment. Resulting delta-front facies are homogeneous, sand-rich, and record significant sediment redistribution and reworking by waves which reduces facies diversity.

The progressive changes from fluvial- to wave-dominated delta-front facies in seaward- to landward-stepping short-term cycles reflect the sensitivity of delta-front profiles to the balance between ocean waves and currents and sediment discharge from distributary channels. Changes in offshore platform slope and delta-front profiles reflect variations in sediment flux. Wright and Coleman (1973) showed that high-flux, river-dominated deltas have low-gradient slopes, whereas low-flux, wave-dominated deltas have steeper depositional slopes. Changes in delta morphology are accompanied by changes in smaller scale geomorphic elements preserved in the delta. These changes produce different facies mosaics and facies proportions within similar water-depth facies associations of different short-term stratigraphic cycles.

Tidal deposits formed during base-level rise are common above shoreface strata irrespective of cycle stacking pattern. However, seaward-stepping cycles show the most complex facies mosaic reflecting a high proportion of interdistributary bay-fill strata. Tidal deposits record delta-front reworking and compose interdistributary bay fills. Significantly, tidal deposits are best developed in SC-3 recording the intermediate-term turnaround from base-level fall to rise and are enhanced in all landward-stepping cycles (Figure 11). Hence, tidal deposits in shallow-marine successions provide important recognition criteria for base-level rise at all scales.

The shoreface facies tracts of Ferron progradational/aggradational units contain good examples of the two types of facies differentiation. The first type — stratigraphic control on the types of geomorphic elements that occupy a geomorphic environment — is manifest by the conspicuous change from fluvial- to wave-dominated deltas in the transition from seaward- to landward-stepping stratigraphic cycles. The only control on the change in delta morphology we can detect is the change in A/S conditions that accompany the stacking pattern of small-scale stratigraphic cycles. Different delta types do not appear to have resulted from changes in climate, drainage-basin size, discharge, fetch, shelf width, water depth, tectonic regime, or other control. Instead, the flux of sediment to the delta front varied as a function of differential sediment storage in the coastal-plain facies tract during changing A/S regimes. As the sediment flux and composition changed, so did the relative balance between fluvial input and marine reworking and cannibalization, and different delta morphologies resulted. The other type of facies differentiation — a change in degree of preservation of original geomorphic elements — is exemplified by conspicuous sedimentologic differences in the facies that compose the shoreface facies tracts of seaward- and landward-stepping stratigraphic cycles. Increased facies diversity and degree of preservation is typical of seaward-stepping cycles, whereas amalgamation, cannibalization and low facies diversity is typical of shoreface deposits of landward-stepping cycles. Again, this change in the facies and architecture of shoreface deposits is attributed to changing A/S conditions that control the proportions and completeness of original geomorphic elements that are preserved.

FACIES DIFFERENTIATION IN THE COASTAL-PLAIN FACIES TRACT

Coastal-plain strata contain the same facies in all short-term stratigraphic cycles, regardless of position in the stacking pattern. However, the proportions of facies, geometry and size of architectural elements, and degree of preservation of geomorphic elements change regularly with the stacking pattern. Stratigraphic cycles in the coastal-plain facies tract contain alternating organic-poor, sand-rich facies (distributary-channel and crevasse-splay/crevasse-channel sandstones) recording base-level fall, and organic-rich, sandstone-poor facies (paludal and floodplain mudstones, carbonaceous shales and coal) recording base-level rise. These base-level changes produce cyclic coal-sandstone successions. Although these facies coexist as laterally equivalent deposits in both halves of a base-level cycle, the proportion is modulated by position within a cycle.

The coastal-plain facies tract of landward-stepping cycles contains greater sandstone and total sediment volumes but lower sandstone to mudstone ratios than the coastal-plain facies tract of seaward-stepping cycles. This tendency is progressive through the stacking pattern.

Coastal-plain strata of seaward-stepping cycles thin upward, and have progressively increasing sandstone-to-shale ratios, decreasing facies diversity, deeper chan-

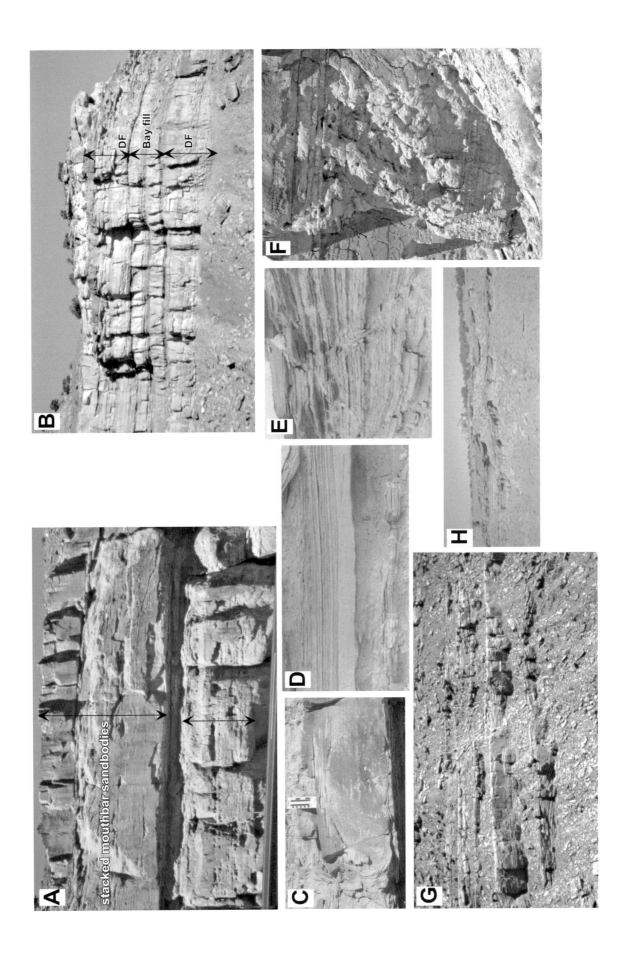

Figure 10. Examples of facies and stratigraphic architecture from the shoreface facies tract of seaward-stepping, low-A/S, short-term cycles. (A) Delta-front and distributary mouthbar deposits of SC2 from the I-70 roadcut. Bar-front and bar-crest facies of distributary mouthbar deposits are well exposed in this roadcut oriented oblique to progradation. The 1.5-m-thick (4.9-ft) bar-front deposits are characterized by thinly bedded, laminated to combined-flow, ripple-laminated sandstone with disseminated organic fines and form the thin recessive zone separating the two massive sandstones. Bar-front deposits overlie delta-front sandstones up to 10 m (32 ft) thick, and underlie bar-crest facies up to 15 m (4.9 ft) thick. Bar-crest facies are characterized by amalgamated, unidirectional trough and planar-tabular cross-stratified sandstone. Bar-crest facies at this locality contain low-angle inclined clinoforms. The thinner tabular sandstone capping the cliff is another distributary mouthbar sandstone. Stacked mouthbar deposits record autogenic subdelta lobe switching. (B) Upward-coarsening delta-front sandstone that is overlain by heterolithic interdistributary-bay deposits of SC2 from Dry Wash. (C) Close-up of shallow-water sediment gravity flow in 48-cm-thick (19-in.) sandstone bed from SC2 at Miller Canyon. Graded, structureless sandstone (Bouma A) at base is overlain by horizontally laminated sandstone (Bouma B), with small scale hummocky and swaley cross-stratification at top. (D) Fully preserved hummock with waning flow cap (parallel lamination, symmetrical aggradational wave ripples, asymmetrical aggradational wave ripples) from SC3 at Dutch Flat. (E). Very well-preserved hummocks with waning flow caps, interbedded burrowed mudstones, and beds with various symmetrical and asymmetrical wave ripples from SC3 at Dutch Flat. (F) Soft-sediment deformation in the lower delta front of SC2 at the I-70 roadcut. Compaction of the underlying mud-rich, SC1 delta front may have contributed to the extensive soft-sediment deformation and sand-rich delta front of SC2 at this locality. (G) Listric normal growth faults in shoreface of SC1 in Muddy Creek Canyon. (H) Soft-sediment synclinal, trough-like depression with bounding anticlinal ridges formed by slump block from SC3 shoreface loading the shelf mudstones of SC2 from Dutch Flat.

nel incisions, laterally extensive erosion surfaces, and thin aggradational paleosols and coal seams. Thin, laterally extensive coals commonly cap seaward-stepping short-term cycles. Coastal-plain strata of vertically stacked cycles contain laterally expanded multistory and multilateral distributary-channelbelt sandstones which interfinger with heterolithic crevasse splay and crevasse channel deposits. Vertically stacked cycles also contain higher proportions of mudstone and carbonaceous mudstone, poorly developed coals, the coarsest sediment fraction, and approximately equal channelbelt-to-floodplain volume ratios. Coastal-plain strata of landward-stepping units contain a high channelbelt-to-floodplain ratio, channelbelt sandstones interfinger with a high proportion of crevasse-channel/crevasse-splay complexes, and amalgamated coal beds up to 10 m (32 ft) thick. Thick pods of locally developed coal seams commonly flank channelbelt sandstones of landward-stepping short-term cycles.

Distributary channels of approximately the same scale, morphology and bank-full dimension fed the shorefaces in all cycles. Yet many sedimentologic attributes of distributary channelbelt sandstones are conspicuously different in cycles at different positions in the stacking pattern (Figures 12 and 13). In distributary channelbelt sandstones, the composition and thickness of lag deposits, the types and geometries of channel sandstone bodies, the degree of preservation and proportion of original bedforms and macroforms, and the diversity of facies record changes in A/S regime and concomitant sediment volume partitioning (Figure 14).

Distributary channelbelt sandstones in all cycles have a grossly similar cross-sectional geometry and a similar internal progression of facies changes. They have a characteristic "funnel" or "longhorn steer" cross-sectional shape (Figure 13). At the base, steep-sided chan-

nelbelt margins are narrow (the nose of the steer); the channelbelt margins expand 4 to 10 times in width toward the top and have low-gradient margins (the "horns" of the steer). The progression of facies attributes is from highly interconnected, amalgamated, cannibalized, laterally restricted, vertically stacked sandstone bodies at the base, to expanded, more open framework and more fully preserved, laterally stacked sandstone bodies toward the top. Even though channelbelt sandstones in all cycles share this basic motif, several sedimentologic changes record the changes in A/S and the sediment volume partitioning that accompany the short- and intermediate-term base-level cycles.

Distributary channelbelt sandstones of high A/S, landward-stepping cycles are typically 15–25 m (49–82 ft) thick and 1–1.5 km (0.6–0.9 mi) wide. They contain a diverse assemblage of moderately interconnected, cut-and-fill, low-sinuosity, high-sinuosity and abandonment-fill macroforms 5–20 m (16–66 ft) thick (Figure 15). Macroforms often are separated by mud drapes or lags. Mud-matrix-supported, mud-boulder-intraclast lag deposits on channel and macroform scour bases are up to 1 m (3 ft) thick and laterally continuous (hundreds of meters). Internal accretion and reactivation surfaces of equal length record lateral and downstream barform migration. Bedforms, cut-and-fill macroforms, and accretionary macroforms are well preserved (low amalgamation and little cannibalization). Facies diversity within macroforms is high, including thick and thin sets of trough cross-stratification, planar-tabular stratification, horizontal lamination, convolute lamination, and other structures indicative of fluidization, meter-scale fully preserved straight-crested dunes climbing the backs of larger barforms, and thick sets of ripple and climbing ripple lamination (Figure 16).

By contrast, channelbelt sandstones of low A/S, sea-

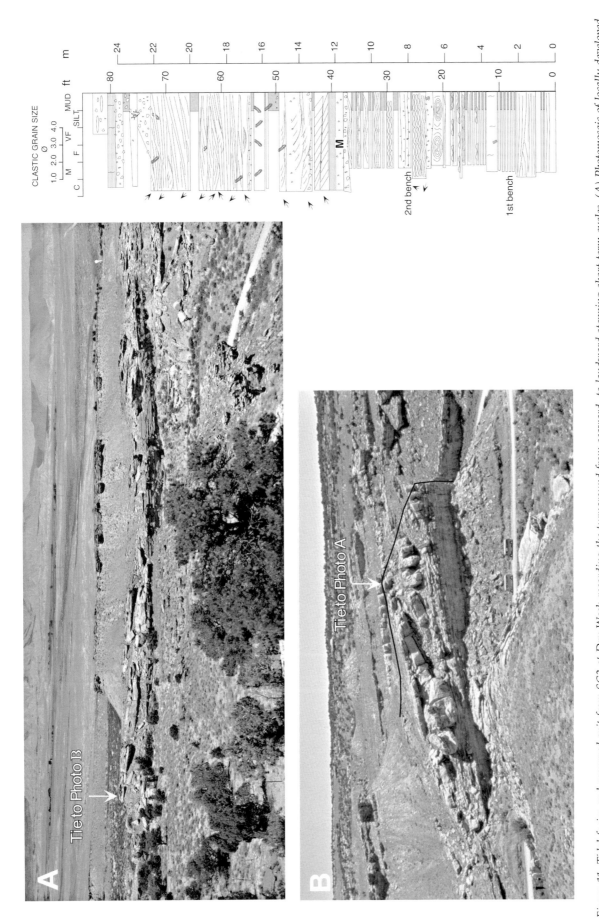

Figure 11. Tidal facies and recurved spit from SC3 at Dry Wash recording the turnaround from seaward- to landward-stepping short-term cycles. (A) Photomosaic of locally developed transgressive, tidal-channel-inlet facies in SC3 at Dry Wash along a 600-m-long (2000-ft) canyon cut oriented to depositional dip. Exposures comprise a series of 14 offlapping 5–10-m-thick (16–32 ft), imbricated, accretionary sandstone lenses that are encased in burrowed mudstones and that often contain one to several thickly bedded, unidirectional sandwaves. (B) Transverse cross-sectional view looking northwest of laterally accreting, imbricated sandstone lenses from SC3 tidal inlet at Dry Wash. At the thickest portion of the drum-stick-shaped sandstone lens, the weight of the sandstone lens has deformed underlying SC 3 delta-front sandstones into a broad synform.

Distributary Channelbelt Architecture of Landward-Stepping Cycle

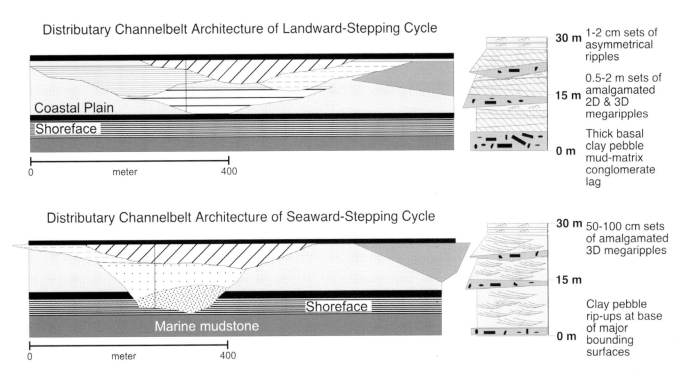

1-2 cm sets of asymmetrical ripples

0.5-2 m sets of amalgamated 2D & 3D megaripples

Thick basal clay pebble mud-matrix conglomerate lag

Distributary Channelbelt Architecture of Seaward-Stepping Cycle

50-100 cm sets of amalgamated 3D megaripples

Clay pebble rip-ups at base of major bounding surfaces

Figure 12. Diagram summarizing changes in distributary channelbelt architecture of landward- and seaward-stepping short-term cycles.

Figure 13. Examples showing variable distributary channelbelt architecture of landward- and seaward-stepping short-term cycles. (A) Outcrop photo of moderately interconnected sandstone lenses of distributary channelbelt from landward-stepping SC7 at Emery mine. (B) Distributary channelbelt sandbodies of seaward and landward-stepping cycles. SC2 at I-70 roadcut. Bars on scale are 3 m.

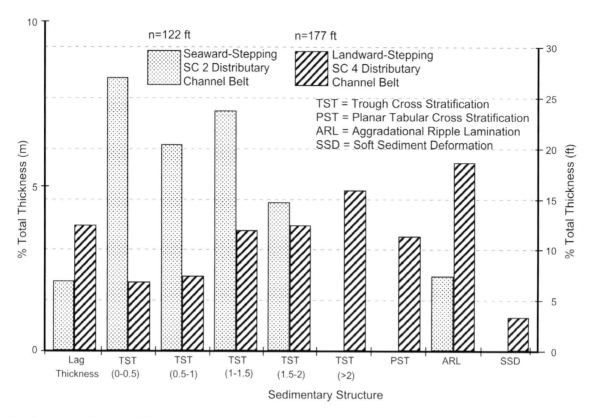

Figure 14. Plot showing diversity difference of sedimentary structures from distributary channelbelt deposits of landward- and seaward-stepping short-term cycles. The increased proportion of higher-energy and more amalgamated, cannibalized structures in distributary channelbelt deposits of seaward-stepping, short-term cycles reflects decreased preservation of the upper part of channel fills.

ward-stepping cycles are of comparable thickness but only a few hundred meters wide. They contain a low diversity of highly amalgamated cut-and-fill and low-sinuosity, erosive-based macroforms 5–10 m (16–32 ft) thick (Figure 17). Bedforms and macroforms are strongly cannibalized and amalgamated, resulting in very low facies diversity (typically >95% by volume of thin sets of amalgamated and top-truncated, trough cross-stratified sandstone). Channel and macroform scour bases are occasionally and discontinuously overlain by sand-matrix-supported, mud-pebble-intraclast lags 2–20 cm (0.8–8 in.) thick.

These channels do not extend beyond their delta, nor do they incise below associated delta-front deposits (one notable exception is the SC5 distributary channelbelt at Cedar Ridge), and the scale of incision conforms to macroforms composing the channelbelt. These are attributes of freely migrating channels constructing a channelbelt unimpeded by valley confinement. The size of these larger channelforms may lead to the misinterpretation of them as valley fills and highlights the limitation of size as a criterion for valley fill recognition. In this case, the increased size of Ferron distributaries of seaward-stepping cycles reflects higher delta subsidence promoted by the higher sediment volumes and more rapid progradation (Morgan, 1973). Higher delta subsi-

dence is also reflected by growth faults, large slumps, and pervasive soft-sediment deformation in the shoreface facies tract of seaward-stepping cycles.

The coastal-plain facies tracts of Ferron progradational/aggradational units contain good examples of facies differentiation produced by differences in degree of preservation of geomorphic elements under varying A/S regimes. In high-accommodation, landward-stepping stratigraphic cycles, coastal-plain facies are diverse, reflecting good preservation of multiple, diverse geomorphic elements (Figure 18). Many of the original bedforms and barforms of distributary channels are fully or nearly fully preserved. By contrast, the diversity of coastal-plain facies in low-accommodation, seaward-stepping stratigraphic cycles is very low, reflecting preservation of only those geomorphic elements that are most easily preserved. The original bedforms and barforms of distributary channels are intensely cannibalized and amalgamated (Figure 12).

FACIES MODELS IN A STRATIGRAPHIC CONTEXT

Facies models summarize the facies associations presumed indicative and characteristic of particular sedimentary environments. They are constructed through

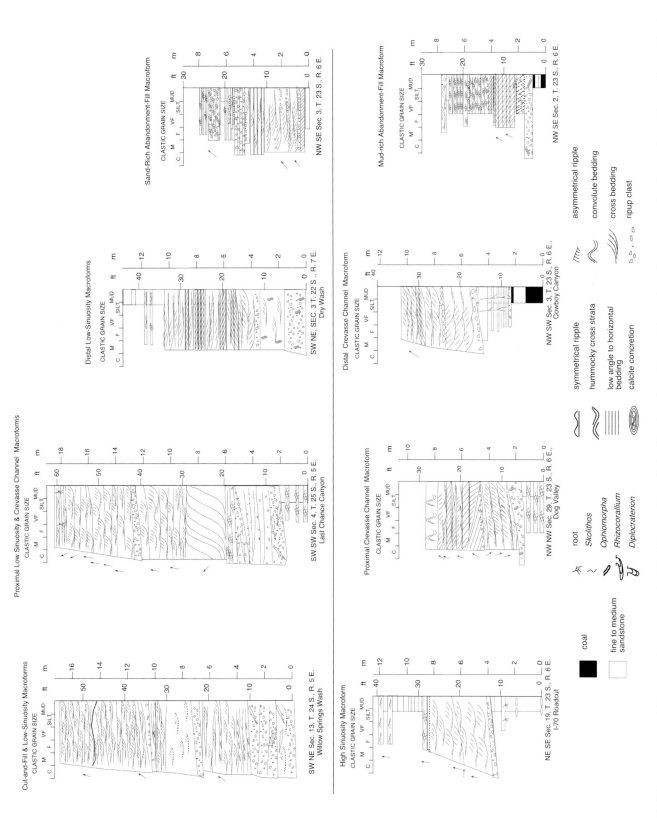

Figure 15. Diagram summarizing macroforms composing distributary channelbelt of landward- and seaward-stepping, short-term cycles. Seaward-stepping cycles are dominated by cut-and-fill macroforms, whereas landward-stepping cycles contain a diverse assemblage of macroform types.

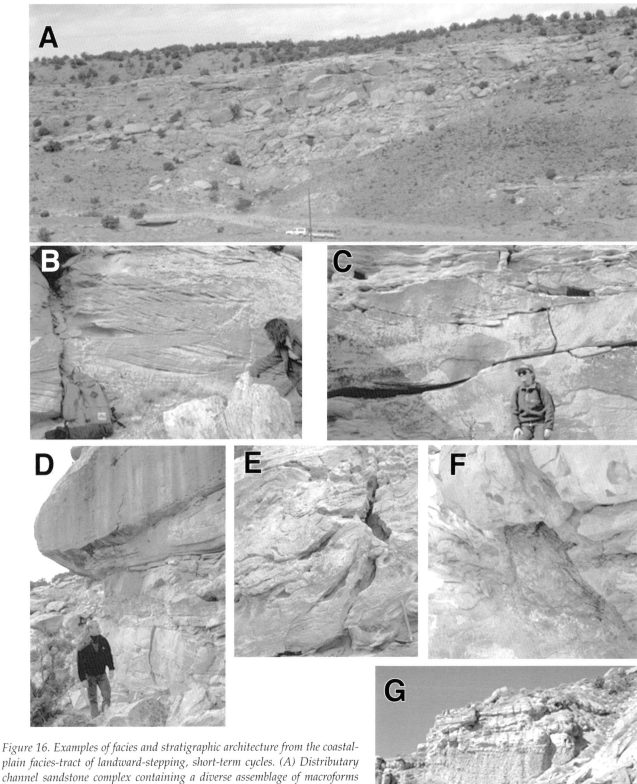

Figure 16. Examples of facies and stratigraphic architecture from the coastal-plain facies-tract of landward-stepping, short-term cycles. (A) Distributary channel sandstone complex containing a diverse assemblage of macroforms separated by meter-scale mudstone drapes and lags. From SC6 at Dog Valley. (B) Amalgamated trough cross-stratified sandstone from the base of the distributary channel (nose of the longhorn steer) of Figure 16A. Backpack rests on channel scour base. (C) Meter-scale climbing sandwave (analogous to climbing ripples) from near the top of the distributary channel sandstone at Dog Valley. (D) Mudstone lag (at person's head) separating overlying low-sinuosity macroform from top of the longhorn steer's nose at Dog Valley. (E) Close-up of convolute bedding in soft-sediment-deformed sandstone of SC5 channel at Ivie Creek. High shear stress on partially liquefied bedforms results in dewatering and development of these soft-sediment deformation features. (F) Detail of thick (meter-scale) boulder/cobble mudstone ripup-clast lag at channel base. (G) Abandonment fill macroform of SC4 along Coal Cliffs.

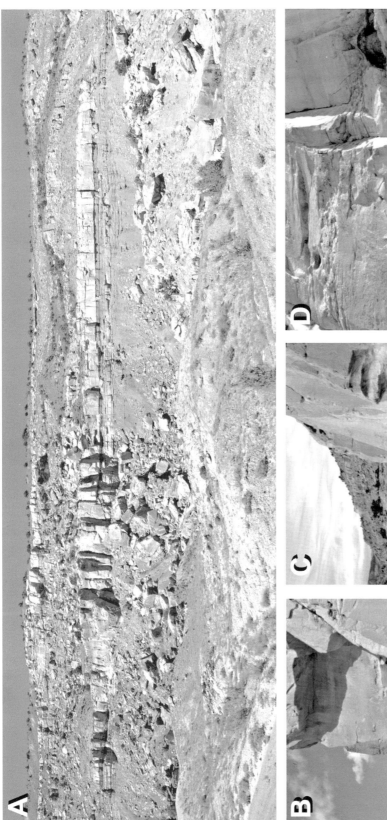

Figure 17. Examples of facies and stratigraphic architecture from the coastal-plain facies-tract of seaward-stepping, short-term cycles. (A) Outcrop photomosaic of transverse, cross-sectional view of 25-m-thick (82-ft) and 280-m-wide (919-ft), distributary channelbelt from seaward-stepping SC2 at Willow Springs Wash. The lower portion of the channelbelt consists of four cut-and-fill macroforms that are overlain by laterally expanded sandstones that consist of four to five low-sinuosity and crevasse-channel macroforms. (B) Amalgamated, 10–20-cm-thick (4–8-in.) sets of trough cross-stratified sandstone from the distributary channelbelt in seaward-stepping SC2 at Willow Springs Wash. (C) View looking west of erosional basal contact of highly amalgamated distributary channelbelt in seaward-stepping SC2 at Willow Springs Wash. Basal contacts form highly irregular, concave-upward surface with up to 5 m (16 ft) of local relief where incised in less resistant, intraformational interdistributary-bay deposits of SC1. (D) Lateral pinch-outs of cut-and-fill sandstone lenses against channel-base bounding surface of seaward-stepping SC2 channelbelt at Willow Springs Wash.

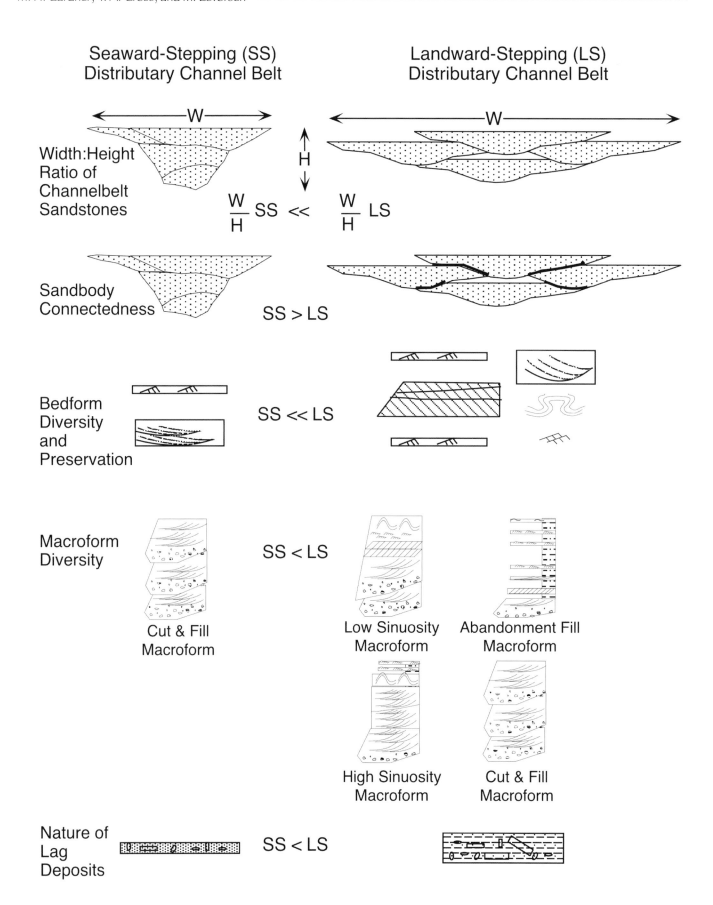

Figure 18. Summary of sedimentologic and stratigraphic attributes of upper Ferron distributary channelbelts that change as a function of short-term cycle stacking pattern.

synthesis, reduction and simplification of observations from multiple specific examples to abstract the "essence," or the essential facies elements of a particular environment, from the "noise," or variations from whatever is perceived as the norm (Walker, 1984, 1990). The only requirement for selection of examples is that they must be a product of a particular geomorphic environment. Facies models are constructed with the presumption that the preserved stratigraphic record of an environment is similar to, and a composite of, all geomorphic elements in that environment. Accordingly, the geomorphic elements in an environment are preserved in the same ratios as facies in strata. This presumption requires that the mosaic of geomorphic elements that form the patchwork quilt on the Earth's surface at an instant in time aggrade in place to form a stratigraphic facies mosaic of identical complexity and areal distribution. If facies associations and successions representing a single depositional environment are collected from different stratigraphic cycles, or different halves of the same base-level cycle, the resulting facies model of that depositional environment is derived from a mixture of unrelated elements that may never have coexisted (Figure 18).

Facies models are specific to each environment; they do not mix or merge the facies of laterally linked environments. A facies model for laterally linked braidplain-lake-alluvial fan environments does not exist, but each environment has its own models. Environments identified as geomorphically distinct have separate facies models attached to them. Multiple facies models exist for different river morphologies (e.g., braided, meandering, anastomosed) and for different delta morphologies (wave-, fluvial-, and tide-dominated).

Facies models are static. Facies attributes, associations, and successions are collapsed from multiple examples into a single geomorphic and sedimentologic character set presumed to exist at an instant in time. Facies models are constructed with the presumption that the stratigraphic products (preserved depositional remnants) of a particular depositional environment are similar from place to place and from time to time. Facies models do not recognize that geomorphic elements at the same positions on a depositional profile may differ in different A/S regimes, even in the two halves of a single base-level cycle. Nor do they recognize that given identical geomorphic elements in a particular environment, changes in degree of preservation (volume and proportion) of those elements will create major differences in the facies observed in the stratigraphic record. Although facies differentiation has not been a focus of most sedimentologic and stratigraphic studies, it has been hinted at, noted, suggested, or described from a variety of environments in a handful of papers (e.g., Barrell, 1912; Van Siclen, 1958; Curtis, 1970; MacKenzie, 1972; Bridge and Leeder, 1979; Brett and Baird, 1986; Gal-

loway, 1986; Cross, 1988; Carter et al., 1991; Borer and Harris, 1991; Boyd et al., 1992; Cross et al., 1993; Sonnenfeld and Cross, 1993; Cant, 1995; Mellere and Steel, 1995; Gardner et al., 1995, Kerans and Fitchen, 1995; Talling et al., 1995; Ramon and Cross, 1997; Cross, 2000).

None of these characteristics of facies models employs four-dimensional stratigraphic appreciation for the accumulation of sediment. Facies models ignore the fact that sediments accumulate during the migration of laterally linked environments. The mosaic of geomorphic environments does not aggrade in place producing a stratigraphic product that closely resembles the geomorphic parent.

This study demonstrates that stratigraphic processes influence the types of geomorphic elements which compose an environment, as well as the proportion and degree of preservation of elements which enter the stratigraphic record. Changing A/S conditions during base-level cycles control stacking patterns and sediment volume partitioning. The latter contributes to the two types of facies differentiation discussed previously. The next generation of facies models should be constructed from a stratigraphic perspective in which there is a continuum of transitional forms of facies associations and successions that are different products of the same parent.

CONCLUSIONS

At the scale of the upper Ferron Sandstone clastic wedge, sediment accumulated in different paleogeographic positions through time form a series of stratigraphic cycles arranged in a seaward-stepping, vertically stacked, and landward-stepping stacking pattern (Figure 19). Geographic partitioning of sediment volumes at this scale is related to the changes in A/S conditions during an intermediate-term, base-level cycle.

Within each stratigraphic cycle, sediment was partitioned into depositional environments in different volumes and ratios. Superposition of the two scales of base-level cycles causes systematic changes in sediment volume partitioning through time. The total sediment volume and total sandstone in the shoreface facies tract decreases regularly from seaward- to landward-stepping stacking patterns. The proportion of marine-to-nonmarine sandstone also decreases. This demonstrates increasing sediment storage in uphill continental environments during the transition from seaward- to landward-stepping stacking patterns (Figure 20).

One product of the changing A/S regime and sediment volume partitioning is a change in delta morphology. Deltas in seaward-stepping cycles are fluvial dominated, whereas those in landward-stepping cycles are wave-dominated. The change in delta morphology is related to the change in sediment storage capacity uphill in continental environments. This is an illustration of

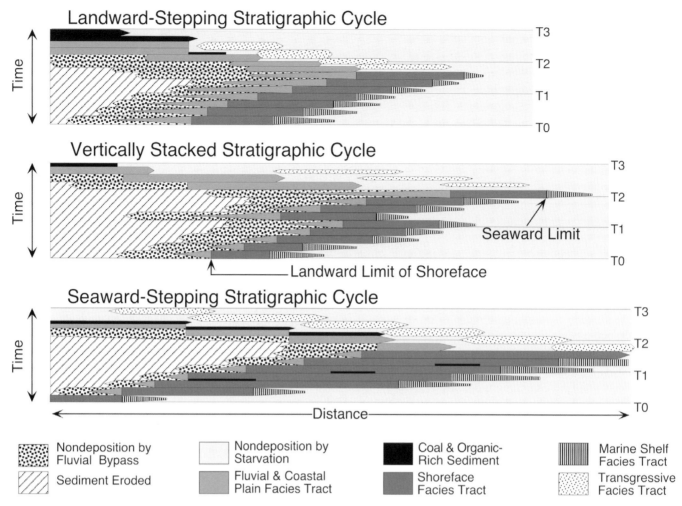

Figure 19. Wheeler (time-space) diagram showing distribution of three time-space domains within facies tracts that compose seaward-stepping, vertically stacked, and landward-stepping stratigraphic cycles of the Ferron sequence. This diagram relates changes in the stacking pattern, stratal geometry, sediment volume distributions, and facies arrangements to variations in accommodation within base-level cycles.

geomorphic facies differentiation, where stratigraphic processes control the types of geomorphic elements that occupy a particular depositional environment.

Sediment volume partitioning at the scale of short-term stratigraphic cycles also affects the rate of net sediment accumulation in different environments, and reflects the balance between rates of sediment addition and rates of sediment reworking and cannibalization. Changes in A/S conditions during a short-term cycle control or modulate the degree of cannibalization and amalgamation of geomorphologic elements that compose an environment. Variations in facies diversity, facies associations and successions, lithologic heterogeneity, and petrophysical properties within a facies tract are manifestations of changing A/S conditions. Shorefaces of high-accommodation, landward-stepping cycles comprise homogeneous, wave-dominated sandstones, whereas shorefaces of low-accommodation, seaward-stepping cycles are lithologically heterogeneous and river-dominated containing diverse facies and well-preserved original geomorphic elements. Tidal deposits

are present within all stratigraphic cycles, are most common immediately above the shoreface facies tract, but show the most complex mosaic in seaward-stepping cycles reflecting the increased proportion of interdistributary bay-fill strata. Distributary channelbelt sandstones of high A/S, landward-stepping cycles are composed of high-diversity, well-preserved macroforms and bedforms, whereas those of seaward-stepping cycles are composed of strongly cannibalized, amalgamated, low-diversity macroforms and bedforms. The latter channelforms show increased size and may be misinterpreted as valley fills recording unconformity development under conditions of no accommodation. These examples illustrate the other type of preservational facies differentiation, where stratigraphic processes control the proportions and ratios of original geomorphic elements that are preserved. It is important to emphasize that facies differentiation describes the tendency for specific facies associations, proportions and degree of preservation within a facies tract. It is not an absolute attribute of the facies tract composing a specific stratigraphic cycle. For

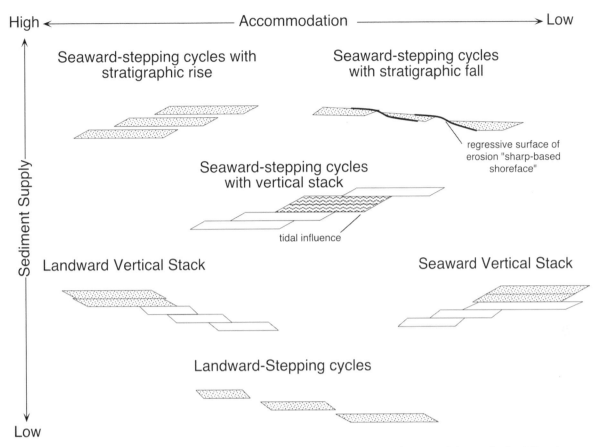

Figure 20. Illustration of variability of stacking patterns of the shoreface facies tract as a function of changes in the A/S ratio.

example, wave-generated sedimentary structures occur within river-dominated as well as wave-dominated deltas and reflect the three-dimensional geomorphology and variability in processes operating within a depositional system.

Sediment volume partitioning and facies differentiation in different A/S regimes create radically different facies constituents, associations, and successions from the same geomorphic environment. The stratigraphic products are transitional forms along a continuum from high to low A/S conditions for each facies tracts. Existing facies models assume the products of geomorphic environments are similar regardless of time, place, and condition of accumulation. Facies models are insensitive to stratigraphic controls on facies associations and successions. Moreover, most are incorrectly constructed from observations of facies elements that never coexisted. If facies models are to be useful for stratigraphic prediction, they must be calibrated to A/S conditions that drive volumetric partitioning and facies differentiation. A new generation of stratigraphically sensitive facies models is required and they need to be placed into an A/S context.

Systematic changes in numerous stratigraphic and sedimentologic attributes emphasize the well-ordered behavior of the stratigraphic process-response system,

and demonstrate systematic linkages among attributes of all scales and many types. These attributes are complementary records of multiple, interdependent processes which may be analyzed from the simple perspectives of conservation laws and stratigraphic base level. The systematic organization of disparate and diverse data types is the basis for robust stratigraphic prediction. From knowledge of attributes at one scale, attributes of other types and scales are predictable.

ACKNOWLEDGMENTS

This paper reports part of the Ph.D. research of the first author. Financial support was provided by the Industrial Associates of the Genetic Stratigraphy Research Program administered by T. A. Cross, the American Chemical Society Petroleum Research Fund, the American Association of Petroleum Geologists grants-in-aid program, the SEPM Donald Smith Research grant, the Sigma Xi grants-in-aid program, the U.S. Geological Survey Branch of Coal Resources, and the Utah Geological Survey. We thank these organizations, companies, and societies for their support. Helpful reviews by Ron Steel and John Cater are gratefully acknowledged.

REFERENCES CITED

Arnott, R. W. C., 1995, The parasequence definition — Are transgressive deposits inadequately addressed?: Journal of Sedimentary Research, v. B65, p. 1–6.

Armstrong, R. L., 1968, Sevier orogenic belt in Nevada and Utah: Geological Society of America Bulletin, v. 79, p. 429–458.

Barrell, J., 1912, Criteria for the recognition of ancient delta deposits: Geological Society of America Bulletin, v. 23, p. 377–446.

—1917, Rhythms and the measurement of geologic time: Geological Society of America Bulletin, v. 28, p. 745–904.

Barton, M. D., 1994, Outcrop characterization of architecture and permeability structure within fluvial-deltaic sandstones, Cretaceous Ferron Sandstone, Utah: Ph.D. Dissertation, University of Texas, Austin, 255 p.

Borer, J. M., and P. M. Harris, 1991, Depositional facies and cyclicity in the Yates Formation, Permian Basin — implications for reservoir heterogeneity: AAPG Bulletin, v. 75, p. 726–779.

Boyd, R., R. Dalrymple, and B. A. Zaitlin, 1992, Classification of clastic coastal depositional environments: Sedimentary Geology, v. 80, p. 139–150.

Brett, C. E., and G. C. Baird, 1986, Symmetrical and upward shallowing cycles in the middle Devonian of New York State and their implications for the punctuated aggradational hypothesis: Paleoceanography, v. 1, p. 431–445.

Bridge, J. S., and M. R. Leeder, 1979, A simulation model of alluvial stratigraphy: Sedimentology, v. 26, p. 617–644.

Busch, D. A., 1959, Prospecting for stratigraphic traps: AAPG Bulletin, v. 43, p. 2829–2843.

—1971, Genetic units in delta prospecting: AAPG Bulletin, v. 55, no. 8, p. 1137–1154.

Cant, D., 1995, Sequence stratigraphic analysis of individual depositional successions — effects of marine/nonmarine sediment partitioning and longitudinal sediment transport, Mannville Group, Alberta Foreland Basin, Canada: AAPG Bulletin, v. 79, no. 5, p. 749–762.

Carter, R. M., S. T. Abbott, C. S. Fulthorpe, D. W. Haywick, and R. A. Henderson, 1991, Application of global sea-level and sequence-stratigraphic models in Southern Hemisphere Neogene strata from New Zealand, in D. I .M. Macdonald, ed., Sedimentation, tectonics and eustasy: International Association of Sedimentologists Special Publication 12, p. 41–65.

Collom, C. J., 1991, High-resolution stratigraphic and paleoenvironmental analysis of the Turnonian-Coniacian Stage boundary interval (Late Cretaceous) in the lower Fort Hays Member, Niobrara Formation, Colorado and New Mexico: M.S. Thesis, Brigham Young University, Provo, 371 p.

Cotter, E., 1975, Deltaic deposits in the Upper Cretaceous Ferron Sandstone, Utah, in M. L. S. Broussard, ed., Deltas, models for exploration: Houston Geological Society, p. 471–484.

Cross, T. A., 1988, Controls on coal distribution in transgressive-regressive cycles, Upper Cretaceous, Western Interior, U.S.A., in C. K. Wilgus, B. S. Hasting, C. G. Kendall, St. C., H. W. Posamentier, C. A. Ross, and J. C. Van Wagoner, eds., Sea-level changes — an integrated approach: Society for Sedimentary Geology (SEPM), Special Publication 42, p. 371–380.

—2000, Stratigraphic controls on reservoir attributes in continental strata: Earth Science Frontiers, v. 7, p. 322–350.

Cross, T. A., and P. W. Homewood, 1998, Amanz Gressly's role in founding modern stratigraphy: Geological Society of America Bulletin, v. 109, p. 1617–1630.

Cross, T. A., and M. A. Lessenger, 1998, Sediment volume partitioning — rational for stratigraphic model evaluation and high-resolution stratigraphic correlation, in F. M. Gradstein, K. O. Sandvik, and N. J. Milton, eds., Sequence stratigraphy — concepts and applications: Norwegian Petroleum Society Special Publication 8, p. 171–195.

—1999, Construction and application of a stratigraphic inverse model, in J. W. Harbaugh, W. L. Watney, E. C. Rankey, R. Slingerland, R. H. Goldstein, and E. K. Franseen, eds, Numerical experiments in stratigraphy — recent advances in stratigraphic and sedimentologic computer simulations: Society for Sedimentary Geology (SEPM) Special Publication 62, p. 69–83.

—2001, Method for predicting asratigraphy: U.S. Patent, 6246963 B1.

Cross, T. A., M. R. Baker, M. A. Chapin, M. S. Clark, M. H. Gardner, M. S. Hanson, M. A. Lessenger, L. D. Little, K. J. McDonough, M. D. Sonnenfeld, D. W. Valasek, M. R. Williams, and D. N. Witter, 1993, Applications of high-resolution sequence stratigraphy to reservoir analysis, in R. Eschard and B. Doligez, eds., Reservoir characterization from outcrop investigations: Proceedings of the 7th Exploration and Production Research Conference, Paris, Technip, p. 11–33.

Crowley, K. D., 1983, Large-scale bed configurations (macroforms) Platte River Basin, Colorado and Nebraska: Geological Society of America Bulletin, v. 94, p. 117–133.

Cowles, H. C., 1901, Physiographic ecology of Chicago and vicinity: Botanical Gazette, v. 31, p. 73–108, 145–182.

Curtis, D. M., 1970, Miocene deltaic sedimentation, Louisiana Gulf Coast, in J. P. Morgan, ed., Deltaic sedimentation — modern and ancient: Society for

Sedimentary Geology (SEPM) Special Publication 15, p. 293–308.

Davis, W. M., 1902, Baselevel, grade and peneplain: Journal of Geology, v. 10, p. 77–111.

Elder, W. P., 1985, Biotic patterns across the Cenomanian Turonian extinction boundary near Pueblo, Colorado, *in* L. M. Pratt, E. G. Kauffman, and F. B. Zelt, eds., Fine-grained deposits and biofacies of the Cretaceous Western Interior Seaway — evidence of cyclic sedimentary processes: Society for Sedimentary Geology (SEPM), Rocky Mountain Section Field Trip Guidebook No. 4, p. 157–169.

Frazier, D. E., 1974, Depositional episodes — their relationship to the Quaternary stratigraphic framework in the northwestern portion of the Gulf Basin: University of Texas at Austin, Bureau of Economic Geology Geological Circular 74–1, 28 p.

Galloway, W. E., 1986, Reservoir facies architecture of microtidal barrier systems: AAPG Bulletin, v. 70, p. 787–808.

Gardner, M. H., 1993, Sequence stratigraphy and facies architecture of the Upper Cretaceous Ferron Sandstone Member of the Mancos Shale, east-central Utah: Ph.D. dissertation, Colorado School of Mines, Golden, 528 p.

—1995a, The stratigraphic hierarchy and tectonic history of Mid-Cretaceous foreland basin of central Utah: *in* S. Dorobek and J. Ross, eds., Stratigraphic evolution of foreland basins: Society for Sedimentary Geology (SEPM) Special Publication 52, p. 283–301.

—1995b, Tectonic and eustatic controls on the stratal architecture of Mid-Cretaceous stratigraphic sequences, central Western Interior foreland basin of North America, *in* S. Dorobek and J. Ross, eds., Stratigraphic evolution of foreland basins: Society for Sedimentary Geology (SEPM) Special Publication 52, p. 283–303.

—1995c, Stratigraphic cross section showing facies relationships in the Upper Cretaceous Ferron Sandstone, central Utah (plate): *in* S. Dorobek and J. Ross, eds., Stratigraphic evolution of foreland basins: Society for Sedimentary Geology (SEPM) Special Publication 52.

Gardner, M. H., and T. A, Cross, 1994, Middle Cretaceous paleogeography of Utah, *in* M. V. Caputo, J. A. Peterson, and K. J. Franczyk, eds., Mesozoic systems of the Rocky Mountain Region, United States: Society for Sedimentary Geology (SEPM), Rocky Mountain Section Special Publication, p. 471–502.

Gardner, M. H., B. J. Willis, and I. N. Widya Dharmasamadhi, 1995, Outcrop-based reservoir characterization of low accommodation/sediment supply fluvial deltaic sandbodies, *in* Reservoir characterization — integration of geology, geophysics and reservoir engineering: Special Publication of the Technology Research Center, JNOC., No. 5, p. 17–54.

Grabau, A. W., 1924, Principles of stratigraphy: Dover, Toronto, Ontario, 1185 p.

Hayes, C. W., 1899, Physiography of the Chattanooga District in Tennessee, Georgia, and Alabama: U.S. Geological Survey, Nineteenth Annual Report, pt. 2, p. 1–58.

Hancock, J. M., and E. G. Kauffman, 1979, The great transgressions of the Late Cretaceous: Geological Society of London Journal, v. 136, p. 175–186.

Haq, B. U., J. Hardenbol, and P. R. Vail, 1988, Mesozoic and Cenozoic chronostratigraphy and eustatic cycles: *in* C. K. Wilgus, B. S. Hasting, C. G. St. C. Kendall, H. W. Posamentier, C. A. Ross, and J. C. Van Wagoner, eds., Sea level changes — an integrated approach: Society for Sedimentary Geology (SEPM) Special Publication 42, p. 71–109.

Kauffman, E. G., 1977, Geological and biological overview — Western Interior Cretaceous Basin, *in* Cretaceous facies, faunas and paleoenvironments across the Western Interior Basin: Mountain Geologist, v. 14, p. 75–99.

Kauffman, E. G., W. A. Cobban, and D. L. Eicher, 1976, Albian through lower Coniacian strata, biostratigraphy and principal events, Western Interior United States: Annales Du Museum D'Histoire Naturelle De Nice-Tome IV, 51 p.

Kerans, C., and W. M. Fitchen, 1995, Sequence hierarchy and facies architecture of a carbonate-ramp system — San Andres Formation of Algerita Escarpment and Western Guadalupe Mountains, West Texas and New Mexico: University of Texas Bureau of Economic Geology Report of Investigations No. 235, 86 p.

Krumbein, W. C., and L. L. Sloss, 1951, Stratigraphy and sedimentation: W. H. Freeman and Co., San Francisco, 497 p.

MacKenzie, D. B., 1972, Primary stratigraphic traps in sandstones, *in* R. E. King, ed., Stratigraphic oil and gas fields: AAPG Memoir 16, p. 47–63.

Mackin, J. H., 1948, Concept of the graded river: Geological Society of America Bulletin, v. 59, p. 463–512.

McDonough, K. J., and T. A. Cross, 1991, Late Cretaceous sea level from a paleoshoreline: Journal of Geophysical Research, v. 96, B4, p. 6591–6607.

Mellere D., and R. J. Steel, 1995, Variability of lowstand wedges and their distinction from forced-regressive wedges in the Mesa Verde Group, southeast Wyoming: Geology, v. 23, v. 9, p. 803–806.

Molenaar, C. M., and W. A. Cobban, 1991, Middle Cretaceous stratigraphy on south and east sides of the Uinta Basin, northeastern Utah and northwestern Colorado: U.S. Geological Survey Bulletin 1787-P, 34 p.

Muto, T., and R. J. Steel, 2000, The accommodation concept in sequence stratigraphy — some dimensional problems and possible redefinition: Sedimentary Geology, v. 130, p. 1–10.

Morgan, J. P., 1973, Impact of subsidence and erosion on Louisiana coastal marshes and estuaries, *in* R. H. Chabreck, ed., Coastal marsh and estuary management: Louisiana State University Division of Continuing Education, p. 217–233.

Obradovich, J. D., 1991, A revised Cenomanian–Turonian time scale based on studies from the Western Interior United States (abs.): Geological Society of America, Abstracts with Programs, A296 p.

Posamentier, H. W., and P. R. Vail, 1988, Eustatic controls on clastic deposition, II — sequence and systems tract models, *in* C. K. Wilgus, B. S. Hasting, C. G. St. C. Kendall, H. W. Posamentier, C. A. Ross, and J. C. Van Wagoner, eds., Sea level changes — an integrated approach: Society for Sedimentary Geology (SEPM) Special Publication 42, p. 125–155.

Powell, J. W., 1875, Exploration of the Colorado River of the West and its tributaries: Washington, D.C., Smithsonian Institute, 291 p.

Ramon, J. C., and T. A. Cross, 1997, Characterization and prediction of reservoir architecture and petrophysical properties in fluvial channel sandstones, Middle Magdalena Basin, Colombia: Ciencia, Tecnologia y Futuro, v. 1, no. 3, p. 19–46.

Rice, W. N., 1897, Revised textbook of geology (by J. D. Dana): New York, American Book Co., 482 p.

Ryer, T. A., 1981, Deltaic coals of Ferron Sandstone Member of Mancos Shale — predictive model for Cretaceous coal-bearing strata of Western Interior: AAPG Bulletin, v. 65, p. 2323–2340.

—1983, Transgressive-regressive cycles and the occurrence of coal in some Upper Cretaceous strata of Utah: Geology, v. 11, p. 207–210.

Ryer, T. A., and P. B. Anderson, 1995, Parasequence sets, parasequences, facies distributions, and depositional history of the Upper Cretaceous Ferron deltaic clastic wedge, central Utah (abs.): AAPG Bulletin 79, p. 924.

Schumm, S. A., 1993, River response to base-level change — implications for sequence stratigraphy: Journal of Geology, v. 101, p. 279–294.

Sloss, L. L., 1962, Stratigraphic models in exploration: AAPG Bulletin, v. 49, p. 1050–1057.

Stalkup, F. I., and W. J. Ebanks, Jr., 1986, Permeability variation in a sandstone barrier island–tidal delta complex, Ferron Sandstone (Lower Cretaceous), central Utah: Society of Petroleum Engineers Paper 15532, p. 1–8.

Sahagian, D., and M. Jones, 1993, Quantified Middle Jurassic to Paleocene eustatic variations based on Russian Platform stratigraphy — stage level resolution: Geological Society of America Bulletin, v. 105, p. 1109–1118.

Schlanger, S. O., M. A. Arthur, H. C. Jenkyns, and P. A. Scholle, 1986, The Cenomanian–Turonian oceanic anoxic event I — stratigraphy and distribution of organic carbon-rich beds and the marine excursion, *in* J. Brooks and A. J. Flebt, eds., Marine petroleum sources rocks: London Geological Society Special Publication 24, p. 347–375.

Sonnenfeld, M. D., and T. A. Cross, 1993, Volumetric partitioning and facies differentiation within the Permian Upper San Andres Formation of Last Chance Canyon, Guadalupe Mountains, New Mexico, *in* R. G. Loucks and J. F. Sarg, eds., Carbonate sequence stratigraphy — recent developments and applications: AAPG Memoir 57, p. 435–474.

Speiker, E. M., 1949, Sedimentary facies and associated diastrophism in the Upper Cretaceous of central and eastern Utah: Geological Society of America Memoir 39, p. 55–82.

Talling, P. J., T. F. Lawton, D. W. Burbank, and R. S. Hobbs, 1995, Evolution of latest Cretaceous-Eocene nonmarine deposystems in the Axhandle piggyback basin of central Utah: Geological Society of America Bulletin, v. 107, p. 297–315.

Van Siclen, D. C., 1958, Depositional topography — examples and theory: AAPG Bulletin, v. 42, p. 1897–1913.

Van Wagoner, J. C., R. M. Mitchum, K. M. Campion, and V. D. Rahmanian, 1990, Siliciclastic sequence stratigraphy: AAPG Methods in Exploration Series, No. 7, 55 p.

Walker, R. G., 1984, General introduction — facies, facies sequences and facies models: Geosciences Canada Reprint Series 1, (2nd ed.), p. 1–11.

—1990, Facies modeling and sequence stratigraphy: Journal of Sedimentary Research, v. 60, p. 777–786.

Wheeler, H. E, 1959, Stratigraphic units in space and time: American Journal of Science, v. 257, p. 692–706.

—1964, Baselevel, lithosphere surface, and time stratigraphy: Geological Society of America Bulletin, v. 75, p. 599–610.

—1966, Baselevel transit cycles, *in* Symposium on cyclic stratigraphy: Kansas Geological Survey Bulletin 169, v. 75, p. 623–630.

Willis, B., 1895, The Northern Appalachians: National Geographic Monograph, p. 169–202.

Wright, L. D., and J. M. Coleman, 1973, Variations in morphology of major river deltas as functions of ocean wave and river discharge regimes: AAPG Bulletin, v. 57, p. 370–398.

Analog for Fluvial-Deltaic Reservoir Modeling: Ferron Sandstone of Utah
AAPG Studies in Geology 50
T. C. Chidsey, Jr., R. D. Adams, and T. H. Morris, editors

High-Resolution Depositional Sequence Stratigraphy of the Upper Ferron Sandstone Last Chance Delta: An Application of Coal-Zone Stratigraphy

James R. Garrison, Jr.[1,2] and T. C.V. van den Bergh[1,3]

ABSTRACT

The deposition of the late Turonian-early Coniacian Upper Ferron Sandstone Last Chance Delta occurred during a long slow relative rise in sea level, interrupted only by three minor 4th-order relative falls in sea level. An analysis of stratigraphic, geometric, and architectural data, for both near-marine and non-marine facies indicate that Last Chance Delta architecture was controlled by changes in sediment supply. The systems tract style and parasequence set stacking pattern within depositional sequences reflect the relationship of the rate of sedimentation to the rate of relative change in sea level.

The Upper Ferron Sandstone Last Chance Delta was deposited along the western margin of the Cretaceous Western Interior Seaway between 90.3–88.6 Ma, as a wave-modified, river-dominated fluvial-deltaic system. Detailed stratigraphy and quantitative cross sections based on volcanic ash layer correlations and coal-zone stratigraphy have been used to delineate the depositional sequence stratigraphy of the Last Chance Delta clastic wedge exposed in Castle Valley of east-central Utah. Relative sea level and local subsidence curves, and sedimentation rates have been determined from analysis of these cross sections.

Within the Last Chance Delta, at least 42 parasequences organized into 14 parasequence sets form four 4th-order depositional sequences (denoted FS1 through FS4). In the non-marine to transitional near-marine facies associations, the upper boundaries of parasequence sets, when not coincident with sequence boundaries or transgressive ravinement surfaces, are coal zones. Laterally extensive unconformities, with 20–30 m (66–98 ft) of erosional relief, locally mark the lower boundaries of sequences FS2, FS3, and FS4. These unconformities are interpreted as type-1 sequence boundaries that record basinward shifts of paleoshorelines by up to 3–7 km (2–4 mi) and the development of incised-valley systems ranging up to 6–10 km (4–6 mi) wide.

The lower boundary of sequence FS1 is a correlative conformity immediately below a condensed section within the underlying Tununk Shale. The upper boundary of FS4 is a correlative conformity

[1]The Ferron Group Consultants, Emery, Utah
[2]Present address: Colorado Plateau Field Institute, Price, Utah
[3]Present address: SGS Minerals Services, Huntington, Utah

stratigraphically above a concretion-bearing condensed section a few meters above the uppermost sandstones of sequence FS4. FS1 is a 4th-order highstand sequence of the 3rd-order Hyatti Composite Sequence. FS2 and FS3 are the progradational (early lowstand) to aggradational (late lowstand) sequences of the Ferron Composite Sequence; FS4 is the transgressive sequence. Lowstand sequence FS2 contains well-defined forced regression deposits and higher-order sequence boundaries. The highstand deposits of the Ferron Composite Sequence are represented, within Castle Valley, only by offshore marine shale deposits.

INTRODUCTION

The concepts of depositional sequence stratigraphy (e.g., Posementier and Vail, 1988; Posementier et al., 1988; Van Wagoner et al., 1990; and Van Wagoner, 1995) have improved the ability of the geoscientist to make correlations in complex lithologic settings (e.g., fluvial-deltaic and shoreline depositional systems). Depositional sequence stratigraphic concepts provide a framework by which chronostratigraphic correlations can be made instead of, the traditional and many times erroneous, lithostratigraphic correlations. These chronostratigraphic correlations have improved the ability of the geoscientist to make accurate subsurface correlations (e.g., see Van Wagoner et al., 1990, for examples). While, in general, greatly improving our ability to make subsurface correlations, in many situations, it is difficult to apply these concepts due to lack of well control and/or outcrop analogs. In non-marine settings, these correlations are particularly difficult, even in outcrops, where the correlative surfaces to unconformities and marine-flooding surfaces are difficult to recognize and shoaling-upward depositional cycles are not manifest. The lack of sufficient data and the difficulty in recognition of significant surfaces in these non-marine rocks severely limits our ability to define the basic depositional sequence stratigraphic building blocks (e.g., the parasequence and parasequence set).

In fluvial-deltaic systems, coal seams are ubiquitous, although in many places discontinuous, lithostratigraphic horizons, especially in the non-marine upper delta-plain and lower alluvial-plain facies associations. These coal seams are intuitively excellent stratigraphic correlation horizons because they are easily recognizable in outcrop and core, and many times on logs, and they tend to be laterally extensive, and generally chronostratigraphic markers. Therefore, there is an intuitive tendency to connect together these compact, lithologically unique beds in outcrop and subsurface correlation exercises, although the exact rules applying to such correlations have not been quantified.

There appears to be a genetic relationship between the geometries of major coal beds and the geometries of the associated transitional near-marine and near-marine sediments in fluvial-deltaic deposits. Numerous authors have proposed theories to utilize these relationships in lithostratigraphic, genetic sequence stratigraphic, depositional sequence stratigraphic, and/or regression/transgression sequence stratigraphic correlations (e.g., see Ryer, 1981; Cross, 1988; Aitken and Flint, 1994, 1995; Hamilton and Tadros, 1994; Kamola and Van Wagoner, 1995; van den Bergh, 1995; Garrison and van den Bergh, 1996, 1997; and Garrison, 2004).

Garrison and van den Bergh (1997) and Garrison (2004) found that within the informally named Last Chance Delta, the upper boundaries of parasequence sets, when not coincident with depositional sequence boundaries or transgressive ravinement surfaces, are coal zones, defined as coal seams and their laterally equivalent, genetically related lithologies. This observation has allowed for a detailed depositional sequence stratigraphy of the Upper Ferron Sandstone Last Chance Delta to be worked out, even in the non-marine facies associations.

This paper outlines the depositional sequence stratigraphy of the Upper Ferron Sandstone Last Chance Delta and attempts to illustrate the use of coal zone and volcanic ash layer correlations in sequence stratigraphy, and in the use of coal zones in the recognition of significant surfaces (i.e., correlative conformities), particularly in the non-marine facies association. Using molluscan faunal geochronology and the regional cross sections generated from this study, we have developed a subsidence curve and a relative sea-level curve, and estimates of sedimentation rates. These estimates are in turn used to further interpret the evolution of the Last Chance Delta.

GENERAL STRATIGRAPHY OF THE FERRON SANDSTONE

Middle Turonian to late Santonian sedimentation along the western margin of the Cretaceous Western Interior Seaway, at the present location in Utah, was controlled by two near-contemporaneous, major fluvial-deltaic complexes (informally named): (1) the middle Turonian to late Turonian age Vernal deltaic complex, located at the present-day position of north-central Utah, and (2) the middle Turonian to late Santonian age Southern Utah Deltaic Complex, located at the present-day position of east-central Utah (Garrison, 2004). Both the Vernal Deltaic Complex and the Southern Utah Deltaic Complex contain fluvial-deltaic rocks assigned to the Ferron Sandstone Member of the Mancos Shale.

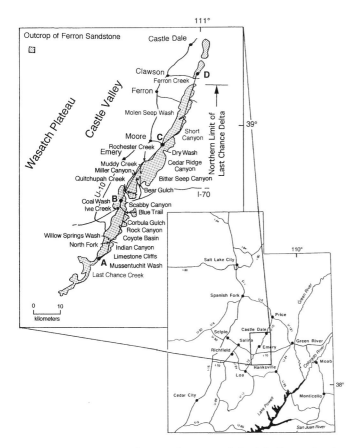

Figure 1. Location map showing the outcrop belt of the Ferron Sandstone in the southern Castle Valley (after Cotter, 1975). Line A-B-C-D represents the line of the cross section shown in Figure 3.

The Ferron Sandstone extends from the Henry Mountains region to Castle Valley to the Uinta Basin of southeastern, east-central, and northeastern Utah, respectively. Early workers suggested that the middle Turonian Ferron Sandstone deposits of northeastern Utah represented a series of deltaic deposits along the western margin of the Cretaceous Western Interior Seaway. The westerly and northwesterly sourced Vernal deltaic complex of north-central Utah (Cotter, 1975; Ryer and McPhillips, 1983; and Bunnel and Holberg, 1991) appears to have been composed of two major deltaic systems, one prograding southeastward in the vicinity of present-day Price, Utah, and another prograding eastward into the Uinta Basin, in the vicinity of Vernal, Utah (Gardner, 1995a). Hale (1972) described a southwesterly sourced, upper Turonian to lower Coniacian, Ferron fluvial-deltaic system, which he called the Last Chance Delta, located at the present-day position of the Castle Valley (Cotter, 1975; Ryer and McPhillips, 1983; Garrison and van den Bergh, 1997; Garrison and van den Bergh, this volume; and Garrison, 2004). Fluvial-deltaic deposits of the middle Turonian Ferron Sandstone, in the Henry Mountains region have been described by Peterson and Ryder (1975), Uresk (1978), and Hill (1982). These deposits have been informally called the Notom Delta (Hill, 1982). We consider both the Last Chance

Delta and the Notom Delta to be part of the much larger Turonian-Santonian Southern Utah Deltaic Complex (also see Garrison, 2004).

The Ferron Sandstone, which crops out in the Castle Valley (Figure 1 and Plate I [all plates are found on the CD-ROM), can be divided into two distinct clastic wedges. The lower portion of the Ferron is referred to as the Lower Ferron Sandstone (Ryer and McPhillips, 1983). The Lower Ferron has been considered by many workers to be shelf sandstone reworked from the Price lobe of the Vernal Delta (e.g., see Garrison, 2004, for a discussion of regional correlations). The upper portion of the Ferron, named the Upper Ferron Sandstone (Ryer and McPhillips, 1983), is the younger, thicker, wave-modified, river-dominated deltaic system, called the Last Chance Delta by Hale (1972). In this paper, only the upper clastic wedge of the Ferron known as the Upper Ferron Sandstone Last Chance Delta or the Upper Ferron Sandstone (Ryer and McPhillips, 1983) will be addressed, unless otherwise noted.

Along the outcrop belt, the Blue Gate Shale Member of the Mancos Shale overlies the Upper Ferron Sandstone, which in turn overlies the Tununk Shale Member of the Mancos Shale containing the distal facies of the Lower Ferron Sandstone. The Last Chance Delta is exposed along the outcrop belt generally parallel to the deltaic progradational direction. There is almost a complete 90-km-long (56-mi) dip section exposure of the deltaic complex. The width of the outcrops of the Ferron perpendicular to the trend of the outcrop belt is generally less than about 5 km (3 mi), thus allowing very little opportunity to examine the deltaic deposits in the strike direction (Figure 1).

PREVIOUS WORKS IN FERRON SAND-STONE SEQUENCE STRATIGRAPHY

Ryer and McPhillips (1983) recognized that the fluvial-deltaic Ferron Sandstone was composed of sediments deposited during a series of transgressions and regressions of the Cretaceous shoreline. Ryer (1981, 1982) and Ryer and McPhillips (1983) described five deltaic cycles within the Last Chance deltaic complex of the southern Upper Ferron Sandstone outcrop belt. These delta-front sandstones were referred to as sandstones 1 through 5. Ryer (1981, 1982) noted that the delta-front sandstone 2 had both a seaward and landward limit displaced farther seaward than delta-front sandstone 1 and that delta-front sandstones 3 through 5 each have seaward limits displaced successively landward. In a subsurface well-log study, Ryer and McPhillips (1983) identified delta-front sandstones 1 through 5 in the subsurface beneath Castle Valley and the Wasatch Plateau. Subsequent work by Ryer (1991) and Gardner (1992, 1993, 1995b) led to the identification of two more, stratigraphically higher, delta cycles in the outcrop belt, referred to as as 6 and 7; both have seaward limits displaced successively landward. The Ferron was subdivided into seven major deltaic events (Gardner, 1992, 1993, 1995b). Based on well-log analysis, Gardner also suggested that an eighth delta-front sandstone may occur in the subsurface farther south near Paradise Valley. Gardner's eighth, subsurface, deltaic unit cannot be verified and will not be discussed, in detail, in this paper. Recent works have subsequently delineated an additional eighth delta-front sandstone in the outcrop belt (Anderson et al., 1997; Garrison and van den Bergh, 1997).

Gardner (1992, 1993, 1995b) examined the deltaic events of the Ferron Sandstone and placed them into a mixed genetic sequence and depositional sequence stratigraphic framework. Gardner recognized the upper portion of the Ferron to be a type-2, 3rd-order depositional sequence, which originally was called the "Ferron sequence" (Gardner, 1992, 1993). Gardner later changed the name of this sequence to the "Ferronensis sequence" (Gardner, 1995a, 1995b) to reflect the occurrence of the ammonite *Scaphites ferronensis*. The southwesterly derived deltaic deposits of the 3rd-order Ferron sequence are separated by a type-2 sequence boundary from the underlying 3rd-order "Hyatti sequence," which contains the ammonite *Prionocyclus hyatti* (Gardner, 1992). According to Gardner (1993, 1995a), the Hyatti sequence has its sediment source from the northwest. Gardner (1993, 1995b) places the upper Hyatti sequence boundary at the base of the Upper Ferron above the Hyatti condensed section. Within the outcrop belt, the lower Hyatti sequence boundary lies within the lower part of the Clawson sandstone unit. The Hyatti sequence is underlain by the "Woollgari sequence" that contains the ammonite *Collignoniceras woollgari* (Gardner, 1993, 1995a; Garrison, in preparation).

Ryer (1994) suggested that the fluvial-deltaic sandstones of the lower Ferron Sandstone (i.e., specifically, the Hyatti sequence) probably formed in response to a dominantly eustatic sea-level change. Leithold (1994) also suggested that the Woollgari and Hyatti sequences (i.e., Lower Ferron Sandstone and its correlative siltstones and shales of the Tununk Shale Member of the Mancos Shale) were formed as a result of a high frequency (i.e., 3rd-order) eustatic sea-level change superimposed on the tectonically induced 2nd-order Greenhorn sea-level cycle. Leithold (1994) also suggested that the Lower Ferron probably formed during the regressive phase of the Greenhorn sea-level cycle (i.e., at the point of maximum sea-level regression). These 3rd-order cycles are postulated to be on the order of 400,000 years in duration (Leithold, 1994).

Ryer (1994) postulated that the Upper Ferron Sandstone (i.e., the Ferron sequence) probably formed in response to an increase in sediment supply associated with increased tectonic activity in the Sevier orogenic belt, during a period of slow sea-level rise (i.e. during a 3rd-order slow sea-level rise). Gardner (1992, 1993) had earlier documented the general decrease in the volume of sediments deposited during each successive delta cycle and discussed the architectural ramifications of the change in the ratio of sedimentation rate to the rate of generation of accommodation space. It was speculated that the 3rd-order sea-level rise was relatively constant, while the rate of sedimentation progressively decreased during the deposition of the Last Chance Delta clastic wedge. Garrison and van den Bergh (1997) later confirmed and expanded the hypothesis of Gardner (1992, 1993).

Gardner (1992, 1993) described the individual deltaic events, within the 3rd-order Ferron sequence, as genetic sequences (Galloway, 1989), and later as stratigraphic cycles (Gardner, 1995b). Recent workers have described them as depositional parasequence sets (e.g., Anderson and Ryer, 1995; Ryer and Anderson, 1995; van den Bergh, 1995; Garrison and van den Bergh, 1996, 1997; Anderson et al., 1997; Garrison, 2004). After examining the seaward and landward limits of the delta-front complexes, Gardner (1993) recognized that: (1) genetic sequences 1 through 3 successively stepped seaward recording an overall regressive event, (2) genetic sequences 4 through 7 recorded a relative transgression, (3) genetic sequences 4 through 5 were aggradational, and (4) genetic sequences 6 through 7 back-stepped sharply landward. Although the delta cycles were described as depositional parasequence sets, the stacking pattern of the Ferron delta cycles, recognized by Gardner (1993), was later confirmed by Ryer and Anderson (1995). Leithold (1994) hypothesized that the high-frequency Greenhorn events (i.e., the parasequence sets and parasequences) are probably associated with local autocyclic events, although it was acknowledged that many of these events appeared to be more basin-wide in extent and thus, could be associated with climatic changes or Milankovitch cycles. van den Bergh and Sprague (1995) postulated that some of the Ferron parasequence sets could likely be grouped into high-frequency, 4th-order depositional sequences within Gardner's 3rd-order Ferron sequence. Garrison (2004) examined the chronostratigraphy of the Turonian to Santonian clastic wedges exposed along the Utah segment of the Western Interior Seaway and concluded that both sedimentation rate and 3rd- and 4th-order relative changes in sea level were controlled by tectonic activity in the Cordillera.

Garrison (2004) documented the depositional sequence stratigraphy of the Upper Ferron Sandstone Last Chance Delta and made regional chronostratigraphic correlations between the Upper Ferron of the Last Chance Delta in Castle Valley, the Turonian-Coniacian rocks exposed in the Henry Mountains region (i.e., Ferron Sandstone Notom Delta), and in the Kaiparowits Plateau ("A" delta of the Straight Cliffs Formation).

Based on the correlation of coal zones, volcanic ash correlations, and detailed stratigraphic mapping, Garrison and van den Bergh (1996, 1997) documented, within the Last Chance clastic wedge, the existence of 14 parasequence sets (i.e., including five single hierarchically equivalent parasequences) which form four high-frequency, 4th-order depositional sequences. Garrison (2004) suggests that the three stratigraphically lowest parasequence sets, within the Last Chance Delta, actually belong to the underlying 3rd-order Hyatti sequence, defined by Gardner (1992, 1993). The detailed depositional sequence stratigraphy of the Last Chance Delta is the subject of this paper.

Garrison (2004) redefines the boundaries of the Woollgari and Hyatti genetic sequences of Gardner (1995a) and the Ferron genetic sequence of Gardner (1992) to make them depositional sequences, instead of genetic sequences. Since the Last Chance Delta has been shown to contain high-order depositional sequences, Garrison (2004) elevates the Woollgari, Hyatti, and Ferron sequences to composite sequences (e.g., see Van Wagoner, 1995, for the definition and characteristics of a composite sequence). The Hyatti composite sequence not only contains the lower part of the Last Chance Delta, but also the fluvial-deltaic rocks of the Ferron Sandstone Notom Delta exposed in the Henry Mountains region. The Ferron composite sequence is overlain by marine rocks that are correlative to the 3rd-order "A" sequence exposed in the Straight Cliffs of the Kaiparowits Plateau (Garrison, 2004).

DEFINITIONS, TERMINOLOGY, AND NOMENCLATURE

Fluvial-Deltaic Systems

In this study, the term "fluvial-deltaic system" is broadly used to describe a depositional system in which there are genetically related coastal, non-marine fluvial-dominated depositional facies and near-marine shoreline depositional facies, both originally deposited by the same coastal plain fluvial systems. This broad definition does not require that there be a shoreline protuberance (i.e., as in the strict definition of a delta [e.g., as defined in Elliott, 1986]), since the extent to which a shoreline protuberance develops, especially in tectonically active areas, is very sensitive to sedimentation rate, amount of wave energy, and rate of relative rise in sea level.

Fluvial-deltaic systems consist of a non-marine "delta-plain facies association" and a "near-marine facies association." In this study, the non-marine delta-plain association is considered to be composed of fluvial channel belts and their associated, and laterally correlative, floodplain facies (e.g., levee deposits, crevasse-splay deposits, overbank mudstones, and lateral swamp and mire facies), and interdistributary bay and lagoonal facies. Near-marine deposits are considered to be sand-rich delta-front deposits that have been subjected to various degrees of wave and tidal destruction and reworking (i.e., these deposits include not only delta-front sandstones and mudstones, but also shoreface deposits produced by wave reworking of delta-front deposits). Marine facies are those deposits formed under deeper water marine conditions, below wave base, where shelfal and basinal processes dominate (e.g., prodelta and offshore deposits).

In this study, fluvial-deltaic deposits are qualitatively classified according to their degree of wave modification. Although this classification is a continuum, three broad classes are defined: river-dominated, wave-reworked, and wave-dominated. This classification is based on the sedimentological and ichnological characteristics of the near-marine sandstones (Table 1).

The near-marine sandstones of river-dominated and wave-reworked fluvial-deltaic systems are separated into stream-mouth bar facies (SMB), proximal delta-front facies (pDF), distal delta-front facies (dDF), and prodelta facies (PD). Near-marine sandstones of wave-dominated deltaic systems are divided into upper shoreface facies (USF), middle shoreface facies (MSF), lower shoreface facies (LSF), and offshore facies (OS). The general criteria for these divisions are shown in Table 1.

Coals and Coal Zones of the Delta-Plain Facies Association

In this study, a coal seam is defined as a laterally continuous bed (i.e., layer) of coal. Coal seams commonly split laterally into multiple seams, separated by delta-plain deposits. These coal seams may be interlayered with carbonaceous shales and mudstones and/or other organic-poor clastic rocks. For the purposes of discussions in this paper, a coal zone is defined as consisting of one or more genetically related coal seams and their laterally correlative, genetically related delta-plain mudstones, siltstones, carbonaceous shales, altered volcanic ash layers, and paleosols. Coal zones may be split by major intervals of delta-plain sediments (i.e., fluvial channels, crevasse splays, and delta-plain mudstones). These interlayered delta-plain intervals represent a hiatus in mire development and are thus not included within the coal zone. This broad, genetic definition implies that, in a facies perspective, the coal zone consists of a central peat-mire facies and its marginal, terrigenously influenced subfacies, as well as any foreign air-fall volcanic ash layers deposited into the mire.

A coal nomenclature was first developed for Ferron Sandstone coals of the Emery coal field by Lupton (1916), in which he described 13 coal beds in the Upper Ferron and denoted them as the A, B, C, ..., and M coal

Table 1. Qualitative characteristics for classification of near-marine sandstones from the Upper Ferron Sandstone Last Chance Delta.

Delta Class	River-dominated			Wave-reworked			Wave-dominated		
Facies	dDf	pDf	SMB	dDf	pDf	SMB	LSF	MSF	USF
% Shale	10-30%	5-10%	<1%	5-20%	1-10%	<1%	10-15%	1-10%	<1%
Bed Thickness	0.5-20 cm	2-50 cm	10-50 cm	5-20 cm	5-30 cm	5-20 cm	0.5-15 cm	5-30 cm	5-30 cm
Bedset Thickness	10-80 cm	10-150 cm	10-150 cm	10-50 cm	20-60 cm	50-100 cm	10-30 cm	30-100 cm	40-200 cm
Dominant Cross-stratification Type or Bedform	horizontal ripples massive	clinoform toes horizontal ripples minor hummocks	troughs ripples	horizontal massive	ripples horizontal minor hummocks minor swales	troughs ripples	hummocks horizontal ripples	hummocks ripples swales	troughs planar tabular horizontal
Burrowing	moderate	rare to slight	rare	moderate	moderate to high	slight	moderate to high	moderate to high	slight
Dominant Ichnofossil	*Skolithos Planolites Arenicolites*	*Thalassinoides Ophiomorpha*	*Ophiomorpha Thalassinoides Skolithos*	*Skolithos Arenicolites*	*Ophiomorpha*	*Ophiomorpha*	*Skolithos Arenicolites Thalassinoides Ophiomorpha*	*Ophiomorpha*	*Ophiomorpha*

seams. This nomenclature has been retained by Garrison and van den Bergh (1996, 1997) and in this study. We have identified and correlated seven of the coal seams, described by Lupton (1916) and have expanded them into coal zones. These seven coal zones and one additional lower coal zone are denoted as the Sub-A, A, C, G, H, I, J, and M coal zones. In addition, a widespread rooted horizon, located stratigraphically between the C coal zone and the G coal zone has been given the informal designation of "E Rooted Zone." When major coal zones split, they are given a letter based on their general coal zone affiliation (e.g., C coal zone) and a subscripted number denoting their stratigraphic position (e.g., C_1, C_2, and C_3 coal zones). The larger the subscripted number, the higher the stratigraphic position of the split within the overall coal zone.

Depositional Sequence Stratigraphy

The depositional sequence stratigraphic interpretations, presented in this paper, utilize the terminology and concepts presented by Posamentier et al. (1988), Posamentier and Vail (1988), Van Wagoner et al. (1990), and Van Wagoner (1995). For detailed discussions of the terminology and the application of depositional sequence stratigraphic concepts, the reader is referred to the definitive work of Van Wagoner et al. (1990) and Van Wagoner (1995).

In this study, the depositional sequence represents a relatively conformable succession of genetically related strata bounded by unconformities or their correlative

conformities. Parasequences and parasequence sets are the building blocks of the depositional sequence. A parasequence is a relatively conformable succession of related bed and bedsets bounded by marine-flooding surfaces or their correlative surfaces. A parasequence set is a set of genetically related parasequences forming a distinctive stacking pattern and bounded by marine-flooding surfaces or their correlative surfaces. The depositional sequence is subdivided into systems tracts. Systems tracts are contemporaneous depositional systems composed of parasequences and one or more parasequence sets. Lowstand, transgressive, and highstand systems tracts are recognized within depositional sequences. Within the depositional sequence, systems tracts are defined based on stratal properties (e.g., see Van Wagoner, 1995).

Recognition of Sequence Stratigraphic Stratal Elements

Parasequences and parasequence boundaries: Following the principles outlined in Van Wagoner et al. (1990) and Van Wagoner (1995), in this study, parasequences are recognized by two or more of the following criteria: (1) the existence of shoaling-upward depositional patterns in the near-marine facies association, (2) evidence of an substantial increase in water depth across the upper boundary of the near-marine to transitional marine facies association of the parasequence (i.e., a marine-flooding surface), (3) evidence of marine transgression at the upper boundary of the parasequence, or (4) evidence of stratigraphic rise between the initial paleoshoreline position of a parasequence, relative to the final paleo-shoreline position of the immediately underlying parasequence. The upper boundary of a parasequence, in the near-marine facies association, is marked by a marine-flooding surface that indicates an abrupt increase in water depth (Van Wagoner et al., 1990; Van Wagoner, 1995). There is no time duration implied in the formation of this boundary. The only restriction is that water depth increases during a hiatus in shoreline sedimentation. This upper surface may or may not exhibit transgressive ravinement and may or may not be overlain by a transgressive lag deposit.

Parasequence set: The stacking pattern of parasequences within a parasequence set is either progradational, aggradational, or retrogradational. These parasequence-stacking patterns reflect the general relationship of the rate of sedimentation to the rate of relative rise in sea level. In a progradational-stacking pattern, each successively younger parasequence has the landward pinchout of its near-marine sandstone displaced seaward relative to the landward pinchout of the near-marine sandstone of the underlying parasequence (i.e., the paleoshorelines associated with successive parasequences exhibit significant seaward migration). In an aggradational-stacking pattern, the landward pinchout of the near-marine sandstone in each successively younger parasequence is located generally in the same areal position relative to the landward pinchout of the near-marine sandstone of the underlying parasequence (i.e., to a 1st-order approximation, the paleoshorelines associated with successive parasequences do not exhibit significant landward or seaward migration). In a retrogradational-stacking pattern, each successively younger parasequence has the landward pinchout of its near-marine sandstone displaced landward relative to the landward pinchout of the near-marine sandstone of the underlying parasequence (i.e., the paleoshorelines associated with successive parasequences exhibit significant landward migration).

Since the parasequence sets within the Upper Ferron Sandstone Last Chance Delta exhibit a stacking pattern, we have defined an "external stacking pattern" nomenclature to describe qualitatively the general 1st-order relative stacking of parasequence sets. The external parasequence set stacking patterns can be either progradational, aggradational, or retrogradational. These parasequence set stacking patterns also reflect the general relationship of the rate of sedimentation to the rate of relative rise in sea level. If a parasequence set is assigned a progradational external stacking pattern, the landward pinchout of its oldest near-marine sandstone is displaced seaward relative to the landward pinchout of the oldest near-marine sandstone of the underlying parasequence set. These parasequence sets generally have

internal progradational-stacking patterns. If a parasequence set is assigned an aggradational external stacking pattern, the landward pinchout of its oldest near-marine sandstone is located generally in the same areal position relative to the landward pinchout of the oldest near-marine sandstone of the underlying parasequence set (i.e., to a 1st-order approximation, the paleoshorelines associated with successive initial parasequences of the sets do not exhibit significant landward or seaward migration). These parasequence sets generally have internally aggradational-stacking patterns. If a parasequence set is assigned a retrogradational external stacking pattern the landward pinchout of its youngest near-marine sandstone is displaced landward relative to the landward pinchout of the youngest near-marine sandstone of the underlying parasequence set. These parasequence sets generally have an internally retrogradational-stacking pattern.

Depositional sequence: Following the principles outlined in Van Wagoner et al. (1990) and Van Wagoner (1995), in this study, depositional sequences are recognized by (1) the presence of widespread bounding erosional unconformities and their correlative surfaces (i.e., depositional sequence boundaries), (2) a significant basinward shift in depositional facies across the depositional sequence boundaries, and (3) the organization of internal stratal patterns that delineate systems tracts.

Systems tracts: In this paper, systems tracts within depositional sequences are defined based on stratal properties, not the inferred position of strata on a relative sea-level curve. The lowstand systems tract (LST) is defined as those strata lying between the sequence boundary, at the base of the sequence, and the first widespread flooding surface, which forms the lower boundary of the transgressive systems tract (TST). The LST may contain not only valley fill fluvial deposits and progradational to aggradational deltaic parasequences, but estuarine and tidally influenced facies as well (Van Wagoner, 1995). The TST is defined as the strata bounded by the first major flooding (or transgressive) surface and the base of the progradational parasequences of the highstand systems tract (HST). Parasequences within the TST are retrogradationally stacked. The HST is defined as those strata lying between the downlap surface at the top of the TST and the overlying depositional sequence boundary. The HST contains aggradational to progradational parasequences.

Nomenclature

The original delta cycle and/or genetic sequence and/or depositional sequence nomenclature (e.g., cycles 1 through 7, genetic sequences 1 through 7, and parasequence sets 1 through 7) is so prevalent in the literature (e.g., Ryer, 1991; Gardner, 1993, 1994a; Ryer and Anderson, 1995; Garrison and van den Bergh, 1996, 1997; and Anderson et al., 1997), that we have chosen to retain this numeric scheme in this study, although more than seven events have been identified. For example, multiple parasequences identified within the original delta cycle 2 will be consecutively denoted as Parasequences 2a, 2b, 2c, etc. ("Parasequence" denoted as upper case). Furthermore, multiple parasequence sets identified within delta cycle 2 will be consecutively denoted as Parasequence Sets 2A, 2B, 2C, etc. ("Parasequence Set" denoted as upper case; on the cross sections in Plates II through V, parasequence sets are denoted as PSS2A, etc.). This convention will retain the connection to the original Ferron nomenclature and make discussions of the newer depositional sequence stratigraphy easier. The only exception to this scheme is that the oldest identified parasequence in Parasequence Set 1B is given the non-sequential designation of Parasequence 1z. This was done, in part, because its entire dip length had not yet been quantified in our studies when the designations were assigned (e.g., Garrison and van den Bergh, 1996, 1997).

We have attempted to retain the names of the 3rd-order depositional and/or genetic sequences applied to the near-marine rocks of the Ferron Sandstone of Castle Valley and the Henry Mountains area by Gardner (1995a). Each of these sequences contains a distinct molluscan faunal assemblage, as described below. Fourth-order depositional sequences are only given letter designations (e.g., FS1, FS2, etc., for 4th-order sequences within the Ferron Sandstone Last Chance Delta). Sequence boundaries are given the sequential designations of SB1, SB2, etc. For example, Sequence FS1 is bound below and above by Sequence Boundaries SB1 and SB2, respectively.

Geochronology and Chronostratigraphy

The chronostratigraphy of the Last Chance Delta discussed in this paper is based on the biostratigraphic correlation charts of Molenaar and Cobban (1991), Kauffman et al. (1993), and correlation charts compiled by Gardner (1995a) and Shanley and McCabe (1995). Figure 2 presents a compilation of these biostratigraphic correlation charts for the Upper Cretaceous from the middle Turonian to the upper Santonian.

The biostratigraphic correlation chart, shown in Figure 2, is calibrated using the ^{40}Ar/^{39}Ar isotopic data presented by Obradovich (1993). These isotopic age data represent the isotopic ages for sanidine crystals extracted from bentonites associated with molluscan faunal collection sites. For the Upper Cretaceous Western Interior biozones *Prionocyclus hyatti* to *Desmoscaphites bassleri*, the typical sanidine age date errors, at the 95% confidence level, range from about 0.28 m.y. to 0.72 m.y. The isotopic age data of Obradovich (1993) suggests that the average biozone duration, within the late Turonian to late Santonian, is approximately 390,000 years.

In order to facilitate regional chronostratigraphic correlations, the names applied to the depositional sequences described in this paper are based on the occurrence of distinct ammonite and inoceramid faunal assemblages. The Woollgari composite sequence has the widespread occurrence of the ammonite *Collignoniceras woollgari*, and has other molluscan fauna older than *Prionocyclus percarinatus* and younger than *Mammities nodosoides*. The Hyatti composite sequence, named for the widespread occurrence of *Prionocyclus hyatti*, spans the Western Interior biozones from *Prionocyclus percarinatus/Inoceramus flaccidusi* to *Prionocyclus macombi/Inoceramus dimidius dimidius*. The Ferron composite sequence generally spans the Western Interior biozones from *Prionocyclus wyomingensis/Inoceramus dimidius* to *Forresteria blancoi/Cremnoceramus deformis*. The "A" Sequence spans the Western Interior biozones from *Scaphites ventricosi* to *Desmoscaphites bassleri*.

METHODOLOGY

This depositional sequence stratigraphic study of the Upper Ferron Sandstone Last Chance Delta of Castle Valley represents a detailed field sedimentological and stratigraphic study of the entire Ferron Sandstone outcrop belt from just north of Last Chance Creek to Ferron Creek (Figure 1 and Plate I). The data sets used in the study include detailed sedimentological measured sections, geometric sections, core descriptions, and interpretations of photomosaics that represent complete photomosaic coverage of the outcrop belt.

Photomosaic interpretations were used to record regional correlations between the measured section positions. In all cases, photomosaic interpretations and correlations were field checked to ensure accuracy. Correlations were further verified using coal zones and volcanic ash layers as stratigraphic markers. In total, seven coal zones and 13 distinct volcanic ash layers were identified and correlated within the study area.

The regional cross sections (Plates II through IV) were generated from these data sets. A 63.5-km-long (39.4-mi) cross section generally parallel to the depositional dip of the Last Chance Delta and an intersecting 3 km (2 mi) strike cross section in Willow Springs Wash were generated. In addition, core descriptions of several cores taken from the Ferron Sandstone were used to augment the measured section and photomosaic data when generating the regional cross sections.

FERRON SANDSTONE DEPOSITIONAL SEQUENCE STRATIGRAPHY — STRATAL ELEMENTS AND PATTERNS

The sedimentology and general stratigraphy of the Upper Ferron Sandstone Last Chance Delta has been extensively described in the literature (Cleavinger, 1974; Cotter, 1975; Ryer, 1981, 1982; Thompson, 1985; Thompson et al., 1986; Howe, 1989; Barton, 1994; Anderson et al., 1997; Garrison and van den Bergh, 1997; Mattson, 1997; Corbeanu et al., 2001). For details of the sedimentology of the Upper Ferron Last Chance Delta, the reader is referred to these works.

In this paper, we focus mainly on the correlation and definition of stratal elements within the Upper Ferron Sandstone Last Chance Delta, in the context of depositional sequence stratigraphy. Sedimentological data will only be presented and discussed when necessary to illustrate stratal properties and stratal element boundaries.

Since many authors have presented correlations and unit definitions within the Upper Ferron Sandstone Last Chance Delta (e.g., Gardner, 1993; Anderson et al., 1997; Garrison and van den Bergh, 1997), this paper will focus on and present, in detail, stratigraphic correlation data and stratal interpretations that are pertinent to outlining the depositional sequence stratal elements of the Upper Ferron Sandstone Last Chance Delta. These data and interpretations are summarized in Table 2, but will be discussed below in more detail to facilitate comparisons to other works (e.g., Gardner, 1995b and Anderson et al., 1997). In order to establish, for the reader, the nature of the sequence stratal elements prior to the presentation of the high-frequency depositional sequence stratigraphy of the Last Chance Delta, the definition and field recognition of the Last Chance Delta parasequences, parasequence sets, and major correlative surfaces will be outlined first, before a discussion of the depositional sequences and systems tracts is presented.

The detailed field studies (Plate I) and cross sections (Plates II through V) have delineated 42 parasequences which compose 14 parasequence sets within the Upper Ferron Sandstone. Table 3 contains the geometry, stratal patterns, and geochronology of these parasequence sets. Table 4 contains the geometry, stratal patterns, and geochronology of these parasequences. The single Parasequences 2e, 4a, 4b, 5c, and 6a have been given hierarchical equivalence to a parasequence set because: (1) they occur stratigraphically between well-defined parasequence sets, (2) they represent a change in stacking pattern between the underlying and overlying parasequences, and/or (3) they are separated from an underlying or overlying parasequence or parasequence set by an erosional unconformity. These parasequences have been designated as Parasequence Sets 2B, 4A, 4B, 5B, and 6, respectively. Parasequence Sets 3 through 8 are represented along the Ferron outcrop belt as dominantly delta-plain facies associations (e.g., fluvial channel belts, crevasse splays, coal zones, and delta-plain mudstones and siltstones).

South of Mussentuchit Wash toward Last Chance Creek, Gardner (1995b) and Anderson et al. (1997) described a series of parasequences, within what may be defined as an additional parasequence set, lying stratigraphically below Parasequence Set 1 of the original study of Garrison and van den Bergh (1997). From subsequent reconnaissance studies we have identified these units, in Last Chance Creek, as an additional parasequence set consisting of at least three parasequences. Since our detailed study did not extend much farther south than Mussentuchit Wash, these parasequences will only be discussed briefly in this paper. This original Parasequence Set 1 of Garrison and van den Bergh (1997) has been denoted as Parasequence Sets 1B and the additional parasequence set, identified at Last Chance Creek, is denoted as Parasequence Sets 1A.

Within the exposed Ferron Sandstone Last Chance clastic wedge, the 14 parasequence sets form four high-frequency, 4th-order depositional sequences (denoted FS1 through 4) (Figure 3, Plates II through IV). The lowest 4th-order sequence FS1 consists of four parasequence sets (denoted as Parasequence Sets 1A, 1B, 2A, and 2B). FS2 also consists of three parasequence sets (denoted as Parasequence Sets 2C, 2D, 3, and 4A). FS3 consists of two parasequence sets (denoted 4B and 5A) and FS4 consists of four parasequence sets (denoted as Parasequence Sets 5B, 6, 7, and 8). In general, the 4th-order depositional sequences consist of progradational and/or aggradational parasequence sets (Table 2), with the upper-most, highstand parasequence set lying stratigraphically above a transgressive lag deposit.

In the non-marine to transitional near-marine facies associations, the upper boundaries of the parasequence sets, when not coincident with sequence boundaries or transgressive ravinement surfaces, are coal zones and their correlative conformities. The sequence boundary erosional unconformities that form the upper or lower boundaries of some parasequence sets have correlative surfaces within coal zones, in the interfluvial areas.

Progradational Parasequence Sets

In the context of the overall Last Chance Delta clastic wedge architecture and the larger scale 3rd-order Ferron composite sequence, Ferron Parasequence Sets 1A, 1B, 2A through 2D, and 3 have an external progradational parasequence stacking pattern. The landward pinchout of the near-marine sandstones (i.e., location of paleoshorelines) of each younger parasequence set is progressive farther down depositional dip. Internally, within each of these parasequence sets, the parasequences themselves exhibit an internal progradational-stacking pattern. Unless coincident with a sequence boundary, coal zones occur at or near the top of these depositional parasequence sets, with the thickest portion of the coal zones near the landward pinchout of the near-marine delta-front sandstones of the depositional parasequence sets. The Sub-A coal zone caps Ferron Parasequence Set 1B; a composite A coal zone caps

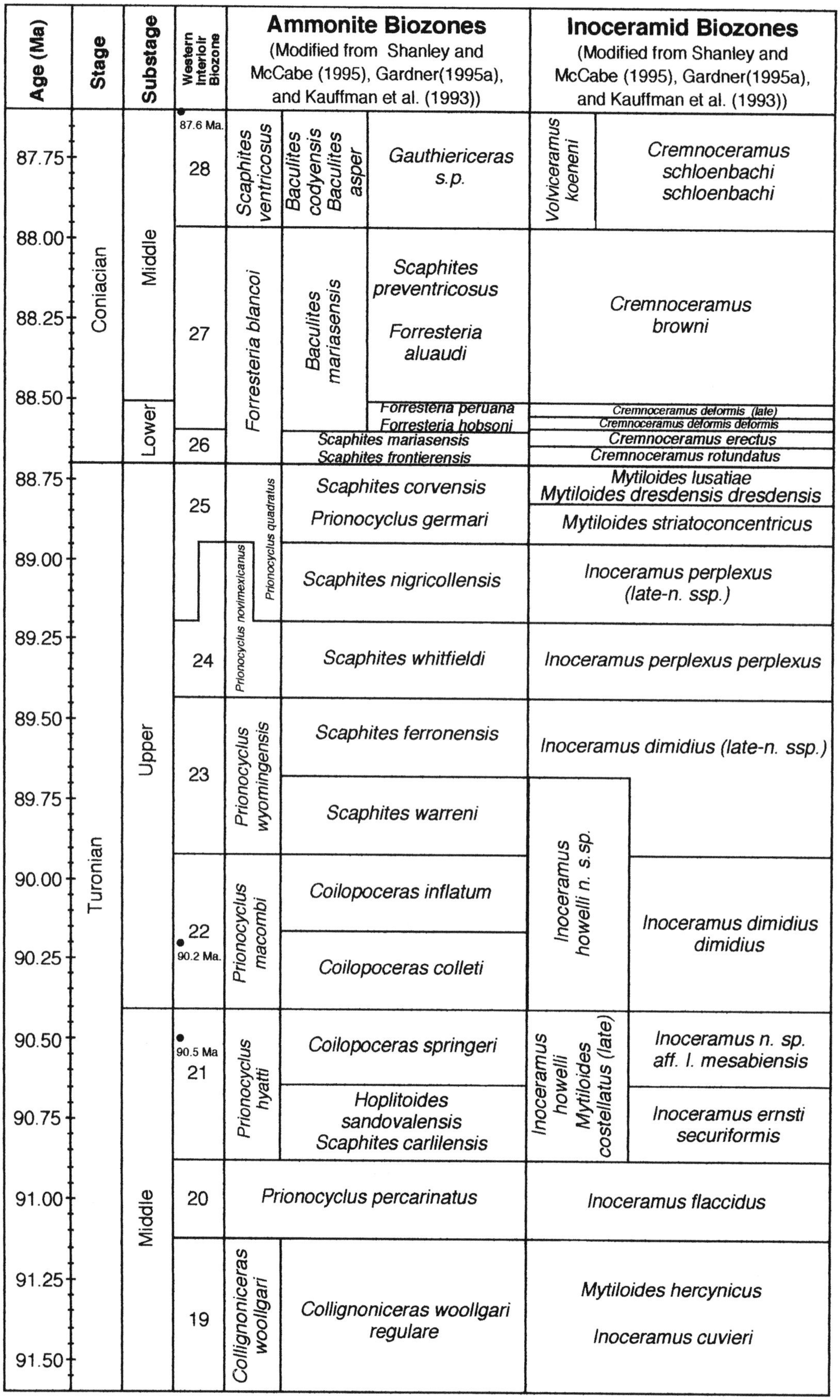

Age (Ma)	Stage	Substage	Western Interior Biozone	Ammonite Biozones (Modified from Shanley and McCabe (1995), Gardner(1995a), and Kauffman et al. (1993))		Inoceramid Biozones (Modified from Shanley and McCabe (1995), Gardner(1995a), and Kauffman et al. (1993))
87.75 — • 87.6 Ma.	Coniacian	Middle	28	*Scaphites ventricosus*	*Baculites codyensis Baculites asper*	*Gauthiericeras s.p.*	*Volviceramus koeneni*	*Cremnoceramus schloenbachi schloenbachi*
88.00 — 88.25 —			27	*Forresteria blancoi*	*Baculites mariasensis*	*Scaphites preventricosus* / *Forresteria aluaudi*		*Cremnoceramus browni*
88.50 —		Lower	26			*Forresteria peruana* / *Forresteria hobsoni* / *Scaphites mariasensis* / *Scaphites frontierensis*		*Cremnoceramus deformis (late)* / *Cremnoceramus deformis deformis* / *Cremnoceramus erectus* / *Cremnoceramus rotundatus*
88.75 —	Turonian	Upper	25	*Prionocyclus quadratus*	*Scaphites corvensis* / *Prionocyclus germari*		*Mytiloides lusatiae* / *Mytiloides dresdensis dresdensis* / *Mytiloides striatoconcentricus*
89.00 —				*Prionocyclus novimexicanus*	*Scaphites nigricollensis*		*Inoceramus perplexus (late-n. ssp.)*
89.25 — 89.50 —			24		*Scaphites whitfieldi*		*Inoceramus perplexus perplexus*
89.75 —			23	*Prionocyclus wyomingensis*	*Scaphites ferronensis*		*Inoceramus dimidius (late-n. ssp.)*
90.00 —					*Scaphites warreni*	*Inoceramus howelli n. s. sp.*	
90.25 — • 90.2 Ma.			22	*Prionocyclus macombi*	*Coilopoceras inflatum* / *Coilopoceras colleti*		*Inoceramus dimidius dimidius*
90.50 — • 90.5 Ma			21	*Prionocyclus hyatti*	*Coilopoceras springeri*	*Inoceramus howelli* / *Mytiloides costellatus (late)*	*Inoceramus n. sp. aff. I. mesabiensis*
90.75 —					*Hoplitoides sandovalensis* / *Scaphites carlilensis*		*Inoceramus ernsti securiformis*
91.00 —		Middle	20		*Prionocyclus percarinatus*		*Inoceramus flaccidus*
91.25 — 91.50 —			19	*Collignoniceras woollgari*	*Collignoniceras woollgari regulare*		*Mytiloides hercynicus* / *Inoceramus cuvieri*

Figure 2. Biostratigraphic correlation chart for middle Turonian to Campanian strata of the Western Interior. Adapted from correlation charts presented by Gardner (1995a), Kauffman et al. (1993), and Shanley and McCabe (1995). The • symbol denotes calibration age dates from Obradovich (1993).

Age (Ma)	Stage	Substage	Western Interior Biozone	Ammonite Biozones (Modified from Shanley and McCabe (1995) and Kauffman et al., 1993))	Inoceramid Biozones (Modified from Shanley and McCabe (1995) and Kauffman et al. (1993))
83.75 – 84.00	Santonian	Upper	34 • 83.9 Ma.	Desmoscaphites bassleri	Sphenoceramus digitatus: Sphenoceramus lundbreckensis: Sphenoceramus lobatus
84.25 – 84.75	Santonian	Upper	33	Desmoscaphites erdmanni	Endocostea simpsoni: Endocostea baltica s.s.: Sphenoceramus cf. Sphenoceramus patootensis
85.00 – 85.25	Santonian	Middle	32	Clioscaphites choteauensis	Sphenoceramus pachti pachti / Cordiceramus n. sp. (quadrate): Cordiceramus n. sp. aff Cordiceramus cordiformis
85.50 – 85.75	Santonian	Middle	31	Clioscaphites vermiformis	Sphenoceramus pachti pachti
86.00	Santonian	Lower	30	Clioscaphites saxitonianus	Cladoceramus undulatoplicatus / Platyceramus cycloides
86.25 – 86.50	Santonian	Lower	30	Clioscaphites saxitonianus	Cladoceramus undulatoplicatus / Cladoceramus undulatoplicatus s.s. Mytiloides n. sp. cf. Mytiloides stantoni
86.75	Coniacian	Upper	86.9 Ma • 29 / Scaphites depressus / Baculites codyensis Baculites asper	Peroniceras sp.	Magadiceramus subquadratus subquadratus / Magadiceramus austinensis / Magadiceramus subquadratus complicatus
87.00	Coniacian	Upper	29	Phlycticrioceras oregonense	Magadiceramus soukupi. Magadiceramus subquadratus crenelatus Magadiceramus sp. cf. Magadiceramus austinensis
87.25	Coniacian	Upper	Scaphites ventricosus		Mytiloides stantoni / Mytiloides n. sp. aff. Mytiloides stantoni
87.50	Coniacian	Middle	28 • 87.6 Ma	Gauthiericeras sp.	Volviceramus koeneni / Cremnoceramus wandereri Cremnoceramus inconstans (late) / Cremnoceramus schloenbachi schloenbachi

Figure 2 continued.

135

Table 2. Depositional sequence stratigraphy of the Upper Ferron Sandstone Last Chance Delta.

Parasequence*	Parasequence Set*	Internal Stacking Pattern	Delta Class	4th-order Sequence	4th-order Systems Tract**	3rd-order Sequence	3rd-order Systems Tract
1v	1A	Aggradational	River-dominated	FS1	HST	Hyatti Composite	Early-HST
1w	1A	Aggradational	River-dominated	FS1	HST	Hyatti Composite	Early-HST
1x	1A	Aggradational	River-dominated	FS1	HST	Hyatti Composite	Early-HST
1y	1A	Aggradational	River-dominated	FS1	HST	Hyatti Composite	Early-HST
1z	1B	Progradational	River-dominated	FS1	HST	Hyatti Composite	Late-HST
1a	1B	Progradational	River-dominated	FS1	HST	Hyatti Composite	Late-HST
1b	1B	Progradational	River-dominated	FS1	HST	Hyatti Composite	Late-HST
1c	1B	Progradational	River-dominated	FS1	HST	Hyatti Composite	Late-HST
1d	1B	Progradational	River-dominated	FS1	HST	Hyatti Composite	Late-HST
1e	1B	Progradational	River-dominated	FS1	HST	Hyatti Composite	Late-HST
1f	1B	Progradational	River-dominated	FS1	HST	Hyatti Composite	Late-HST
1g	1B	Progradational	River-dominated	FS1	HST	Hyatti Composite	Late-HST
1h	1B	Progradational	River-dominated	FS1	HST	Hyatti Composite	Late-HST
1i	1B	Progradational	River-dominated	FS1	HST	Hyatti Composite	Late-HST
1j	1B	Progradational	River-dominated	FS1	HST	Hyatti Composite	Late-HST
1k	1B	Progradational	River-dominated	FS1	HST	Hyatti Composite	Late-HST
2a	2A	Progradational	Wave-reworked	FS1	HST	Hyatti Composite	Late-HST
2b	2A	Progradational	Wave-reworked	FS1	HST	Hyatti Composite	Late-HST
2c	2A	Progradational	Wave-reworked	FS1	HST	Hyatti Composite	Late-HST
2d	2A	Progradational	Wave-reworked	FS1	HST	Hyatti Composite	Late-HST
2e	2B[+]	single parasequence	Wave-reworked	FS1	HST	Hyatti Composite	Late-HST
2f	2C	Progradational	Wave-reworked	FS2	LST	Ferron Composite	Early-LST
2g	2C	Progradational	Wave-reworked	FS2	LST	Ferron Composite	Early-LST
2h	2C	Progradational	Wave-reworked	FS2	LST	Ferron Composite	Early-LST
2i	2C	Progradational	Wave-reworked	FS2	LST	Ferron Composite	Early-LST
2j	2C	Progradational	Wave-reworked	FS2	LST	Ferron Composite	Early-LST
2k	2D	Progradational	Wave-reworked	FS2	LST	Ferron Composite	Early-LST
2l	2D	Progradational	Wave-reworked	FS2	LST	Ferron Composite	Early-LST
2m	2D	Progradational	Wave-reworked	FS2	LST	Ferron Composite	Early-LST
3a	3	Progradational	Wave-reworked	FS2	TST	Ferron Composite	Early-LST
3b	3	Progradational	Wave-reworked	FS2	TST	Ferron Composite	Early-LST
3c	3	Progradational	Wave-reworked	FS2	TST	Ferron Composite	Early-LST
4a	4A[+]	single parasequence	Wave-dominated	FS2	HST	Ferron Composite	Early-LST
4b	4B[+]	Aggradational	Wave-dominated	FS3	LST	Ferron Composite	Late-LST
5a	5A	Aggradational	Wave-dominated	FS3	HST	Ferron Composite	Late-LST
5b	5A	Aggradational	Wave-dominated	FS3	HST	Ferron Composite	Late-LST
5c	5B[+]	single parasequence	Wave-dominated	FS4	LST	Ferron Composite	TST
6a	6[+]	single parasequence	Wave-dominated	FS4	TST	Ferron Composite	TST
7a	7	Retrogradational	Wave-dominated	FS4	TST	Ferron Composite	TST
7b	7	Retrogradational	Wave-dominated	FS4	TST	Ferron Composite	TST
7c	7	Retrogradational	Wave-dominated	FS4	TST	Ferron Composite	TST
7d	7	Retrogradational	Wave-dominated	FS4	TST	Ferron Composite	TST
8a	8	Retrogradational	Wave-dominated	FS4	TST	Ferron Composite	TST
8b	8	Retrogradational	Wave-dominated	FS4	TST	Ferron Composite	TST

*Number refers to the delta cycle number assigned by Ryer (1981) and Gardner (1993).

**LST = Lowstand Systems Tract, TST = Transgressive Systems Tract, HST = Highstand Systems Tract.

+Based on their stratigraphic stacking and stratal relationships, these single parasequences are given hierarchical equivalence to parasequence sets.

Table 3. Parameters describing the parasequence sets within the Upper Ferron Sandstone Last Chance Delta.

Parasequence Set	Internal Stacking	External Stacking	Section	Position	Age (m.y.)	Bayline* (km)	Shoreline* (km)	Seaward Limit* (km)	Seaward Step** (km)	Stratigraphic Rise (m)	Progradation (km)	Length (km)	Maximum Thickness (m)	Length/ Thickness Aspect Ratio	Duration (years)
PSS1A	Aggradational	Progradational	Dip	Bottom	90.3000										
				Top	90.2144										
PSS1B	Progradational	Progradational	Dip	Bottom	90.2143	1.3	1.3	2.9		69.1	23.2	24.1	29.5	817	85610
				Top	90.1287	24.5	24.5	25.5		76.2					
PSS2A	Progradational	Progradational	Dip	Bottom	90.1286	8.9	8.9	12.4	7.6	52.7	22.5	27.3	31.9	856	68468
				Top	90.0601	31.4	31.4	36.2		4.7					
PSS2B	single PS	Progradational	Dip	Bottom	90.0600	25.2	25.2	29.4	16.3	6.7	1.8	6.2	9.7	634	14142
				Top	90.0458	27.0	27.0	31.3		-16.1					
PSS2C	Progradational	Progradational	Dip	Bottom	90.0428	28.0	29.7	32.1	4.6	24.9	14.6	17.6	24.8	709	85610
				Top	89.9572	42.6	44.3	45.9		6.2					
PSS2D	Progradational	Progradational	Dip	Bottom	89.9571	39.4	42.2	45.8	12.4	20.8	6.9	21.3	27.6	772	85610
				Top	89.8715	42.3	49.0	58.2		5.5					
PSS3	Progradational	Progradational	Strike Oblique	Bottom	89.8714	32.2	32.2	41.1	-10.0	8.4	9.5	31.3	21.5	1457	85610
				Top	89.7858	41.7	41.7	63.5		6.0					
PSS4A	single PS	Aggradational	Dip	Bottom	89.7857	25.6	25.6	27.7	-6.6	12.1	8.6	15.0	27.2	552	82710
				Top	89.7030	34.2	34.2	40.6		-9.4					
PSS4B	single PS	Aggradational	Dip	Bottom	89.7000	29.4	32.4	40.9	6.8	6.7	9.5	11.4	17.9	634	249900
				Top	89.4501	41.9	41.9	43.7		5.0					
PSS5A	Aggradational	Aggradational	Dip	Bottom	89.4500	29.9	29.9	37.4	-2.4	23.8	6.3	11.8	27.5	428	247000
				Top	89.2030	36.3	36.3	41.7		-8.4					
PSS5B	single PS	Aggradational	Dip	Bottom	89.2000	34.1	34.1	38.7	4.1	14.1	5.1	7.5	16.1	466	149900
				Top	89.0501	39.2	39.2	41.6		3.7					
PSS6	single PS	Retrogradational	Dip	Bottom	89.0500	32.1	32.1	34.7	-2.0	7.4	1.8	2.8	10.4	267	149900
				Top	88.9001	33.9	33.9	34.9		18.0					
PSS7	Retrogradational	Retrogradational	Dip	Bottom	88.9000	31.9	31.9	33.1	-0.2	26.7	-4.6	7.1	9.1	779	149900
				Top	88.7501	21.3	27.4	29.7		0.7					
PSS8	Retrogradational	Retrogradational	Dip	Bottom	88.7500	23.4	23.4	27.2	-8.6	6.4	0.7	5.9	3.6	1628	150000
				Top	88.6000	24.1	24.1	24.7							

*Positions determined relative to arbitrary position at regional dip cross section point A; significant errors may occur in the strike oblique sections.

**Seaward step is the change in initial paleoshoreline position relative to the initial paleoshoreline position of the underlying parasequence; significant errors may occur in the strike oblique sections.

Table 4. Parameters describing the parasequences within the Upper Ferron Sandstone Last Chance Delta.

Parasequence	Parasequence Set	Position	Section	Age (m.y.)	Bayline* (km)	Shoreline* (km)	Seaward Limit* (km)	Stratigraphic Rise (m)	Progradation (km)	Seaward Step** (km)	Shoreline Step (km)	Length (km)	Thickness (meters)	Aspect Ratio (Length/thickness)	Duration (yr)
PS0a	PSS1A	Bottom	Dip	90.3000											
		Top		90.2715											
PS0b	PSS1A	Bottom	Dip	90.2714											
		Top		90.2429											
PS0c	PSS1A	Bottom	Dip	90.2428											
		Top		90.2144											
PS1z	PSS1B	Bottom	Dip	90.2143											
		Top		90.2072											
PS1a	PSS1B	Bottom	Dip	90.2071	1.3	1.3	2.9	7.7							
		Top		90.2001	4.6	4.6	6.4	2.7	3.3		3.3	5.1	27.5	186	7042
PS1b	PSS1B	Bottom	Dip	90.2000	2.7	2.7	3.6	7.4			-2.0				
		Top		90.1929	7.0	7.0	8.9	2.3	4.4	1.3	4.4	6.3	22.2	283	7042
PS1c	PSS1B	Bottom	Dip	90.1928	5.6	5.6	7.8	1.3			-1.5				
		Top		90.1858	10.9	10.9	14.2	5.4	5.3	2.9	5.3	8.7	12.8	682	7042
PS1d	PSS1B	Bottom	Dip	90.1857	10.5	10.5	15.3	4.4			-0.4				
		Top		90.1786	14.0	14.0	18.0	4.7	3.5	4.9	3.5	7.5	16.8	446	7042
PS1e	PSS1B	Bottom	Strike	90.1785	16.6	16.6	17.5	0.0			2.7				
		Top		90.1715	18.1	18.1	19.5	6.0	1.5	6.2	1.5	2.9	22.5	129	7042
PS1f	PSS1B	Bottom	Strike	90.1714	18.3	18.3	18.6	0.0			0.2				
		Top		90.1644	19.8	19.8	22.0	5.4	1.5	1.7	1.5	3.6	16.8	216	7042
PS1g	PSS1B	Bottom	Strike	90.1643	19.7	19.7	21.8	-4.7			-0.1				
		Top		90.1572	22.3	22.3	22.4	4.7	2.6	1.4	2.6	2.7	9.4	289	7042
PS1h	PSS1B	Bottom	Dip	90.1571	21.8	21.8	22.9	2.4			-0.5				
		Top		90.1501	23.2	23.2	24.4	5.0	1.4	2.1	1.4	2.6	13.8	187	7042
PS1i	PSS1B	Bottom	Strike	90.1500	23.9	23.9	25.7	0.0			0.7				
		Top		90.1429	24.2	24.2	26.4	3.0	0.4	2.1	0.4	2.6	13.1	195	7042
PS1j	PSS1B	Bottom	Dip	90.1428	18.4	21.8	22.1	10.1			-2.4				
		Top		90.1358	23.4	23.4	24.8	0.0	1.6	-2.1	1.6	3.0	5.7	523	7042
PS1k	PSS1B	Bottom	Dip	90.1357	22.9	22.9	24.5	1.3			-0.5				
		Top		90.1287	24.5	24.5	25.5	0.0	1.7	1.1	1.7	2.6	6.7	386	7042
PS2a	PSS2A	Bottom	Dip	90.1286	8.9	8.9	12.4	7.1			-15.6				
		Top		90.1115	20.7	20.7	26.8	30.9	11.8	-14.0	11.8	17.9	42.0	427	17042
PS2b	PSS2A	Bottom	Dip	90.1114	12.4	12.4	24.3	0.7			-8.3				
		Top		90.0944	23.6	23.6	27.5	9.4	11.1	3.5	11.1	15.0	18.5	814	17042

Table 4 continued.

Parasequence	Parasequence Set	Position	Section	Age (m.y.)	Bayline* (km)	Shoreline* (km)	Seaward Limit* (km)	Stratigraphic Rise (m)	Progradation (km)	Seaward Step** (km)	Shoreline Step (km)	Length (km)	Thickness (meters)	Aspect Ratio (Length/ thickness)	Duration (yr)
PS2c	PSS2A	Bottom	Dip	90.0943	19.5	19.7	20.4	0.7		7.2	-3.9				
		Top		90.0772	24.6	24.6	32.7	4.7	4.9		4.9	13.1	8.4	1557	17042
PS2d	PSS2A	Bottom	Dip	90.0771	22.0	22.0	26.7	1.3		2.3	-2.6				
		Top		90.0601	31.4	31.4	36.2	5.0	9.4		9.4	14.2	22.3	636	17042
PS2e	PSS2B	Bottom	Dip	90.0600	25.2	25.2	29.4	4.7		3.2	-6.3				
		Top		90.0458	27.0	27.0	31.3	6.7	1.8		1.8	6.2	9.7	634	14142
PS2f	PSS2C	Bottom	Dip	90.0428	28.0	29.7	32.1	-16.1		4.6	2.7				
		Top		90.0258	31.8	31.8	36.4	5.7	2.1		2.1	6.7	18.9	352	17042
PS2g	PSS2C	Bottom	Dip	90.0257	28.0	28.3	32.2	1.0		-1.4	-3.6				
		Top		90.0087	36.7	36.7	38.1	5.0	8.4		8.4	9.8	17.5	562	17042
PS2h	PSS2C	Bottom	Dip	90.0086	37.1	37.1	38.3	-12.1		8.8	0.4				
		Top		89.9915	39.9	39.9	40.3	6.0	2.8		2.8	3.2	7.8	415	17042
PS2i	PSS2C	Bottom	Dip	89.9914	35.4	35.4	40.5	6.0		-1.7	-4.5				
		Top		89.9744	42.4	42.4	42.9	7.7	7.1		7.1	7.5	14.8	506	17042
PS2j	PSS2C	Bottom	Dip	89.9743	41.9	42.4	43.9	3.0		7.1	0.0				
		Top		89.9572	42.6	44.3	45.9	2.4	1.9		1.9	3.5	12.7	272	17042
PS2k	PSS2D	Bottom	Dip	89.9571	39.4	42.2	45.8	6.2		-0.3	-2.1				
		Top		89.9287	47.7	47.7	50.8	3.9	5.5		5.5	8.6	10.0	864	28470
PS2l	PSS2D	Bottom	Dip	89.9286	25.5	42.2	51.4	7.3		0.1	-5.4				
		Top		89.9001	48.8	48.8	63.4	0.0	6.6		6.6	21.2	10.3	2058	28470
PS2m	PSS2D	Bottom	Dip	89.9000	41.7	42.3	52.1	2.6		0.0	-6.6				
		Top		89.8715	42.3	49.0	58.2	7.0	6.8		6.8	16.0	9.7	1646	28470
PS3a	PSS3	Bottom	Strike oblique	89.8714	32.2	32.2	41.1	5.5		-10.3	-16.8				
		Top		89.8430	32.9	32.9	41.1	0.0	0.7		0.7	8.9	8.1	1109	28470
PS3b	PSS3	Bottom	Strike oblique	89.8429	32.6	32.6	43.4	1.0		0.5	-0.2				
		Top		89.8144	43.5	43.5	63.5	5.0	10.9		10.9	30.9	21.1	1461	28470
PS3c	PSS3	Bottom	Strike oblique	89.8143	24.5	35.5	51.5	2.4		2.9	-7.9				
		Top		89.7858	41.7	41.7	63.5	0.0	6.2		6.2	28.0	2.7	10283	28470
PS4a	PSS4A	Bottom	Dip	89.7857	25.6	25.6	27.7	6.0		-7.1	-16.1				
		Top		89.7030	34.2	34.2	40.6	12.1	8.6		8.6	15.0	27.2	552	82710
PS4b	PSS4B	Bottom	Dip	89.7000	29.4	32.4	40.9	-9.4		6.8	-1.8				
		Top		89.4501	41.9	41.9	43.7	6.7	9.5		9.5	11.4	17.9	634	249900

Table 4 continued.

Parasequence	Parasequence Set	Position	Section	Age (m.y.)	Bayline* (km)	Shoreline* (km)	Seaward Limit* (km)	Stratigraphic Rise (m)	Progradation (km)	Seaward Step** (km)	Shoreline Step (km)	Length (km)	Thickness (meters)	Aspect Ratio (Length/thickness)	Duration (yr)
PS5a	PSS5A	Bottom	Dip	89.4500	29.9	29.9	37.4	5.0		-2.4	-11.9				
		Top	Dip	89.3251	36.1	36.1	40.5	4.7	6.1		6.1	10.6	15.6	677	124900
PS5b	PSS5A	Bottom	Dip	89.3250	30.9	30.9	35.2	15.5		0.9	-5.2				
		Top	Dip	89.2030	36.3	36.3	41.7	3.6	5.4		5.4	10.9	15.4	703	122000
PS5c	PSS5B	Bottom	Dip	89.2000	34.1	34.1	38.7	-8.4		3.2	-2.2				
		Top	Dip	89.0501	39.2	39.2	41.6	14.1	5.1		5.1	7.5	16.1	466	149900
PS6	PSS6	Bottom	Dip	89.0500	32.1	32.1	34.7	3.7		-2.0	-7.1				
		Top	Dip	88.9001	33.9	33.9	34.9	7.4	1.8		1.8	2.8	10.4	267	149900
PS7a	PSS7	Bottom	Dip	88.9000	31.9	31.9	33.1	18.0		-0.2	-2.0				
		Top	Dip	88.8626	32.9	32.9	33.2	3.4	0.9		0.9	1.3	3.4	375	37400
PS7b	PSS7	Bottom	Dip	88.8625	30.2	30.2	32.8	1.7		-1.7	-2.6				
		Top	Dip	88.8251	31.9	31.9	33.0	1.7	1.7		1.7	2.8	1.8	1536	37400
PS7c	PSS7	Bottom	Dip	88.8250	29.0	29.0	29.9	8.7		-1.2	-3.0				
		Top	Dip	88.7876	30.1	30.1	32.1	4.2	1.1		1.1	3.1	8.5	366	37400
PS7d	PSS7	Bottom	Dip	88.7875	26.1	26.1	29.6	3.4		-2.9	-4.0				
		Top	Dip	88.7501	27.4	27.4	29.7	3.7	1.3		1.3	3.6	3.7	967	37400
PS8a	PSS8	Bottom	Dip	88.7500	23.4	23.4	27.2	0.7		-2.8	-4.0				
		Top	Dip	88.6751	26.9	26.9	27.2	2.9	3.5		3.5	3.9	3.6	1075	74900
PS8b	PSS8	Bottom	Dip	88.6750	21.2	21.3	24.5	1.7		-2.0	-5.6				
		Top	Dip	88.6000	24.1	24.1	24.7	1.9	2.8		2.8	3.4	3.5	966	75000

*Positions determined relative to arbitrary position at regional dip cross-section point A; significant errors in these positions may occur in the strike oblique cross-sections.
**Seaward step is the change in initial paleoshoreline position relative to the initial paleoshoreline position of the underlying parasequence; significant errors may occur in strike oblique cross-sections.

Depositional Sequence Stratigraphy of the Upper Ferron Sandstone Last Chance Delta

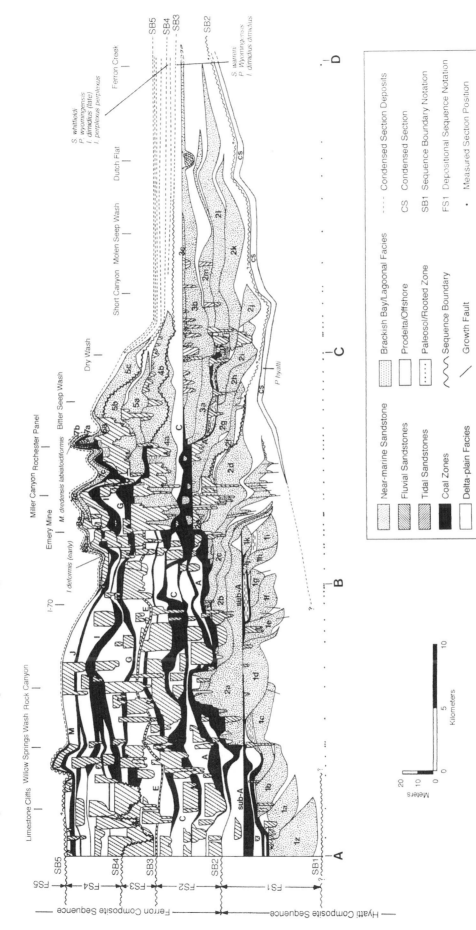

Figure 3. Cross section of the Upper Ferron Sandstone Last Chance Delta clastic wedge generalized from the regional cross sections in Plates II through IV in the CD-ROM. The datums are the same as in Plates II through IV.

Figure 4. Upper Ferron Sandstone Parasequence Set 1B at Ivie Creek/I-70 outcrop.

Parasequence Sets 2C and 2D; the C coal zone caps Parasequence Set 3. It is common for these coal zones to split near the top of the depositional parasequence sets. The top of Ferron Parasequence Set 2B is not capped by a coal zone, but instead the upper boundary is an erosional unconformity that has been identified as 4th-order depositional sequence boundary SB2. The nature and extent of this boundary will be discussed later.

The southern outcrop exposures of river-dominated Parasequence Sets 1A and 1B, south of Last Chance Creek, are poor, making the identification and analysis of the oldest parts of these parasequence sets difficult. For the purposes of this paper, the discussions of these parasequence sets and resulting depositional sequence stratigraphic analysis will be restricted to those outcrops north of Last Chance Creek. The cross sections in Figure 3 and Plate II begin just north of Last Chance Creek and just south of Mussentuchit Wash and will not include the southern exposures of Parasequence Set 1A.

Parasequence Set 2A contains four wave-reworked fluvial-deltaic parasequences (denoted 2a through 2d), and exhibits an internal progradational parasequence stacking pattern (Figure 3, Plates II through IV). The near-marine sandstones of Parasequence Set 2A extend about 27.3 km (17 mi) in the dip direction from Willow Springs Wash to Bitter Seep Wash. Parasequence Set 2B contains only Parasequence 2e that is given hierarchical equivalence to a parasequence set. Parasequence 2e is retrogradational relative to underlying Parasequence 2d. Parasequence 2e is severely truncated by fluvial erosion, such that the true vertical and lateral extent of Parasequence Set 2B cannot be ascertained. The erosional remnants of Parasequence Set 2B extend from north Quitchupah Creek to the northern reaches of Muddy

Creek Canyon. Parasequence Set 2C contains five wave-reworked parasequences (denoted 2f through 2j), some of which exhibit progradational/erosional stratigraphic relationships suggesting shoreline erosion produced by a forced regression. The near-marine sandstones of Parasequence Set 2C extend about 17.9 km (11.1 mi) in the dip direction from just north of Miller Canyon northeastward to north of Dry Wash. Parasequence Set 2D contains three wave-reworked parasequences (denoted 2k through 2m). The near-marine sandstones of Parasequence Set 2D extend about 21.3 km (13.2 mi) in the dip direction from just north of Dry Wash northeastward to Ferron Creek.

River-Dominated Parasequence Set 1A

Ferron Parasequence Set 1A contains at least three wave-reworked deltaic parasequences (denoted as Parasequences 1v, 1x, and 1y). The near-marine sandstone facies of these parasequences exhibit an aggradational to slightly progradational-stacking pattern (Gardner, 1995b). The landward limits of these near-marine parasequences are not exposed. The seaward pinchouts of these near-marine sandstones occur at Last Chance Creek. The upper bounding surface of the exposed portion of Parasequence Set 1A is a marine-flooding surface. Overlying progradational Parasequence Set 1B downlaps onto this marine-flooding surface.

The near-marine parasequences in Ferron Parasequence Set 1A have clinoformal geometry and exhibit both vertical and lateral facies changes from (1) reworked stream-mouth bar to (2) reworked proximal and distal-delta front to (3) prodelta. The delta-plain facies association is not exposed. The upper bounding surfaces of these parasequences are marine-flooding surfaces.

142

River-Dominated Parasequence Set 1B

Sedimentologic characteristics: The river-dominated near-marine facies of the parasequences in Ferron Parasequences Set 1B exhibit both vertical and lateral facies changes from (1) stream-mouth bar to (2) proximal delta front to (3) distal-delta front to (4) prodelta (Figures 4 and 5). The delta-front deposits are shale-rich, ranging from 10–30% shale (van den Bergh, 1995) (Table 1). The delta-front succession generally fines upward from silt-stone to fine-grained sandstone. Horizontal and ripple cross-stratification are the dominant bed form type. Distal prodelta/shelf slumping, indicating instability as a result of rapid deposition, is common. Contorted bedding, flame structures, escape burrows, growth faults, and the cannibalizing of distal stream-mouth bar by proximal stream-mouth bar (or distributary) channels also suggests an overall rapid rate of deposition (van den Bergh, 1995). In general, Parasequence Set 1B delta-front deposits exhibit little to mild evidence of wave influence. They locally may exhibit poorly developed hummocky stratification and occasionally rare bi-directional ripple stratification.

South of Ivie Creek at Interstate 70 (I-70), distributary channels and sand-poor, delta-plain facies associations occur in parasequences of Ferron Parasequence Set 1B. Fluvial channel belts form the major sandstone lithofacies. Floodplain overbank mudstones and siltstones, crevasse-splay deposits, peat-mire facies, and brackish-water interdistributary bay deposits make up the balance of the delta-plain to transitional marine facies association. Fluvial channel belts are generally laterally restricted and multi-storied with channel-fill elements stacked vertically within the channel-belt boundaries (Garrison and van den Bergh, 1997). Channel-fill elements usually have erosional tops and incompletely preserved vertical profiles. The lower channel-fill elements frequently have bounding surfaces delineated by thick, clay-pebble lag deposits, although clay-pebble lag zones, composed of clay pebbles floating in a sand matrix, are also common. These lower channel-fill elements are generally fine upward from medium-grained sandstones to fine-grained sandstones. They generally have large-scale trough cross-stratified beds, greater than 15 cm (6 in.) thick, near their bases, which change upward into faint large troughs and massive sandstones. The trough cross-stratification decreases in size upward. The upper channel-fill elements are dominated by lateral accretion sedimentary surfaces, structures, and bed forms. The bounding surfaces between these channel-fill elements are frequently defined by bedding surfaces between major bed-form domains. These channel-fill elements exhibit a general overall fining-upward grain-size trend from medium-grained to fine-grained sandstones. Where channel-fill element tops are preserved, ripple and climbing ripple cross-stratification is common.

Stratal properties: Ferron Parasequence Set 1B contains 12 progradationally stacked, river-dominated fluvial-deltaic parasequences (denoted as Parasequences 1z, 1a through 1k) (Figure 3, Plates II and III). Within Parasequence Set 1B, the landward pinchout (i.e., the initial paleoshoreline position) of the near-marine sandstone facies of each successively younger parasequence steps seaward by an average of about 1–5 km (0.6–3 mi), relative to the landward pinchout of the near-marine facies of the immediately underlying parasequence. The stacked package of near-marine sandstones of Parasequence Set 1B extends at least 27 km (17 mi) in the dip direction from near Last Chance Creek northeastward to just north of Bear Gulch (Plate I).

In the southern part of the outcrop belt, the Sub-A coal zone marks the top of Parasequence Set 1B. The top of this coal zone is the non-marine correlative conformity to the marine-flooding surface at the top of the more distal near-marine facies of Parasequence Set 1B. In the Limestone Cliffs, the Sub-A coal zone is approximately 15 m (45 ft) thick and splits into two components denoted Sub-A_1 and Sub-A_2. The Sub-A_1 coal zone disappears in Coyote Basin and the Sub-A_2 coal zone disappears in Corbula Gulch. A coal zone stratigraphically equivalent to the Sub-A_2 reappears in Blue Trail Canyon and splits into two components near I-70, where they are designated the Sub-A_3 and Sub-A_4 (Figure 4). The Sub-A_3 and Sub-A_4 section of the Sub-A coal zone extends some 8.5 km (5.3 mi), from Blue Trail Canyon to North Quitchupah Creek. The total dip length for the Sub-A coal zone is at least 27 km (17 mi). In the Limestone Cliffs, the Sub-A_2 coal zone contains a very well-preserved volcanic ash layer. This Sub-A_2 ash layer has not been identified farther downdip to the northeast.

The upper bounding surfaces of the near-marine facies of Parasequences 1z, 1a, 1b, 1c, and 1d are either locally marine-flooding surfaces or the correlative conformity lying above the Sub-A coal zone. Parasequence 1a through 1d have landward pinchouts at Mussentuchit Wash, South Limestone Cliffs, Indian Canyon, and Coyote Basin, respectively. Parasequences 1a through 1d, each progressively steps seaward approximately 1.3 km (0.8 mi), 2.9 km (1.8 mi), and 4.9 km (3.0 mi), respectively (Table 4). The landward pinchouts of the near-marine sandstone facies of Parasequences 1a, 1b, 1c, and 1d exhibit 7.7 m (25.3 ft), 7.4 m (24.3 ft), 1.3 m (4.3 ft), and 4.4 m (14.4 ft) of stratigraphic rise, respectively, relative to the final paleoshoreline of the underlying parasequence. The near-marine facies of Parasequences 1a, 1b, 1c, and 1d are approximately 5.1 km (3.2 mi), 6.3 km (3.9 mi), 8.7 km (5.4 mi), and 7.5 km (4.7 mi), in dip length, respectively, and maximum thicknesses are 28 m (92 ft), 22 m (72 ft), 13 m (43 ft), and 17 m (56 ft), respectively.

The upper bounding surfaces of the near-marine facies of Parasequences 1e, 1f, 1g, and 1i are also either locally marine-flooding surfaces or the correlative con-

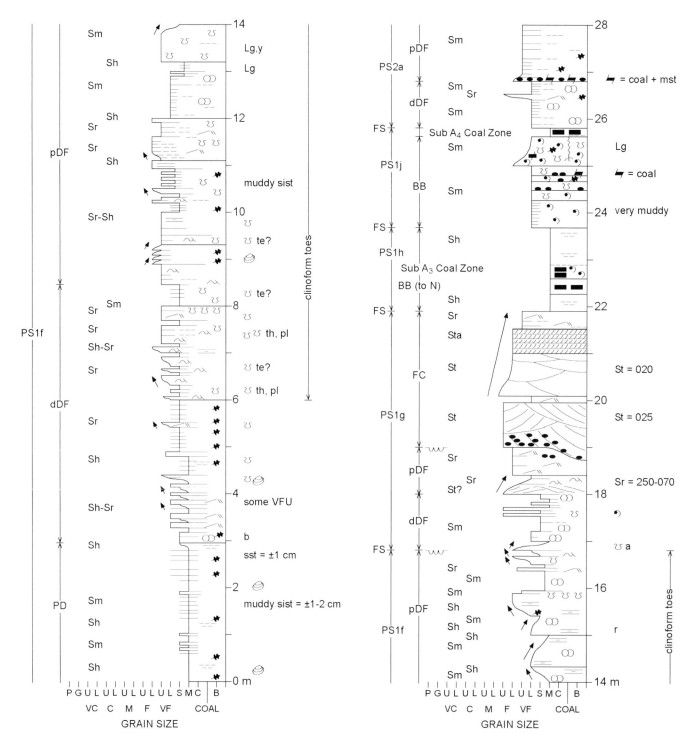

Figure 5. A portion of measured section GV5 at the Ivie Creek/I-70 roadcut showing Upper Ferron Sandstone Parasequence Set 1B.

formity lying above the Sub-A coal zone. These parasequences represent the strike cross sections of a series of laterally shifting delta lobes and therefore the landward pinchouts of their near-marine sandstone facies exhibit no stratigraphic rise relative to the final paleoshoreline of the underlying parasequence. The outcrop belt cuts these parasequences in a strike-oblique direction. Parasequences 1e, 1f, 1g, and 1i have southeastern landward pinchouts at Blue Trail Canyon, Scabby Canyon, I-

70, and near Cowboy Mine Canyon, respectively. Parasequences 1e, 1f, 1g, and 1i are approximately 2.9 km (1.8 mi), 3.7 km (2.3 mi), 2.7 km (1.7 mi), and 2.6 km (1.6 mi) in strike width, respectively, and the maximum thicknesses are 22 m (72 ft), 17 m (56 ft), 9 m (30 ft), and 13 m (43 ft), respectively.

The upper bounding surfaces of the near-marine facies of Parasequences 1h, 1j, and 1k are again, either locally marine-flooding surfaces or the correlative con-

Explanation

Sedimentary Structures and Detailed Rock Descriptions

Sh ---	Horizontal lamination in mudstone
Sh ≡	Horizontal stratification in sandstone
Sr ∿	Wave ripples
Sr ⟋	Current ripples
St	Trough cross-stratification
St$_{La}$	Large-scale trough cross-stratification
St$_S$	Small-scale trough cross-stratification
St = 090	Paleocurrent direction
Shu	Hummocky trough cross-stratification
Ssw	Swaley trough cross-stratification
Sm	Massive bedding
Sher	Herringbone trough cross-stratification
\\\\\\\	Wedge-tabular cross-stratification
↑ML-FU	Fining-upward on lamina scale
≈	Contorted bedding
∞	Bumpy weathering
�	Mudstone rip-up clasts
∿∿	Scour surface
⌣	Flutes and grooves
◑	Septarian nodules
○ ○ ○	Extrabasinal pebbles
● ● ●	Intrabasinal pebbles

Sequence Stratigraphy and Depositional Facies

SB	Depositional sequence boundary
PSS	Parasequence set
PS	Parasequence
MFS	Maximum flooding surface
FS	Flooding surface
TL	Transgressive lag
TRS	Transgressive ravinement surface
SMB	Stream-mouth bar
DP	Delta-plain facies
FC	Fluvial channel
DC	Distributary channel
PD	Prodelta
pDF	Proximal delta front
dDF	Distal delta front
BB	Brackish-bay facies
USF	Upper shoreface
MSF	Middle shoreface
LSF	Lower shoreface
OS	Offshore

Ichnofossils and Other Biostratigraphic Information

➘	Oysters
◑	Brackish bay fauna
℧	Burrow (unidentified)
℧ H	Horizontal burrow
℧ esp	Escape burrow
℧ sk	*Skolithos*
℧ te	*Teichichnus*
℧ th	*Thalassinoides*
℧ oph	*Ophiomorpha*
℧ pl	*Planolites*
℧ ch	*Chondrites*
⅄	Roots
⬳	Logs and trees
⬳ +t	Log with *Teredolites*
⊝	Bivalve

General Rock Description

sst	Sandstone
sist	Siltstone
▨	Altered volcanic ash layer
■	Coal
➖➖	Almost coal
mst ---	Mudstone
▬ ▬	Carbonaceous shale
✶	Carbonaceous plant fragments
▬	Coal fragment
C	Clay grain size
M	Mudstone grain size
S	Silt grain size
VFL	Very fine lower grain size
VFU	Very fine upper grain size
FL	Fine lower grain size
FU	Fine upper grain size
ML	Medium lower grain size
MU	Medium upper grain size
CL	Coarse lower grain size
CU	Coarse upper grain size
G	Granule grain size
P	Pebble grain size
B	Altered volcanic ash deposit (including tonsteins and bentonites)
➘	Coarsening-upward grain size
➚	Fining-upward grain size

b	Brown	y	Yellow
bl	Black	gr	Green
g	Gray	L	Light color
r	Red	d	Dark color

Figure 5 continued.

formity lying above the Sub-A coal zone. The landward pinchouts of the near-marine sandstone facies of Parasequences 1h, 1j, and 1k exhibit 2.4 m (7.9 ft), 10.1 m (33.2 ft), and 1.3 m (4.3 ft) of stratigraphic rise, respectively, relative to the final paleoshoreline of the underlying parasequence. The dip lengths of 1h and 1k are both approximately 2.6 km (1.6 mi) and maximum thicknesses are 14 m (46 ft) and 7 m (23 ft), respectively.

Interpretation: In general, Parasequence Set 1B delta-front deposits are extremely river-dominated and exhibit little to mild evidence of wave influence. High sedimentation rates are implied by the sedimentary structures, such as contorted bedding, flame structures, escape burrows, and by the frequent development of growth faults. The geometry and size of the delta lobes, the development of steep clinoforms, and the existence of multiple progradation directions also suggest rapid sedimentation rates accompanied by frequent local avulsions and delta lobe abandonment. The reconstructed geometry of the fluvial-deltaic system suggests an overall finger-like, birdsfoot-type river-dominated delta.

Parasequences 1a through 1d represent a series of small delta lobes which prograded seaward 3.3 km (2 mi), 4.4 km (2.7 mi), 5.3 km (3.3 mi), and 3.5 km (2.2 mi), respectively, in a northeasterly direction (025°-040° azimuth). The outcrop belt is approximately along the depositional dip direction. Parasequence 1c is a composite delta lobe with two to three preserved mouth bars.

Parasequences 1e, 1f, 1g, and 1i formed in response to the northwesterly (310°-335° azimuth) progradation of very small, river-dominated sub-delta lobes. Parasequence 1f appears to be a composite delta with two preserved mouth bars. Parasequence 1e, 1f, 1g, and 1i represent four laterally shifting delta lobes. Each lobe represents a slight northward shift in deltaic deposition relative to the underlying parasequence.

Parasequences 1h and 1k represent small delta lobes that prograded seaward in a northeasterly direction 1.4 and 1.7 km (0.9 and 1.1 mi), respectively. Parasequence 1k pinches out, landward, into a split of the Sub-A coal zone at I-70. Parasequence 1h pinches out, landward in the Ivie Creek amphitheater.

Parasequence 1j, occurs within a split in the Sub-A coal zone, and is represented in the outcrop belt as a brackish-water bay mudstone, to the south near I-70. To the north in Quitchupah Creek Canyon, Parasequence 1j becomes a small 3-km-long (2-mi), 6-m-thick (20-ft) delta-front sandstone body.

Sedimentological Characteristics of Wave-Reworked Parasequence Sets 2A, 2B, 2C, and 2D

The near-marine parasequences in Ferron Parasequence Set 2A, 2B, 2C and 2D all exhibit both vertical and lateral facies changes from (1) reworked stream-mouth bar to (2) reworked proximal and distal-delta

Figure 6. Upper Ferron Sandstone Parasequence Set 2A at the I-70 roadcut.

front to (3) prodelta (Figures 6 and 7). In general, reworked stream-mouth bar and delta-front deposits of Parasequence Sets 2A through 2C, are very sand-rich and exhibit evidence of moderate wave influence (Table 1). Parasequence Set 2D exhibits a slightly higher degree of wave modification, compared to Parasequence Sets 2A through 2C. The wave-reworked stream-mouth bar deposits are fine- to medium-grained sandstones, frequently poorly burrowed (ichnofossil *Skolithos*), and exhibit well-developed trough cross-stratification, and, less frequently, herringbone cross-stratification, suggesting the development of an incipient shoreface profile; horizontal and ripple cross-stratification are also common. The reworked delta-front deposits generally fine upward from very fine- to fine-grained sandstone and are moderately burrowed (ichnofossil *Skolithos* and *Cruziana*). They exhibit minor hummocky and swaley cross-stratification, planar stratification, and bi-directional ripple stratification.

South of Willow Springs Wash, a well-developed delta-plain facies association is present. Fluvial channel-belt facies form the dominant sandstone lithofacies of the delta-plain facies association. As observed in Parasequence Set 1B, floodplain overbank mudstone and siltstone deposits, crevasse-splay siltstone and sandstone deposits, and peat-mire facies make up the balance of the delta-plain to transitional marine facies association. Channel-fill elements in this facies generally fine upward from medium- to fine-grained sandstone and exhibit a vertical bed-form succession from large trough cross-stratification to small trough cross-stratification to wedge tabular and planar tabular cross-stratification. Where channel-fill element tops are preserved, ripple and climbing ripple cross-stratification is common.

Amalgamated fluvial deposits occur locally within Parasequence Set 2C. These amalgamated fluvial deposits are sedimentologically very distinct from the isolated, laterally restricted, multi-storied fluvial channel-belts deposits within the delta-plain facies association.

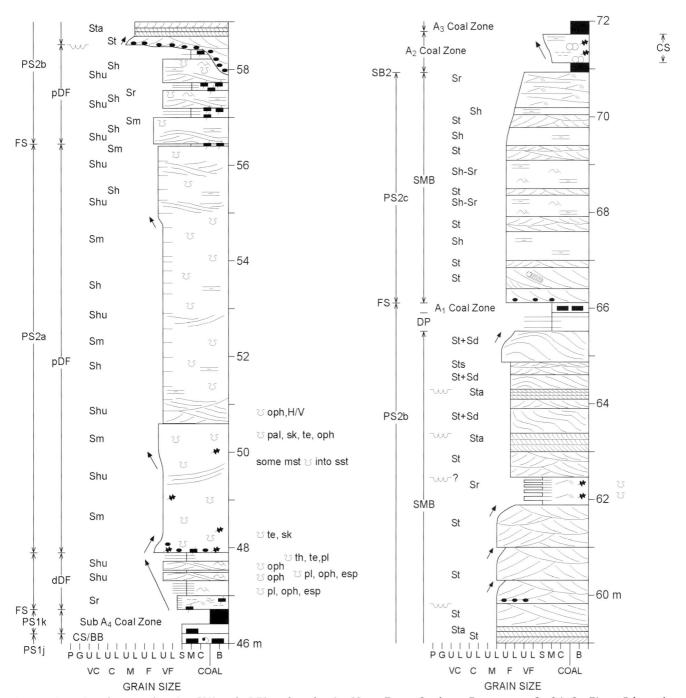

Figure 7. A portion of measured section GV3 at the I-70 roadcut showing Upper Ferron Sandstone Parasequence Set 2A. See Figure 5 for explanation.

From North Quitchupah Creek Canyon to the South Goosenecks of Muddy Creek Canyon, Parasequence Set 2C consists of a major fluvial channel-belt complex that incises up to 25 m (82 ft) deep into underlying Parasequence Sets 2A and 2B (Figures 8 and 9). South of I-70, this incision is represented as an erosional unconformity, with up to 30 m (98 ft) of topographic relief, overlain by tidally influenced fluvial sandstones, siltstones, and mudstones containing brackish-water faunas. The northern fluvial deposits appear to be amalgamated braided channel belts that exhibit complex vertical and

lateral stacking of channel-fill deposits (Garrison, 2004). They obtain a thickness of 30 m (98 ft) (Figure 8).

These amalgamated braided channel-belt deposits exhibit a multi-storied, internally scoured vertical stacking of amalgamated braided channel belts 6–8 m (20–26 ft) thick, composed of trough cross-stratified, medium-grained to fine-grained sandstone. Individual channel belts may contain a thin abandonment phase consisting of ripple cross-stratified, very fine grained sandstone channel barforms. These channel-belt barforms have scour bases and range in thickness from 1.5–2 m (4.9–6.6

Figure 8. Tri-Canyon area showing the erosional unconformity at the base of Upper Ferron Sandstone, Parasequence Set 2C fluvial valley fill and the incision into the near-marine deposits of Parasequence Sets 2A and 2B.

ft). The barforms have a sigmoidal shape in the dip direction and are channel-form in cross section. Locally, these deposits contain heterolithic channel-fill consisting of very fine to fine-grained, trough cross-stratified to rippled sandstones interlayered with laminated mudstone containing very thin lamina of very fine grained sandstone. The sandstone layers range in thickness from 10–60 cm (4–24 in.) and the mudstone layers range from 5–35 cm (2–14 in.) in thickness. Trough cross-beds commonly contain climbing ripples on their surfaces. The sandstones frequently contain burrows (ichnofossils *Scoyenia* and *Skolithos)* and fragments of coaly material.

Locally within Parasequence Sets 2C and 2D, there is evidence of tidal reworking and the development of tidally influenced fluvial deposits. A small tidal-channel complex has also been documented in Parasequence 2a in Coyote Basin (Plate II). These tidally influenced fluvial channel belts consist of sandstone and heterolithic channel-fill deposits containing brackish-water molluscan faunas (e.g., oysters) and are frequently burrowed by marine ichnofossils *Ophiomorpha, Thalassinoides, Skolithos*, and *Teichichnus*. Locally, cross-stratification indicates bimodal current directions. In the Miller Canyon area, tidally influenced fluvial channel deposits contain small tidal bundles interlayered with thin mud drapes.

In Parasequence Set 2C, there are heterolithic, locally carbonaceous, bay/lagoonal deposits that contain a brackish-water molluscan fauna. These bay/lagoonal deposits occur adjacent to tidally influenced fluvial and distributary channel-belt sandstones. The bay deposits consist of mudstone, interlayered silty mudstone and siltstone, and massive to poorly cross-stratified, intensely burrowed, fine-grained sandstone containing very

thin lamina of mudstone and siltstone. The sandstone layers are burrowed, horizontal to ripple laminated, and exhibit syndepositional deformation. The burrowing in the sandstone is generally so intense that the mudstone and siltstone lamina have been reworked into the sandstone by the burrowing. The sandstone facies contains fragments of coal, wood (containing *Teredolites* burrows), and oysters.

In the area between Bear Gulch and Miller Canyon, small flood-tidal delta deposits occur in the A_2-A_3 coal zone (Figure 9). The best exposed of these flood-tidal delta deposits occurs in the Tri-Canyon area north of Bear Gulch. This flood-tidal deltaic deposit is a lenticular sandstone body containing small trough, planar and wedge tabular, and herringbone cross-stratification, and frequently the ichnofossils *Ophiomorpha* and *Thalassinoides*. In the Dry Wash amphitheater, washover fan deposits, interfingered with lagoonal facies within the A_2-A_3 coal zone, can be traced seaward into the reworked stream-mouth bar/upper shoreface facies of Parasequences 2k-2m of Parasequence Set 2D. These washover fan deposits are wedge-shaped sandstone bodies. They are commonly intensely burrowed, with little of the original sedimentary structures preserved, although in the seaward extents of these fans, well-developed trough and planar tabular cross-stratification occurs. These washover fan deposits contain the ichnofossils *Ophiomorpha* and *Thalassinoides*, and are frequently rich in carbonaceous material.

Wave-Reworked Parasequence Set 2A

Stratal properties: Parasequences 2a, 2b, 2c, and 2d have their landward pinchouts in Willow Springs Wash, Rock Canyon, I-70, and South Quitchupah Creek Canyon,

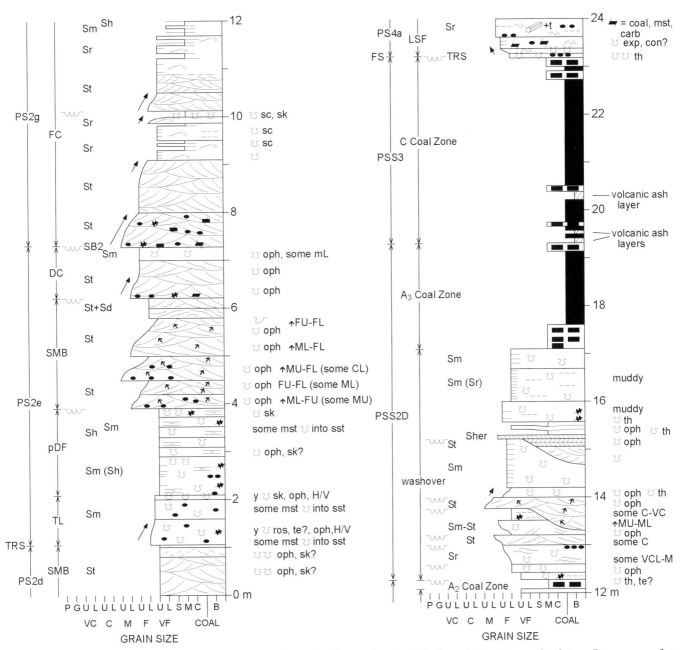

Figure 9. Measured section GV8a in Bear Gulch showing the erosional unconformity at the base of Upper Ferron Sandstone, Parasequence Set 2C fluvial valley fill and the incision into the near-marine deposits of Parasequence Sets 2A and 2B. See Figure 5 for explanation.

respectively. The near-marine facies of Parasequences 2a, 2b, 2c, and 2d are approximately 17.9 km (11.1 mi), 15.0 km (9.3 mi), 13.1 km (8.1 mi), and 14.2 km (8.8 mi) in dip length, respectively and maximum thicknesses are 42 m (138 ft), 18 m (59 ft), 8 m (26 ft), and 22 m (72 ft), respectively. The landward pinchouts of the near-marine facies of Parasequences 2b, 2c, and 2d step seaward 3.5 km (2.2 mi), 7.2 km (4.5 mi), and 2.7 km (1.7 mi), respectively, relative to the landward pinchout of the near-marine facies of the immediately underlying parasequence. The landward pinchouts of the near-marine sandstone facies of Parasequences 2a, 2b, 2c, and 2d exhibit 7.1 m (23.3 ft), 0.7 m (2.2 ft), 0.7 m (2.2 ft), and 1.3 m (4.3 ft) of strati-

graphic rise, respectively, relative to the final paleo-shoreline position of the underlying parasequence.

The upper bounding surface of the near-marine facies of Parasequence 2a is marine-flooding surface, north of Rock Canyon. South of Rock Canyon, the upper bounding surface of the near-marine facies of Parasequence 2a is the sequence boundary erosional unconformity at the base of Parasequence Set 2C (denoted as SB2 in Plate II).

The upper bounding surface of the near-marine facies of Parasequence 2b is marine-flooding surface, north of I-70. Between I-70 and Scabby Canyon, the upper bounding surface is the correlative conformity at the top of the A_1 coal zone. South of Scabby Canyon, the upper bounding surface of the near-marine facies of Parasequence 2b is the sequence boundary erosional unconformity at the base of Parasequence Set 2C.

The upper bounding surface of the near-marine facies of Parasequence 2c is a marine-flooding surface, north of Quitchupah Creek. Between I-70 and Quitchupah Creek, the upper bounding surface is a correlative conformity near the base of the A_2 coal zone. South of I-70, the delta-plain facies association of Parasequence 2c has been removed by the sequence boundary erosional unconformity at the base of Parasequence Set 2C.

North of Rochester Creek, the upper bounding surface of the near-marine facies of Parasequence 2d is a marine-flooding surface. Between Bear Gulch and Rochester Creek, the upper bounding surface of the near-marine facies of Parasequences 2d is a transgressive ravinement surface overlain by a thick transgressive lag deposit (Plate III). Between I-70 and Quitchupah Creek, the upper bounding surface is a correlative conformity near the base of the A_2 coal zone. South of I-70, the delta-plain facies association of Parasequence 2d has been removed by the sequence boundary erosional unconformity at the base of Parasequence Set 2C. Between the Tri-Canyon area of Miller Canyon and the South Goosenecks of Muddy Creek Canyon, the upper bounding surface of Parasequence 2d is the sequence boundary erosional unconformity at the base of Parasequence Set 2C.

South of Willow Springs Wash, the correlative conformity to the marine-flooding surface at the top of Parasequences 2a through 2d cannot be traced into the exposed delta-plain facies associations.

Interpretation: In general, Parasequence Set 2A delta-front deposits are river-dominated, but exhibit moderate evidence of wave influence. The reconstructed geometry of these fluvial-deltaic systems suggests an overall lobate wave-reworked, river-dominated delta system. Preserved sand thicknesses, moderate shoreline progradation distances, and stratigraphic rises during shoreline progradation suggest high sedimentation rates for these deltaic systems.

Parasequences 2a and 2b appear to represent large wave influenced, river-dominated delta lobes that prograded seaward in a northeasterly direction. South of I-70, Parasequence 2a prograded northeast (34° azimuth) (van den Bergh, 1995). In the vicinity of I-70 and Ivie Creek, Parasequence 2a prograded northeast (25° azimuth). Parasequence 2a appears to have prograded seaward at least 11.8 km (7.3 mi). Parasequence 2b is actually a composite delta lobe with a more easterly progradation direction. In the vicinity of Scabby Canyon and I-70, Parasequence 2b has a northeast (45° azimuth) progradational direction. The near-marine facies of Parasequence 2b prograded seaward at least 11.1 km (6.9 mi).

Near-marine Parasequence 2c appears to represent a relatively small delta lobe and a shoreline progradation to the northeast (25° azimuth) of approximately 4.9 km (3.0 mi). The delta-front sandstones contain well-developed growth faults from Bear Gulch to Muddy Creek Canyon, suggesting rapid sedimentation onto an unstable substrate. Parasequence 2c exhibits only a slight amount of wave reworking, relative to other parasequences in Parasequence Set 2A. The small amount of stratigraphic rise (i.e., delta-plain aggradation) exhibited by Parasequence 2c (Table 4) during shoreline progradation suggests rapid progradation into relatively shallow water.

Parasequence 2d represents a major delta lobe that prograded to the northeast (50° azimuth) approximately 9.4 km (5.8 mi). Parasequence 2d appears to contain at least four preserved mouth-bar complexes. The overall geometry and the sedimentary characteristics of the near-marine deposits of Parasequence 2d suggest a high sedimentation rate and a moderate degree of wave reworking of the delta-front deposits.

Wave-Reworked Parasequence Set 2B

Parasequence Set 2B is represented by the erosional remnants of a single parasequence (i.e., Parasequence 2e). Parasequence 2e has its landward pinchout, in Quitchupah Creek Canyon, about 3.2 km (2.0 mi) seaward from the pinch-out of Parasequence 2d. The overall preserved length of Parasequence 2e is 6.2 km (3.9 mi). In Bear Gulch, the base of Parasequence 2e is a highly bioturbated transgressive lag deposit resting on a transgressive ravinement surface (Figure 9). This transgressive lag deposit can be traced over 3 km (1.9 mi) northward into Muddy Creek Canyon, where it goes into the subsurface. Parasequence 2e has a very small stream-mouth-bar deposit extending from south of Anderson Ranch to just north of Bear Gulch. In Muddy Creek Canyon, Parasequence 2e has distal delta-front and prodelta facies lying stratigraphically above the transgressive lag and the stream-mouth-bar deposits of Parasequence 2d, suggesting a moderate landward migration of the initial paleoshoreline relative to Parasequence 2d (Figure 9). Regional mapping indicates that the initial paleoshoreline of Parasequence 2e is approximately 6.3 km (3.9 mi) landward of the final paleoshoreline position of Parasequence 2d. Based on these observations, Parasequence 2e, which appears to be retrogradational relative to Parasequence 2d, has been assigned to the stratigraphically higher retrogradational Parasequence Set 2B. Due to severe degrees of erosional truncation by overlying fluvial deposits, it cannot be determined whether additional parasequences existed in Parasequence Set 2B, above Parasequence 2e, prior to Parasequence Set 2C fluvial erosion and the development of Sequence Boundary SB2.

Wave-Reworked Parasequence Set 2C

Stratal properties: The sand-rich wave-modified, river-dominated, near-marine facies of Parasequence Set 2C, represented by Parasequences 2f and 2g, have their landward pinchouts in South Muddy Creek Canyon and Muddy Creek Canyon, respectively (Plates I and III). Parasequences 2f and 2g are slightly retrogradationally stacked, with the initial paleoshoreline of Parasequence 2g 0.7 km (0.4 mi) landward of the initial paleoshoreline of Parasequence 2f. The near-marine facies of Parasequences 2f and 2g, are approximately 6.7 km (4.2 mi) and 9.8 km (6.1 mi) in dip length, respectively, and have maximum thicknesses of 19 m (62 ft) and 17 m (56 ft), respectively. The initial paleoshoreline of Parasequence 2g exhibits about 1 m (3 ft) of stratigraphic rise relative to final paleoshoreline of Parasequence 2f.

The upper bounding surface of the near-marine facies of Parasequence 2f is a marine-flooding surface, north of Rochester Creek. Between Rochester Creek and the North Goosenecks of Muddy Creek Canyon, the upper bounding surface of the near-marine facies of Parasequence 2f is a transgressive ravinement surface overlain by a thin transgressive lag. From the North Goosenecks of Muddy Creek Canyon to the Tri-Canyon area of Miller Canyon, the upper boundary of Parasequence 2f is a flooding surface overlain by brackish-bay deposits. Farther south, the correlative surface cannot be traced into the amalgamated fluvial deposits of Parasequence Set 2C. North of Bitter Seep Wash, the upper bounding surface of the near-marine facies of Parasequence 2g is an erosional unconformity overlain by sharp-based, wave-reworked delta-front/shoreface deposits. From Bitter Seep Wash to Rochester Creek, the upper bounding surface of Parasequence 2g is a flooding surface overlying the thin A_2 coal zone. From Rochester Creek to Muddy Creek Canyon, the upper surface is an erosional unconformity at the base of a small incised-valley system (Garrison, 2004). South of Muddy Creek Canyon, the upper surface is a correlative conformity within the A_2 coal zone.

The sand-rich, wave-modified, river-dominated, near-marine facies of Parasequence Set 2C, represented by Parasequences 2h, 2i, and 2j, have their landward pinchouts north of Bitter Seep Wash, and south of Bitter Seep Wash, respectively. Parasequence 2h represents a seaward step of 8.8 km (5.5 mi), while Parasequence 2i is stepped landward 1.7 km (1.1 mi) relative to the underlying Parasequence 2h. The near-marine facies of 2h, 2i, and 2j are approximately 3.2 km (2.0 mi), 7.5 km (4.7 mi), and 3.5 km (2.2 mi) in dip length, respectively, and maximum thicknesses are 8 m (26 ft), 15 m (49 ft), and 13 m (43 ft), respectively. The landward pinchouts of the near-marine sandstone facies of Parasequences 2i and 2j exhibit 6.0 m (20 ft) and 3.0 m (10 ft) of stratigraphic rise, respectively, relative to the final paleoshoreline position of the underlying parasequence.

The upper bounding surface of the near-marine facies of Parasequence 2h is a marine-flooding surface. North of North Cedar Ridge Canyon, the upper bounding surface of the near-marine facies of Parasequence 2i is also a marine-flooding surface. From North Cedar Ridge Canyon to Bitter Seep Wash, the upper boundary of Parasequence 2i is a marine-flooding surface at the top of the thin A_2 coal zone. South of Bitter Seep Wash, the upper boundary of Parasequence 2i is a correlative conformity at the top of the A_2 coal zone. North of Castle Peak, the upper bounding surface of the near-marine facies of Parasequence 2j is a marine-flooding surface. South of North Cedar Ridge Canyon, the upper bounding surface of Parasequence 2j is a correlative conformity at the top of A_2 coal zone.

Interpretation: From North Quitchupah Creek Canyon to the South Goosenecks of Muddy Creek Canyon, Parasequence Set 2C consists of a major fluvial channel-belt complex that incises up to 25 m (82 ft) deep into underlying Parasequence Sets 2A and 2B (Figures 8 and 9). South of I-70, this incision is represented as an erosional unconformity, with up to 30 m (98 ft) of topographic relief, overlain by tidally influenced fluvial sandstones, siltstones, and mudstones, containing brackish-water faunas. The northern fluvial deposits are amalgamated channel belts that exhibit complex vertical and lateral stacking of braided stream channel-fill deposits. These fluvial incisions are interpreted to represent the development of two small incised-valley systems and the development of a tripartite valley-fill facies association (see Dalrymple et al., 1994; Zaitlin et al., 1994, for a discussion of the tripartite classification of valley-fill deposits). The paleovalley system appears to bifurcate west of the outcrop belt and be expressed at the inferred paleoshoreline as two valleys. The incised-valley system is oriented approximately 60°-70° azimuth. At the paleoshoreline, the northern incised valley is approximately 6 km (4 mi) wide and the southern valley is approximately 10 km (6 mi) wide (Garrison, 2004). Down dip, the northern valley-fill amalgamated fluvial deposits feed more isolated channel belts that probably represent distributary channel-belt deposits. As these valley-fill deposits are traced toward the near-marine deposits at the paleoshoreline, there is clear, well-developed influence by near-marine and tidal processes in distributary-channel deposits. In addition, as much as 1–2 km (0.6–1.2 mi) inland from the paleoshoreline, inter-fluvial incised-valley areas consist of heterolithic, locally carbonaceous, brackish-bay/lagoonal deposits containing a brackish-water molluscan fauna. These deposits have been interpreted to represent estuarine conditions. West of the exposed incised-valley deposits in Muddy Creek Canyon, cores have penetrated subsurface rock units that are interpreted as bay-head delta deposits (Garrison, 2004).

Detailed three-dimensional cross sections have been constructed through the Parasequence Set 2C incised-valley fill between North Quitchupah Creek Canyon, near the Emery Mine and the South Goosenecks of Muddy Creek Canyon (Garrison, 2004). These cross sections suggest that the inner segment (Zaitlin et al., 1994) of the incised-valley fill contains fluvial-fill facies representing initial amalgamated lowstand braided stream fluvial facies overlain by highstand fluvial deposits; transgressive deposits were not identified. The middle segment consists of amalgamated lowstand braided stream fluvial facies overlain by brackish-bay and bay-head delta transgressive deposits, which are in turn overlain by highstand-fluvial and interfluvial-bay facies. The outer segment of the incised-valley fill consists of small lowstand wave-reworked, delta-front sand bodies (Parasequence 2f) overlain by transgressive estuarine bay facies; locally the underlying lowstand facies are truncated by wave ravinement; the HST consists of wave-reworked stream-mouth-bar facies (Parasequence 2g).

From Rochester Creek to Muddy Creek Canyon, a small 5th-order incised-valley system developed into the reworked delta-front deposits of Parasequence 2g (Plate III). This small incised-valley system supplied sediment to the small deltas that produced the near-marine deposits of Parasequences 2h, 2i, and 2j (Plate III).

The amalgamated fluvial incised-valley deposits are interpreted to be associated with the development of an erosional unconformity representing a 4th-order sequence boundary (here denoted as SB2) (Figure 3, Plates II through IV). This erosional unconformity is considered to be a depositional sequence boundary because: (1) it represents a basinward shift in depositional facies, (2) sub-regional extent, (3) the internal consistency of systems tracts above and below the unconformity, (4) the erosional incision suggests a base-level fall of at least 16 m (52 ft), and (5) the nature of the inferred incised-valley and estuarine depositional systems immediately overlying the unconformity. These fluvial systems mark a basinward shift of at least 3 km (2 mi) in the paleoshoreline and a base-level fall of about 16 m (52 ft). Parasequence Set 2C incised-valley erosion cuts completely through underlying Parasequence 2e and down into the distal delta-front deposits of Parasequence 2d (Plate III). The areal extent of the erosional unconformity is estimated to be at least 250 km^2 (96 mi^2). The systems tracts immediately above the erosional unconformity have been interpreted to represent well-defined 5th-order lowstand, transgressive, and highstand incised-valley/estuarine deposits (Garrison, 2004). Fourth-order depositional sequence boundary SB2 is coincident with the 3rd-order depositional sequence boundary at the base of the Ferron Composite Sequence (Garrison, 2004). These boundaries will be discussed in more detail later.

Parasequences 2g, 2h, and 2i appear to exhibit stratigraphic relationships commonly associated with forced regressions (Posementier et al., 1992). Parasequences 2h and 2i were both deposited stratigraphically below the final shoreline position of the underlying Parasequence 2g. Along the Molen Reef outcrop belt near Bitter Seep Wash, the position of the near-marine facies of Parasequence 2h reflect a 12 m (39 ft) stratigraphic drop in paleoshoreline position and subsequent erosion into the reworked distal delta-front deposits of Parasequence 2g. The initial paleoshoreline of Parasequence 2i is 1.7 km (1.1 mi) landward of the initial paleoshoreline of the underlying Parasequence 2h. Parasequence 2i appears to have been deposited conformably on Parasequence 2h and has its stratigraphic top at the same stratigraphic level as the top of Parasequence 2g, reflecting a stratigraphic rise in shoreline position and complete highstand infilling of the initial erosional topography developed on Parasequence 2g. The cross section in Plate III suggests that, between the North Goosenecks of Muddy Creek and the Muddy Creek Graveyard, a second incised-valley system developed, incising into Parasequence 2g shoreline deposits, and supplied sediment to Parasequences 2h and 2i. This situation suggests a 5th-order fall in base level following the initial fall that resulted in the development of the erosional unconformity associated with the erosional lower boundary of Parasequence Set 2C (i.e., 3rd- and 4th-order Sequence Boundary SB2).

Wave-Reworked Parasequence Set 2D

Stratal properties: Parasequence Set 2D exhibits a progradational external stacking pattern and a slightly progradational to aggradational internal parasequence stacking pattern. The sand-rich wave-reworked, near-marine facies of Parasequences 2k, 2l, and 2m, all have their landward pinchouts in the Dry Wash amphitheater area. The near-marine facies of Parasequences 2k, 2l, and 2m are approximately 8.6 km (5.3 mi), 21.2 km (13.2 mi), and 16.0 km (9.9 mi) in dip length, respectively, and maximum thicknesses are 10 m (33 ft), 10 m (33 ft), and 9.7 m (31.8 ft), respectively. The landward pinchouts of the near-marine sandstone facies of Parasequences 2k, 2l, and 2m exhibit 6.2 m (20.3 ft), 7.3 m (23.9 ft), and 2.6 m (8.5 ft) of stratigraphic rise, respectively, relative to the final paleoshoreline position of the underlying parasequence.

The upper bounding surfaces of the near-marine facies of Parasequences 2k and 2l are marine-flooding surfaces, north of the Dry Wash amphitheater. From the Dry Wash amphitheater to Rochester Creek, the upper boundary of Parasequences 2k is a correlative surface within brackish-bay deposits. Farther south, the correlative surface extends into the A$_3$ coal zone. South of Dry Wash, the upper boundary of Parasequence 2l is a correlative surface within the A$_3$ coal zone.

Locally, the reworked stream-mouth bar/upper shoreface and reworked proximal delta-front/middle shoreface facies of Parasequence 2m have been removed by a transgressive ravinement surface and are overlain by a transgressive lag deposit from Dry Wash to Ferron Creek. Comparisons of the geometries of facies and the thickness of Parasequence 2m, relative to other parasequences in Parasequence Set 2D, suggest that up to 5 m (16 ft) of erosion may have occurred during the transgressive ravinement event. This transgressive lag can be traced over 24 km (15 mi) from just south of Dry Wash to north of Ferron Creek. South of Dry Wash, the upper boundary of Parasequence 2m is a correlative surface at the top of the A_3 coal zone.

Interpretation: The near-marine facies of Parasequences 2k, 2l, and 2m appear to represent wave-reworked deltaic deposits. The degree of wave-modification is significant in some areas and a shoreface profile is developed. The paleoshoreline of these parasequences never transgressed farther southwest than Dry Wash. During the development of Parasequences 2k, 2l, and 2m, the paleoshorelines prograded seaward 5.5 km (3.4 mi), 6.6 km (4.1 mi), and 6.8 km (4.2 mi), respectively.

Locally, in the Miller Canyon and Dry Wash areas, the upper part of Parasequence Set 2D contains small flood-tidal delta and washover fan deposits (Plates III and IV). These deposits are intimately associated with the A3 coal zone and appear to be related to the extensive lagoonal facies landward of the paleoshorelines of Parasequences 2k through 2m. The most extensive lagoonal facies appears to be associated with Parasequence 2l. Regional mapping suggests that the bayline of Parasequence 2l was over 16 km (10 mi) landward of the paleoshoreline.

The A Coal Zone and Its Relationship to Parasequence Sets 2A Through 2D

The A coal zone extends over 40 km (25 mi), from the Limestone Cliffs to just north of Dry Wash. Figures 10 and 11 show the general character of the A coal zone in Cowboy Mine Canyon. South of Blue Trail Canyon, the A coal zone consists of two splits denoted A_2 and A_4. In Blue Trail Canyon, the A coal zone splits into three components denoted A_1, A_2, and A_4; the stratigraphically lower A_1 coal zone disappears near I-70; the A_2 and A_4 coal zones continue into Quitchupah Creek Canyon where they become one coal zone denoted as the A coal zone. In the region from just south of Bear Gulch to Muddy Creek, the A coal zone lies apparently conformably below the C coal zone. Volcanic ash layers in the A coal zone provide data to assign splits of the A coal zone to parasequence sets. Through most of the outcrop belt, Parasequence Set 2C is capped by the A_2 split of the A coal zone (Figure 3, Plates II and III). The A_2 split of the

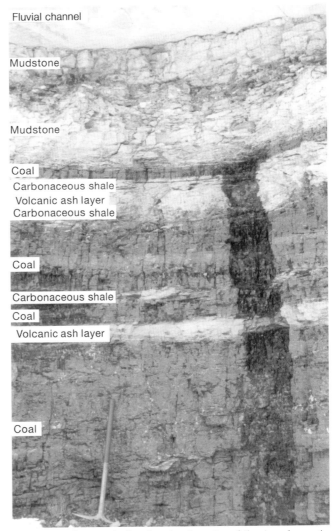

Figure 10. Photograph showing the character of the A coal zone in Cowboy Mine Canyon.

A coal zone contains an altered volcanic ash layer that can be traced over 26 km (16 mi) from the Limestone Cliffs to the Muddy Creek amphitheater. North of Dry Wash, Parasequence Set 2D is overlain by the A_3 split of the A coal zone (Figure 3, Plates III and IV). The volcanic ash layer associated with the A_3 split of the A coal zone can be traced 35 km (22 mi) from Indian Canyon to north of Dry Wash. The A_4 coal zone from south of Blue Trail Canyon to Quitchupah Creek appears to be a stratigraphically higher split of the A_3 coal zone. The A_1 split of the A coal zone appears to be related to the underlying Parasequence Set 2A. Most of the coal deposited at the top of Parasequence Sets 2A and 2B was probably subsequently removed during the development of the erosional unconformity at the base of Parasequence Set 2C.

Wave-Reworked Parasequence Set 3

Parasequence Set 3 contains three wave-modified, river-dominated parasequences, denoted Parasequences

153

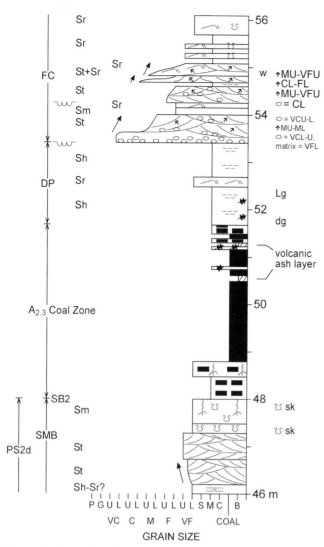

Figure 11. A portion of measured section GV10 in Cowboy Mine Canyon showing the character of the A coal zone. See Figure 5 for explanation.

3a, 3b, and 3c, that exhibit a slightly progradational internal stacking pattern (Figure 3, Plates III and IV). The near-marine sandstones of Parasequence Set 3 extend about 22 km (14 mi) along the outcrop belt. The internal stacking pattern is difficult to evaluate because the outcrop belt cuts Parasequence Set 3 oblique to depositional dip. The landward pinchout of Parasequence Set 3 appears to have stepped seaward only about 6–7 km (4 mi) relative to the bayline position of underlying Parasequence Set 2D, although this is difficult to determine because of the radically different progradation direction of Parasequence 3, relative to other Last Chance Delta parasequence sets.

South of Bitter Seep Wash, the upper boundary of Parasequence Set 3 is the top of the C coal zone. Figure 12 shows the character of the C coal zone at the I-70 roadcut. The C coal zone extends over 35 km (22 mi) from the Limestone Cliffs to its seaward pinchout in

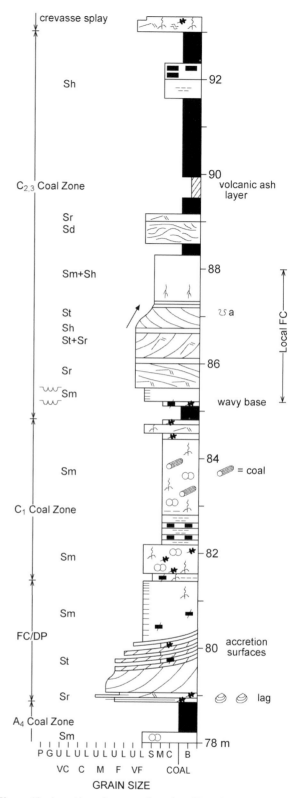

Figure 12. A portion of measured section GV3 showing the character of the C coal zone at the I-70 roadcut. See Figure 5 for explanation.

Cedar Ridge Canyon. In Willow Springs Wash, the C coal zone splits into two components denoted C_2 and C_4; the C_4 coal zone appears to not be present farther downdip than north Quitchupah Creek; the C_2 coal zone locally splits into two components near Rock Canyon, denoted C_2 and C_3, but merge at Blue Trail Canyon and splits again at Ivie Creek; the C_2 and C_3 coal zones merge again just south of Bear Gulch (Plates II and III). The C coal zone has multiple volcanic ash and tonstein layers. The C_3 coal zone ash layer can be traced from Ivie Creek to the pinchout of the C_3 coal zone in north Quitchupah Creek. The thick tonstein in the C_2 coal zone can be traced from Rock Canyon northward to northern portions of Muddy Creek, near Rochester Creek (Figure 12). Numerous smaller tonsteins and ash layers can also be correlated from north of I-70 to Muddy Creek.

Sedimentological characteristics: The near-marine parasequences in Ferron Parasequence Set 3 exhibit both vertical and lateral facies changes from (1) wave-reworked stream-mouth bar, commonly preserved as upper shoreface deposits to (2) wave-reworked delta-front deposits, most commonly preserved as middle and lower shoreface deposits to (3) prodelta deposits (Table 1) (Figure 13). The wave-reworked stream-mouth-bar deposits are fine-grained sandstones and are frequently slightly burrowed (ichnofossil *Skolithos*). These deposits exhibit well-developed, small-scale trough cross-stratification and, less frequently, herringbone cross-stratification, suggesting the development of an incipient shoreface profile. The reworked delta-front deposits generally fine upward from very fine to fine-grained sandstone and are moderately burrowed (ichnofossil *Skolithos* and *Cruziana*). They exhibit hummocky and swaley cross-stratification, planar stratification, and bi-directional ripple-stratification (Figure 13).

In the southern half of the outcrop belt, Parasequence Set 3 is represented by a non-marine, delta-plain facies association. Fluvial channel-belt facies form the dominant sandstone lithofacies of the delta-plain facies association. As observed in the other parasequence sets, the delta-plain to transitional marine facies association consists of floodplain overbank, crevasse-splay, and peat-mire deposits.

From Cedar Ridge Canyon to Castle Peak, north of Dry Wash, Parasequence Set 3 contains many accretionary tidal-channel and abandoned tidal-inlet deposits (Figures 13 and 14). The tidal channel and tidal inlets in Parasequence Set 3 have axes (i.e., axis azimuths 80–100°) that are oriented generally perpendicular to the paleoshoreline of Parasequence 3b. The best exposure of a tidal channel is on the county road, just east of the western entrance to the canyon of Dry Wash. This Parasequence 3b tidal channel, locally, completely scours through the underlying reworked stream-mouth-bar/upper shoreface facies. This tidal channel contains sigmoidal accretionary sandstone wedges interlayered with bioturbated carbonaceous mudstone and shale and fine- to very fine grained sandstone and muddy sandstone containing coal fragments. The accretion direction

of the channel has an azimuth 350°. The accretionary sandstones contain both small-scale planar tabular and trough cross-stratification bound by the inclined accretion surfaces, abundant oyster fragments, and the ichnofossil *Ophiomorpha*. The heterolithic interlayers contain the ichnofossils *Ophiomorpha*, *Teichichnus*, and *Thalassinoides*. North of Dry Wash, the abandoned tidal inlets are commonly filled with heterolithic carbonaceous sandstone and carbonaceous mudstone. These abandoned tidal-inlet deposits also contain the ichnofossil *Ophiomorpha*.

Stratal properties: The landward pinchouts of the near-marine sandstones of Parasequences 3a and 3b are in the Molen amphitheater, but due to the orientation of the paleoshoreline, the landward pinchouts project to the cross section of Garrison and van den Bergh (1997) near Rochester Creek. The near-marine sandstone of Parasequence 3c has its landward pinchout just north of Dry Wash. Along the cross section of Garrison and van den Bergh (1997), Parasequences 3a, 3b, and 3c have near-marine facies that are 8.9 km (5.5 mi), 30.9 km (19.2 mi), and 30 km (19 mi) in oblique dip length, respectively. Parasequences 3a and 3b have maximum thicknesses of 8 m (26 ft) and 21 m (69 ft), respectively. Parasequence 3a steps landward 14 km (9 mi) relative to the youngest parasequence in Parasequence Set 2C. Parasequence 3b steps seaward less than 1 km (0.6 mi) relative to Parasequence 3a.

The upper bounding surfaces of the near-marine facies of Parasequences 3b is a marine-flooding surface, north of Bitter Seep Wash. Farther south, the upper boundary of Parasequence 3b is a correlative surface at the top of the C coal zone.

Parasequence 3a and 3b appear to exhibit the erosional stratigraphic relationship commonly associated with forced regressions (Posementier et al., 1992). Parasequence 3b appears to have been deposited stratigraphically below the final shoreline position of the underlying Parasequence 3a. Along the Molen Reef outcrop belt near the Miller Canyon area, the upper boundary of the near-marine facies of Parasequence 3a reflects approximately 5 m (16 ft) of erosion down into the reworked stream-mouth-bar deposits of Parasequence 3a. Parasequence 3b is deposited onto this wave-ravinement surface. Farther south, the upper boundary of Parasequence 3a is a correlative conformity within the C coal zone.

Parasequence 3c is severely truncated by a transgressive ravinement surface and is overlain by a transgressive lag deposit. Most of the reworked stream-mouth-bar deposits (including upper shoreface deposits) and reworked proximal delta-front deposits (including middle shoreface deposits) have been removed by the transgressive ravinement event. Parasequence 3c has a preserved thickness of only 2.7 m (8.9 ft).

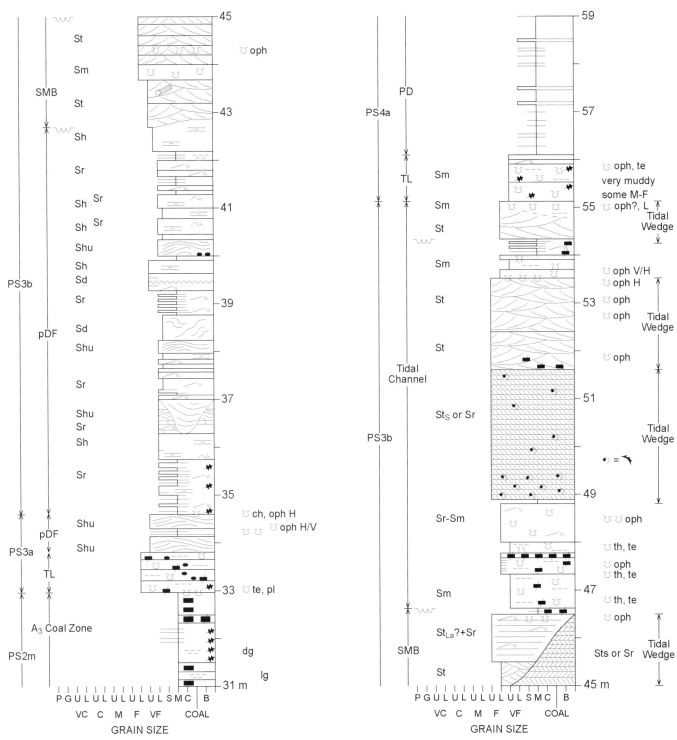

Figure 13. A portion of measured section GV22 showing the tidal-inlet deposits of Upper Ferron Sandstone, Parasequence 3b in Dry Wash, cutting into the near-marine deposits. See Figure 5 for explanation.

Regional data suggests that 5–10 m (16–33 ft) of ravinement may have occurred. The transgressive lag deposit lying on this transgressive ravinement surface can be traced over 38 km (24 mi) from the Tri-Canyon area to north of Ferron Creek. In the Tri-Canyon/Miller Canyon area, the transgressive ravinement removes over 1 m (3 ft) of coal and lies at the top of the thick altered volcanic ash (tonstein) layer within the C_2 coal zone. At Ferron Creek, the volcanic ash layer lies 0.75 m (2.5 ft) above the ravinement surface, within the top 20 cm (8 in.) of the transgressive lag deposit. These relationships indicate the timing of the ravinement event and the amount of non-marine deposition occurring during the transgression.

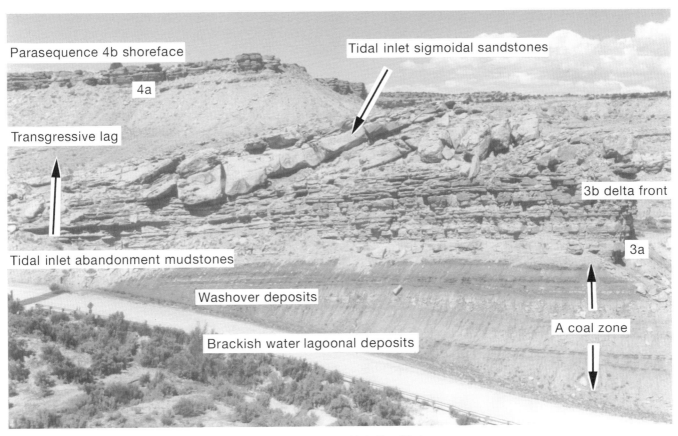

Figure 14. Tidal-inlet deposits of Upper Ferron Sandstone, Parasequence 3b in Dry Wash.

Interpretation: Parasequences 3a and 3b represent wave-modified delta lobes with a general east-north-eastern progradational direction (75° azimuth). This progradation direction is more east oriented than that of underlying and overlying parasequence sets that have a 25°-50° progradation direction. The reasons for this change in paleoshoreline orientation are speculative. Thicknesses and facies patterns suggest deposition of the near-marine facies of Parasequences 3a and 3b into relatively shallow water.

The degree of wave-modification is greater than that exhibited by Parasequence Sets 2A through 2C, but similar to that exhibited by Parasequence Set 2D. The widespread development of tidal inlets and tidal channels suggest the occurrence of an extensive lagoonal facies landward of the paleoshorelines (i.e., out of the plane of the outcrop belt). The degree of wave modification and the presence of the lagoonal facies suggest a well-developed, wave-dominated shoreline strand adjacent to the main fluvial network of the deltaic system.

Major surfaces bounding and internal to Parasequence Set 3 suggest high order base-level fluctuations are recorded. The extensive transgressive ravinement surfaces, overlain by thick transgressive lag deposits, occur below and above Parasequence Set 3. These transgressive ravinement surfaces and the presence of a sur-

face of shoreface erosion between Parasequences 3a and 3b suggest that Parasequence Set 3 was deposited during a period of rapid base-level fluctuation. The pattern of base-level fall and rise, and the changes in the general paleoshoreline positions suggest that Parasequence Set 3 can be divided into a series of 5th-order systems tracts reflecting these rapid, although minor, base-level changes. Parasequence 3a represents a 5th-order HST. The sharp-based shoreface erosion at the top of Parasequence 3a suggests that Parasequence 3b represents a LST. The transgressive lag deposits below and above Parasequence Set 3 represent preserved TST deposits.

Aggradational Parasequence Sets

In the context of the total Last Chance clastic wedge architecture and the larger scale 3rd-order Ferron Composite Sequence, Ferron Parasequence Sets 4A, 4B, 5A, and 5B exhibit general external aggradational stacking (Figure 3, Plates III and IV). In the southern 25 km (16 mi) of the outcrop belt, Parasequence Sets 4A, 4B, 5A, and 5B are represented by non-marine delta-plain facies associations composed of delta-plain mudstones, large distributary channel belts, crevasse-splay sandstones, over-bank deposits, and carbonaceous shales and siltstones. Coal zones occur at or near the tops of Parasequence Sets 4B and 5B, with the thickest portion of the coal zones occurring near the landward pinchout of the near-marine delta-front sandstones. Parasequence Sets 4B and 5B are capped by the G and I coal zones, respectively. A very laterally restricted coal occurs locally within Parasequence Set 5A. This coal seam is denoted as the H coal zone. The near-marine facies of both Parasequence Set 4A and Parasequence Set 5A are sub-regionally bounded above by erosional unconformities that have been identified as 4th-order depositional sequence boundaries SB3 and SB4, respectively. The nature and extent of these boundaries will be discussed later.

Parasequence Sets 4A and 4B

Parasequence Sets 4A and 4B each appear to contain only one parasequence (Figure 3, Plates III and IV). These two parasequences (denoted 4a and 4b) are given the hierarchical equivalence to parasequence sets, due to their stratigraphic and systems tract relationships to overlying and underlying parasequence sets. Parasequence 4a is back-stepped 7 km (4 mi) landward of the initial paleoshoreline of Parasequence 3b, in the underlying Parasequence Set 3, placing Parasequence 4a delta-front, prodelta, and transgressive lag deposits stratigraphically on Parasequence 3b (and 3c) delta-plain facies. Parasequence 4a is separated from overlying Parasequence 4b by an extensive sub-regional erosional unconformity. These relationships suggest that Parasequence 4a is hierarchically equivalent to a highstand parasequence set. Parasequence 4b is bounded above by the G coal zone and below by the sub-regional unconformity. Therefore, by analogy to other Upper Ferron Sandstone parasequence sets, that are not bounded above by sequence boundaries but capped by coal zones, Parasequence 4b is also considered hierarchically equivalent to a lowstand parasequence set.

Sedimentological characteristics: The near-marine facies of the single parasequence in each Parasequence Sets 4A and 4B represent wave-dominated, wave-reworked stream-mouth-bar and delta-front deposits. The near-marine facies of both parasequences exhibit vertical and lateral facies changes from (1) reworked stream-mouth bar, preserved as upper shoreface deposits to (2) reworked proximal and distal delta front, preserved as middle and lower shoreface deposits to (3) prodelta/offshore deposits (Table 1). The upper shoreface deposits are of fine-grained to medium-grained, trough cross-stratified sandstones. Locally, horizontal to ripple stratified bedsets are preserved at the top of the facies. Rare *Ophiomorpha* burrows occur near the top of the facies. The middle and lower shoreface deposits are very fine grained, bioturbated sandstones containing hummocky, planar, and ripple cross-stratification and frequently the ichnofossils *Ophiomorpha*,

Teichichnus, *Chondrites*, *Cylindrichnus*, *Arenicolites*, *Diplocraterion*, *Rosselia*, *Asterosoma*, and *Thalassinoides* (ichnofacies *Skolithos* and *Cruziana*). In many areas, Ophiomorpha bioturbation has destroyed most of the stratification within the middle shoreface deposits.

South of Anderson Ranch, a well-developed delta-plain facies association is present with fluvial channel-belt facies forming the dominant sandstone lithofacies of the delta-plain facies association (see previous descriptions of these types of deposits). Channel-fill elements generally have well-preserved vertical profiles that generally fine upward from coarse- or medium-grained sandstone, at the base, to very fine grained sandstone at the top. They exhibit a vertical bed form succession from large trough cross-stratification to small trough cross-stratification to ripple and climbing ripple cross-stratification. Locally, the tops of channel-fill elements are very intensely burrowed (i.e., *Planolites*, *Skolithos*, and *Scoyenia*). In some areas, channel-fill elements may be separated by massive mudstones. Near the top of Ferron Parasequence Set 4B, coarse grained sandstones occur in fluvial channel belts. These sandstones contain extrabasinal pebbles of quartzite, chert, sandstone, siltstone, volcanic-derived lithics, and large K-feldspar crystals. These clasts can be up to 1.5 cm by 0.8 cm (0.6 by 0.3 in.), averaging 8 mm by 4 mm. These well rounded to subangular extrabasinal clasts occur along the base of channels, beds and troughs (Garrison, 2004). Along the base of the channels and sporadically along bedset surfaces, angular intrabasinal mud rip-ups up to 1.8 cm by 0.8 cm (0.7 by 0.3 in.) occur; the average is 1.5 cm by 1.8 cm (0.6 by 0.7 in.). This occurrence of abundant coarse-grained material in fluvial and distributary channels is a unique occurrence among all of the Ferron parasequence sets.

Amalgamated fluvial deposits occur locally within Parasequence Set 4B. These deposits are sedimentologically very distinct from the laterally accretionary, multi-storied fluvial channel belts deposits within the delta-plain facies association. In Miller Canyon and southward into Bear Gulch, Parasequence 4B consists of a major fluvial channel-belt complex that incises up to 30 m (98 ft) into the underlying parasequence 4a of Parasequence Set 4A (Figures 15 and 16). These deposits are amalgamated channel belts that exhibit a multi-storied, internally scoured vertical stacking of channel-fill elements. They are generally exposed in vertical cliffs along Miller Canyon, obtaining a thickness of almost 30 m (90 ft). Sandstone-rich braided stream deposits consist of amalgamated braided channel belts 4–8 m (13–26 ft) thick, composed of trough cross-stratified, coarse-grained to fine-grained sandstone barforms (Figure 16). These channel-belt barforms have scour bases and range in thickness from 2.3–5 m (7.5–16 ft). The barforms have a sigmoidal shape in the dip direction and are channel-form in cross section. They consist of trough cross-beds

Figure 15. Bear Gulch showing the erosional unconformity at the base of Upper Ferron Sandstone, Parasequence Set 4B fluvial valley fill and the incision into the near-marine deposits of Parasequence Set 4A.

and fine upward from medium-grained to fine-grained sandstone and from coarse-grained to medium-grained sandstone. In some barforms, ripples are found climbing up the trough cross-stratification. Trough cross-stratification fines upward on the lamina scale from medium-grained to fine-grained sandstone. Extrabasinal material ranging in size from very coarse grained sand to pebbles up to 0.5 cm (0.2 in.) in diameter are found along cross-bed boundaries, and also floating within the sandstone bedsets. Locally, fine carbonaceous and coaly material can be found along lamina set and bedset boundaries. Locally, these amalgamated deposits contain heterolithic channel-fill facies. These deposits consist of very fine to fine-grained, rippled sandstones interlayered with laminated carbonaceous mudstone and mudstone containing very thin lamina of siltstone. The sandstone layers range in thickness from 2 cm (0.8 in.) to 2.7 m (8.9 ft) and the mudstone layers range from 25 cm (10 in.) to 2.6 m (8.5 ft) in thickness. Ripple bedsets range in thickness from 2 cm (0.8 in.) to 2 m (7 ft). The sandstones frequently contain burrows, fragments of coaly material, and clay clasts; rare lags of pelecypod fragments also occur.

Locally within Parasequence Set 4B, there is evidence of tidal reworking and the development of tidally

influenced fluvial deposits. These fluvial channel-belt sandstones exhibit evidence marine influence. These channel belts commonly pinchout laterally into interdistributary, brackish-bay facies. The upper heterolithic portions and the margins of the channel belts contain abundant marine burrows (e.g., *Ophiomorpha*). The channel-belt sandstones contain floating oyster fragments, coal fragments, and lags of woody material and rounded clay clasts. These deposits consist of fine-grained to coarse-grained trough and ripple cross-stratified sandstone. The trough bedsets frequently fine upward in grain size. Trough cross-beds often have ripples on their surfaces. Frequently near the tops of these fluvial channel belts, there are heterolithic in-fills consisting of carbonaceous mudstone containing lamina of fine-grained sandstone, interlayered with trough, planar tabular, and ripple cross-stratified medium-grained sandstone and bioturbated fine-grained carbonaceous sandstone with mudstone lamina. These fine-grained and medium-grained sandstones are burrowed by *Ophiomorpha* and *Thalassinoides*, and commonly contain floating grains of coarse sand.

Brackish-bay fill deposits are locally present in the Parasequence Set 4B. These bay deposits consist of mud-

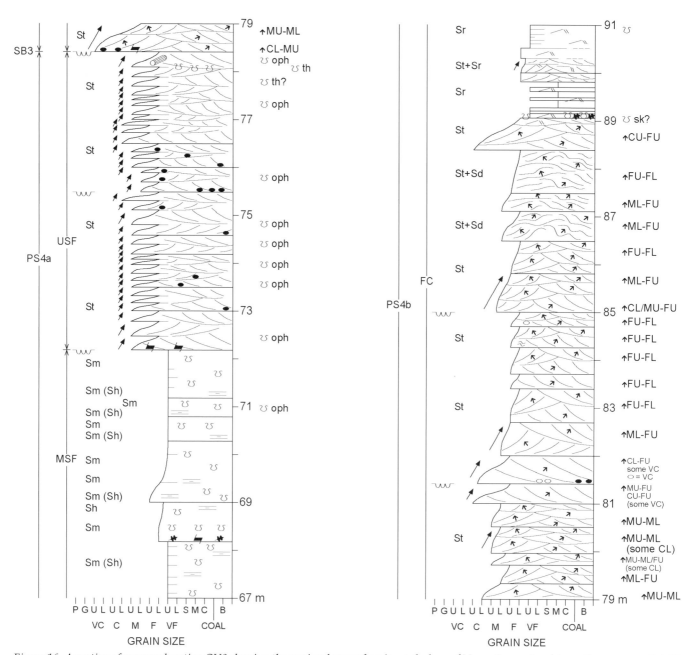

Figure 16. A portion of measured section GV8 showing the erosional unconformity at the base of Upper Ferron Sandstone, Parasequence Set 4B fluvial valley fill and the incision into the near-marine deposits of Parasequence Set 4A. See Figure 5 for explanation.

stone, carbonaceous mudstone, and carbonaceous siltstone, interlayered with thin layers of intensely burrowed, fine-grained ripple cross-stratified to flaser-bedded sandstone. These flaser-bedded sandstones average 5 cm (2 in.) in thickness and are burrowed. Locally, crevasse-splay sandstones interfinger with the bay mudstone. The crevasse-splay sandstone facies contains coal fragments, wood fragments, and a brackish-water pelecypod fauna. The sandstones are thin (< 2 m [7 ft] thick), range from very fine grained to coarse grained, and contain extrabasinal pebbles, up to 1 cm (0.4 in.) in diameter, and oyster fragments. They are burrowed by *Planolites*, *Ophiomorpha*, and *Thalassinoides*.

Stratal properties: Near-marine sandstones of the Parasequence 4a of Parasequence Set 4A are 15 km (9 mi) in dip length, and the maximum thickness is approximately 27 m (89 ft). Near-marine facies of Parasequences 4b of Parasequence Set 4B is 11.4 km (7.1 mi) in dip length with a maximum thickness of approximately 18 m (59 ft). The landward pinchout of the near-marine facies of Parasequence 4a shifts landward approximately 10 km (6 mi) relative to the paleoshoreline of underlying Parasequence 3c. The landward pinchout of the near-marine facies of Parasequence Set 4B shifts basinward some 7 km (4 mi) relative to the landward pinchout of the near-marine sandstones of Parasequence Set 4A.

160

Parasequence 4a exhibits 6 m (20 ft) of stratigraphic rise relative to the final shoreline position of underlying Parasequence 3b.

North of Bear Gulch, the base of Parasequence 4a of Parasequence Set 4A is a transgressive lag deposit on a transgressive ravinement surface (Plate III). South of Bitter Seep Wash, this transgressive lag deposit reworks the upper part of the C coal zone; to the north, it reworks the top of Parasequence Set 3. It exhibits extreme bioturbation (ichnofossils *Ophiomorpha* and *Thalassinoides*) and contains coal and clay rip-up clasts. This deposit can be traced seaward, along the outcrop belt, over 38 km (24 mi) to north of Ferron Creek. Between Anderson Ranch and the South Goosenecks of Muddy Creek, the upper boundary of the near-marine sandstones of Parasequence Set 4A is an erosional unconformity at the base of a small incised-valley system and exhibits as much as 30 m (98 ft) of relief. In the non-marine part of the Parasequence Set 4A, south of Anderson Ranch, Parasequence 4a is capped by a very laterally extensive rooted zone, designated the E Rooted Zone (Plates II and III). This rooted zone can be traced from just north of Last Chance Creek northward to just south of Anderson Ranch, where its base is correlative with the erosional unconformity between Parasequence Set 4A and 4B that crops out in Bear Gulch and Miller Canyon.

South of Anderson Ranch, the base of Parasequence 4b of Parasequence Set 4B is represented by a correlative conformity at the base of the E Rooted Zone. Between Anderson Ranch and Rochester Creek, Parasequence Set 4B is represented by amalgamated braided fluvial incised-valley fill deposits lying on the erosional unconformity at the top of the near-marine sandstones of Parasequence Set 4A. North of Rochester Creek, the near-marine deposits of Parasequence 4b lie conformable on top of the near-marine deposits of Parasequence 4a of Parasequence Set 4B. North of Bitter Seep Wash, Ferron Parasequence Set 4B has been truncated by a transgressive ravinement surface overlain by the transgressive lag deposits of Parasequence 5a (see Plates III and IV). These lag deposits are best developed in the area between Bitter Seep Canyon and Cedar Ridge Canyon, where locally they may be up to 1 m (3 ft) in thickness. In the area near Dry Wash, the reworked stream-mouth-bar facies has been almost completely removed, exposing the reworked proximal delta-front deposits, which have been locally incised by the basal sections of distributary channels. It appears from an analysis of the regional cross sections (Plates III and IV) that as much as 5–10 m (16–33 ft) of material may have been removed during ravinement. From Rochester Creek to the Limestone Cliffs, the upper boundary of Parasequence Set 4B is the top of the G coal zone. The top of the G coal zone is correlative to the transgressive ravinement surface at the top of Parasequence Set 4B.

South of Indian Canyon, in the Limestone Cliffs, the G coal zone is completely scoured out by fluvial channel belts and is not present farther south than Mussentuchit Wash. Curiously, the G coal zone does not split within the study area. The G coal zone extends over 27 km (17 mi) from its erosional limit in the Limestone Cliffs to its seaward limit north of Bitter Seep Canyon. Figure 17 shows a portion of a measured section through the G coal zone in Miller Canyon.

Interpretation: The near-marine facies of Parasequences 4a, and 4b appear to represent extensively wave-reworked deltaic deposits. The degree of wave-modification is significant in most areas and a shoreface profile is fully developed. Paleogeographic reconstructions, based on facies distributions and fluvial paleocurrent directions, indicate a northeasterly progradation of the paleoshoreline. The paleoshoreline of Parasequence 4a never transgressed farther southwest than Anderson Ranch. The transgression of this shoreline was accompanied by the development of a transgressive ravinement surface and an extensive transgressive lag deposit. Parasequence Set 4A has a highstand position relative to Parasequence Set 3. The paleoshoreline position did not move substantially during the base-level fall and the valley incision at the beginning of deposition of Parasequence 4b.

The amalgamated braided fluvial deposits within Parasequence Set 4b appear to represent the development of a minor incised-valley fill and development of a valley-fill facies association (Garrison, 2004). The incised-valley system appears to bifurcate in the area southwest of Quitchupah Creek, north of I-70, such that in the Miller Canyon/Muddy Creek area, the incised-valley system is composed of two distinct, parallel incised-valley fill sequences. Both incised-valley segments are up to 30 m (98 ft) in depth. The incised-valley system is oriented approximately 60° and has a floor with substantial topographic relief. Near the mouth of the valleys, the north and south branches of the valley system are approximately 2 km (1 mi) and 5 km (3 mi) wide, respectively. Garrison (2004) has documented the detailed internal stratigraphy of the valley-fill sequence. The incised-valley deposits exhibit a very poorly developed tripartite segmentation and contain only the 4th-order lowstand and TSTs, within the incised-valley fill deposits; the highstand deposits lie stratigraphically above the incised valley. As the amalgamated, braided fluvial incised-valley fill deposits are traced toward the near-marine deposits at the paleoshoreline, there is significant influence by marine processes. As much as 3 km (2 mi) inland from the paleoshoreline, inter-fluvial areas contain heterolithic, locally carbonaceous, brackish-bay/lagoonal deposits with brackish fauna. Fluvial and distributary channel-belt sandstones are burrowed by marine ichnofossils *Ophiomorpha* and *Thalassinoides*. Flu-

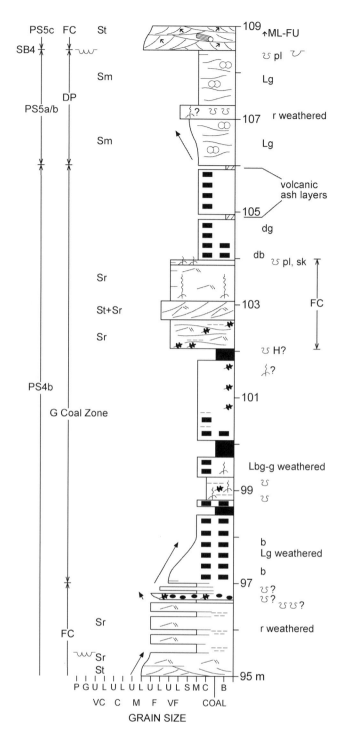

Figure 17. A portion of measured section GV6 in Miller Canyon showing the character of the G coal zone. See Figure 5 for explanation.

thin brackish-water estuarine facies that occur locally at the top of the valley fill. Trangressive lag deposits overlie this thin estuarine facies, and represent the TST. The 4th-order HST is represented by wave-modified shoreface sandstones of Parasequence 5a that overlie and have "over-filled" across the top of the older Parasequence 4b incised-valley fill deposits.

The amalgamated fluvial incised-valley deposits are interpreted to be associated with the development of an erosional unconformity representing a 4th-order sequence boundary (here denoted as SB3) (Figure 3, Plates II and IV). This erosional unconformity is considered to represent a depositional sequence boundary because: (1) sub-regional extent, (2) the internal consistency of systems tracts above and below the unconformity, (3) the magnitude of the erosional incision suggests a base-level fall of about 10 m (30 ft), (4) the lateral extent of the correlative E Rooted Zone indicates a widespread depositional hiatus synchronous with, and following, valley incision and base-level fall, and (5) the nature of the inferred incised-valley and estuarine depositional systems immediately overlying the unconformity. Parasequence Set 4B incised-valley erodes down through the prodelta deposits of Parasequence 4a (Plate III). The areal extent of the erosional unconformity is estimated to be at least 75–100 km^2 (29–39 mi^2) (Garrison, 2004). The systems tracts immediately above the erosional unconformity have been interpreted to represent well-defined 4th-order lowstand and transgressive incised-valley/estuarine deposits (Garrison, 2004).

Parasequence Sets 5A and 5B

Parasequence Set 5A is a slightly progradational to aggradational parasequence set containing two parasequences (denoted as Parasequences 5a and 5b) (Figure 3, Plates III and IV). Parasequence Set 5B contains a single parasequence (denoted Parasequence 5c). Parasequence 5c is given the hierarchical equivalence to a parasequence set, due to their stratigraphic and systems tract relationships to overlying and underlying parasequence sets. From the South Goosenecks of Muddy Creek Canyon to Picture Flats, the top of Parasequence Set 5A has from 10–20 m (33–66 ft) of erosional relief developed, as amalgamated fluvial channel belts of Parasequence 5c scour down into underlying Parasequence 5b (Figures 18 and 19). The upper boundary of Parasequence 5c is the I coal zone. Therefore, by analogy to other Upper Ferron Sandstone parasequence sets, that are not bounded above by sequence boundaries but capped by coal zones, Parasequence 5c is also considered hierarchically equivalent to a lowstand parasequence set.

The landward pinchout of the near-marine facies of Parasequence Set 5A is about 2.4 km (1.5 mi) landward of the position of the landward pinchout of Parasequence Set 4B. Parasequence Set 5B has the landward

vial and shoreline aggradation allowed the incised valley to be filled completely with the 4th-order, low-stand fluvial-fill facies of Parasequence Set 4B. The incised-valley fluvial system supplied sediment to the lowstand, wave-dominated delta system represented by the near-marine facies of Parasequence 4b. The 4th-order latest LST, within these incised-valley fills, is represented by

Figure 18. Muddy Creek Canyon showing the erosional unconformity at the base of Upper Ferron Sandstone, Parasequence Set 5B fluvial valley fill and the incision into the near-marine deposits of Parasequence Set 5A.

pinchout of its near-marine sandstones displaced seaward by 3.2 km (2 mi).

Sedimentological characteristics: The near-marine facies of the parasequences in each Parasequence Sets 5A and 5B represent wave-dominated, wave-reworked stream-mouth bar and delta-front deposits. The near-marine facies exhibit vertical and lateral facies changes from (1) reworked stream-mouth bar, preserved as upper shoreface deposits to (2) reworked proximal and distal delta front, preserved as middle and lower shoreface deposits to (3) prodelta/offshore deposits (Table 1). The upper shoreface deposits are of fine-grained to coarse-grained trough cross-stratified sandstones. Large low-angle, fine-grained to coarse-grained trough cross-stratified sandstones, near the base, fine upward into fine-grained to medium-grained trough cross-stratified sandstone, near the top. Locally, horizontal to ripple stratified bedsets are preserved at the top of the facies (Figure 19). The upper shoreface deposits frequently contain *Ophiomorpha* burrows. The middle and lower shoreface deposits consist of very fine grained bioturbated, horizontal bedded and ripple-stratified sandstone interlayered with siltstone, near the base, to hummocky cross-stratified sandstone at higher levels (Figure 19). These facies commonly contain the ichnofossils *Ophiomorpha* and *Thalassinoides*, and less commonly *Skolithos*. In many areas, bioturbation has destroyed most of the stratification within the middle shoreface deposits.

South of Miller Canyon, a typical well-developed delta-plain facies association is present. The delta-plain facies association consists of fluvial channel belts interlayered with a floodplain facies association consisting of overbank mudstone and siltstone deposits, crevasse-splay siltstone and sandstone deposits, and peat-mire deposits. Fluvial channel belts are similar to those preserved in Parasequences Sets 4A and 4B.

Amalgamated fluvial deposits occur locally within Parasequence Set 5B, also similar to those preserved in Parasequences Sets 4A and 4B. From Muddy Creek amphitheater to just north of Picture Flats, Parasequence Set 5B consists of a major fluvial channel-belt complex that incises up to 20 m (66 ft) into the underlying parasequence 5b of Parasequence Set 5A (Figures 18 and 19, Plate III). These deposits are generally exposed in vertical cliffs along the northern end of Muddy Creek Canyon, where they obtain a thickness of almost 20 m (66 ft).

The Parasequence 5c fluvial channel belts within 2 km (1 mi) of the near-marine deposits, of Parasequence 5c, exhibit marine influence (Garrison, 2004). These channel belts are fine-grained to medium-grained trough cross-stratified sandstone channel-fill elements interlayered with heterolithic channel-fill elements. The sandstone channel-fill elements are composed of trough cross-beds that fine upward. The bases of some channel-fill elements contain lags composed of clay clasts, mud rip-up clasts, and woody material burrowed by *Teredolites*. The channel-fill elements range in thickness from 3.2–6.7 m (10.5–22.0 ft). The heterolithic channel belts consist of accretionary beds composed of very fine grained, *Ophiomorpha* burrowed, ripple cross-stratified sandstone interlayered with carbonaceous siltstone.

In the area between Rochester Creek and the North Goosenecks of Muddy Creek, a back-barrier lagoonal and mire facies association is exposed. This facies association is composed of carbonaceous mudstone and car-

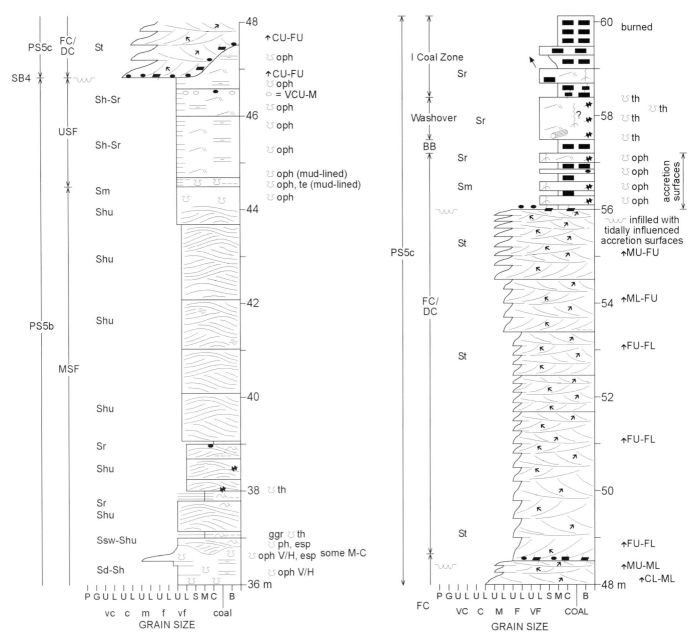

Figure 19. A portion of measured section GV19 showing the erosional unconformity at the base of Upper Ferron Sandstone, Parasequence Set 5B fluvial valley fill and the incision into the near-marine deposits of Parasequence Set 5A. See Figure 5 for explanation.

bonaceous siltstone, and carbonaceous fine-grained sandstone interlayered with thin, very fine grained to medium-grained bioturbated washover fan sandstones. Occasionally, thin discontinuous coal seams occur interlayered with the landward portion of the lagoonal facies. The carbonaceous facies of the lagoonal deposits consists of 50–70 cm (20–28 in.) thick coarsening upward cycles of mudstone, siltstone, and very fine grained sandstone. The carbonaceous sandstones are horizontal laminated to ripple cross-stratified and bioturbated. Occasionally, the carbonaceous mudstones contain very fine grained sandstone lamina. The carbonaceous lithofacies is commonly rooted. The washover fan sandstone facies consists of coarsening upward cycles of fine-grained to medium-grained bioturbated, ripple cross-

stratified sandstone, interlayered with very thin lamina of mudstone. The washover fans are carbonaceous, intensely burrowed by *Ophiomorpha* and *Thalassinoides*, and occasionally rooted. The washover fan sandstone facies is frequently interlayered with the coal seams of the I coal zone. These seams range in thickness from 10 cm (4 in.) to 1.1 m (3.6 ft).

Stratal Properties: The landward pinchout of the near-marine facies of Parasequence 5a is located about 2.4 km (1.5 mi) landward of the position of the landward pinchout of underlying Parasequence 4b. The landward pinchout of Parasequence 5b is truncated by amalgamated braided fluvial channels of Parasequence Set 5B, but can be located to within 1 km (0.6 mi). It is estimated that the landward pinchout of the near-marine facies of Parasequence 5b steps seaward about 0.9 km (0.6 mi), relative to the landward pinchout of the near-marine facies of underlying Parasequences 5a. The landward pinchout of the near-marine facies of Parasequence 5c steps seaward 3.2 km (2.0 mi), relative to the landward pinchout of the near-marine facies of underlying Parasequences 5b. The initial paleoshorelines of Parasequences 5a and 5b exhibit about 5.0 m (16.4 ft) and 15.5 m (50.9 ft) of stratigraphic rise, respectively. Parasequence 5c represents an 8.4 m (27.6 ft) base-level fall relative to the final paleoshoreline position of Parasequence 5b.

Near-marine facies of the Parasequences 5a and 5b of Parasequence Set 5A are approximately 10.6 km (6.6 mi) and 10.8 km (6.7 mi) in dip length, respectively, and maximum thicknesses are approximately 16 m (52 ft) and 15 m (49 ft), respectively. The estimated overall dip length of Parasequence Set 5A is approximately 11.8 km (7.3 mi) with a maximum thickness of approximately 33 m (108 ft). The near-marine facies of the Parasequence 5c, in Parasequence Set 5B, is 7.5 km (4.7 mi) in dip length and has a maximum thickness of 16 m (52 ft).

North of the Rochester Creek, the lower boundary of the near-marine facies of Parasequence 5a is a marine-flooding surface on top of underlying Parasequence 4b. From Rochester Creek to the North Goosenecks of Muddy Creek Canyon, the lower part of Parasequence 5a is represented by a transgressive lag lying on a transgressive ravinement surface on top of Parasequence 4b. Farther south, the lower boundary is a correlative conformity to the marine-flooding surface, located at the top of the G coal zone. North of the North Goosenecks of Muddy Creek Canyon, the upper boundary of Parasequence 5a is a marine-flooding surface. To the south, the upper boundary lies within the delta-plain facies association and cannot be delineated with any certainty.

North of the Rochester Creek, the lower boundary of the near-marine facies of Parasequence 5b is a marine-flooding surface on top of underlying Parasequence 5a (Plates III and IV). In Picture Flats and in the Muddy Creek amphitheater, the lower boundary is a marine-flooding surface, located at the top of the localized H coal zone at the top of Parasequence 5a. South of Muddy Creek amphitheater, the lower boundary lies within the delta-plain facies association and cannot be delineated with any certainty. From south of Bitter Seep Canyon to Miller Canyon, the top of Parasequence 5b is an erosional unconformity lying at the base of the amalgamated braided fluvial deposits of Parasequence Set 5B. This erosional unconformity has been interpreted as being the base of a small incised-valley system and as representing a depositional sequence boundary. Farther south, this erosional unconformity is correlative with a conformable surface at or near the base of the I coal zone. Farther north, the erosional unconformity is correlative with a marine-flooding surface.

North of Picture Flats, the upper and lower boundaries of Parasequence 5c are marine-flooding surfaces. From south of Bitter Seep Canyon to Miller Canyon, the base of Parasequence 5c is an erosional unconformity lying at the base of the amalgamated braided fluvial deposits interpreted as incised-valley fill deposits. Farther south, the lower boundary of Parasequence 5c is a correlative surface at or near the base of the I coal zone. From south of Bitter Seep Canyon to Rochester Panel, the upper boundary of Parasequence 5c is a marine-flooding surface beneath the near-marine facies of Parasequence 6a. Farther south, the upper surface is a correlative surface that is located at the top of the I coal zone.

The I coal zone caps Ferron Parasequence Set 5B. South of Indian Canyon, in the Limestone Cliffs, the I coal zone is completely scoured out by fluvial channel belts and does not occur farther south than Mussentuchit Wash. South of Coyote Basin, the I coal zone splits into three zones, denoted as the I_1, I_2, and I_3 coal zones. Figure 20 shows a portion of a measured section through the I coal zone in Bear Gulch. In the Muddy Creek amphitheater and in Picture Flats, a thin, laterally restricted intra-parasequence set coal seam, denoted as the H coal zone, occurs at the top of Parasequence 5a. Although in some areas of the Last Chance Delta, coal zone splits can be locally correlated with the tops of parasequences, this is the only clearly intra-parasequence set coal zone that has been identified in the study area.

Interpretation: The near-marine facies of Parasequences 5a, 5b, and 5c appear to represent extensively wave-reworked deltaic deposits. The degree of wave-modification is significant in most areas and a shoreface profile is full developed. Paleogeographic reconstructions (Garrison, 2004), based on facies distributions and the inferred direction of fluvial sediment transport, suggest a northeasterly progradation of the shoreline. The paleoshoreline of Parasequence 5a never transgressed, across the top of underlying Parasequence Set 4B, farther southwest than the Muddy Creek amphitheater. The transgression of this paleoshoreline was accompanied by the development of a transgressive ravinement surface and the deposition of a thick transgressive lag. Parasequence Set 5A has a highstand position relative to underlying Parasequence Set 4B. Parasequences 5a and 5b, of Parasequence Set 5A, represent aggradational highstand deposits that overfill across the top of the underlying incised-valley of Parasequence Set 4B. Parasequence 5c represents a small lowstand wave-dominated deltaic deposit fed by the incised-valley fluvial system. The establishment of the incised valley and the subsequent initiation of Parasequence 5c shoreline deposition was accompanied by only a 3.2 km (2 mi) seaward shift in the paleoshoreline position during the 8.4 m (27.6 ft) of base-level fall.

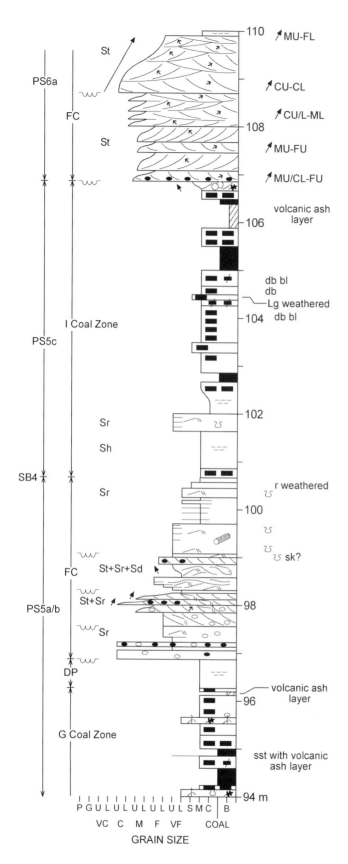

Figure 20. A portion of measured section GV8 in Bear Gulch showing the character of the I coal zone. See Figure 5 for explanation.

Similar to the amalgamated fluvial deposits of Parasequence Set 2C and 4B, the amalgamated fluvial deposits of Parasequence Set 5B appear to represent the development of a small, incised valley and the development of a valley-fill facies association. These deposits are dominantly amalgamated channel belts that exhibit complex vertical and lateral stacking of channel-fill deposits. Down dip these amalgamated braided fluvial deposits feed more isolated channel belts that probably represent distributary channel-belt deposits. As these valley-fill deposits are traced toward the near-marine deposits at the paleoshoreline, there is evidence of influence by marine processes. As much as 2–3 km (1–2 mi) inland from the paleoshoreline, inter-fluvial areas contain heterolithic, locally carbonaceous, brackish-bay/lagoonal deposits and fluvial channel-belt sandstones are burrowed by marine ichnofossils *Ophiomorpha, Teichichnus,* and *Thalassinoides.* Garrison (2004) has described the detailed internal stratigraphy of the valley-fill sequence. The incised-valley system does not exhibit a tripartite segmentation, but contains only the 4th-order LST, within the incised-valley fill deposits; the transgressive deposits lie stratigraphically above the incised valley. Fluvial and shoreline aggradation allowed the incised valley to be filled completely with the 4th-order, low-stand fluvial-fill facies of Parasequence Set 5B. Thin brackish-water estuarine facies occur locally at the top of the fluvial incised-valley fill facies. These lowstand estuarine facies locally contain washover fan deposits. The 4th-order TST is represented by the wave-modified shoreface sandstones of Parasequence Sets 6, 7, and 8 that overly and have "over-filled" across the top of the older Parasequence Set 5B incised-valley fill deposits.

The amalgamated fluvial incised-valley deposits are interpreted to be associated with the development of an erosional unconformity representing a 4th-order sequence boundary (here denoted as SB4) (Figure 3, Plates II through IV). This erosional unconformity is considered to represent a depositional sequence boundary because: (1) sub-regional extent, (2) the internal consistency of systems tracts above and below the unconformity, (3) the magnitude of the erosional incision suggesting a base-level fall of about 8 m (26 ft), and (4) the nature of the inferred incised valley and estuarine depositional systems immediately overlying the unconformity. Parasequence Set 5B incised-valley erosion erodes down through the middle shoreface deposits of Parasequence 5b and into the upper shoreface deposits of Parasequence 5a (Plate III). The areal extent of the erosional unconformity is estimated to be at least 75–100 km² (29–39 mi²). The systems tracts immediately above the erosional unconformity have been interpreted to represent well-defined 4th-order lowstand incised-valley/estuarine deposits.

Retrogradational Parasequence Sets

Ferron Parasequence Sets 6, 7, and 8 all exhibit an internal and external retrogradational-stacking pattern (Figure 3, Plates II and III). Parasequence Set 7 contains four retrogradational wave-reworked deltaic parasequences (denoted 7a, 7b, 7c, and 7d) and Parasequence 8 contain two retrogradational wave-reworked deltaic parasequences (denoted 8a and 8b). Between south Quitchupah Creek and the Emery Mine, Parasequence Set 8 is separated from Parasequence Set 7 by a transgressive lag deposit containing oyster fragments and rounded extrabasinal pebbles (1–10 cm [0.4–4 in.] in length). At Coal Wash and in the area between Willow Springs Wash and Indian Canyon, the boundary between Parasequence Set 7 and Parasequence Set 8 is the top of the M coal zone. Parasequence Set 6 contains a single parasequence. This parasequence (denoted 6a) is given the hierarchical equivalence to a parasequence set, due to its stratigraphic relationship to overlying and underlying parasequence sets and lithofacies. Parasequence 6a is back-stepped 2 km (1 mi) landward of the initial paleoshoreline of Parasequence 5c, in the underlying Parasequence Set 5B, placing Parasequence 6a delta-front deposits stratigraphically on the transgressive surface at the top of Parasequence 5c lowstand incised-valley fill deposits. The J coal zone forms the upper delta-plain lithofacies of Parasequence 6a. Parasequence 6a is separated from overlying retrogradational Parasequence Set 7 by a marine-flooding surface above the J coal zone. These relationships suggest that Parasequence 6a is hierarchically equivalent to a parasequence set.

Throughout most of the outcrop belt, Parasequence Sets 6, 7, and 8 are represented by a delta-plain facies association composed of wide distributary channel belts, crevasse-splay deposits, and floodplain mudstones, siltstones, and carbonaceous shales and siltstones. Near-marine sandstone facies are restricted to a relative small area of the outcrop belt. Near-marine facies of Parasequence Sets 6, 7, and 8 are approximately 2.8 km (1.7 mi), 7.1 km (4.4 mi), and 6.0 km (3.7 mi), in dip length, respectively (Table 3). The thicknesses of the near-marine facies of the Parasequence Sets 6, 7, and 8 are 10 m (33 ft), 9 m (30 ft), and 7 m (23 ft), respectively. The near-marine facies of Parasequence Set 6 extends from Rochester Panel to Bitter Seep Canyon. The near-marine facies of Parasequence Set 7 extends from Christiansen Wash to Rochester Panel. The near-marine facies of Parasequence Set 8 extends from Coal Wash to just south of Miller Canyon (Plates II and III). The landward pinchout of the near-marine facies of Parasequence Set 7 is 0.2 km (0.1 mi) landward of the landward pinchout of the near-marine facies of Parasequence Set 6. The landward pinchout of the near-marine facies of Parasequence Set 8 is 8.6 km (5.3 mi) landward of the landward pinchout of the near-marine facies of Parasequence Set 7.

Sedimentological Characteristics

The near-marine facies of the parasequences in each Parasequence Sets 6, 7, and 8 represent a wave-dominated, wave-reworked delta-front deposits. They all exhibit both vertical and lateral facies changes from (1) reworked stream-mouth bar, preserved as upper shoreface deposits to (2) reworked delta front, preserved as middle and lower shoreface deposits to (3) prodelta. Figures 21 and 22 show measured sections through Parasequence Sets 6, 7, and 8. The upper shoreface facies consists of very fine grained to medium-grained trough, low-angle trough, and planar tabular cross-stratified sandstones. Cross-stratification and bedsets fine upward. The upper shoreface sandstones frequently contain the ichnofossils *Ophiomorpha* and *Arenicolites*. In some areas, rounded extrabasinal pebbles and oyster fragments are found floating within the upper shoreface sandstones of Parasequence Set 8. The middle shoreface deposits consist of a succession of very fine grained, ripple and hummocky cross-stratified sandstone. The sandstones contain *Ophiomorpha* and *Thalassinoides* burrows. The lower shoreface deposits consist of very fine grained horizontal to ripple cross-stratified sandstones interlayered with thin siltstone and mudstone layers. Small-scale hummocky cross-stratification occurs in some sandstone layers near the top of the facies. Mudstone and siltstone layers range in thickness from 2–20 cm (0.8–8 in.). *Ophiomorpha* and *Thalassinoides* burrows are common in the lower shoreface facies.

South of Muddy Creek Canyon, a well-developed delta-plain facies association is present in Parasequences 6 and 7. In the area of Coal Wash and in the area between Willow Springs Wash and Indian Canyon, a very thin Parasequence Set 8 delta-plain facies association is exposed consisting of the same types of deposits in the other parasequence sets. Fluvial channel belts are laterally extensive and sheet-like with channel-fill elements generally stacked vertically within the channel-belt boundaries (Garrison and van den Bergh, 1997; van den Bergh and Garrison, this volume). Channel-fill elements generally have erosional tops and incompletely preserved vertical sedimentary profiles; soft-sediment deformation is common. The channel belts consist of medium- to coarse-grained sandstones at the base, to very fine grained sandstones at the top. The vertical sedimentary structure profile ranges from trough cross-stratification at the base, to locally small-scale planar tabular cross-stratification, and to ripple stratification at the top (Garrison and van den Bergh, 1997; van den Bergh and Garrison, this volume). Contorted bedding is very common in the upper and middle parts of the channel belts. There are abundant tree logs and plant fragments, and locally, abundant angular mud rip-ups and subangular to well rounded intra-basinal mud clasts along the base of the channel belts. Locally, extrabasinal

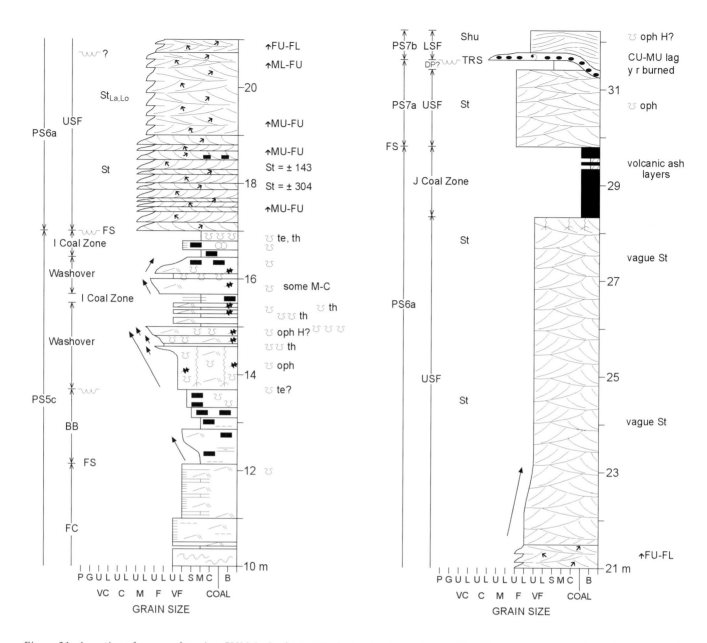

Figure 21. A portion of measured section GV20 in Rochester Creek showing Upper Ferron Sandstone, Parasequence Set 6. See Figure 5 for explanation.

pebbles, up to 5 mm in diameter, occur as lags in a coarse-grained matrix along the bases of channel belts. In some areas, trough cross-stratified beds may be interlayered with muddy siltstones (van den Bergh, 1995). Locally, the tops of channel belts are burrowed by the ichnofossil *Planolites*.

Stratal Properties

Parasequence 6a of Parasequence Set 6 is 2.8 km (1.7 mi) in dip length and has a maximum thickness of 10.4 m (34.1 ft). Parasequences 7a, 7b, 7c, and 7d of Parasequence Set 7 are 1.3 km (0.8 mi), 2.8 km (1.7 mi), 3.1 km (1.9 mi), and 3.6 km (2.2 mi) in dip length, and have maximum thicknesses of 3.4 m (11.2 ft), 1.8 m (5.9 ft), 8.5 m

(27.9 ft), and 3.7 m (12.1 ft), respectively. Parasequences 8a and 8b of Parasequence Set 8 are 3.9 km (2.4 mi) and 3.4 km (2.1 mi) in dip length and have maximum thicknesses of 3.6 m (11.8 ft) and 3.5 m (11.5 ft), respectively.

All of the parasequences in Parasequence Sets 6, 7, and 8 exhibit evidence of significant shoreline retrogradation. The landward pinchout of the near-marine facies of Parasequence 6a steps landward 2.0 km (1.2 mi), relative to landward pinchout of the near-marine facies Parasequence 5c. This transgression at the top of Parasequence 5c was accompanied by a stratigraphic rise of 3.7 m (12.1 ft). The landward pinchout of the near-marine facies of Parasequence 7a steps landward 0.2 km (0.1 mi), relative to landward pinchout of the near-marine

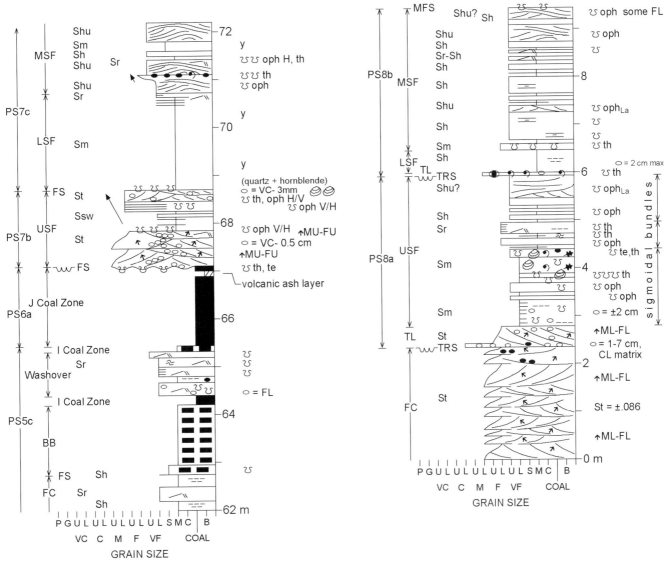

Figure 22. A portion of measured sections showing the Upper Ferron Sandstone, Parasequence Set 7 (GV18 in Muddy Creek Canyon) (left) and Parasequence Set 8 (GV4 in Walker Flat) (right). See Figure 5 for explanation.

facies Parasequence Set 6. This transgression at the top of Parasequence 6a was accompanied by a stratigraphic rise of 18 m (59 ft). The J coal zone caps Ferron Parasequence Set 6 (Figure 23). Figure 24 shows a portion of a measured section showing the character of the J coal zones in Willow Springs Wash. Locally, the base of Parasequence Set 7, resting on the J coal zone, contains the trace fossils *Teichichnus, Thalassinoides, Planolites,* and *Rhizocorallium* (ichnofacies *Glossifungites*). Within Parasequence Set 7 all of the parasequences exhibit a retrogradational stacking. The landward pinchout of the near-marine facies of Parasequence 7b is 1.7 km (1.1 mi) landward of the landward pinchout of the near-marine facies of Parasequence 7a. The transgression across the top of Parasequence 7a was accompanied by 1.7 m (5.6 ft) of stratigraphic rise. The landward pinchout of the near-marine facies of Parasequence 7c is 1.2 km (0.7 mi) landward of the landward pinchout of the near-marine

facies of Parasequence 7b. The transgression across the top of Parasequence 7b was accompanied by 8.7 m (28.5 ft) of stratigraphic rise. The landward pinchout of the near-marine facies of Parasequence 7d is 2.9 km (1.8 mi) landward of the landward pinchout of the near-marine facies of Parasequence 7c. The transgression across the top of Parasequence 7c was accompanied by 3.4 m (11.2 ft) of stratigraphic rise. Locally, the M coal zone caps Parasequence Set 7. The M coal zone is only exposed from North Fork Canyon to Willow Springs Wash and from Ivie Creek to south Quitchupah Creek Canyon. Figure 24 shows a portion of a measured section showing the character of the M coal zones in Willow Springs Wash. In the area North of Mussentuchit Wash in the Limestone Cliffs and in the area from Rock Canyon to Corbula Gulch, the top of Parasequence Set 7 is represented only as a burrowed transgressive/flooding surface; the burrows are marine *Thalassinoides*. Parase-

169

Figure 23. Photograph from the western end of Willow Springs Wash showing the character of the J$_2$ coal zone at the top of Upper Ferron Sandstone, Parasequence Set 6 (after van den Bergh, 1995).

quence Set 8 is a retrogradational parasequence set with the near-marine facies of Parasequence 8b back-stepped landward 2.0 km (1.2 mi), relative to Parasequence 8a. The transgression at the top of Parasequence 8a was accompanied by 1.7 m (5.6 ft) of stratigraphic rise. Locally, a thin lag of rounded extrabasinal pebbles and oyster fragments form a lag separating Parasequences 8a and 8b. In the area near Coal Wash, Parasequence Set 8 is locally capped by a lag of oyster fragments and a thin shale interval containing septarian concretion nodules. South of Coal Wash, in areas where the M coal zone is absent, it is not possible to apportion the exposed non-marine facies at the top of the Ferron Sandstone to either Parasequence Sets 7 or 8. In these areas, the upper bounding surface of the non-marine facies is a marine-flooding surface burrowed by marine ichnofossils.

Interpretation

The near-marine facies of Parasequence Sets 6, 7, and 8 represent a series of small wave-reworked delta-front deposits forming a TST. The initial paleoshoreline of Parasequence 6a of Parasequence Set 6 represents a transgression of 7.1 km (1.1 mi) from the final paleoshoreline of the underlying lowstand deposits of Parasequence Set 5B. The initial paleoshoreline of Para-

sequence Set 7 represents a transgression of 2.0 km (1.2 mi) relative to the final paleoshoreline of Parasequence Set 6. The initial shoreline of Parasequence Set 8 represents a transgression of 4.0 km (2.5 mi) from the final paleoshoreline of Parasequence Set 7. This TST represents at total transgression of 18 km (11 mi) and is punctuated by a major maximum marine-flooding surface that can be traced along the entire length of the outcrop belt. Within the outcrop belt, no highstand deposits have been found upon this maximum flooding surface.

The small deltaic systems represented by these parasequence sets appear to have a north-northeast progradation direction (28° azimuth). In Parasequence Sets 6 and 7, a well-developed progradational mire facies occurs near the top of the delta-plain facies association. Each progressively younger deltaic system (i.e., parasequence set) becomes smaller in size and represents a thinner total accumulation of near-marine sandstone (Table 3), suggesting a declining sedimentation rate associated with the transgression.

Depositional Sequences

Gardner (1992, 1993) described the Upper Ferron Sandstone Last Chance Delta to be a type-2 3rd-order depositional sequence. Gardner suggested that the Fer-

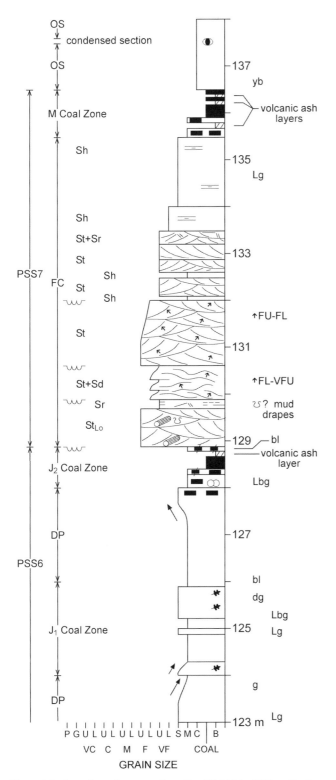

Figure 24. A portion of measured section VMS6 in Willow Springs Wash showing the character of the J and M coal zones. See Figure 5 for explanation.

two sequences is below a condensed section deposit lying upon a widespread bentonitic volcanic ash layer. This volcanic ash layer was assigned an age of 90.5 Ma by Obradovich (1993). The condensed section contains the ammonite *Prionocyclus hyatti*. These condensed section deposits lie stratigraphically above the Clawson and Washboard Sandstone units of Cotter (1975) that are encased in the upper part of the Tununk Shale.

Gardner placed the upper boundary of the Ferron Sequence at the top of the uppermost sandstones of the Last Chance Delta. Garrison and van den Bergh (1997) described a widespread concretion-bearing horizon a few meters above the Upper Ferron Sandstone, which they interpreted as a condensed section deposit. This deposit may be similar in origin to the condensed section containing the ammonite *Prionocyclus hyatti*. The inoceramid *Inoceramus deformis* (early) (Western Interior Biozone *Forresteria blancoi/Cremnoceramus deformis* [early]) has been identified in condensed section deposits lying above Parasequence Set 8 sandstones, just south of the Emery Mine (Figure 3) (Gardner, 1993). The inoceramid *Mytiloides dresdensis labiatoidiformis* (Western Interior Biozone *Prionocyclus quadratus/Mytiloides dresdensis*) was identified in marine deposits lying above Parasequence Set 6 deposits (Ryer, 1981), near Rochester Panel (Figure 3). The geochronology and molluscan fauna indicate that Gardner's Ferron Sequence (i.e., the Last Chance Delta) was deposited sometime between 90.5 and 88.6 Ma.

Gardner (1992) interpreted the Ferron Sequence as a type-2 sequence based on (1) the absence of significant erosional truncation, (2) the absence of recognized incised-valley deposits, and (3) inconsistent systems tract development. As discussed above, the detailed stratigraphic work we have delineated three major erosional unconformities produced by major fluvial events that record incised-valley development within the Last Chance Delta. These three unconformities have been interpreted to be type-1 depositional sequence boundaries that record significant base-level falls and basinward shifts in depositional facies and superimpose amalgamated braided fluvial and estuarine incised-valley fill deposits on near-marine sandstones (i.e., Ferron Sequence Boundaries SB2, SB3, and SB4 in Plates II and III). Based on regional biostratigraphic correlations and interpreted systems tracts, Garrison (2004) suggested that Ferron Sequence Boundary SB2 is the 3rd-order, type-1 depositional sequence boundary that separates the Hyatti and Ferron Sequences. The identification of higher-order sequence boundaries in both the Ferron Sequence of the Upper Ferron Sandstone Last Chance Delta in Castle Valley, and the Hyatti Sequence Ferron Sandstone Notom Delta in the Henry Mountains region (Garrison, 2004), indicates that the Hyatti and Ferron Sequences of Gardner (1993, 1995a) are composite sequences.

ron Sequence is separated from the underlying 3rd-order Hyatti Sequence by a type-2 sequence boundary. Gardner (1993) indicated that the boundary between these

Systems Tracts

Within the Last Chance Delta, the 14 parasequence sets, described above, form four 4th-order depositional sequences (denoted as Ferron Sequences FS1 through FS4) (Figure 3, Plates II through IV). The lowest 4th-order Ferron Sequence FS1 contains the Parasequence Sets 1A, 1B, 2A, and 2B. Ferron Sequence FS2 consists of the Parasequence Sets 2C, 2D, 3, and 4A. Ferron Sequence FS3 contains the Parasequence Sets 4B and 5A. Ferron Sequence FS4 consists of Parasequence Sets 5B, 6, 7, and 8. Based on stratal properties it is possible to assign of each of these parasequence sets and the 3rd-order and 4th-order depositional sequences to depositional systems tracts.

Third-Order Depositional Sequences

An Upper Turonian to Lower Santonian regional depositional sequence stratigraphic synthesis of the western margin of Utah portion of the Western Interior Basin is presented in Garrison (2004). In that model the 3rd-order Hyatti Composite Sequence and the 3rd-order Ferron Composite Sequence represent the lowstand and transgressive sequences of the Southern Utah Deltaic Complex 2nd-order composite sequence, respectively. The Ferron Composite Sequence contains two major transgressive ravinement surfaces (i.e., within Ferron Sequences FS2 and FS3) and retrogradational parasequence sets (i.e., within Ferron Sequence FS4) attesting to its overall transgressive nature. The lowstand Hyatti Composite Sequence is dominated by highly progradational parasequence sets (Garrison, 2004).

The regional depositional sequence stratigraphic synthesis places Ferron Sequence FS1 as the uppermost sequence in the 3rd-order Hyatti Composite Sequence and assigns Ferron Sequences FS2, FS3, and FS4 to the 3rd-order Ferron Sequence. Ferron Sequence FS1 has a highstand position within the 3rd-order Hyatti Composite Sequence. Ferron Sequences FS2 and FS3 are the 4th-order progradational early lowstand and aggradational late lowstand sequences of the 3rd-order Ferron Composite Sequence, respectively. Ferron Sequence FS4 is the 4th-order transgressive depositional sequence of the Ferron Composite Sequence. Within the Castle Valley study area, the 4th-order highstand sequence of the Ferron Composite Sequence (denoted as Ferron Sequence FS5 in Figure 3) is represented only by marine shale and condensed section sediments. The 4th-order highstand sequence of the Ferron Composite Sequence is represented by the "A" Sandstone in the Kaiparowits Plateau (i.e., the lower part of the "A" Delta of the Straight Cliffs Formation) (see Garrison, 2004).

Fourth-Order Depositional Sequences

Sequence FS1: Upper Ferron Sandstone 4th-order Sequence FS1 is composed of Parasequence Sets 1A, 1B,

2A, and 2B. Examining the stratal characteristics of the parasequence sets within Ferron Sequence FS1, exposed in Castle Valley, suggests that Parasequence Set 1A represents the aggradational early LST and that Parasequence Sets 1B and 2A represent the highly progradational, late LST of Ferron Sequence FS1 (i.e., scenario one). In this scenario, Parasequence Set 2B, composed of the erosional remnants of a single retrogradational parasequence, would represent the HST. The TST of Ferron Sequence FS1 is represented only by the development of a ravinement surface at the top of Parasequence Set 2A and a transgressive lag deposit occurring at the base of Parasequence Set 2B. When the regional correlations and 3rd-order stratal units recognized by Garrison (2004) are examined, it is possible to suggest an alternative interpretation (i.e., scenario two). The case could be made that Parasequence Set 1A represents the early HST, and Parasequence Sets 1B and 2A represent the late HST of a 4th-order sequence that includes the upper portion of the Ferron Sandstone Notom Delta exposed in the Henry Mountains region. In this second scenario, the TST would be represented by the transgressive lag at the top of the Ferron Sandstone exposed in the Henry Mountains region and below the Last Chance Delta in Castle Valley. In scenario one, the location of the HST of the Notom Delta is problematic. In scenario two, the apparently retrogradational erosional remnants of Parasequence Set 2B must represent a minor 5th-order transgressive event. On balance, scenario two appears to best fit the overall regional chronostratigraphy and depositional sequence stratigraphic model, and to explain the overall systems tract development. In the following discussion, the second scenario will be assumed.

In the Last Chance Delta, most of the Ferron Sequence FS1 sediments were deposited in a highly progradational, highstand, wave-reworked river-dominated deltaic system. This deltaic system has near-marine facies with a preserved cumulative dip length of 34.9 km (21.7 mi). Over 268 m (879 ft) of sand-rich sediment was deposited during progradation. The maximum preserved stratigraphic thickness is 54 m (177 ft). The transgressive condensed section deposit at the base of the Last Chance Delta in Castle Valley extends for over 43 km (27 mi) and reaches a thickness of over 3 m (10 ft). The high-order transgressive lag deposit between Parasequence Sets 2A and 2B is only 1.5 m (4.9 ft) thick, and is only exposed between Bear Gulch and Rochester Panel, a distance of only 5.7 km (3.5 mi). The preserved 5th-order highstand sediments of Parasequence Set 2B are of limited extend, having a dip length of only 6.1 km (3.8 mi) and a preserved thickness of less than 10 m (33 ft). At the northern extent of the outcrop belt at Ferron Creek, Ferron Sequence FS1 is only represented by condensed section deposits. The middle part of the black shale condensed section at Ferron Creek is assigned to FS1.

In the outcrop belt, north of Mussentuchit Wash, the lower boundary of Sequence FS1 is a correlative conformity lying below the condensed section deposits described by Gardner (1993) that contain the ammonite *Prionocyclus hyatti*. In the Miller Canyon area and at I-70, this condensed section deposit contains abundant calcareous concretions. In Miller Canyon, this deposit is underlain by a thin, bioturbated transgressive lag that rests upon a ravinement surface at the top of the Washboard Sandstone unit. This lag deposit represents the earliest transgressive deposits of Ferron Sequence FS1.

The upper boundary SB2 of Sequence FS1 is a type-1 boundary. South of Willow Springs Wash and from south Blue Trail Canyon to north Quitchupah Creek, the upper boundary of FS1 lies mainly within an inter-fluvial area and is represented as a correlative conformity lying below the A_2 coal zone. From Willow Springs Wash to south Blue Trail Canyon and from north Quitchupah Creek to the South Goosenecks of Muddy Creek Canyon, the upper boundary is represented by an erosional unconformity at the base of an incised-valley system. North of the South Goosenecks of Muddy Creek Canyon, the upper boundary of Sequence FS1 is a correlative conformity; the delta-front deposits of Ferron Sequence FS2 lie conformably on the delta-front deposits of Ferron Sequence FS1.

Based on molluscan faunal data within Castle Valley, volcanic ash correlations and geochronology, and regional biostratigraphic correlations, Garrison (2004) suggests that within the Last Chance Delta, the HST of Ferron Sequence FS1 was deposited between 90.3 Ma and 90.0 Ma (Western Interior Biozone *Prionocyclus macombi/Inoceramus dimidius dimidius*). Including the lowstand deposits of the upper part of the Notom Delta in the Henry Mountains region, Garrison (2004) speculated that 4th-order Ferron Sequence FS1 was deposited between 90.5 Ma and 90.0 Ma.

Sequence FS2: Upper Ferron Sandstone 4th-order Sequence FS2 is composed of Parasequence Sets 2C, 2D, 3, and 4A. Parasequence Sets 2C and 2D represent the highly progradational LST of 4th-order Ferron Sequence FS2 and Parasequence Set 4A, composed of the erosional remnants of a single parasequence, is the HST. The TST of Sequence FS2 is represented, in part, by ravinement surfaces and transgressive lag deposits that occur at the bases of Parasequence Sets 3 and 4A. The occurrence of Parasequence Set 3, lying stratigraphically above the first major flooding surface and between two major transgressive ravinement surfaces, suggests that it also belongs to the TST of FS2, although it exhibits an aggradational to slightly progradational internal stacking pattern instead of a retrogradational internal stacking pattern generally associated with parasequence sets within a TST.

Most of the FS2 sediments were deposited as the highly progradational LST, represented by Parasequence Sets 2C and 2D, which have a cumulative dip length of 34.4 km (21.4 mi) and represent a cumulative thickness of almost 102 m (335 ft) of sediment deposited during progradation, resulting in a total preserved stratigraphic thickness of approximately 34 m (112 ft). The preserved HST sediments of Parasequence Set 4A are locally truncated by the development of the erosional unconformity of SB3 bounding Sequence FS2, although locally it has a cumulative preserved thickness of 27 m (89 ft) and a total dip length of 15 km (9 mi).

The 2-m-thick (7-ft) transgressive lag TST deposit at the base of Parasequence Set 4a of Sequence FS2, representing a major regional transgression, is exposed between the Tri-Canyon area and Ferron Creek, a distance of over 42.4 km (26.3 mi). The 1-m-thick (3-ft) transgressive lag TST deposit at the base of Parasequence Set 3 of Sequence FS2, also representing a major regional transgression, is exposed from south of Dry Wash to Ferron Creek, a distance of over 24.4 km (15.2 mi). Parasequence Set 3 lying stratigraphically between these two major transgressive ravinement surfaces represents over 40 m (131 ft) of sediment deposited, although due to ravinement only a preserved stratigraphic thickness of 21 m (70 ft). The atypical aggradational to progradational internal stacking pattern of TST Parasequence Set 3 is probably the result of the high sedimentation rate during the highly progradational early stage of the development of the Last Chance Delta. This will be addressed in more detail later.

The upper boundary SB3 of Sequence FS2 is a type-1 depositional sequence boundary. South of Scabby Canyon, south of I-70, the upper boundary of FS2 is a correlative conformity lying below the E Rooted Zone. From Scabby Canyon to Rochester Creek, the upper boundary is an erosional unconformity. This unconformity represents up to 32 m (105 ft) of erosion by fluvial channels of Parasequence Set 4B of overlying Sequence FS3. North of Rochester Creek, the upper boundary of Sequence FS2 is a correlative conformity with delta-front deposits of Sequence FS3 lying conformably on delta-front deposits of Sequence FS2. At the northern extent of the outcrop belt at Ferron Creek, the upper boundary of FS2 lies at the base of a thin bioturbated sandstone lag (?) deposit (Figure 25).

The lower boundary of Sequence FS2 is a type-1 depositional sequence boundary, described above as SB2. From Willow Springs Wash to South Blue Trail Canyon and from North Quitchupah Creek to the South Goosenecks of Muddy Creek Canyon, the lower boundary SB2 is an erosional unconformity representing up to 25–30 m (82–98 ft) of incised-valley erosion. This unconformity is overlain by the amalgamated braided fluvial channels and estuarine incised-valley fill facies of

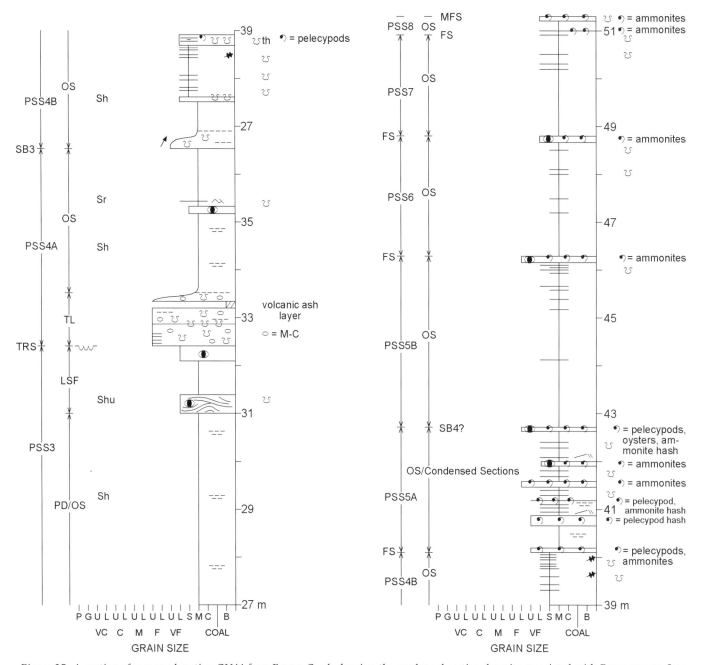

Figure 25. A portion of measured section GV44 from Ferron Creek showing the condensed section deposits associated with Parasequence Sets 4A, 4B, 5A, 5B, 6, 7, and 8. See Figure 5 for explanation.

Parasequence Set 2C. South of Willow Springs Wash and between Blue Trail Canyon and North Quitchupah Creek, the Sequence Boundary SB2 lies mainly within an inter-fluvial area and is represented as a correlative conformity lying below the A$_2$ coal zone. North of the Muddy Creek amphitheater, the lower boundary of Sequence FS2 is a correlative conformity that places delta-front deposits of Parasequence 2f conformably upon Parasequence 2d. North of Bitter Seep Wash to Ferron Creek, the correlative conformity at the base of FS2 lies within the black shale condensed section deposits

stratigraphically above the Washboard Sandstone unit. Fourth-order depositional sequence boundary SB2 is coincident with the 3rd-order depositional sequence boundary at the base of the Ferron composite sequence (Garrison, 2004).

At the northern extent of the outcrop belt at Ferron Creek, Gardner (1993) reported the ammonites *Scaphites warreni,* and *Prionocyclus wyomingensis,* and the inoceramid *Inoceramus dimidius dimidius* from within the deposits that we have assigned to Sequence FS2. A younger molluscan fauna has been recovered from overlying deposits. Based on the occurrence of these molluscan fauna, we suggest that, within the Last Chance Delta, Ferron Sequence FS2 was deposited between 90.0 Ma and 89.7 Ma (Western Interior Biozone *Scaphites warreni/Inoceramus dimidius* [late]).

Sequence FS3: Upper Ferron Sandstone 4th-order Sequence FS3 is composed of Parasequence Sets 4B and 5A. Parasequence Set 4B represents the progradational LST of Sequence FS3 and Parasequence Set 5A represents an aggradational HST. The TST of Sequence FS3 is represented only by a ravinement surface and a transgressive lag deposit occurring at the base of Parasequence Set 5A. The progradational LST, represented by Parasequence Set 4B has a cumulative dip length of 11.4 km (7.1 mi) and a stratigraphic thickness of 18 m (59 ft) of sediment deposited during progradation. The preserved highstand sediments of Parasequence Set 5A are locally truncated by the development of the erosional unconformity of SB4 bounding Sequence FS3. Parasequence Set 5A has an estimated total dip length of 11.8 km (7.3 mi) and represents at least 31 m (102 ft) of sediment deposited during aggradation. The total preserved maximum stratigraphic thickness is 28 m (92 ft). The TST of Sequence FS3 is represented only by a localized ravinement surface and a laterally restricted transgressive lag deposit. The restricted 1-m-thick (3-ft) transgressive lag deposit is exposed between the North Goosenecks of Muddy Creek and Rochester Panel, a distance of only 1.1 km (0.7 mi) and between South Dry Wash and Castle Rock, a distance of only 1.9 km (1.2 mi).

The upper boundary SB4 of Sequence FS3 is a type-1 depositional sequence boundary. South of Miller Canyon, the upper boundary of FS3 is a correlative conformity lying below the I_1 coal zone. From Miller Canyon to north of Picture Flats, the upper boundary is an erosional unconformity, representing 10–20 m (33–66 ft) of erosion into Parasequence Set 5A by amalgamated, braided fluvial incised-valley systems of Parasequence Set 5B. North of Bitter Seep Canyon, the upper boundary of Sequence FS3 is a correlative conformity lying somewhere below the concretion-bearing condensed section reported by Garrison and van den Bergh (1997). In the Dry Wash Tank area, north of Dry Wash, this surface may be correlative with a thin, bioturbated condensed section/lag deposit containing extrabasinal pebbles and shark teeth located 2 m (7 ft) above the distal delta-front deposits of Parasequence 5b of Parasequence Set 5A. At Ferron Creek, the upper boundary of FS3 lies at the top of a concretion-bearing horizon containing ammonites and shell hash.

The lower boundary of Sequence FS3 is a type-1 depositional sequence boundary, described above as SB3. South of Scabby Canyon, the lower boundary of FS3 is a correlative conformity lying below the E Rooted Zone.

From Scabby Canyon to Rochester Creek, the lower boundary of FS3 is an erosional unconformity representing as much as 30 m (98 ft) of incised-valley erosion by the 4th-order, lowstand, amalgamated braided fluvial system. North of Rochester Creek, the lower boundary of Sequence FS3 is a correlative conformity. At the northern extent of the outcrop belt near Ferron Creek, the lower boundary of Sequence FS3 lies at the base of a thin bioturbated sandstone lag deposit (Figure 25).

At the northern extent of the outcrop belt at Ferron Creek, Sequence FS3 is represented by 6 m (20 ft) of silty mudstone containing thin (1 cm [0.4 in.]) very fine grained sandstone layers and thin lag deposits (up to 20 cm [8 in.] thick), containing ammonites and oyster and pelecypod shell hash. Gardner (1993) reported the ammonoideae *Scaphites whitfieldi* and *Prionocyclus wyomingensis* and the inoceramid *Inoceramus dimidius (late-n.ssp.)* and *Inoceramus perplexus perplexus* within the thin lag deposits that we have assigned to Sequence FS3. Based on the occurrence of these molluscan fauna, within the Last Chance Delta, Ferron Sequence FS3 was deposited between 89.7 Ma and 89.2 Ma (i.e., spanning Western Interior Biozones *Scaphites ferronensis/ Inoceramus dimidius* to *Scaphites whitfieldi/Inoceramus perplexus perplexus*).

Sequence FS4: Upper Ferron Sandstone 4th-order Sequence FS4 is composed of Parasequence Sets 5B, 6, 7, and 8. Parasequence Set 5B represents the progradational LST of Sequence FS4 and Parasequence Sets 6, 7, and 8 represent a retrogradational TST. The HST is represented in the Castle Valley by marine shale deposits of the Blue Gate Shale. The upper depositional sequence boundary of FS4 lies above a trangressive systems tract, concretion-bearing, condensed section deposit, located a few meters above the upper most sandstones of the Last Chance Delta. Garrison (2004) suggests that the Ferron Sequence FS4 HST is represented by the "A" Sandstone exposed in the Kaiparowits Plateau (i.e., the lower part of the "A" Delta). The lowstand near-marine sandstone deposit has a dip length of 7.5 km (4.7 mi) and represents 16 m (52 ft) of sediment deposited during progradation. The TST deposits are exposed between Coal Wash and Bitter Seep Canyon, a distance of over 14.2 km (8.8 mi) and represent a total of 35 m (115 ft) of sediment deposited during progradation, resulting in a total preserved stratigraphic thickness of approximately 22 m (72 ft).

The lower boundary of Sequence FS4 is a type-1 depositional sequence boundary, described above as SB4. South of Miller Canyon, this boundary is a correlative conformity lying below the I_1 coal zone. From Miller Canyon to north of Picture Flats, the boundary is represented as an erosional unconformity at the base of the Parasequence Set 5B incised valley. North of Bitter Seep

Canyon, SB4 is a correlative conformity lying somewhere below the concretion-bearing condensed section reported by Garrison and van den Bergh (1997). North of Dry Wash, this surface may be correlative with the thin, bioturbated condensed section/lag deposit containing extrabasinal pebbles and shark teeth. At Ferron Creek, the lower boundary of FS4 lies at the top of a concretion-bearing horizon containing ammonites and shell hash.

In the study area, the upper boundary of Sequence FS4 TST deposits appears to be a correlative conformity/maximum flooding surface boundary. Locally, there is evidence of transgressive ravinement. In the area between Rock Canyon and Corbula Gulch and in the Limestone Cliffs north of Mussentuchit Wash, the upper boundary of the TST is a flooding surface, delineated by marine ichnofossils in the upper surfaces of fluvial deposits of Parasequence Sets 7 and 8. In the area of Coal Wash, the sandstones of Parasequence Set 8 (i.e., the youngest sediments of Sequence FS4) are overlain by a transgressive lag deposit consisting of oyster hash. This deposit is overlain by a concretion-bearing condensed section horizon.

In the study area, SB4 at the top of Sequence FS4 is a correlative conformity lying somewhere above the concretion-bearing condensed section reported by Garrison and van den Bergh (1997). North of Dry Wash, this surface is a correlative surface above the thin, bioturbated condensed section/lag deposit, described above. From the Willow Springs Wash to Ferron Creek, the condensed section is composed of concretion-bearing deposits located a few meters above the outcrops of the uppermost sandstones of the Last Chance Delta. These deposits consist of scattered septarian nodule, calcareous concretions. From the area between Ivie Creek and the Emery Mine portal, Gardner (1993) reported the occurrence of *Inoceramus deformis deformis* within the thin condensed section deposits. The inoceramid *Mytiloides dresdensis labiatoidiformis* was identified in marine deposits lying above Parasequence Set 6 (Ryer, 1981), near Rochester Panel.

At the northern extent of the outcrop belt at Ferron Creek, Sequence FS4 is represented by 8.7 m (28.5 ft) of silty mudstone containing thin (1 cm [0.4 in.]) very fine sandstone layers and thin lag deposits (up to 10 cm [4 in.] thick) containing ammonites. Gardner (1993) reported the ammonite *Scaphites ferronensis* within the thin lag deposits that we have assigned to Sequence FS4.

Based on the occurrence of molluscan fauna within the Last Chance Delta in Castle Valley, the near-marine facies of Ferron Sequence FS4 were deposited between 89.2 Ma and 88.6 Ma (i.e., spanning Western Interior Biozones *Prionocyclus quadratus/Inoceramus perplexus* [late] to *Scaphites marianensis/ Cremnoceramus erectus*). Including the highstand deposits represented by the "A" Sandstone of the Kaiparowits Plateau, the deposition of Fer-

ron Sequence FS4 spanned the interval between 89.2 Ma and 88.0 Ma (Garrison, in preparation).

Fifth-Order Depositional Sequences

The erosional unconformity at the base of Ferron Sequence FS2 is interpreted to represent the development of two small incised-valley systems. These incised-valleys contain a tripartite valley-fill facies association. Within the incised-valley fill succession, the sediments have been interpreted to represent well-defined 5th-order lowstand, transgressive, and highstand incised-valley/estuarine deposits (Plate III). The inner segment of the incised-valley fill contains fluvial fill facies representing initial amalgamated, lowstand braided stream fluvial facies overlain by transgressive and highstand fluvial deposits. The middle segment consists of amalgamated, lowstand braided stream fluvial facies overlain by brackish-bay and bay-head delta transgressive deposits (i.e., located only in the subsurface), which are in turn overlain by highstand fluvial and interfluvial bay facies. The outer segment of the incised-valley fill consists of small lowstand, wave-reworked, delta-front sand bodies (Parasequence 2f) overlain by transgressive estuarine bay facies; locally the underlying lowstand facies are truncated by wave ravinement and the HST consists of wave-reworked stream-mouth-bar facies (Parasequence 2g).

In addition to the 5th-order systems tracts, described above, that fill the incised-valley system above sequence boundary SB2, there are additional 5th-order surfaces and stratal geometries represented within Parasequence Sets 2C and 2D that suggest additional 5th-order forced regressions (Posementier et al., 1992). During continued regression and development of Parasequence Sets 2C and 2D, there appears to have been periodic erosion of shoreface sandstones during the base-level fall and the subsequent progradation of shoreline sediments (Plates III and IV). This is particularly well developed in Parasequence Set 2C, in the Bitter Seep Wash and Cedar Ridge Canyon areas of the Molen Reef, south of Dry Wash, where progradation of Parasequences 2h and 2i was accompanied by the erosion of up to 12 m (39 ft) of underlying parasequence 2g. This erosion surface can be described as a 5th-order sequence boundary. The small, amalgamated fluvial system at the northern end of Muddy Creek Canyon that incised a valley into the highstand deposits of Parasequence 2g is thought to source the near-marine shoreline of Parasequences 2h and 2i. Parasequence 2h and 2i represent the 5th-order lowstand and HSTs, respectively.

Within Parasequence Set 3 there are similar stratal surfaces and geometries that suggest continued 5th-order forced regressions even during the overall 4th-order transgression when Ferron Sequence FS2 developed. In the Molen amphitheater area of the Molen Reef, the progradation of Parasequence 3b is accompanied by the local erosion of up to 5 m (16 ft) of the underlying shoreface sandstones of Parasequence 3a. This surface of erosion, within the TST, can also be described as an additional 5th-order sequence boundary within Sequence FS2.

THE DEPOSITIONAL HISTORY AND STRATAL PATTERNS OF THE HYATTI AND FERRON COMPOSITE SEQUENCES

Figure 26 presents a schematic composite cross section through the Southern Utah Deltaic Complex containing the 3rd-order Hyatti and Ferron Composite Sequences (Garrison, 2004). The three non-contemporaneous fluvial-deltaic systems are projected into the plane of the cross section. The vertical scale of the Straight Cliffs Formation of the Kaiparowits Plateau has been exaggerated in order to illustrate the geochronology of the internal correlation surfaces. For the scales of these three systems, the reader is referred to the cross sections presented by Garrison (2004).

Initial deposition of the Southern Utah Deltaic Complex (i.e., the beginning of the deposition of the Hyatti Composite Sequence) began in the Kaiparowits Plateau region prior to about 90.7 Ma (i.e., earliest part of Western Interior Biozone *Prionocyclus Hyatti*) with the deposition of the highly progradational shoreface sandstones of the upper part of the Tropic Shale and the lower part of the Tibbet Canyon Member of the Straight Cliffs Formation. In the Kaiparowits region, Leithold (1994) described a basinward progradation of sandy heterolithic facies in the Tropic Shale suggesting a type-1 depositional sequence boundary with an age of 92.0 Ma. These Tibbet Canyon shoreface deposits represent a rapidly prograding highstand shoreline. A high-order base-level fall and a major basinward shift in depositional facies occurred at about 90.7 Ma, as deposition of the deltaic shoreline sandstones shifted seaward to the Henry Mountains region. Deposition at the newly established Henry Mountains shoreline position is referred to as the Ferron Sandstone Notom Delta. Initial deposition of the Ferron Sandstone Notom Delta began around 90.7 Ma. The lower part of the Ferron Sandstone of the Notom Delta represents the establishment of the LST of Notom Sequence NS1 at the new shoreline position. Notom Sequence NS1 represents a high-frequency, lowstand sequence within the Hyatti Composite Sequence. Lowstand Notom Sequence NS1 is probably represented in the Kaiparowits as a sediment by-pass surface.

At approximately 90.5 Ma, a 4th-order base-level fall that resulted in a 10 km (6 mi) basinward shift in Notom Delta shoreline position and the development of a small

incised-valley estuarine system representing up to 30 m (98 ft) of erosion. The erosional unconformity, at the base of this newly established lowstand system, is represented by Notom Sequence Boundary SB2. In the Kaiparowits region, there was deposition of the fluvial-estuarine facies association of the upper Tibbet Canyon Member (i.e., the lower part of the Tibbet Sequence). The Notom Sequence NS2 lowstand deltaic sedimentation halted, near 90.3 Ma, as marine flooding and a major regional avulsion shifted the depocenter landward 20 km (12 mi) and north-northwestward along the coastline 60 km (37 mi), to the present-day position of the southern Castle Valley. The Notom Delta was abandoned and became the site of 15–30 m (49–98 ft) of transgressive ravinement (i.e., the development of a condensed TST), followed by condensed section and marine shelf deposition until at least 88.6 Ma. In the Kaiparowits Plateau, this transgressive surface is represented as a correlative conformity at the top of the Tibbet Canyon Member of the Straight Cliffs Formation (Figure 26).

At the new northerly depocenter, Hyatti Composite Sequence highstand deposition was resumed as the Upper Ferron Sandstone Last Chance Delta developed at about 90.3 Ma. The highstand deposits are represented in the Last Chance Delta by Ferron Highstand Sequence FS1 (Figures 26). At about 90.0 Ma, a 3rd-order base-level fall resulted in an initial 3 km (2 mi) basinward shift in paleoshoreline and the development of small incised-valley system representing up to 30 m (98 ft) of erosion. This event represents the initiation of development of the LST of the Ferron Composite Sequence. The development of this type-1 sequence boundary was manifest in the Kaiparowits Plateau as the Calico Sequence Boundary. Ferron Composite Sequence lowstand deltaic deposition is represented in the Last Chance Delta as Ferron Early Lowstand Sequence FS2. Lowstand deposition continued until approximately 89.1 Ma. This lowstand deposition was interrupted by a 4th-order base-level fall at 89.7 Ma, which was accompanied by a basinward shift in paleoshoreline of 7 km (4 mi) and the development of an incised-valley system with up to 32 m (105 ft) of erosional relief. These deposits are represented in the Last Chance Delta as Ferron Late Lowstand Sequence FS3. At 89.2 Ma, a second 4th-order base-level fall shifted the paleoshoreline 3 km (2 mi) seaward and allowed the development of a second incised-valley system which developed 10–20 m (33–66 ft) of erosional relief. At about 89.2 Ma, a Ferron Composite Sequence 3rd-order TST (i.e., Ferron Transgressive Sequence FS4) began to develop. Initial FS4 lowstand deposition continued until 89.1 Ma, at which time retrogressive parasequence sets began to develop and the paleoshoreline progressively stepped landward (Figures 26). In the Kaiparowits Plateau, fluvial-estuarine deposition was occurring during the time of the deposition of the latest lowstand and transgres-

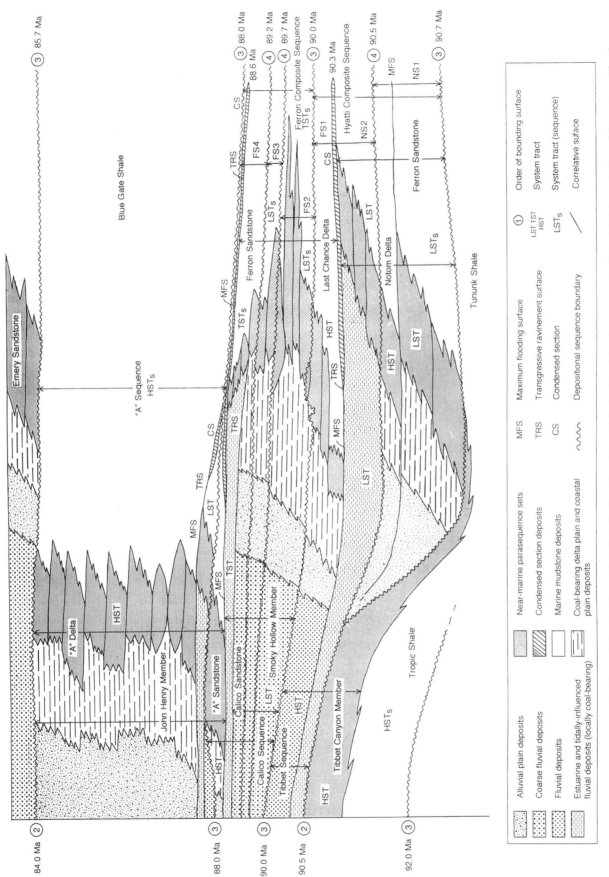

Figure 26. Schematic, composite cross section through the Southern Utah Deltaic Complex along with the depositional sequence stratigraphic interpretation of the system. The three non-contemporaneous deltas are projected into the plane of the cross section. The vertical scale of the Smoky Hollow Member in the Kaiparowits Plateau has been exaggerated to illustrate internal surfaces. Systems tracts are based on depositional sequence internal stratal properties and imply no specific position on a local sea-level curve. The lowstand, transgressive, and highstand systems tracts are denoted as LST, TST, and HST, respectively. Systems tracts of the Hyatti and Ferron Composite Sequences are high-order depositional sequences (denoted as LSTs, TSTs, etc.). Systems tracts within the Ferron Sandstone high-order sequences of Hyatti and Ferron Composite sequences are not shown.

178

sive deposits of the Last Chance Delta. A major marine-flooding event, coupled with a major regional avulsion, ended Last Chance deltaic deposition at about 88.6 Ma. The depocenter shifted landward 80 km (50 mi) and south-southeastward 60 km (37 mi), to the present-day position of the Straight Cliffs. Ferron Composite Sequence TST condensed section and highstand marine shelf deposition continued at the site of the abandoned Last Chance Delta. The maximum flooding surface and transgressive surface at the top of the Upper Ferron Sandstone Last Chance Delta is represented in the Kaiparowits Plateau as a transgressive ravinement surface, and its correlative conformity, at the base of the "A" Sandstone of the John Henry Member of the Straight Cliffs Formation. The depositional sequence boundary at the top of the 3rd-order Ferron Composite Sequence (i.e. Ferron Sequence Boundary SB5) is a correlative conformity lying within the Blue Gate Shale above the top of the Upper Ferron Sandstone.

At the new southerly depocenter, Ferron Composite Sequence highstand deposition was resumed as the John Henry Member "A" Delta developed at about 88.6 Ma. The Ferron Composite Sequence highstand deposits are represented, in the "A" Delta, by the "A" Sandstone (i.e., a parasequence set of the John Henry Member of the Straight Cliffs Formation, Figures 26). At about 88.0 Ma, a 3rd-order base-level fall, marking the end of Ferron Composite Sequence deposition, interrupted "A" Delta deposition. The paleoshoreline shifted basinward and incised-valley and estuarine lowstand depositional systems developed within the John Henry Member. This lowstand deposition began at about 88.0 Ma and marks the beginning of the development of the LST of the "A" Sequence, as defined by Shanley and McCabe (1995). "A" Sequence deposition continued at this southerly position until at least 86.0 Ma.

FACTORS CONTROLLING THE DEVELOPMENT OF THE LAST CHANCE DELTA

Geochronology

The faunal data summarized by Garrison (in preparation) suggest that the Upper Turonian-Lower Coniacian, Upper Ferron Sandstone Last Chance Delta was deposited between approximately 90.3 Ma and 88.6 Ma. Figure 27 shows the depositional sequence stratigraphy and coal zone stratigraphy integrated with the chronostratigraphy and the molluscan faunal data summarized by Garrison (2004) and with the eustatic sea-level curves of Haq et al. (1988). During the deposition of the Last Chance Delta, these eustatic sea-level curves indicate a

slow 2nd-order sea-level fall event, upon which are superimposed portions of two 3rd-order sea-level cycles.

In an attempt to scale the time intervals associated with parasequence sets deposited between biozone time data points, two assumptions were made. First, we assume that, in general, the cyclicity of parasequence deposition is controlled by cyclic processes, such as the Milankovitch cycles suggested by Leithold (1994) for slightly older parasequences in the Tropic and Tununk Shales. Secondly, the deposition of each parasequence, occurring within a specific time interval, is assumed to have occurred over a similar time span.

Based on stratigraphic data, Garrison (2004) estimated that the average length of time for the deposition of each of the Upper Ferron 4th-order Sequences (i.e., FS1 through FS4) would be about 625,000 years. Ferron Sequences FS1, FS2, and FS3 have durations of 500,000, 300,000, and 500,000 years, respectively. The duration of Ferron Sequence FS4 was much longer, lasting 1,200,000 years. The individual depositional events that produced the parasequences within the Upper Ferron Sandstone have durations from 7000 to 150,000 years, averaging about 41,000 years. The parasequence sets within the Upper Ferron Sandstone, have durations of 69,000 to 250,000 years, averaging about 123,000 years. The duration of the Upper Ferron 4th-order depositional sequence events are similar to those of older 3rd-order events within the Tununk Shale (Leithold, 1994). The duration of the depositional episodes that resulted in the Upper Ferron parasequences and parasequence sets are not inconsistent with the 20,000 to 100,000 year duration postulated for 4th-order Milankovitch cycle-driven events in the underlying Tununk Shale (Leithold, 1994).

Relative Changes in Sea Level

Based on the geochronology and the detailed depositional sequence stratigraphy, relative sea- level and subsidence curves for the Last Chance Delta were developed (Figure 28). Relative changes in sea level were estimated from the stratigraphic rise of paleoshorelines during progradation and from estimates of the magnitude of stratigraphic rise in shoreline during transgression and final flooding of parasequences.

The earliest deposition of Ferron Parasequence Set 1A occurred during the beginning of a slow east-central Utah relative rise in sea level about 90.3 Ma. The initial rate of relative rise in sea level, during the initial deposition of Parasequence Set 1B, was approximately 1.3 mm/year. The youngest parasequence set of the outcrop belt, Parasequence Set 8, was deposited in earliest Coniacian time at about 88.7 Ma, while a local relative rise in sea level continued, albeit at a very slow rate. During the final stages of Last Chance Delta deposition, the rate of

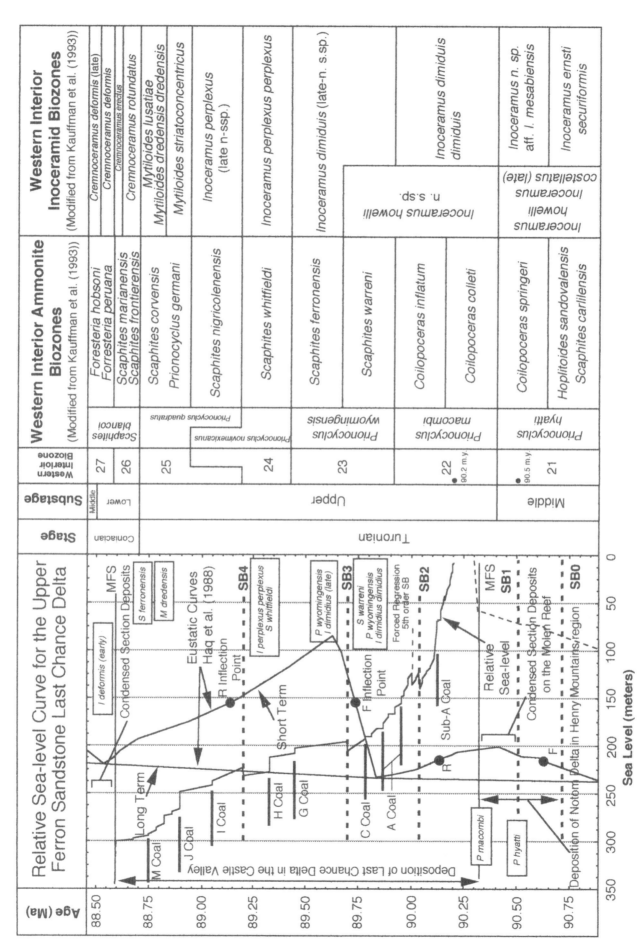

Figure 27. Diagram showing the depositional sequence stratigraphy of the Upper Ferron Sandstone Last Chance Delta, as defined in this paper, integrated with the chronostratigraphy and the faunal data reported by Gardner (1993) and the eustatic sea-level curves of Haq et al. (1988).

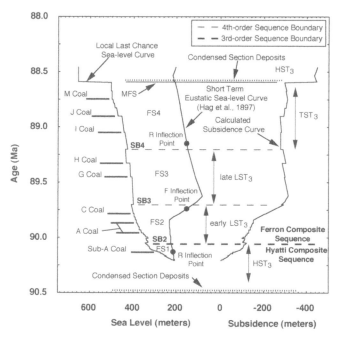

Figure 28. Diagram showing a relative sea-level curve for central Utah and a subsidence curve for the Last Chance Delta developed from the regional cross sections.

relative rise in sea level is estimated to have been as low as 0.05 mm/year. Figure 27 shows that this slow sea-level rise was not a constant rate phenomenon and that periodically this trend of slow sea-level rise was interrupted by small fluctuations, at which time there were short periods of relative fall in sea level or near still stand. This is particularly evident during the deposition of Sequence FS2 (Figure 27).

The Interplay of Tectonics, Subsidence, Relative Sea Level, and Sediment Supply

Based on the regional cross sections (Plates II through IV), the geochronology, and the sea-level and subsidence curves shown in Figure 28, it is possible to speculate on the sedimentation rate, the development of accommodation space, and the timings and magnitude of relative changes in sea level during each of the depositional episodes occurring within the Last Chance Delta clastic wedge. This information, when combined with the regional model for the control and development of fluvial-deltaic shoreline successions along the south-central and east-central Utah portions of the Western Interior Basin, makes it also possible to theorize on the interplay of tectonics, subsidence, relative sea level, and sediment supply during the deposition of the Last Chance Delta.

The relative sea level and subsidence curves for the Last Chance Delta suggest that local subsidence, coupled with the eustatic events, produced a local, slow 3rd-order sea-level rise event in central Utah during the

late Turonian and early Coniacian deposition of the Upper Ferron Sandstone Last Chance Delta. Regional analysis of the Southern Utah Deltaic Complex (i.e., including the Last Chance Delta), indicates that in the rapidly subsiding parts of the Western Interior Basin, such as in the area of the Southern Utah Deltaic Complex, Cordilleran tectonics exerted, at least locally, significant control over basin subsidence and sedimentation (Garrison, 2004). The 3rd-order and 4th-order relative changes in sea level, recorded within the Last Chance Delta and within the entire Southern Utah Deltaic Complex, had a significant tectonic component. This conclusion was based on the correlation of changes in coastal onlap, changes in sedimentation rate, and relative changes in sea level with the timings of basin faulting, Cordilleran volcanism, and thrust faulting (Garrison, 2004).

Effects of Sediment Supply and Accommodation Space on Stratal Elements

Gardner (1992, 1995a) previously demonstrated that within the Upper Ferron Sandstone, the total volume of sediment progressively decreased during each successive depositional episode (i.e., Gardner's genetic sequences). The stratigraphic and architectural data, for both near-marine and non-marine facies, as quantified by Garrison and van den Bergh (1997) also indicate that the over-all Ferron Sandstone Last Chance Delta deposition was controlled by sediment supply.

In this study, vertical sedimentation rate is calculated from depositional facies maximum thicknesses within a near-marine sandstone facies. The rate of development of vertical accommodation space is estimated from the magnitude of the relative changes in sea level occurring during parasequence development. Figure 29 shows the sedimentation rate and the rate of development of accommodation space during the deposition of the Last Chance Delta clastic wedge. Figure 30 shows sedimentation rate and relative sea level as a function of time. Figure 31 shows the relative changes in shoreline (bayline) position (Posementier et al., 1988) and the seaward limit of sandstone deposition as a function of time for the Last Chance Delta. Given a slow, relatively constant, sea-level rise, any changes in sedimentation rate will result in dramatic alterations in the ratio of the sedimentation rate (S) to the rate of development of accommodation space (A) (i.e., the ratio S/A). This process is very well demonstrated by the Upper Ferron Sandstone Last Chance Delta (Figures 29, 30, and 31).

Parasequences and parasequence sets: The high sedimentation rates (>1.5 mm/year), in the early stages of development of the Last Chance Delta resulted in a volume of sediment that exceeded the accommodation space created by the average 0.7 mm/year rate of rela-

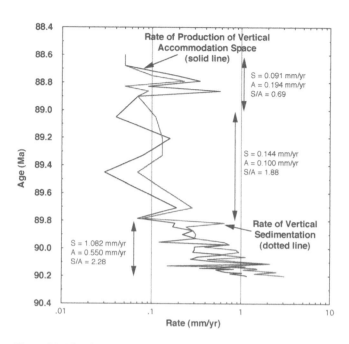

Figure 29. Plot showing the sedimentation rate and the rate of development of accommodation space during the deposition of the Last Chance Delta clastic wedge, as calculated from the geochronology and the regional cross sections.

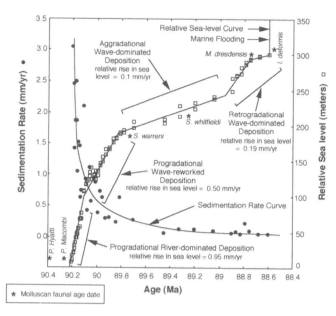

Figure 30. Plot showing interplay of sedimentation rate and relative change in sea level as a function of time during deposition of the Last Chance Delta.

tive rise in sea level. This resulted in deltaic events prograding very rapidly and parasequences stepping seaward at a dramatic rate. This is well illustrated by the extensive seaward progradations of the deltaic components of Parasequence Sets 1, 2A, 2C, 2D, and 3. These parasequence sets have near-marine sandstones with total progradational lengths that average 25 km (16 mi). A decrease in sediment supply during the development of FS3 resulted in a more balanced system, with sedimentation rates of 0.1 mm/year and a relative rise in sea level of only 0.1 mm/year. This resulted in the development of a more aggradational system. The parasequence sets deposited during this phase have near-marine sandstones with a mean total progradational length of only 14 km (8.7 mi). During the final stages of development of the Last Chance Delta, the rate of relative change in sea level was 0.2 mm/year and the 0.1 mm/year sedimentation rate could not keep pace. The increase in accommodation space initiated a transgression such that the final parasequence sets back-stepped landward. The near-marine sandstones of these final parasequence sets have total progradational lengths that average only 5 km (3 mi).

Fourth-order depositional sequences: The internal architecture of the 4th-order depositional Sequences FS1 through FS4 of the Ferron clastic wedge (Plates II through IV), also reflect the progressive change in the ratio of sediment supply to available accommodation space. Fourth-order highstand Sequence FS1 formed during a period in which the sedimentation rate (1.9

mm/year) was much greater than available accommodation space, created by an average relative rise in sea level of 0.9 mm/year. It has an internal architecture dominated by progradational parasequence sets. Fourth-order highstand Sequence FS2 formed during a period when the sedimentation rate was initially high (1.1 mm/year) and the rate of relative rise in sea level was 0.3 mm/year (i.e., S/A>1), but later became more balanced as the sedimentation rate dropped to 0.3 mm/year. It has an internal architecture consisting of two progradational lowstand parasequence sets overlain by a slightly progradational parasequence set that is stratigraphically between two transgressive ravinement surfaces and lags (i.e., a TST). This TST does not contain a retrogradational parasequence set. The TST is in turn overlain by a well-developed highstand parasequence set. Fourth-order, late-lowstand Sequence FS3 developed during a period when sediment supply (i.e. S = 0.1 mm/year) was balanced with the rate of development of accommodation space (i.e. A = 0.1 mm/year). Its internal architecture is dominated by aggradational parasequences. The oldest parasequence set consists of a single parasequence overlain by a transgressive ravinement surface and lag deposit. The highstand parasequence set is slightly progradational to aggradational. Fourth-order transgressive Sequence FS4 developed during a period when the sediment supply (S = 0.1 mm/year) could not keep up with the 0.2 mm/year average relative rise in sea level. Its internal architecture is dominated by retrogradational parasequence sets. The oldest parasequence set consists of a single parasequence whose external geometry is consistent with that of older aggra-

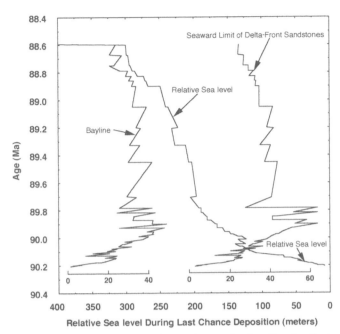

Figure 31. Plot showing the relative changes in shoreline (bayline) position (Posementier et al., 1988) and the seaward limit of sandstone deposition as a function of time for the Last Chance Delta.

dational parasequences. The TST is represented by retrogradational parasequence sets. A transgressive lag is only locally developed within the TST. The upper part of the TST consists of a thin condensed section lying a meter or so above the uppermost sandstones. In the study area, the HST is represented by marine shale.

Incised-valley fill deposits: Based on the regional cross sections, it can be concluded that the nature of the incised-valley fill deposits progressively change with time. The lithologies and internal organization of the incised-valley fill facies and the development of either 4th-order or 5th-order systems tracts within the incised-valley fills are controlled by the sediment supply/ accommodation space systematics.

The oldest incised-valley system occurring at the base of sequence FS2, within the lower, highly progradational part of the Last Chance Delta clastic wedge, was filled during a time when sediment supply far outpaced the slow 3rd-order relative rise in sea level. Moderate sediment by-pass allowed this incised-valley fill to develop elements of 5th-order lowstand, transgressive, and HSTs. The inner segment of the incised-valley fill contains fluvial fill facies representing initial, amalgamated, lowstand braided stream fluvial facies overlain by highstand fluvial deposits. The middle segment consists of amalgamated, lowstand braided stream fluvial facies overlain by brackish-bay and bay-head delta transgressive deposits, which are in turn overlain by highstand fluvial and interfluvial bay facies. The outer segment of the incised-valley fill consists of small, lowstand wave-reworked delta-front sand bodies overlain

by transgressive estuarine bay facies; the HST consists of wave-reworked stream-mouth-bar facies.

The incised-valley system occurring at the base of sequence FS3 was filled during a period when sedimentation rate was balanced with the rate of relative rise in sea level. Fluvial and shoreline aggradation occurring during this period allowed the incised valley to be filled completely with 4th-order, lowstand amalgamated braided fluvial fill facies. The 4th-order TST is represented by thin brackish-water estuarine facies occurring locally at the top of the valley fill. The 4th-order HST is represented by wave-modified shoreface sandstones that "over-filled" across the top of the older incised-valley fill deposits.

The incised-valley system occurring at the base of sequence FS4 was filled during a period when sedimentation rate was slightly less than the rate of relative rise in sea level. Fluvial and shoreline aggradation occurring during this period allowed the incised valley to be filled completely with 4th-order, lowstand, amalgamated braided fluvial fill facies. The 4th-order early-TST is represented by thin brackish-water estuarine facies occurring locally at the top of the valley fill. The sedimentation rate was outpaced by the relative rise in sea level and the 4th-order late-TST, represented by thin retrogradational wave-modified shoreface parasequences, "over-filled" across the top of the older incised-valley fill deposits.

Subsidence, Regional Avulsion, and Sedimentation Rate

The systematics surrounding the abandonment of the Last Chance Delta provide insight into the interplay of subsidence, regional avulsion, depocenter shifts, and changes in sedimentation rate. The sedimentation rate of the Last Chance Delta progressively decreased as the delta evolved, from an initial 3–4 mm/year, until in the final stages of the delta's development sedimentation rates were incredibly low (<0.1 mm/year). Upon abandonment of the delta and the shift of the depocenter to the site of the HST of Ferron Sequence FS4 at the base of the "A" Delta in Kaiparowits Plateau, the deltaic sedimentation rate was again restored to a rate comparable to that existing early in the development of the Last Chance Delta. This suggests that the regional avulsion was correlative with an increase in sedimentation rate. The cause of the increase in sedimentation rate is still problematic. The marine flooding of the Last Chance Delta, after abandonment, and the resulting ravinement and initiation of marine condensed section deposition would imply that delta abandonment was nearly contemporaneous with a major relative rise in sea level. A plot of sedimentation rate versus rate of relative change in sea level for the Last Chance Delta suggests that sedimentation rate is strongly correlated with relative rise in

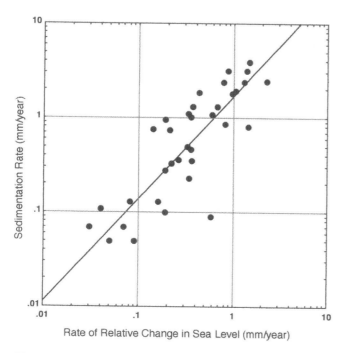

Figure 32. Plot showing the change in sedimentation rate versus the rate of relative change in sea level for the Last Chance Delta.

sea level (Figure 32). Rate of relative sea-level rise slowly decreased during the evolution of the Last Chance Delta and sedimentation rate declined as well. It can be argued that the final abandonment and flooding of the Last Chance Delta indicates an increase in regional subsidence concurrent with an increase in sedimentation rate. The tectonic regime of this Cretaceous foreland basin makes episodic increases in subsidence likely. In fact, the abandonment and flooding of the Notom Delta of the Henry Mountains and the concurrent shift of the depocenter northward to the Last Chance Delta was also accompanied by an increase in sedimentation rate.

Effects of Tectonics and Eustasy on Relative Changes in Sea Level

Garrison (2004) examines the relative timing of relative changes in sea level and sedimentation rate, as recorded in the depositional sequence stratigraphy of the Southern Utah Deltaic Complex, relative to tectono-magmatic events, within the Cordillera and the Western Interior Basin, and speculates about mechanisms for local subsidence and uplift that result in relative changes in sea level. The timing of chronostratigraphic events, within the Southern Utah Deltaic Complex, is compared with the timing of basin subsidence events, igneous activity, and periods of increased basin tectonism in the Sevier thrust belt of the Cretaceous Cordillera and adjacent Western Interior Basin. In the discussion below we will summarize the data relevant to the evolution of the Last Chance Delta.

Episodic volcanism occurred in the Cordillera during the time span when the Hyatti and Ferron Composite Sequences were deposited. A major period of volcanism began at about 90.7 Ma and continued until about 88.6 Ma. Peak volcanism occurred at about 90.3 Ma and 88.7 Ma. This volcanism roughly corresponds to a time of increased active vertical uplift of basement blocks in the fore bulge, hinge zone, and other parts of the foreland basin (Kauffman and Caldwell, 1993). This period of volcanism and tectonism coincides with the deposition of the 3rd-order, lowstand Hyatti Composite Sequence and the 3rd-order, transgressive Ferron Composite Sequence of the 2nd-order composite sequence of the Southern Utah Deltaic Complex (Garrison, 2004). The abandonment and transgressive ravinement of the Notom Delta, within the Hyatti Composite Sequence, corresponds to the beginning of the extensive period of regional basinal subsidence, suggested by Kauffman and Caldwell (1993). The period of quiescence in basin tectonism, following this volcanic event, coincides with the deposition of the 4th-order transgressive Ferron Sequence FS4, within the transgressive 3rd-order Ferron Composite Sequence, and the subsequent abandonment and flooding of the Last Chance Delta.

During Turonian to early Campanian time, tectonic activity in the Sevier orogenic belt of the Cordillera was restricted to the Pavant 2 thrust event from about 88.8 Ma to 84.0 Ma (Schwans and Campion, 1997). The deposition of the 3rd-order, lowstand Hyatti Composite Sequence and the 3rd-order, transgressive Ferron Composite Sequence occurred during the tectonically quiet period preceding the initiation of the Pavant 2 event. The development of high-frequency 4th-order and 5th-order depositional sequences within the Last Chance Delta and 4th-order depositional sequences within the Notom Delta also characterizes the period of tectonic quiescence in the Cordillera (Garrison, 2004). The maximum flooding and abandonment of the Last Chance Delta coincided with the beginning of the Pavant 2 thrust activity at about 88.8 Ma.

The two major transgressions (at 90.3 Ma and 88.6 Ma), which correspond to the abandonment, flooding, and transgressive ravinement of the Notom and Last Chance Deltas, respectively, immediately follow periods of increased or renewed volcanic activity in the Cordillera and generally coincide with periods of reduced tectonic activity or tectonic quiescence in the foreland. The formation of the 3rd-order and 4th-order depositional sequence boundaries (i.e., the development of erosional unconformities and major basinward shifts in paleoshorelines) at 90.7 Ma, 90.5 Ma, 90.0 Ma, 89.7 Ma, 89.2 Ma, and 88.0 Ma correspond to the times of renewed or intensified volcanic activity in the Cordillera and to times of increased basin tectonism.

The relative importance of eustasy and tectonically induced basin subsidence and basin uplift in influencing variations in relative changes in sea level in foreland basins have been discussed extensively (e.g., Kauffman and Caldwell, 1993; Beaumont et al.,1993; Ryer, 1994). Beaumont et al. (1993) argued that the partitioning of relative changes in sea level in foreland basins between eustasy and tectonic subsidence is problematic and ambiguous at best, since the sedimentation patterns necessary to deconvolve these two effects lie beyond the influences of tectonic processes responsible for basin subsidence. Nevertheless, the data from the Last Chance Delta can be used in an attempt to weigh these two processes in evaluating the sedimentation patterns within the Utah sector of Western Interior Basin.

Garrison (2004) suggests that the middle Turonian to upper Santonian high-frequency depositional sequence stratigraphic events within the Southern Utah Deltaic Complex that are correlative with Cordilleran volcanism and Cordilleran and Basin tectonism indicate that 3rd-order and 4th-order changes in the rate of relative change in sea level were synchronous with tectonic activity. This correlation further suggests that the rate and stratal geometry of local foreland basin sedimentation were controlled by Cordilleran tectonics. The correlation of regional avulsions and depocenter shifts with increases in basin subsidence further argues that changes in subsidence rate (i.e., relative changes in sea level) and in sedimentation rate are coupled, since thrust fault activity and thrust sheet stacking control both sediment supply and basin subsidence.

Therefore, we conclude that in the rapidly subsiding parts of the Western Interior Basin the Cordilleran tectonics did exert, at least locally, significant control over basin subsidence and sedimentation. From the summary presented above, it is clear that the 3rd-order and 4th-order relative changes in sea level, recorded within the Southern Utah Deltaic Complex, do have a significant tectonic component. The extent to which eustasy affected stratal geometry and architecture is only speculative at this time, although the general lack of correlations between 2nd-order relative changes in sea level and tectono-magmatic events in the Cordillera suggest that the longer-term, 2nd-order relative sea-level curve may have a significant eustatic component.

COAL ZONES AND VOLCANIC ASH LAYERS AS STRATIGRAPHIC TOOLS

This study of the depositional sequence stratigraphy of the Upper Ferron Sandstone Last Chance Delta was conducted, in part, to determine the uses of and limitations of coal zones, as defined above, in depositional sequence stratigraphic studies. It can be concluded, from this study, that coal zones are very useful in defining the basic building blocks (i.e., stratal elements) of deposi-

tional sequences. Coal zones, especially those containing volcanic ash layers, are powerful tools for making regional correlations within depositional sequences from marine facies associations, where conventional shoaling upward cycles and flooding surfaces are easily recognizable, landward into non-marine facies associations, where significant surfaces are more elusive.

Coal Zones in Regional Correlations and Depositional Sequence Stratigraphy

The regional stratigraphic mapping of the Last Chance Delta indicates that the upper boundaries of depositional parasequence sets, when not coincident with sequence boundaries or transgressive ravinement surfaces, are the tops of coal zones. For example, the Sub-A, A_2 and A_3, C, G, I, J, and M coal zones occur at the tops of Parasequence Sets 1B, 2C, 2D, 3, 4B, 5B, 6 and 7, respectively. Garrison et al. (2004) concludes that the Last Chance Delta coals formed in progradational back-barrier/lagoonal environments. This conclusion is based on the observation that the seaward limits of coal zones lie stratigraphically above near-marine reworked mouth-bar and shoreface facies, indicating mire progradation across near-marine facies. In addition, these coals are commonly associated with washover fan facies. Mire progradation and subsequent stratigraphic rise of the paleoshoreline facilitate the development of laterally extensive mires with internal aggradation.

Coal zones are frequently removed or partially eroded by transgressive ravinement (e.g., the ravinement of the C coal zone by the transgressive ravinement surface at the base of Parasequence Set 4A) or erosional unconformities associated with depositional sequence boundaries (e.g., the removal of a portion of the A_1 coal zone above Parasequence Set 2A by depositional sequence boundary SB2 and the erosion of coal zones at the tops of Parasequence Sets 4A and 5A by depositional sequence boundaries SB3 and SB4) (Plates II and III). Within interfluvial areas of the delta-plain facies association, it is difficult to determine whether subaerial exposure associated with a base-level fall and the subsequent development of a depositional sequence boundary has removed a coal zone.

Coal zones frequently split into multiple sub-zones. These sub-zones can be separated by fluvial channel-belt sandstones, crevasse splays, landward pinchouts of near-marine sandstones, or by delta-plain mudstones and siltstones. In many cases, in the delta-plain facies association, these splits are related to the existence of multiple parasequence sets that are only recognized seaward in the near-marine facies association (e.g., the A_2 and A_3 splits of the A coal zone associated with Parasequence Sets 2C and 2D) (Plates II and III). Care must be exercised in correlating coal splits or sub-zones, in the absence of volcanic ash layers. For example, in this

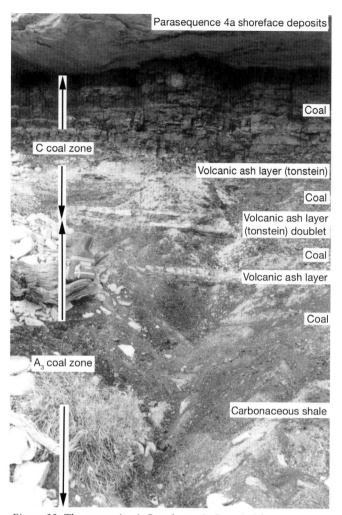

Parasequence 4a shoreface deposits

Coal

C coal zone

Volcanic ash layer (tonstein)

Coal

Volcanic ash layer (tonstein) doublet

Coal

Volcanic ash layer

Coal

A₃ coal zone

Carbonaceous shale

Figure 33. The composite A-C coal zone in Bear Gulch.

study, in the absence of volcanic ash layers, it would have been impossible to reliably determine that the A and C coal zones were stacked in the Bear Gulch-Miller Canyon area (Figures 33 and 34, Plate III) and that the A_2 and A_3 splits of the A coal zone were associated with Parasequence Sets 2C and 2D (Plate III), respectively.

Although minor coal zone splits may locally cap a single parasequence (e.g., the Sub-A_1-A_2 coal zone splits above and below Parasequence 1d in the Coyote Basin-Rock Canyon area), rarely do coal zones cap single parasequences within a parasequence set. In the Last Chance Delta, the H coal zone capping Parasequence 5a within Parasequence Set 5A is the only documented example of a coal zone forming the top of a genetically related parasequence, internally within a parasequence set (Plate III).

It is likely that in the non-marine portions of the outcrop belt, coal zones may occasionally represent peat deposited during the span of several parasequences or parasequence sets, but this is generally impossible to determine simply by field inspection, in the absence of volcanic ash layers. For example, the composite A-C coal

zone in the Bear Gulch-Miller Canyon area (Plate III) suggests that the precursor peat accumulation was stable geographically for at least the length of time required for the deposition of Parasequence Sets 2D and 3 (i.e., the time for the deposition of at least six parasequences).

Volcanic Ash Layers in Regional Correlations

Volcanic ash layers are extremely important in verifying coal zone correlations, particularly if a coal zone contains splits. Volcanic ash layers are reliable only in the parts of the fluvial-deltaic environment favorable for their preservation. Landward and seaward of near-marine pinchouts of parasequence sets, preservation potential of volcanic ash layers decreases dramatically. This phenomenon is well illustrated by the seaward and landward ends of the C coal zone. For example, the southern most 12 km (8 mi) of the C coal zone, where the C coal zone occurs in the non-marine portion of the outcrop belt, is devoid of volcanic ash layers. From Rock Canyon to I-70 only a single tonstein is present in the C coal zone. In Bear Gulch four volcanic ash layers are assigned to the C coal zone, while in Miller Canyon three volcanic ash layers occur. About 4 km (3 mi) north of Miller Canyon, near the landward pinchout of associated Parasequence Set 3, the C coal zone still contains two volcanic ash layers. This is approximately 20 km (12 mi) landward of the seaward pinchout of Parasequence Set 3. At some distance landward of the paleoshorelines (e.g., in the upper-most delta plain and alluvial plain), the probability that an ash layer will be reworked by terrigeneous incursions into peat swamps increased dramatically.

Volcanic ash layers that are generally well preserved throughout most of the fluvial-deltaic facies associations can be subject to erosion at their seaward ends by transgressive ravinement. The thickest ash layer horizon in the C coal zone can be traced over 51 km (32 mi), although locally it is eroded or reworked by transgressive ravinement (e.g., in the region from Bitter Seep Wash to just south of Ferron Creek [Plate III]).

Datums for Regional Correlations

Although, in general, while coal zones are time transgressive, they are laterally extensive and make good datums for regional correlations. Volcanic ash layers are laterally extensive and represent compact chronostratigraphic horizons, which make good datums for regional correlations. However, using coal zones and volcanic ash layers as datums do not go without some problems. First, rarely does a coal zone extend throughout a study area. Secondly, the frequent splits of coal zones into sub-zones, makes choosing a coal zone as a datum somewhat difficult. A coal zone chosen as a

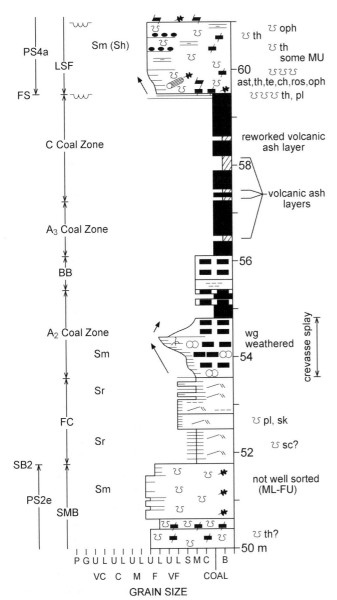

Figure 34. A portion of measured section GV8 in Bear Gulch showing the character of the composite A-C coal zone. See Figure 5 for explanation.

ravinement surface at the top of the stratigraphic horizon correlative with the top of the C coal zone.

Conceptually, choosing a laterally extensive volcanic ash layer as a datum is ideal (e.g., see Gardner, 1993). It gives the datum chronostratigraphic significance. The problems associated with using volcanic ash layers as datums in non-marine and near-marine facies associations arise from the variable preservation potential of ash layers. Ash layers, in upper delta-plain settings, are frequently reworked by fluvial and crevasse processes. In the near-marine setting, they are frequently subject to transgressive ravinement events.

In summary, care should be used when making regional correlations based on coal zones and volcanic ash layers as datums. When possible, it is most advantageous to choose a coal zone, with a minimal number of splits and which contains at least one volcanic ash layer over much of its length.

DEFINING DEPOSITIONAL SEQUENCE STRATIGRAPHIC BUILDING BLOCKS IN NON-MARINE FACIES ASSOCIATIONS

Parasequences and Parasequence Sets

The recognition of parasequences and parasequence sets in near-marine and proximal near-marine settings is well documented (e.g., see Van Wagoner et al., 1990), but their recognition in the non-marine delta-plain and lower alluvial-plain settings remains problematic. The recognition of depositional sequences, which are bounded by erosional unconformities, is also well documented in marginal near-marine settings, but their recognition in inter-fluvial areas of the delta-plain and lower alluvial-plain facies associations (i.e., outside paleovalleys) remain also problematic. The coal zone and depositional sequence stratigraphy of the Upper Ferron Sandstone Last Chance Delta has provided insight into the recognition of the basic building blocks of the depositional sequence in non-marine facies associations (i.e., delta-plain and lower alluvial-plain settings).

We were not able to recognize individual parasequences within the non-marine facies associations of the Last Chance Delta. The only exception to this general conclusion occurs when a parasequence is contained within a minor split in a coal zone. For example, the non-marine facies of Parasequence 1d in Willow Springs Wash was delineated where it occurs within the Sub-A_1 and Sub-A_2 splits of the Sub-A coal zone and could be correlated with its correlative near-marine facies 2 km (1 mi) seaward near Rock Canyon (Plate II). These observations are in marked contrast to the study of Kamola and Van Wagoner (1995), in which the coal seams of the Upper Cretaceous (Campanian) Spring Canyon Member of the Blackhawk Formation appear to mark tops of sin-

datum should be one that contains as few sub-zone splits as possible and one that the stratigraphic thickness between splits is minimal.

The datums chosen for the regional cross sections shown in Plates II, III, and IV serve to illustrate these observations and limitations. The top of the Sub-A coal zone was chosen as datum in the southern part of the study area, south of North Quitchupah Creek, because of its lack of major laterally extensive splits. North of the seaward pinchout of the Sub-A coal zone at North Quitchupah Creek, the base of the composite A-C coal zone was chosen for the same reason. At Rochester Creek, this composite coal zone dramatically splits into two distinct coal zones separated by a major stratigraphic interval, so the datum was redefined as the gle parasequences in near-marine to transitional non-marine facies associations. In the Last Chance Delta, we conclude that a coal zone represents a time transgressive, progradational mire accumulation deposited progressively farther seaward during the development of each successive parasequence in a progradational parasequence set. Therefore, a coal zone or coal seam would form the top of a single parasequence only in the most landward portion of a parasequence set where the initial near-marine parasequence is overlain, for the most part, by the progradational mire accumulations of younger near-marine parasequences.

Last Chance Delta parasequence sets are easily delineated in the non-marine section because the upper boundaries of parasequence sets, when not coincident with sequence boundaries, are coal zones or rooted horizons and paleosols. In many cases, care must be taken to use the uppermost coal zone split when defining the top of a coal zone that forms a parasequence set boundary because minor splits of coal zones, in the delta-plain facies, may simply be due to minor fluvial and crevasse incursions into precursor peat mires.

Depositional Sequence Boundaries

The recognition of sequence boundaries, in non-marine facies associations, can be very difficult in many situations. The recognition of depositional sequence boundaries, represented by erosional unconformities such as the base of a paleovalley fill facies association, is well documented (e.g., in this study, the paleovalley fill sequences of Parasequence Sets 2C, 4B, and 5B rest on type-1 depositional sequence boundaries). Within the Last Chance Delta, the depositional sequence boundaries, for the most part, occur in the interfluvial areas of delta-plain facies association. They are represented either by rooted horizons and paleosols (e.g., the E Rooted Zone at the top of Parasequence Set 4A south of North Quitchupah Creek) (Plate II) or as correlative conformities stratigraphically associated within coal zones (e.g., sequence boundary FS2 represented as a correlative conformity near the base of the A_2 coal zone). Generally, in outcrops, the nature of the poor exposures of interfluvial delta-plain and alluvial-plain mudstones, siltstones, and coal zones, hinder the recognition of these correlative conformities, such that they can usually only be located to within a few meters.

Provided that base-level drops are substantial, sequence boundaries within the lower alluvial plain may be represented by erosional unconformities manifested as the fluvial incision of a channel-belt sandstone complex into an underlying older channel-belt sandstone complex (e.g., the stacked channel complex in the Limestone Cliffs where fluvial channel belts of Parasequence Set 5B erode down into inter-fluvial facies of Parasequence Set 5A and fluvial channel-belt sandstones

of Parasequence Set 4B, resulting in the formation of the erosional unconformity designated as sequence boundary SB4). These situations can usually be recognized by the development of an abnormally thick succession of fluvial deposits. In these instances, care must be taken to trace these alluvial plain fluvial erosional unconformities seaward into regional correlative conformities or paleovalley fill associations before assigning sequence boundary significance to them.

EFFECTS OF TRANSGRESSIVE RAVINEMENT SURFACES ON CORRELATIONS IN NEAR-MARINE FACIES ASSOCIATIONS

Within the Last Chance Delta clastic wedge, transgressive ravinement surfaces such as those at the tops of Parasequence Sets 2D, 3, and 4B complicate regional correlations because of the degree of truncation of near-marine facies occurring during transgressive events. It is estimated that transgressive ravinement events at the tops of Parasequence Sets 2D, 3, and 4B may have each removed as much as 5–10 m (16–33 ft) of sediment. This material is frequently preserved as a transgressive lag deposit, which in outcrop, rarely represents the total thickness of sediment removed by ravinement. The truncated parasequences frequently do not exhibit the preservation of a normal facies succession.

The net stratigraphic effect of transgressive ravinement is to truncate the vertical shoaling-upward facies succession within a parasequence to produce an apparent abnormal thickening and thinning of the parasequence and to juxtapose vertical depositional facies in a confusing manner. Examples of these architectural anomalies are found in the Upper Ferron Sandstone Last Chance Delta. For instance, in the region north of Dry Wash, the vertical facies successions of both Parasequences 2m and 3c are truncated by transgressive ravinement. North of Red Hole Draw, as much as 75% of Parasequence 2m has been removed by transgressive ravinement, such that almost all of the reworked proximal delta-front deposits are missing and the parasequence is represented only by the reworked distal delta-front facies. North of Dry Wash Tanks, as much as 50–80% of Parasequence 3c has been removed by transgressive ravinement, such that most of the reworked stream-mouth bar and all of the distal delta-front deposits are missing and the parasequence is represented dominantly by reworked proximal delta-front facies. In the region between Bitter Seep Wash and Castle Rock, Parasequence 4b has been truncated by transgressive ravinement, removing most of the upper shoreface deposits and the upper portions of coarse- to medium-grained distributary channel deposits. This ravinement of Parasequence 4b leaves middle and lower shoreface deposits incised by the lower parts of distributary channels, which at first examination appears to be an apparent facies juxtaposition.

In making regional correlations, care should be taken to evaluate apparent anomalous depositional facies stacking patterns and thicknesses for evidence of possible transgressive ravinement surfaces. Transgressive lag deposits should signal the potential for parasequence facies truncation, thinning, and juxtaposition. These ravinement surfaces make the recognition of parasequence and parasequence set boundaries more difficult, particularly in outcrop studies, where the outcrop preservation of transgressive lags may have been hindered by surface erosion. If these transgressive ravinement surfaces are not recognized, errors will result in correlations and in definition of parasequences making the application of an outcrop analog to subsurface correlations more complicated.

ACKNOWLEDGMENTS

The Ferron Sandstone depositional sequence stratigraphy research of the Ferron Group Consultants, L.L.C., was supported, in part, by grants from Arco Exploration and Production Technology, Anadarko Petroleum, Amoco Production Company, British Petroleum Exploration (Alaska), Chevron Overseas Petroleum, Live Earth Products, Norsk-Hydro ASA, Phillips Petroleum Company, Shell E&P Technology, Texaco, U.S.A., and Union Pacific Resources in 1997, 1998, and 1999. Field discussions of the depositional sequence stratigraphy of the Ferron Sandstone with J. P. Bhattacharya, J. M. Boyles, K. Soegaard, C. Jenkins, M. Scheihing, C. J. O'Byrne, and K. W. Shanley proved invaluable. Field discussions of the Ferron coals with C. E. Barker, S. S. Crowley, and B. S. Pierce are gratefully acknowledged. Discussions of molluscan faunal biostratigraphic correlations of the Ferron Sandstone and other Late Cretaceous rocks of the Western Interior Basin with J. I. Kirkland and W. A. Cobban are gratefully acknowledged. D. Triplehorn provided insight into volcanic ash recognition and petrography. Discussions with J. Thomas of Amoco always provided timely wisdom. P. B. Anderson kindly provided the core description for UGS/DOE Well #3 and Ivie Creek outcrop measured section PA1. The Ivie Creek photomosaic used in this paper is courtesy of T. A. Ryer. J. M. Boyles of Union Pacific Resources and K. W. Shanley of Amoco provided graphics and reprographics support during the compilation phase of this project. M. Boyles of Phillips Petroleum Company provided the color versions of the cross sections on the CD-ROM. We thank Jim Parker, Utah Geological Survey, for drafting many of the figures. P. K. Link, E. R. Gustason, and A. Pulham reviewed an early interpretive manuscript, from which this manuscript evolved. Critical reviews by J. B. Thomas and P. J. Varney are gratefully acknowledged.

REFERENCES CITED

Aitken, J. F., and S. S. Flint, 1994, High frequency sequences and the nature of incised valley fills in fluvial systems of the Breathitt Group (Pennsylvanian), Appalachian foreland basin, eastern Kentucky, in R. Dalrymple, R. Boyd, and B. Zaitlin, eds., Incised valley systems: origin and sedimentary sequences: Society for Sedimentary Geology (SEPM) Special Publication 51, p. 353–368.

—1995, The application of high resolution sequence stratigraphy to fluvial systems: a case study from the Upper Carboniferous Breathitt Group, eastern Kentucky, U.S.A: Sedimentology, v. 42, p. 3–30.

Anderson, P. B., and T. A. Ryer, 1995, Proposed revisions to parasequence set nomenclature of the Upper Cretaceous Ferron Sandstone Member of the Mancos Shale, Central, Utah (abs.): AAPG Bulletin, v. 79, p. 914–915.

Anderson, P. B., T. C. Chidsey, Jr., and T. A. Ryer, 1997, Fluvial sedimentation and stratigraphy of the Ferron Sandstone, in P. K. Link and B. J. Kowallis, eds., Mesozoic to Recent geology of Utah: Brigham Young University Geology Studies, v. 42, pt. 2, p. 135–154.

Barton, M. D., 1994, Outcrop characterization of architecture and permeability structure in fluvial-deltaic sandstones, Cretaceous Ferron Sandstone, Utah: Ph.D. Dissertation, University of Texas, Austin, 259 p.

Beaumont, C., G. M. Quinlan, and G. S. Stockmal, 1993, The evolution of the Western Interior Basin: causes, consequences and unsolved problems, in W. G. E. Caldwell and E. G. Kauffman, eds., Evolution of the Western Interior Basin: Geological Association of Canada Special Paper 39, p. 97–117.

Cleavinger, H. B., 1974, Paleoenvironments of deposition of the Upper Cretaceous Ferron Sandstone near Emery, Emery County, Utah: Brigham Young University Geology Studies, v. 21, pt. 1, p. 247–274.

Corbeanu R. M., K. Soegaard, R. B. Szerbiak, J. B. Thurmond, G. A. McMechan, D. Wang, S. Snelgrove, C. B. Forster, and A. Menitove, 2001, Detailed internal architecture of a fluvial channel sandstone determined from outcrop, cores, and 3-D ground-penetrating radar: example from the Middle Cretaceous Ferron Sandstone, east-central Utah: AAPG Bulletin, v. 85, p. 1583–1608.

Cotter, E., 1975, Late Cretaceous sedimentation in a low energy coastal zone: the Ferron Sandstone of Utah: Journal of Sedimentary Petrology, v. 45, p. 669–685.

Cross, T. A., 1988, Controls on coal distribution in transgressive-regressive cycles, Upper Cretaceous, Western Interior, U.S.A., in C. K. Wilson , B. S. Hastings,

C. G. St. C. Kendall, H. W. Posamentier, C. A. Ross, and J. C. Van Wagoner, eds., Sea-level changes: an integrated approach: Society for Sedimentary Geology (SEPM) Special Publication 42, p. 371–380.

Dalrymple, R. W., R. Boyd, and B. A. Zaitlin, 1994, History of research, valley types and internal organization of incised-valley systems: introduction to the volume, *in* R. Dalrymple, R. Boyd, and B. Zaitlin, eds., Incised valley systems: origin and sedimentary sequences: Society for Sedimentary Geology (SEPM) Special Publication 51, p. 3–10.

Elliott, T., 1986, Deltas, *in* H. G. Reading, ed., Sedimentary environments and facies: Oxford, Blackwell Scientific Publications, p. 113–154.

Galloway, W. E., 1989, Genetic stratigraphic sequences in basin analysis, I. architecture and genesis of flooding-surface bounded depositional units: AAPG Bulletin, v. 73, p. 125–142.

Gardner, M. H., 1992, Sequence stratigraphy of the Ferron Sandstone, east-central Utah, *in* N. Tyler, M. D. Barton, and R. S. Fisher, eds., Architecture and permeability structure of fluvial-deltaic sandstones: a field guide to selected outcrops of the Ferron Sandstone, east-central Utah: University of Texas, Austin, Bureau of Economic Geology Guidebook, p. 1–12.

—1993, Sequence stratigraphy of the Ferron Sandstone (Turonian) of east-central Utah: Ph.D. Dissertation, Colorado School of Mines, Golden, 406 p.

—1995a, Tectonic and eustatic controls on the stratal architecture of Mid-Cretaceous stratigraphic sequences, central Western Interior foreland basin of North America, *in* S. L. Dorobek and G. M. Ross, eds., Stratigraphic evolution of foreland basins: Society for Sedimentary Geology (SEPM) Special Publication 52, p. 243–282.

—1995b, The stratigraphic hierarchy and tectonic history of the Mid-Cretaceous foreland basin of central Utah, *in* S. L. Dorobek and G. M. Ross, eds., Stratigraphic evolution of foreland basins: Society for Sedimentary Geology (SEPM) Special Publication 52, p. 283–303.

Garrison, J. R., Jr., 2004, A field guide to the outcrops of the Upper Ferron Sandstone Last Chance Delta of east-central Utah: AAPG Field Trip Guidebook.

Garrison, J. R., Jr., and T. C. V. van den Bergh, 1996, Coal zone stratigraphy — a new tool for high-resolution depositional sequence stratigraphy in near-marine to non-marine fluvial-deltaic facies associations: a case study from the Ferron Sandstone, east-central Utah: AAPG Rocky Mountain Section Meeting, expanded abstracts volume, Montana Geological Society, p. 31–36.

—1997, Coal zone and high-resolution depositional sequence stratigraphy of the Upper Ferron Sandstone, *in* P. K. Link and B. J. Kowallis, eds., Mesozoic to Recent geology of Utah: Brigham Young University Geology Studies, v. 42, pt. 2, p. 160–178.

Haq, B. U., J. Hardenbol, and P. R. Vail, 1988, Mesozoic and Cenozoic chronostratigraphy and eustatic cycles, *in* C. K Wilson, B. S. Hastings, C. G. St. C. Kendall, H. W. Posamentier, C. A. Ross, and J. C. Van Wagoner, eds., Sea-level changes: an integrated approach: Society for Sedimentary Geology (SEPM) Special Publication 42, p. 39–45.

Hale, L. A., 1972, Depositional history of the Ferron Formation, central Utah, *in* J. L. Baer and E. Callaghan, eds., Plateau-Basin and Range transition zone, central Utah: Utah Geological Association Publication 2, p. 29–40.

Hamilton, D. S., and N. Z. Tadros, 1994, Utility of coal seams as genetic stratigraphic sequence boundaries in nonmarine basins: an example from the Gunnedah Basin, Australia: AAPG Bulletin, v. 78, p. 267–286.

Hill, R. B., 1982, Depositional environments of the Upper Cretaceous Ferron Sandstone south of Notom, Wayne County, Utah: Brigham Young University Geology Studies, v. 29, p. 59–83.

Howe, D. M., 1989, Comparison of surface and subsurface vertical sequences in the Cretaceous Ferron Sandstone, Utah: Ph.D. Dissertation, University of Nevada, Reno, 191 p.

Kamola, D. L., and J. C. Van Wagoner, 1995, Stratigraphy and facies architecture of parasequences, *in* J. C. Van Wagoner and G. T. Bertram, eds., Sequence stratigraphy of foreland basin deposits: AAPG Memoir 64, p. 27–54.

Kauffman, E. G., and W. G. E. Caldwell, 1993, The Western Interior Basin in space and time, *in* W. G. E. Caldwell and E. G. Kauffman, eds., Evolution of the Western Interior Basin: Geological Association of Canada Special Paper 39, p. 1–30.

Kauffman, E. G., B. B. Sageman, J. I. Kirkland, W. P. Elder, P. J. Harries, and T. Villamil, 1993, Molluscan biostratigraphy of the Cretaceous Western Interior Basin, North America, *in* W. G. E. Caldwell and E. G. Kauffman, eds., Evolution of the Western Interior Basin: Geological Association of Canada Special Paper 39, p. 397–434.

Leithold, E. L., 1994, Stratigraphical architecture at the muddy margin of the Cretaceous Western Interior Seaway, southern Utah: Sedimentology, v. 41, p. 521–542.

Lupton, C. T., 1916, Geology and coal resources of Castle Valley in Carbon, Emery, and Sevier Counties, Utah: U.S. Geological Survey Bulletin, v. 628, p. 1– 88.

Mattson, A. 1997, Characterization, facies relationships, and architectural framework in a fluvial-deltaic sandstone: Cretaceous Ferron Sandstone, central Utah: M.S. thesis, University of Utah, Salt Lake City, 174 p.

Molenaar, C. M., and W. A. Cobban, 1991, Middle Cretaceous stratigraphy on the south side of the Uinta Basin, east-central Utah, in T. C. Chidsey, Jr., ed., Geology of east-central Utah: Utah Geological Association Publication 19, p. 29–43.

Obradovich, J. D., 1993, A Cretaceous time scale, in W. G. E. Caldwell and E. G. Kauffman, eds., Evolution of the Western Interior Basin: Geological Association of Canada Special Paper 39, p. 379–396.

Peterson, F., and R. T. Ryder, 1975, Cretaceous rocks in the Henry Mountains region, Utah and their relation to neighboring regions, in J. E. Fassett and S. A. Wengerd, eds., Canyonlands Country: Four Corners Geological Society Guidebook, 8th Field Conference, p. 167–189.

Posamentier, H. W., and P. R. Vail, 1988, Eustatic controls on clastic deposition II — sequence and systems tract models, in C. K Wilson, B. S. Hastings, C. G. St. C. Kendall, H. W. Posamentier, C. A. Ross, and J. C. Van Wagoner, eds., Sea-level changes: an integrated approach: Society for Sedimentary Geology (SEPM) Special Publication 42, p. 125–154.

Posamentier, H. W., Jervey, M. T., and Vail, P. R., 1988, Eustatic controls on clastic deposition I — conceptual framework, in C. K Wilson, B. S. Hastings, C. G. St. C. Kendall, H. W. Posamentier, C. A. Ross, and J. C. Van Wagoner, eds., Sea-level changes: an integrated approach: Society for Sedimentary Geology (SEPM) Special Publication 42, p. 109–124.

Ryer, T. A., 1981, Deltaic coals of Ferron Sandstone Member of Mancos Shale: predictive model for Cretaceous coal-bearing strata of Western Interior: AAPG Bulletin, v. 65, p. 2323–2340.

—1982, Cross section of the Ferron Sandstone Member of the Mancos Shale in the Emery coal field, Emery and Sevier Counties, central Utah: U.S. Geological Survey Map MF-1357, vertical scale : 1 inch = 30 meters, horizontal scale: 1 inch = 1.25 kilometers.

—1991, Stratigraphy, facies, and depositional history of the Ferron Sandstone in the canyon of Muddy Creek, east-central Utah, in T. C. Chidsey, ed., Geology of east-central Utah: Utah Geological Association Publication 19, p. 45–54.

—1994, Interplay of tectonics, eustasy, and sedimentation in the formation of Mid-Cretaceous clastic wedges, central and northern Rocky Mountain regions: Rocky Mountain Association of Geologists Unconformity Controls Symposium, p. 35–44.

Ryer, T. A., and P. B. Anderson, 1995, Parasequence sets, parasequences, facies distributions, and depositional history of the Upper Cretaceous Ferron deltaic clastic wedge, central Utah (abs.): AAPG Bulletin, v. 79, p. 924.

Ryer, T. A. and McPhillips, M., 1983, Early Late Cretaceous paleogeography of east-central Utah, in M. W. Reynolds and E. D. Dolly, eds., Mesozoic paleogeography of west-central United States: Society of Economic Paleontologist and Mineralogists, Rocky Mountain Section Paleogeographic Symposium 2, p. 253–272.

Schwans, P., and K. M. Campion, 1997, Sequence architecture and stacking patterns in the Cretaceous foreland basin, Utah: tectonism versus eustasy, in P. K. Link and B. J. Kowallis, eds., Mesozoic to Recent geology of Utah: Brigham Young University Geology Studies, v. 42, pt. 2, p. 105–134.

Shanley, K. W., and P. J. McCabe, 1995, Sequence stratigraphy of Turonian-Santonian strata, Kaiparowits Plateau, southern Utah, U.S.A.: implications for regional correlation and foreland basin evolution, in J. C. Van Wagoner and G. T. Bertram, eds., Sequence stratigraphy of foreland basin deposits: AAPG Memoir 64, p. 103–136.

Thompson, S. L., 1985, Ferron Sandstone Member of the Mancos Shale: a Turonian mixed-energy deltaic system, M. A. thesis, University of Texas, Austin, 164 p.

Thompson S. L., C. R. Ossian, and A. J. Scott, 1986, Lithofacies, inferred processes, and log response characteristics of shelf and shoreface sandstones, Ferron Sandstone, central Utah, in T. F. Moslow and E. G. Rhodes, eds., Modern and ancient shelf clastics — a core workshop: Society of Sedimentary Geology (SEPM) Core Workshop 9, p. 325–361.

Uresk, J., 1978, Sedimentary environment of the Cretaceous Ferron Sandstone near Caineville, Utah: Brigham Young University Geology Studies, v. 26, pt. 2, p. 81–100.

van den Bergh, T. C. V., 1995, Facies architecture and sedimentology of the Ferron Sandstone Member of the Mancos Shale, Willow Springs Wash, east-central Utah: M. S. thesis, University of Wisconsin, Madison, 255 p.

Van Wagoner, J. C., 1995, Overview of sequence stratigraphy of foreland basin deposits: terminology, summary of papers, and glossary of sequence stratigraphy, in J. C. Van Wagoner and G. T. Bertram, eds., Sequence stratigraphy of foreland basin deposits: AAPG Memoir 64, p. ix-xxi.

Van Wagoner, J. C., R. M. Mitchum, K. M. Campion, and V. D. Rahmanian, 1990, Siliciclastic sequence stratigraphy in well-logs, cores, and outcrop: AAPG Methods in Exploration Series, v. 7, 55 p.

Zaitlin, B. A., R. W. Dalrymple, and R. Boyd, 1994, The stratigraphic organization of incised-valley systems associated with relative sea-level change, *in* R. Dalrymple, R. Boyd, and B. Zaitlin, eds., Incised valley systems — origin and sedimentary sequences: Society for Sedimentary Geology (SEPM) Special Publication 51, p. 45–60.

Analog for Fluvial-Deltaic Reservoir Modeling: Ferron Sandstone of Utah
AAPG Studies in Geology 50
T. C. Chidsey, Jr., R. D. Adams, and T. H. Morris, editors

Stratigraphic Architecture of Fluvial-Deltaic Sandstones from the Ferron Sandstone Outcrop, East-Central Utah

Mark D. Barton[1], Edward S. Angle[2], and Noel Tyler[3]

ABSTRACT

The Cretaceous Ferron Sandstone is a fluvial-deltaic system that is superbly exposed along the western flank of the San Rafael Swell in east-central Utah. The Ferron consists of fluvial, near shore-zone, and shallow-marine strata that were deposited along the active margin of an evolving foreland basin. The 180-m-thick (590-ft) Ferron Sandstone forms an east- to northeast-thinning clastic wedge that is bounded by marine strata and pinches out over the distance of 40 km (25 mi). Numerous local transgressive intervals further subdivide the Ferron into 10–20-m (30–60-ft) thick successions of fluvial and shallow-marine strata that are similar to the commonly used 'parasequence.'

Along the Molen Reef escarpment, cliffs of 100 m (300 ft) in height and 20 km (12 mi) in length provide a nearly complete dip view through the entire thickness of the Ferron Sandstone. A detailed stratigraphic framework is constructed for the exposure by tracing beds between closely spaced measured sections in the field and from photomosaics.

The stratigraphic framework documents abrupt lateral facies changes across five surfaces interpreted as unconformities. The surfaces are characterized by incision (up to 30 m [90 ft]) of fluvial and estuarine strata into shallow-marine strata and an abrupt basinward shift of coastal-plain and near-shorezone strata. The unconformities are onlapped by coastal-plain and near-shorezone strata associated with sets of aggradational-to-retrogradational parasequences. Accompanying the change in parasequence stacking pattern are distinct changes in the facies of near-shorezone strata. Those associated with the abrupt basinward shifts are relatively mud-rich, display rapid lateral changes in facies, and are dominated by sediment-gravity-flow processes. In contrast, near-shorezone strata associated with the aggradational and retrogradational parasequences are relatively sand-rich, display minimal changes in facies laterally, and are dominated by wave and storm processes.

The incision of fluvial strata and abrupt basinward shifts in environments are interpreted to reflect falls in relative sea level related to minor tectonic and/or eustatic events. Variations in near-shorezone deposits, between those dominated by sediment gravity flows and those dominated by wave or storm reworking, are interpreted to reflect changes in the ratio of sediment supply and accommodation.

[1]Shell International Exploration and Production Inc., Houston, Texas;
[2]Texas Water Development Board, Austin, Texas; [3]ARC Group, LLC, Leander, Texas

INTRODUCTION

Fluvial-deltaic reservoirs typically display a complex internal architecture that plays a dominant role in limiting the volumes of oil and gas that can be produced during conventional recovery operations. Oil recovery efficiencies from mature fluvial-deltaic reservoirs in Texas are typically less than 40% of the oil in place (Tyler, 1988). A key to effective development is the early identification of those reservoirs that may be internally heterogeneous and poorly drained from those that may be internally homogeneous and well drained.

Fluvial-deltaic deposits are influenced by a number of variables that include fluvial, wave, and tidal processes, as well as tectonic, eustatic, and climatic forces. Because the ratio of these controlling variables may shift over time, the depositional character of discrete but juxtaposed progradational pulses may be radically different. In other words, a depositional unit dominated by fluvial processes with large lateral variations in reservoir quality may directly overlie a depositional unit dominated by wave processes that displays much lower lateral variability in reservoir quality.

To help understand and predict the depositional variability that exists in fluvial-deltaic sandstones, exposures were studied of the Cretaceous Ferron Sandstone, east-central Utah. Facies and key surfaces were mapped along an exceptionally well exposed, but largely unstudied portion of the Ferron Sandstone outcrop. The goals of this study are to document the lithologic character and stacking pattern of parasequences and relate changes in lithofacies and stratal geometry to stratigraphic position and depositional processes.

Geologic Setting

The Cretaceous Ferron Sandstone Member of the Mancos Shale exposed in east-central Utah (Figure 1) consists of fluvial, deltaic, and marine deposits that formed during a widespread regression of the Western Interior Seaway (Hale, 1972; Cotter 1975). The strata accumulated along the active margin of a foreland basin as sediments from the Sevier thrust belt were shed eastward (Ryer and McPhillips, 1983). Consequently, the Ferron Sandstone is generally considered to have been deposited under conditions of high sediment supply and accommodation (Ryer and Lovekin, 1986). The Ferron Sandstone forms an east- to northeast-thinning clastic wedge that is as much as 180 m (540 ft) in thickness (Figure 2). It is bounded at the base (Tununk Shale Member of the Mancos Shale) and the top (Blue Gate Shale Member of the Mancos Shale) by marine strata associated with two regional transgressions (Lupton, 1916).

The Ferron Sandstone is well exposed along a northwest-dipping, northeast-trending escarpment referred to as the Molen Reef and Coal Cliffs (Figure 3). The escarpment extends for a distance of 100 km (62 mi) and parallels the dominant direction of shoreline progradation and sediment transport that was to the northeast (Katich, 1953; Davis 1954). Small side canyons that cut the escarpment provide limited views parallel to the depositional strike of the system. The quality of the exposures offers the opportunity to document regional stratigraphic relationships with a high degree of detail and accuracy.

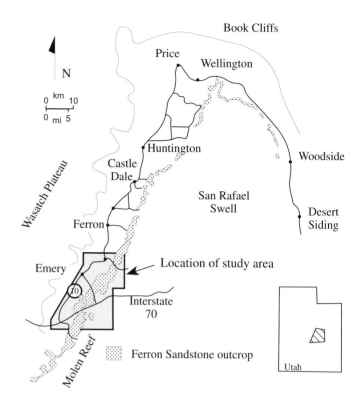

Figure 1. Location map of the Ferron Sandstone outcrop and field study area, east-central Utah.

ment extends for a distance of 100 km (62 mi) and parallels the dominant direction of shoreline progradation and sediment transport that was to the northeast (Katich, 1953; Davis 1954). Small side canyons that cut the escarpment provide limited views parallel to the depositional strike of the system. The quality of the exposures offers the opportunity to document regional stratigraphic relationships with a high degree of detail and accuracy.

Ryer (1981) documented that the Ferron Sandstone is composed of up to seven sandstone tongues (Figure 4). Each sandstone tongue is composed of a delta-front succession overlain by a coal and bounded by marine mudstones. The sandstone tongues range from 15–30 m (45–90 ft) in thickness and extend basinward 5–50 km (3–31 mi). Initial sandstone tongues are widespread and step basinward. Later sandstone tongues are less widespread and stack vertically or step landward. Subsequent studies (Ryer, 1993) recognized that the sandstone tongues could be further divided by localized transgressive flooding surfaces into upward-shoaling and upward-coarsening successions that are similar to the commonly used "parasequence" defined by Van Wagoner et al. (1988, 1990).

An important aspect of Ferron facies architecture is the link between delta-front facies associations and their stratigraphic position. Gardner (1993) showed that the seaward-stepping sandstone tongues identified by Ryer

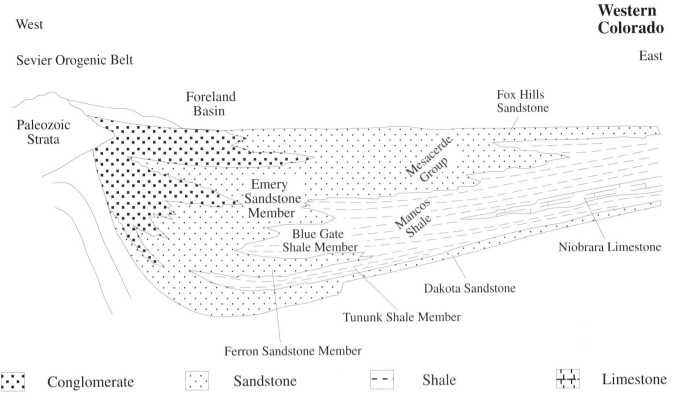

Figure 2. Cross section of Cretaceous strata in central Utah (modified from Armstrong, 1968). The Ferron Sandstone is a fluvial-deltaic system deposited under conditions of high sediment supply and accommodation.

Figure 3. Photograph of the Ferron Sandstone looking south from the Molen Reef. The trend of the escarpment parallels the dominant direction of shoreline progradation which is towards the viewer. Small side canyons that cut the escarpment provide limited views parallel to depositional strike of the Ferron system.

(1981) consist largely of river-dominated shoreline deposits, where as, landward-stepping sandstone tongues consist largely of wave-dominated shoreline deposits. River-dominated shoreline deposits were interpreted to have formed along a highly embayed coastline where sediment fed from distributary channels was deposited and preserved as discrete lobes. In contrast, wave-dominated shoreline deposits were interpreted to

have accumulated along a coastline where sediment from distributary channels was reworked and redistributed by waves and storms into a relatively broad, sandy, strike-aligned shoreline.

The link between delta-front facies and their stratigraphic position is interpreted to record the stratigraphic response of the Ferron system to changes in sediment supply and accommodation (Gardner, 1993). Seaward-

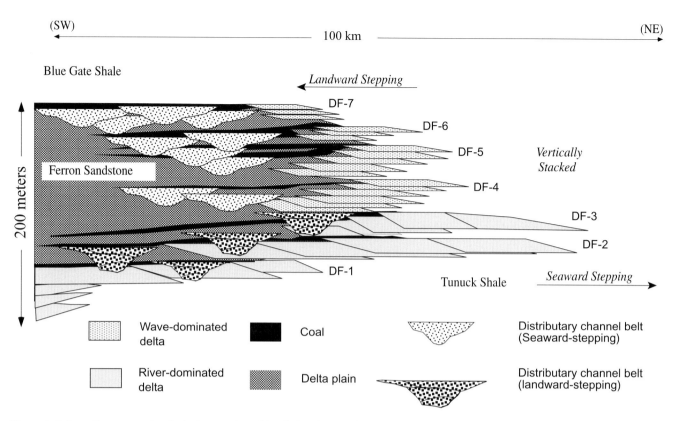

Figure 4. Schematic cross section depicting stratigraphic relationships in the Ferron Sandstone, east-central, Utah (based on Ryer, 1981; modified by Gardner, 1993). Delta-front units (DF) number 1 through 7 in ascending stratigraphic order.

stepping sandstone tongues are deposited during a period of relatively low accommodation in which most of the sediment passed through the coastal-plain facies tract and deposited in the shallow-marine facies tract. A fluvial-, or river-dominated, shoreline develops as waves and storms have insufficient time to rework the incoming sediment. By comparison, the landward-stepping sandstone tongues are deposited during a period of relatively high accommodation in which a much greater proportion of the incoming sediment is deposited in the coastal-plain facies tract and a smaller amount of sediment bypassed to the shallow-marine facies tract. Waves and storms have sufficient time to rework the incoming sediment into a wave-dominated shoreline.

Data and Methods

Exposures of the Ferron Sandstone were examined north of Interstate Highway 70 (I-70) and south of the town of Moore (Figure 5). Data consist of measured descriptive lithologic logs and photomosaics that provide complete coverage of the nearly continuous outcrops. The outcrops were photographed from a helicopter using of a large format camera to maximize resolution of outcrop images, and rock climbing equipment to gain access to the vertical exposures. Stratigraphic cross sections were constructed by mapping the distribution of facies and tracing key bounding surfaces

between measured logs. The laterally continuous exposures allowed for correlations between measured sections with a high degree of accuracy.

Key bounding surfaces mapped in the field include marine flooding surfaces and their correlative nonmarine equivalents, transgressive ravinement surfaces, and erosional onlap surfaces (unconformities). Facies were grouped into associations that include offshore marine/prodelta, lower shoreface/lower delta front, middle shoreface/middle delta front, upper shoreface/upper delta front, transgressive shallow marine, sand-rich fluvial, mud-rich fluvial or estuarine, coastal plain, and coal. Near-shorezone strata were further differentiated into wave dominated and river dominated. Wave-dominated delta fronts are interpreted to have been formed by wave- and storm-generated currents. They display (1) clean, well-sorted sandstones; (2) an abundance of hummocky cross-stratification in lower-shoreface sandstones; (3) an abundance of multidirectional cross-stratification in upper-shoreface sandstones; and (4) and parallel to subparallel lamination in foreshore sandstones. Burrowing tends to be common to intense throughout the succession with the most abundant trace fossils being *Ophiomorpha* and *Thalassinoides*.

Fluvial-dominated delta fronts are interpreted to have been deposited at the mouths of distributary channels by unidirectional traction currents and sediment

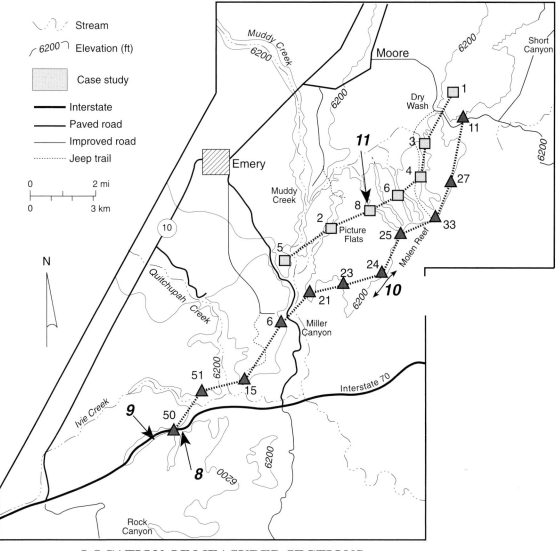

**LOCATION OF MEASURED SECTIONS
FERRON SANDSTONE**
···□··· **Landward stepping**
···▲··· **Seaward stepping**
10 - Location of photograph

Figure 5. Index map showing location of measured logs (indicated by number) used in stratigraphic cross sections and photographs. Data were assembled into two cross sections, one through the seaward-stepping portion, and one through the landward-stepping portion, of the Ferron system. The cross sections parallel the depositional axis of the Ferron system in which sediment transport was towards the northeast.

gravity flows. They exhibit an abundance of beds with waning flow sequences and unidirectional traction features. Many beds have erosive bases, are graded, and display a succession of stratification types similar to the Bouma Tabcd divisions. The sandstone beds are often inclined and when traced up dip tend to thicken and merge with amalgamated sets of unidirectional, trough cross-stratified sandstones. Thin mudstone interbeds are common throughout the interval and burrowing is generally sparse to absent. Locally, soft-sediment deformation and gravity remobilization features, including rotated slump blocks and growth faults are common.

STRATIGRAPHIC ARCHITECTURE

The stratigraphic architecture of the Ferron clastic wedge is divided into two parts informally referred to as "lower" (Figure 6) and "upper"(Figure 7). The lower part of the Ferron Sandstone was measured from a condensed section at the base of the Ferron up to a laterally extensive bentonite within the C coal that can be traced throughout the study area. The upper Ferron Sandstone was measured from the C-coal bentonite to the base of the overlying Blue Gate Shale Member. Location of lithologic logs used in the cross sections are shown in Figure 5.

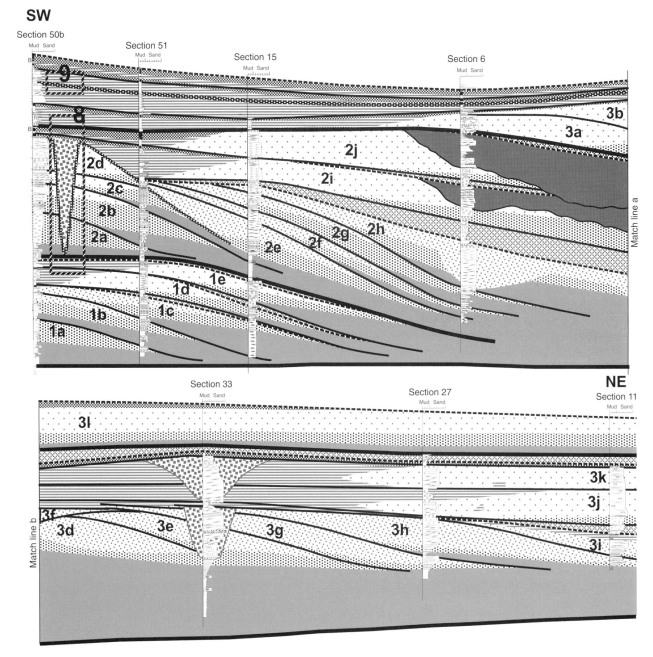

Figure 6. Southwest-northeast cross section showing stratigraphic relationships in the lower or seaward-stepping portion of the Ferron Sandstone between Interstate 70 and Dry Wash.

Lower Ferron

In the lower portion of the Ferron Sandstone the overall depositional style is one of progradation with the shoreline migrating from I-70 to Dry Wash a distance of about 24 km (15 mi) (Figure 6). The general style of progradation is episodic rather then uniform displaying several abrupt landward and basinward shifts of the shoreline. There are three main episodes of shoreline progradation and retrogradation. Lithosomes that are represented by these episodes are informally referred to as cycles and numbered 1 through 3 in ascending strati-

graphic order. They are bounded by relatively widespread or major marine flooding surfaces. The cycles vary in thickness from 20–30 m (60–90 ft) thick and extend basinward 15–30 km (9–19 mi).

Parasequence Architecture

Within each cycle, relative local or minor transgressive flooding surfaces extend several kilometers along depositional dip and subdivide the cycle into parasequences. Parasequences are relatively conformable successions of marine and nonmarine strata bounded by marine flooding surfaces and their correlative nonma-

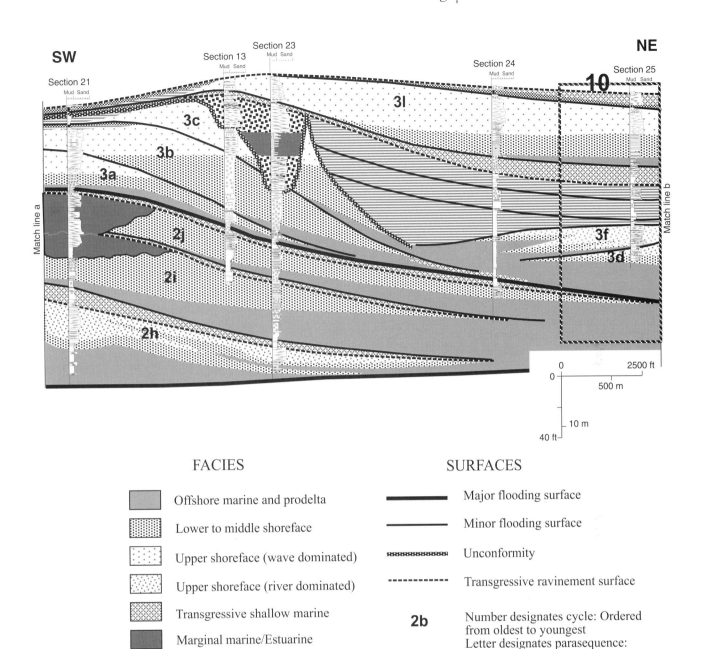

FACIES

Offshore marine and prodelta

Lower to middle shoreface

Upper shoreface (wave dominated)

Upper shoreface (river dominated)

Transgressive shallow marine

Marginal marine/Estuarine

Marginal marine/Coastal plain

Coal

Fluvial

SURFACES

Major flooding surface

Minor flooding surface

Unconformity

Transgressive ravinement surface

2b   Number designates cycle: Ordered from oldest to youngest
Letter designates parasequence: Ordered from oldest to youngest

9   Location of photo Number designates figure

Figure 6 continued.

rine equivalents (Van Wagoner et al., 1990). Individual parasequences are 5–15 m (15–45 ft) thick and extend parallel to depositional dip 1–10 km (0.6–6 mi). Each parasequence has a regressive marine portion that consists from the base up of offshore marine, lower shoreface, upper shoreface, and foreshore strata. The regressive portion records a gradual decrease in water depth associated with a discrete episode of shoreline progradation. In a seaward direction, the regressive marine portion progressively thins, fines, and becomes indistinguishable from offshore marine mudstones. In a landward direction the regressive marine portion of the parasequence is replaced by nonmarine and marginal-marine strata. The transition from marine to nonmarine strata occurs abruptly, over the distance of several hundred meters, or in a gradual intertonguing fashion that takes place over the distance of several thousand meters. The nonmarine portion of the parasequence typically

199

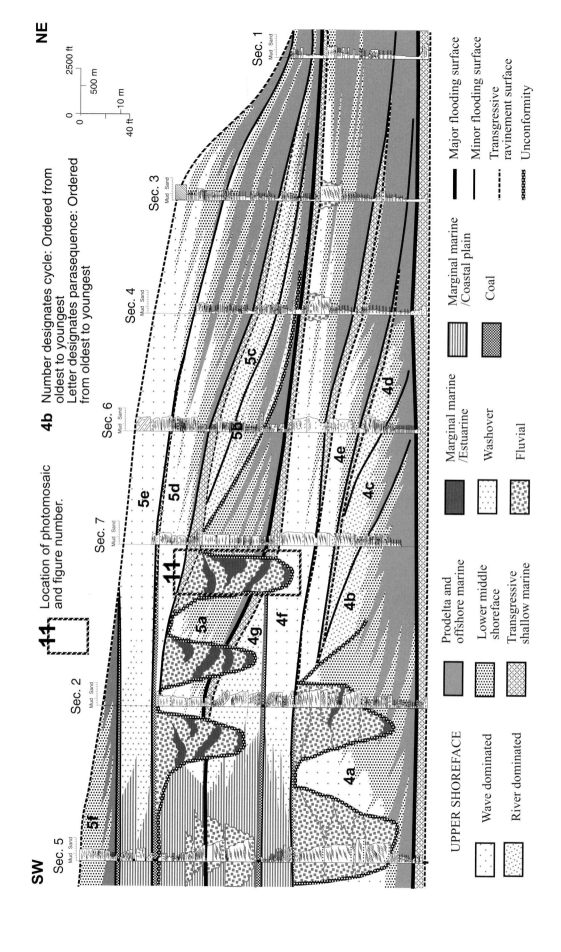

Figure 7. Southwest-northeast cross section showing stratigraphic relationships in the upper Ferron Sandstone between Miller Canyon and Dry Wash.

consists of carboneous shales, coals, and lenticular sandstone channels.

Some of the parasequences described in the Ferron Sandstone differ from the conventional definition in that the regressive marine portion is truncated by a low-relief erosion surface and overlain by a relatively thin, upward-fining shallow-marine succession. The upward-fining succession consists from the base up of highly burrowed, coarse-grained sandstones containing abundant clay and coal clasts; fine-grained, hummocky cross-stratified sandstones; and marine mudstones. The erosive-based, upward-fining succession is interpreted to record marine erosion followed by deposition during a gradual transgression of the shoreline. The coarse-grained sandstones represent a lag deposit formed by marine reworking of the underlying deposits. The hummocky cross-stratified sandstones and marine mudstones were deposited below fair weather wave base, then storm wave base, as the transgression continued and water depth increased.

Parasequence Stacking Patterns

Parasequences in the lower Ferron Sandstone stack to form progradational, aggradational, and retrogradational sets (Figure 6). The organization within each depositional cycle is similar. From oldest to youngest, the parasequences initially stack to form a normal progradational set that is followed by an offlapping to strongly progradational set and concluded with an aggradational to retrogradational set. The transition from one style of stacking to another is abrupt.

The normal progradational set (parasequences 2a, 2b, 2c, 2d, 3a, 3b, 3c) is composed of 3 to 4 parasequences that display a moderate degree of stratigraphic climb (Figure 6). Individual parasequences are relatively thick and widespread, ranging from 7-15 m (21-45 ft) in thickness, and extending 2-5 km (1-3 mi) landward into coastal-plain deposits. The basal contact is transitional with offshore marine mudstones.

Widespread erosion surfaces displaying a high degree of erosional relief (up to 30 m [90 ft]) overlie each normal progradational parasequence set. Locally, the normal parasequence set is completely eroded and replaced by lenticular sandstone bodies that are up to 25 m (75 ft) thick and 1000 m (3000 ft) across (Figure 8). The bodies are a composite of smaller channel-form sandstone bodies that are 3–10 m (9–30 ft) in thickness. Most of the channel-form bodies have an erosional lag of siderite nodules and clay clasts at the base, and display an upward-fining grain-size profile composed of unidirectional, cross-stratified and ripple laminated sandstones. In areas where the normal parasequence set has not been highly eroded the surface is heavily stained with iron oxides and disrupted by rootlets and clay slickensides (Figure 9).

Progradational to offlapping sets of parasequences (parasequences 2e, 2f, 2g, 2h, 3d, 3e, 3f, 3g, 3h, 3i) that display little to no degree of stratigraphic climb overlie and are offset basinward from the normal progradational sets. Individual parasequences are relatively thin, on the order of 5–10 m (15–30 ft), and laterally they rarely extend for more than a kilometer along depositional dip (Figure 10). The basal contact is often abrupt, with upper delta-front deposits erosively overlying offshore marine mudstones.

Aggradational to retrogradational parasequences (parasequences 1d, 1e, 2i, 2j, 3j, 3k, 3l) overlie the strongly progradational parasequence set. In cycles 1 and 2 the parasequences stack vertically while in cycle 3 they step landward. Similar to the normal progradational set individual parasequences are thick, widespread bodies that are composed largely of wave-dominated shoreface deposits (see Figure 10). In a landward direction the parasequences terminate by onlap onto the erosion surfaces that cap the normal progradational sets. Transgressive shallow-marine deposits at the base of the aggradational and retrogradational parasequences are typically thick and well developed.

Upper Ferron

In the upper portion of the Ferron Sandstone the overall depositional style is one of aggradation with the position of the shoreline showing no overall migration. There are two main episodes of shoreline progradation and retreat. The lithosomes represented by the episodes are informally referred to as cycles 4 and 5 in ascending stratigraphic order (Figure 7). The cycles are thicker than in the lower Ferron, 40–45 m (120–135 ft), and less widespread, extending basinward 10–12 km (6–7 mi).

The organization of parasequences is similar to the lower Ferron Sandstone with there being sets of normal progradational, to strongly progradational, and then to aggradational to retrogradational parasequences. However, there are differences in the relative proportions of each type of set. In the lower Ferron there are subequal amounts of normal progradational, strongly progradational, and aggradational parasequences. In contrast, the upper Ferron consists largely of aggradational and retrogradational parasequences (parasequences 4e, 4f, 5d, 5e, 5f). Strongly progradational parasequences (parasequences 4b, 4c, 4d, 5b, 5c) are present but poorly developed. Facies relationships show the same relationship to stacking pattern with near-shore strata of normal progradational, aggradational, and retrogradational sets composed of wave-dominated shoreface deposits and strongly progradational sets of river-dominated delta-front deposits.

In the upper Ferron Sandstone a much greater portion of the initial normal progradational parasequences set has been eroded and replaced by valley fill deposits.

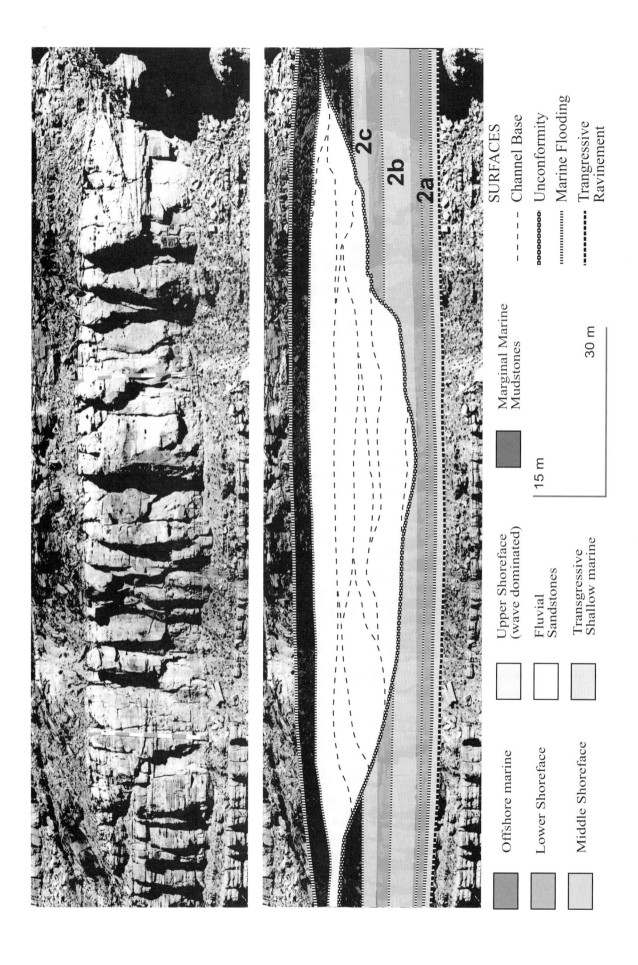

Figure 8. Photomosaic of valley fill cutting out delta-front deposits within the lower Ferron Sandstone. Valley fill consists of vertically and laterally stacked channels. Location of photo shown on Figures 5 and 6.

Figure 9. Photograph shows a soil horizon with well-developed rootlets, pedogenic clay slickenslides, and iron oxide staining. The horizon correlates with an unconformity in cycle 3 and separates coastal deposits that accumulated during a normal regression from coastal-plain deposits that accumulated during transgression. Location of photograph shown on Figures 5 and 6.

The valleys are up to 25 m (75 ft) thick and several kilometers in width (Figure 11). Similar to the lower Ferron they are a composite of smaller channels that are 3–8 m (9–24 ft) thick. In contrast, valley fills associated with the upper Ferron contain sub-equal amounts of sandstone and mudstone. Shell fragments and burrowing are common in the mudstones and sandstones display unidirectional as well as bi-directional trough cross-stratification suggesting the deposits were influenced by tidal and shallow-marine processes.

STRATIGRAPHIC RESPONSE TO CHANGES IN ACCOMMODATION AND SEDIMENT SUPPLY

In the Ferron Sandstone parasequences are organized into five regressive-to-transgressive cycles. Each cycle consists of a succession of progradational, offlapping, aggradational, and retrogradational sets. A widespread erosion surface separates the normal progradational set of parasequences from the overlying aggradational and retrogradational sets in each cycle. Accompanying the shifts in parasequence stacking patterns are changes in delta-front facies associations. The delta-front facies in normal progradational, aggradational, and ret-

rogradational parasequences are interpreted to have formed along a wave-dominated coastlines where sediment from feeding distributary channels was reworked and redistributed by waves and currents into broad, sandy, strike-aligned shoreline deposits. In contrast, the delta-front facies association of strongly progradational parasequences is interpreted to have accumulated along a highly embayed, river-dominated coastline where sediment fed from distributary channels was deposited and preserved as sandy lobes without subsequent reworking. The observations are similar to those made by Gardner (1993) who demonstrated a link between aggradational and progradational portions of the Ferron clastic wedge and the type of delta-front deposits preserved. In this study the link between progradational style and the type of delta-front deposits preserved is also observed to occur at the scale of individual parasequences.

The bodies that incise into the normal progradational parasequence sets are interpreted as valley fills that were eroded during relative falls in sea level and filled during relative rises. They are up to 30 m (90 ft) thick and expressed on outcrop as a cluster of channel-form bodies bounded by an unconformity that is incised through multiple shallow marine parasequences. The channel-form bodies, which are much smaller (3–8 m [9–24 ft]), are interpreted as bars and channel fills that

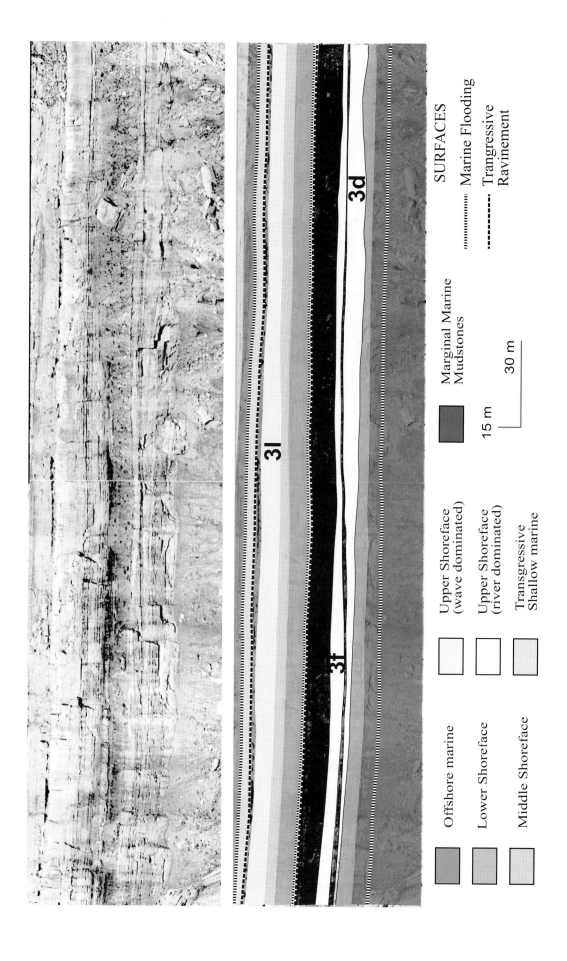

Figure 10. River-dominated shoreline deposits are associated with the strongly progradational part of cycle 3 (parasequences 3d and 3f), marginal marine deposits with the aggradational part of cycle 3, and wave-dominated shoreline deposits with the retrogradational portion of cycle 3 (parasequence 3l). Location of the photograph is shown on Figures 5 and 6.

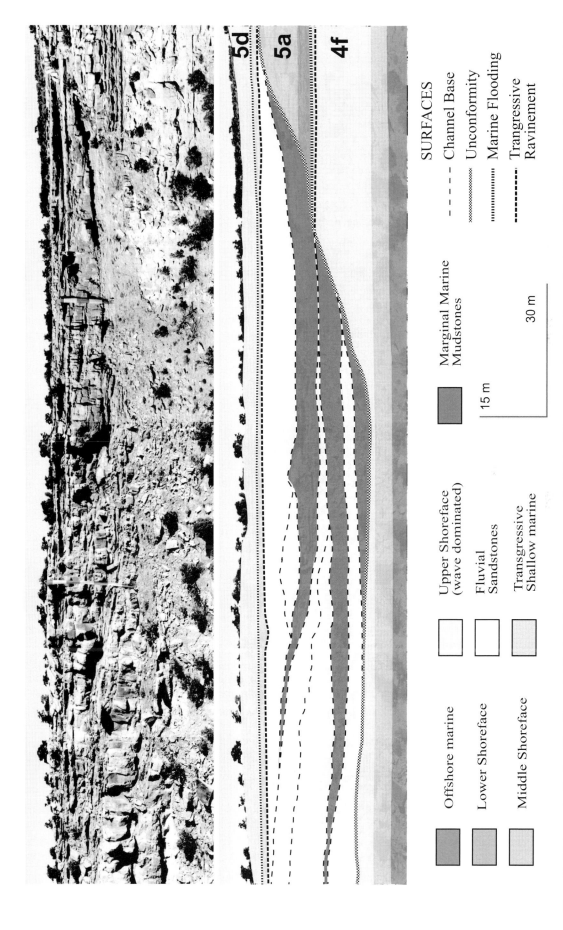

Figure 11. Photomosaic of valley fill cutting out delta-front deposits (parasequences 4f and 5a) within upper portion of the Ferron Sandstone. Valley fill consists of vertically and laterally stacked channels interstratified with marginal-marine mudstones. Location of photo shown on Figures 5 and 7.

aggraded within the larger valley. In the lower Ferron Sandstone the channels, which consist largely of unidirectional trough cross-stratified sandstone, result solely from fluvial processes. In contrast, within the upper Ferron valleys the presence of burrowed mudstones, shell fragments, and bi-directional cross-stratification suggest the channels were influenced by tidal currents or storm surges that forced marine waters up the channel in addition to fluvial processes. The presence of rootlets, clay slickenslides, and iron oxide staining in areas not highly eroded are interpreted as pedogenic features that formed in interfluve areas that were exposed but not eroded during a drop in relative sea level.

The interpreted stratigraphic response of the Ferron system to changes in sediment supply and accommodation is shown in Figure 12. Within each depositional cycle, the regressive to transgressive pattern of sedimentation punctuated by an episode of widespread erosion is interpreted to record a cycle of relative sea-level rise-fall-rise. The progradational parasequence set likely records a rise in relative sea level in which rates of deposition were greater than rates of accommodation. The widespread erosion surface and offlapping parasequence set likely records a fall in relative sea level. The aggradational and progradational parasequence set likely records a rise in relative sea level in which rates of deposition were nearly equal to rates of accommodation. The retrogradational parasequence set likely records a rise in relative sea level in which rates of deposition were less than rates of accommodation.

During a fall in relative sea level the previous progradational highstand shoreline was apparently cannibalized and sediment was bypassed into a shallow-marine lowstand environment. Fluvial systems were incised, interfluve areas were subaerially exposed, and the shoreline abruptly forced basinward. A river-dominated coastline developed as the rate of sediment supply exceeded the ability of ocean waves and storms to redistribute it. As the system expanded or as rates of relative sea level rose, rates of accommodation likely increased to the point that they nearly equaled rates of sediment supply. Ocean waves and currents were more capable of redistributing the incoming sediment, and the river-dominated shoreline was transformed into a wave-dominated shoreline. Landward of the shoreline, the antecedent topography of the previous highstand shoreline was onlapped by coastal-plain and marginal-marine deposits. Farther up depositional dip fluvial incision ceased and incised valleys were filled with a complex assemblage of facies ranging from fluvial to estuarine.

As relative sea level continued to rise the antecedent topography was eventually overstepped. At this point, the rate accommodation landward of the shoreline apparently abruptly increased and the system was drowned and transgressed. During transgression shore-line sediments, eroded by storms and waves, were transported landward and deposited in back barrier environments or carried offshore. Episodes of progradation during the transgression may reflect minor still stands in sea level, abundant sediment supply, or even brief forced regressions. A new cycle of deposition was evidently initiated when a balance between rates of sediment supply and accommodation was reached. At this point, the transgression ceased and an aggradational to progradational barrier or wave-dominated shoreline system was reestablished and maintained prior to another fall in relative sea level.

UPPER VS LOWER FERRON

Within the Ferron system there are progressive changes vertically in cycle thickness, parasequence stacking pattern, facies proportions, and valley-fill architecture. A schematic cross section summarizes the differences (Figure 13). Cycle thickness gradually increases from 25 m (75 ft) in cycle 2, to 45 m (135 ft) in cycle 5. The increase in thickness is largely the result of increased development and/or preservation of the transgressive portion of each cycle. The transgressive or retrogradational portion of the cycle increases from 0 m in cycle 1, to 3 m (9 ft) in cycle 2, to 12 m (36 ft) in cycle 3, to 20 m (60) in cycle 4, and 25 m (75 ft) cycle 5. Offlapping parasequence sets are well developed in cycles 1, 2, and 3 and poorly developed in cycles 4 and 5. In contrast, retrogradational parasequence sets are well developed in cycles 4 and 5, moderately developed in cycle 3, and absent in cycles 1 and 2. The proportion of coastal-plain to delta-front deposits increases from a small fraction in cycles 1 and 2 to subequal amounts in cycles 4 and 5. Valley fills in the lower Ferron Sandstone have a narrow shoestring geometry composed largely of sand-rich fluvial deposits. In contrast, valley fills in the upper portion of the Ferron Sandstone are preserved as broad ribbons composed of a mixture of sand-rich fluvial and mud-rich marginal marine (estuarine) deposits.

The progressive changes in thickness, parasequence stacking pattern, facies proportion, and valley fill architecture suggest that during Ferron time the ratio of sediment supply to accommodation progressively decreased. Cycles became thicker and more compressed, aggradational and retrogradational parasequence sets increase in number, offlapping parasequences decrease in number, and valleys show progressively greater marine influence because a greater percentage of the sediment is stored within the coastal-plain facies tract and less bypassed to the shallow-marine environment. The change in the sediment supply to accommodation ratio may reflect a decrease in the amount of sediment supplied from the Sevier thrust belt and/or an increase in subsidence due to tectonic loading associated with the Sevier thrust belt.

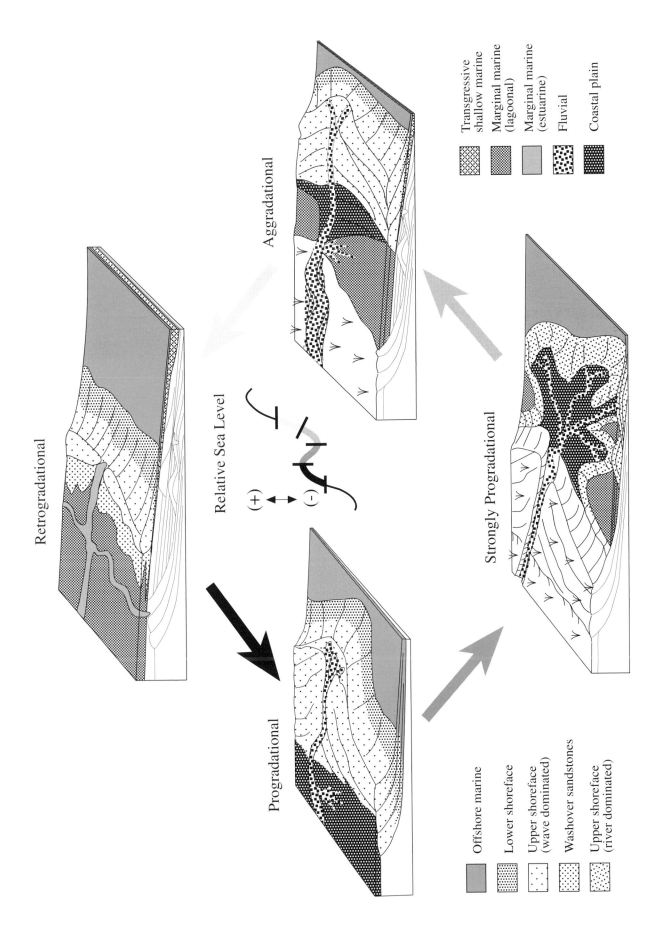

Retrogradational

Aggradational

Relative Sea Level

(+) ↔ (−)

Strongly Progradational

Progradational

Offshore marine

Lower shoreface

Upper shoreface (wave dominated)

Washover sandstones

Upper shoreface (river dominated)

Transgressive shallow marine

Marginal marine (lagoonal)

Marginal marine (estuarine)

Fluvial

Coastal plain

Figure 12. Diagram illustrating changes in shoreline morphology and facies with changes in parasequence stacking and accommodation or relative sea level.

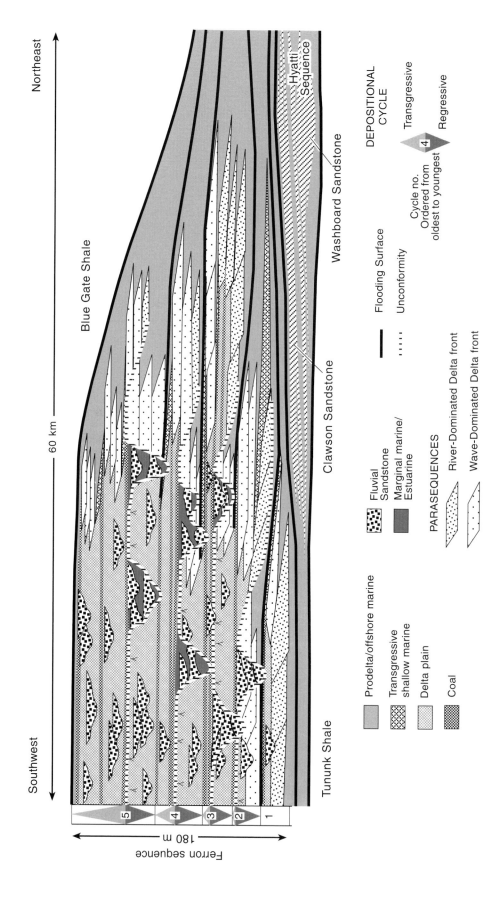

Figure 13. Schematic cross section depicting stratigraphic relationships of the Ferron Sandstone in east-central Utah, based on the study (compare with Figure 4).

SUMMARY

Marine flooding surfaces divide the Ferron Sandstone into numerous parasequences. Individual parasequences are 5–15 m (15–45 ft) in thickness and range from 1–10 km (0.6–6 mi) in length along depositional dip. They are interpreted to have been deposited during multiple episodes of shoreline progradation followed by marine flooding.

Correlations suggest the parasequences are organized into regressive to transgressive cycles each of which is punctuated by an unconformity. Each cycle begins with a normal progradational parasequence set, followed by a strongly progradational parasequence set, and concludes with an aggradational to retrogradational parasequence set. An unconformity overlies the initial progradational set of parasequences within each cycle and is onlapped by a set of aggradational to retrogradational parasequences. The unconformities are marked by incised fluvial systems, subaerial exposure of marine and marginal-marine strata, and an abrupt basinward shift in facies. A total of five cycles, numbered 1 through 5 in ascending stratigraphic order were recognized.

Facies variability can be linked to stratigraphic position. The marine portion of strongly progradational parasequence sets consists predominantly of river-dominated delta-front deposits, whereas the marine portion of normal progradational, aggradational, and retrogradational parasequence sets consists largely of wave-dominated shoreface deposits. In a similar fashion, variability in valley fill deposits is also linked to stratigraphic position. Incised valleys associated with seaward-stepping cycles are filled largely with fluvial sandstones. In contrast, incised valleys associated with landward-stepping cycles filled with a heterogeneous mix of fluvial sandstones and marginal marine mudstones.

ACKNOWLEDGMENTS

This paper represents part of a Ph.D. study conducted at the University of Texas at Austin under the supervision of Dr. William Fisher and Noel Tyler. Financial support for this project was provided by the Gas Research Institute under contract no. 5089-260-1902. The project was conducted at the Bureau of Economic Geology at the University of Texas in Austin under the supervision of Noel Tyler and Steve Fisher. The paper benefited from discussions with Brian Willis, Janok Bhattacharya, Michael Gardner, and Paul Knox. Illustrations were assisted by the cartography department at the Bureau of Economic Geology under the supervision of Joel Lardon. We thank Lee F. Krystinik and Jeffrey A. May for their careful reviews and constructive criticisms of the manuscript.

REFERENCES CITED

Armstrong, R. L., 1968, Sevier orogenic belt in Nevada and Utah: Geological Society of America Bulletin, v. 79, p. 429–458.

Cotter, E., 1975, Deltaic deposits in the Upper Cretaceous Ferron Sandstone, Utah, in M. L. S. Broussard, ed., Deltas, models for exploration: Houston Geological Society, p. 471–484.

Davis, L. J., 1954, Stratigraphy of the Ferron Sandstone: Intermountain Association of Petroleum Geology Fifth Annual Field Guidebook, p. 55–58.

Gardner, M. H., 1993, Sequence stratigraphy and facies architecture of Upper Cretaceous Ferron Sandstone Member of the Mancos Shale, east-central Utah: Ph.D. dissertation T-3975, Colorado School of Mines Colorado, Golden, 528 p.

Hale, L. A., 1972, Depositional history of the Ferron Sandstone, central Utah, in J. L. Baer and E. Callaghan, eds., Plateau-basin and range transition zone: Utah Geologic Association Publication No. 2, p. 115–138.

Katich, P. R, 1953, Source direction of the Ferron Sandstone: AAPG Bulletin, v. 37, no. 4, p. 858–862.

Lupton, C. T., 1916, Geology and coal resources of Castle Valley in Carbon, Emery, and Sevier Counties, Utah: U.S. Geological Survey Bulletin 628, 88 p.

Ryer, T. A., 1981, Deltaic coals of the Ferron Sandstone Member of the Mancos Shale — predictive model for Cretaceous coal-bearing strata of the Western Interior: AAPG Bulletin, v. 65, no. 11, p. 2323–2340.

—1993, The autochthonous component of cyclicity in shoreline deposits of the Upper Cretaceous Ferron Sandstone, central Utah (abs.): AAPG Bulletin, v. 77, p. 175.

Ryer, T. A., and J. R. Lovekin, 1986, The Upper Cretaceous Vernal delta of Utah — depositional or paleotectonic feature? in J. A. Peterson, ed., Paleotectonics and sedimentation in the Rocky Mountain region, United States: AAPG Memoir 41, p. 497–510.

Ryer, T. A., and Maureen McPhillips, 1983, Early Late Cretaceous paleogeography of east-central Utah, in M. W. Reynolds, ed., Mesozoic paleogeography of west-central United States: Society of Economic Paleontologists and Mineralogists, Rocky Mountain Paleogeography Symposium 2, p. 253–272.

Tyler, Noel, 1988, New oil from old oil fields: Geotimes, July, p. 8–10.

Van Wagoner, J. C., H. W. Posamentier, R. M. Mitchum, P. R. Vail, J. F. Sarg, T. S. Loutit, and J. Hardenbol, 1988, An overview of the fundamentals of sequence stratigraphy and key definitions, *in* C. K. Wilgus, B. S. Hastings, C. G. St. C. Kendall, H. W. Posamentier, C. A. Ross, and J. C. Van Wagoner, eds., Sea-level changes — an integrated approach: Society of Economic Paleontologists and Mineralogists Special Publication No. 42, p. 39–45.

Van Wagoner, J. C., R. C. Mitchum, K. M. Campion, and V. D. Rahmanian, 1990, Siliciclastic sequence stratigraphy: AAPG Methods in Exploration Series, No. 7, p. 55.

Analog for Fluvial-Deltaic Reservoir Modeling: Ferron Sandstone of Utah
AAPG Studies in Geology 50
T. C. Chidsey, Jr., R. D. Adams, and T. H. Morris, editors

Regional Stratigraphy of the Ferron Sandstone

Paul B. Anderson[1] and Thomas A. Ryer[2]

ABSTRACT

The Ferron Sandstone Member of the Mancos Shale is divided informally into upper and lower units. The upper Ferron has been divided by earlier workers into seven or eight delta-front units. These units correspond to parasequence sets. We suggest one additional delta-front or parasequence set, bringing the total to nine for the upper Ferron. The opportunity to study many of these stratal units in detail on outcrop has led to the further division into numerous parasequences. These parasequences often contain distinct, mappable genetically related packages of beds which are clearly not bounded by marine-flooding surfaces and have hence been labeled bedsets. Nineteen parasequences and four bedsets are named and described along with several undivided parasequence sets. The landward and seaward pinchouts of the nearshore marine facies of most of these units are mapped, enhancing our ability to predict geometries of associated reservoir facies. Parasequences tend to follow an evolution of delta types from initial regression to maximum regression. This evolution begins with a wave-dominated shoreline, passing through a transitional wave-modified shoreline, and typically ending with a fluvial-dominated shoreline. Parasequences in seaward-stepping parasequence sets have an average dip length of nearshore marine facies of 4.3 mi (6.9 km) and average maximum thickness of 55 ft (17 m). Although less detailed information was gathered on the aggradational parasequences, their average dip length of nearshore marine facies is 6 mi (10 km) and average maximum thickness of 56 ft (17 m). The best reservoir facies are found in the wave-dominated deltas and distributary channels. Wave dominated delta facies are consistently found in the initial regression of each parasequence, and range in shoreline strike orientation from N60W to N5E. Ambiguities in the definitions of sequence stratigraphic units and differing application of those definitions have lead to a variety of stratigraphic schemes from different workers on the same Ferron outcrops.

[1]Consulting Geologist, Salt Lake City, Utah
[2]The ARIES Group, Inc., Katy, Texas

INTRODUCTION

The Ferron Sandstone was originally defined as a member of the Mancos Shale by Lupton (1914, 1916). Ryer (1981, 1991) (see Ryer, this volume) recognized a series of seven "delta-front units;" an eighth was designated by Gardner (1993) (see Ryer, this volume) and a ninth is added here. We refer to these units as Kf-Last Chance and Kf-1 through Kf-8 (Figure 1). Had the Ferron been originally described as a formation, these nine units would constitute members. Should the various members of the Mancos Shale in east-central Utah, the Ferron among them, be elevated to formation status in the future, these nine units should be elevated to member status.

SEQUENCE STRATIGRAPHY

Sequences

We follow Gardner's (1993, 1995) interpretation and attribute the Ferron Sandstone to two sequences. The lower Ferron lies at the top of Gardner's Hyatti sequence and consists of shelf sandstones (Washboard and Clawson units); the upper Ferron constitutes Gardner's Ferronensis sequence and consists of fluvio-deltaics (Figure 1). However, we place the thick marine sandstone at the mouth of Last Chance Canyon in the Ferronensis se-

quence instead of the Hyatti. We do this based on the presence of a rather dark and organic-rich shale zone which consistently lies above the upper shelf sandstone of the Hyatti sequence along the outcrop and appears to either be equivalent to the lowest Ferronensis marine sandstone at Last Chance Creek or possibly underlies it.

Parasequences Sets

The "upper" Ferron Sandstone (not including the underlying Washboard and Clawson units) contains nine parasequence sets and are designated based on (1) genetically related parasequences within the set, (2) recognition of major flooding surfaces which bound the sets reflecting a change of stacking pattern in the parasequence sets, and (3) the coupling of parasequence sets with major coal zones. The parasequence sets are designated in Figure 1.

Parasequences

The parasequence is a smaller-scale depositional unit recognized on the basis of marine flooding surfaces. Van Wagoner et al. (1987) define the parasequence as "a relatively conformable succession of genetically related beds or bedsets bounded by flooding surfaces or their correlative surfaces." They go on to define the marine-flooding surface (now generally shortened to flooding surface) as "a surface separating younger from older

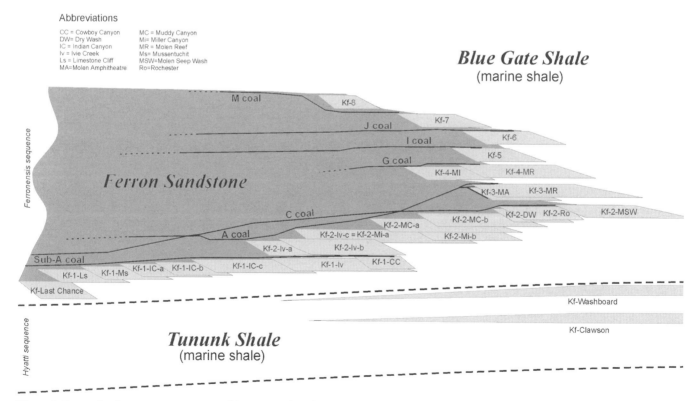

Figure 1. Ferron Sandstone parasequence stacking pattern based on nearshore marine facies. Each parallelogram represents a parasequence. The horizontal edges of each parallelogram represents the landward and seaward edges of the nearshore marine sandstone associated with each parasequence. Sequence boundaries are modified from Gardner (1995) and are short, black dashed lines.

strata across which there is evidence of an abrupt increase in water depth." "Marine-flooding surface" is a replacement term for the familiar "transgressive surface." Swift (1967) defined the "ravinement surface," another synonym that is commonly used to describe this same surface in the literature.

The Ferron Sandstone contains many marine-flooding surfaces and most who have studied the Ferron (but not all) agree on where they are located. In theory, the parasequence includes all strata lying between two marine-flooding surfaces and the correlative surfaces in adjacent continental and offshore marine environments. In our experience in the Ferron the correlative surfaces can very rarely be recognized. As a consequence, our Ferron parasequences are, for all practical purposes, recognized on the basis of bodies of predominantly sandy strata deposited in the nearshore environment.

Bedsets

The Ferron shoreline progradation is not the result of a single type of deltaic deposition but a complex mix of a variety of styles. Some of these styles are recognizable on outcrop as distinct and mappable genetic units but unrelated to marine-flooding surfaces. We reject the use of parasequence to describe this rock package because of the lack of clear evidence for a flooding surface. We instead borrow the next term in Van Wagoner et. al.'s (1990) hierarchical scheme below a parasequence and call these units bedsets.

In our view, many parasequences identified in the literature have been based on the assumption that a change in sedimentation pattern is always related to a change in water depth or a marine-flooding event. Hence, when stacked coarsening-upward rock packages are found in a distal position to a contemporaneous beach, each cycle is presumed to represent a change in water depth or a flooding event. When these "flooding surfaces" are traced up depositional dip they are often lost in about the middle shoreface. Multiple "flooding surfaces" are related to a single documented landward pinchout of the beach into coastal-plain facies (a confirmed marine-flooding surface). We interpret many of these individual sub-parasequence cycles or bedsets as events related to changes in sedimentation patterns, not relative sea-level change. We have often identified distinct coarsening upward cycles that when traced landward simply feather to a common stratigraphic horizon with adjacent rock packages above and below, show no intertonguing with coastal-plain or marginal marine facies, and no stratigraphic rise (typically associated with progradation in the Ferron Sandstone). We have interpreted these distinctive sub-parasequence cycles as bedsets, related to changes in sedimentation patterns, not relative sea-level changes.

A change from higher energy (sandy) nearshore facies to lower energy (muddy) facies is not always associated with a change in water depth or "flooding event." The Indian Canyon area (a side canyon of Willow Springs Wash) offers an excellent example of a change in shoreline energy, and hence facies, which is unrelated to a change in relative sea level (specifically the change from Kf-1-IC-c to Kf-1-IC-c[d], see Dewey and Morris, this volume). Some of our bedsets may be parasequences, but until the bounding surfaces can be shown to be marine-flooding events the bedset designation is more appropriate.

Shoreline Types

Sub-parasequence packages of rocks or bedsets can generally be classified on the basis of the three shoreline types: wave-dominated, wave-modified, and fluvial-dominated. The facies of these shoreline types are described in Ryer and Anderson (this volume). Figures 2 through 5 show progradation of a hypothetical shoreline. Figure 6 displays the symbols used to represent the three shoreline bedset types. Figure 7 shows the common arrangement of the three basic shoreline bedset types within the parasequences of a hypothetical parasequence set. Figure 7 emphasizes one of our principal conclusions: initial progradation of each parasequence (or bedset) of the Ferron Sandstone consists of wave-dominated shoreline deposits.

NOMENCLATURE SCHEME

We have adopted a scheme (Figure 8) for naming stratigraphic units in the Ferron Sandstone. "Kf" is an abbreviation for the Cretaceous-aged Ferron Sandstone Member of the Mancos Shale. The first, second, third, and so forth (Kf-LC, Kf-1, Kf-2, Kf-3, ...) of nine stratigraphic units are identified on the basis of laterally traceable, cliff-forming units composed of shoreline sandstone facies. They are generally the "delta-front units" of Ryer (1981), the "genetic units" of Gardner (1993), and the "stratigraphic cycles" of Barton (1994). Some are equivalent to parasequence sets, whereas others include more than one parasequence set. We refer to them by the very general term "shoreline units" and they are designated by a number or letters after the first dash in the abbreviation. Kf-Clawson and Kf-Washboard are shelf sandstone units in the lower Ferron. Kf-Last Chance is a unit of uncertain origin that pre-dates Kf-1 and probably post-dates Kf-Washboard. We include it in the upper Ferron, as discussed previously.

At the parasequence level of our hierarchical scheme we add another dash behind the abbreviation that designates the shoreline unit or parasequence set followed by the parasequence name, for example Kf-1-{parase-

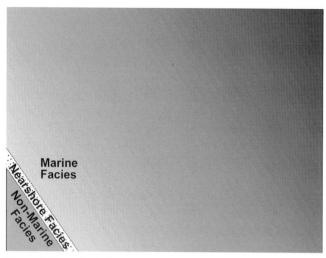

Figure 2. Hypothetical prograding strand plain.

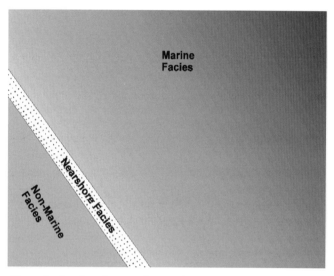

Figure 3. Continued progradation of the strand plain. Addition of sediment is by longshore drift, presumably from a deltaic source.

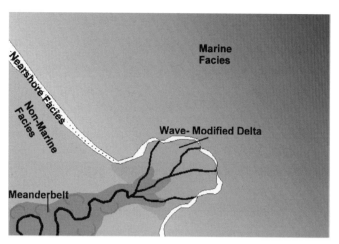

Figure 4. Progradation of a wave-modified delta following avulsion of a river system to the segment of coast shown.

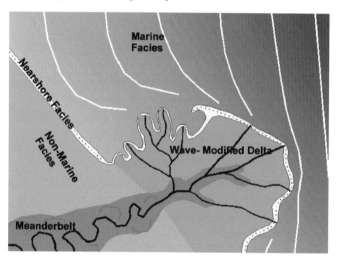

Figure 5. Continued progradation of delta. The white lines represent the crests of waves approaching from the east. The northwestern side of the delta experiences lower wave energies and takes on a fluvial-dominated form, whereas the eastern and southeastern sides have a wave-modified form.

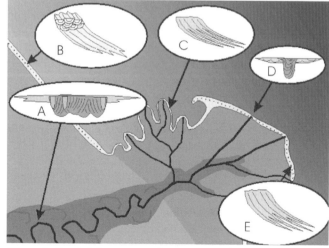

Figure 6. Symbols/patterns used to represent the various sandstone body types in Figures 7 and 10. (A) meanderbelts, (B) wave-dominated deltaic deposits, (C) fluvial-dominated deltaic deposits, (D) distributary channels, and (E) wave-modified deltaic deposits.

quence name or abbreviation}. The Kf-1-Iv and Kf-2-MC-a are parasequence designations. As discussed earlier, a parasequence must have clear evidence of a marine-flooding surface. The best possible evidence of this is the existence of a landward pinchout, where shoreface sandstone tongues out into continental facies. The Ferron Sandstone includes many excellent examples of this phenomenon (such as Kf-1-Limestone Cliffs, Kf-1 Indian Canyon-b, Kf-1-Indian Canyon-c, Kf-1-Ivie Creek, Kf-2-Ivie Creek-a and -c, Kf-2-Muddy Canyon-a and -b, Kf-2-Dry Wash, Kf-3, and Kf-6).

A parasequence is named for a type area. A type area, like Indian Canyon, can be the type area for more than one parasequence, in which case parasequences are labeled sequentially with a lower case alphabetic character. There is a problem with designating stratigraphic units in the Ferron Sandstone by means of type areas: there are simply not enough names to go around. We have restricted our names to features that are labeled on

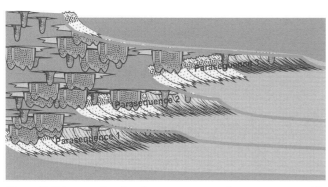

Figure 7. Diagram showing the arrangement of parasequences within a forward-stepping parasequence set and, within the parasequences, the arrangement of bedsets. Alluvial-plain deposits are in dark gray and offshore marine shale is shown in light gray. Medium gray lines separating parasequences are marine-flooding surfaces. The oldest and youngest of the four shown are major flooding surfaces that bound the parasequence set.

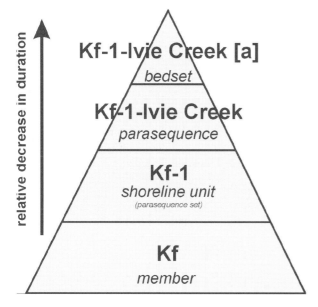

Figure 8. Hierarchical scheme of stratigraphic units. Italicized text is the type of stratigraphic unit. The shoreline unit is similar to a parasequence set, but does not rigorously follow published definitions.

the U.S. Geological Survey 7.5 minute quadrangle maps. In some areas, we have identified many more parasequence-level units than we can find unique names for. This necessitates the a, b, c suffix convention. In one case we have two names for the same parasequence: Kf-2-Ivie Creek-c and Kf-2-Miller Canyon-a. Detailed work was completed in both the Muddy Creek Canyon area (Kf-2-Miller Canyon-a) and Ivie Creek area (Kf-2-Ivie Creek-c) before the final regional correlations were complete. Hence, this parasequence carries one name in the southern area and another name in the northern area. The Muddy Creek Canyon work pre-dates the Ivie Creek work and in the future studies we recommend abandoning the term Kf-2-Iv-c.

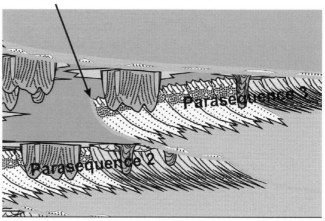

Figure 9. Landward pinchout marks point where shoreline sandstone body pinches out laterally into coastal-plain or delta-plain deposits. The pinchout is underlain by the most landward part of the transgressive surface of erosion.

At the bedset level of our hierarchical scheme we attach brackets behind the abbreviation that designates the parasequence with a bedset designation, that is Kf-1-Iv[bedset name or abbreviation]. When a bedset is named, a type area is used. A type area, like Ivie Creek, can be the type area for more than one bedset, in which case, bedsets in the same shoreline unit are labeled sequentially. Presently, some parasequences consist of a single bedset. We have only formally named four bedsets, however, numerous bedset bounding surfaces are recognized and mapped (see Anderson et al., 2003).

LANDWARD PINCHOUTS

From the perspective of sequence stratigraphic analysis, the landward pinchout of a shoreline sandstone body (shown diagrammatically in Figure 9) is a distinctive and very important feature. Examples of Ferron landward pinchouts on outcrop are shown in Figures 10 through 14. In each of the examples shown, the shoreline sandstone body pinches out into delta-plain or coastal-plain strata. In the examples shown in Figures 10 and 12 through 14, the landward pinchouts are both underlain and overlain by coal and/or carbonaceous shale. In all four examples, there can be no doubt that a relative rise of sea level accompanied the transgression that preceded progradation of the wave-dominated shoreline. The transgressive surface of erosion at the base of each shoreline sandstone body is the most landward part of a marine-flooding surface. Each marine-flooding surface marks a parasequence boundary.

Many landward pinchouts were subsequently eroded and replaced by channel belt deposits (for example parasequence 2 in Figure 7). Only the more seaward parts of the marine-flooding surfaces remain. In such cases, marine-flooding surfaces must be recognized on the basis

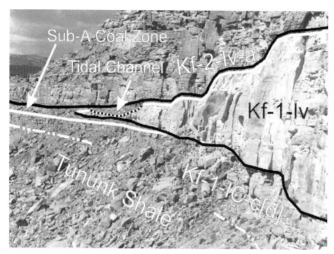

Figure 10. Landward pinchout of parasequence Kf-1-Iv on the south-facing cliffs east of Coyote Basin. A small tidal channel filled with muddy sandstone and oyster shells has eroded part of the wave-dominated shoreline sandstone body of Kf-1-Iv near the pinchout.

Figure 11. Landward pinchout of parasequence Kf-2-Rochester on the northern side of Dry Wash. A fluvial channel belt cuts the wave-dominated shoreline sandstone body near the pinchout and is outlined in the upper right of the photo. This situation is similar to that shown in Figure 9.

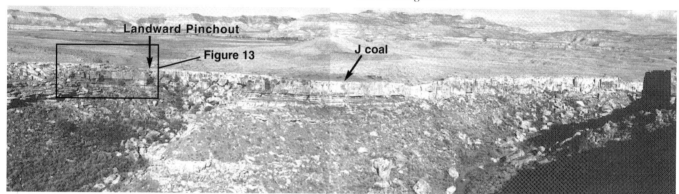

Figure 12. Landward pinchout of Kf-6 in the uppermost part of Muddy Creek Canyon. The white to rusty tan weathering, cliff-forming sandstone was deposited on a strongly wave-dominated shoreline. Carbonaceous shales beneath the transgressive surface of erosion at the base of Kf-6 represent the I coal zone (see Figure 1); the bed of coal apparent in the middle of the photo that immediately overlies Kf-6 is the J coal bed. A bed of muddy, burrowed sandstone lies between the I and J coal zone to the south and represents a lagoon that formed during the pre-Kf-6 transgression. The J coal is directly overlain by another transgressive surface of erosion with a thin transgressive lag, which is, in turn, overlain by the Blue Gate Shale.

Figure 13. Closer view of the landward pinchout of Kf-6 shown in Figure 12.

Figure 14. The uppermost sandstone is the Kf-2-Iv-c and is rapidly thinning toward the pinchout, which occurs near road level. Carbonaceous-rich sediments from the A coal zone (see Figure 1) enclose the landward pinchout of this marine sandstone.

of an increase in water depth represented by the deposits above the surfaces relative to the deposits beneath the surfaces. In cases where the increase in water depth is great (such as marine shale overlying coal), marine-flooding surfaces can be recognized with certainty. Where the increase is small (such as a marine shale overlying lower shoreface deposits) there is some uncertainty; as demonstrated in our study of the Kf-1 deposits in Indian Canyon, lithologic changes of this type can be produced by purely autocyclic processes that do not involve deepening of water (see Dewey and Morris, this volume).

The transgressions that ended each of the nine "delta-front units" and preceded the next are represented by major marine-flooding surfaces. Ryer (1981, 1982a, 1982b) mapped the positions of the pinchouts associated with the major marine-flooding surfaces, although we have reinterpreted the orientations of some of the pinchouts (Figure 15). In the following section, we document the bedsets, parasequences, and shoreline units (parasequence sets) that we recognize in the Ferron Sandstone.

FERRON STRATIGRAPHY

Our analysis of Ferron stratigraphy made extensive use of oblique aerial and ground-based photomosaics (Anderson et al., 2003). Our extensive detailed field effort lead to the naming of many geographic features. These new names appear in this paper in quotation marks and can be found on Plate 1, CD-ROM. Our work began with the definition of parasequences within the shoreline sandstone units, each of which was assigned a name based on a type area. The next smaller mappable and genetically related unit is the bedset. We have taken a conservative approach in defining bedsets and parasequences: others would possibly choose to subdivide further or differently.

Choosing a bounding surface between adjacent bedsets from one or another of the three shoreline types sometimes had to be done in a somewhat arbitrary fashion, since the change from one type of shoreline to another may be gradational. Numerous transitions, particularly between the wave-dominated and wave-modified shoreline deposits, are observed without a distinct bedset bounding surface. Therefore, every change in shoreline type is not divided by a bedset bounding surface. In some instances (Kf-1-Ivie Creek[a] to Kf-1-Ivie Creek[c], see Anderson et al., figure 7, this volume) adjacent bedsets are assigned to the same shoreline type. For river-dominated shorelines, the bedsets probably represent different delta lobes; for wave-dominated shorelines, the bedsets probably represent either different wave-energy levels, different rates of sediment supply, and/or accommodation space.

We located all of the recognizable landward pinchouts of shoreline sandstone bodies and identified the marine-flooding surfaces associated with them. We identified marine-flooding surfaces that cannot be traced to a landward pinchout owing to subsequent fluvial erosion of the underlying shoreline strata at the pinchout. These two types of rock packages were designated parasequences.

It must be emphasized that the parasequences we identified encompass only the shoreline and nearshore facies of the parasequence and not the continental and offshore marine facies. With only a few possible exceptions, we did not attempt to trace correlative surfaces for any significant distance landward in the continental deposits nor seaward into prodelta shale.

The Appendix (on CD-ROM) sets forth the complete description of our Ferron stratigraphy. Geographical features used to delineate the locations of these units are shown on Plate 1, CD-ROM. A detailed, diagrammatic regional cross section is provided as Plate 2, CD-ROM. Complete photomosaics coverage of the study area is available on CD-ROM from the Utah Geological Survey (Anderson et al., 2003). Numerous references are made to different facies in the Appendix. A definition of these facies can be found in Ryer and Anderson (this volume).

ANALYSIS

Applicability of Sequence Stratigraphic Terminology

The application of sequence stratigraphic terminology in this study was driven by two factors: (1) the hope that its models and terms would enhance our understanding of the packages of Ferron rocks and increase our ability to predict the occurrence and distribution of reservoir-quality facies, and (2) its popularity at this time. At the outset of the project the regional stratigraphic focus was on subdividing Ryer's (1981) previous "deltaic units" into the next logical sequence stratigraphic unit, the parasequence. The key in defining parasequences is the recognition of marine-flooding surfaces. As discussed in the Introduction, Van Wagoner et al. (1990) define these surfaces as "A surface separating younger from older strata across which there is evidence of an abrupt increase in water depth." The "law of superposition" makes almost any bedding surface qualify for the first part of the definition, so it is the "evidence of an abrupt increase in water depth" which becomes the important element of establishing a marine-flooding surface or parasequence boundary. It is not clear why a gradual change in water depth is not equally as important to recognize and should not also receive equal status in the hierarchy.

Early in the study numerous coarsening-upward and bed-thickening-upward successions of marine rocks were recognized, and a marine-flooding surface was

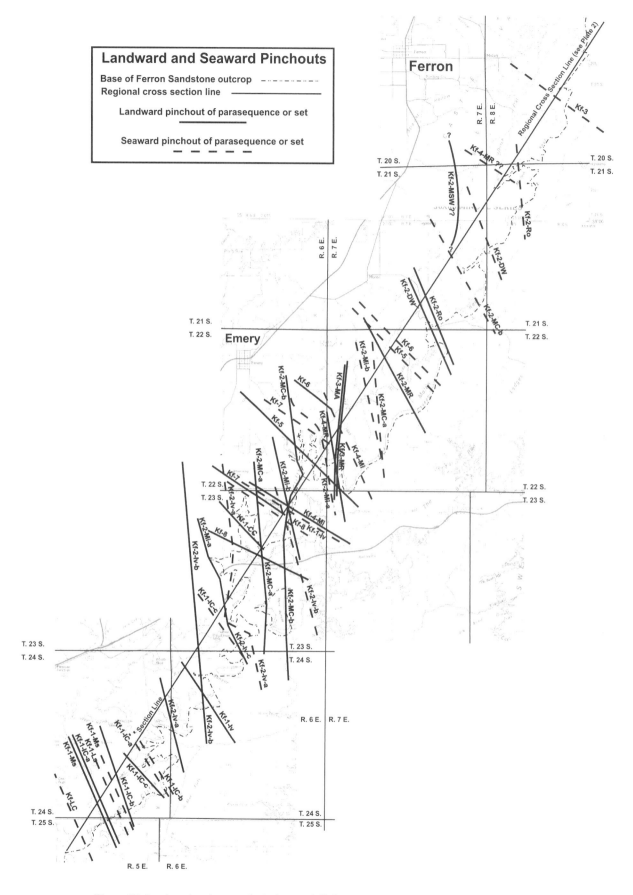

Figure 15. Landward and seaward pinchouts of all the parasequences identified in the study area.

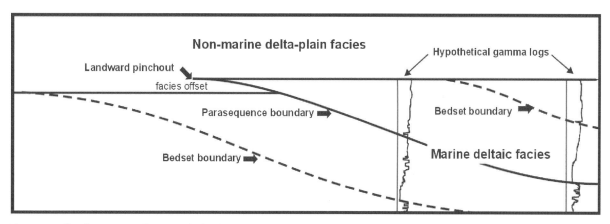

Figure 16. In more distal positions relative to the landward pinchout of nearshore marine facies, the top of stacked packages of coarsening-upward successions are easily mistaken for flooding surfaces. These surfaces were designated bedset boundaries because when traced to the marine/non-marine contact no facies offset into the delta plain could found.

placed at the top of many of these packages when they were overlain by another similar succession (Figure 16). As these units were traced laterally, some of these "flooding surfaces" merged with the top of the marine facies with no indication of a vertical offset of the marine into the overlying non-marine or delta-plain facies. This strongly suggests there was no relative change in sea level associated with these important bedding surfaces. The rocks in these packages bounded by this bedding surface are genetically related to an event which has differentiated them from rocks above and below.

It seems possible to have a "parasequence boundary" in the distal portion of genetically related beds but no such boundary (by definition) in the proximal portion of the same chronostratigraphic unit at the beach. Considering the higher compaction ratio of mud to sand, it is likely that the more distal and muddy portions of a delta would compact more than the proximal and sandy facies. No guidance is given for this situation in Van Wagoner's definitions and based on his work in the Book Cliffs (Van Wagoner, 1995) he has chosen to view these minor offshore changes in lithology as always reflecting a "flooding event." We see this as an inherent ambiguity in the application of sequence stratigraphy in high-resolution stratigraphic analysis which creates a variety of interpretations for the same stratigraphic interval.

When examining rocks only in the distal portion of the system, it appears hopeless to differentiate a change in sand-shale lithology related to changes in sediment supply or minor compaction-related subsidence, and one of sufficient magnitude to affect sea level at the beach. This fact limits the predictive power of the sequence stratigraphic model in correlating each "parasequence" identified in more distal facies of deltas with corresponding flooding events which affect sedimentation at the beach.

We are aware of two other attempts to apply sequence stratigraphic analysis (terminology) to the Fer-

ron Sandstone: Garrison and van den Bergh (1996) and Knox and Barton (1999). Many of the interpretations are similar, but there remain many differences. Regardless of who is right or wrong, the noted differences reflect the utility of sequence stratigraphy when applied by various workers. The terms we use in stratigraphy are useful if they enable us to clearly communicate with each other. We see room for improvement by more clearly defining terms.

Progradation Styles

The early progradations of the Ferron Sandstone (Kf-LC [Figure 17], Kf-1-Ls, Kf-1-Ms) are shorter in length from landward pinchout to seaward pinchout, and the internal inclination of beds are steeper than in later seaward-stepping parasequences. The cause of this geometry is not clear, but may be related to more rapid subsidence in the foreland basin locally, creating the steeper inclinations into the basin.

All initial progradations of the Ferron parasequences are wave-dominated (Figure 18). With the onset of initial transgression, the sea inundates the low delta-plain and the shoreline moves landward. With this landward transgression of the shoreline, the mouths of rivers are initially flooded and sediment is trapped in the old channel systems. During base-level "turn around" and transgression the principal source of sediment at the shoreline is from eroding delta plain sediments. As waves rework old delta-plain sediments, new strandline-type deposits are formed.

Progression from one type of delta (wave-dominated, wave-modified, or fluvial-dominated) to another often, but not always, follows a predictable order. Often this change is abrupt and is identified as a bedset, other times the change is gradational. A change from one progradational style to another does not necessary correspond to a new parasequence (Figure 19). The Ferron shorelines prograded as a complex of contemporaneous

Figure 17. View of the full sandstone thickness of Kf-LC just north of Last Chance Creek, along the Limestone Cliffs. Note the thin transition facies at the base of the sandstone and contact with the underlying Tununk Shale. The top of the unit is overlain by a split of the Sub-A coal zone (see Figure 1).

Figure 18. Wave-dominated facies within 0.25 mi of the landward pinchout of Kf-2-MC-b. Note chunks of coal below ledge. The A-C coal lies directly on top of the planar-bedded foreshore at the top of the ledge (not visible). The hammer lies on trough cross-bedded upper shoreface.

Figure 19. The lower cliff-forming sandstone is parasequence wave-dominated Kf-2-Mi-b on the east side of Muddy Creek Canyon, just north of the first meander above the confluence with Miller Canyon. Note the increase in shoreface sandstone thickness in the upper bedset. This change in deposition is related to an autocyclic event and not to a relative change in sea level. The Kf-2-MC a and Kf-2-MC-b contain a full range of wave-dominated to fluvial-dominated facies. See Appendix for detailed descriptions.

and diachronous intertonguing delta-types. Bhattacharya (1998) describes Danube River delta as a modern example of various deltaic facies associated with one river along a "wave-influenced" coast. This is a good analogy for portions of the Ferron where a wide variety of facies and architectures are observed within a relatively short distance along the paleoshoreline.

REGIONAL RESERVOIRS

Regional stratigraphic study has resulted in a better understanding of the geometries of delta progradation as observed in one general orientation to the Ferron Sandstone impulse of sediments. The outcrop of the Ferron serves as a generally depositional-dip oriented cross section through the depositional system (Figure 15 and Plate 2, CD-ROM). Using the delineation of the landward and seaward pinchouts in Figure 15, Table 1 was constructed. Measured sections and photomosaic coverage of the outcrop provided information on the thickness of parasequences found in Table 1. Kf-1 and Kf-2 have the greatest dip lengths of all the parasequence sets. The average maximum thickness of the nearshore facies of parasequences in these two sets is 55 ft (17 m) with the average dip length of the nearshore facies 4.3 mi (6.9 km). Most of this thickness is sandstone and the upper quarter to third is often good quality reservoir rock. In the aggradational parasequence sets, Kf-3 (Figure 20) to Kf-6 we have poorer data on the breakout of parasequences and an average maximum thickness of the nearshore facies is not as meaningful, but where full

Table 1. Parasequence and parasequence set nearshore marine facies geometries and thickness.

Parasequence *indicates a set	Depositional-dip Length (mi)	Maximum Thickness (ft)	Shoreline Orientation	Identified Bedsets
Kf-Last Chance*	>2.0	200	?	
Kf-1*	15.2	-	-	
Kf-1-Limestone Cliffs	>2.4	90	-	
Kf-1-Mussentuchit	1.0	60	N33W?	
Kf-1-Indian Canyon-a	2.4	55	N38W?	
Kf-1-Indian Canyon-b	1.8	50-60	N29W?	
Kf-1-Indian Canyon-c	5.8	60	N43W	1
Kf-1-Ivie Creek	7.6	70	N35W?	6
Kf-1-Cowboy Canyon	?	35	N42W	
Kf-2*	>30.4	-	-	
Kf-2-Ivie Creek-a	3.8	50	N15W?	
Kf-2-Ivie Creek-b	2.8	40	N4W	
Kf-2-Miller Creek-a	5.3	55	N17W	
Kf-2-Miller Creek-b	4.0	60	N13W?	2
Kf-2-Muddy Canyon-a	5.0	56	N5W	
Kf-2-Muddy Canyon-b	9.1	50	N2W	1
Kf-2-Dry Wash	3.3	40	N23W?	
Kf-2-Rochester	4.3	40	N22W?	
Kf-2-Molen Seep Wash	?	40	?	2?
Kf-3*	15.8	-	-	
Kf-3-Molen Amphitheater	**?	15	N5E?	
Kf-3-Molen Reef	15.8**	55	N5E?	
Kf-4*	14.6	-	-	
Kf-4-Miller Creek	3.6	85	N59W?	
Kf-4-Molen Reef	11.3	75	N8W?	
Kf-5*	5.2	45	N46W?	
Kf-6*	3.6	30	N52W	
Kf-7*	3.1	20	N57W?	
Kf-8*	1.7	15	N63W?	

**Additional study needed to verify one or more parasequences.

development of the facies were available the average maximum thickness is about 56 ft (17 m) — a bit thicker than the seaward-stepping parasequences. The average dip length for the aggradational parasequences nearshore facies is 6 mi (10 km). This may be high if more parasequences lie within these sets. The landward-stepping parasequences are definitely thinner. We did not divide these parasequence sets into parasequences, if further study merits such division the average maximum thickness will be less than the present 22.5 ft (6.9 m). These average maximum thicknesses correlate well with low accommodation at the shoreface during seaward stepping, moderate to high accommodation during aggradation, and rapid relative sea-level rise and

Figure 20. View to the west of the landward pinchout of parasequence Kf-3-MA. Note the dark upper split of the C coal under the
Kf-3-MA just seaward of the landward pinchout. The landward pinchout of the Kf-3-MR is located about where this photograph was
taken and Kf-4-MR landward pinchout is just out of view in the
upper right.

Figure 21. Alluvial-plain section of Ferron along the Limestone
Cliffs. Sandy fluvial rocks make up a high percentage of the section.
The lowest sandstone cliff is marine nearshore deposits in Kf-1
parasequence set with the Tununk Shale at the bottom of the photograph. Landslide debris in the bottom left foreground.

transgression during landward stepping. There appears
to be no correlation between parasequences' dip length
and thickness.

One important conclusion related to regional reservoirs is the recognition that all initial progradations
begin as wave-dominated. From associated studies
(Mattson and Chan, this volume, Forster et al., this volume, and Barton et al., this volume) we know that the
wave influenced facies generally have better porosity
and permeability than the fluvial-dominated facies,
therefore, the landward pinchouts of parasequences
make good reservoir rocks. The trends of these reservoirs are shoreline parallel.

Collectively, seaward-stepping parasequence sets
have much greater dip length (>40 mi [64 km]) than do
the vertically aggrading and landward-stepping parasequence sets (Table 1). A significant portion of the volume
of rock within any seaward-stepping parasequence set
consists of channel and meanderbelt facies which have
incised older or relatively contemporary shoreface
deposits. These channel facies are some of the best reservoirs in the system (Barton, 1994, and Forster et al., this
volume). Their elongation direction relative to shoreline
is orthogonal.

Alluvial-plain deposits consist of a high percentage
of sandstone. This is qualitatively obvious when examining the photomosaics along the Limestone Cliffs (Figure 21). Shale drapes and plugs can create a reservoir
which is highly compartmentalized, but individual
meanderbelt systems in the alluvial-plain facies have
continuity of several miles, based on exposures along
the Limestone Cliffs.

Landward-stepping parasequence sets are much

smaller in volume of sand because of the relatively short
distance they prograded before the continuing rise of sea
level overtook sediment supply and the shoreline
pushed farther landward (Figure 1). Compared to seaward-stepping units, their dip length is short. Landward-stepping units offer the attraction of having an
excellent seal of offshore muds stratigraphically above.
These units are generally wave-dominated, providing
good reservoir sandstones and are not commonly
incised by channel systems which could act as conduit to
leak hydrocarbons into the delta-plain and alluvial-plain
deposits found landward of these deposits.

SUMMARY

The Ferron Sandstone Member of the Mancos Shale
can be divided into nine parasequence sets. Close examination of outcrops of several Ferron parasequence sets
has lead to further subdivision of chronostratigraphic
units into parasequences. Where bounding surfaces cannot be confirmed as marine-flooding surfaces but contain rock packages of genetically related beds, the term
bedset is applied. Several such bedsets have been
named. Our proposed nomenclature is an alphanumeric
scheme in which the first letter designates the geologic
age and the second is a one letter abbreviation for the
unit name (Kf for Cretaceous Ferron), followed by a
dash. The next alphanumeric sequence is for the parasequence set name followed by a dash and the parasequence name (Kf-1-Ivie Creek, for parasequence set 1
and the parasequence named Ivie Creek). Nineteen
parasequences are named. A genetically related package
of beds, bounded by a mappable surface which is clearly not a marine-flooding surface is a bedset and attached
at the end of the alphanumeric string in brackets (Kf-1-
Iv[a] for the bedset "a" of the Ivie Creek parasequence).
Only four bedsets are named, but many others recognized. Coals in the section retain the names assigned by
Lupton (1916).

Progradation of Ferron shorelines begins as wave-
dominated, often changing to a wave modified, and then

to fluvial-dominated. In some instances these changes in delta-type are mappable as distinct bedsets. Some of the best reservoir rocks are associated with the shallow nearshore facies of a wave-dominated coast and the distributary channels of the fluvial-dominated coast. The first is shoreline parallel and the latter, normal to oblique to the shoreline. Shoreline orientation, based on the landward pinchout of nearshore facies, varied from about N60W to N5E. The nearshore marine facies of seaward-stepping parasequences average 55 ft (17 m) maximum thickness and have an average depositional dip length of 4.3 mi (6.9 km). The detail of study on the aggradational parasequences sets make their statistics more suspect, but indicate an average maximum thickness of 56 ft (17 m) and dip length of 6 mi (10 km) of the nearshore facies.

ACKNOWLEDGMENTS

This paper was supported by the Utah Geological Survey (UGS), contract numbers 94-2488 and 94-2341, as part of a project entitled *Geological and Petrophysical Characterization of the Ferron Sandstone for 3-D Simulation of a Fluvial-Deltaic Reservoir*, M. Lee Allison and Thomas C. Chidsey, Jr., Principal Investigators. The project was partially funded by the U.S. Department of Energy (DOE) under the Geoscience/Engineering Reservoir Characterization Program of the DOE National Petroleum Technology Office, Tulsa, Oklahoma, contract number DE-AC22 93BC14896. The Contracting Office Representative was Robert Lemmon.

A special thanks to Thomas C. Chidsey, Jr., Utah Geological Survey, for his excellent help on the outcrop and his continued support, guidance, and management of the DOE project. Mark Barton and Ed Angle provided helpful insight and stimulating discussions in the early stages of the work.

We thank the Mobil Exploration/Producing Technical Center and Amoco Production Company for their technical and financial support of the project which was conducted between 1993 and 1998. We also thank the following individuals and organizations for their technical contributions at the time this work was performed: R.L. Bon, Brigitte Hucka, D.A. Sprinkel, D.E. Tabet, Kevin McClure, K.A. Waite, S.N. Sommer, R. D. Adams, and M. D. Laine of the UGS, Salt Lake City, Utah; and M.A. Chan of the University of Utah, Salt Lake City, Utah.

Jim Parker (UGS) drafted figures and Cheryl Gustin (UGS) prepared the manuscript. Finally, we thank David E. Tabet and Rex D. Cole for their careful reviews and constructive criticism of the manuscript.

REFERENCES CITED

Anderson, P. B., and T. A. Ryer, 1995, Proposed revisions to parasequence-set nomenclature of the Upper Cretaceous Ferron Sandstone Member of the Mancos Shale, central Uta (abs.): AAPG Bulletin, v. 79, no. 6, p. 914–915.

Anderson, P. B., K. McClure, T. C. Chidsey, Jr., T. A. Ryer, T. H. Morris, J. A. Dewey, Jr., and R. D. Adams, 2003, Interpreted regional photomosaics and cross section, Cretaceous Ferron Sandstone, east-central Utah: Utah Geological Survey Open-File Report 412, compact disk.

Barton, M. D., 1994, Outcrop characterization of architecture and permeability structure in fluvial-deltaic sandstones, Cretaceous Ferron Sandstone, Utah: Ph.D. dissertation, University of Texas, Austin, 255 p.

Bhattacharya, J. P., 1998, Wave-influenced deltas (abs.): AAPG Annual Convention Extended Abstracts, v. I, p. A-64.

Campbell, C. V., 1967, Lamina, lamina-set, bed and bedset: Sedimentology, v. 8, no. 1, p.

Cotter, Edward, 1975, Deltaic deposits in the Upper Cretaceous Ferron Sandstone, Utah, *in* M. L. S. Broussard, ed., Deltas, models for exploration: Houston Geological Society, p. 471 484.

—1976, The role of deltas in the evolution of the Ferron Sandstone and its coals: Brigham Young University Geology Studies, v. 22, pt. 3, p. 15–41.

Gardner, M. H., 1993, Sequence stratigraphy and facies architecture of the Upper Cretaceous Ferron Sandstone Member of the Mancos Shale, east-central Utah: Ph.D. dissertation T 3975, Colorado School of Mines, Golden, 528 p.

—1995, Tectonic and eustatic controls on the stratal architecture of mid-Cretaceous stratigraphic sequences, central Western Interior foreland basin of North America, in Stratigraphic evolution of foreland basins: Society for Sedimentary Geology (SEPM) Special Publication No. 52, p. 243–281.

Garrison, J. R., Jr., and T. C. V. van den Bergh, 1996, Coal zone stratigraphy — a new tool for high resolution depositional sequence stratigraphy in near-marine fluvial-deltaic facies — a case study from the Ferron Sandstone, east-central Utah: AAPG Rocky Mountain Section Meeting, expanded abstracts volume, Montana Geological Society, p. 31–36.

Garrison, J. R., Jr., T. C. V. van den Bergh, C. E. Barker, and D. E. Tabet, 1997, Depositional sequence stratigraphy and architecture of the Cretaceous Ferron Sandstone: implications for coal and coal bed methane resources — a field excursion, *in* P. K. Link and B. J. Kowallis, eds., Mesozoic to Recent geology of Utah: Provo, Brigham Young University Geology Studies, v. 42, pt. 2, p. 155–202.

Gustason, E. R., T. A. Ryer, and P. B. Anderson, 1993, Integration of outcrop and subsurface information for geological model building in fluvial-deltaic reservoirs: unpublished Utah Geological Survey/AAPG Rocky Mountain Section Core Workshop Manual, September 16, 1993, 24 p.

Kamola, D. L., and J. C. Van Wagoner, 1995, Stratigraphy and facies architecture of parasequences with examples from the Spring Canyon Member, Blackhawk Formation, Utah, *in* J. C. Van Wagoner and G. T. Bertram, eds., Sequence stratigraphy of foreland basin deposits: AAPG Memoir 64, p.

Knox, P. R., and M. D. Barton, 1999, Predicting interwell heterogeneity in fluvial-deltaic reservoirs — effects of progressive architecture variation through a depositional cycle from outcrop and subsurface observations, *in* R. A. Schatzinger and J. F. Jordan, eds., Reservoir characterization — recent advances: AAPG Memoir 71, p. 57–72.

Lupton, C. T., 1914, Oil and gas near Green River, Grand County, Utah: U.S. Geological Survey Bulletin 541, p. 115–133.

—1916, Geology and coal resources of Castle Valley in Carbon, Emery, and Sevier Counties, Utah: U.S. Geological Survey Bulletin 628, 88 p.

Mitchum, R. M., Jr., P. R. Vail, and S. Thompson III, 1977, Seismic stratigraphy and global sea level changes, Part 2: the depositional sequence as a basic unit for stratigraphic analysis — global cycles of relative changes of sea level, *in* C. E. Payton, ed., Seismic stratigraphy- applications to hydrocarbon exploration: AAPG Memoir 26, p. 53–62.

Nix, T. L., 1999, Detailed outcrop analysis of a growth fault system and associated structural and sedimentologic elements, Ferron Sandstone, Utah: M.S. thesis, Brigham Young University, Provo, 48 p.

Ryer, T. A., 1981, Deltaic coals of Ferron Sandstone Member of Mancos Shale — predictive model for Cretaceous coal-bearing strata of Western Interior: AAPG Bulletin, v. 65, no. 11, p. 2323–2340.

—1982a, Possible eustatic control on the location of Utah Cretaceous coal fields, *in* K. D. Gurgel, ed., Proceedings — 5th ROMOCO symposium on the geology of Rocky Mountain coal: Utah Geological and Mineralogical Survey Bulletin 118, p. 89–93.

—1982b, Cross section of the Ferron Sandstone Member of the Mancos Shale in the Emery coal field, Emery and Sevier Counties, central Utah: U.S. Geological Survey Map MF-1357, vertical scale: 1 inch = 30 meters, horizontal scale: 1 inch = 1.25 kilometers.

—1991, Stratigraphy, facies, and depositional history of the Ferron Sandstone in the canyon of Muddy Creek, east-central Utah, *in* T. C. Chidsey, Jr., ed., Geology of east-central Utah: Utah Geological Association Publication 19, p. 45–54.

Ryer, T. A., and Maureen McPhillips, 1983, Early Late Cretaceous paleogeography of east-central Utah, *in* M. W. Reynolds and E. D. Dolly, eds., Mesozoic paleogeography of the west central United States: Society of Economic Paleontologists and Mineralogists, Rocky Mountain Paleogeography Symposium 2, p. 253–272.

Swift, D. J. P., 1967, Coastal erosion and transgressive stratigraphy: Journal of Geology, v. 76, no. 4, p. 444–456.

Van Wagoner, J. C., 1995, Overview of sequence stratigraphy of foreland basin deposits- terminology, summary of papers, and glossary of sequence stratigraphy, *in* J. C. Van Wagoner and G. T. Bertram, eds., Sequence stratigraphy of foreland basin deposits: AAPG Memoir 64, p. ix-xxi.

Van Wagoner, J. C., R. M. Mitchum, H. W. Posamentier, and P. R. Vail, 1987, An overview of sequence stratigraphy and key definitions, *in* A. W. Bally, ed., Atlas of seismic stratigraphy: AAPG Studies in Geology 27, v. 1, p. 11–14.

Van Wagoner, J. C., R. M. Mitchum, K. M. Campion, and V. D. Rahmanian, 1990, Siliciclastic sequences, stratigraphy in well logs, cores and outcrops: AAPG Methods in Exploration, no. 7, 55 p.

General Geology of the Ferron Sandstone

"Wild men armed," Charles T. Lupton second from the left, circa 1910. Photograph courtesy of the family of C. T. Lupton.

Analog for Fluvial-Deltaic Reservoir Modeling: Ferron Sandstone of Utah
AAPG Studies in Geology 50
T. C. Chidsey, Jr., R. D. Adams, and T. H. Morris, editors

Petrophysics of the Cretaceous Ferron Sandstone, Central Utah

Richard D. Jarrard[1], Carl H. Sondergeld[2],
Marjorie A. Chan[1], and Stephanie N. Erickson[3]

ABSTRACT

The fluvial-deltaic sandstones of the Cretaceous Ferron Sandstone, Utah, provide an opportunity to document and compare petrophysical properties of outcrop and subsurface rocks. We find that the processes that generate outcrop exposures — uplift, erosion, and exhumation — can overprint patterns of velocity, porosity, and permeability developed in the subsurface. Burial to depths of 3000–3400 m (9800–11,100 ft), with associated compaction and carbonate cementation, was followed by uplift, which exhumed different portions of the Ferron Sandstone by 0 to >3400 m. A complex diagenetic history culminated with the development of secondary intergranular porosity by carbonate dissolution during exhumation, because of increasing groundwater flux at depths shallower than ~2 km (7000 ft) subsurface. Velocity logs show velocity decreases larger than expected from porosity increase; we attribute the excess to presence of microcracks. Outcrop plugs exhibit even higher porosity and lower velocity than shallow logs, probably because of enhanced leaching of carbonate cement. Ferron core-plug and log velocity responses to this secondary porosity are comparable to that of primary intergranular porosity, but these samples lack permeability anisotropy and sensitivity of velocity and permeability to clay content, both of which are typical of primary porosity. Ferron Sandstone permeability is very closely related to porosity, and therefore exhumation increases permeability by porosity enhancement. The influence of grain size on porosity and permeability persists after both initial compaction/cementation, and subsequent exhumation and secondary porosity development. Consequently, Ferron outcrop stratigraphy can provide useful clues to fluid-flow patterns in other deltaic formations, despite its complex diagenetic history.

[1]Department of Geology and Geophysics, University of Utah, Salt Lake City, Utah
[2]Mewbourne School of Petroleum and Geological Engineering, University of Oklahoma, Tulsa, Oklahoma
[3]ConocoPhillips Alaska, Inc., Anchorage, Alaska

INTRODUCTION

Subsurface geological interpretations, whether based on geophysical data or on drilling, can be guided by case studies that combine stratigraphic concepts with links between geological and geophysical variables. A team of academic, government, oil-industry, and consultant researchers, led by the Utah Geological Survey (UGS), selected the Cretaceous Ferron Sandstone for a multidisciplinary investigation of the interactions between stratigraphy, petrophysics, fluid flow, and burial history in a fluvial-deltaic system.

This study focuses on the interactions between diagenesis and petrophysics of the Ferron Sandstone, by examining the reliability of a common assumption: that geophysical properties and relationships observed on outcrop are representative of those operating in the subsurface. In particular, the process that generates outcrop exposures — exhumation resulting from uplift and erosion — may overprint the original patterns of velocity, porosity, and permeability developed in the subsurface. Petrophysical overprints may result either from diagenetic processes or from mechanical rebound — the expansion of rocks due to pressure release. In contrast to most previous studies of diagenetic influences on petrophysical properties, the emphasis here is on dissolution rather than on cementation. Furthermore, previous research on rebound has concentrated mainly on engineering and soils applications (Terzaghi, 1965) and on cores (Hamilton, 1976), rather than on in situ sedimentary rocks.

GEOLOGIC SETTING OF THE FERRON SANDSTONE

The Cretaceous inland sea, which extended north to the Arctic Ocean and south to the Gulf of Mexico, had a western margin in the region that is now eastern Utah (Williams and Stelck, 1975). The Sevier orogeny produced mountains in western Utah (Armstrong, 1968), and networks of rivers carried the rapidly eroded sediments eastward to the inland epeiric sea. During the Late Turonian, a suite of deltaic sediments known as the Ferron Sandstone was deposited along this margin of the inland sea. These deltaic sediments graded eastward into marine muds and westward into coals and alluvial sediments (Ryer, 1981; Ryer and McPhillips, 1983). The Ferron Sandstone is a member of the more laterally and temporally extensive Mancos Shale formation. More detailed aspects of the Ferron stratigraphy and sedimentology are presented in Anderson and Ryer (this volume) and Anderson et al. (this volume).

The northeastward progradation of Ferron sedimentation occurred in pulses, producing at least seven depositional cycles, or parasequence sets. These parasequence sets are numbered in ascending stratigraphic

order: Kf-1, Kf-2, ..., Kf-7 (Ryer, 1991; Gardner et al., 1995 a, b). The stratigraphic architecture of each parasequence set was controlled by a combination of rate of transgression or regression of the inland sea and rate of deposition (Ryer and Anderson, 1995). Slow sea-level rise and large sediment supply generated widespread units with little stratigraphic climbing, whereas rapid sea-level rise plus reduced sediment supply developed more localized sandstones with substantial stratigraphic climbing (Ryer, 1983, 1991). Some cycles are wave dominated and consist of tabular, laterally homogeneous fine-, medium- and coarse-grained sands. Others are river dominated and form deltaic lobes of interbedded mudstone, siltstone and sandstone (Ryer and McPhillips, 1983; Gardner et al., 1995 a, b). Our study focuses on the river-dominated Kf-1 parasequence set and overlying wave-dominated Kf-2 (see Anderson et al., this volume).

Cretaceous marine and marginal-marine deposition in central and eastern Utah was followed by Paleogene fluvial and lacustrine deposition of the Wasatch and Green River formations. Based on vitrinite reflectance data, surface and subsurface Ferron coals are similar to each other in maturity, suggesting relatively uniform depth of maximum burial (Tabet et al., 1995; D. A. Sprinkel, personal communication, 1996). A regional cross section (Sprinkel, 1994) indicates that the Ferron Sandstone was uniformly buried to a depth of about 3000–3400 m (9800–11,100 ft). During the Laramide orogeny, a broad uplift associated with the San Rafael Swell raised the region containing the eastern portion of the Ferron Sandstone. The westernmost Ferron Sandstone was not uplifted significantly, but it did undergo east-west compressional development of folds and fractures. Finally, beginning about 10 Ma, the Colorado Plateau uplift raised central and eastern Utah and adjacent regions by 2000–3000 m (6600–9800 ft) (Hintze, 1988), triggering intensive erosional downcutting. Sprinkel (1993) provides a geohistory diagram summarizing the burial and exhumation history of the portion of western Wasatch Plateau containing the Ferron Sandstone.

The result of this geologic history is that the Ferron Sandstone today exhibits an immense range in degree of exhumation. Along its northern and western extent the Ferron is still buried by 3000–3400 m (9800–11,100 ft) of sedimentary rocks, about the same as the previous maximum burial depth of the entire Ferron Sandstone. Overburden thickness decreases rapidly to the east and south, and the Ferron Sandstone crops out over a broad region (Figure 1).

METHODS

Data acquisition mainly involved petrophysical and petrographic analysis of outcrop core plugs from the Ivie Creek study area (Figure 1). This dataset was augment-

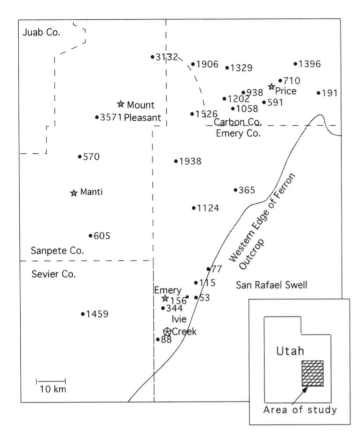

Figure 1. Map of east-central Utah showing Ferron Sandstone locations considered in this study. Solid dots are wells with both sonic and density logs from the Ferron; adjacent numbers are average depths of the Ferron data, in meters. Stars are towns. Starred pentagon is the Ferron outcrop at Ivie Creek, where core plugs were obtained.

ed by well logs and core plugs from five drill cores at Ivie Creek, and by well logs from 24 widely spaced wells penetrating the Ferron Sandstone.

Regional Ferron Well Logs

Coal and petroleum companies have drilled more than 500 wells through the Ferron Sandstone (Tripp, 1989; Hucka et al., 1995). Most of these wells have density logs, but few have sonic logs, and only 31 wells contain both sonic and density logs through the Ferron Sandstone. Of these, we discarded six because of neutron/density crossover or other evidence (Tripp, 1989) of gas presence. These gas sandstones were avoided because gas causes density-based porosity to be overestimated and velocity to plummet. We discarded one additional well, plus portions of several other wells, because of poor or dubious log quality. Thus, 24 wells (Figure 1) survived these selection and rejection criteria: all appear to contain good quality sonic and density logs representative of water-saturated Ferron sandstones.

Wells penetrating the Ferron interval typically encounter 17–323 m (56–1060 ft) of alternating sand-

stone, shaley sandstone, shale, and coal (Ryer and McPhillips, 1983). For each of the 24 wells, we identified sandstone or slightly shaley sandstone beds based on gamma-ray log cutoffs, and then picked average sonic traveltime and density for 2–20 (avg. 7.5) beds or relatively homogeneous portions of beds. Sonic traveltime (μs/ft) was converted to velocity (km/s). To convert density to porosity, we used an estimate of grain density provided by core-plug petrophysical data.

Core Plugs and Drill Holes from the Ivie Creek Study Area

Outcrop exposures offer a cost-effective opportunity to examine lateral stratigraphic and petrophysical variations at high resolution. Concepts and patterns based on these high-resolution studies can then be applied to actual exploration plays, where data coverage is more limited, and interpolation is common. At Ivie Creek (Figures 1 and 2), river incisions provide excellent exposures of the lowest two units of the Ferron in vertical to near-vertical cliffs. The lowest unit, Kf-1, is a river-dominated deltaic lobe with steeply prograding beds of sandstone, siltstone, and shale. This unit is overlain by Kf-2, a more homogeneous and sandier tabular, wave-dominated parasequence set.

The UGS drilled five wells on the plateau adjacent to Ivie Creek (Figure 2), at distances of 100 to 400 m (300–1300 ft) from the outcrops, to delineate three-dimensional (3-D) Ferron stratigraphy more thoroughly than is possible from outcrops alone. The Kf-1 and Kf-2 sandstones were cored at four wells, and we obtained logs of caliper, density, and gamma ray at all five wells. The Kf-1 and Kf-2 units lie above the water table at all wells, but the lower portions of two holes (IC-3 and IC-11) could be saturated so that sonic logging was feasible. For cores from three drill holes, we used the U.S. Geological Survey (USGS) continuous-core system to measure velocity of whole-core samples, plus continuous-core magnetic susceptibility and gamma-gamma density.

We took 722 oriented core plugs from the Ferron outcrop at Ivie Creek (Figure 2) using a portable rock drill with water as coolant. The core samples were 3–8 cm long and 2.5 cm in diameter. Sampling was conducted along more than a dozen near-vertical traverses, some accessible on foot and others by rock-climbing gear. A vertical sample spacing of 2–4 m (7–13 ft) was usually chosen, in order to sample all major lithologic units. We did not, however, sample shale units on outcrop because they were too weathered. In contrast, shale units in the Ivie Creek wells were unweathered and could be sampled. We took 86 core plugs from UGS Ivie Creek drill holes IC-3, IC-9, and IC-11. We also undertook a database search for any Ferron cores from much deeper wells. We identified two suitable wells with cores at

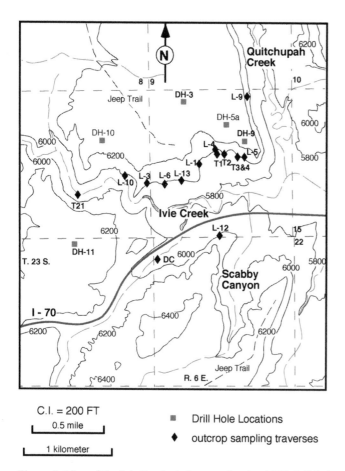

Figure 2. Map of the Ivie Creek study area showing UGS drill-hole locations and outcrop sampling traverses.

USGS Denver Core Repository and took 28 core-plug samples. Nearly all core-plug samples — from both outcrops and drill core — were drilled horizontally. However, at 79 sample locations a 3-D set of three samples (horizontal, vertical, and 45° diagonal) was collected for detection of petrophysical anisotropy.

Of these plug samples, 471 were analyzed using the Geoscience Evaluation Module (Sondergeld and Rai, 1993a, b) at Amoco Production Company. For each sample, P-wave and two S-wave velocities were measured for both dry and water-saturated states, at the following pressures: 1.72, 3.45, 5.17, 6.90, 10.3, 13.8, 20.7, 27.6, and 34.5 MPa. Porosity, matrix density, and bulk density were measured by helium porosimeter and mercury immersion. Pycnometer porosity, or Boyle's Law porosity, is calculated from grain density and dry density. Saturated porosity is calculated from saturated and dry densities. Because saturated porosity considers only pore space accessible to water, it is slightly lower than pycnometer porosity (maximum difference 2 porosity units at high porosities) and is more appropriate for our subsequent consideration of seismic response of saturated rocks. Porosity was also measured along with air permeability at a pressure of 6.9 MPa. Magnetic susceptibil-

ity was measured on a 1 KHz susceptibility bridge. Transmission infrared spectroscopy gave approximate concentrations of the following 14 mineral components: quartz, feldspars (orthoclase, oligoclase, albite, and intermediate plagioclase), carbonates (calcite, dolomite, siderite), clays (illite, kaolinite, chlorite, and smectite), pyrite, and anhydrite.

DIAGENESIS AND POROSITY OF THE FERRON SANDSTONE

Petrography

Two prior petrographic studies of the Ferron Sandstone were undertaken (Fisher et al., 1993; Barton, 1994). They noted compositional differences between distributary channels and delta-front deposits. The distributary channel sandstones were not so well sorted, contained more feldspar, and were classified as quartz-rich arkoses. The delta-front deposits ranged from reworked, distributary mouth-bar quartz arenites, to delta-front sub-arkoses and arkoses. As is commonly expected, more reworked sediments contained fewer feldspar and lithic fragments. This study corroborates these previous observations and considers differences between parasequences Kf-1 and Kf-2, other lithofacies, and mini-permeameter data (Mattson and Chan, 1996; Snelgrove et al., 1996; Mattson, 1997).

Thirty-eight thin sections were prepared from a suite of samples encompassing the dominant lithofacies in the Kf-1 and Kf-2 sequences at Ivie Creek. Thin sections were not made of the mudstones and low-permeability lithofacies, because of difficulties in distinguishing the mineralogy of fine-grained sediments in thin section.

For 18 thin sections from outcrop plugs of fluvial-dominated Kf-1-Iv[a] bedset (Anderson et al., this volume), the following grain parameters were estimated (Table 1): 14 framework grain compositions, two types of matrix, six types of cement, and porosity. The textural aspects of grain size, rounding, sorting, and contacts were also estimated. For subsequent samples (Table 2), the focus was placed on the examination of the major categories of grain size, grain type, cement, and porosity, and on the potential relationships to various sedimentary structures, lithofacies or parasequence set(s) (both fluvial-dominated Kf-1-Iv[a] bedset and wave-modified Kf-2 parasequences). In Table 2, the categories of grains, cement, and porosity total 100%. For the detailed examinations of Table 1, individual parameters within each category are normalized to sum to 100% of that particular category. Note that "matrix" is treated as a grain type and the sum of its subcategories (clay protomatrix and pseudomatrix; Dickinson, 1970) equals the "matrix" value. Because all values on the table are visual estimates, they have 10–20% margin of error.

Table 1. Compositional estimates for selected permeability plug thin sections. Ferron samples are from the Kf-1 in the Ivie Creek case study area and can be keyed to studies of Snelgrove et al. (1996) and Mattson (1997). Texture abbreviations: (1) grain size: vf = very fine-grained sand, f = fine-grained sand, m = medium-grained sand, and cg = coarse-grained sand; (2) rounding: r = rounded, sr = subrounded, sa = subangular, a = angular; and (3) sorting: poor = poorly sorted, mod = moderately sorted, and well = well sorted. Permeability measurements from Snelgrove et al. (1996). Sedimentary structure abbreviations: hor = horizontally laminated, ripple = ripple cross-laminated, lac = low-angle cross-stratified, and tro = trough cross-stratified. Facies code abbreviations: cp = clinoform proximal, cm = clinoform medial, and cd = clinoform distal.

SAMPLES	1	2	3	4	5	6	7	8	9	10	11	12	13	14	15	16	17	18
	T1-101	T2-50	T2-73	T2-78	T2-159	T2-161	T3-30	T3-149	T4-54	T9-72	T9-75	T9-82	T9-86	T12-41	T12-47	T14-8	T14-14	T14-19
GRAINS (100%)	88	83	90	75	77	82	85	88	80	84	77	83	85	85	87	88	85	85
Qtz- monoxlln	60	35	30	36	52	60	53	43	46	35	35	45	55	40	50	52	30	55
Qtz- polyxlln	7	3	6	5	3	7	12	12	10	25	20	5	7	10	6	2	20	2
Kspar/microcline	5	6	2	2	3	5	15	19	7	10	8	10	7	10	5	5	10	5
Plag	2	1	1	2	1	1	2	2	6	10	8	5	5	10	5	5	10	5
Clay altered rock frags	7	6	2	5	3	5	6	6	10	10	10	10	5	10	10	10	20	10
CO3	7	40	5	20	10	7	4	10	5	5	12	12	7	15	10	10	6	10
Muscovite		2	3	2	3	1	1	2	1	1	2	2	1	1	1		1	
Biotite	2		7	7		1		2	1	1	2	1	1					
Chlorite			2		3	2	1		1	2		2	2			1	1	
Glauconite	1		1			2			1		1	2		1				
Heavy mins	1			1					2									
Iron Opaques	1		2		3	2		2	2			1				1		
Organic matter			10	2	8	1										1		
Other/Unidentifable		6			3	3						2			3	3	3	3
MATRIX	7	1	29	18	8	3	6	2	10	1	2	5	10	3	10	10	2	10
protomatrix (clay)	2	1	16	18	8	3	3	2	2		2	5	2	3	4	4	2	4
pseudomatrix	5		13				3		8	1			8		6	6		6
CEMENT (100%)	10	15	8	10	15	15	12	10	8	8	8	12	10	10	10	10	8	10
carbonate/calcite	15	30	28	15	20	10	40	28	10	10	25	10	10	20	18	20	15	20
dolomite rhombs	45	65	40	80	40	40	28	40	50	20	25	30	28	40	50	50	20	48
qtz overgrowths	15	2	2	3	5	10	2	2	5	10	10	5	2	5	2	10	5	2
Iron (stain)	10	3	30	2	2	25	15	20	15	10	10	25	30	5	20	10	30	20
gypsum (late vein)	5				23													
kaolinte porefill/repl.	10				10	15	15	10	20	50	30	30	30	30	10	10	30	10
POROSITY (100%)	2	2	2	15	23	25	15	20	15	10	10	25	30	5	20	10	7	5
primary	50	50	50	15	25	50	25	50	15	25	25	25	25	25	25	25	25	25
secondary	50	50	50	85	75	50	75	50	85	75	75	75	75	75	75	75	75	75
TEXTURES																		
grain size	vf-f	vf-f	slt-vf	slt-vf	slt-vf	vf-f	f-crse	vf-med	f-med	vf-f	vf-m	vf	vf-m	vf-f	slt-m	vf-f	f-m	vf-f
rounding	r,sr,sa	r,sr,sa	sa,a	sr,sa	sa,a	sr,sa,a	sr,sa,a	sr,sa,a	sr,sa,a	r,sr,sa	r,sr,sa	sr,sa	sr,sa	sr,sa	sr,sa	sr,sa	sr,sa	sr,sa
sorting	mod-well	mod	poor	mod	mod	mod-well	mod	poor	mod	poor	poor	mod	poor	poor	poor	mod-well	mod	mod-well
COMMENTS																		
Facies Code	cd	cm	cm	cm	cd	cd	cd	cd	cp	cp	cp	cp	cp	cp	cp	cp	cp	cp
Sed Structure	hor	hor	ripple	ripple	ripple	ripple	tro	horiz	tro	lac	lac	hor	hor	tro	tro	lac	lac	hor
Permeability, md	2.2	2.1	36.4	2.3	12.7	2	3.4	3.4	31.7	565.4	74	96.3	8.6	16	4.8	9.2	101	6.5

Table 2. Summary porosity estimates with permeability measurements for drill hole (DH) samples and permeability plugs of Table 1 (T samples). Kf numbers indicate the parasequence set (fluvial-dominated Kf-1 vs. wave-modified Kf-2) with more detailed units from measured sections (Mattson, 1997). Asterisk (*) indicates units outside of the Kf-1 units. Facies codes are listed in Table 1 and also include additional facies: dmb = distributary-mouth bar, dic = distributary channel, msf = middle shoreface, and ldf = lower delta-front. Grain size and sedimentary structure abbreviations are given in Table 1 caption. Additional sedimentary structures are: bio = bioturbated, and hcs = hummocky cross-stratified. Permeability measurements from DH samples are from automated stage-mounted mini-permeameter equipment. Measurements for T samples are mini-permeameter plugs taken in the field (Snelgrove et al., 1996).

Well (IC) or Outcrop Plug (T)	Depth (feet)	Parase-quence Set	Facies Code	Grain Size	Sed. Struc.	Grains (%)	Cmt (%)	Porosity (%)	Perm. (md)
IC-11	312.0	2	dmb	mf	tro	90	7	3	18.3
IC-11	321.6	2	dmb	vf	bio	90	2	8	87.3
IC-11	332.0	2b	msf	fg	bio	90	3	7	25.1
IC-11	339.4	2	dic	fg	hor	80	5	15	20.2
IC-11	342.8	2a	dic	mfg	tro low	80	5	15	109.8
IC-11	345.2	2	dic	mfg	hor	85	5	10	23.2
IC-11	348.8	2	dic	cg	tro	85	5	10	36.9
IC-11	351.3	2	dic	cg	tro	85	3	12	146.3
IC-11	356.3	2	dic	cg	tro	85	5	10	72.0
IC-11	360.9	2a	msf	fvf	hor	85	5	10	39.7
IC-3	285.5	2	bay	fvf	rip	95	3	2	
IC-3	297.3	2	msf	f	hcs?	85	7	8	6.1
IC-3	306.5	2	msf	fvf	rip	90	2	8	14.0
IC-3	323.7	2	msf	fvf	hcs?	90	3	7	11.9
IC-3	362.4	*	bay	fvf	rip	85	12	3	
IC-3	376.0	*	ldf	m	tro	85	10	5	
IC-9	221.5	1	ldf	mc	tro	90	5	5	
IC-9	231.6	1	cp	fvf	hor	90	5	5	
IC-9	255.6	1	cd	fvf	hor	90	5	5	3.4
IC-9	262.0	1	cd	f	hor	90	5	5	4.3
T1	101.0	1a	cd	fvf		88	10	2	2.2
T2	50.0	1a	?	fvf	hor	83	15	2	2.1
T2	73.0	1a	cm	fvf	ripple	90	8	2	36.4
T2	78.0	1a	cm	fvf	ripple	75	10	15	2.3
T2	159.0	1a	cd	fvf	ripple	77	15	8	12.7
T2	161.0	1a	cd	fvf	ripple	82	15	3	2.0
T3	30.0	1	?	m	tro	85	12	3	3.4
T3	149.0	1a	cd	f	hor	88	10	2	3.4
T4	54.0	1a	cp	f	lac	80	8	12	31.7
T4	114.0	1a	cp	f	hor-lac	85	10	5	0.6
T4	117.0	1a	cp	f	hor-lac	85	10	5	0.8
T9	72.0	1a	cp	m	lac	84	8	8	565.4
T9	75.0	1a	cp	m	lac	77	8	15	74.0
T9	82.0	1a	cp	f	lac	83	12	5	96.3
T9	86.0	1a	cp	f	lac	85	10	5	8.6
T12	41.0	1a	cp	f	tro	85	10	5	16.0
T12	47.0	1a	cp	f	tro	87	10	3	4.8
T14	8.0	1a	cp	mf	lac	88	10	2	9.2
T14	14.0	1a	cp	mf	lac	85	8	7	10.1
T14	19.0	1a	cp	f	hor	85	10	5	6.5

These thin-section examinations (Table 1) indicate that all Ferron samples fall within a broad category of quartz arenites to quartzo-feldspathic arenites. Compositionally, these sandstones are classified as submature to mature. Framework grains consist of monocrystalline quartz, polycrystalline quartz, potassium feldspar (microcline), plagioclase, carbonate- and clay-altered rock fragments, muscovite, biotite, chlorite, glauconite, heavy minerals, iron opaque minerals, organic matter, other indistinguishable accessory grains, and matrix (either proto- or pseudomatrix), in approximate order of decreasing abundance. Cement types comprise calcite cement, dolomite rhombs, quartz overgrowths, iron oxide cement, kaolinite pore fill and replacement, and gypsum vein fill (a late, possibly surface-weathering, phenomenon).

The sandstone samples range from very fine- to medium-grained with some silt and coarse sand fractions. Most samples are poorly to moderately well sorted, and grains exhibit varying degrees of rounding. Kf-2 shoreface sandstones (hummocky cross-stratified, horizontal laminated, or burrowed facies) tend to be the "cleanest." Kf-1 samples are typically "dirtier" and contain more lithic fragments and altered grains with a higher degree of compaction, likely because of rapid sedimentation fed by fluvial-dominated deltas.

Porosity values are variable, averaging 5–10% and sometimes as high as 15%. Although discrimination between primary and secondary porosity is difficult, little primary porosity appears to remain in any of the samples. Many samples have undergone substantial reduction of primary porosity by compaction. Compacted rock fragments and point to concavo-convex grain contacts are common, indicating moderate to strong compaction. Much original porosity was likely occluded by early carbonate cement, and later dissolution of those same carbonate phases (calcite, dolomite, and possibly siderite) created secondary porosity in the same intergranular spaces. Evidence of secondary porosity development is shown by partial to complete dissolution of carbonate minerals showing corroded edges, dissolved-out grain interiors, and some over-sized pores. In some cases, secondary porosity only partly dissolves edges of carbonate cements (e.g., along cement-grain boundaries or in some unstable grain interiors) and still leaves some remnant carbonate cements. Where secondary porosity is more pervasive, surviving cements consist primarily of quartz and kaolinite, with some interspersed minor carbonate and/or iron oxide cement (Figure 3).

Ferron sandstone diagenesis is complex, in part due to compaction, alteration of grains, and a number of successive cement types. Quartz overgrowths generally reflect the earliest diagenetic stage, displayed as cements in optical continuity with the host monocrystalline quartz grains. Clay (kaolinite) pore-filling and replace-

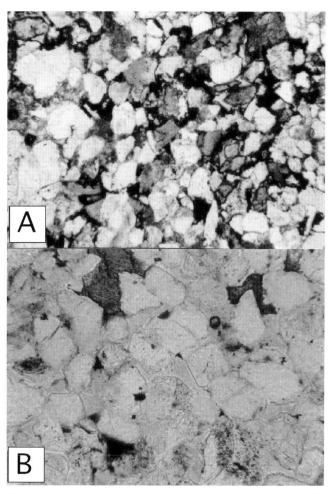

Figure 3. Polarized-light photomicrographs of fluvial-dominated Kf-1 sandstone. (A) Sample T4 117, showing localized secondary porosity development (porosity shown in black) from dissolution of cements; field of view is approximately 1.9 mm across. (B) Sample DH11 321.6, showing more uniform secondary porosity development (corresponding to higher permeability, see Table 2) from dissolution of cements; field of view is approximately 0.8 mm across.

ment follows the quartz overgrowths. This early diagenetic stage is typically followed by a middle diagenetic stage of calcite cement and dolomite rhombs. Carbonate cements can be partly distinguished based on dual stain (alizarin red-S and potassium ferricyanide) of the thin section, which shows calcite (pink stain) with some minor ferroan calcite (purple), and dolomite rhombs (colorless). Unstable lithic or feldspathic grains commonly are altered to clays and/or replaced by carbonates. In some cases, dolomite appears to replace and "cut into" both clay cement and some quartz overgrowths. Late diagenetic stages include iron/hematite staining. Final stages of diagenesis are dissolution of these same carbonate phases and development of secondary porosity (Figures 3 and 4) and some microfractures.

Based on thin-section petrography, it is difficult to fully evaluate the effect of weathering. For the most part, the outcrop plugs have permeability and porosity values

Figure 4. Photomicrographs of secondary porosity development in carbonate cements from sample T4-117. (A) Nearly "floating" center dolomite rhomb (center of cross hairs) in pore space. Pore space edges (enhanced by the dark iron cement) look corroded, where carbonate cements were likely dissolved to form the secondary porosity (pores shown in black). Example is shown in plane-polarized light, where q = quartz grains. (B) Central carbonate cement shows a dissolved and corroded secondary pore development (center dark spot shown by arrow). Example is shown in cross-nicols (cross-polarized light), where q = quartz grains.

comparable to samples from adjacent drill holes. However, identical beds from outcrop and from unweathered drill-hole cores have not been sampled and compared.

Surprisingly, some samples with relatively high permeability are very fine-grained sandstones with fairly low porosity, as visually noted from the thin sections (e.g., T2-73). Generally those fine-grained samples do have relatively little cement (<5%) compared with other samples (5–15%). Another possible explanation is that localized porosity, perhaps along some bedding/parting planes or where organic matter is present, may locally produce unusually high permeability. Porosity localiza-

tion is visible in thin section (e.g., T9-82), although the interconnection of pore space is unknown in the third dimension. The two outcrop samples with lowest permeability (T4-117 and T4-114) have pervasive iron oxide cement, yet there exists visible porosity in thin section (Figure 3); thus the cause of their very low permeability is not evident in thin sections. Microporosity in clay is a factor that needs further evaluation and could be a contributor to permeability even where visible porosity appears low.

Petrographic examinations provide several generalizations concerning relationships among grain size, sorting, depositional energy, and lithofacies. Coarser-grained samples (generally fine to coarse-grained) possess more abundant and generally larger pores. Coarse-grained sandstones also tend to be more quartz rich and are relatively "cleaner," with less matrix. Organic matter may form small-scale permeability barriers, and coarse-grained samples with better porosity generally contain less organic matter. Fine-grained samples commonly contain more organic matter, reduced porosity, and more concavo-convex grain contacts.

Overall, it appears that reservoir potential is well correlated to facies. Coarse-grained distributary channel, proximal clinoform, burrowed bar, and shoreface deposits likely contain the highest permeability and porosity. Wave-modified (reworked) Kf-2 sandstones are generally better sorted and cleaner and may have higher porosity and good permeability in the coarser sand intervals. Fluvial-dominated deposition, as in the Kf-1 samples, implies less sorting and more rapid sedimentation fed by rivers. These Kf-1 samples therefore exhibit a corresponding decrease in both porosity and permeability. Although diagenetic overprint is pervasive and complex, the original intergranular pore spaces occluded by carbonate cements are the same sites of later dissolution and secondary porosity.

Roche (1999) reports a complementary petrographic study of the Ferron Sandstone in the Coyote Basin (south of the Ivie Creek study area). Thin section macroporosities there are 1.6 — 18.4% in outcrop, and 0.4 — 14.8% in core samples. Roche (1999) notes porosity reduction by calcite cement, with increasing porosity of outcrop samples due to dissolution of calcite cement. The author observes that bulk porosity in some sandstones is contained within interstitial clays, which lower effective porosity. The study also confirms that thin section macroporosity correlates with Hassler cell permeability in core samples.

Ferron Sandstone diagenesis is similar to that of the Frontier Formation, another suite of fluvial-deltaic sediments deposited on the western margin of the Cretaceous inland sea. Frontier Sandstone diagenesis has been examined for samples from about 3900 m (13,000 ft) subsurface in the Spearhead Ranch Field, Wyoming (Tillman and Almon, 1979), and about 3400 m (11,000 ft) sub-surface in Moxa Arch, Wyoming (Stonecipher et al., 1984; Stonecipher and Diedrich, 1993). These studies emphasize the importance of facies, and particularly sandstone maturity, to variations in diagenesis both within and between wells. At Spearhead Ranch, in contrast, Frontier Sandstone porosity is almost entirely primary, and diagenetic history has been strongly affected by early chlorite pore linings (Tillman and Almon, 1979).

Our observations of Ferron diagenesis at Ivie Creek show that the Ferron diagenetic history was generally similar to the Frontier diagenesis at Moxa Arch. The main difference is that early carbonate cementation and replacement were more pervasive in the Ferron, perhaps inhibiting the later-stage quartz cementation that was more pervasive in the Frontier diagenetic history. In both the Moxa Arch Frontier and the Ivie Creek Ferron, later fluid flow caused leaching of feldspars and especially carbonate, mainly in the coarsest-grained sediments, with minor kaolinite pore lining.

Controls on Porosity: Evidence from Ivie Creek Core Plugs

Ivie Creek core-plug porosity averages 13.0% (std. dev. 4.0%, range 1.7–20.9%). Cross-plots of porosity versus mineral abundance show, surprisingly, no significant correlation between porosity and clay content, at least among these low-clay (<20%) samples. Similarly, porosity is not correlative with abundances of other minerals, with one notable exception: carbonate cement. Increasing carbonate content (the sum of calcite, dolomite, and siderite abundances, based on infrared spectroscopy) is strongly associated with porosity reduction (Figure 5A). Carbonate-free samples have porosity values of approximately 12–19%, probably reflecting the porosity of these samples at the time that carbonate cementation initiated. The carbonate content of the low-porosity samples is not 12–19% as expected for simple pore filling, but considerably higher. Thin sections show that carbonates here are present not merely as pore filling but also as feldspar replacements. Therefore, carbonate content can exceed primary porosity.

A correlation between porosity and carbonate content can be either primary (reflecting early carbonate precipitation) or secondary (reflecting earlier carbonate filling of pores followed by later carbonate dissolution). The latter effect is indicated by petrographic observations where highest-porosity samples possess significant secondary intergranular porosity (Figure 5B).

The effects of cementation and dissolution on Ferron porosity impact the conversion between bulk density and porosity. Bulk density is a function of mineral and fluid components:

$$\rho_b = \rho_f \phi + \rho_{ma}(1-\phi) \tag{1}$$

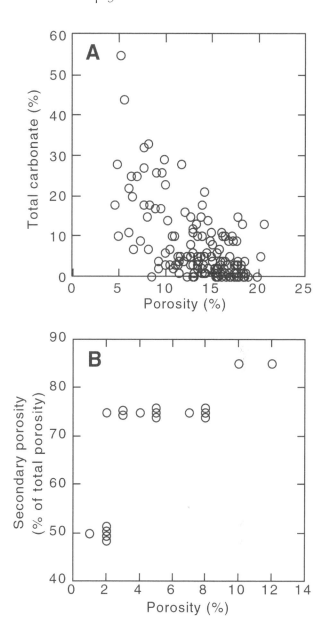

Figure 5. (A) Cross-plot of core-based analysis showing that total carbonate content increases with decreasing porosity. (B) Cross-plot of total porosity versus percentage of the porosity that is secondary, based on thin sections. These two patterns suggest that porosity reduction was caused mainly by carbonate cementation, and the current high porosity is mainly secondary porosity generated by carbonate dissolution.

where ϕ = fractional porosity, ρ_f = fluid density, and ρ_{ma} = matrix (or grain) density. For sandstones and slightly shaley sandstones such as these, a single matrix density of 2.65, corresponding to quartz, is usually appropriate for all porosities. Carbonates, however, have higher matrix density (2.71 for calcite, 2.88 for dolomite, and 3.9 for siderite). Consequently, a plot of matrix density versus porosity shows a matrix density of 2.70±0.05 for the lowest-porosity (carbonate-rich, Figure 5A) rocks, dropping to the 2.65 value of quartz at porosity values of approximately 18% (carbonate-free, Figure 5A). Linear regression indicates that the dependence of matrix density on porosity is $\rho_{ma} = 2.719 - 0.406\phi$ ($R = 0.62$). Combining this equation with (1) above, and using a value of 1.02 g/cc for ρ_f, gives

$$\rho_b = 2.719 - 2.105\phi + 0.406\phi^2, \text{ or}$$
$$\phi = 2.592 - 1.232 \cdot \text{sqrt}(0.0154 + 1.624\rho_b) \quad (2)$$

To convert density log response in Ferron sandstones to porosity, equation 2 is more appropriate than equation 1.

Effects of Exhumation on Ferron Porosity

The 179 paired Ferron log picks of velocity and porosity are plotted as a function of depth in Figures 6A and 6B. Two features are evident on these plots: (1) heterogeneous porosity and velocity at each depth, and (2) gradual porosity increase and velocity drop with decreasing depth (increasing exhumation). The wide ranges of porosity and velocity are an advantage in defining velocity/porosity relationships, but these ranges partially obscure the changes in these values with depth. Depth-dependent variations are more evident on plots of three-well averages of porosity and velocity, showing 95% confidence limits (Figures 6C and 6D).

Only two deep wells survived our rejection criteria. The deeper well (~3600 m [12,000 ft]) exhibits porosity values of 2–4%, and the shallower well (~3100 m [10,000 ft]) possesses porosity values of 4–8% (Figure 6A). These differences reflect heterogeneity within wells and do not necessarily suggest porosity increase associated with initial exhumation. Porosity measurements for the two wells indicate that pre-exhumation porosity in the Ferron was 4±4%. Shallower than 1800 m (6000 ft), where the abundance of well data is much higher, the systematic increase in porosity with continuing exhumation is more reliably indicated.

Porosity increase accelerates as the Ferron is progressively exhumed, leading to a porosity/depth trend that is approximately logarithmic (Figure 6C). By the time the Ferron reaches near-surface depths (20–140 m [70–460 ft]), average porosity increases to 9.3±2.4% (95% confidence limits). Outcrop samples are even more porous than log-based well porosity: average porosity of 162 Ivie Creek outcrop samples is 13.0±0.6%, and 90% of the 59 outcrop samples from nearby Muddy Creek Canyon exhibit porosity values within the range of 12–20% (Miller et al., 1993a).

Many authors have emphasized the influence of provenance and depositional environment on sandstone diagenesis and on development of secondary porosity (e.g., Hayes, 1979; Stonecipher et al., 1984). Could the regional variations in Ferron porosity be caused by lateral changes in lithology? The general sedimentary pattern of fining and thinning eastward (Ryer and McPhillips, 1983; Tripp, 1989) could cause higher primary porosity in more deeply buried western sediments than in shallower, eastern sediments, but the present-day porosity distribution exhibits the opposite pattern.

The increase in Ferron porosity is not correlative with subsurface variations in fracture development (Tripp, 1991, 1993a, b). Nor do fractures explain the magnitude of this porosity increase: macrofracture and microfracture porosity is seldom more than 1% (Bourbié et al., 1987). Porosity increase is inevitable because of the elastic expansion of pores due to pressure release (e.g., Domenico and Mifflin, 1965), but this unloading-curve effect is far too small to account for observed porosity changes.

Secondary Porosity in the Ferron Sandstone

Carbonate dissolution can account for the presence and magnitude of depth-dependent porosity increases in the Ferron. The petrographic and petrophysical studies discussed above demonstrate that outcrop samples host secondary porosity that results from carbonate dissolution. The depth dependence of Ferron porosity implies a relationship between exhumation and dissolution of carbonate cement. This effect may reflect a direct influence of exhumation on calcite solubility or an indirect link between exhumation and fluid flow. In either case, determining the origin of Ferron secondary porosity is an essential prerequisite to identifying implications for secondary porosity of other exhumed sandstones.

Once thought to be relatively rare, secondary intergranular porosity has increasingly been recognized as being comparable to primary porosity among sandstones globally (Schmidt and McDonald, 1979). Secondary porosity commonly develops below the depths at which primary porosity is destroyed, as a consequence of kerogen breakdown at temperatures of >100°C (Schmidt and McDonald, 1979; Surdam et al., 1984). Maturation generates carbon dioxide (e.g., Hunt, 1979) and organic carboxylic acids, both of which can induce carbonate dissolution, perhaps the most common cause of secondary porosity development (Schmidt and McDonald, 1979). Organic acids may also induce silicate dissolution (Hayes, 1979; Surdam et al., 1984), an equally important process among immature sandstones (Hayes, 1979).

In contrast to maturation-induced secondary porosity, development of exhumation-related secondary porosity is less predictable. Calcite solubility is controlled mainly by pH, temperature, carbon dioxide fugacity, and salinity (Garrels and Christ, 1965). Although the effects of exhumation on temperature and pressure can be determined accurately, associated

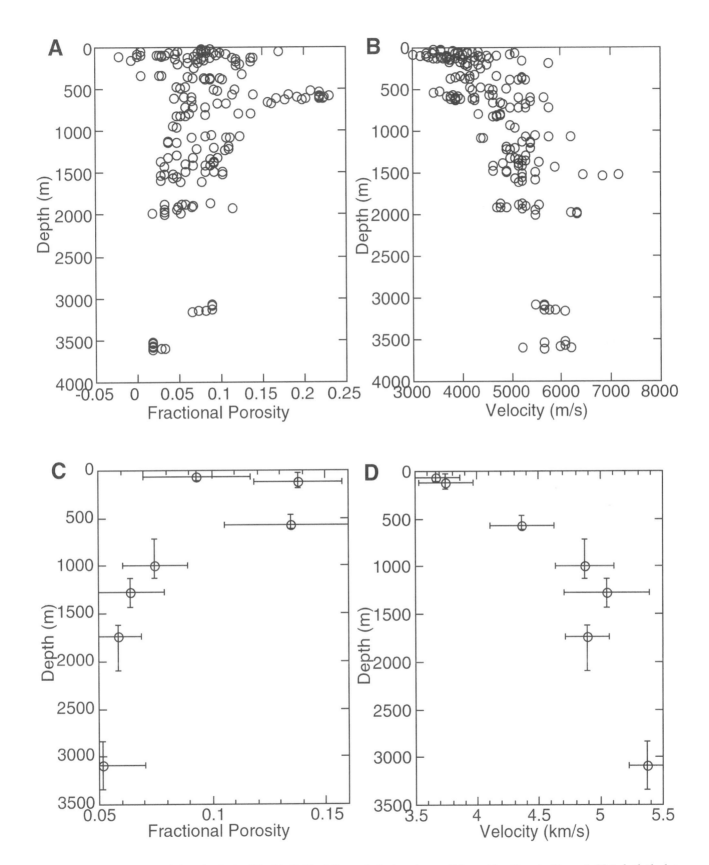

Figure 6. Fractional porosity (A) and velocity (B) of individual Ferron beds, based on well logs at locations in Figure 1. Note both the heterogeneity at a given depth and the systematic changes with depth. Variation of average Ferron porosity (C) and velocity (D) with depth, based on three-well adjacent-depth averaging of data from (A) and (B), respectively.

changes in pH, carbon dioxide fugacity, or salinity cannot. This problem of predicting carbonate dissolution is likely to be most acute in regions where hydrocarbon generation is accompanied by production of carbon dioxide and organic acids — the very regions where we are most likely to have well data.

Cementation and dissolution rates depend on both calcite solubility and fluid flow rate (Maxwell, 1960). For significant dissolution to occur within a formation, substantial fluid flux is required; ion diffusion is insufficient (Hayes, 1979; Bjørlykke, 1984). This flux can come from dewatering of adjacent shales (Hayes, 1979), smectite/illite transition (Surdam et al., 1984), and/or meteoric water (Bjørlykke, 1984). Galloway (1984) emphasized the impact of hydrologic regime — either meteoric, compactional, or thermobaric — on diagenetic history. The author noted that basins commonly undergo a temporal evolution from one hydrologic regime to another, with associated diagenetic changes. A common hydrologic evolution is from early compaction flush to later meteoric flush (Harrison and Tempel, 1993). For example, the diagenetic history of foreland basins, such as Western Canada Basin and the Appalachian Basin, has been strongly affected by early compaction or tectonic-compression expulsion of syndepositional waters, followed by later orogenic uplift drive of meteoric waters (Harrison and Tempel, 1993; Longstaffe, 1993). The meteoric influence in foreland basins is not necessarily confined to late-stage diagenesis, however. Early meteoric diagenesis is inferred for both the Western Canada Basin (Longstaffe, 1993) and Wyoming's Green River Basin (Stonecipher and Diedrich, 1993).

The Ferron Sandstone, like some other foreland-basin sandstones, demonstrates the impact of meteoric flux on its late-stage diagenetic history. Flow rates in regional groundwater-flow systems nearly universally decrease rapidly with subsurface depth (Freeze and Witherspoon, 1967), compatible with the strong depth dependence that we observe for carbonate dissolution in the Ferron Sandstone. The maximum depth of meteoric leaching depends on many factors: potential gradients, permeability, and the rates of recharge, flow, reactions, and solubility change (Schwartz and Longstaffe, 1988; Bjørlykke, 1994). The 2-km (7000 ft) depth of Ferron secondary porosity enhancement is compatible with the 3-km (10,000 ft) maximum depth of meteoric flow in other regions with comparable head (Toth, 1980).

The modern hydrologic regime of the Ferron Sandstone is dominated by recharge in the Wasatch Plateau and Book Cliffs at the western and northern margins, respectively, of the study area (Figure 1), with discharge near the Ferron outcrop (Lines and Morrissey, 1983). Topography slopes down to the east and southeast of these recharge regions, in contrast to stratigraphic dips to the west and northwest. Consequently, meteoric flux

is not simply downslope in the subsurface. Confined by relatively impermeable, bracketing shales, groundwater moves up-dip along Ferron sandstone beds toward the eastern outcrops (Lines and Morrissey, 1983). The Ferron hydrologic analysis of Lines and Morrissey (1983) is confined to relatively shallow depths (<500 m, [<1600 ft]) and to the south-central portion of Figure 1. The patterns of secondary porosity development seen in our study suggest that upward flow along the Ferron beds is pervasive and that this meteoric flow extends to a subsurface depth of about ~2 km (~7000 ft).

This upward flux of meteoric waters into discharge areas is orders of magnitude faster than the rate of exhumation. For example, Lines and Morrissey (1983) report a carbon-14 age of 28,000 years for a fluid sample from 450 m (1480 ft) depth, whereas only about 6 m (20 ft) of erosion would occur in this time period if the erosion rate were steady-state at 2 km (6600 ft) per 10 m.y. Thus upward meteoric flux, not exhumation, probably generates an increase in carbonate solubility responsible for late-stage carbonate dissolution within the Ferron Sandstone. Upward flow of meteoric water can be an extremely effective mechanism for development of secondary porosity, because of the combination of high flow rates (compared to other types of fluid flow) and the cooling-induced increase in calcite solubility (Bjørlykke, 1994).

Ground-water flow at Ferron outcrops contrasts with the large-scale meteoric flux, which is upward and eastward within Ferron Sandstone beds. At Ferron outcrops such as Ivie Creek, rainwater percolates downward through the vadose zone to the water table and is discharged at deeply incised canyons (Lines and Morrissey, 1983). Near and above the water table, carbon dioxide buildup from decaying organic matter can enhance carbonate leaching (Bjørlykke, 1994), and this mechanism probably accounts for the porosity increases among outcrop samples that we previously noted. Although Muddy Creek Canyon and Ivie Creek outcrop samples represent different Ferron parasequences and partly different lithofacies, porosity values of carbonate-free samples are remarkably similar: 12–20% (Miller et al., 1993a, b) and 12–19%, respectively.

P-WAVE VELOCITY OF THE FERRON SANDSTONE

Controls on Velocity: Evidence from Ivie Creek Core Plugs

In a classic series of papers, Biot developed a general theoretical description for the mechanical response of fluid-filled porous materials (Biot, 1956a, b). In the low frequency limit, the formalism reduces to the Gassmann equation (Gassmann, 1951). Porosity is incorporated in this equation both explicitly and implicitly in the porous frame properties. It is through such a formulation and bulk density that porosity can be seen to have a dominant control on velocity.

Recognizing this porosity dominance, Wyllie et al. (1956) introduced an empirical "time-average" relationship between porosity and velocity that has been used extensively, particularly to calculate porosity from velocity logs. Other empirical relations (e.g., Raymer et al., 1980; Erickson and Jarrard, 1998) have been developed for high-porosity rock. Common to these empirical models are assumptions of isotropy, invariant mineralogy, 100% brine saturation, normal pressures, and normal cementation. However, Ferron cementation is representative of rocks with secondary rather than primary porosity. Pressure dependence is also a potential concern: overpressured zones are absent, but overburden pressures have plummeted during exhumation.

While the measured change in porosity from atmospheric pressure to 6.9 MPa is only about 0.6% for our Ivie Creek samples, such a change can have a profound influence on velocity if it represents crack volume. Figure 7 confirms the presence of microcracks as manifest in the strong pressure dependence of velocity. We attribute generation of these cracks to the 3-km exhumation of Ferron sandstones at Ivie Creek.

Core-plug compressional velocity at the highest measured pressure (34.5 MPa) is plotted as a function of porosity in Figure 8D. This pressure is comparable to the pressure at 3000 m (9800 ft), the maximum depth to which the Ferron Sandstone has been buried. Also plotted in Figure 8D is the time-average equation (Wyllie et al., 1956) for sandstone (V_{ma} = 5950 m/s). This relationship fits these data moderately well, as does the revised equation by Raymer et al. (1980), not shown. For carbonate-rich samples, higher velocity is expected and observed. Samples with >25% carbonate have velocity values compatible with a typical limestone matrix velocity of about 6450 m/s.

The velocity of shaley sandstones is better represented by a function of both porosity and clay content than by the time-average equation (Castagna et al., 1985; Han et al., 1986). Clay minerals have a double effect on velocity. The first is increased microporosity and the second is a reduction in frame moduli if the clays are grain-supporting. For the Ferron samples, however, the shaley-sand relationships are no more successful than the time-average equation in predicting velocity. Furthermore, velocity residuals from the time-average equation (observed velocity minus that predicted from porosity based on the time-average equation) are not correlated with clay content. We conclude that the effect of clay content on velocity can be neglected for these low-clay (<5–10%) Ferron sandstones and moderately shaley sandstones.

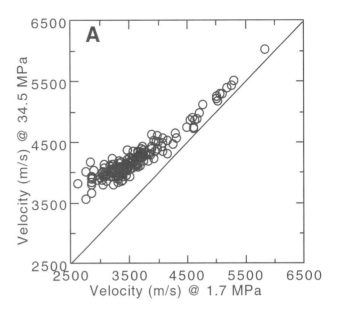

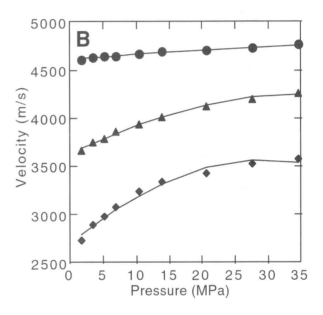

Figure 7. (A) Comparison of the lowest-pressure and highest-pressure measurements of compressional-wave velocity of core plugs. (B) Pressure dependence of velocity for three representative samples. All sample velocity measurements are sensitive to pressure change due to closing of microcracks, and this effect is most conspicuous in the lower-velocity samples.

Effects of Exhumation on Ferron P-wave Velocity

Wherever exhumation causes development of secondary porosity, velocity is expected to decrease because velocity is dependent on porosity. Figures 6B and 6D confirm that Ferron velocity does exhibit a strong depth dependence. Indeed, the velocity drop at the shallowest depths appears to be even stronger than the associated porosity increase (Figure 6C).

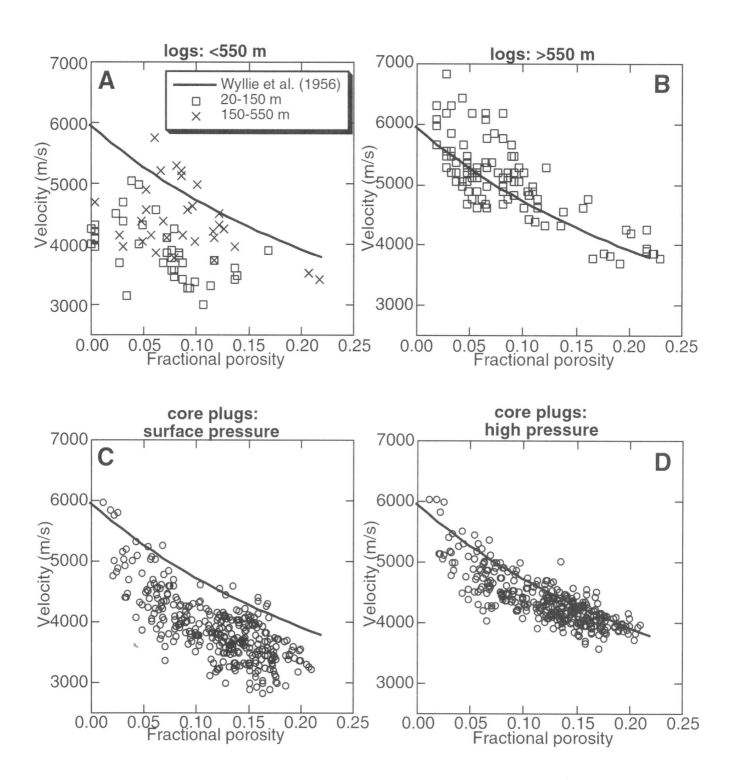

Figure 8. Comparison of Ferron velocity/porosity patterns, based on both well logs (A and B) and core plugs (C and D). Ferron log-based velocity and porosity from Figure 6 show systematic changes of the velocity/porosity relationship with sample depth. Data from deeper than 550 m (B) are consistent with the empirical relationship of Wyllie et al. (1956), whereas shallower data (A) are systematically lower in velocity than one would expect based on their porosity. The shallowest log data in A, from 0–150 m (squares), are even lower in velocity/porosity pattern than data from 150–550 m (crosses). Low-pressure (1.7 MPa) measurements on outcrop core plugs (C) exhibit a similar velocity/porosity relationship to that of log data from about the same pressure (A), whereas high-pressure measurements on these core plugs (D) have a velocity/porosity pattern that is similar to deeper (>550 m) logs (B).

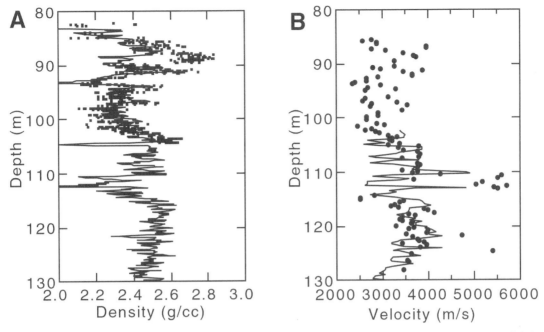

Figure 9. Comparison of log (line) and core (dots) measurements for density (A) and velocity (B) at Ivie Creek drill hole IC-3.

Figure 8 shows that observed velocity variations are mostly attributable to variations in porosity. Log data from depths greater than approximately 550 m (1800 ft) (Figure 8B) fit the Wyllie et al. (1956) time-average relationship for sandstone matrix fairly well. Several points with near-zero porosity fall far above the time-average prediction for sandstone; corresponding velocity values imply high carbonate content.

For Ferron sandstones that have been exhumed to within 550 m (1800 ft) of ground level, porosity is no longer the only control on velocity; it is not even the main control. Figure 8A shows that the entire velocity/porosity relationship is systematically offset toward lower velocity for these shallow-burial data. Most points — and particularly those with low porosity — are offset by 1000±500 m/s below the time-average predictions. For the depth interval 0–150 m (0–500 ft), the offset is even larger (Figure 8A): all are below the time-average predictions, averaging about 1500 m/s too low, and most have velocity values less than those observed at 150–550 m (500–1800 ft).

These systematic depth-dependent changes in the porosity/velocity relationship indicate that exhumation-induced pressure release decreases velocity. Decrease in frame bulk modulus, which is strongly pressure dependent, is responsible. As increased pressure closes microcracks, reduction in total porosity is trivial (e.g., Han et al., 1986). Initial microcrack porosities of <0.5% are sufficient to cause pressure-dependent velocity variations of 5–50% (O'Connell and Budiansky, 1974), indicating that the primary effect of pressure on velocity is through cracks, not porosity or density (Walsh, 1965; Nur and Murphy, 1981; Bourbié et al., 1987).

The measured pressure dependence of plug P-wave velocity (Figure 7) can account for the observed log-based changes in velocity and in the velocity/porosity relationships. Figure 8 compares the velocity/porosity pattern for lowest-pressure core-plug measurements (Figure 8C) to that for shallow (<550 m [<1800 ft]) log data (Figure 8A). Log velocity for 150–550 m (500–1800 ft) is comparable to core-plug velocity for the same porosity; core-plug pressure of 1.72 MPa is comparable to or lower than log pressures of 1.7–6.2 MPa for these depths. Both are systematically lower than the Wyllie et al. (1956) time-average relationship, but we have already seen that pressure increase can close microcracks and remove this discrepancy (Figure 8D). The shallowest log data (20–140 m [665–460 ft]) are even lower in velocity than the 150–550 m (500–1800 ft) log (Figure 8A) and 1.72 MPa core-plug data (Figure 8C); core-plug measurements do not extend to pressures low enough for comparison to these shallowest log data. The similar systematic change in velocity/porosity relationship for log data (versus depth) and core-plug data (versus pressure) implies that the two are equally sensitive to closing of microcracks.

Ferron log and core velocity responses can also be compared within a single drill hole, at IC-3. This drill hole has continuous core, with >99% recovery, throughout the Kf-1 and Kf-2 parasequence sets of the Ferron. These cores were run through the USGS gamma-gamma density logger, measuring density every 2 cm along core. The equivalent downhole interval was logged with a gamma-gamma density tool. The two density datasets agree (Figure 9A). At this well we also took 113 horizontal measurements of whole core velocity using transducers with a frequency of 500 KHz, the same as for plug velocity measurements and much higher than the frequency of about 25 KHz employed in sonic logging. For part of the same depth interval, we were able to retain water in the hole long enough to run sonic logs. Rock velocity for air-filled pores (cores) is expected to be lower than that for water-filled pores (logs); for Ferron core plugs, this difference is 300–500 m/s (Gokturkler, 1996). Log and whole-core velocity data for drill hole IC-3 show good agreement (Figure 9B).

Log analysts often generalize that sonic logs do not recognize secondary porosity, particularly secondary porosity in carbonates (e.g., Schlumberger, 1972, 1989; Ellis, 1987; Winkler and Murphy, 1995), and that one can therefore estimate the amount of secondary porosity (ϕ_{sec}) by comparing sonic (ϕ_{son}) and neutron-density (ϕ_{nd}) porosity: $\phi_{sec} = \phi_{nd} - \phi_{son}$. These generalizations assume that secondary porosity consists of either fractures or vugs. Sonic logs record only the fastest of many compressional-wave paths between source and receiver. For vertically traveling waves in a vertically fractured rock, waves paralleling fractures will arrive sooner than waves from within fractures. Similarly, waves traversing isolated large vugs arrive later than waves whose paths encounter no vugs. Furthermore, spherical vugs are expected to reduce velocity less than an equal volume of smaller aspect-ratio cracks (O'Connell and Budiansky, 1974; O'Connell, 1984).

Among sandstones, a more common type of secondary porosity than either large-scale fractures or vugs is intergranular secondary porosity (Schmidt and McDonald, 1979) like that encountered by the Ferron logs. Both Ferron core-plug velocity measurements and sonic logs respond to secondary porosity in the same manner as to primary intergranular porosity, based on the fits to the time-average equation (Figure 8). This agreement suggests that the pore shapes of primary and secondary intergranular porosity are sufficiently similar to generate the same velocity responses.

IMPLICATIONS FOR ESTIMATING THE AMOUNT OF EXHUMATION FROM VELOCITY OR POROSITY CHANGES WITH DEPTH

Normal mechanical compaction and consolidation of siliciclastic sediments cause porosity to decrease and velocity to increase with increasing depth. If lithology and pore pressure are relatively uniform laterally, then this process generates a characteristic pattern — valid throughout a region — of porosity and velocity change with depth. Departures from this characteristic pattern may indicate overpressures (Hottmann and Johnson, 1965), lithologic changes (Schlumberger, 1989; Japsen, 1993), or exhumation (Magara, 1976).

Various procedures have been proposed for using velocity/depth or porosity/depth patterns to estimate the amount of exhumation at a well. All assume that uplift does not modify porosity or velocity. The Ferron sandstones provide a striking illustration of the problem: all major velocity and porosity changes with depth (Figure 6) result from processes — pressure release and dissolution of carbonate cement — ignored by traditional exhumation estimation techniques. Both processes accelerate as beds are brought closer to the ground surface, causing substantial underestimation of amount of erosion and exhumation. Focusing on shales does not avoid the problem; shales also undergo carbonate cementation and dissolution, as well as pressure-dependent velocity. Furthermore, exhumation of impermeable shales may generate abnormally low fluid pressures (Neuzil and Pollock, 1983), presumably resulting in velocity change.

PERMEABILITY OF THE FERRON SANDSTONE

Permeability studies presented here utilize core-plug measurements. A complementary suite of permeability data has been obtained from the Ivie Creek area using mini-permeameter transects (Snelgrove et al., 1996, 1998; Chidsey, 2001; Mattson and Chan, this volume). The mini-permeameter data were used to determine the statistical structure of the spatially variable permeability field within delta fronts (Mattson and Chan, this volume), to investigate how geological processes control the spatial distribution of permeability (Snelgrove et al., 1996, 1998; Chidsey, 2001; Mattson and Chan, this volume), and to estimate relative permeability values used in single-phase reservoir models (Forster et al., this volume).

Controls on Permeability: Evidence from Ivie Creek Core Plugs

Most sedimentary rocks exhibit a linear relationship between porosity and the logarithm of permeability, but that relationship varies so much among formations and among regions that it must be locally determined (Nelson, 1994). Ferron Sandstone permeability is very closely related to porosity, with a relationship (Figure 10) that is compatible with other high-porosity sedimentary rocks. This pattern is much stronger than we had anticipated based on the Texas Bureau of Economic Geology (BEG) measurements (Miller et al., 1993a) or Ivie Creek porosity/mini-permeameter data (Chidsey, 2001), and it extends far below the 2-md threshold for useful mini-permeameter measurements. This relationship means that permeability can be estimated for all core-plug samples for which porosity is measured. Furthermore, density logs from Ivie Creek drill holes can provide contin-

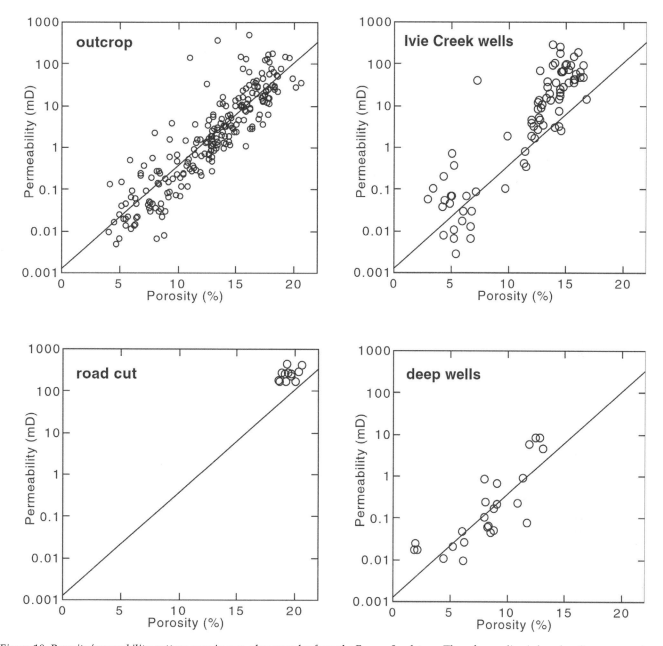

Figure 10. Porosity/permeability pattern seen in core-plug samples from the Ferron Sandstone. The reference line is based on linear regression of the 3.44 MPa outcrop data.

uous records of down-hole permeability variations, complementing results from outcrop mini-permeameter transects (Chidsey, 2001).

Effects of Exhumation and Outcrop Exposure on Ferron Permeability

Dissolution associated with exhumation is expected to increase permeability. Figure 10 shows that the observed permeability variations are mostly attributable to variations in porosity. Similar patterns are observed for outcrop samples, shallow subsurface samples from Ivie Creek, and deep subsurface samples. This corre-

spondence indicates that the primary impact of exhumation on permeability is through porosity enhancement. Outcrop exposure does not diagenetically modify the pattern, as evidenced by the similarity among three groups of samples (Figure 10): mature natural outcrops, a road cut less than a few decades old, and wells (both shallow and deep).

Comparison of atmospheric-pressure porosity measurements to those made at 6.9 MPa shows that atmospheric-pressure porosity is systematically higher than the 6.9 MPa data by up to 1%. Thus, atmospheric-pressure measurements of permeability, such as the BEG data and all mini-permeameter data, are probably increased only slightly (relative to 6.9 MPa measurements) by microcrack opening.

Ferron Sandstone permeability (k) is related to fractional porosity (ϕ) by:

$$\ln(k) = -6.657 + 56.6 \cdot \phi \quad (R = 0.921) \qquad (3)$$

This relationship is based on measurements of outcrop core plugs, excluding four anomalous points. Analysis of residuals shows, remarkably, no significant correlation with clay content. This observation does not imply that shaley samples have the same permeability as sandy ones; indeed, shaley parasequence set Kf-1 is much lower in porosity and permeability than sandy parasequence set Kf-2. It indicates, however, that the relationship between clay content and permeability is entirely indirect: shaley samples have lower porosity and therefore lower permeability.

Not all Ferron Sandstone beds display porosity increase associated with exhumation (Figure 6A). Even at near-surface depths, a few beds exhibit porosity values comparable to those at 2900–3400 m (9500–11,100 ft). The standard deviation of porosities increases with exhumation, indicating perhaps a heterogeneity of the process that generates this porosity increase. Hayes (1979) and Bjørlykke (1984) have predicted that leaching might concentrate secondary porosity in beds that have the highest primary porosity, because fluid flux is localized within the most permeable (coarsest grained) beds. Ferron outcrop studies at the Ivie Creek and other areas indicate that grain size is positively correlated with permeability (Barton and Tyler, 1991; Mattson and Chan, 1996, this volume). Our outcrop petrophysical data show that permeability is strongly dependent on porosity (R = 0.92; Figure 10), and thin sections demonstrate that porosity is largely secondary (Figures 3 and 4). These observations imply that grain size has influenced the entire history of Ferron porosity evolution: (1) early mechanical compaction in coarse-grained sandstones created higher porosity and permeability than in fine-grained sediments; (2) carbonate precipitation reduced porosity and permeability but did not obscure this grain-size dependence; and (3) dissolution enhanced this grain-size dependence because fluid flow and associated dissolution were concentrated in the most permeable, coarsest-grained sediments. Consequently, Ferron outcrop stratigraphy can provide useful clues to fluid-flow patterns in other deltaic formations, despite its complex diagenetic history.

Permeability Anisotropy

Anisotropy is a variable that needs consideration in fluid-flow modeling. Anisotropy is also of broader petrophysical concern because it can be caused by either mineral alignment (for example, platy, subhorizontal clay minerals) or microfractures (usually one set of vertical microfractures), because both affect fluid flow, and because large-scale anisotropy can be estimated by seismic methods. Based on the abundance of finely laminated sediments, particularly in parasequence set Kf-1, substantial permeability anisotropy might be expected at the scale of individual core plugs.

Our core-plug petrophysical measurements included 38 3-D sample sets, each comprising adjacent horizontal, vertical, and 45° diagonal measurements. Figure 11 compares permeability within these sets. As expected, vertical permeability is lower than horizontal permeability, but this difference is surprisingly small: usually a factor of two or less, in contrast to the more than four orders of magnitude variation in permeability attributable to porosity variations (Figure 10). This magnitude of anisotropy is comparable to the impact on permeability of changing porosity by only about 1%. Indeed, companion samples often vary in porosity by this much due to local heterogeneity. Porosity variations among samples of an individual 3-D set can be corrected for, by normalizing each set of permeability data to an equivalent effective porosity, but this correction accomplishes only minor improvement. Diagonal permeability is not significantly different than horizontal permeability.

This core-plug permeability anisotropy is much less than is often determined for shaley sands from other regions, probably because Ferron permeability is associated with leaching-induced intergranular secondary porosity rather than compaction-induced primary porosity. Ferron permeability, like Ferron velocity, lacks the sensitivity to clay content that is characteristic of primary porosity. The small amount of plug-scale permeability anisotropy does not, however, imply that permeability anisotropy can be neglected in reservoir modeling. Layering-induced permeability anisotropy at the parasequence level is demonstrated by reservoir modeling studies of the Ivie Creek study area (Forster et al., this volume).

CONCLUSIONS

This petrophysical study challenges the reliability of a common assumption: that geophysical properties and relationships observed in outcrop are representative of those in the deep subsurface. We find that processes that generate outcrop exposures — uplift, erosion, and exhumation — can overprint patterns of velocity, porosity, and permeability developed in the subsurface. This conclusion is based primarily on a suite of measurements of the Ferron Sandstone that includes the following: petrophysical analysis of 471 outcrop plug samples from Ivie Creek, evaluation of subsurface (0–3400 m [0–11,000 ft]) velocity and porosity for 179 sandstone beds from 24 wells or drill holes, comparison of whole-core and log measurements of velocity and density at one drill hole, and determination of porosity, mineralo-

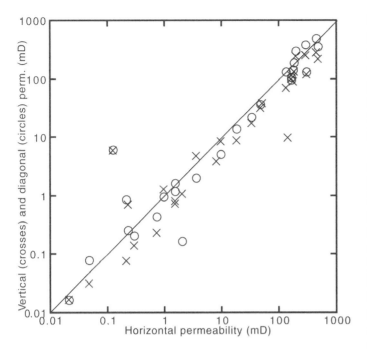

Figure 11. Relationship between horizontal permeability and both vertical (crosses) and diagonal (open circles) permeability, based on 3-D sets of adjacent core-plug samples.

gy, and diagenetic patterns from both thin sections and core-plug analysis.

Thin-section examinations showed that little, if any, primary porosity appears to remain in any of the Ferron outcrop samples. A complex diagenetic history culminated with development of substantial secondary porosity as a result of carbonate dissolution. This secondary porosity is intergranular, not vugular or crack porosity. Petrophysical measurements confirm that decreasing carbonate content is strongly associated with porosity increase. Surprisingly, clay-mineral content has only a minor direct correlation with petrophysical properties.

Ferron porosity in deep wells indicates that pre-exhumation porosity was 4±4%. Increase in porosity accelerated as the Ferron Sandstone was progressively exhumed, resulting in development of 9.3±2.4% porosity at near-surface depths and 13.0±0.6% for outcrop samples. This pattern of secondary porosity development differs from the diagenetic pattern commonly observed in deeply buried, but not exhumed, sandstones — growth of secondary porosity below the depths at which primary porosity is destroyed. Ferron Sandstone diagenetic history is similar to portions of the Frontier Formation at Moxa Arch, Wyoming. Stonecipher et al. (1984) and Stonecipher and Diedrich (1993) linked the Frontier leaching to both fluid flow and hydrocarbon generation, whereas we interpret the acceleration of Ferron leaching with exhumation as meteoric leaching.

The Ferron Sandstone, like some other sedimentary rocks in foreland basins, demonstrates the impact of meteoric flux on late-stage diagenesis. Upward meteoric

flux provides a combination of high flow rates and a cooling-induced increased calcite solubility that is probably responsible for late-stage carbonate dissolution within the Ferron Sandstone. The patterns of secondary porosity development observed in our study suggest that up-dip flow along the Ferron beds is pervasive, and that this meteoric flow extends to depths of about 2 km (~7000 ft).

Wherever exhumation causes development of secondary porosity, velocity decrease is also expected because velocity is primarily dependent on porosity. Depth-dependent changes in both velocity and porosity are observed in Ferron well logs. Cross-plots indicate that the velocity drop at depths of <550 m (<1800 ft) is even stronger than expected from the associated porosity drop. Most of these points are offset by 1000±500 m/s below predictions based on deeper data. For the depth interval 0–150 m (0–550 ft), the decrease is even larger: all data are ~1500 m/s below time-average predictions and most have velocity values less than those observed at 150–550 m (550–1800 ft). Core-plug measurements indicate that pressure dependence of velocity accounts for these log-based observations of changes in velocity and in the velocity/porosity relationship. The lowest-pressure, core-plug measurements exhibit a velocity/porosity pattern similar to that for shallow (<550 m [<1800 ft]) log data, and the highest-pressure core measurements are consistent with deep log data. We conclude that pressure decrease associated with exhumation has opened microcracks.

Log analysts often generalize that sonic logs do not respond to secondary porosity, particularly vugular porosity in carbonates and fracture porosity. Among sandstones, however, the most common type of secondary porosity is that encountered by the Ferron logs — secondary intergranular porosity created by dissolution of early carbonate cement. Ferron core-plug velocity measurements and sonic logs respond to this secondary porosity in the same manner as for primary intergranular porosity.

Velocity/depth or porosity/depth patterns are sometimes used to estimate the amount of exhumation at a well. All techniques for estimating exhumation from velocity/depth or porosity/depth patterns assume that present velocity and porosity are functions only of previous maximum burial depth. However, Ferron sandstones demonstrate that erosion can sometimes modify porosity and velocity dramatically: all major velocity and porosity changes with depth result from processes — pressure release and dissolution of carbonate cement — ignored by traditional exhumation estimation techniques.

Ferron Sandstone permeability is very closely related to porosity, as illustrated by the linear relationship between porosity and the logarithm of permeability. The primary impact of exhumation on permeability is through porosity enhancement; the porosity/permeability relationship is not changed. Permeability anisotropy is unusually low in the Ferron Sandstone, probably because permeability here results from leaching rather than mechanical compaction. Grain size has affected the entire Ferron diagenetic history: the coarsest sands retained highest permeability during compaction and cementation, and they experienced the greatest permeability enhancement during late-stage expansion of secondary porosity. Consequently, Ferron outcrop stratigraphy can provide useful clues to fluid-flow patterns in other deltaic formations, despite its complex diagenetic history. At Ivie Creek, the wave-dominated Kf-2 parasequence set exhibits much higher porosity and permeability than the river-dominated deltaic Kf-1 parasequence set. Within each parasequence set, reservoir potential is well correlated to lithofacies.

ACKNOWLEDGMENTS

We thank fellow members of the Ferron Project for the team effort that made collection and analysis of these data fruitful. We particularly wish to recognize the leadership role, both scientific and technical, of the Utah Geological Survey (UGS). We thank Craig B. Forster and Stephen H. Snelgrove for their collaboration in field work. R. Kayen provided whole-core measurements from one Ivie Creek well. We thank David E. Tabet (UGS), Barbara A. Marin (TerraTech), and S. Robert Bereskin (Tesseract Corp.) for constructive reviews.

This project was funded through the Utah Geological Survey (contracts 94-2113 and 94-2976) under the Geoscience Engineering Reservoir Characterization Program of the U.S. Department of Energy National Petroleum Technology Office, contract DE-AC22-93BC14896, with additional technical and financial support from Amoco Production Company (now BP) and the Mobil Exploration/Production Technology Center (now part of the ExxonMobil Company).

REFERENCES CITED

Armstrong, R. L., 1968, Sevier orogenic belt in Nevada and Utah: Geological Society America Bulletin, v. 79, no, 4, p. 429–458.

Barton, M. D., 1994, Outcrop characterization of architecture and permeability structure in fluvial-deltaic sandstones, Cretaceous Ferron Sandstone, Utah: Ph.D. dissertation, University of Texas at Austin, Austin, 259 p.

Barton, M. D., and N. Tyler, 1991, Quantification of permeability structure in distributary-channel deposits, Ferron Sandstone, Utah, in T. C. Chidsey, Jr., ed., Geology of east-central Utah: Utah Geological Association Publication 19, p. 273–281.

Biot, M. A., 1956a, Theory of propagation of elastic waves in a fluid saturated porous solid — I. low frequency range: Journal of Acoustical Society of America, v. 28, p. 168–178.

—1956b, Theory of propagation of elastic waves in a fluid saturated porous solid — II. high frequency range: Journal Acoustical Society of America, v. 28, p. 179–191.

Bjørlykke, K., 1984, Formation of secondary porosity: how important is it? in D. A. McDonald and R. C. Surdam, eds., Clastic diagenesis: AAPG Memoir 37, p. 277–286.

—1994, Fluid-flow processes and diagenesis in sedimentary basins, in J. Parnell, ed., Geofluids — origin, migration and evolution of fluids in sedimentary basins: Geological Society of America Special Publication 78, p. 127–140.

Bourbié, T., O. Coussy, and B. Zinszner, 1987, Acoustics of porous media: Paris, Educational Technology, 334 p.

Castagna, J. P., M. L. Batzle, and R. L. Eastwood, 1985, Relationships between compressional-wave and shear-wave velocities in clastic silicate rocks: Geophysics, v. 50, p. 571–581.

Chidsey, T. C., Jr. (compiler and editor), 2001, Geological and petrophysical characterization of the Ferron Sandstone for 3-D simulation of a fluvial-deltaic reservoir, Final Report: U.S. Department. of Energy, DOE/BC/14896-24, 471 p.

Dickinson, W. R., 1970, Interpreting detrital modes of graywacke and arkose: Journal Sedimentary Petrology, v. 40, p. 695–707.

Domenico, P. A., and M. D. Mifflin, 1965, Water from low-permeability sediments and land subsidence: Water Resources Research, v. 1, p. 563–576.

Ellis, D. V., 1987, Well logging for earth scientists: New York, Elsevier, 532 p.

Erickson, S. N., and R. D. Jarrard, 1998, Velocity-porosity relationships for water-saturated siliciclastic sediments: Journal Geophysical Research, v. 103, p. 30385–30406.

Fisher, R. S., M. D. Barton, and N. Tyler, 1993, Quantifying reservoir heterogeneity through outcrop characterization — 2. architecture, lithology, and permeability distribution of a seaward-stepping fluvial-deltaic sequence, Ferron Sandstone (Cretaceous), central Utah: Topical Report GRI-93-0023 for Gas Research Institute, 83 p.

Freeze, R. A., and P. A. Witherspoon, 1967, Theoretical analysis of regional groundwater flow — 2. effect of water-table configuration and subsurface permeability variation: Water Resources Research, v. 3, p. 623–634.

Galloway, W. E., 1984, Hydrogeologic regimes of sandstone diagenesis, *in* D. A. McDonald and R. C. Surdam, eds., Clastic diagenesis: AAPG Memoir 37, p. 3–13.

Gardner, M. H., 1995 a, Tectonic and eustatic controls on the stratal architecture of mid-Cretaceous stratigraphic sequences, central Western Interior foreland basin of North America, *in* S. L. Dorobek and G. M. Ross, eds., Stratigraphic evolution of foreland basins: SEPM Special Publication No. 52, p. 243–281.

—1995 b, The stratigraphic hierarchy and tectonic history of the mid-Cretaceous foreland basin of central Utah, *in* S. L. Dorobek and G. M. Ross, eds., Stratigraphic evolution of foreland basins: Society of Economic Paleontology and Mineralogy Special Publication No. 52, p. 283–303.

Garrels, R. M., and C. L. Christ, 1965, Solutions, minerals, and equilibria: New York, Harper and Row, 450 p.

Gassmann, F., 1951, Elastic waves through a packing of spheres: Geophysics, v. 16, p. 673–685.

Gokturkler, G., 1996, Seismic imaging of the Ferron Sandstone, Utah: Master's thesis, University of Utah, Salt Lake City, 113 p.

Hamilton, E. L., 1971, Elastic properties of marine sediments: Journal Geophysical Research, v. 76, p. 579–604.

Hamilton, E. L., 1976, Variations of density and porosity with depth in deep-sea sediments: Journal Sedimentary Petrology, v. 46, p. 280–300.

Han, D., A. Nur, and D. Morgan, 1986, Effects of porosity and clay content on wave velocities in sandstones: Geophysics, v. 51, p. 2093–2107.

Harrison, W. J., and R. N. Tempel, 1993, Diagenetic pathways in sedimentary basins, *in* A. D. Horbury and A. G. Robinson, eds., Diagenesis and basin development: AAPG Studies in Geology 36, p. 69–86.

Hayes, J. B., 1979, Sandstone diagenesis — the hole truth, *in* P. A. Scholle and P. R. Schluger, eds., Aspects of diagenesis: SEPM Special Publication 26, p. 127–139.

Hintze, L. F., 1988, Geologic history of Utah: Brigham Young University Geology Studies, Special Publication 7, 202 p.

Hottmann, C. E., and R. K. Johnson, 1965, Estimation of formation pressures from log-derived shale properties: Journal Petroleum Technology, v. 17, p. 717–722.

Hucka, B. P., S. N. Sommer, D. A. Sprinkel, and D. E. Tabet, 1995, Ferron Sandstone drill-hole database, Ferron Creek to Last Chance Creek, Emery and Sevier Counties, Utah: Utah Geological Survey Open-File Report 317, 1130 p.

Hunt, J. M., 1979, Petroleum geochemistry and geology: San Francisco, W.H. Freeman, 617 p.

Japsen, P., 1993, Influence of lithology and Neogene uplift on seismic velocities in Denmark — implications for depth conversion of maps: AAPG Bulletin, v. 77, p. 194–211.

Lines, G. C., and D. J. Morrissey, 1983, Hydrology of the Ferron Sandstone aquifer and effects of proposed surface-coal mining in Castle Valley, Utah: U.S. Geological Survey Water-Supply Paper 2195, 39 p.

Longstaffe, F. J., 1993, Meteoric water and sandstone diagenesis in the Western Canada sedimentary basin, *in* A. D. Horbury and A. G. Robinson, eds., Diagenesis and basin development: AAPG Studies in Geology 36, p. 49–68.

Magara, K., 1976, Thickness of removed sedimentary rocks, paleopore pressure, and paleotemperature, southwestern part of Western Canada Basin: AAPG Bulletin, v. 60, p. 554–565.

Mattson, A., 1997, Characterization, facies relationships, and architectural framework in a fluvial-deltaic sandstone, Cretaceous Ferron Sandstone, central Utah: Master's thesis, University of Utah, Salt Lake City, 174 p.

Mattson, A., and M. A. Chan, 1996, Facies relationships and statistical measures for reservoir heterogeneities in the Cretaceous Ferron Sandstone, central Utah (abs.): AAPG Annual Convention, Abstracts with Program, v. 5, p. A93.

Maxwell, J. C., 1960, Experiments on compaction and cementation of sand: Geological Society America Memoir 79, p. 105–132.

Miller, M. A., J. Holder, H. Yang, Y. Jamal, K. E. Gray, R. S. Fisher, and N. Tyler, 1993a, Petrophysical and petrographic properties of the Ferron Sandstone: Topical Report GRI-93/0219 for Gas Research Institute, 242 p.

Miller, M. A., H. Yang, J. T, Holder, M. J. Jounus, K. E. Gray, and R. S. Fisher, 1993b, Petrophysical properties of Ferron Sandstone during two-phase flow — implications for heterogeneity in Tertiary deltaic reservoirs: Society Petroleum Engineers Paper SPE 26489, p. 757–762.

Nelson, P. H., 1994, Permeability-porosity relationships in sedimentary rocks: Log Analyst, v. 35, no. 3, p. 38–62.

Neuzil, C. E., and D. W. Pollock, 1983, Erosional unloading and fluid pressures in hydraulically "tight" rocks: Journal of Geology, v. 91, p. 179–193.

Nur, A., 1971, Effects of stress on velocity anisotropy in rocks with cracks: Journal Geophysical Research, v. 76, p. 2022–2034.

Nur, A., and W. Murphy, 1981, Wave velocities and attenuation in porous media with fluids: Proceedings Fourth International Conference on Continuum Models of Discrete Systems, Stockholm, p. 311–327.

O' Connell, R. J., 1984, A viscoelastic model of anelasticity of fluid saturated porous rocks, in D. L. Johnson and P. N. Sen, eds., Physics and chemistry of porous media: New York, American Institute of Physics, p. 166–175.

O'Connell, R. J. and B. Budiansky, 1974, Seismic velocities in dry and saturated cracked solids: Journal Geophysical Research, v. 79, p. 5412–5426.

Raymer, L. L., E. R. Hunt, and J. S. Gardner, 1980, An improved sonic transit time-to-porosity transform: Transactions SPWLA 21st Annual Logging Symposium, p. P1–P13.

Roche, K. N., 1999, A high-resolution petrographic study of the Cretaceous Ferron Sandstone, Coyote Basin, Utah — integrating petrology and petrophysics: Master's thesis, University of New Mexico, Albuquerque, 110 p.

Ryer, T. A., 1981, Deltaic coals of Ferron Sandstone Member of Mancos Shale — predictive model for Cretaceous coal-bearing strata of western interior: AAPG Bulletin, v. 65, p. 2323–2340.

—1983, Transgressive-regressive cycles and the occurrence of coal in some Upper Cretaceous strata of Utah: Geology, v. 11, p. 207–210.

—1991, Stratigraphy, facies, and depositional history of the Ferron Sandstone in the canyon of Muddy Creek, east-central Utah, in T. C. Chidsey, Jr., ed., Geology of east-central Utah: Utah Geological Association Publication 19, p. 45–54.

Ryer, T. A., and P. B. Anderson, 1995, Parasequence sets, parasequences, facies distributions, and depositional history of the upper Cretaceous Ferron deltaic clastic wedge, Utah (abs.): AAPG Bulletin, v. 79, p. 924.

Ryer, T. A., and M. McPhillips, 1983, Early Late Cretaceous paleogeography of east-central Utah, in M. W. Reynolds and E. D. Dolly, eds., Mesozoic paleogeography of west-central United States: Rocky Mountain Section SEPM Paleogeography Symposium 2, p. 253–272.

Schlumberger, 1972, Log interpretation volume I — principles: New York, Schlumberger Ltd., 113 p.

—1989, Log interpretation principles/applications: Houston, Schlumberger Educational Services, 116 p.

Schmidt, V., and D. A. McDonald, 1979, The role of secondary porosity in the course of sandstone diagenesis, in P. A. Scholle and P. R. Schluger, eds., Aspects of diagenesis: SEPM Special Publication 26, p. 175–207.

Schwartz, F. W., and F. J. Longstaffe, 1988, Ground water and clastic diagenesis, in W. Back, J. S. Rosenshein, and P. R. Seaber, eds., Hydrogeology — geology of North America: Geological Society of America, v. O-2, p. 413–434.

Snelgrove, S. H., C. B. Forster, A. Mattson, and M. A. Chan, 1996, Impact of lithofacies architecture and distribution on fluid flow — examples from Cretaceous Ferron Sandstone (abs.): AAPG Annual Convention, Abstracts with Program, v. 5, p. A132.

Snelgrove, S. H., G. McMechan, R. B. Szerbiak, D. Wang, R. Corbeanu, K. Soegaard, J. Thurmond, C. B. Forster, L. Crossey, and K. Roche, 1998, Integrated 3-D characterization and flow simulation studies in a fluvial channel, Ferron Sandstone, east-central Utah (abs.): AAPG Annual Convention, Abstracts with Program, v. 2, p. A617.

Sondergeld, C. H., and C. S. Rai, 1993a, A new concept in quantitative core characterization: The Leading Edge, July, p. 774–779.

—1993b, A new exploration tool — quantitative core characterization: PAGEOPH, v. 141, p. 249–268.

Sprinkel, D. A., 1993, Ferron Sandstone, in C. A. Hjellming, ed., Atlas of major Rocky Mountain gas reservoirs: New Mexico Bureau of Mines and Mineral Resources, p. 89.

—1994, Stratigraphic and time-stratigraphic cross sections — a north-south transect from near the Uinta Mountain axis across the Basin and Range transition zone to the western margin of the San Rafael Swell, Utah: U.S. Geological Survey Miscellaneous Investigations Series, I-2184-D, 31 p., 2 plates.

Stonecipher, S. A., and R. P. Diedrich, 1993, Petrographic differentiation of fluvial and tidally influenced estuarine channels in Second Frontier sandstones, Moxa Arch area, Wyoming, in B. Strook and S. Andrew, eds., Guidebook Wyoming Geological Association, Jubilee Anniversary Field Conference, Casper Wyoming, August 14–19, 1993: Wyoming Geological Association, p. 181–200.

Stonecipher, S. A., R. D. Winn, Jr., and M. G. Bishop, 1984, Diagenesis of the Frontier Formation, Moxa Arch — a function of sandstone geometry, texture and composition, and fluid flux, in D. A. McDonald and R. C. Surdam, eds., Clastic diagenesis: AAPG Memoir 37, p. 289–316.

Surdam, R. C., S. W. Boese, and L. J. Crossey, 1984, The chemistry of secondary porosity, in D. A. McDonald and R. C. Surdam, eds., Clastic diagenesis: AAPG Memoir 37, p. 127–149.

Tabet, D. E., B. P. Hucka, and S. N. Sommer, 1995, Maps of total Ferron coal, depth to the top, and vitrinite reflectance for the Ferron Sandstone Member of the

Mancos Shale, central Utah: Utah Geological Survey Open-File Report 329, 3 plates, 1:250,000.

Terzaghi, K., 1965, Theoretical soil mechanics: New York, John Wiley, 510 p.

Tillman, R. W., and W. R. Almon, 1979, Diagenesis of Frontier Formation offshore bar sandstones, Spearhead Ranch field, Wyoming, *in* P. A. Scholle and P. R. Schluger, eds., Aspects of diagenesis: SEPM Special Publication 26, p. 337–378.

Toth, J., 1980, Cross-formational flow of groundwater — a mechanism for transport and accumulation of petroleum (the generalized hydraulic theory of petroleum migration), *in* W. H. Roberts III and R. J. Cordell, eds., Problems in petroleum migration: AAPG Studies in Geology, v. 10, p. 121–167.

Tripp, C. N., 1989, A hydrocarbon exploration model for the Cretaceous Ferron Sandstone Member of the Mancos Shale, and the Dakota Group in the Wasatch Plateau and Castle Valley of east-central Utah, with emphasis on post-1980 subsurface data: Utah Geological Survey Open-File Report 160, 81 p.

—1991, Ferron oil and gas field, Emery County, Utah, *in* T. C. Chidsey, Jr., ed., Geology of east-central Utah: Utah Geological Association Publication 19, p. 265–272.

—1993a, Clear Creek, *in* B. G. Hill and S. R. Bereskin, eds., Oil and gas fields of Utah: Utah Geological Association Publication 22, non-paginated.

—1993b, Flat Canyon/Indian Creek/East Mountain, *in* B. G. Hill and S. R. Bereskin, eds., Oil and gas fields of Utah: Utah Geological Association Publication 22, non-paginated.

Walsh, J. B., 1965, The effect of cracks on the compressibility of rock: Journal Geophysical Research, v. 70, p. 381–389.

Williams, G. D., and C. R. Stelck, 1975, Speculations on the Cretaceous paleogeography of North America, *in* W. G. E. Caldwell, ed., The Cretaceous system in the western interior of North America: Geological Association of Canada Special Paper 13, p. 1–20.

Winkler, K. W., and W. F. Murphy III, 1995, Acoustic velocity and attenuation in porous rocks, *in* T. J. Ahrens, ed., Rock physics and phase relations: a handbook of physical constants: Washington DC, American Geophysical Union, p. 20–34.

Wyllie, M. R. J., A. R. Gregory, and L. W. Gardner, 1956, Elastic wave velocities in heterogeneous and porous media: Geophysics, v. 21, p. 41–70.

Moving field camp, circa 1910. Photograph courtesy of the family of C. T. Lupton.

Analog for Fluvial-Deltaic Reservoir Modeling: Ferron Sandstone of Utah
AAPG Studies in Geology 50
T. C. Chidsey, Jr., R. D. Adams, and T. H. Morris, editors

Facies and Permeability Relationships for Wave-Modified and Fluvial-Dominated Deposits of the Cretaceous Ferron Sandstone, Central Utah

Ann Mattson[1] and Marjorie A. Chan[1]

ABSTRACT

Two parasequence intervals in the Cretaceous Ferron Sandstone Member of the Mancos Shale exhibit contrasting facies and permeability relationships that can be used as models for reservoir characterization. Stratigraphic sections, permeability transects, photomosaic panels, and drill-hole core data are combined to establish relationships between sedimentary features and permeability, and to create a three-dimensional (3-D) framework. The fluvial-dominated bedset of one parasequence locally exhibits seaward-dipping clinoform geometries. In contrast, the wave-modified parasequence set contains tabular facies that range from foreshore to prodelta environments. A comparison of these two depositional types shows the fluvial-dominated deposits contains: (1) higher percent shale, (2) lower degree of bioturbation, (3) finer average grain size in reservoir facies, (4) overall lower permeability values, and (5) low-permeability bounding layers between bedforms.

Various sedimentary features are used to summarize and characterize potential reservoir units. Facies, sedimentary structure, and average grain size each are related to permeability in approximately log-normal distributions. Permeability values range from 0.5 to 1121 md. The geometric mean permeability for the clinoform facies ranges from 8.3 md in proximal sandstones to 1.5 md in distal sandstones. The geometric mean permeability for the wave-modified facies range from 146 md in the foreshore to 0.95 md in the transgressive deposits.

A 3-D clinoform facies model of the fluvial-dominated bedset is constructed to preserve the prograding clinoform geometries and facies relationships. The 3-D facies model combined with permeability distribution provides input to detailed 2-D and 3-D fluid flow simulations. This study links comprehensive field studies to reservoir characterization and can serve as an analog for other fluvial-deltaic reservoirs with fluvial and or wave influence.

[1]Department of Geology and Geophysics, University of Utah, Salt Lake City, Utah

INTRODUCTION

Outcrop analogs of petroleum reservoirs have been studied for several decades in an attempt to better understand the influence of interwell-scale (tens to hundreds of meters) permeability structures on petroleum production. These analog studies cover a variety of clastic depositional settings (e.g., Fisher et al., 1993a, b; Begg et al., 1994; Doyle and Sweet, 1995; summary in Liu et al., 1996; Robinson and McCabe, 1997; Dutton and Willis, 1998; Eschard et al., 1998) and typically include high-resolution stratigraphic mapping, systematic analysis of bedform geometries, detailed permeability/porosity measurements, and geophysical surveys.

Cretaceous fluvial-deltaic exposures of the Western Interior have long been recognized as excellent analogs to subsurface fluvial-deltaic systems such as the Alaskan North Slope, the Gulf Coast, and the Rocky Mountain region, due to similarities of size, extent, and architecture. Fluvial-deltaic reservoirs are significant exploration targets because of their volumetrically large sand accumulations, proximity to source rocks, syndepositional structural traps, and active subsidence to aid maturation (Coleman and Wright, 1975; Fisher et al., 1969; Wescott, 1992). The Ferron Sandstone has been the subject of numerous geologic studies since the early 1900s (Ryer, this volume), and more recently has been the site of reservoir analog studies (Anderson et al., this volume). The Ferron Sandstone lends itself to integrated field analysis and reservoir simulation because of its vertical and lateral exposures.

This paper is focused on a subset of the larger regional study area and compares two separate parasequence intervals within the Ferron Sandstone that exhibit contrasting depositional styles. Our goal is to establish relationships between sedimentary features and permeability, and construct a three-dimensional (3-D) framework. We accomplish these tasks by integrating photomosaic interpretations, detailed stratigraphic sections, and permeability transect data in a spatial framework. In wave-modified deposits, a parasequence set of three parasequences was chosen for study because the stacked parasequences showed similar lithologies and facies types in an overall progradational package. In the fluvial-dominated deposits a smaller, individual bedset from a parasequence was chosen for study because of its unusual clinoform geometries and lateral extent. Other bedsets and parasequences within the fluvial-dominated parasequence set are thinner and less extensive.

The first portion of this paper summarizes the characteristics of both a wave-modified parasequence set and a fluvial-dominated bedset. The characterization is followed by a discussion of the geologic controls on permeability and a comparison of the reservoir facies for each depositional style. Finally, we construct facies cross sections of the fluvial-dominated bedset and build a 3-D facies model that can be used as input to reservoir simulation modeling.

FERRON SANDSTONE: IVIE CREEK CASE STUDY

The Ferron Sandstone is a clastic member of the Upper Cretaceous (Turonian, Coniacian) Mancos Shale and is exposed for over 70 km (43 mi.) along the flanks of the San Rafael Swell in central Utah (Figure 1A). The fluvial-deltaic Ferron Sandstone was deposited along the western margin of the Cretaceous Western Interior Seaway in a rapidly subsiding foreland basin (Figure 1B). The basin trends northeast-southwest and is bounded to the west by the Pavant thrust (Gardner, 1995a, b). Paleozoic strata and granitic basement of the Sevier orogenic belt sourced the Ferron deltas (Ryer, 1981). Stratigraphically, the Ferron Sandstone is underlain by the Tununk Shale and overlain by the Blue Gate Shale, also members of the Mancos Shale.

The Ivie Creek case study (Figure 2) represents a small area chosen for characterization because of the well-exposed facies on extensive cliff-faces combined with a drill-hole core study conducted immediately behind the outcrop (Anderson et al., this volume). This combination provides an excellent 3-D view typical of a small oil field (thousands of meters) and interwell (hundreds of meters) scales. This paper focuses on and the wave-modified Kf-2 parasequence set and the fluvial-dominated Kf-1-Iv[a] bedset from the Kf-1 parasequence set (Anderson and Ryer, this volume; Anderson et al., this volume). These deposits are exposed in outcrops along Quitchupah Creek, Scabby Canyon, Ivie Creek, and the Interstate 70 road-cut (Figure 2). The stratigraphic units are named "Kf" for Cretaceous Ferron followed by parasequence set number (e.g., "1" = oldest set), type locale (e.g., "Iv" for Ivie Creek), and individual parasequence or bedset letter [e.g., "a" = oldest parasequence or bedset]. Two different scales of investigation are implemented based on the length scales of the different facies architecture: the wave-dominated Kf-2 at a small oil field-scale termed the "local model" (2.25 x 2.75 km [1.4 x 1.7 mi]); and the fluvial-dominated Kf-1-Iv[a] at an interwell-scale termed the "detail model" (610 x 610 m [2000 x 2000 ft]).

Field Data

Sedimentary features of bed thickness, grain size, induration, sedimentary structure (genetically similar structures are grouped together), and lithotype were measured in 28 outcrop and four drill-hole core sections (Anderson et al., this volume). Facies designations were based on the field description of these sedimentary fea-

(a) Ferron Sandstone outcrop map, central Utah

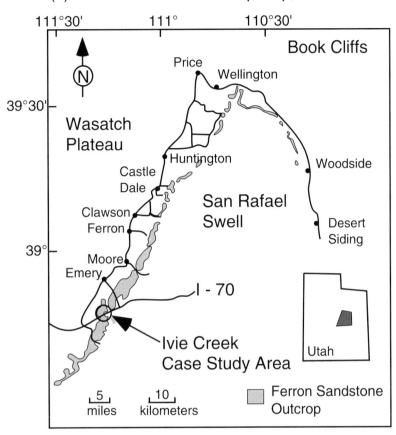

(b) Stratigraphic cross section of Cretaceous foreland basin across Utah, including Ferron Sandstone

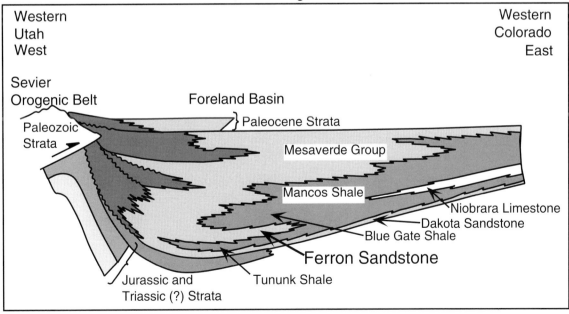

Figure 1. (A) Ferron Sandstone outcrop map, central Utah; (B) Stratigraphic cross section of Cretaceous foreland basin across Utah, including Ferron Sandstone (after Armstrong, 1968; and Ryer, 1981). See Anderson et al. and Anderson and Ryer (this volume) for the detailed stratigraphy of the Ivie Creek parasequence sets.

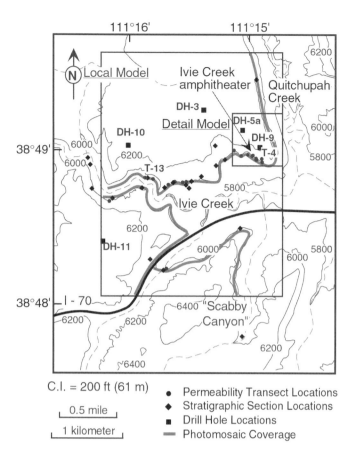

Figure 2. Database map of Ivie Creek case-study area showing locations of permeability transects, stratigraphic sections, drill hole, and photomosaics. Transect / stratigraphic sections T-4 and T-13 shown in Figure 3.

tures. Thirteen stratigraphic sections measured 144 m (472 ft) of the fluvial-dominated Kf-1-Iv, and 19 stratigraphic sections measured 384 m (1253 ft) of the wave-modified Kf-2. Digitally processed cliff-face photos were merged to form photomosaic panels (Anderson et al., this volume). These photomosaics provided a framework to relate geologic and permeability data, and to interpret facies distributions and geometries. The two-dimensional (2-D) architectural geometry of the parasequence intervals was interpreted from the photomosaic panels, and the 3-D architectures were derived from the photomosaic panels combined with drill-hole core sections.

Mini-permeameter studies of outcrop core plugs (co-located with stratigraphic sections) and drill-hole cores (taken behind the cliff face) were used to characterize the facies for the two depositional environments (Anderson et al., this volume; Jarrard et al., this volume). These data were used to quantify the relationship between various sedimentary features and permeability and to establish spatial distributions of permeability (Figure 2). The position and permeability measurement for each transect was superimposed on the architecture of each parase-

quence interval and the permeability values were correlated to facies and sedimentary features. The data set included 18 vertical and three lateral permeability outcrop transects and four drill-hole core transects. Thirteen outcrop transects clustered within the detail model area primarily sample the Kf-1-Iv[a], and eight vertical outcrop transects widely scattered in the local model primarily sample parasequences of the Kf-2.

Wave-Modified Parasequence Set: Kf-2

Geologic Parameters

The wave-modified Kf-2 parasequence set is composed of three progradational parasequences in which 14 facies are identified. A typical stratigraphic section for the Kf-2 parasequence set in Figure 3A (indexed to Figure 2) illustrates the typical upward shoaling of marine facies from prodelta to foreshore. This vertical shoaling transition is characterized by increases in sandstone to shale ratio, average grain size, bed size, and degree of cementation. All facies exhibit a fairly high degree of bioturbation (~ 50%). Sedimentary structures range from horizontal and ripple cross-lamination in the prodelta to trough cross-stratification in the upper shoreface. The wave-dominated deposits typically exhibit tabular bedding and laterally continuous facies. The geologic parameters of the wave-modified Kf-2 parasequences and facies are summarized in Figure 4 and Tables 1 and 2.

Permeability Data

Permeability was measured in eight facies of the wave-modified, Kf-2 parasequence set: transgressive, lower shoreface, middle shoreface, upper shoreface, foreshore, distal bar, distributary mouthbar, and bayfill. Permeability values measured in these sampled facies range from 0.34–1121 md (Table 3). The foreshore facies individually has the highest permeabilities (up to 1121 md), but represents a small fraction of the total volume. The volumetrically most important reservoir facies are the distributary mouthbar and upper shoreface. These reservoir facies have the highest overall permeability values (up to 852 md), the highest percent of sandstone (~100%), the coarsest average grain size (medium-fine to fine), the highest percent of trough cross-stratification (60–80%), and the lowest degree of bioturbation (12–50%). A relatively high degree of sorting due to higher depositional energy is inferred from textural observations in the reservoir facies. The coefficient of variation (the ratio of standard deviation to mean) of permeability for the geometric and arithmetic means is low for these reservoir facies, indicating a lower degree of variability relative to the mean (Table 3).

A coefficient of variation greater than one indicates the presence of erratic sample values that may have a significant impact on the final estimates (Isaaks and Srivas-

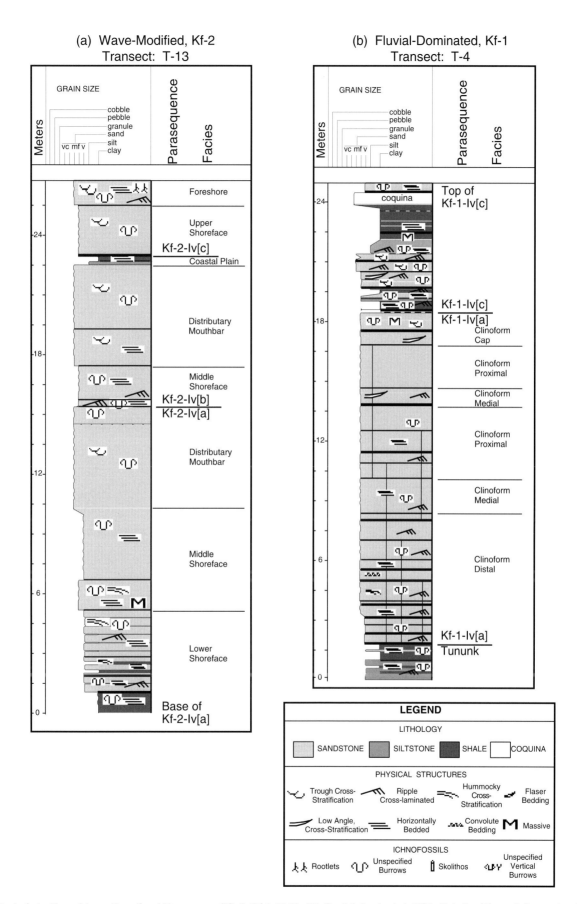

Figure 3. Typical stratigraphic sections for: (A) wave-modified, Kf-2: T-13; (B) fluvial-dominated, Kf-1: T-4. See Figure 2 for section locations (after Anderson et al., this volume).

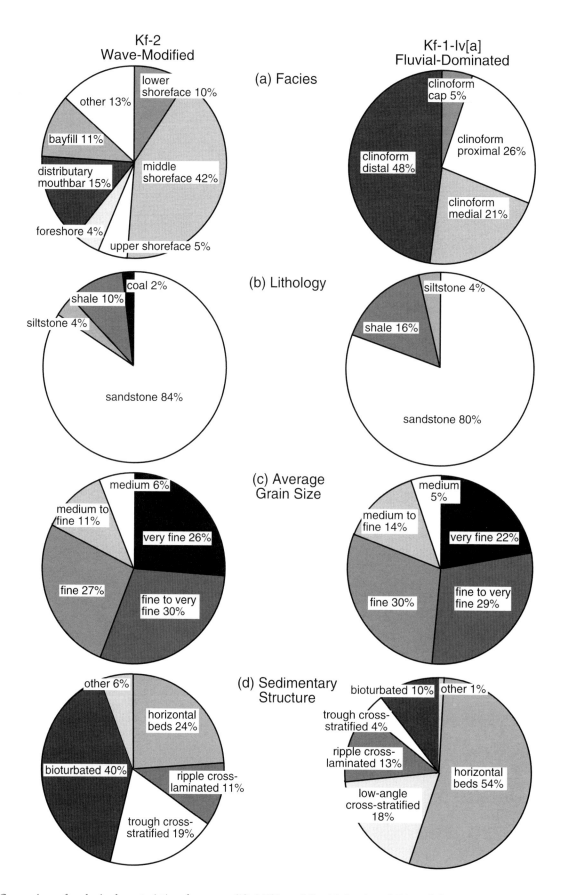

Figure 4. Comparison of geologic characteristics of wave-modified Kf-2, and fluvial-dominated Kf-1-Iv[a]: (A) facies, (B) lithology, (C) average grain size, and (D) sedimentary structure.

Table 1. Wave-modified, Kf-2 parasequence summary.

Property *	Kf-2	%	Kf-2-Iv[a]	%	Kf-2-Iv[b]	%	Kf-2-Iv[c]	%
Facies	Middle shoreface	42	Middle shoreface	54	Middle shoreface	53	Bayfill	46
	Distributary mouthbar	15	Lower shoreface	18	Distributary mouthbar	22	Upper shoreface	18
	Bayfill	11	Distributary mouthbar	17			Foreshore	12
	Lower shoreface	10						
	Upper shoreface	5						
Lithology	Sandstone	84	Sandstone	94	Sandstone	88	Sandstone	61
	Shale	10	Shale	5	Shale	9	Shale	23
							Siltstone	9
Grain Size-dominant	Fine to very fine-grained	30	Very fine-grained	38	Fine-grained	37	Fine-grained	37
Sedimentary Structures	Bioturbated	40	Bioturbated	51	Bioturbated	35	Horizontal beds	33
	Horizontal beds	24	Horizontal beds	20	Horizontal beds	23	Bioturbated	28
	Trough cross-stratified	19	Trough cross-stratified	18	Trough cross-stratified	22	Trough cross-stratified	17
	Ripple cross-laminated	11			Ripple cross-laminated	15	Ripple cross-laminated	15
Bed Size-dominant	Thick	24	Thick to medium	30	Thick to medium	26	Thin to laminated	29
Induration-dominant	Well to moderately cemented	28	Moderately cemented	29	Well to moderately cemented	32	Friable	29

* All properties cutoff at 5%.
Note: Properties of this table are based on Anderson et al., this volume.

tava, 1989). In this study, the coefficient of variation for each facies type is greater than one for the arithmetic mean, but is closer to one for the geometric mean (Table 3). The arithmetic mean is more sensitive to erratic values than is the geometric mean. In permeability studies the geometric mean is often considered a more robust measure of mean permeability.

A cross-plot of geometric mean permeability versus percent sandstone (estimated in the field) shows a logarithmic increase in permeability with a linear increase in percent sandstone (Figure 5A). Regression of geometric mean permeability with respect to percent sandstone has a coefficient of determination (R^2) of 0.23, indicating only a very weak correlation. For example, an R^2 of 0.23 implies that the model only explains 23% of the variation in the data. The error bar associated with sandstone percentage is assumed to be a field estimation error of no more than 10%. The error bar associated with geometric mean permeability is the 95% confidence interval for the data set. This value is calculated by finding the log of the geometric mean and the standard deviation of the log of the individual permeability measurements, applying a 95% confidence test, and then taking the inverse log (Conover, 1980).

Bulk permeability measurements are presented in terms of geometric mean rather than arithmetic or harmonic mean, because the geometric mean more accurately reflects the bulk permeability of a sample when fluid flow and sedimentary bedding are arbitrarily oriented (Isaaks and Srivastava, 1989). The effective permeability for fluids flowing through sedimentary beds is estimated by: (1) the arithmetic mean for flow parallel to bedding, (2) the harmonic mean for flow perpendicular to bedding, or (3) the geometric mean for flow in a random orientation relative to bedding or for flow through complex facies distributions.

Cumulative probability of permeability is plotted versus: (1) facies (Figure 6A); (2) average grain size (Figure 6C); (3) sedimentary structure (Figure 7A); and (4) parasequence (Figure 8A). All four plots are relatively straight lines on log-normal plots. Two step functions appear at ~ 0.5 and ~ 2.0 md, and seem to coincide with the resolution limits of the two permeability instruments used in the analysis (Jarrard et al., this volume). The linear relationships shown on these plots support the general assumption that permeability measurements as grouped by geologic facies are log-normally distributed (Barton, 1994; Rhea et al., 1994; and Liu et al., 1996). These curves show an increase in average permeability as: (1) facies become more proximal, (2) average grain size increases, and (3) depositional energy as inferred from sedimentary structures increases. The plots of permeability for individual parasequences (Figures 8A and 8C-E) demonstrate an increase in permeability from the oldest parasequence, Kf-2-Iv[a], to the youngest, Kf-2-Iv[c] that corresponds to progradation within the Kf-2.

Table 2. Wave-modified, Kf-2 facies summary.

Facies	%	Lithology *	%	Grain Size Sandstone-dominant	%	Sedimentary Structures	%	Bed Size-% dominant		Induration-dominant	%
Prodelta	0.5	Sandstone Shale	50 50	Very fine-grained	100	Horizontal beds Ripple cross-laminated	50 50	Thin to laminated	100	Friable	100
Transgressive	2	Sandstone Siltstone	76 20	Very fine-grained	50	Bioturbated Horizontal beds	87 8	Medium to thin	38	Moderately cemented to friable	53
Lower Shoreface	9.5	Sandstone Shale	70 27	Very fine-grained	70	Bioturbated Horizontal beds Ripple cross-laminated	55 25 13	Medium to thin	47	Moderately cemented to friable	39
Middle Shoreface	42	Sandstone	98	Fine to very fine-grained	40	Bioturbated Ripple cross-laminated Horizontal beds	51 13 6	Medium to thin	24	Moderately cemented	33
Upper Shoreface	5	Sandstone	98	Fine-grained	50	Trough cross-stratified Bioturbated Ripple cross-laminated	59 24 12	Medium	37	Well to moderately cemented	46
Foreshore	4	Sandstone	99	Fine to very fine-grained	46	Bioturbated Horizontal beds Ripple cross-laminated	50 32 8	Medium	35	Well to moderately cemented	39
Backshore	0.2	Sandstone	100	Medium to fine-grained	100	Horizontal beds Bioturbated Ripple cross-laminated	55 25 20	Medium to thin	100	Well to moderately cemented	100
Distal Bar	0.8	Sandstone Siltstone	50 50	Fine-grained	100	Bioturbated Irregular laminae Trough cross-stratified	70 15 10			Well to moderately cemented	100
Distributary Mouthbar	15	Sandstone	99	Medium to fine-grained	50	Trough cross-stratified Bioturbated Horizontal beds	78 12 5	Thick to medium	62	Well to moderately cemented	50
Distributary Channel	2	Sandstone	100	Medium-grained	98	Trough cross-stratified	97	Very thick	51	Well cemented	100
Meander Belt	1	Sandstone	97	Medium to fine-grained	41	Trough cross-stratified Ripple cross-laminated Horizontal beds	38 24 24	Thick to medium	40	Well to moderately cemented	40
Bayfill	11	Sandstone Shale Siltstone	42 41 17	Fine to very fine-grained	39	Horizontal beds Bioturbated Ripple cross-laminated Irregular laminae	39 32 18 7	Thin to laminated	54	Friable	36
Bayhead Delta	2	Sandstone Siltstone	88 10	Fine-grained	64	Bioturbated Trough cross-stratified Horizontal beds	52 20 19	Medium	39	Moderately cemented to friable	33
Coastal Plain	3	Sandstone Shale Coal	20 61 11	Fine-grained	63	Horizontal beds Bioturbated Ripple cross-laminated	59 30 7	Thin to laminated	38	Friable	58
Swamp	2	Coal Shale	76 20	None		Horizontal beds Irregular laminae Bioturbated	80 10 10	Thick to medium	31	Friable	82

* All properties cutoff at 5%.

Table 3. Kf-2 permeability summary.

Permeability (md)	All Facies	Trans-gressive	Lower Shoreface	Middle Shoreface	Upper Shoreface	Foreshore	Distal Bar	Distributary Mouthbar	Bayfill
Arithmetic Mean	34.10	1.24	3.33	13.25	115.1	219.1	13.75	97.95	7.98
Coefficient of Variation	2.20	1.01	1.50	2.42	1.12	1.00	1.27	0.86	1.32
Geometric Mean	7.05	0.95	1.61	4.59	57.15	146.3	7.85	68.35	4.69
Coefficient of Variation	0.87	1.99	1.91	0.85	0.07	0.02	0.36	0.04	0.60
Harmonic Mean	2.25	0.81	1.02	2.25	20.06	76.77	5.02	37.13	2.85
Coefficient of Variation	31.74	1.63	5.37	15.06	7.99	3.41	3.91	2.79	4.14
Median	5.60	0.75	1.00	4.13	66.76	174.1	9.38	78.45	5.24
Range	0.34-1121.0	0.40-7.4	0.34 -27.2	0.46-598.5	2.12-625.2	5.10-1121.0	1.95-72.0	1.97-851.8	0.55-51.4
Count	2469	109	299	1438	145	48	31	369	30

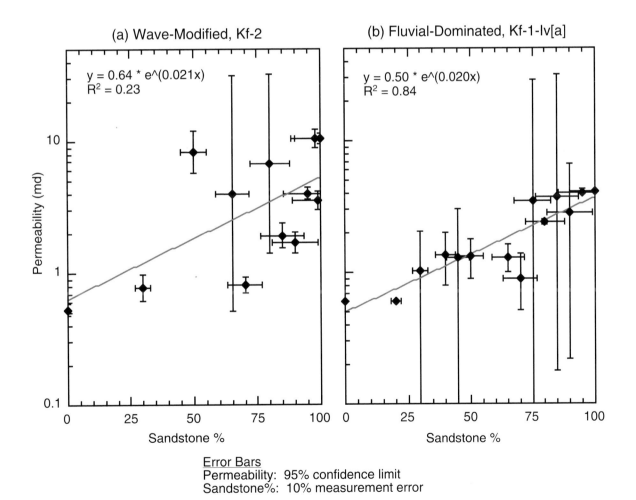

Figure 5. Percent sandstone vs. permeability: (A) wave-modified Kf-2, and (B) fluvial-dominated Kf-1-Iv[a].

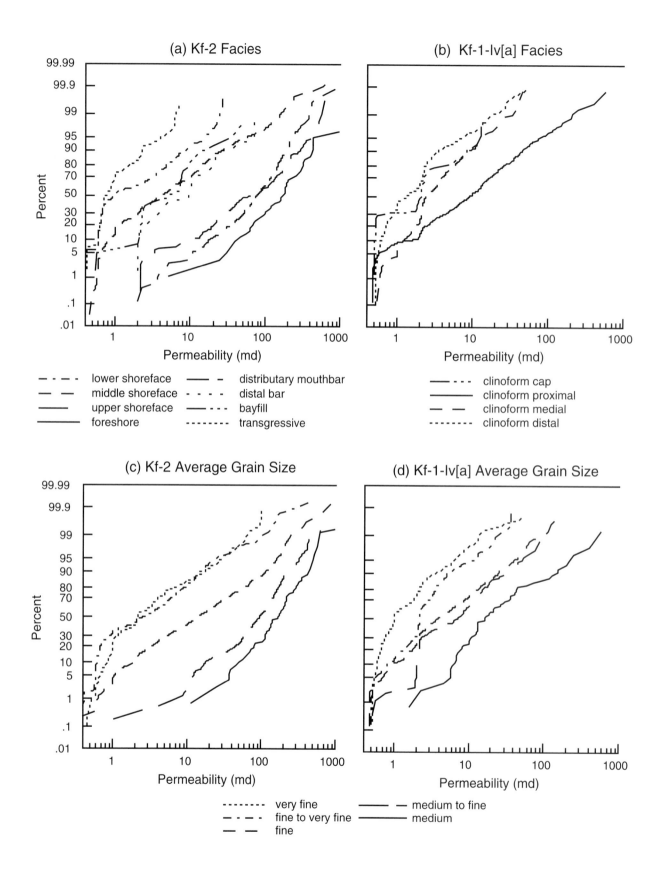

Figure 6. Wave-modified Kf-2, and fluvial-dominated Kf-1-Iv[a]: facies vs. permeability (A and B), and average grain size vs. permeability (C and D).

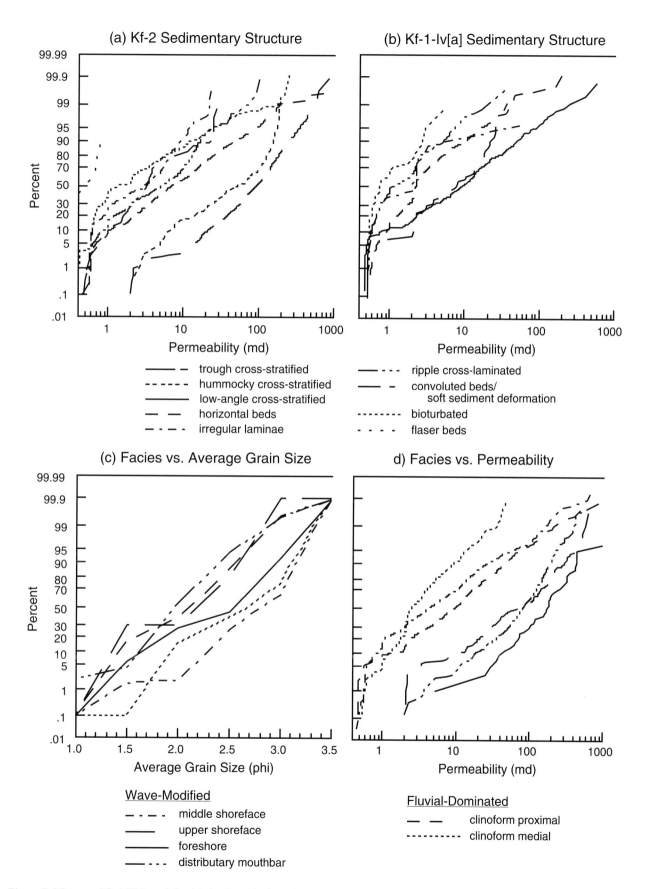

Figure 7. Wave-modified Kf-2, and fluvial-dominated Kf-1-Iv[a]: sedimentary structure vs. permeability (A and B); and reservoir facies vs. average grain size (C) and permeability (D).

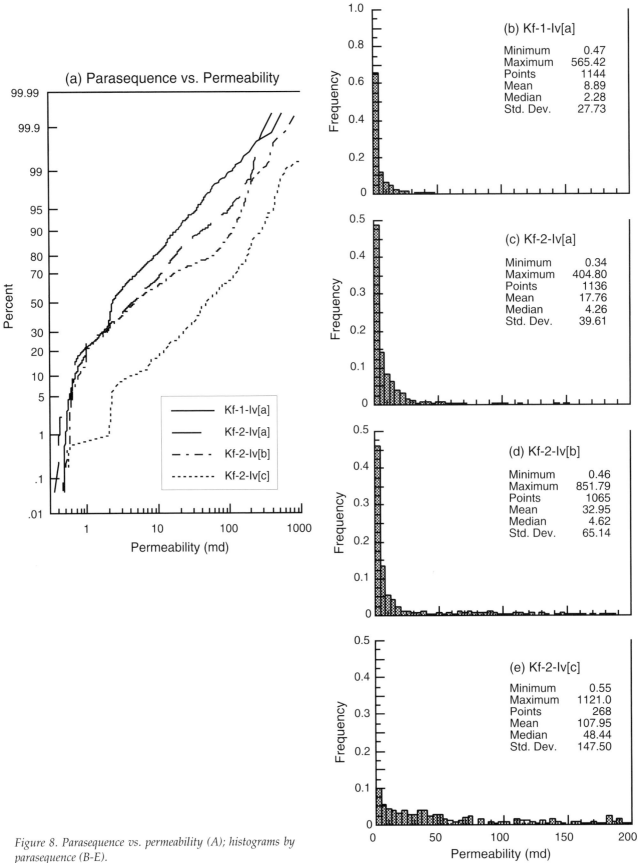

Figure 8. Parasequence vs. permeability (A); histograms by parasequence (B-E).

(a) Isopach of Fluvial-Dominated, Kf-1-Iv[a]

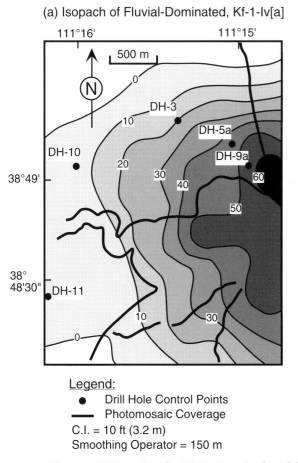

(b) Location of Detail Model Cross-Sections

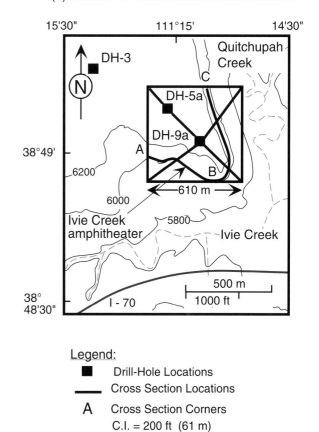

Legend:
● Drill Hole Control Points
— Photomosaic Coverage
C.I. = 10 ft (3.2 m)
Smoothing Operator = 150 m

Legend:
■ Drill-Hole Locations
— Cross Section Locations
A Cross Section Corners
C.I. = 200 ft (61 m)

Figure 9. (A) Isopach of fluvial-dominated, Kf-1-Iv[a]; (B) location of cross sections for construction of 3-D detail model.

Fluvial-Dominated Bedset: Kf-1-Iv[a]

Geologic Parameters

The Kf-1-Iv[a] is the oldest bedset within the Kf-1 parasequence set and is the focus of the 610 x 610 m (2000 x 2000 ft) detail model. The Kf-1-Iv[a] is a fluvial-dominated deltaic lobe that thins and becomes more distal toward the northwest (Figure 9A). A typical stratigraphic section for this bedset in Figure 3 (indexed to Figure 2) illustrates the clinoform facies. Clinoform bodies are characterized by apparent dip inclination angles ranging from ~ 28° in the proximal facies to ~ 0° in the distal facies. Facies are assigned within each clinoform body on the basis of average grain size, sedimentary structure, lithology, bed size, angle of inclination, and position within a clinoform body (Figures 4 and 10, Tables 4 and 5, Anderson et al., this volume). Similar attributes are grouped to distinguish four clinoform facies: proximal, medial, distal, and cap. The down-dip trend from clinoform proximal to clinoform distal facies represents a decrease in depositional energy within an individual clinoform body. This down-dip trend is characterized by decreases in: sandstone to shale ratio, average grain size, bed thickness, inclination angle between bedsets, and low-angle, cross-stratified beds. The clino-

form cap facies represents a laterally continuous zone of reworked sediment that truncates the tops of the clinoform bedforms.

Thin, recessive beds observed between clinoform bodies are thought to be a result of shale drapes (bounding surfaces) that were deposited during lower energy depositional periods (Anderson et al., this volume). These recessive beds typically increase in thickness and continuity distally, but are typically absent proximally. In the clinoform proximal facies, these drapes were either not deposited or were scoured away when the next clinoform body was deposited. In 2-D reservoir simulations these shale drape units are treated as baffles and/or barriers to hydrocarbon flow (Fisher et al., 1992; Forster et al., this volume). One goal of 2-D simulation is to observe the impact of these bounding layers on fluid flow.

Low-angle, cross-stratification (inclined at < 10–14° in the up-dip direction to the bedform-bounding surface) in fine- to medium-grained sandstones is common within the Kf-1-Iv[a]. These structures are primarily found in the clinoform proximal facies and are probably related to the development of a fair-weather, swash-bar along the prograding delta front. Similar anti-dune features have been noted in coarse-grained deltas of southern Italy (Massari, 1996), but Froude numbers for water

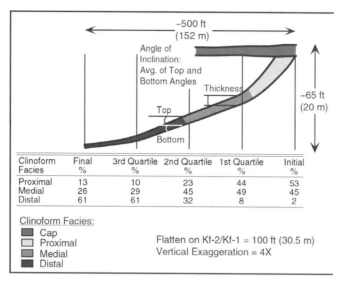

Figure 10. Typical clinoform characteristics, and measurements in a fluvial-dominated bedset.

Table 4. Fluvial-dominated, Kf-1-Iv[a] clinoform facies summary.

Facies	%	Lithology	%	Grain Size Sandstone-dominant	%	Sedimentary Structures	%	Bed Size-dominant	%	Induration-dominant	%
All Facies		Sandstone Shale	80 16	Fine-grained	30	Horizontal beds Low-angle cross-stratified Ripple cross-laminated Bioturbated	54 18 13 10	Thin to laminated	33	Well to moderately cemented	31
Clinoform Cap	5	Sandstone	99	Fine to very fine-grained	49	Horizontal beds Low-angle cross-stratified Bioturbated Trough cross-stratified	30 28 24 15	Thick	35	Well cemented	60
Clinoform Proximal	26	Sandstone	100	Fine-grained	49	Low-angle cross-stratified Horizontal beds Trough cross-stratified	60 26 10	Thick to medium	28	Well to moderately cemented	68
Clinoform Medial	21	Sandstone Shale	94 5	Fine to very fine-grained	36	Horizontal beds Ripple cross-laminated Bioturbated	80 7 6	Medium	33	Well cemented	38
Clinoform Distal	48	Sandstone Shale Siltstone	62 31 7	Very fine-grained	39	Horizontal beds Ripple cross-laminated Bioturbated	59 21 16	Thin to laminated	49	Moderately cemented to friable	37

* All properties cutoff at 5%.

depths of a few tens of meters, as estimated for the Ferron clinoforms, do not appear to be high enough to generate anti-dunes. Thus, the exact origin of these features is unknown, but they do represent a higher energy portion of the system.

Architectural Parameters

The 2-D deterministic framework for the Kf-1-Iv[a] bedset was generated from two photomosaic panels (Ivie Creek amphitheater and Quitchupah Canyon, see

Table 5. Kf-1-Iv[a] clinoform geometry summary.

Facies	Inclination Angle (deg)	Clinoform Thickness (m)	Characteristic Length (m)
Clinoform Cap	0 to 2		
Clinoform Proximal	10.0 ± 0.8	1.28 ± 0.21	25.8 ± 8.8
Clinoform Medial	6.0 ± 0.6	0.95 ± 0.09	45.3 ± 10.9
Clinoform Distal	2.1 ± 0.3	0.73 ± 0.09	71.7 ± 25.8

Table 6. Kf-1-Iv[a] permeability summary.

Permeability (md)	All Facies	Clinoform Proximal	Clinoform Medial	Clinoform Distal	Clinoform Cap
Arithmetic Mean	8.89	21.34	4.38	2.30	3.00
Coefficient of Variation	3.12	2.19	1.25	1.75	1.16
Geometric Mean	3.13	8.30	3.00	1.48	1.78
Coefficient of Variation	1.11	0.48	0.73	1.49	1.57
Harmonic Mean	1.77	3.30	2.30	1.16	1.15
Coefficient of Variation	16.12	15.19	2.54	3.60	3.45
Median	2.28	9.20	2.42	1.37	2.05
Range	0.47 - 565.42	0.47 - 565.42	0.55 - 46.40	0.51 - 55.58	0.50 - 13.43
Count	1144	357	331	419	37

Anderson et al., 2003) that intersect at ~ 50°. The linework interpretation for each panel was based on recessive units (bounding layers) observed on the photomosaic panels (Anderson et al., this volume). These bounding layers are used to define individual clinoform bodies. For the architectural analysis a clinoform body was defined in one of three ways. (1) The bounding layers enclose a body on the photomosaic interpretation. (2) The bounding layer could no longer be interpreted on the photomosaic panel and therefore the clinoform body was terminated. (3) The bounding layers for the distal end of the clinoform body became subparallel, and the body was terminated when the bedform thickness dropped below 0.6 m (2 ft). This photomosaic clinoform interpretation was flattened on the parasequence set boundary (Kf-2 / Kf-1 = 30.5 m [100 ft]) to provide a datum for quantitative analysis. For 2-D reservoir simulation, clinoform bodies must be fully enclosed, so this initial framework interpretation was later modified prior to modeling.

Along each panel, apparent inclination angle, bedform thickness and length, and clinoform facies were measured at the beginning, end, and quartile points of each bedform (Figure 10, Table 5). Thicknesses were measured from vertical, and apparent inclination angles were measured from horizontal. The photomosaic interpretation for Quitchupah Canyon is less detailed than for Ivie Creek, and effectively grouped several clinoform

bodies together. Hence, inclination angles may be compared between the two panels, but bedform thickness and length measurements are not comparable.

In the architectural analysis, four main relationships emerge. (1) The initial apparent dip inclination angle measured in Quitchupah Canyon (~ 14°) is steeper than in Ivie Creek (~ 11°). This implies that Quitchupah Canyon is closer to being a "true dip section" with respect to deposition of the clinoform bodies than Ivie Creek. (2) The apparent angle of inclination decreases along the length of the bedform, and the angle decreases as the facies become more distal. (3) The average clinoform thickness decreases along a bedform. (4) Finally, position within a clinoform bedform may be used to predict the facies present (Figure 10).

Permeability Data

Permeability values for the Kf-1-Iv[a] bedset are generally quite low compared to permeabilities of reservoir facies in many producing fields, due to the very fine-grained nature of the Ferron outcrops (Table 6). The highest Kf-1-Iv[a] permeability values (< 50 md) occur in the clinoform proximal facies. This proximal facies (geometric mean = 8.3 md) is considered the most important reservoir facies within the fluvial-dominated Kf-1-Iv[a] because it has the highest depositional energy, coarsest average grain size, and lowest shale percentage of the clinoform facies. The coefficient of variation of perme-

Table 7. Sedimentary features passing the Lilliefors Log-Normality Test for the 0.95 and 0.99 quantiles.

Property	K f-1-Iv [a]	Quantile	Kf-2	Quantile
Facies	Clinoform cap	0.99	Distal bar	0.95
	Clinoform proximal	0.95	Foreshore	0.95
Grain Size	Medium-grained	0.99	Fine-grained	0.99
Sedimentary Structure	Irregular laminae	0.95	Flaser beds	0.95
	Trough cross-stratified	0.95	Horizontal beds	0.95
	Low-angle cross-stratified	0.95		

ability in the clinoform proximal and medial facies is about half that of the clinoform distal and cap facies, indicating a lower degree of variability relative to the mean.

A cross-plot of geometric mean permeability versus percent sandstone (estimated in the field) is similar to the plot obtained for the Kf-2 (Figure 5B). Regression of permeability mean with respect to percent sandstone has a coefficient of determination (R^2) of 0.81, indicating a strong relationship between percent sandstone and permeability. By comparison, the coefficient of determination measured for the Kf-2 parasequence set indicates only a very weak relationship between percent sandstone and permeability ($R^2 = 0.23$).

Permeability is related to depositional energy, and is inferred from the sedimentary structure, average grain size, and lithology. Low-angle, cross-stratified beds have the highest permeability (geometric mean = 7.8 md) and ripple cross-laminated beds the lowest (geometric mean = 1.43 md). Correspondingly, medium-grained sandstones have the highest permeability (geometric mean = 21.0 md) and very fine-grained sandstones the lowest (geometric mean = 1.27 md). In general, permeability decreases with increasing shale percentage and therefore with decreasing depositional energy.

Cumulative probability plots of permeability versus clinoform facies (Figure 6B), average grain size (Figure 6D), and sedimentary structure (Figure 7B) mimic those obtained for the Kf-2. The curves also appear to exhibit the same log-normal behavior and step-function increases at ~ 0.5 md and ~ 2.0 md as observed in the Kf-2 plots. The cumulative curves and histogram (Figures 8A and 8B) of the Kf-1-Iv[a] are shifted to slightly lower permeability values, reflecting the finer average grain size and implied lower relative degree of sorting of this bedset, in comparison to the Kf-2 parasequence set.

Geologic Controls on Permeability

Permeability from the Ivie Creek database was evaluated for log-normality by both visual inspection of the distribution curves and the Lilliefors test (Conover, 1980). The Lilliefors test was chosen because it is considered a robust test for log-normality and it can be used to evaluate large data sets. The method computes a test statistic based on the number of data points and the maximum difference between the empirical distribution and a standard log-normal distribution. Each data set was measured at the 0.95 quantile (level of significance, ($\alpha = 0.05$), a more rigorous standard, and the 0.99 quantile ($\alpha = 0.01$), a less rigorous standard. The level of significance is the maximum probability that the data cannot be described by a log-normal distribution. The results of this analysis (Mattson, 1997, Table 7) indicate that cumulative probability curves for permeability versus facies, average grain size, and sedimentary structure may be approximated by a log-normal distribution, at least for data above the 0.5 and 2.0 md resolution limits of the mini-permeameter instruments (Figures 6A-D and 7A-B). These log-normal approximations can then be used as input to subsequent reservoir simulation modeling of the fluvial-deltaic facies.

In the preceding data analysis, we have shown that three of the geologic parameters measured in the field, average grain size, lithology, and sedimentary structure, show the strongest correlation with permeability. Permeability appears to be ultimately controlled by the energy of the depositional setting. Low energy, fine-grained deposits yield low permeability values. In contrast, the higher energy, coarse-grained deposits (e.g., 100% sandstone, trough cross-stratified units) correspond to the highest permeability values.

Other workers have demonstrated that average grain size is the primary control on permeability, with sorting and lithology as secondary controls (e.g., Barton, 1994; Doyle and Sweet 1995; Liu, et al., 1996; Dutton and Willis, 1998). Barton (1994) further hypothesized that 90% of permeability variation is due to four factors (in decreasing order of importance): (1) grain size, (2) sorting, (3) mineralogical composition, and (4) porosity. If average grain size were the only control, no variation in permeability would be predicted for different sedimentary structures with a given average grain size (assuming a single lithotype). However, for a given average grain size, high-energy sedimentary structures (e.g., trough cross-stratification) have higher average permeabilities than do lower energy sedimentary structures (e.g., ripple cross-lamination) for the same grain size and lithology, probably due to sorting (Mattson, 1997). Thus, sedimentary structure should be considered along with the grain size in modeling studies.

Lithology also acts as a significant control on permeability. An increase in siltstone or shale percentage corresponds to a decrease in permeability (Figures 5A and 5B). Again, this phenomenon relates to depositional energy. If the inferred depositional energy is high, the unit tends to be better sorted, has a higher percentage of sandstone, and has a higher average permeability. If the inferred depositional energy is variable, the unit is poorly sorted, more heterolithic, and has a lower average permeability.

These relationships between geologic parameters and permeability are dependent on the assumption that diagenesis has not obliterated or destroyed the primary grain size, fabric, and sorting (Dutton and Willis, 1998). In initial petrographic studies from a U.S. Department of Energy/Utah Geological Survey Ferron project (Chidsey, 2001; Jarrard et al., this volume), it appears that permeability is still a function of primary fabrics and textures. Even where secondary porosity is developed, original cements that filled primary porosity have dissolved, thus preserving the relationship of permeability and relative energy.

Comparison of Fluvial-Dominated and Wave-Modified Intervals

Marked differences in the depositional conditions of the fluvial-dominated Kf-1-Iv[a] bedset and the wave-modified Kf-2 parasequence set influence the range and distribution of permeability. The fluvial-dominated Kf-1-Iv[a] exhibits lower overall permeability values (geometric mean = 3.1 md). In contrast, wave action reworked the delta front of the Kf-2 to sort the sediments and improve the overall permeability of the deposits (geometric mean = 7.1 md). Increased sorting also has the effect of decreasing the variability of the permeability values and increasing the average grain size of the reservoir facies. The variability in geologic parameters and permeability within a facies or individual bedform can quantify the heterogeneity of these fluvial-deltaic deposits.

The primary potential reservoir facies are the clinoform proximal (geometric mean = 8.3 md) facies in the Kf-1-Iv[a] and the distributary mouthbar (geometric mean = 68 md) and upper shoreface (geometric mean = 57 md) facies in the Kf-2. Secondary reservoir facies include the clinoform medial (geometric mean = 3 md) facies of the Kf-1-Iv[a] and the middle shoreface (geometric mean = 5 md) and foreshore (geometric mean = 146 md) facies of the Kf-2. A cumulative grain-size plot for the reservoir facies (Figure 7C) shows similar curves for the clinoform proximal and upper shoreface facies,

but the permeability values for the clinoform proximal are shifted to slightly lower values (Figure 7D). Likewise, the clinoform medial and middle shoreface grain-size curves are similar, but the clinoform medial facies exhibits slightly lower permeability values. This result reinforces the hypothesis that average grain size is not the sole influence on permeability; sorting and textural maturity are also significant factors. The wave modification of the Kf-2 likely has the effect of improving sorting and increasing the percentage of sandstone in the facies that in turn raises the permeability.

The overall shape and distribution of cumulative probability curves of permeability versus geologic parameters (e.g., facies) shown in these Ferron outcrop studies can be applied to, analogous subsurface reservoirs, although the absolute values should be evaluated for each individual case (Barton, 1994). The suite of permeability curves would likely shift depending on the average grain size, lithologic composition, and diagenesis of the reservoir facies.

LITHOFACIES CROSS SECTIONS AND PERMEABILITY OBSERVATIONS

Scaled photomosaic panels provided a deterministic framework for two apparent-dip, lithofacies cross sections for the fluvial-dominated Kf-1-Iv[a] (Figures 11A and 11B). These cross sections were based on ten stratigraphic sections along the Ivie Creek amphitheater and one stratigraphic section along Quitchupah Canyon. The clinoform facies were interpreted from observations of lithology, sedimentary structure, average grain size, apparent inclination angle, and position within a clinoform body. Permeability was not used in the facies definition process, but the permeability transects were shown relative to the architectural framework in Figure 11A. Average grain size and sedimentary structures measured in the stratigraphic sections were combined with the interpreted lithofacies cross section for the Ivie Creek amphitheater to create cross sections of these attributes (Figures 12A and 12B).

The southeastern portion of the lithofacies sections contains the highest percentage of clinoform proximal facies. The clinoform facies prograde toward the north and west into medial and distal facies. Visual comparison of the permeability transects to the lithofacies section (Figure 11A), shows lateral permeability variations within each facies. Permeability in clinoform proximal facies can vary by more than an order of magnitude over a lateral extent of 30.5 m (100 ft). In contrast, permeabilities in clinoform medial and distal facies vary by about a factor of two over a comparable distance.

Permeability correlation length, in the vertical direction, can be modeled with an aggregate semivariogram. A semivariogram measures spatial information in either

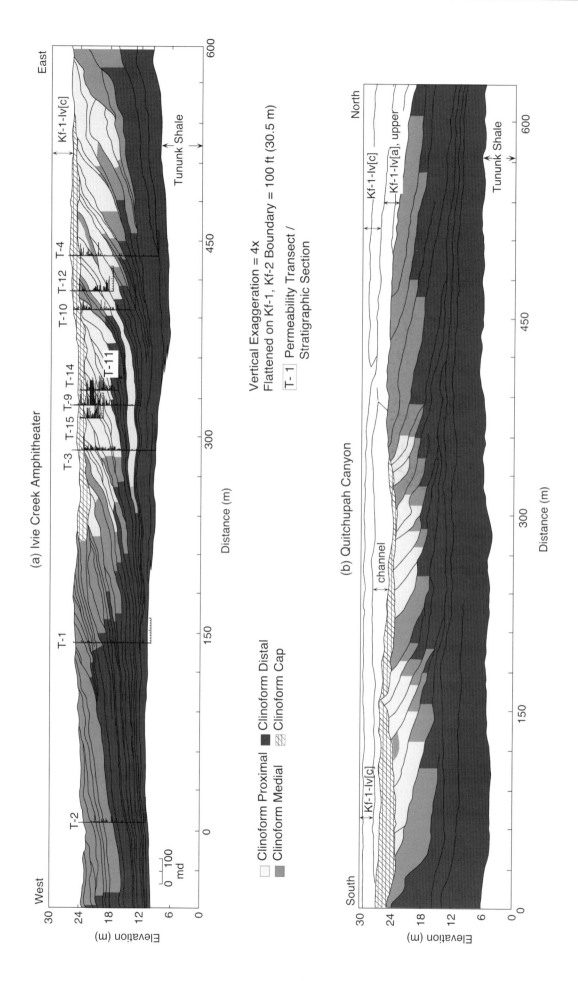

(a) Ivie Creek Amphitheater

Vertical Exaggeration = 4x
Flattened on Kf-1, Kf-2 Boundary = 100 ft (30.5 m)

T-1 Permeability Transect / Stratigraphic Section

Clinoform Proximal ▪ Clinoform Distal
▨ Clinoform Medial ▨ Clinoform Cap

(b) Quitchupah Canyon

Figure 11. Kf-1-Iv[a] clinoform facies distributions: (A) Ivie Creek amphitheater, and (B) Quitchupah Canyon. Horizontal distances relative to 3-D detail model (see Figures 14 and 15).

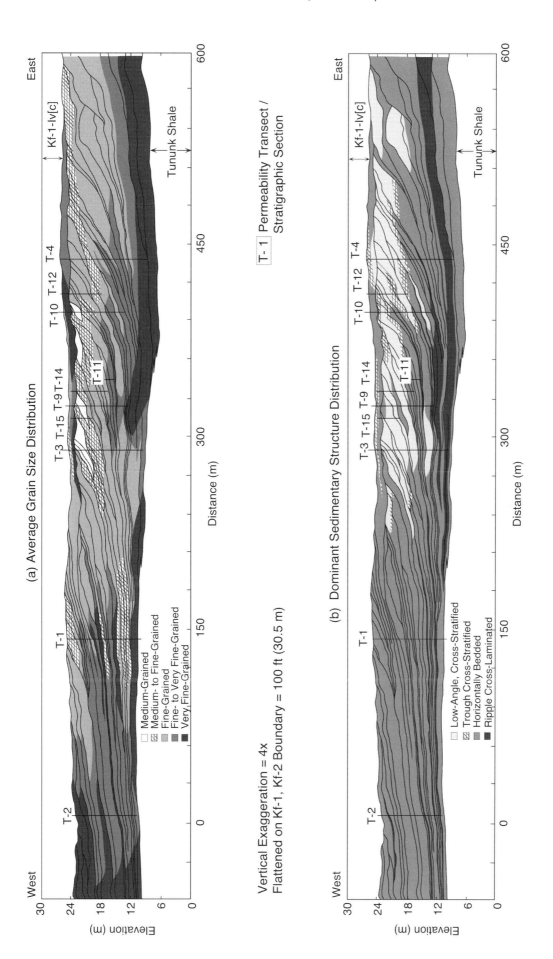

Figure 12. Kf-1-Iv[a] lithofacies maps, Ivie Creek amphitheater: (A) average grain size, and (B) dominant sedimentary structure (see Figure 11 for permeability transects). Horizontal distances relative to 3-D detail model (see Figures 14 and 15).

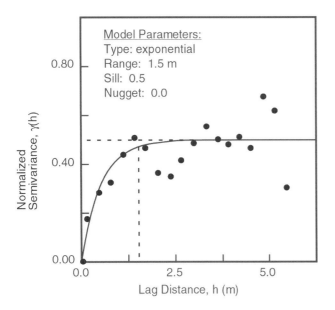

Figure 13. Vertical aggregate semivariogram of permeability for the fluvial-dominated, Kf-1-Iv[a] bedset.

an omni-directional or a uni-directional sense based on the offset or lag between data points (Isaaks and Strivastava, 1989).

$$\gamma(h) = [1/(2N)] \Sigma (x_{i+h} - x_i)^2 \qquad (1)$$

The semivariance (1) for a given lag distance is $\gamma(h)$, where h is the lag distance, N is the number of lags, and x is a data point. The results are plotted with semivariance as a function of lag and are fitted with a model curve (Figure 13). The range is the lag distance over which data points correlate and the sill is the semivariance value where the variogram flattens out. The sill is roughly equal to the variance of the dataset. A third parameter, the nugget, is the y-axis intercept and measures variability at small offsets. The semivariogram in Figure 13 references all permeability transects for the Kf-1-Iv[a] bedset to a datum on the top of the Kf-1 parasequence set and then computes one experimental variogram. An exponential model is fit to the experimental data and indicates a vertical correlation length of 1.5 m (5 ft), a sill of 0.5, and a nugget of 0.0.

Three-Dimensional Clinoform Facies Model Construction

The detail model area of the fluvial-dominated Kf-1-Iv[a] bedset was defined as 610 x 610 m (2000 x 2000 ft) to simulate interwell scale facies variations as input to reservoir simulation (Figure 9B). Eight lithofacies cross sections were interpreted for this area and provided control for constructing the 3-D clinoform facies model. These sections included: two outcrop sections along Ivie Creek and Quitchupah Canyon, four cross sections around the perimeter of the model domain, and two diagonal cross sections through drill holes 5 and 9a.

Each of the vertical, lithofacies cross sections was initially discretized into 1.2 x 1.2 x 1.2 m (4 x 4 x 4 ft) blocks. Then five laterally adjacent blocks were grouped into 1.2 x 6.1 x 6.1 m (4 x 20 x 20 ft) blocks. In discretization, if a block contained more than 50% of a given lithofacies type, then that lithofacies was assigned to the whole block. Thin beds are lost when they are split between blocks and do not comprise 50% of either block (Fogg, 1989). Blocks, 1.2 m (4 ft) thick, were chosen to minimize the loss of thin beds and maximize number of blocks in the proposed model domain. The discretized cross sections for Ivie Creek and Quitchupah Canyon (Figures 14A and 14B) can be compared to the non-discretized sections (Figure 11). Although the individual clinoform bodies are not preserved in discretization, the overall progradational nature of the facies distribution is preserved. The block thickness (1.2 m [4 ft]) seems reasonable since this thickness is approximately the correlation range (1.5 m [5 ft]) of the vertical aggregate permeability semivariogram.

Facies data from the vertical sections were transferred to horizontal slices through the model domain. Facies interpolated between cross section control points were based on the depositional hypothesis that arcuate lobes of sediment were sourced from the east-southeast as shown in the Kf-1-Iv[a] isopach (Figure 9A; Anderson et al., this volume). Slices above and below each interpreted slice were checked for internal consistency, resulting in a coherent model volume.

Three-Dimensional Clinoform Facies Model Interpretation

Two layers from the model, 24.4 m (80 ft) and 18.3 m (60 ft) elevation, display the complexity of the facies distribution (Figures 15A and 15B). The model is most complex at ~ 20 m (65 ft) elevation, with the complexity decreasing toward the top (dominantly proximal and cap facies) and bottom (dominantly distal facies) of the model. The complexity of the interfingering facies is a function of rapid lateral variations within the clinoform bodies.

The 3-D facies model exhibits arcuate geometries that interfinger in a complex pattern. During production, fluids tend to flow toward the higher permeability facies found primarily at the higher elevations in the southeast corner of the model volume. The location of the high permeability facies implies that water injection wells should be located in the northwest, northeast, and southwest corners. A production well should be located in the southeast corner. Flow would be oriented up-dip along the clinoform bodies.

The resulting facies framework provides a realistic geologic model that can be used as an input to reservoir simulation. The relationship between facies and permeability (discussed earlier) establishes the framework to

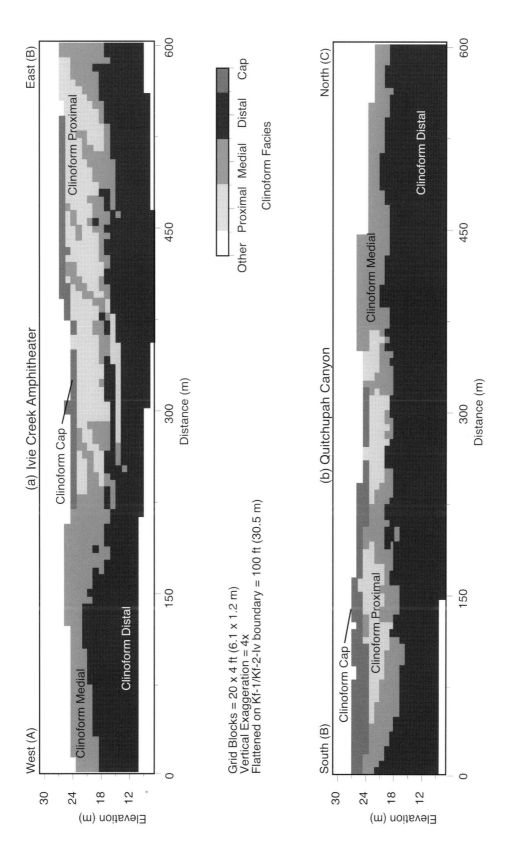

Figure 14. Kf-1-Tv[a], cross sections from 3-D facies model: (A) Ivie Creek amphitheater, and (B) Quitchupah Canyon. See Figure 15 for horizontal slices from model volume.

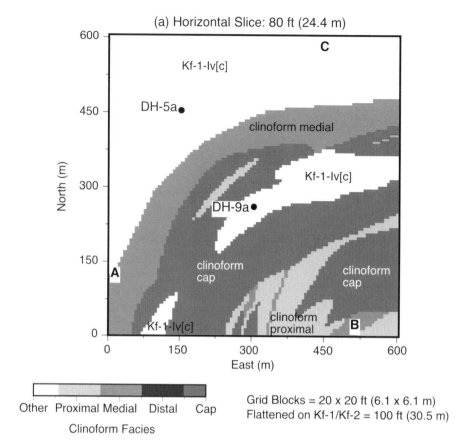

(a) Horizontal Slice: 80 ft (24.4 m)

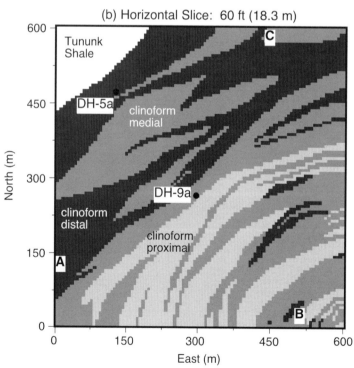

(b) Horizontal Slice: 60 ft (18.3 m)

Grid Blocks = 20 x 20 ft (6.1 x 6.1 m)
Flattened on Kf-1/Kf-2 = 100 ft (30.5 m)

Other Proximal Medial Distal Cap
Clinoform Facies

Figure 15. Kf-1-Iv[a], horizontal slices from 3-D facies model: (A) 24.4 m elevation, and (B) 18.3 m elevation. These elevations are relative to 30.5 m elevation at the top of the Kf-1 parasequence set. Points A, B, and C refer to locations on cross sections (see Figure 14).

create permeability structures to use in populating a geologic/facies model. Some measuring errors (e.g., scaling the photomosaics and locating the database in x/y space) can affect the model framework construction. Modeling uncertainties, such as kriging the parasequence boundaries and using a deterministic approach, result in less definite outcomes. However, potential model errors probably comprise less than 10–20% variability or error in the final simulation results. This detail geologic model incorporates photomosaic panels, stratigraphic sections, and drill-hole core data, and provides more input control to a reservoir simulation model than simply relying on isolated drill-hole core data. Although a high degree of detail may not be necessary for accurate reservoir predictions, an in-depth image of the subsurface provides guidelines for scaling up dimensions in reservoir simulations. Reservoir simulations are carried out in Forster et al. (this volume).

SUMMARY AND IMPLICATIONS FOR RESERVOIR SIMULATION

The evaluation of geologic parameters for two fluvial-deltaic depositional styles of the Cretaceous Ferron Sandstone indicates physical relationships that can be quantified for reservoir characterization. This study focused on two different scales of interpretation: the internal structure of a bedform (i.e., average grain size and sedimentary structure) and the geometry of a bedform (i.e., clinoform dimensions). Permeability shows a strong correlation to lithology, average grain size, and sorting as inferred from sedimentary structure. Although the rocks have undergone diagenesis, the original depositional relationships are preserved because secondary porosity developed where earlier cements were dissolved (Jarrard et al., this volume).

Tabular delta-front facies and geometries in medium- to fine-grained sandstones characterize the wave-modified Kf-2 parasequence set (geometric mean = 7.1 md). Potential reservoir facies are distributary mouth-bar, upper shoreface, foreshore, and middle shoreface. Higher permeability values in these facies are interpreted to be a function of strong wave energy and reworking that produced better sorting and coarser grain sizes.

The fluvial-dominated Kf-1-Iv[a] bedset is characterized by fine-grained sandstones divided into four clinoform facies: proximal, medial, distal, and cap. All the permeability measurements taken together have a geometric mean permeability of 3.1 md. Proximal and medial facies represent the best potential reservoir facies where low-angle, cross-stratification, and horizontal beds are the prominent preserved sedimentary structures.

Permeability can be approximated by a log-normal distribution with respect to facies, sedimentary structure, and average grain size. These relationships establish probability distribution functions that can be input into reservoir simulation. The probability functions could be applied to analogous delta systems, although the absolute values of the curves should be individually evaluated for each case study.

Clinoforms of the fluvial-dominated Kf-1-Iv[a] appear to produce partitioned reservoir facies that may be difficult to model and evaluate. Thin shale beds which commonly occur between clinoform bodies in the distal facies, would act as baffles and/or barriers to hydrocarbon production. In contrast, wave action in the Kf-2 produced tabular beds that form more simple targets for hydrocarbon production. Thus, an understanding of the depositional conditions, processes, resulting geometries, and sorting of each deltaic system can be quantified in a model designed to optimize hydrocarbon recovery.

The Cretaceous Ferron outcrop and drill-hole core data set provides an excellent opportunity to quantify facies relationships and statistical measures of heterogeneity. This paper provides a basis for evaluating the relationships between geological parameters and permeability for other fluvial-deltaic parasequences. The synthesis of the large database, including both field measurements (e.g., lithology, facies, average grain size, and sedimentary structures) and lab measurements (e.g., permeability and microscopy), can be used to evaluate permeability. Permeability control from average grain size and relative sorting is an important function of depositional facies and processes and should be built into reservoir simulation models.

ACKNOWLEDGMENTS

We gratefully acknowledge the support of the numerous investigators, participants and partners in this comprehensive, integrated Ferron project, especially Roy D. Adams, Paul B. Anderson, Thomas C. Chidsey, Jr., Craig B. Forster, Thomas A. Ryer, and Stephen H. Snelgrove. This project was contracted and managed by the Utah Geological Survey (contract no. 94-2517), and funded through the U.S. Department of Energy, National Petroleum Technology Office, Tulsa, Oklahoma (contract no. DE-AC22-93BC14896). Mobil Exploration and Producing Technical Center supplied the electronic miniprobe permeameters (EMP). (RC)[2], Reservoir Characterization and Research Consulting provided geostatistical programs used for part of the data analysis. Forrest M. Terrell, Wexpro, Salt Lake City, Utah, and Thomas Brill, Utah Department of Natural Resources, Salt Lake City, Utah, provided thoughtful reviews that improved this manuscript.

REFERENCES CITED

Anderson, P. B., K. McClure, T. C. Chidsey, Jr., T. A. Ryer, T. H. Morris, T. A. Dewey, Jr., and R. D. Adams, 2003, Interpreted regional photomosaics and cross section, Cretaceous Ferron Sandstone, east-central Utah: Utah Geological Survey Open-File Report 412, compact disc.

Armstrong, R. L., 1968, Sevier orogenic belt in Nevada and Utah: Geological Society of America Bulletin, v. 79, p. 429–458.

Barton, M. D., 1994, Outcrop characterization of architecture and permeability structure in fluvial deltaic sandstones, Cretaceous Ferron Sandstone, Utah: Ph.D. dissertation, University of Texas at Austin, Austin, 259 p.

Begg, S. H., A. Kay, E. R. Gustason, and P. F. Angert, 1994, Characterization of complex fluvial deltaic reservoir for simulation: 69th Annual Technical Conference and Exhibition of the Society of Petroleum Engineers, SPE Paper 28398, p. 375–384.

Chidsey, T. C., Jr., compiler and editor, 2001, Geological and petrophysical characterization of the Ferron Sandstone for 3-D simulation of a fluvial-deltaic reservoir — final report: U.S. Department of Energy, DOE/BC/4896-24, 471 p.

Coleman, J. M., and L. D. Wright, 1975, Modern river deltas — variability of processes and sand bodies, in M. L. Broussard, ed., Deltas: Houston Geologic Society, p. 99–150.

Conover, W. J., 1980, Practical nonparametric statistics: New York, John Wiley & Sons, 493 p.

Doyle, J. D., and M. L. Sweet, 1995, Three-dimensional distribution of lithofacies, bounding surfaces, porosity, and permeability in a fluvial sandstone-Gypsy Sandstone of Northern Oklahoma: AAPG Bulletin, v. 79, p. 70–96.

Dutton, S. P., and B. J. Willis, 1998, Comparison of outcrop and subsurface sandstone permeability distribution, lower Cretaceous Fall River Formation, South Dakota and Wyoming: Journal of Sedimentary Research, v. 68, no. 5, p. 890–900.

Eschard, R., P. Lemouzy, C. Bacchiana, G. De´saubliaux, J. Parpant, and B. Smart, 1998, Combining sequence stratigraphy, geostatistical simulations, and production data for modeling a fluvial reservoir in the Chaunoy field (Triassic, France): AAPG Bulletin, v. 82, p. 545–568.

Fisher, R. S., M. D. Barton, and N. Tyler, 1993a, Quantifying reservoir heterogeneity through outcrop characterization — 1. architecture, lithology, and permeability distribution of a landward-stepping fluvial-deltaic sequence, Ferron Sandstone (Cretaceous), central Utah: Topical Report GRI-93-0022 for the Gas Research Institute, 132 p.

Fisher, R. S., M. D. Barton, and N. Tyler, 1993b, Quantifying reservoir heterogeneity through outcrop characterization — 2. architecture, lithology, and permeability distribution of a seaward-stepping fluvial-deltaic sequence, Ferron Sandstone (Cretaceous), central Utah: Topical Report GRI-93-0023 for the Gas Research Institute, 83 p.

Fisher, R. S., N. Tyler, M. D. Barton, M. A. Miller, K. Sepehrnoori, J. Holder, K. E. Gray, 1992, Quantification of flow unit and bounding element properties and geometries, Ferron Sandstone, Utah — implications for heterogeneity in Gulf Coast Tertiary deltaic reservoirs: Annual Report for the Gas Research Institute, GRI-92-0072, 160 p.

Fisher, W. L., L. F. Brown, Jr., A. J. Scott, J. H. McGowen, 1969, Delta systems in the exploration for oil and gas: University of Texas, Austin, Bureau of Economic Geology, 78 p.

Fogg, G. E., 1989, Stochastic analysis of aquifer interconnectedness: Wilcox Group, Trawick area, east Texas: Bureau of Economic Geology, The University of Texas at Austin, Report of Investigations No. 189, 68 p.

Gardner, M. H., 1995a, Tectonic and eustatic controls on the stratal architecture of mid-Cretaceous stratigraphic sequences, central Western Interior foreland basin of North America, in Stratigraphic evolution of foreland basins: SEPM Special Publication No. 52, p. 243–281.

—1995b, The stratigraphic hierarchy and tectonic history of the mid-Cretaceous foreland basin of central Utah, in Stratigraphic evolution of foreland basins: SEPM Special Publication No. 52, p. 283–303.

Isaaks, E. H., and R. M. Srivastava, 1989, An introduction to applied geostatistics: New York, Oxford University Press, 561 p.

Liu, K., P. Boult, S. Painter, and L. Paterson, 1996, Outcrop analog for sandy braided stream reservoirs — permeability patterns in the Triassic Hawkesbury Sandstone, Sydney basin, Australia: AAPG Bulletin, v. 80, p. 1850–1866.

Massari, F., 1996, Upper-flow-regime stratification types on steep-face, coarse-grained, Gilbert type progradational wedges (Pleistocene, Southern Italy): Journal of Sedimentary Research, v. 66, p. 364–375.

Mattson, A., 1997, Characterization, facies relationships, and architectural framework in a fluvial-deltaic sandstone, Cretaceous Ferron Sandstone, central Utah: Master's thesis, University of Utah, Salt Lake City, 174 p.

Rhea, L., M. Person, G. de Marsily, E. Ledoux, and A. Galli, 1994, Geostatistical models of secondary oil migration within heterogeneous carrier beds: a theoretical example: AAPG Bulletin, v. 78, p. 1679–1691.

Robinson, J. W., and P. J. McCabe, 1997, Sandstone-body and shale-body dimensions in a braided fluvial system — Salt Wash Sandstone Member (Morrison Formation), Garfield County, Utah: AAPG Bulletin, v. 81, p. 1267–1291.

Ryer, T. A., 1981, Deltaic coals of Ferron Sandstone Member of Mancos Shale — predictive model for Cretaceous coal-bearing strata of Western Interior: AAPG Bulletin, v. 65, p. 2323–2340.

Wescott, W. A., 1992, Deltaic provinces a major focus of world-wide exploration efforts: Oil and Gas Journal, June 15, p. 52–55.

C. T. Lupton's U. S. Geological Survey field camp near Castle Dale, Utah, circa 1910. Photograph courtesy of the family of C. T. Lupton.

Outcrop Case Studies

"Flood caused by one hour's rain. Stream bed absolutely dry before rain began" near Emery, Utah, circa 1910. Photograph courtesy of the family of C. T. Lupton.

Analog for Fluvial-Deltaic Reservoir Modeling: Ferron Sandstone of Utah
AAPG Studies in Geology 50
T. C. Chidsey, Jr., R. D. Adams, and T. H. Morris, editors

Sedimentology and Structure of Growth Faults at the Base of the Ferron Sandstone Member Along Muddy Creek, Utah

Janok P. Bhattacharya[1] and Russell K. Davies[2]

ABSTRACT

This paper describes normal growth faults at the base of the Ferron Sandstone exposed along the highly accessible walls of Muddy Creek Canyon in central Utah. Although there have been several studies of growth faults in outcrops this is the first that integrates detailed sedimentological measured sections with fault kinematics and section restorations. We measured 20 sedimentological sections and interpreted a photomosaic covering approximately 200 m (550 ft) lateral distance. The outcrop is oriented parallel to depositional dip and perpendicular to the general strike of the faults.

Distinctive pre-growth, growth, and post-growth strata indicate a highly river-dominated crevasse delta, that prograded northwest into a large embayment of the Ferron shoreline. The growth section comprises medium- to large-scale cross stratified sandstones deposited as upstream and downstream accreting mouth bars in the proximal delta front. Deposition of mouth bar sands initiates faults. Because depositional loci rapidly shift, there is no systematic landward or bayward migration of fault patterns. During later evolution of the delta, foundering of fault blocks creates an uneven sea-floor topography that is smoothed over by the last stage of deltaic progradation.

Faults occur within less than 10 m (30 ft) water depths in soft, wet sediment. Detailed examination of the fault zones shows that deformation was largely by soft-sediment mechanisms, such as grain rolling and by lubrication of liquefied muds, causing shale smears. Mechanical attenuation of thin beds occurs by displacement across multiple closely spaced small throw faults.

Analogous river-dominated deltaic subsurface reservoirs may be compartmentalized by growth faults, even in shallow-water, intracratonic, or shelf-perched highstand deltas. Reservoir compartmentalization would occur where thicker homogenous growth sandstones are placed against the muddy pre-growth strata and where faults are shale-smeared, and thus potentially sealing.

[1]Geosciences Department, University of Texas at Dallas, Richardson, Texas
[2]Rock Deformation Research, U.S.A. Inc., McKinney, Texas

INTRODUCTION

Synsedimentary normal faults, or growth faults, associated with deltas are involved in the formation of major traps for oil and gas reservoirs and they may isolate compartments in subsurface hydrocarbon reservoirs or aquifers (Busch, 1975; Galloway et al., 1982; Bishop et al., 1995; Diegel et al., 1995). Subsurface studies of growth faults typically are based on seismic data, which are useful for mapping and describing regional-scale geologic architecture of growth faults, but not as useful for imaging finer-scaled details. Well log and core data invariably alias smaller scale structures. Outcrops, in contrast, can provide complete information about the lateral variability of growth-faulted strata and fault geometries at a range of scales. Because of the self-similarity of many structural styles at different scales, knowledge about growth faults at the outcrop scale may be applied to regional-scale growth fault systems.

Although there have been several studies of growth faults in outcrops (Brown et al., 1973; Edwards, 1976; Rider, 1978; Elliot and Lapido, 1981), there are no studies that integrate detailed sedimentological measured sections with fault kinematics and section restorations. The best-studied examples include beautifully exposed, but rather inaccessible Carboniferous-age growth strata along the sea-cliffs of Western Ireland (Rider, 1978; Elliot and Lapido, 1981) and Triassic and Cretaceous sea-cliff exposures along the Norwegian coast (Edwards, 1976; Nemec et al., 1988).

This paper describes the detailed sedimentology and structure associated with normal growth faults at the base of the Ferron Sandstone exposed along the highly accessible walls of Muddy Creek Canyon in central Utah (Figures 1 and 2). A prior study of these faults suggested that fault displacement was accommodated by movement of underlying mobile prodelta muds (Nix, 1999; Morris and Nix, 2000). This earlier study interpreted the fault style and broad stratigraphic relationships from photomosaics, but did not incorporate any detailed sedimentological measured sections. Our study provides a more detailed description of the interrelationship between the stratigraphy and structure.

Regional Stratigraphy and Paleogeography

The Ferron Sandstone is a fluvio-deltaic clastic wedge deposited near the end of the Sevier Orogeny into a rapidly and asymmetrically subsiding foreland basin that rimmed the western margin of the Late Cretaceous (Turonian) seaway in central Utah (Gardner, 1995; Gardner, this volume). The Ferron consists of seven regressive-transgressive stratigraphic cycles, each bounded by a flooding surface and associated coals (Ryer, 1984; Gardner, 1995; Anderson and Ryer, this volume). In out-

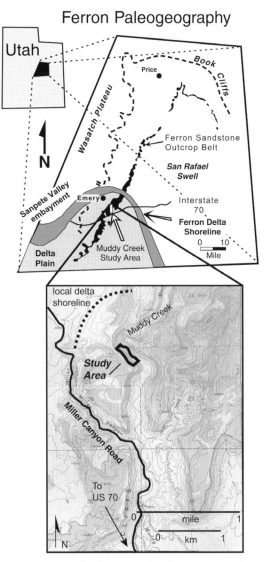

Figure 1. (A) Regional paleogeographic base map. In the Muddy Creek study area the delta lobe prograded locally to the northwest into the Sanpete Valley embayment (from Ryer and McPhillips, 1983). (B) Topographic inset with location of photomosaic along south cliff face of Muddy Creek as well as orientation of shoreline associated with northwest prograding bay-fill, crevasse delta lobe.

crop, the clastic wedges are exposed in sandstone cliffs above slope-forming mudstones (Figure 2).

Regionally, the Ferron is mapped as a large lobate body that prograded northwest, north, and northeast forming a large western embayment (the Sanpete Valley embayment in Figure 1) of the Cretaceous seaway (Ryer and McPhillips, 1983). Facies descriptions (see "stratigraphy and facies" section below) show that this bay experienced diminished wave activity compared to most of the Ferron Sandstone and was infilled with river-dominated delta lobes (Figure 3). This study focuses on one river-dominated delta lobe that built-out towards the northwest at a high angle to the regional progradation direction. This lobe may represent a large bay-fill crevasse delta associated with avulsion of the Ferron dis-

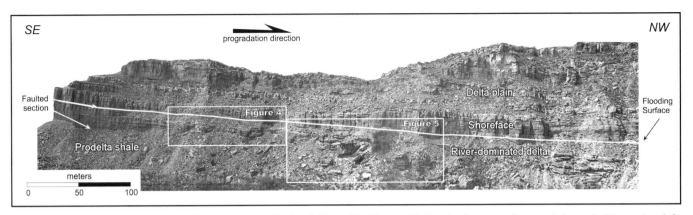

Figure 2. Panoramic photo of south cliff-face along Muddy Creek (located in Figure 1B) showing location of interpreted panels (Figure 4 — left; Figure 5 — right). Northwest dipping beds in unfaulted strata at the southeast end of the outcrop reflect inclined delta front sandstone beds. Closeup of this area is shown in Figure 8. Color version on CD-ROM.

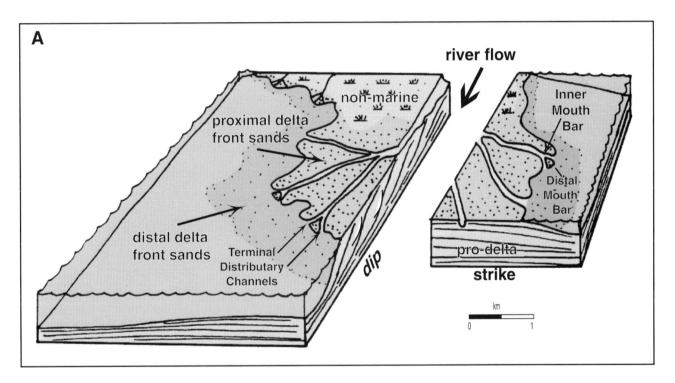

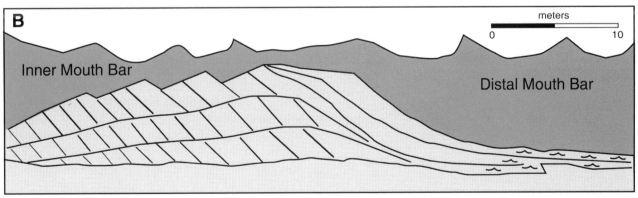

Figure 3. (A) Block diagram of a river-dominated, shoal-water delta showing position of inner and distal (outer) mouth bar deposits associated with the delta front. Lobate geometry reflects river dominance. (B) Schematic cross section of a distributary mouth bar, based on the Ferron example. Inner-mouth bar is characterized by movement of subaqueous dune-scale bedforms. Decreasing flow velocities cause upstream accretion of the bar. At the distal end of the bar, sands flow down the bar foreset and are locally reworked by waves. Color version on CD-ROM.

tributary channel system into this low area (see Moiola et al., this volume). We use the terms proximal and distal to refer to the southeast and northwest position associated with the landward and bayward progradation of this crevasse delta, rather than the more general southwest to northeast trend associated with progradation of the entire Ferron clastic wedge.

The lower portion of the Ferron has been associated with a minor drop of sea level within "short-term stratigraphic cycle 1" at the base of the "Ferronensis Sequence" of Gardner (1995). This sea-level drop may have forced the deposition of delta front and prodelta mudstones onto highly bioturbated older "shelf" mudstones of the underlying Tununk Shale Member of the Mancos Shale. Listric normal faults sole into this shale and growth of the section across the faults seems to occur exclusively within the delta front sands. It is the faults in this lower section that were mapped and are described in this paper.

Study Area

The mapped set of growth faults are well exposed on the southwest face of a 50 m (160 ft) high cliff along Muddy Creek canyon (Figures 1 and 2) and cover a 200 m (650 ft) lateral section near the base of the cliff. The faults are restricted to a 15–20 m (50–65 ft) section of prograding delta sands above the mud-rich pro-delta Tununk shale and below a flooding surface that separates the faulted delta sands from the thick overlying unfaulted strata of Ferron stratigraphic cycle 2. Faults are also exposed across the valley in the northeast cliff-face and along other meanders in the Muddy Creek Canyon and in adjacent creeks. This study focuses only on the faults in the southwest cliff of Muddy Creek.

Although the faults form a continuous set of structures across the face of the outcrop, for convenience we separate the description and discussion into a distal and proximal fault exposure. The proximal fault exposure extends from the southeast, unfaulted landward extension of the delta complex towards the central northwest bayward prograding and faulted section (Figure 2). The distal fault exposure extends from the center of the mapped fault set to the most bayward-mapped exposure (Figure 2). The exposure in both the proximal and distal fault sets is excellent over the thick clastic growth sections, but poor or covered in the underlying heterolithic more mud-rich section.

Methodology

We measured 20 sedimentological sections and interpreted a photomosaic covering the 200 m (650 ft) lateral distance over the exposed outcrop (Figures 4 and 5). Paleocurrent data and fault strike and dip data show that the cliff face studied is oriented parallel to the depositional dip of the delta lobe and perpendicular to the general strike of the growth faults (Figure 6).

Integrating facies and structure, we documented offset of specific beds across the faults. Distinctive facies geometry allowed us to determine the pre-growth, growth, and post-growth stratigraphy. These distinctive beds are characterized by specific grain sizes, sedimentary structures and stacking order, and are designated with different colors (see compact disc) in the measured sections (Figures 4C and 5C) and structural interpretation (Figures 4B and 5B). For example, at 6 m (20 ft) in Section 1 (designated with an A in Figure 4C and colored yellow in digital version) there is a distinctive bed, about 25 cm (10 in.) thick, that consists of current-rippled sandstone (base of Figure 7). This ripple bed is overlain by mudstones followed by a flat-stratified sandstone, which is overlain by about 10 cm (4 in.) of deformed sandstone, and mudstone (Figures 4C and 7). These distinctive beds, along with others (e.g. bed B) below, could be identified and correlated throughout most of the outcrop face, as illustrated in Figure 4, although it was more difficult to find these beds in the distal fault set because of poorer exposure of the pre-growth section (Figure 5). The correlations and interpretations were used as the basis to constrain the structural geometry and models for restoration of each major fault.

STRATIGRAPHY AND FACIES

Across the outcrop the distinctive pre, growth, and post-growth strata were described (Figures 4 and 5). The pre-growth section is a mud-rich heterolithic section interpreted as delta front mudstones and sandstones, which contrasts with the growth section of nearly homogeneous interbedded bars and delta front sandstones and shallow distributary channels and bars that are expanded across the faults. A transgressive flooding surface separates the growth section from the overlying strata of Ferron stratigraphic cycle 2 (Gardner, 1995).

At the broadest mapped scale, the delta front sandstones form an upward coarsening facies succession (Figures 4C and 5C) suggesting progradation. Bayward-dipping sandstones in the unfaulted correlative growth succession at the SE end of the outcrop, as shown in an oblique photo (Figure 8), are interpreted as foreset strata associated with the prograding delta lobe. The thickness of the upwards coarsening succession (Figure 9) and height of the clinoform beds approximates water depth, and suggests the pre-growth strata were deposited in a maximum of about 25 m (80 ft) water depth. The homogenous growth sands lie at the top of the growth section, following infilling of this space, and were thus likely deposited in less than a few meters water depth. The stratigraphy and facies are described and interpreted in detail below.

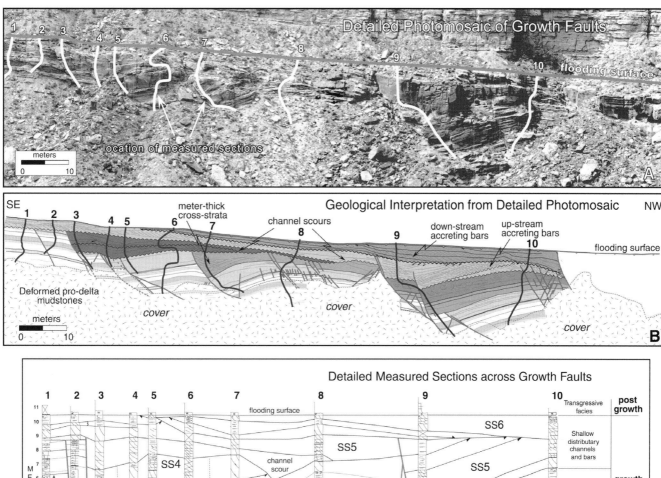

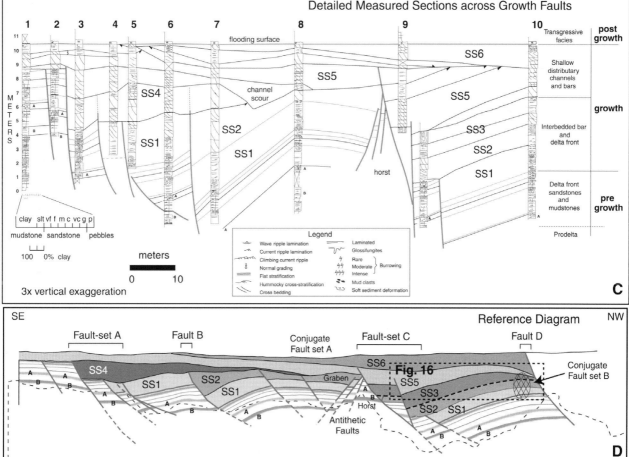

Figure 4. Cross section and interpretation of proximal exposure of growth faults along Muddy Creek showing: (A) detailed photomosaic, (B) geological interpretation of structure, (C) detailed measured sections, and (D) reference diagram. Lettered beds A and B (color version on CD-ROM) are matched to the sands in the measured section and show offset on faults. The growth interval consists of upstream and downstream accreting cross-bedded sandstones deposited in shallow distributary channels and proximal distributary mouth bars. The relative ages of these sands are indicated with numbers SS1 to SS6 from oldest to youngest, respectively. Location of panel shown in Figure 2.

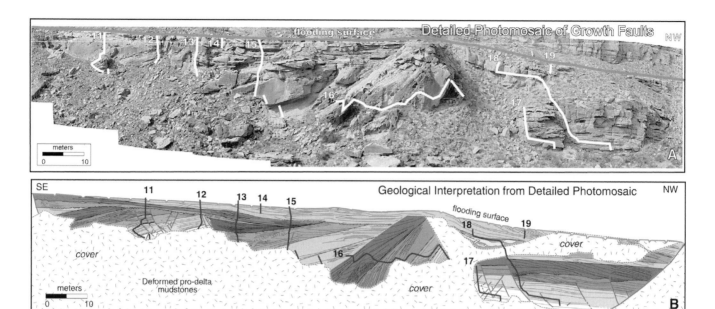

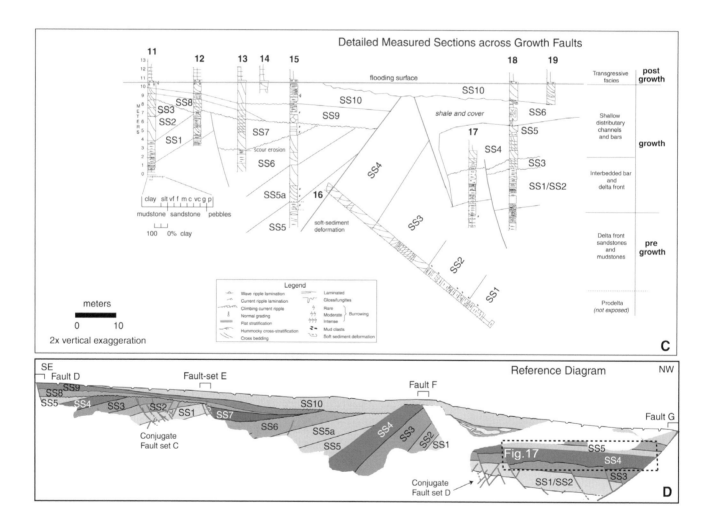

Figure 5. Cross section and interpretation of distal exposure of growth faults along Muddy Creek showing: (A) detailed photomosaic, (B) geological interpretation of structure, (C) detailed measured sections, and (D) reference diagram. Shaded beds (color version on CD-ROM) are matched to the sands in the measured section and show offset on faults. The growth interval consists of upstream and downstream accreting cross-bedded sandstones deposited in shallow distributary channels and proximal distributary mouth bars and larger-scale cross-stratified sandstones interpreted as distal-mouth bar foreset beds. The relative ages of these sands are indicated with numbers SS6 to SS10 from oldest to youngest, respectively. Location of panel shown in Figure 2.

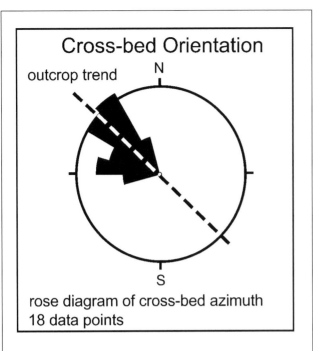

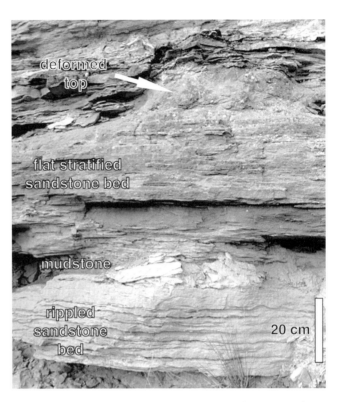

Figure 7. Photograph of key marker beds within the pre-growth section that were correlated across growth faults (see Figure 4). Lower bed consists of ripple-cross laminated sandstones followed by mudstones. Overlying sandstone consists of flat-stratified and deformed strata. Color version on CD-ROM.

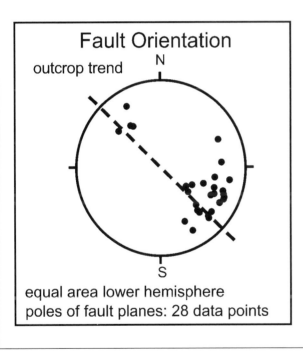

Figure 6. Stereonets of paleocurrents and poles to the planes of faults show that the Ferron delta lobe prograded to the northwest, parallel to the cliff face, and that the exposure is nearly perpendicular to the fault strike which affords a nearly perfect dip view.

Figure 8. Oblique view of top-truncated, inclined delta front sandstones at southeast end of outcrop. Color version on CD-ROM.

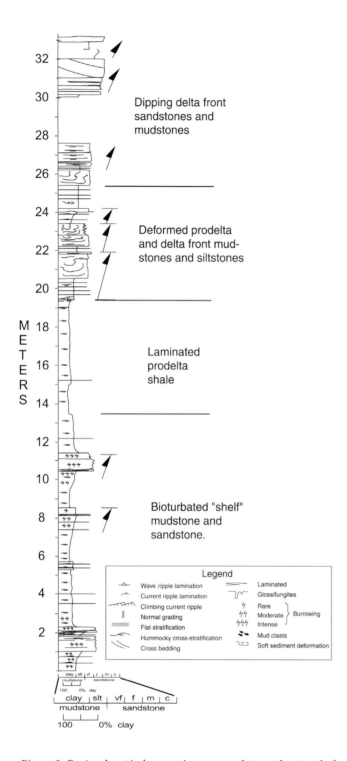

Pre-Growth Strata

Description

Pre-growth strata are sandstones interbedded with shales in Figures 4 and 5 (colored green, red and yellow in the digital version). The faults sole into the underlying facies, which are covered by rubble within the mapped outcrop, but are well exposed along the cliffs about 300 m (980 ft) to the south in a landward direction (Figure 9). There, the section consists of about 6 m (20 ft) of laminated silty mudstones with rare, very-fine grained current-ripple cross-laminated sandstone beds and ironstone nodules (Figure 10). This passes up into about 7 m (23 ft) of deformed interbedded mudstones and sandstones showing interstratal recumbent folding and loading (Figure 11). This deformed facies passes into interbedded decimeter-thick sandstones and mudstones (Figure 12). Sandstone beds are sharp-based, normally graded and show Bouma-like sequences passing from massive to horizontally stratified and capped by climbing current-ripple cross-lamination (Figure 13). Burrows are few but *Planolites*, *Skolithos*, and *Arenicolites* occur throughout the succession (Figures 12, 13, and 14) with rare small diameter (< 0.5 cm) *Thallasinoides*, *Ophiomorpha*, and *Rosselia*. Coalified plant debris is common on parting planes and rare larger woody clasts are bored by marine *Teredolites* (Figure 15).

Figure 10. Laminated mudstones of the pre-growth section. Note lack of burrows and numerous current-rippled sandstone inter-beds. Lack of burrowing reflects fluvial influence and rapid deposition of prodelta sediments. Photo taken at about 20 m (60 ft) in Figure 9. Color version on CD-ROM.

Figure 9. Regional vertical succession measured at southeast end of south cliff along Muddy Creek. Bioturbated mudstone and sandstones, below 13 m (39 ft) are interpreted as being deposited on an open marine shelf. Laminated un-bioturbated mudstones, above 13 m (39 ft), reflect increasing fluvial influence and formation of the Sanpete Valley embayment. Thick zones of deformed strata reflect instability of rapidly deposited prodelta and delta front sediments. The transgressive surface is at 32 m (96 ft). Thickness of the section above the bioturbated shelf facies suggests that the maximum water depth of the bay into which the delta prograded was about 25 m (75 ft). Color version on CD-ROM.

Figure 11. *Recumbent fold in interbedded sandstones and mudstones of the distal delta front. Photo taken from strata shown at 22 m (66 ft) in Figure 8. Color version on CD-ROM.*

Figure 12. *Sharp-based, flat-stratified sandstones with burrowed tops, interbedded with burrowed mudstones. Color version on CD-ROM.*

Figure 13. *Photo of graded, flat-stratified, to current rippled sandstones, with thin mudstone interbeds. These are interpreted to form as delta-front turbidites in the pre-growth strata and reflect fluvial dominance. Color version on CD-ROM.*

Figure 14. *Skolithos trace fossil within stratified sandstones.*

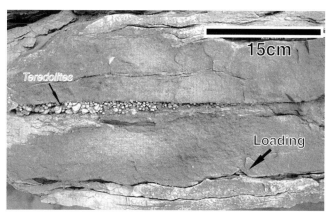

Figure 15. *Teredolites bored log forms "grapelike" bubbly looking unit at the base of upper sandstone bed. Coaly material surrounds the casts of the burrows. Sandstones are structureless to flat-laminated to current rippled. Note flame-structures and load casts at the base of the lower sandstone bed. Color version on CD-ROM.*

Interpretation

The low diversity, low abundance, and small diameter of trace fossils suggests a stressed and probably brackish-water environment that was likely proximal to a river mouth (Moslow and Pemberton, 1988; Bhattacharya and Walker, 1992; Gingras et al., 1998). The climbing ripples, lack of burrowing, and abundant load casts and soft-sediment deformation suggest high sedimentation rates that are typical of deposition in a river-dominated prodelta environment (Moslow and Pemberton, 1988; Bhattacharya and Walker, 1992; Gingras et al., 1998). The loading indicates that the prodelta muds were less dense than overlying more-dense sandstones. Density differences result from the fact that rapidly deposited muds typically have much higher porosity than overlying sands (Rider, 1978). The folding in the muds is interpreted to indicate flow of the early-deposited mudstones and we hypothesize that this basal layer accommodated the displacement at the base of the faults cutting the younger section, consistent with the interpretation of Nix (1999) and Morris and Nix (2000). Dewatering features, such as pipes and flame-structures associated with small-scale loading (Figure 15), indicate that the prodelta sediments were waterlogged at the time of deformation and had not experienced significant compaction.

Growth Strata

Description

The growth section forms as a series of 10 offlapping sandstone wedges, labeled SS1-SS10 from oldest to youngest, respectively, in Figures 4 and 5. These wedges consist mostly of fine- to medium-grained cross-bedded

sandstones, 2–9 m (7–30 ft) thick, with meter-scale truncation and erosion between cross sets. Paleocurrent directions are strongly unimodal and indicate flow to the northwest (Figure 6). Cross-bedding is also locally organized into climbing co-sets (e.g. between sections 9 and 10, Figure 4 and expanded version in Figure 16). Locally, within individual fault blocks, cross-sets decrease in thickness from 2 m (7 ft) to a few decimeters away from the faults (Figure 16).

At the scale of the entire exposure, the offlapping organization of the growth sands also shows that cross-sets step seaward. In the most distal portion of SS4 (Figures 5, 17) similar medium-scale cross beds pass seaward into a single thick set of inclined strata. These thick single cross-stratified sandstones have a distinctly sigmoidal shape (Figure 18). Much less mudstone was observed in the growth section compared to the pre-growth section, although the sigmoidal cross-stratified sandstones can ultimately be traced into heterolithic facies in more distal exposures of the delta lobe. As an example, SS1 sandstone at the base of measured sections 4 and 5, in Figure 4C correlate with the more heterolithic strata of SS1 in measured section 10 at the base of the growth section.

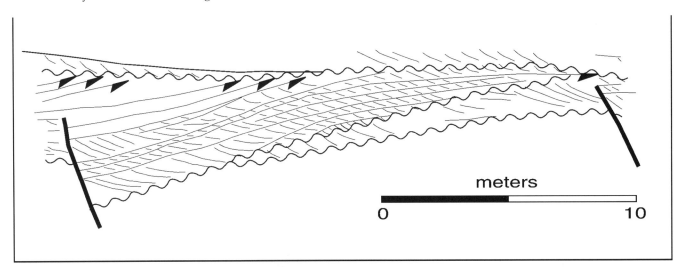

Figure 16. Upstream accreting cross-strata interpreted to form on the inner margin of a distributary-mouth bar. Closeup of bedding diagram of SS5, between measured sections 9 and 10 in Figure 4B.

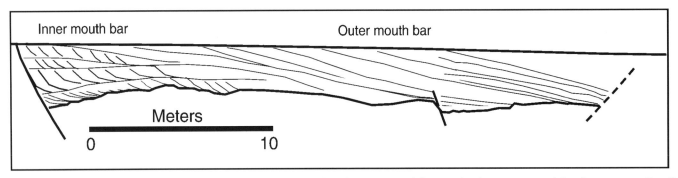

Figure 17. Bedding diagram showing dune-scale cross-strata, formed in the inner-mouth bar, passing into outer mouth-bar foreset strata. Detail is of SS4, immediately northwest of measured section 18 in Figure 5B.

Figure 18. Large-scale sigmoidal cross-strata, interpreted to form as foresets on the prograding distal, outer-mouth bar. Color version on CD-ROM.

In places, the larger scale sigmoidal strata are draped by mudstones, as seen in SS10 at 1 m in section 19 (Figure 5C). These cross-stratal units also locally show 2–3 m-thick (6–9 ft) upward coarsening successions. In places a gradational contact with underlying mudstones is seen (e.g. section 15 in Figure 19). Locally the large-scale cross-strata pass from massive medium-grained sandstones, such as from 8–9 m (24–36 ft) in section 13 (Figure 19) bayward into finer-grained ripple cross-laminated sandstones interbedded with thin muds (section 15, Figure 19). The ripple cross-laminations are organized into spectacular near-vertical climbing sets (Figure 20) suggesting very high sediment aggradation. Some of the ripples look symmetrical in plan view indicating some wave-activity during deposition.

Locally growth sandstones also show extensive soft-sediment deformation and dewatering structures, such

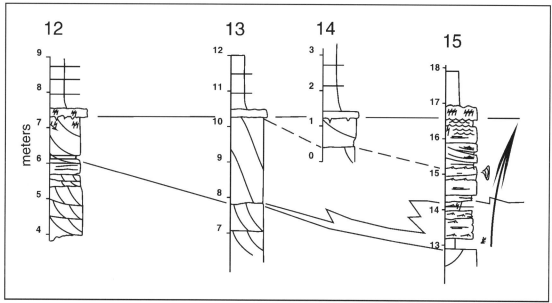

Figure 19. Closeup of SS11 in sections 12–14 (Figure 5) showing passage from sandstone into muddier strata. Two-m (6-ft) thick sets of inclined strata form as foresets on the prograding distal margin of a mouth bar. Distal deposits in section 15 show near-vertical climbing ripples and display a coarsening-upward facies succession.

Figure 20. Vertically climbing symmetrical ripples formed in a distal-mouth bar. Note edge of the rock hammer for scale at right side of photo.

as dish structures and pipes (see sandstone SS3, between sections 9 and 10, Figure 4C).

Interpretation

The medium-scale cross-strata represent shallow-water, migrating subaqeous dunes. The truncation associated with the dunes suggests that they formed within shallow (a few meters deep) distributary channels in the proximal delta front (channel scours are labeled in Figures 4B and C). Decreased preservation of cross-strata within the hanging walls suggests that the faults were moving during deposition of the cross-bedded sands. The climbing cross-sets possibly indicate high sand aggradation rates and upstream accretion of bars. Alternatively, cross-sets might step landward as rotation on the hanging wall increases towards the fault. The landward-stepping and seaward-stepping cross-strata show that both upstream and downstream accretion of sands has occurred, as is seen in modern river-dominated deltas (Van Heerden and Roberts, 1988). The organization of cross-strata suggests that fault movement and sand deposition were roughly synchronous and reflects an intimate interplay between sedimentation and the formation of faults.

The larger sigmoidal features are interpreted as the topset, foreset and bottomsets of prograding distributary-mouth bars, rather than the smaller-scale dune bedforms within distributary channels (Figures 18, 3B). The thickness of the mouth bars likely approximates water depth and suggests depths of a few meters. The smaller-scale cross-strata are interpreted to lie at the inner margin of the delta front within the region confined by the most distal, "terminal" distributary channels (Figure 3). The large-scale sigmoidal cross-strata represent the region of unconfined flow at the bayward (distal) margin of a distributary-mouth bar and can be identified throughout the cross sections (Figures 4, 5).

In the distal fault exposure (Figure 5), units SS8 to SS10 contain a greater proportion of prodelta mud and prograde over a complex topography caused by foundering of the underlying growth strata. This mud may have been injected along Fault E, but there is also evidence that prodelta muds were deposited along with sandstones SS8-SS10. This style of deformation and sedimentation is somewhat more similar to the style of synsedimentary faulting described by Nemec et al. (1988) and by Pulham (1993) in which failure of the delta front and proximal slope occurs first forming a complex seafloor topography that is then later infilled or "healed" with prodelta and delta front sediments.

The thickness of these younger offlapping, upward-coarsening facies successions, also likely approximates water depth, and suggest the submarine topography was only on the scale of a few meters to 10 m (30 ft) deep. The soft-sediment deformation and estimation of water

Figure 21. Bioturbated coarse sandstone with Skolithos and Ophiomorpha burrows, overlying cross-stratified sandstones. This facies is interpreted to mark the transgressive lag across the top of the drowned delta lobe and can be traced across the top of the cliff. Color version on CD-ROM.

depths from facies shows that all of the growth faulting occurred in less than about 10 m (30 ft) water depth, while sediments were still waterlogged and before significant compaction had occurred.

Post-Growth Strata

The overall growth succession is truncated by a decimeter-thick bed of bioturbated sandstone containing abundant centimeter-diameter Ophiomorpha and Skolithos burrows (Figures 4, 5, and 21). This bed is in turn overlain by marine shales at the base of the next upward-coarsening succession in stratigraphic cycle 2 of Gardner (1995). No mapped faults persist above this bioturbated unit showing that growth faulting ceased with the transgression of the lobe and deposition of this thin sandstone.

STRUCTURE

The predominant structural style across the entire mapped outcrop is seaward (northwest) dipping listric normal faults with extensive growth of the strata into the hanging wall of the faults. The faults curve gently along their length as dip decreases and terminate in a mud-rich prodelta section underlying the prograding sands. The upper tips of the faults terminate near or below the overlying flooding surface that separates the faulted growth section from the unfaulted overlying strata of Ferron stratigraphic cycle 2. The upper portions of the faults dip at an angle of approximately 70°. The basal fault terminations are covered, however, preventing a complete description of their lower geometry.

A set of regularly spaced individual faults and fault sets separate the growth sections along the exposure. These are labeled as Fault or Fault-set A to F in Figures 4

290

and 5. A fault-set is two or more closely spaced faults that together bound an extended growth section in their composite hanging wall.

From southeast to northwest, or from a landward to bayward direction, the facies architecture shows a thickening of the nearly homogeneous cross-bedded sandstones across the outcrop. Clearly, the thickening is structurally controlled with dramatic changes across the faults, but the thinner more proximal growth section (Figure 4) exposes the underlying heterolithic pre-growth section, which is nearly completely covered in the distal location. The unique facies within the pre-growth section are more easily correlated and restored across the faults in the thinner proximal mapped exposure (Figure 4) than in the thick cross-stratified sands that comprise nearly the entire distal mapped exposure (Figure 5).

Proximal Exposure

The proximal exposure is divided into three major fault blocks separated by Fault-set A, Fault B, and Fault-set C (Figure 4D). The pre-growth section is exposed well enough across the proximal exposure to measure the throw across the faults. The composite throw, totaled across all of the faults in the proximal exposure, is approximately 13 m (40 ft), although the throw across any individual fault is at most a few meters. This section is folded as it is displaced along the curved growth faults. The shallowest prograding sandstones overlying the strongly folded section are less strongly deformed and they show a distinct, offlapping, prograding stratal geometry. The pre-growth section, as described above, is a heterolithic section of thinly bedded sand and shales (Figures 10 through 13). The growth section is a nearly homogeneous section of cross-bedded sandstones. Shales occur in the growth section only locally as discussed above.

Fault-Set A

The first major growth section in a landward direction in the proximal exposure is bound by a set of three faults (Fault-set A) across a zone 10 m (30 ft) wide. The faults step down into the section from a proximal to distal position. The most proximal and shallowest fault in the set offsets a wedge of cross-bedded sandstone (SS4) that is thickest against the fault and thins in a bayward position, pinching-out landward of Fault-set C (Figure 4). The upper fault tip terminates at the bed surface separating SS4 from the overlying prograding sand (SS5) that thickens across Fault-set C in a more distal location. The gentle curvature of the underlying bedding surface suggests that the deeper section rode passively with the stratigraphic growth against this fault.

The other two more distal faults in Fault-set A are older faults with a shallower dip than the proximal fault in the set indicating that they have also rotated with stratigraphic growth and displacement across the youngest fault in the set. This fault pair terminates at the bed surface between SS1 and SS4. Although there is large differential offset across these faults deeper in the section, the mapped displacement across the base of SS1 shallow in the section is small across the more distal fault in the set. The two distal segments in Fault-set A most likely represent the overlapping tip regions of two contemporaneous faults forming a transfer zone or relay ramp (Peacock and Sanderson, 1991, 1994) across which throw is transferred. The measured throw across Fault-set A in the pre-growth section is 8 m (24 ft).

Fault B

A single fault (Fault B) cuts the section approximately 15 m (45 ft) seaward from the outermost fault in Fault-set A (Figure 4). The fault is strongly curved along its length, which is consistent with the strong curvature of the beds in the hanging wall. Fault B terminates against the base of SS4 like the set of faults in the hanging wall of Fault-set A, but cuts a thicker cross-bedded sand growth section than the previous set. The cross-bedded sand (SS2) comprises the growth fill in the hanging wall of Fault B.

In the distal hanging wall of Fault B (the proximal footwall of Fault-set C) is a set of conjugate faults (Conjugate Fault-set A; Figure 4) bounding a graben. These faults most likely formed in response to the bending of the layers (McClay, and Ellis, 1987; McClay and Scott, 1991; Brewer and Groshong, 1993). The beds thicken into the hanging wall of the antithetic faults dipping to the southeast and bounding the graben to the northwest. These antithetic faults, together with the synthetic faults in Fault-set C, bound a complementary horst to the graben (Figure 4D). The thickening of the section across the antithetic faults suggests that they were active at the same time as Fault B. The synthetic faults in the conjugate fault set are more numerous than the antithetic faults and comprise tens of closely spaced faults with lengths from several centimeters up to 10 m (30 ft) and throws of millimeters to 30 cm (12 in.) (Figure 4B).

Fault-Set C

Fault-set C is the most distal fault set in the proximal exposure and comprises three faults that step down into the section from a proximal to distal location as in Fault-set A. The most proximal fault in the set beheads the antithetic faults bounding the graben. The shallow prograding sand, SS5, expands across this fault, which terminates at the base of the youngest prograding sand in the proximal exposure, SS6. The central fault in the set connects along the deeper trace of the proximal fault in the set. The most distal fault in the set terminates at the upper bed surface of SS1 similar to the deep faults in Fault-set A.

The distal hanging wall of Fault-set C has a set of conjugate faults, which we also interpret as related to bending (Conjugate Fault-set B, Figure 4D). In this case, however, the lateral extent of the faults in the set is narrow and the conjugate fault pairs are nested up through the section with no growth across the faults.

Distal Exposure

The distal mapped exposure (Figure 5) is markedly different from the proximal exposure (Figure 4) in several important ways. The pre-growth section is nearly completely covered with only the thick cross-bedded growth section exposed. This limits the accuracy of the fault restoration and estimates of throw. In addition, most of the major growth faults are covered with only local exposure of the fault tips. We infer the presence of the faults by the dip of the beds, but the poor exposure prevents as detailed an interpretation as in the more proximal exposure. The bed dip increases across the entire mapped outcrop and dips are greatest in this distal exposure.

Fault D

The first or most proximal fault in the interpreted section in Figure 5 is labeled Fault D. This is the fault at the distal end of the hanging wall in Fault-set C (Figure 4). Only the upper 5 m (15 ft) of the fault is exposed and although mapped as a single fault at the scale of the map, in detail it is a narrow zone approximately 1–2 m (3–6 ft) wide of thin fault segments and deformed sand. Massive shallow beds of growth strata (SS8 and SS9) immediately adjacent to the fault in the hanging wall are only slightly rotated towards the fault and form large-scale cross-bedded sandstones that thin seaward. These young prograding sands dip gently bayward in their more distal position. Their offlapping stratal geometry is similar to the younger prograding sands (SS5 and SS6) in the proximal section, but correlation across the fault is difficult.

Beneath the gently dipping prograding sands, SS8 and SS9, is a thick section of nearly uniformly thick cross-bedded sands (SS1 to SS5). The beds are rotated and truncated by an erosional surface across which lies the prograding section. The facies of the rotated beds is indicative of the growth sections, but their down-dip termination towards fault D is covered. Near the very base of the section about 1 m (3 ft) of interbedded shales and sandstone is exposed (sections 11 and 12, Figure 5C), which we correlate with the pre-growth section described earlier.

Although the section is incomplete because of erosion and cover, the exposed beds have a nearly uniform thickness of approximately 10 m (30 ft). This thickness is equivalent to the growth section close to the faults comprising Fault-set C and we interpret these as the same beds (SS1 to SS5).

A conjugate set of faults (Conjugate Fault-set C, Figure 5D) cut these thick beds in a distal location of the hanging wall of Fault D just below the erosional surface. The conjugate fault pairs in the set form a 10-m-wide (30-ft) graben that narrows deeper in the section to the intersection of the fault pairs. The lower terminations of the outer faults bounding the graben are covered. These conjugate faults most likely formed in response to the bending of the layers (McClay, and Ellis, 1987; McClay and Scott, 1991; Brewer and Groshong, 1993).

Fault-Set E

At the distal ends of the thick dipping cross-bedded sands in the hanging wall of Fault D are a set of faults, which are exposed only along their upper extent (Fault-set E, Figure 5). This set of faults extends horizontally over a zone approximately 2 m (6 ft) wide. We interpret these faults to be the upper terminations of a listric growth fault set bounding the next bayward section of thick cross-bedded dipping sands. Between Fault-set E and Fault F is the thickest growth section exposed in the outcrop. The rotated beds in the most distal part of the hanging wall of Fault-set E have dips up to 45° southeast. This strongly rotated fault block acts as a buttress to the overlying prograding sediments.

Because the faults bounding this growth section are covered, we are uncertain of the deeper fault geometry. It is likely, based on the structural style in the proximal section and on our restorations, that this extensive growth section developed across several sub-parallel listric faults.

The shallowest growth section completely truncated by Fault-set E, SS7, thickens on the downthrown side of the exposed fault trace and thins onto the dipping hanging wall over a distance of approximately 10 m (30 ft). Onlap of individual sigmoidal cross-strata against the fault, suggest that a distributary-mouth bar prograded into the actively growing space. The base of the prograding sands overlying this wedged section is slightly offset by the fault. The youngest prograding sands overlying this section (SS9 and SS10) are thickened over the hanging wall, but are not faulted indicating some differential compaction in the hanging wall of the fault.

Beneath this youngest section of prograding sands, the cross-bedded sands in the hanging wall of Fault-set E maintain a nearly uniform thickness across their length. This implies that the sands were deposited over a broad surface that was not strongly dipping, which is very different from the youngest sands with a prominent wedge shape.

The thick steeply dipping cross-bedded sands that terminate against Fault F at the distal end of the hanging wall in Fault-set E are the oldest growth section in this fault block. A heterolithic section exposed at the base of the sands (Figure 5C; section 16) we interpret as pre-growth. We therefore, correlate the thick steeply dipping sands, which overlie the pre-growth section with SS1 to SS4 exposed in the hanging wall of Fault D and Fault-set C. The trailing edge of these sands in the hanging wall of Fault-set E are sharply truncated by Fault F, which is a distinct planar surface in outcrop.

The thick sand beds above the SS1 to SS4 section are slightly discordant with the underlying section. The older beds in the section (SS5), onlap the upper bed surface of section SS4, but without lateral thinning of individual beds in the section. The upper most beds in the thick section are truncated across their thickness by an erosional surface separating the older dipping section from the younger prograding sands SS9 and SS10.

We interpret the bedding surface separating SS4 from the overlying discordant beds of SS5 as a bed-parallel fault. Slip on the fault would account for the onlapping relationships. The youngest prograding sands, SS10, thicken across the southeast dipping beds of SS4 lending credence to the interpretation that this bed surface acts as a fault displacing the beds antithetic to the primary growth orientation along the bayward dipping faults.

Fault F

The most distal mapped section in the outcrop is a flat lying section of heterolithic sands and shales overlain by cross-bedded sands with an anomalous, shale section near the top. This section is unique in the outcrop because the beds are not rotated across Fault F as in the style observed in other bayward dipping faults. The style of unrotated beds suggests that Fault F is not a curved listric fault, but is planar. As we show in the section on fault restoration, we interpret this as a fault rotated from a southeast dip to a northwest dip.

The correlation across this distal exposure to the other mapped sections is difficult because of erosional unconformities through the section. We associate the heterolithic beds at the base of the section with the heterolithics at the base of the steeply dipping beds in the footwall of Fault F. The cross-bedded sands above these heterolithics are thinner than the cross-bedded sands comprising the hanging wall of Fault-set E.

Above the cross-bedded sands capped by SS5 in the flat lying section is a shale-rich section 3–5 m (9–15 ft) thick. This shale is anomalous in the younger growth section, which is mostly homogeneous cross-bedded sandstones across the outcrop. The total thickness of the shale in this fault block is uncertain because much of the area is covered. Adjacent to Fault F, however, a thin

exposed sand layer is folded and contorted within the shale section. The deformation of these sands most likely formed due to gravity collapse with progradation across the buttressed high of the rotated distal footwall of Fault E onto the steep dip of Fault F.

The anomalous shale shallow in the clastic section may have been injected along the fault plane of Fault F or be the distal muddy influx from a prograding sand. Nix (1999) and Morris and Nix (2000) hypothesize that the shale is diapiric and sourced from the underlying prodelta muds. One possible mechanism for the emplacement of muds shallow in the section from a deeper source is overpressure of the muds and injection along the fault plane. The steeply dipping rotated beds truncated by Fault F lie above a regional dip as demarcated by the base of the youngest prograding sand. This sand thickens on either side of the buttressed fault block. The differential relief across the block suggests that the shale beneath the block is fully compacted, unlike the prodelta muds on either side, which compact slowly to accommodate the growth section. The rotation and bayward displacement of this block may have overpressured the muds driving them up the fault plane. Trenching along the fault plane seems to support this hypothesis although it remains inconclusive as to whether the trenched muds are in place.

The alternative explanation for the shallow muds is that they are the distal muds of a prograding sand. This hypothesis may be supported by the evidence of muds in SS9 measured in section 15 (Figure 5C). These muds may have prograded across the rotated beds into the more distal section. Our correlation of the beds across the buttressed block does not agree with this interpretation, but the correlation of beds in this distal location is arguable. The exposures are still too limited to support one hypothesis more strongly over the other.

At the base of this distal section adjacent to Fault F is a set of conjugate faults labeled Conjugate Fault-set D. Although these faults are not in a present-day folded distal hanging wall as the other conjugate fault sets, our restorations show that they are most likely an older set that formed in the curved footwall of the nascent Fault-set E.

The exposed limit of the mapped cliff-face extends approximately 50 m (150 ft) beyond the cross-sections described. Much of this unmapped exposure is covered but at the most distal end is a partly exposed thick section of cross-bedded sands with a small counter-clockwise rotation apparently in response to displacement on a fault covered beneath the rock rubble.

FAULT MECHANISMS

As we have shown in our description, no growth faults extend above the flooding surface separating the growth section from the overlying Ferron stratigraphic

293

Figure 23. Faulted contact between the cross-bedded sandstones of the growth section on the right and the heterolithic pre-growth section to the left. The fault is a zone of thin seams interpreted to be local disaggregation and mixing of the grains.

Figure 22. Sand and shale smear within a narrow fault zone separating beds in the heterolithic pre-growth section. Color version on CD-ROM.

cycles. The facies interpretation suggests that the cross-stratified growth section was never deposited in water more than a few meters deep. These observations and the obvious growth of the section across active faults is proof that the faults nucleated and remained active during only shallow burial that scales with the maximum depth to which the faults extend, which we expect to be less than 30 m (90 ft).

The active fault mechanisms at these depths are restricted to deformation in soft-sediments such as shale and sand smear and disaggregation zone mixing or grain-rolling in narrow zones without concomitant grain-crushing (Fisher and Knipe, 1998). In the following discussion we describe some of the detailed fault mechanisms observed in the outcrop.

Shale and Sand Smear

In shallow-marine siliciclastic sediments, shale smear along fault planes is a common mechanism described in surface outcrops (Lindsay et al., 1993) and sub-surface reservoirs (Berg and Avery, 1995; Yielding et al., 1997). Shale smeared along the fault plane from a source layer is commonly interpreted to be a sealing mechanism between hydrocarbon-filled sands juxtaposed across faults in oil and gas reservoirs. Less commonly described is sand smeared concomitantly with the shales along the fault plane. Because of the soft state of the sediments deformed across the growth faults, sand and shale are commonly observed smeared along the planes of faults in the Ferron outcrop.

Mud-rich fault smear occurs where the faults are exposed cutting the pre-growth section of sands and shales. Figure 22 shows the lowermost section of a fault segment along Fault-set A (Figure 4) showing sand and shale entrained along the fault plane. Although the beds adjacent to the fault would have been unconsolidated during the deformation, the fault forms a narrow zone 5–10 cm (2–4 in.) wide. Within this zone, sand and shale beds are attenuated and smeared. In places, sand wedges within the fault zone are shingled. In most cases, the beds within the fault zone are disconnected from their source layer outside the fault.

Disaggregation Zones

Faults that cut the thick homogeneous sections of cross-bedded sands are often manifested in outcrop by a thin trace darker or lighter in color than the host sand. The fault trace in the sands is most likely a narrow zone, millimeters wide, of grains that roll around their grain boundaries to a more tightly packed configuration without grain crushing. This mechanism has been called disaggregation zone mixing (Fisher and Knipe, 1998). Limited micro-structural analysis of these zones confirms this interpretation. The local compaction of the grains within the fault zone reduces the permeability within the fault zone, but not sufficiently to act as a seal between sands juxtaposed across the fault. Petrophysical studies of these rocks show that the faults may seal if developed in host rocks with clay greater than 15% (Fisher and Knipe, 1998).

Figure 23 shows several closely spaced disaggregation faults in a growth section within a zone 5–10 cm (2–4 in.) wide. The fault zone separates a thick growth

Figure 24. Conjugate fault-set B cutting flat-stratified sands in the hanging wall of Fault-set C. Color version on CD-ROM.

section of cross-bedded sands on the right from the heterolithic pre-growth section to the left, in Figure 23. As in the example of sand and shale smear, the beds adjacent to the fault are not deformed. Within a narrow zone adjacent to the heterolithic section sand and shale is smeared along the fault zone. The trace of the disaggregation faults can be followed up the section into the sand on sand contact, but it is more difficult to follow in outcrop.

The thick sand section in the distal hanging wall of Fault-set C is cut by a conjugate fault (Conjugate Fault-set C; Figure 4). These faults formed later in the deformation after bending and with no growth across their boundaries (McClay and Ellis, 1987; McClay and Scott, 1991; Brewer and Groshong, 1993). The strata in the distal sands are eroded into distinctive bed sets that can be traced across the faults (Figure 24). These faults as in the other examples formed in soft sediments by disaggregation zone processes.

Mixed Mechanisms

Not all of the fault displacement occurs across discrete fault planes or narrow fault traces. In several cases, the displacement is accommodated across closely spaced faults within zones up to 1 m (3 ft) wide. Figure 25A, for example, shows several faults cutting the pre-growth section of thinly bedded sands and shales. The bed offset of 30 m (90 ft) is shared among several closely spaced faults accompanied by a narrow monoclinal bend of the rocks.

In Figure 25B, the displacement is similarly accommodated across several discrete faults in a narrow zone. In this case the deformation style is a conjugate fault set. The faults dipping to the right displaced the section and then they are offset by the fault dipping to the left. Several beds entrained in the left-dipping fault are attenuated during the deformation. Attenuation of the thinner sandstone beds by small local closely spaced faults is a common mechanism in the pre-growth section.

Figure 25. (A) Multiple small displacement faults accommodating a larger composite throw across the zone containing the faults. (B) Multiple faults dipping to the right are offset by a later fault dipping to the left. Both fault sets cut the pre-growth section in the distal hanging wall of Fault B. Color version on CD-ROM.

STRUCTURAL RESTORATIONS

Many models have been proposed for the restoration or back-stripping of structural cross-sections to validate the interpretation and constrain their historical development. One method described is flattening of sequential stratigraphic horizons (Dula, 1991). In this method, particle paths are required to move vertically, and if the layer is curved in the deformed state the line length is shortened in the restored state. If the bed area is not maintained by thickening of the layer during this restoration, the bed area will be reduced in the restored state. An alternative restoration method is to maintain the line length and bed area during the restoration. Restoring a curved surface in this scenario requires bed slip.

More sophisticated methods for "kinematic" restoration have been proposed that assume *a priori* the particle path based on the bed and fault shape. The section is restored along these particle paths. The assumptions inherent in these techniques must erroneously estimate the "true" particle paths. They do, however, constrain the restoration along a path parallel to the fault that results in reasonable restored surfaces through time and preserves the bed length and area.

We restored both the proximal and distal fault interpretation in this study to investigate the fault timing and constrain the interpretation of the fill style and history. For several reasons we used different restoration techniques for the proximal and distal exposures. In the proximal set of faults, we restored the faults using a kinematic approach. The bed shapes were used to predict the fault shale at depth — curved beds result in curved or listric faults, and the beds were restored to their pre-faulted position along the interpreted faults (Dula, 1991). We used ARCO in-house proprietary software for the restoration.

In the more distal fault set, the faults have a more complex style and bed-fill history. Because of the added complexity and because the software available for the previous study was no longer at our disposal we relied on a restoration of line length and area balance for the interpretation.

The kinematic restoration technique displaces the section along paths parallel to the fault and so we often assume that this method is more rigorous. Although neither of the two methods described for restoration accurately describes the particle motion, both restoration techniques provide us with an interpretation of the fault timing and fill history of the normal faults. A more rigorous interpretation would require knowledge of the material properties and boundary conditions. The sequential restorations as described for the proximal and distal fault sets do not correlate as one-to-one time steps because of the progradation of the section over the outcrop distance. The more proximal fault sets have a greater number of restored steps early in the growth history than the more distal fault sets, which have a greater number of steps in the later fault history. Although these restorations constrain the timing of the faults, the difficulty in correlating a proximal growth sand with a distal sand introduces uncertainty into the analysis. The uncertainty is discussed where appropriate for each restoration step.

The restoration of the faults in the proximal fault exposure are referred to steps a to h (Figure 26). We refer the restoration of the faults in the distal exposure to the same numbered steps except include subscripts for the intervening times between the proximal steps (Figure 27). Thus steps labeled g1, g2, and g3 refer to three time steps between the restored step f and h in the proximal set. The hanging wall of Fault-set C is repeated in the restoration of the distal fault set for comparison and reference.

Step A

In Figure 26A, the final restored section is the pre-growth heterolithic sands and shales. The light dashed lines in the figure show the locations of the future faults cutting the section (Figure 26B). A similar restoration is not provided for the distal fault set because the pre-growth section is mostly covered.

Step B

The first faults active in the deformation are Fault-set A and Fault-set C in the proximal mapped section. Two closely spaced fault pairs accommodate the displacement in each fault set. Contemporaneous displacement across synthetic fault pairs is common in normal faults and usually indicates the geometry of overlapping fault tips in map view. These overlaps are known as relay ramps or transfer zones (Peacock and Sanderson, 1991, 1994).

The thick SS1 sand fills the broad hanging walls of the early growth faults. The measured sections show large cross-beds in the sands close to Fault-set A, but a more flat stratified sand across Fault-set C. This observation is consistent with a broad flat hanging wall in Fault-set C that extends across the outcrop. SS1 is interpreted as an unfaulted thick sandstone in the distal mapped outcrop (Figure 27B).

Contemporaneous with the deposition of SS1 across Fault-set A and C, we interpret a set of antithetic faults bounding a horst in the footwall of Fault-set C (Figure 26C). SS1 is either eroded or not deposited over the horst crest. The exposure in the horst is poor, but detailed mapping shows that some of the pre-growth section is mechanically thinned by small-scale faulting over the crest.

Step C

Figure 26C shows continued thickening of the SS1 sands across the bayward dipping growth faults and the antithetic faults bounding the horst. Fault-set A and the antithetic faults define a broad graben in the proximal mapped section.

One interpretation of this early, restored state is that the SS1 sand is a pre-growth, prograding delta sand that is later faulted and eroded. Alternatively, the sand may have been deposited as a hanging wall sand in Fault-set A and then later displaced across Fault-set C and the antithetic faults. This interpretation is consistent with the nearly uniform bed thickness along the length of the outcrop, but is difficult to differentiate from a growth interpretation based on the exposure.

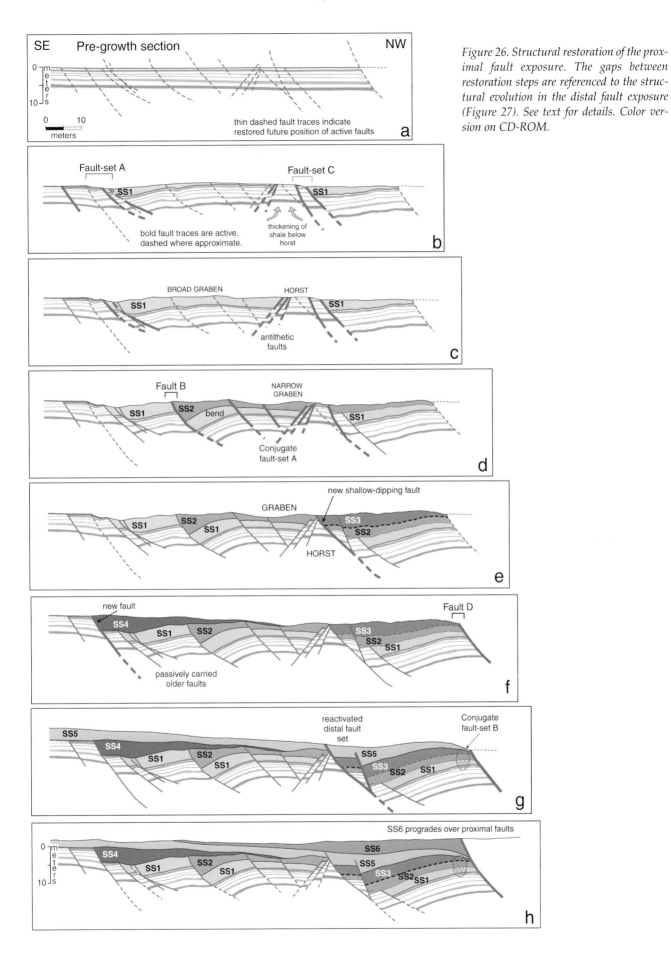

Figure 26. Structural restoration of the proximal fault exposure. The gaps between restoration steps are referenced to the structural evolution in the distal fault exposure (Figure 27). See text for details. Color version on CD-ROM.

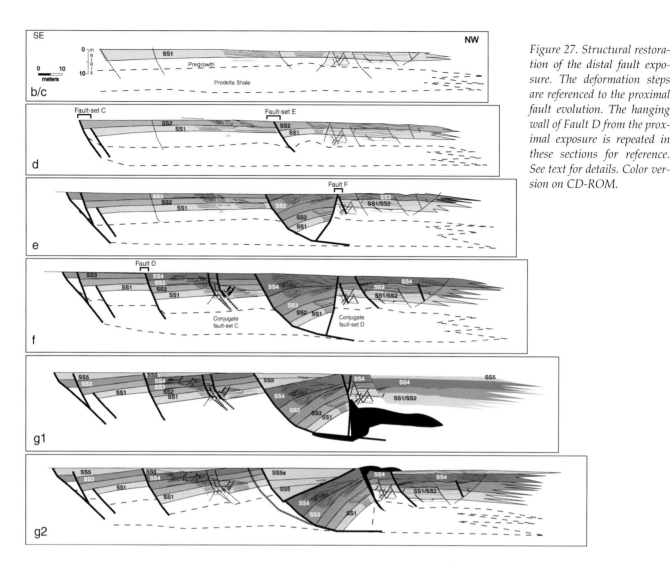

Figure 27. Structural restoration of the distal fault exposure. The deformation steps are referenced to the proximal fault evolution. The hanging wall of Fault D from the proximal exposure is repeated in these sections for reference. See text for details. Color version on CD-ROM.

Step D

At the next step in the deformation (Figure 26D), displacement ceases on Fault-set A and is transferred to Fault B, which cuts the broad graben bounded to the northwest by the antithetic faults of the conjugate fault-set A. The cross-bedded sand SS2 thickens across Fault B and the antithetic faults. Fault B is strongly curved which results in a strongly curved hanging wall. Small throw synthetic faults develop in response to this bending and together with the antithetic faults form the conjugate Fault-set A (McClay and Ellis, 1987; McClay and Scott, 1991; Brewer and Groshong, 1993).

The distal fault in the active growth fault pair of Fault-set C is inactive at this stage, and SS2 thickens across the more proximal fault in the set. The distal set in the pair is buried with deposition and growth across the fault.

At this stage in the deposition, the earliest fault displacement is interpreted in the more distal interpreted exposure with thickening of SS2 across Fault F (Figure 27D).

Steps E and F

During the next step in the deformation in the proximal mapped exposure, the most landward fault in Fault-set C cuts back into the footwall section or horst at a shallow angle beheading and offsetting the antithetic fault-set (Figure 26E). The other faults in Fault-set C are inactive and sand SS3 fills the hanging wall of the active fault.

Figure 26F shows displacement on the most landward fault in Fault-set A in the later step f. This fault cuts the shallowest section of all faults in this local set. We interpret the cross bedded SS4 sands, filling the hanging wall of this fault, as separate from the more distal SS3 sands which show abundant soft-sediment deformation and contorted bedding. This local deformation is unique in the section and, therefore, we interpret the SS3 sand as different from SS4, although it is possible that these sands are related and deposited contemporaneously. The similarity in the cross-bedded growth sands across much of the outcrop precludes a unique description across some fault blocks.

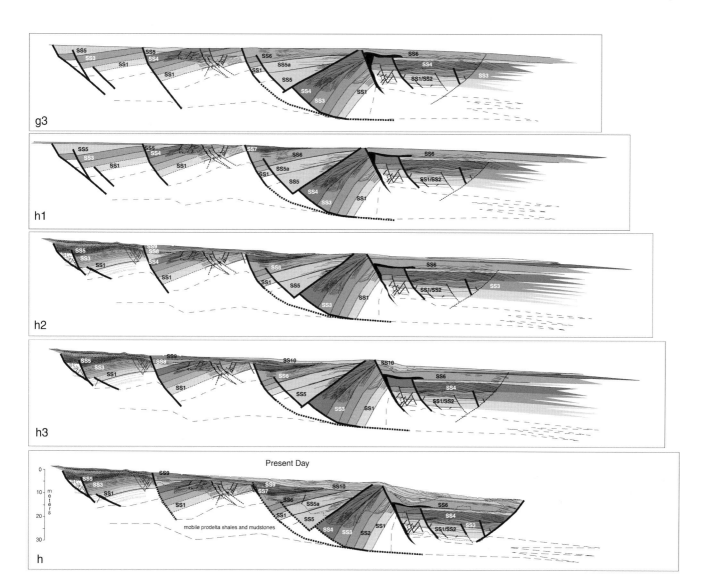

In the distal mapped exposure, the distal equivalent of these sands fills a local graben formed by Fault-set E and Fault F (Figure 27E). At this stage in the deformation Fault F is dipping to the southeast and is antithetic to Fault-set E. The faults together form a conjugate fault pair bounding a wide graben. Deposition across the faults fills the graben with thick sand sections that are only slightly thicker across the northwest dipping fault, which has a greater displacement. Rotation of the beds in the distal graben bound by Fault-set E and Fault F exposes the trailing ends of the beds, which are eroded.

Sand is either not deposited on the most distal flat lying beds in the distal exposure or the beds are uplifted and eroded (Figure 27E). We expect that both of these processes occur limiting the thickness of the sand beds in this distal location.

The first increment of displacement across Fault D at the boundary between the proximal and distal exposure occurs during the deformational step f (Figure 27F) with sand labeled SS4 filling the hanging wall. The sand is numbered sequentially from the underlying section to provide consistency to the distal mapped set of faults, although these sands may correlate with SS3 as discussed above. The bending of the hanging wall section with fault displacement nucleates the Conjugate fault-set C (McClay and Ellis, 1987; McClay and Scott, 1991; Brewer and Groshong, 1993). During this stage in the structural evolution, the SS4 sandstones thicken dramatically across Fault-set E near the center of the distal exposure (Figure 27F).

Step G

The final fault growth in the proximal mapped exposure during step G shows displacement across the two proximal faults comprising Fault-set C and corresponding fill by the prograding SS5 sand (Figure 26G). These faults may have been active continuously throughout the deformation, as explained above, if the SS3 sand was deposited contemporaneously with SS4. The most bayward fault in the set terminates within SS5, but the most landward fault extends to the top of the sand.

Steps g1 to g3 in Figure 27 show the incremental complex deformation in the distal exposure during this stage of the structural evolution. The hanging wall of Fault D rotates with fault displacement accepting new growth sediment from sand SS5 and eroding the uplifted trailing edge of the older sands. The Conjugate fault-set C remains active during this deformation.

The hanging wall of Fault-set E also continues to rotate, but with more dramatic consequences. The fault displacement, responding to the layer extension, creates space in the hanging wall that is filled by sands SS5 to SS6, which are unique to the distal map exposure.

The rotation of the hanging wall reorients Fault F to a northwest dip (Figure 27 g2). The strong rotation and displacement of the deeper beds most likely overpressures the underlying muds at the leading edge of the rotated block and forces the mud up the surface of Fault F. Whether these muds ever reached the surface is questionable.

We expect that the unique geometry of steeply rotated beds extending deeply into the hanging wall of Fault E limits the subsequent motion of these beds. Continued fill within the fault block is accommodated by displacement across the southeast-dipping, bed-parallel surface between SS4 and SS5. This creates intersecting conjugate faults defined by the bed-parallel fault and Fault-set E.

Step H

During the final stages of deformation, the set of offlapping, wedge-shaped prograding sands (SS8-SS10) fill local space created in the hanging walls by offset across the Fault D and Fault-set F (Figure 27 h1), by differential compaction in the hanging wall of the Fault-set F (Figure 27 h2), and by differential relief across the rotated hanging wall of Fault-set E (Figure 27 h3). The sands prograding over the distal edge of the rotated fault block collapse and fold due to the gravity instability created by the slope across Fault F. The muds in this section are due either to the muddy distal toes of the prograding sands or to clays injected up the fault plane or a combination of the two.

In summary, the restorations show a complex growth history with proximal and distal growth fault sets active contemporaneously. The more distal faults in the section, however, have a larger throw than the more proximal faults and therefore a thicker section of cross-bedded sands.

DISCUSSION

Two styles of fault growth fill are present across the mapped exposure. The earlier deposited and more deeply buried medium-scale cross-bedded sandstones (e.g. SS1-SS3, Figure 4, and SS1-SS7, Figure 5) lack mudstone clasts or scarp-collapse breccias, which suggests that topography on the faults was minimal. In the younger prograding sands in the distal exposure (e.g. SS8-SS10, Figure 5), the faulted topography was infilled with offlapping, upward coarsening successions containing some mudstones, suggesting that several meters of topography may have existed. The older growth strata show that deposition of sand was synchronous with fault movement and that the faults were uniformly initiated with the deposition of the cross-bedded sands. An understanding of where deposition of cross-bedded sands occurs in modern delta fronts may give clues as to where and how growth faults nucleate.

Studies of sand deposition in modern shoal-water river-dominated deltas show a highly complex system of bifurcating distributary channels with several orders of channel splitting (Van Heerden and Roberts, 1988). The "terminal" ends of shallow high-order distributary channel are "plugged" by distributary-mouth bars (Figure 3). These mouth bars, in turn, cause sand to be deposited immediately upstream. Eventually, frictional deceleration and instability cause distributary channels to avulse. Depositional loci thus change position at a variety of scales and may shift not only seaward, as the delta progrades, but also landward, as channel plugging causes upstream deposition. Sand may also be deposited laterally as channels migrate or avulse. As demonstrated here, the growth faults initiate in response to deposition of sand in this dynamic proximal delta front area. Because these depositional loci can locally switch and even migrate upstream, associated growth faults show a similar complex pattern of initiation and movement.

The earlier-formed faults are similar to the Namurian deltas described by Rider (1978), who suggested that growth faults form as a natural consequence of delta progradation. Denser sands are deposited over less-dense mobile prodelta muds. Although broader-scale studies of growth faults show that they form in a progressively seaward-stepping fashion, as delta sediments prograde over shelf muds (Evamy et al., 1978; Bruce, 1983), our data shows that in detail, faults are not initiated or formed in such a progressive fashion.

We suspect that some larger-scale growth faults that are infilled with shallow-water facies may initiate in a manner similar to that interpreted here, as suggested by Rider (1978). However, regional-scale deformation in deltas is invariably tied to settings adjacent to a shelf-slope break, such as in the Gulf of Mexico (e.g. Winker, 1982) or Niger delta (Evamy et al., 1978). Large-scale faulting is enhanced by gravity-driven slumping and sliding on the continental slope, and the presence of thick underlying overpressured muds or salt (e.g. Winker, 1982; Winker and Edwards, 1983; Martinsen, 1989; Pulham, 1993). Thinner underlying mobile muds and smaller slopes, such as occur in intracratonic settings or farther inboard on the continental shelf, allow less accommodation for growth faulted strata (c.f.

Figure 28. Sharp contact between pre-growth heterolithic strata and growth sandstones might look like a channel contact in core. Color version on CD-ROM.

Brown, et al., 1973; Tye et al., 1999). Nevertheless, the observations of growth faults formed within a large shallow embayment of the Cretaceous seaway suggests that similar-scale features may be found in other river-dominated deltas deposited in shallow-water, intracratonic or shelf-perched, highstand settings. The main control on development of these growth strata appears to be extremely rapid sedimentation rates associated with highly river-dominated delta processes, rather than proximity to a shelf edge. Tectonic tilting likely formed this embayment. The active tectonic setting also likely caused earthquakes that may have been responsible for liquefying the prodelta mudstones of the Tununk, helping to initiate some of the faulting.

The younger faulting style is somewhat more similar to faults that form on a slope, such as the Cretaceous faults in Spitzbergen, documented by Nemec et al. (1988) and Pulham (1993). Muddy facies above the foundered block bounded by faults F and G (Figure 5) show large-scale deformation features, and may represent material that slumped away from the scarp formed during the movement along Fault F. Slumps naturally form on slopes because of downslope gravitational instability, which is enhanced where slope sediments are rapidly deposited and easily liquefied. Slumping occurs along faults and forms complex sea-floor topography. Deltas subsequently build over these areas and the faulted

topography is filled with delta front turbidites and prodelta mudstones, such as is seen in SS10 (Figure 19). If a larger-scale slope existed, it is likely that fault F would evolve onto a regional-scale type feature.

Although these growth-faults show offset of a few meters, which is well below the scale of features typically imaged by conventional two-dimensional (2D) or three-dimensional (3D) seismic data, fault and fracture patterns often have a similar expression at a range of scales suggesting a self-similar geometry (Tchalenko, 1970). Regional-scale growth faults in areas such as the Niger delta or in the Gulf Coast of the U.S., show similar patterns of growth faulting to those documented here. It is also likely that the seismic-scale faults are associated or are formed of smaller-scale features, such as those documented here, but which may not be well imaged using conventional tools. Growth-faulted strata at a scale similar to that mapped in this study have been described with limited data in other shallow-water, river-dominated fluvial-deltaic reservoirs such as the supergiant Prudhoe Bay field in Alaska (Tye et al., 1999). In these types of fields, the interpretation of complex 3D geometry is incomplete. Use of these outcrop analogs should be considered in placement of horizontal production wells and to explain anomalous production. Local over-thickening of sandstones may provide additional reserves that may be missed in reservoirs delineated on

301

the basis of conventional seismic surveys or widely spaced well logs. Evidence for small-scale synsedimentary growth faulting may be found in dip-meter logs, borehole imaging logs, or in cores, which can be compared to the facies and dip-changes, documented in this outcrop study. As one example, in core, the sharp contact between growth and pre-growth strata across a dipping fault might be misinterpreted as a channel margin (Figure 28). Because these synsedimentary faults form near the surface in soft sediment, features like fracture-filling cement and cataclastic features that commonly characterize late-stage tectonic faults are absent.

CONCLUSIONS

1. Early-formed growth faults initiated with deposition of dense, cross-bedded sands in shallow water (less than a few meters-deep), terminal distributary channels and distributary-mouth bars. Changes in the position of active faults through time reflect shifting depositional loci within the dynamic proximal delta front environment.

2. Later foundering of fault blocks results in an uneven sea-floor topography, which is more passively filled and smoothed over by progradation of shallow water deltas.

3. Growth faulting is accommodated by deformation and movement associated with dense sand loading underlying rapidly deposited, less-dense mobile prodelta shales. These types of shales are typical of those formed in highly river-dominated deltas. The lack of waves and tidal features is consistent with deposition of this portion of the Ferron into a protected embayment.

4. Detailed examination of the fault zones shows that deformation was largely by soft-sediment mechanisms, such as grain rolling and by lubrication of liquefied muds, causing shale smears. Mechanical attenuation of thin beds occurs by displacement across multiple closely spaced small throw faults.

5. Faults do not follow a systematic seaward or landward progression, which reflect the highly variable sedimentation patterns. Locally, faults within closely spaced fault sets step landward as they cut younger section.

6. Although fault blocks are reactivated, individual faults are rarely reactivated. Instead, new faults are formed during subsequent filling of the fault blocks. A single, through-going, long-lived fault was not mapped. Rather, complex sets of faults are active at different times.

7. Our results suggest that analog river-dominated deltaic subsurface reservoirs may be compartmentalized by growth faults, even in shallow-water, intracratonic, or shelf-perched highstand deltas. Reservoir compartmentalization would occur where thicker homogenous growth sandstones are placed against the muddy pre-growth strata and where faults are shale-smeared, and thus potentially sealing.

ACKNOWLEDGMENTS

We thank BP for providing financial support for this field work. Figures were drafted by Lou Bradshaw. We also thank reviewer Allan Driggs and editor Tom Morris who improved the clarity of the manuscript. This is contribution number 964 of the Geosciences Department, University of Texas at Dallas.

REFERENCES CITED

Berg, R. B., and A. H. Avery, 1995, Sealing properties of Tertiary growth faults, Texas Gulf Coast: AAPG Bulletin, v. 79, p. 375–393.

Bhattacharya, J. P., and R. G. Walker, 1992, Deltas, *in* R. G. Walker and N. P. James, eds., Facies models — response to sea-level change: Geological Association of Canada, p. 157–177.

Bishop, D. J., P. G. Buchanan, and C. J. Bishop, 1995, Gravity-driven, thin-skinned extension above Zechstein Group evaporites in the western central North Sea — an application of computer-aided section restoration techniques: Marine and Petroleum Geology, v. 12, p. 115–135.

Brewer, R. C., and R. H. Groshong, Jr., 1993, Restoration of cross-sections above intrusive salt domes: AAPG Bulletin, v. 77, p. 1769–1780.

Brown, L. F., Jr., A. W. Cleaves, II, and A. W. Erxleben, 1973, Pennsylvanian depositional systems in North-Central Texas — a guide for interpreting terrigenous clastic facies in a cratonic basin: Bureau of Economic Geology Guidebook No. 14, University of Texas at Austin, 122 p.

Bruce, C., 1983, Shale tectonics, Texas coastal area growth faults, *in* A. W. Bally, ed., Seismic expression of structural styles: AAPG Studies in Geology No. 15, p. 2.3.1–2.3.1-6.

Busch, D. A., 1975, Influence of growth faulting on sedimentation and prospect evaluation: AAPG Bulletin, v. 59, p. 217–230.

Diegel, F. A., J. F. Karlo, D. C. Schuster, R. C. Shoup, and P. R. Tauvers, 1995, Cenozoic structural evolution and tectono-stratigraphic framework of the northern Gulf Coast continental margin, *in* M. P. A. Jackson, D. G. Roberts, and S. Snelson, eds., Salt tectonics: a global perspective: AAPG Memoir 65, p. 109–151.

Dula, W. F., Jr., 1991, Geometric models of listric normal faults and rollover folds: AAPG Bulletin, v. 75, p. 1609–1625.

Edwards, M. B., 1976, Growth faults in upper Triassic

deltaic sediments, Svalbard: AAPG Bulletin, v. 60, p. 341–355.

Elliot, T., and K.O. Lapido, 1981, Syn-sedimentary gravity slides (growth faults) in the coal measures of South Wales: Nature, v. 291, p. 220–291.

Evamy, D. D., J. Haremboure, P. Kamerling, W. A. Knapp, F. A. Molloy, and P. H. Rowlands, 1978, Hydrocarbon habitat of Tertiary Niger delta: AAPG Bulletin, v. 62, p. 1–39.

Fisher, Q. J., and R. J. Knipe, 1998, Fault sealing processes in siliciclastic sediments, *in* G. Jones, Q. J. Fisher, and R. J. Knipe, eds., Faulting, fault sealing and fluid flow in hydrocarbon reservoirs: Geological Society, London, Special Publication 147, p. 117–134.

Galloway, W. E., D. K. Hobday, and K. Magara, 1982, Frio Formation of Texas Gulf Coastal plain — depositional systems, structural framework and hydrocarbon distribution: AAPG Bulletin, v. 66, p. 649–688.

Gardner, M., 1995, Tectonic and eustatic controls on the stratal architecture of mid-Cretaceous stratigraphic sequences, central western interior foreland basin of North America, *in* S. L. Dorobek and G. M. Ross, eds., Stratigraphic evolution of foreland basins: SEPM Special Publication 52, p. 243–281.

Gingras, M. K., J. A. MacEachern, and S. G. Pemberton, 1998, A comparative analysis of the ichnology of wave- and river-dominated allomembers of the Upper Cretaceous Dunvegan Formation: Bulletin of Canadian Petroleum Geology, v. 46, p. 51–73.

Lindsay, N. G., F. C. Murphy, J. J. Walsh, and J. Watterson, 1993, Outcrop studies of shale smears on fault surfaces: International Association of Sedimentologists Special Publication 15, p. 113–123.

Martinsen, O. J., 1989, Styles of soft-sediment deformation on a Namurian (Carboniferous) delta slope, Western Irish Namurian basin, Ireland, *in* M. K. G. Whateley and K. T. Pickering, eds., Deltas — sites and traps for fossil fuels: Oxford, Blackwell Scientific Publications, Geological Society Special Publication 41, p.167–177.

McClay, K. R., and P. G. Ellis, 1987, Geometries of extensional fault systems developed in model experiments: Geology, v. 15, p. 341–344.

McClay, K. R., and A. D. Scott, 1991, Experimental models of hanging wall deformation in ramp-flat listric extensional fault systems: Tectonophysics, v. 188, p. 85–96.

Moslow, T. F., and S. G. Pemberton, 1988, An integrated approach to the sedimentological analysis of some lower Cretaceous shoreface and delta front sandstone sequences, *in* D. P. James and D. A. Leckie, eds., Sequences, stratigraphy, sedimentology; surface and subsurface: Canadian Society of Petroleum Geologists Memoir 15, p. 373–386.

Morris, T. H., and T. L. Nix, 2000, Structural and sedimentologic elements of an evolving growth fault system: outcrop and photomosaic analysis of the Ferron Sandstone, Utah (abs.): AAPG Annual Convention, Abstracta with Program, v. 9, p. A84.

Nemec, W., R. J. Steel, J. Gjelberg, J. D. Collinson, E. Prestholm, and I. E. Oxnevad, 1988, Anatomy of a collapsed and re-established delta front in Lower Cretaceous of Eastern Spitsbergen — gravitational sliding and sedimentation processes: AAPG Bulletin, v. 72, p. 454–476.

Nix, T. L., 1999, Detailed outcrop analysis of a growth fault system and associated structural and sedimentological elements, Ferron Sandstone, Utah: Master's thesis, Brigham Young University, Provo, 48 p.

Peacock, D. C. P., and D. J. Sanderson, 1991, Displacements, segment linkage and relay ramps in normal fault zones: Journal of Structural Geology, v. 13, p. 721–733.

—1994, Geometry and development of relay ramps in normal fault systems: AAPG Bulletin, v. 78, p. 147–165.

Pulham, A. J., 1993, Variations on slope deposition, Pliocene-Pleistocene, offshore Louisiana, Northeast Gulf of Mexico, *in* P. Weimer and H. W. Posamentier, eds., Siliciclastic sequence stratigraphy, recent developments and applications: AAPG Memoir 58, p. 199–233.

Rider, M. H., 1978, Growth faults in Carboniferous of Western Ireland: AAPG Bulletin, v. 62, p. 2191–2213.

Ryer, T. A., 1984, Transgressive-regressive cycles and the occurrence of coal in some Upper Cretaceous strata of Utah, U.S.A.: International Association of Sedimentologists Special Publication 7, p. 217–227.

Ryer, T. A., and M. McPhillips, 1983, Early Late Cretaceous paleogeography of east-central Utah, *in* M. W. Reynolds and E. D. Dolly, eds., Mesozoic paleogeography of west-central United States Rocky Mountain section: SEPM Symposium 2, p. 253–271.

Tchalenko, J. S., 1970, Similarities between shear zones of different magnitudes: Bulletin of the Geological Society of America, v. 81, p. 1625–1640.

Tye, R. S., J. P. Bhattacharya, J. A. Lorsong, S. T. Sindelar, D. G. Knock, D. D. Puls, and R. A. Levinson, 1999, Geology and stratigraphy of fluvio-deltaic deposits in the Ivishak Formation — applications for development of Prudhoe Bay field, Alaska: AAPG Bulletin, v. 83, 1588–1623.

Van Heerden, I. L., and H. H. Roberts, 1988, Facies development of Atchafalaya delta, Louisiana — a modern bayhead delta: AAPG Bulletin, v. 72, p. 439–453.

Winker, C. D., 1982, Cenozoic shelf margins, northwestern Gulf of Mexico: Transactions — Gulf Coast Association of Geological Societies, v. 32, p. 427–448.

Winker, C. D., and M. B. Edwards, 1983, Unstable progradational clastic shelf margins, *in* D. J. Stanley and G. T. Moore, eds., The shelfbreak — critical interface on continental margins: SEPM Special Publication 33, 139–157.

Yielding, G., G. Freeman, and B. Needham, 1997, Quantitative fault seal prediction: AAPG Bulletin, v. 81, p. 897–917.

Analog for Fluvial-Deltaic Reservoir Modeling: Ferron Sandstone of Utah
AAPG Studies in Geology 50
T. C. Chidsey, Jr., R. D. Adams, and T. H. Morris, editors

Geologic Framework of the Lower Portion of the Ferron Sandstone in the Willow Springs Wash Area, Utah: Facies, Reservoir Continuity, and the Importance of Recognizing Allocyclic and Autocyclic Processes

John A. Dewey, Jr.[1] and Thomas H. Morris[2]

ABSTRACT

Three-dimensional outcrop exposures of the Ferron Sandstone in the Willow Springs Wash area of east-central Utah illustrate the importance of recognizing allocyclic and autocyclic processes in marginal-marine settings. Three chronostratigraphically distinct wave-dominated shoreline sandstone successions display vertical offset at their landward pinchouts. These coarsening upward successions are interpreted to be parasequences produced by allocyclic processes. Fluids trapped within these sandstones may be compartmentalized by this vertical offset. In contrast, a fluvial-dominated coarsening upward succession, interpreted to have been produced by autocyclic delta lobe-switching processes, displays no vertical offset between its landward pinchout and an underlying wave-dominated parasequence. In a landward direction, these two distinct sandstone bodies merge into one and fluids migrating updip would not be compartmentalized. Reservoir continuity is further complicated by younger multi-lateral and multi-story distributary channels that can incise into previously deposited marine parasequences, thereby creating a fluid pathway between these otherwise isolated sandstone bodies.

Within the area of Willow Springs Wash, exposures of the Ferron Sandstone permit detailed examination of both macro-and mega-scale features in three dimensions. Observations of lithology, primary sedimentary structures, bedding, bioturbation, and internal architectural elements allow identification of eleven facies. These facies are grouped into wave-dominated shoreline and fluvial-dominated deltaic depositional systems.

The deposits exposed along Indian Canyon in the Willow Springs Wash area consist of four distinct, mappable depositional units; three deposited along wave-dominated shorelines (depositional units A, B, and C) and one deposited in a lobe of a fluvial-dominated delta (depositional unit D). Vertical offset (or the lack of it) among the landward pinchouts of these individual depositional units helps distinguish between allocyclic and autocyclic parasequences. Although autocyclic deposits pro-

[1]*Anadarko Petroleum Corp., The Woodlands, Texas*
[2]*Department of Geology, Brigham Young University, Provo, Utah*

duced by deltaic lobe switching are recognized by many as parasequences, their macro- and mega-scale features are very distinct relative to marine-dominated allocyclic parasequences.

Misidentification of parasequence type and failure to distinguish between autocyclic and allocyclic changes involved in a rock succession can result in misunderstanding the depositional history, the distribution of related facies, and the quality and extent of potential hydrocarbon reservoirs. Indeed, without recognizing the parasequence type, the utility of parasequence correlation is diminished and optimal reservoir development strategies cannot be accurately formulated. Models based on outcrop analogs such as the Ferron Sandstone in the Willow Springs Wash area can provide insight into the complexities of these reservoir rocks.

INTRODUCTION

This study of the Ferron Sandstone illustrates the need to understand to a high degree the lithofacies and sequence stratigraphy of a mixed wave- and fluvial-dominated shoreface succession in an effort to predict accurately reservoir continuity or the lack of it. Inherent to this understanding is the recognition of allocyclic parasequences relative to autocyclic parasequences. Two important keys to understanding parasequence development in the study area were (1) the details of lithofacies identification and (2) the details of the landward pinchouts of the parasequences. Further, this study illustrates the complexity of reservoir continuity in a mixed wave- and fluvial-dominated delta system even when the broad picture is well understood and thereby illustrates the need for detailed outcrop analog studies.

Location of the Study Area

The Willow Springs Wash area is located on the west flank of the San Rafael Swell, about 15 mi (24 km) south of the town of Emery, Utah (Figure 1). Within the study area, canyons provide three-dimensional exposures of the rock bodies. The canyons of particular interest are Indian Canyon, the North Fork of Indian Canyon, and Willow Springs Wash.

Several locations within the study area are here informally named for convenience of reference (Figure 2). "The Alcove" is a natural alcove found on the south side of the mouth of Indian Canyon. The North Fork of Indian Canyon has two branches at its head. These are here referred to as the "North Branch" and the "South Branch." "The Wall" refers to vertical exposures along the west and northwest sides of Indian Canyon.

Regional Stratigraphic Setting

The regional stratigraphy of the Ferron Sandstone is described by Anderson and Ryer (this volume), Barton et al. (this volume), and Garrison and van der Bergh (this volume), and summarized in Figure 3. Outcrops in the Indian Canyon-Willow Springs Wash area consist of deltaic and marine sandstone and shale of the Kf-1 parasequence set, and overlying deltaic and alluvial-plain deposits associated with the Kf-2 through Kf-8

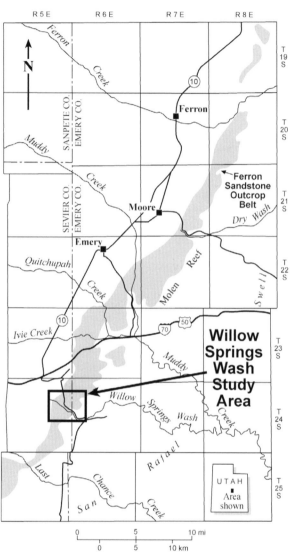

Figure 1. Location map of the Ferron Sandstone outcrop belt (shaded) showing the Willow Springs Wash study area.

parasequence sets. The "Sub-A coal" overlying the Kf-1 and underlying the Kf-2 is primarily a carbonaceous shale, which locally develops into coal. This carbonaceous shale is regionally extensive and serves as an important stratigraphic marker in the area. The landward pinchout of marine sandstone of the Kf-2 is seen on the north side of the mouth of Willow Springs Wash. Kf-2 is in turn overlain by the A coal zone. The Willow Springs Wash study focused on deposits within the Kf-1.

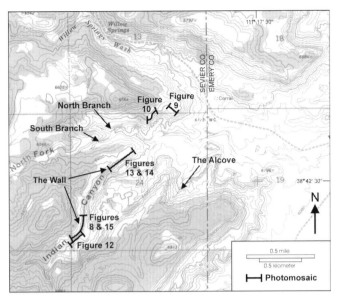

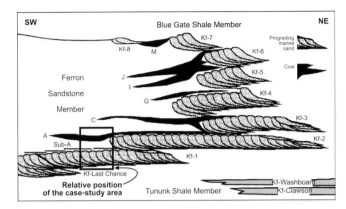

Figure 2. Detailed map of the Willow Springs Wash study area, including the informally named locations. Topographic base taken from U.S. Geological Survey Willow Springs, Utah, 7.5 minute quadrangle map.

Figure 3. Diagrammatic cross section showing relative positions of Ferron parasequence sets with their associated coal zones (black), and the relative stratigraphic position of the Willow Springs Wash study area within the Ferron Sandstone. The diagram is not to scale. Modified from Ryer, 1991.

Methods

Data for the study was derived by measuring stratigraphic sections, mapping photomosaics of canyon walls, measuring orientations of paleoflow indicators, and mapping lithofacies distributions. The measured section locations and the photomosaic coverage are shown in Figure 4. All of the measured sections, paleocurrent data, and interpreted photomosaics are available in Anderson et al. (2003).

LITHOFACIES

Eleven distinct lithofacies comprise the Kf-1 parasequence set in Indian Canyon of the Willow Springs Wash study area. These are distinguished by grain size; type and orientation of sedimentary structures; shape, orientation, and thickness of bedding; bioturbation and ichnofossils; and relationships between bounding surfaces and sedimentary structures (internal architecture). Variation between lithofacies is the direct result of changes in depositional conditions. Herein, lithofacies will be referred to by depositional facies names after the characteristics of each individual facies have been described.

Vertical repetition of several of the lithofacies reflects the repetitive nature of the depositional events that formed the deposits in the Indian Canyon-Willow Springs Wash area. In order to break down the outcrop into workable parts, distinct, mappable depositional units based on these events were identified. These depositional units are analogous to parasequences as defined by Van Wagoner et al. (1990), but differ in the sense that not all are bounded by a marine flooding surface *senso*

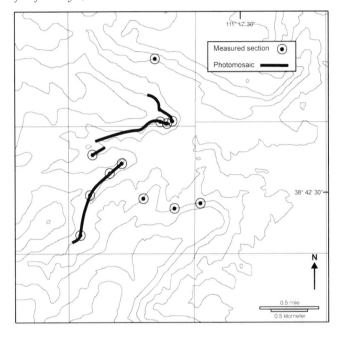

Figure 4. Map showing the locations of measured sections and the extent of photomosaic coverage within the study area. Refer to Anderson et al. (in press) for measured sections and interpreted photomosaics.

stricto. There are four depositional units within the study area. These depositional units are designated A, B, C, and D from base to top. The distribution of lithofacies in different depositional units is shown on lithofacies distribution maps. A sequence stratigraphic interpretation of these depositional units will be discussed later.

In addition, landward and seaward pinchouts of depositional units have been identified in adjacent canyons. Correlation of these landward pinchouts suggests that the shorelines were oriented northwest-southeast during deposition of these units. Land was towards the southwest and the sea was to the northeast. This is in agreement with the observations of Ryer (1981).

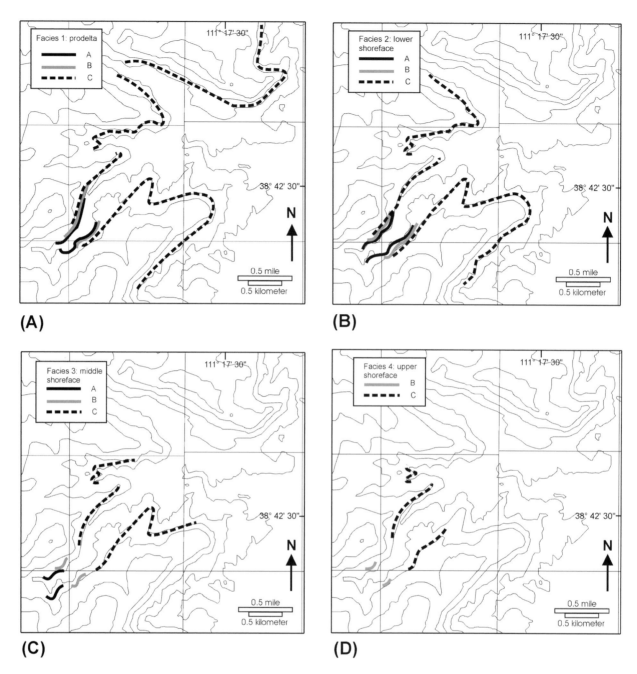

Figure 5. Maps showing extent of outcrops of Lithofacies 1 through 4: (A) Lithofacies 1, (B) Lithofacies 2, (C) Lithofacies 3, and (D) Lithofacies 4. The heavy lines within each map (lines labeled A, B, C, or D) show the occurrence of these lithofacies within different depositional units. Depositional unit A is incompletely exposed. Depositional unit B is poorly developed. These facies are absent within depositional unit D.

Lithofacies 1: Interbedded Siltstone and Shale

Lithofacies 1 consists of interbedded siltstone and shale. Beds are commonly 1–4 in. (3–10 cm) thick. The dominant sedimentary structures are asymmetric ripple lamination, horizontal lamination, and massive bedding. Some of the ripples appear to be starved ripples. Horizontal and vertical burrows of unknown affinity are

common. The extent of this lithofacies within the A, B, and C depositional units is shown in Figure 5A. Internal vertical heterogeneity within this lithofacies is the result of interbedding of siltstone and shale beds, but there is relatively little horizontal variation within the lithofacies (Figure 6).

Lithofacies 1 is interpreted to have been deposited within the prodelta or inner shelf environment. Deposition of siltstone was probably the result of storm currents.

Figure 6. Photomosaic from "the Wall" of Indian Canyon showing the typical succession of Lithofacies 1 through 5. (A) Prodelta siltstone and shale of Lithofacies 1. (B) Hummocky cross-stratified, lower shoreface sandstone and interbedded siltstone of Lithofacies 2. (C) Inclined, middle shoreface sandstone beds of Lithofacies 3. (D) Trough and tabular cross-bedded, upper shoreface/ebb-tidal delta sandstone of Lithofacies 4. (E) Seaward-dipping, foreshore laminae of Lithofacies 5. Markers in the photomosaic are approximately 10 ft (3 m) apart.

Lithofacies 2: Hummocky Cross-Stratified Sandstone and Siltstone

Lithofacies 2 consists of interbedded very fine grained sandstone and muddy siltstone (Figure 6). The siltstones are 1–6-in. (3–15 cm) thick beds and the interbedded sandstones vary in thickness between 4 and 10 in. (10–25 cm). Asymmetric ripple lamination and horizontal lamination are common within the siltstones. Sandstone beds show scoured bases and hummocky or swaley cross-stratification. Tops of individual sandstone beds are ripple-laminated or show rare, low-angle, small-scale trough cross-stratification. Tops of sandstone beds are also bioturbated, with the degree of bioturbation rapidly diminishing downward. Siltstone is commonly bioturbated and shows both horizontal and vertical burrows. The aerial extent of this lithofacies within the different depositional units is shown in Figure 5B. Heterogeneity within this lithofacies results from a combination of vertical grain-size variation due to interbedding and horizontal pinch-and-swell of individual hummocky cross-stratified sandstone beds.

Lithofacies 2 represents deposition within a lower shoreface environment. Hummocky cross-stratification has been interpreted as forming below fair weather wave-base during storm events (Dott and Bourgeois, 1982). Small-scale trough cross-stratification and asymmetric ripple lamination formed as a result of unidirectional currents during the waning of storms. Siltstone deposition and bioturbation represent relatively calm conditions between storms.

Lithofacies 3: Inclined, Medium-Bedded, Planar-Laminated Sandstone

Lithofacies 3 consists of medium-bedded, very fine-grained sandstone with very little siltstone and mudstone (Figure 6). Sandstone beds, 1–2 ft (0.3–0.6 m) thick, are inclined to the northeast (seaward). This lithofacies consists of beds with a well-laminated lower portion and a bioturbated upper portion which, from a distance, appear to be alternating well-indurated and friable sandstones, respectively. The lower portion shows inclined planar-lamination and ripple-lamination (symmetric and asymmetric) with some scour-and-fill trough cross-stratification. The bioturbated portion of these sandstone beds contain abundant vertical burrows assigned to *Cylindrichnus*, *Ophiomorpha*, and *Thalassanoides*. The areal extent of this lithofacies within the various depositional units is shown in Figure 5C. Heterogeneity of this lithofacies results from differences in the amount of bioturbation within individual beds and from the orientation of the beds within the lithofacies. Bedding within this facies downlaps onto the underlying sediments of Lithofacies 2.

Lithofacies 3 formed within the middle shoreface environment. Deposition was episodic, as evidenced by the presence of discrete zones of bioturbation. The absence of siltstone and mudstone in this environment has two possible causes. First, the energy within the system during "normal" conditions was such that mud and silt were probably winnowed out. Second, the frequency and magnitude of storms was such that any mud and silt that may have been deposited during calm conditions was subsequently removed by the next storm.

Lithofacies 4: Thick-Bedded, Trough and Tabular Cross-Stratified Sandstone

Lithofacies 4 consists of very thick beds, 10–15 ft (3–5 m) of fine- to very fine-grained sandstone. Internally, these beds show abundant trough and/or tabular cross-stratification (Figure 6). Within the lower two-thirds of this lithofacies, the cross-stratification appears to be oriented towards the northeast (seaward). Cross-stratification within the upper one-third is oriented towards the southwest (landward) and shows more variability in paleocurrent direction. Bioturbation is also common within this lithofacies. Vertical burrows (*Ophiomorpha* ?) are present, but for the most part burrow types are not recognizable. The lateral extent of this lithofacies within the various depositional units is shown in Figure 5D. This lithofacies appears to be relatively homogeneous when compared to the underlying Lithofacies 3 (middle shoreface). There are relatively few breaks in bedding and no observable grain size variation.

Lithofacies 4 represents deposition within the upper shoreface of a barred shoreline or within an ebb-tidal delta environment. Details of this interpretation will be discussed later.

Lithofacies 5: Thick-Bedded, Inclined Planar-Laminated Sandstone

Lithofacies 5 consists of thick-bedded (3–5 ft [1–1.5 m]), fine-grained sandstone. Internally, the sandstone beds display inclined planar-lamination (Figure 6). The laminae typically dip to the northeast (seaward) but some laminae, though rare, dip to the southwest. The top of this lithofacies also shows minor scour-and-fill with lensoidal geometries and thickness up to 1.5 ft (0.5 m). The top of this lithofacies has been rooted by plants from overlying carbonaceous shales. The lateral extent of this lithofacies within the various depositional units is shown in Figure 7A. As with the upper shoreface, this lithofacies appears to be relatively homogeneous. Lamination within this lithofacies is probably the result of alternating grain-size variation between laminae, however, this variation is not well pronounced in hand sample. The contact between this lithofacies and underlying Lithofacies 4 is commonly gradational.

Lithofacies 5 is interpreted to have formed within a foreshore or beach environment. Rare, landward-dipping laminae probably formed on the landward side of a berm. Channelization at the top of the facies was possibly caused by minor tidal channels.

Lithofacies 6: Sigmoidal-Bedded, Cross-Stratified Sandstone

Lithofacies 6 consists of sigmoidal beds of fine- to medium-grained sandstone (Figure 8). Beds vary in thickness from 0.5–2 ft (0.2–0.6 m). Lithofacies 6 shows scour into underlying deposits of Lithofacies 4. The depth of scour and overall thickness of the lithofacies increases from about 4 ft (1.2 m) where it begins in "The Wall" to greater than 10 ft (3 m) towards the southwest (landward). Individual beds thin and become tangential to the base of the lithofacies, forming a foreset and a bottomset. Comparison of bedding plane orientations with landward pinchout orientation shows that the beds dip to the southwest (oblique to shore). Individual beds are probably separated by fine-grained breaks, but this is hard to verify given the inaccessibility of exposure within the study area. Internally, the beds are trough cross-stratified. Paleocurrent orientations of these troughs show movement of sediment perpendicular to the dip direction of the beds, suggesting that this lithofacies developed by lateral accretion rather than downstream accretion (Miall, 1985). The facies is overlain by foreshore deposits that probably eroded any topset bedding. Some bioturbation is present, but burrow types were not identified. The lateral extent of the lithofacies in depositional unit C is shown in Figure 7B. Heterogeneity within this facies results from the bedding plane breaks between the sigmoidal-shaped beds. The horizontal variation in thickness of the facies is also a source of heterogeneity.

Lithofacies 6 is interpreted to represent deposits within a tidal inlet. Scour resulted from tidal currents passing through the inlet. The inlet was filled by a spit that developed through a combination of longshore drift and tidal currents. Spit development and southeastward inlet migration were followed by development of a foreshore. This lithofacies is differentiated from deposits of meandering fluvial and distributary channels by bioturbation, its position relative to foreshore deposits, and the lack of a spatially related, lensoidal, channel-sandstone body.

Lithofacies 7: Shale with Interbedded Sandstone

Lithofacies 7 consists of silty shale interbedded with very fine-grained sandstone (Figure 9). Sandstone is relatively minor compared to siltstone and shale. Shale beds range from 0.5 to 2 ft (0.2–0.6 m) in thickness. Sandstones are typically thin, 0.5–1 ft (0.2–0.3 m) but do thicken and become bedsets within adjacent (and overlying) deposits of Lithofacies 8. Ripple lamination, horizontal lamination, and massive bedding are common within the siltstones. The sandstones are typically horizontal- or wavy-bedded and ripple-laminated. They are volumetrically minor (less than 10%) compared to the shales. Burrowing is abundant but burrow types were not identified. The lateral extent of the lithofacies in depositional unit D is shown in Figure 7C. The lithofacies is not present within depositional units A, B, or C.

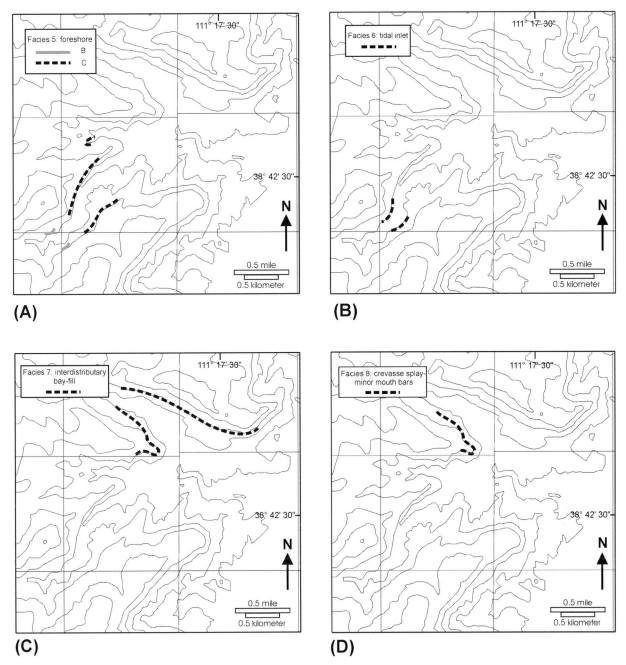

Figure 7. Maps showing the extent of outcrops of (A) Lithofacies 5, (B) Lithofacies 6, (C) Lithofacies 7, and (D) Lithofacies 8. Lithofacies 5 deposits are present in depositional units B and C. Lithofacies 6 is only found in depositional unit C. Lithofacies 7 and 8 are only found in depositional unit D.

Lithofacies 7 is heterogeneous vertically and horizontally as a result of interbedding, and the thinning and pinching out of sandstone beds.

Lithofacies 7 is interpreted as interdistributary bay-fill deposits. These deposits are, in some respects, similar to those of the inner shelf and lower shoreface (Lithofacies 1 and 2), and are distinguished in part by their relation to deposits of distributary channels and crevasse splays. In this respect, the distinction between deposits of Lithofacies 7 and those of Lithofacies 1 and 2 is partly model driven. However, Lithofacies 7 does contain less silt and is more aerially restricted than Lithofacies 1 and

2. While these observations render an identification within a core difficult, correct identification of the facies does allow estimates of the geographic distribution of the facies.

Lithofacies 8: En Echelon Sandstone Bedsets and Interbedded Shales

Lithofacies 8 consists of *en echelon* bedsets of fine- to medium-grained sandstone separated by silty shales (Figure 9). Individual sandstone bedsets start stratigraphically high within the facies and thicken to about 4

Figure 8. Photomosaic from "the Wall" of Indian Canyon showing the relationship between (A) the sigmoidal bedded, tidal-inlet sandstone of Lithofacies 6 and adjacent lithofacies. (B) Lithofacies 2 (lower shoreface). (C) Lithofacies 5 (foreshore). (D) Pronounced channel in the photomosaic has scoured into deposits of Lithofacies 5 and 6, but this channel originated from within the carbonaceous shale deposits overlying the marine sandstone deposits and is not genetically related to the facies described here. Markers in the photomosaic are approximately 10 ft (3 m) apart.

Figure 9. Photomosaic from the mouth of the North Fork of Indian Canyon. (A) Prodelta and lower shoreface deposits of Lithofacies 1 and 2. (B) Interdistributary bay fill of Lithofacies 7. (C) Minor mouth-bar deposits of Lithofacies 8. Markers in the photomosaic are approximately 10 ft (3 m) apart.

ft (1.2 m). At this point they begin to thin and drop stratigraphically beneath subsequently deposited bedsets. The toes of individual bedsets interfinger with underlying interdistributary bay-fill deposits (Lithofacies 7). Sandstone bedsets are isolated from each other by shale and siltstone drapes. These interbedded shales and siltstones range in thickness from less than 1 ft to greater than 2 ft (0.3–0.6 m). Internally, sandstones show a variety of sedimentary structures including trough, tabular, and sigmoidal cross-stratification and planar-lamination. Bedforms appear to be unidirectional and show migration to the northwest. Bedform orientation in con-

junction with northwest-dipping bounding surfaces suggest that these bedsets grew by "downstream" accretion rather than lateral accretion (Miall, 1985). The tops of sandstone bedsets are locally scoured and filled with trough and tabular cross-stratified sandstone. The total areal extent of this lithofacies within depositional unit D is shown in Figure 7D, but it should be noted that the total length of an individual sandstone bedset may only reach 300–400 ft (90–120 m). Heterogeneity within this lithofacies results from isolation of sandstone bedsets within silty shales and shale drapes within sandstone bedsets.

Figure 10. Photomosaic from the north face of the North Fork of Indian Canyon showing the relationship between (A) distributary channel sandstone of Lithofacies 9 and adjacent lithofacies. This channel has scoured into lower and middle shoreface deposits of Lithofacies 2 and 3 (B). Outcrop at (C) is the same exposure as that in Figure 9.

Lithofacies 8 is interpreted to be the deposits of a crevasse-splay system that developed within an interdistributary bay. Individual sandstone bedsets are the deposits of minor distributary mouth bars. Scouring at the tops of these bars was the result of progradation of the minor distributary channels that fed the bars.

Lithofacies 9: Multistory, Lensoidal Sandstone

Lithofacies 9 consists of fine- to medium-grained, multistory, lensoidal sandstone bodies (Figure 10). The multistory set reaches a total thickness of 20–30 ft (6–9 m). The entire set shows scour down to underlying interdistributary bay-fill (Lithofacies 7), crevasse splay (Lithofacies 8) and middle shoreface deposits (Lithofacies 3). Within this multistory sandstone, multiple episodes of scour-and-fill are evident. Each scour-and-fill set varies in thickness from about 3–5 ft (1–1.5 m). Trough cross-stratification is abundant within the scour-and-fill deposits. Thin, fine-grained, ripple-laminated sandstone is common at the top. Extending laterally from the scour-and-fill deposits is a deposit composed of two thick beds of climbing, asymmetric ripple-laminated, very fine to fine-grained sandstone. Ripple-drift direction is away from the multistory channel system. The ripple-laminated deposit has a flat base that appears to have scoured into underlying interdistributary bay-fill sediments (Lithofacies 7). The depth of scour increases towards the multistory scour system. The areal extent of the combined lithofacies is shown in Figure 11A.

Lithofacies 9 is interpreted to be a distributary channel system. The absence of multilateral accretion suggests that the distributary had a straight reach. The thick-bedded, ripple-laminated sandstones probably represent deposits within a sub-aqueous levee associated with the distributary channel and are similar to sub-aqueous levees described by Coleman (1976).

The effect of this facies on heterogeneity within a reservoir is variable. Fine-grained drapes, if present, may act as baffles between and within scour-and-fill lens within this lithofacies. This type of drape has been observed by the authors within other channels in the Ferron Sandstone, but the inaccessibility of the exposure within the North Fork Canyon outcrop makes this difficult to verify here. The fact that this lithofacies exhibits scour into underlying interdistributary bay-fill and middle shoreface deposits suggests that Lithofacies 9 could actually serve to enhance the connectivity of otherwise isolated reservoirs within single or multiple parasequences.

Lithofacies 10: Multilateral, Lensoidal Sandstone

Lithofacies 10 is exposed along the south wall of Indian Canyon. It consists of two parts: laterally accreting deposits and lensoidal channel deposits. The lateral accretion deposits are exposed within "The Alcove" and consist of westward dipping, medium-thick beds of fine- to medium-grained sandstone. Northward-dipping trough cross-stratification and planar-lamination are abundant and confirm the laterally accreting nature of the beds. These lateral accretion sets terminate laterally into deposits of lensoidal sandstone 15–20 ft (5–6 m) thick. These contain fine- to medium-grained sandstone. Like the multistory channel system of Lithofacies 9, this lithofacies shows scour down into middle shoreface deposits of Lithofacies 3. The size of the bedforms and the bedding within the lensoidal bodies decrease

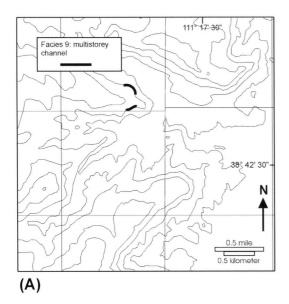

(A)

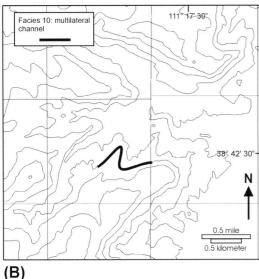

(B)

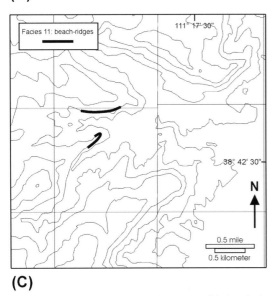

(C)

Figure 11. Maps showing the extent of outcrops of (A) Lithofacies 9, (B) Lithofacies 10, and (C) Lithofacies 11. All of these lithofacies are specific to depositional unit D.

upward. The bases of the lensoidal sandstones show large scale, 2–3 ft (0.6–0.9 m), trough cross-stratification. The tops of lenses contain abundant asymmetric ripple lamination. The location of outcrops of this lithofacies is shown in Figure 11B.

Lithofacies 10 is interpreted to be a distributary channel system. Unlike Lithofacies 9, this channel system was deposited as a meandering channel system rather than a multistory system with a straight reach. The lateral accretion sets of this lithofacies are distinguished from those of Lithofacies 6 (tidal inlet) in that these lateral accretion sets are directly associated with lensoidal channel sandstones. Furthermore, the Lithofacies 6 shows evidence of marine bioturbation.

As with the multistory channel (Lithofacies 9), the effect on reservoir heterogeneity is probably variable. Fine-grained drapes within the lateral accretion sets would act as baffles, however, this lithofacies also scours into underlying deposits and connects isolated rock bodies. This lithofacies is more extensive laterally than Lithofacies 9 and the deposits within the lensoidal sandstone are as thick as the multistory channel deposits.

Lithofacies 11: Planar-Bedded, Cross-Stratified Sandstone and Siltstone

Lithofacies 11 is best preserved at the top of the north wall of North Fork and "North Branch" Canyons. This lithofacies consists of interbedded fine sandstone and muddy siltstone. The lithofacies has three distinct, upward-coarsening bedsets. The bedsets are planar and horizontal, and maintain a relatively constant thickness of about 3 ft (1 m). Local scour-and-fill is present, but rare. These scour-and-full structures are never more than about 1 ft (0.3 m) deep which coincides with the thickness of any given sandstone bed within the lithofacies. Internally, trough and tabular cross-stratification are the most abundant features within the sandstones. Siltstones are commonly ripple-laminated and bioturbation is rare. This lithofacies is laterally extensive in the study area (Figure 11C) and fairly homogeneous horizontally. However, interbedding of sandstone and siltstone would create barriers to vertical fluid flow.

This lithofacies is interpreted to represent deposits of cuspate, strand plain/beach ridges extending from the mouths of rivers along wave-dominated shorelines. Sediment deposited at the mouth of a distributary was reworked by longshore currents and deposited along a strandplain downcurrent from the distributary mouth.

Depositional Systems

Recognition of individual lithofacies and the depositional systems to which they pertain is important for two reasons. First, correct facies identification will allow more accurate prediction of the distribution of adjacent

Table 1. Lithofacies and depositional facies within the study area and their presence within wave-dominated shoreline or fluvial-dominated delta-ic depositional systems.

Lithofacies Number	Lithofacies Name	Depositional Environment	Wave-dominated Shoreline	Fluvial-dominated Delta
1	Interbedded siltstone & shale	Prodelta-inner shelf	X	
2	Hummocky cross-stratified sandstone & siltstone	Lower shoreface	X	
3	Inclined bedded, planar laminated sandstone	Middle shoreface	X	
4	Thick bedded, trough and tabular cross-stratified sandstone	Upper shoreface / ebb-tidal delta	X	
5	Thick-bedded, inclined planar-laminated sandstone	Foreshore	X	
6	Sigmoidal-bedded, cross-stratified sandstone	Tidal inlet	X	
7	Shale & interbedded sandstone	Interdistributary bay-fill		X
8	*En echelon* sandstone & interbedded silty shale	Crevasse splay / minor mouth distributary mouth bar		X
9	Multistory, lensoidal sandstone	Straight reach distributary channel		X
10	Multilaterally accreting, lensoidal sandstone	Meandering distributary channel		X
11	Planar-laminated, cross-stratified sandstone and siltstone	Strandplain / beach ridge		X

facies. Second, correct facies identification will aid in predicting the size, shape, and connectivity of sandstone bodies and the types of barriers to fluid flow between and within these sandstone bodies. This type of knowledge is useful in planning well spacing, and in secondary and tertiary recovery strategies.

The previously described lithofacies can be categorized as components of two major types of depositional systems: (1) wave-dominated shoreline and (2) fluvial-dominated delta (Table 1). Depositional units A, B, and C show deposition within wave-dominated shorelines while depositional unit D represents deposition within a fluvial-dominated delta lobe. Figures 5 through 11 help to illustrate some of the differences in scale of the deposits within these individual systems. From the previous descriptions, it is also apparent that the two depositional systems differ in terms of grain size, bed thickness, bedding orientation, overall lithofacies thickness, and areal extent of the lithofacies. In general, the deposits of wave-dominated shorelines are thicker, cleaner, and more laterally extensive. Deposits of the flu-

vial-dominated delta are thinner, richer in mud, and less laterally extensive than wave-dominated shoreline deposits. These differences come into play when viewed in the context of common well spacings in oil fields.

SEQUENCE STRATIGRAPHY

Sequence stratigraphic concepts provide the geologist with tools to interpret the rock record in a chronostratigraphic context. Rocks can be divided and correlated according to depositional packages that are roughly time equivalent. This approach is a powerful tool for predicting facies distributions. In the case of the Willow Springs Wash area, the rocks are examined within the context of changes of relative sea-level. Viewed in terms of relative sea-level change, depositional variation between autocyclic processes (i.e. stream avulsion and delta lobe switching), and allocyclic processes (i.e. eustatic sea-level change or tectonic subsidence) can be distinguished (Ryer et al., 1995).

The rocks within Indian Canyon can be divided at

(A)

(B)

Figure 12. Uninterpreted (A) and interpreted (B) photomosaics of parasequences Kf-1-IC-a, -b, and -c. Note landward pinchouts of marine sand-stones in parasequences Kf-1-IC-b and -c. Cliff on the right side is approximately 45–50 ft (14–15 m) high. The location is at the southwest end of "The Wall," view is to the northwest. See Figure 2 for location of photomosaic.

the parasequence scale of Van Wagoner et al. (1990). Van Wagoner et al. (1990) defined a parasequence as a "relatively conformable succession of beds and bedsets bounded by marine-flooding surfaces or their correlative surfaces." A marine-flooding surface is defined as a "surface separating younger from older strata across which there is evidence of an abrupt increase in water depth." Marine-flooding surfaces are commonly identified in cores as surfaces separating relatively shallow water or fluvial sandstones below from marine shales above. In well logs, the interpretation of marine-flooding surfaces is based on the occurrence of abrupt decreases in grain size (identified within the constraints of wireline logging tools). Identifying marine-flooding surfaces on the basis of grain size changes, however, is not an entirely valid practice since changes in grain size may be the result of other factors. For example, at the mouth of North Fork Canyon fined-grained, interdistributary bay siltstone and shale of depositional unit D overlie lower and middle shoreface sandstone of depositional unit C (Figure 9). Although this clearly represents an abrupt decrease in grain size, it may not represent a marine flooding surface. Further evidence that this boundary does not represent a marine flooding surface is discussed below.

Within Indian Canyon, landward pinchouts of the various depositional units can be identified and these can be used to understand the history of relative sea-level changes (Figure 12). Vertical offset between two different landward pinchouts suggests that some

amount of change in relative sea level has occurred. If there is no vertical offset between two pinchouts, then no relative sea-level change can be documented and one must call upon autocyclic processes (e.g. delta lobe switching) for deposition of the younger unit. Deposits produced by delta lobe switching are recognized by many as parasequences (Emery and Myers, 1996). In Indian Canyon, however, autocyclic parasequences are architecturally and sedimentologically different than allocyclic parasequences and create more complexity relative to reservoir prediction.

The surfaces between depositional units A and B, and B and C are recognized as ideal marine-flooding surfaces, based on vertical offset between landward pinchouts and the occurrence of marine shales overlying shallower water sandstones (Figure 12). Thus, three parasequences exist due to allocyclic processes. An examination of the pinchout of depositional unit D, however, shows that this unit merges laterally with depositional unit C and no vertical offset of the landward pinchout is present (Figures 13 and 14). The lack of vertical offset between depositional units C and D indicate that the change between units C and D was autocyclic.

Based on the relationships between pinchouts, the rocks in the Indian Canyon-Willow Springs Wash are divided into four parasequences (Figure 15). These are designated from base to top as Kf-1-Indian Canyon-a, -b, -c, and -d (abbreviated as Kf-1-IC-a, -b, -c, and -d; K = Cretaceous and f = Ferron and 1 = parasequence set number 1). Parasequences Kf-1-IC-a, -b and -c consist of

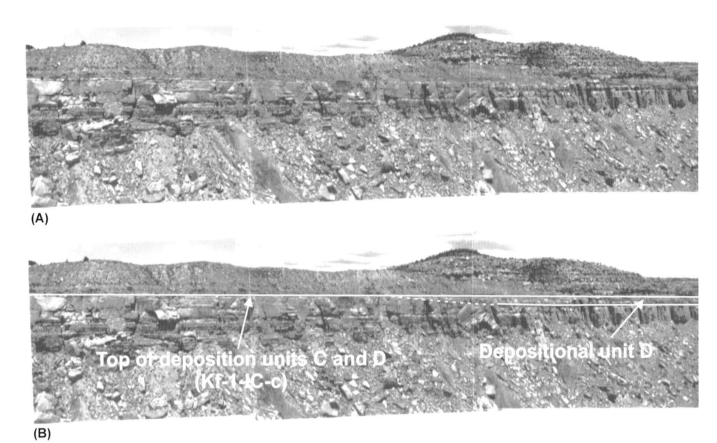

Figure 13. Uninterpreted (A) and interpreted (B) photomosaics showing the merging of the allocyclic parasequence Kf-1-IC-c and autocyclic parasequence Kf-1-IC-d. This exposure is down depositional dip from the exposure in Figure 9. See Figure 2 for location of photomosaic and Figure 14 for details.

ideal parasequences produced by allocyclic processes involving wave-dominated marine shoreface sandstones whereas Kf-1-IC-d represents a local parasequence produced by autocyclic processes involving a fluvial-dominated lobe of the Ferron delta.

DEPOSITIONAL HISTORY AND PALEOGEOGRAPHY

The paleogeography and depositional history of the study area can now be interpreted in the context of parasequence formation.

Kf-1-Indian Canyon-a

Figure 16A shows a paleogeographic reconstruction of the study area during deposition of parasequence Kf-1-IC-a. Outcrops of this parasequence display a coarsening-upward, progradational succession of beds. The depositional environments observable in outcrop (based on recognition of the component lithofacies) include prodelta-inner shelf, lower shoreface, and middle shoreface (Figure 15). This parasequence plunges into the subsurface in Indian Canyon and details of the formation and geometry of this parasequence become spec-

ulative. Based on the available outcrop, however, some inferences about adjacent, unexposed depositional environments can be made.

The lower and middle shoreface of the parasequence are very well developed. By comparison with the same facies in other parasequences, we can infer that the Kf-1-IC-a parasequence also contains well developed upper shoreface and foreshore deposits. The presence of any tidally influenced deposits in this parasequence is not known. The shoreline is inferred to have a northwest-southeast orientation. Due to the wave-dominated nature of the shoreline, the deposits within parasequence Kf-1-IC-a are inferred to be linear. The total dimensions of the parasequence are unknown.

Kf-1-Indian Canyon-b

Figure 16B shows a paleogeographic reconstruction of the study area during deposition of parasequence Kf-1-IC-b. Outcrops of this parasequence display a coarsening-upward, progradational succession of beds and bedsets. Depositional environments observed within this parasequence include the prodelta, a poorly developed lower and middle shoreface, and a foreshore (Figure 15). The areal extent of this parasequence is very restricted

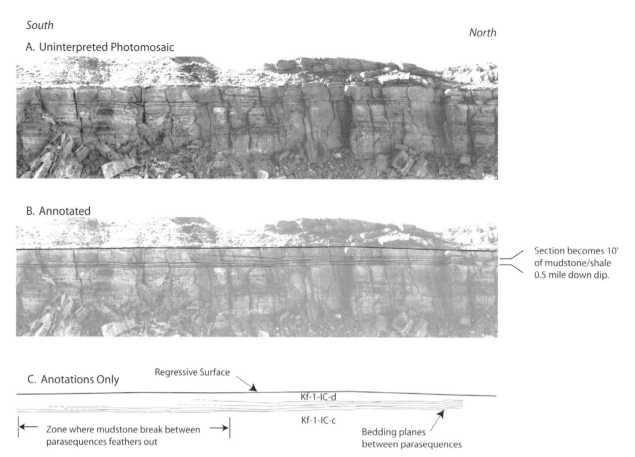

South

North

A. Uninterpreted Photomosaic

B. Annotated

Section becomes 10'
of mudstone/shale
0.5 mile down dip.

C. Anotations Only

Regressive Surface

Kf-1-IC-d

Kf-1-IC-c

Zone where mudstone break between
parasequences feathers out

Bedding planes
between parasequences

Figure 14. Uninterpreted photomosaic (A), annotated photomosaic (B), and annotations only (C) of the area where in Kf-1-IC-c merges with the Kf-1-IC-d. One half mile down depositional slope these muddy bedding plane breaks (between the sandstone bodies) develop into an approximately 10-ft (3-m) thick interval composed of mudstone/shale. Updip these bedding planes pinch out creating a single sandstone body. Note the lack of vertical offset between these two parasequences in the updip (landward) direction.

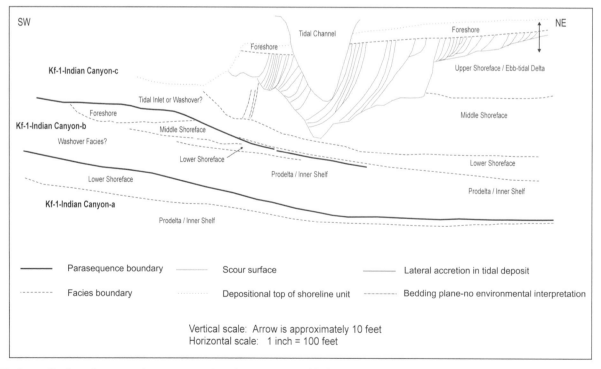

SW

NE

Foreshore

Tidal Channel

Foreshore

Kf-1-Indian Canyon-c

Upper Shoreface / Ebb-tidal Delta

Tidal Inlet or Washover?

Foreshore

Middle Shoreface

Kf-1-Indian Canyon-b

Middle Shoreface

Washover Facies?

Lower Shoreface

Lower Shoreface

Prodelta / Inner Shelf

Lower Shoreface

Prodelta / Inner Shelf

Kf-1-Indian Canyon-a

Prodelta / Inner Shelf

Prodelta / Inner Shelf

——————— Parasequence boundary   .... Scour surface   ——————— Lateral accretion in tidal deposit

-------- Facies boundary   ·········· Depositional top of shoreline unit   --------- Bedding plane-no environmental interpretation

Vertical scale: Arrow is approximately 10 feet
Horizontal scale: 1 inch = 100 feet

Figure 15. Generalized southwest-northeast cross section along a portion of "The Wall" showing parasequences Kf-1-IC-a, -b, and -c, and major depositional facies. See Figure 2 for location of photomosaic. Cross section covers the southwestern part of the photomosaic panel.

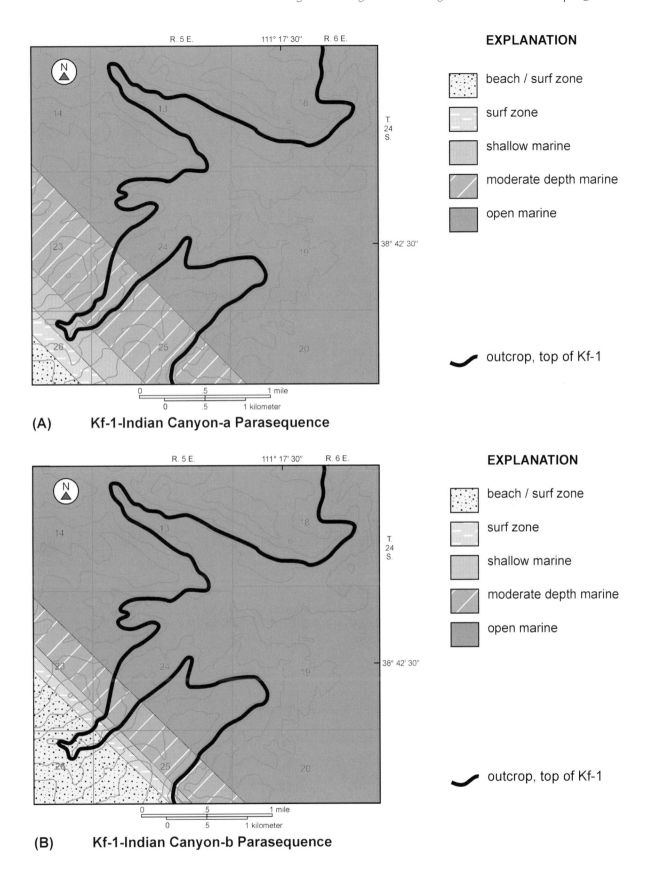

(A)   **Kf-1-Indian Canyon-a Parasequence**

(B)   **Kf-1-Indian Canyon-b Parasequence**

Figure 16. Paleogeographic maps of the study area during deposition of parasequences Kf-1-IC-a and -b. These parasequences are characterized by deposition along wave-dominated shorelines. The narrower facies belts of Kf-1-IC-b are possibly the result of a shoreface with a higher gradient than the shoreface of Kf-1-IC-a.

and probably represents only about 500–600 ft (150–180 m) of progradation. This parasequence is also inferred to have been deposited along a wave-dominated shoreline. As such, it probably had a morphology similar to that of parasequence Kf-1-IC-a. Mapping of the areal extent of lithofacies within this parasequence shows that the facies belts are narrow compared to those of parasequences Kf-1-IC-a and -c. This suggests that the gradient of the shoreface in parasequence Kf-1-IC-b was probably higher than the gradients of shorefaces in parasequences Kf-1-IC-a and -c.

Kf-1-Indian Canyon-c

The Kf-1-IC-c parasequence had a much more complex history than the previous two parasequences and its development will be considered in two stages.

Stage 1. Barrier-Tidal Inlet

Figure 17A is a paleogeographic reconstruction of the shoreline during deposition of Stage 1 of parasequence Kf-1-IC-c. The base of Kf-1-IC-c is seen at its most landward extent where lower and middle shoreface sandstones overlie planar-laminated foreshore sandstone of Kf-1-IC-b. The tidal inlet deposits of Lithofacies 6 and medium-bedded, isolated, and mollusk-bearing sandstone beds located just landward of the landward pinchout suggest that a local embayment existed immediately landward of the shoreline at the beginning of the Kf-1-IC-c parasequence. Southwest-dipping foresets within the tidal inlet suggest that the inlet filled with deposits from a laterally accreting spit. Sediment supply was sufficient to allow progradation to occur. Seaward of the inlet mouth, ebb-tidal delta deposits may have accumulated with paleocurrent indicators showing northeast-directed flow (Figures 15, 17A, and 18 location A). The embayment itself, however, probably did not last long since such lagoons tend to rapidly fill with sediment (Heward, 1981).

Stage 2. Wave-Dominated Shoreline

After continued migration of the tidal inlet, wave action dominated deposition in the study area. Much of the shoreline progradation occurred at this time. As the shoreline prograded and the embayment behind it filled, tidal channels and creeks increased in their importance in distributing the tidal prism (Figure 17B).

Kf-1-Indian Canyon-d

The end of parasequence Kf1-IC-c is marked by the transition of lithofacies resulting from wave-dominated processes to lithofacies resulting from an increasing fluvial influence. The series of events that resulted in this transition are described below.

Stage 1. Strand Plain

At the end of deposition of the Kf-1-IC-c parasequence, a river system avulsed into the area. Initially, the streams emptying into the area were small and the area was still subject to the effects of wave processes. Sediment that was delivered into the basin was redistributed, presumably along cuspate strand lines by longshore currents (Figure 19A). The initial fluvial deposits were reworked by wave action and amalgamated with the uppermost, last-deposited, wave-dominated shoreline sands (Kf-1-IC-c).

Paleocurrent indicators at the head of "North Branch" suggest that sediment derived from the northwest was initially carried by currents towards the southeast (Figure 18, location B). Paleocurrent indicators between the head of "North Branch" and the multistory channel system closer to the mouth of North Fork Canyon show that as the strand plain facies grew eastward, northwest-directed currents became dominant. These paleocurrents indicate a sediment source to the southeast. A distributary system near the south end of the study area could have provided a sediment source and produced the observed paleocurrent pattern. As the distributary system increased in size, the resulting deposits began to reflect the orientation of currents induced by the system. The result, in this case, was a transition from southeast-directed to northwest-directed flow.

Stage 2. Interdistributary Bay

Subsequent to the initiation of fluvial-dominated processes, the area at the mouth of North Fork Canyon and Willow Springs Wash became isolated and protected from the effects of wave energy (Figure 19B). The energy of the water in the area decreased and sedimentation was restricted to the deposition of fine clays during calm conditions and very fine sandstone during storms or flood events. The cause of the isolation is not directly observable within the study area, but several inferences can be made. The uppermost sandstones within the strand plain facies show transport towards the northwest by longshore currents from the proposed distributary system. Isolation of the study area would have to be the result of formation of some type of barrier to these currents located to the southeast.

In this case, isolation may have been the result of progradation of the proposed distributary system or delta lobe. As the distributary mouth prograded, longshore currents could not have carried sediment back into the newly formed interdistributary bay.

This model suggests two things. First, the distributary channels prograded rapidly enough to overcome the effects of wave reworking in the study area. Second, as distributary channels entered the area directly to the

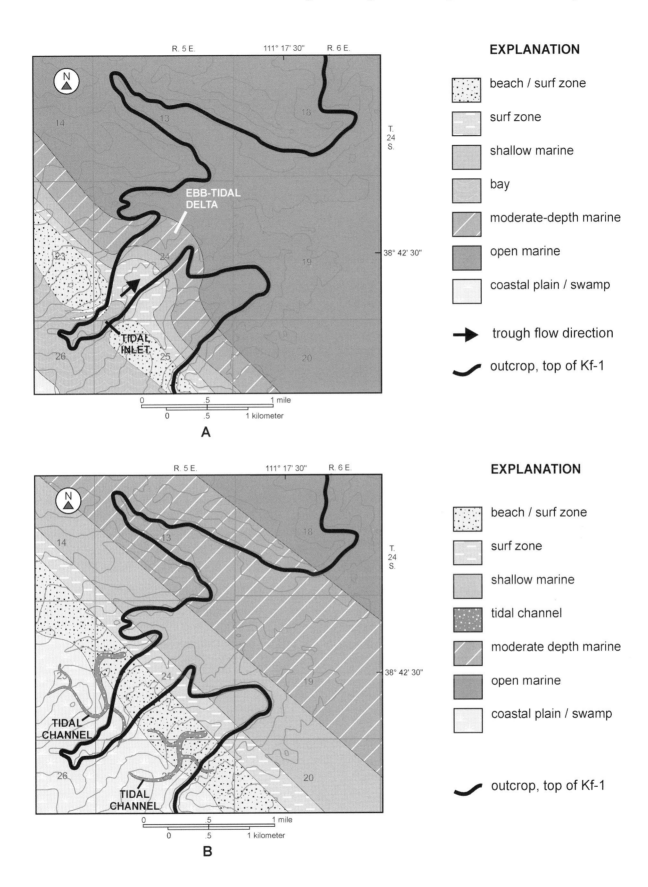

Figure 17. Paleogeographic maps of the study area during deposition of parasequence Kf-1-IC-c. (A) shows tidal inlet and possible ebb-tidal delta deposition during Stage 1. (B) shows wave-dominated deposition after the tidal inlet has migrated to the southeast during Stage 2.

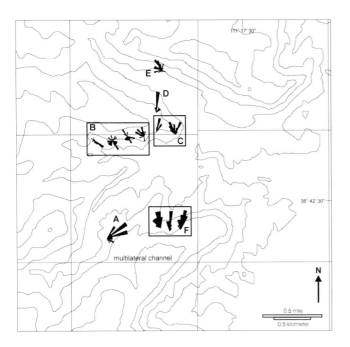

Figure 18. Paleocurrent map of deposits of depositional units C and D. (A) Paleocurrent pattern within the upper shoreface/ebb-tidal delta facies of depositional unit C shows northeast-directed flow. (B) Paleocurrent indicators within the strand plain facies of depositional unit D show a change in the dominant current direction from southeast-directed flow to northwest-directed flow. This was the result of a growing distributary system located to the southeast of the study area. (C) Paleocurrent indicators from the crevasse splay facies (depositional unit D) show northwestward flow into an interdistributary bay. (D) Deposits within the multistory channel facies in Willow Springs Wash show northward flow (depositional unit D). (E) Deposits of depositional unit D show variable current pattern of distal-mouth bar. (F) Deposits within the multilateral channel facies show flow to the north, toward the multistory channel in North Fork Canyon (depositional unit D).

east, the resulting deposits rapidly changed from a cuspate, strand-line morphology to a more elongate, possibly shoestring morphology. The outcrops needed to confirm or deny this, however, not preserved due to regional uplift and erosion.

Stage 3. Crevasse Splay-Minor Mouth Bar

Stage 3 marks the deposition of sandstones at the mouth of North Fork and Willow Springs Wash. Northwest-dipping, *en echelon* bedsets of Lithofacies 8 deposited as minor distributary mouth bars suggest that these sandstone beds were sourced from the east or southeast. Paleocurrent indicators on the ridge separating North Fork and Willow Springs Wash (Figure 18, location C) show north and northwestward flow. These deposits were the result of a crevasse splay developing adjacent to the inferred distributary channels to the east (Figure 20A). Most of the deposits of this crevasse-splay system have been removed by regional uplift and erosion, so it

is impossible to accurately determine the original dimensions of the deposit. We can infer from their absence in the north side of Willow Springs Wash that they did not extend that far north.

Stage 4. Distributary Channel

Finally, a major distributary channel avulsed into the immediate area. Two types of distributary channel deposits are present in the area. On the north side of North Fork Canyon, deposits of a multistory, non-meandering channel are preserved. Deposits of this same non-meandering channel are also present to the north, on the south side of Willow Springs Wash. The deposits in Willow Springs Wash are not as thick as those in North Fork and they tend to be a little wider. On the north side of Willow Springs Wash, the thick channel sandstone accumulations are gone and in their place are the widely spaced deposits of three or four much smaller channels. These smaller channels represent bifurcation of the single, large channel into several smaller channels and distributary mouth bars as the distributary entered the open, distributary bay (Figure 20B).

To the south, along the south and west walls of "The Alcove" and on the south wall of Indian Canyon, the deposits of a meandering channel system are exposed. This channel is not in direct contact with the multistory channel in North Fork and it is not possible to determine the exact relationship between these two deposits. One hypothesis is that the straight-reach system in North Fork is the distal end of the meandering system and that the two existed simultaneously. Paleocurrent data from "The Alcove" and from the south side of Indian Canyon both indicate northward flow towards the North Fork channel (Figure 18F). However, comparison of single, lensoidal sandstone bodies in the Indian Canyon exposures with single, lensoidal sandstone bodies within the North Fork exposures suggests that the North Fork channel wasn't large enough to carry the volume of water moving through the meander belt. This problem would be resolved if there were other multistory channels branching from the meander belt.

Another possibility is that the North Fork channel was deposited prior to the development of the meander belt and that the meander belt eroded previously deposited sediment from the straight-reach system. In this case, we should expect to find deposits of the meander belt or an associated distributary mouth bar system in or near Willow Springs Wash. Presently, we do not find this in the exposed and preserved outcrop, but the absence of these deposits may be the result of the uplift and erosion that exposed the outcrop.

Overlying the bay fill, crevasse splay, and distributary channel deposits, carbonaceous shale and siltstone deposits of the Sub-A coal zone suggest development of peat swamp conditions throughout the area.

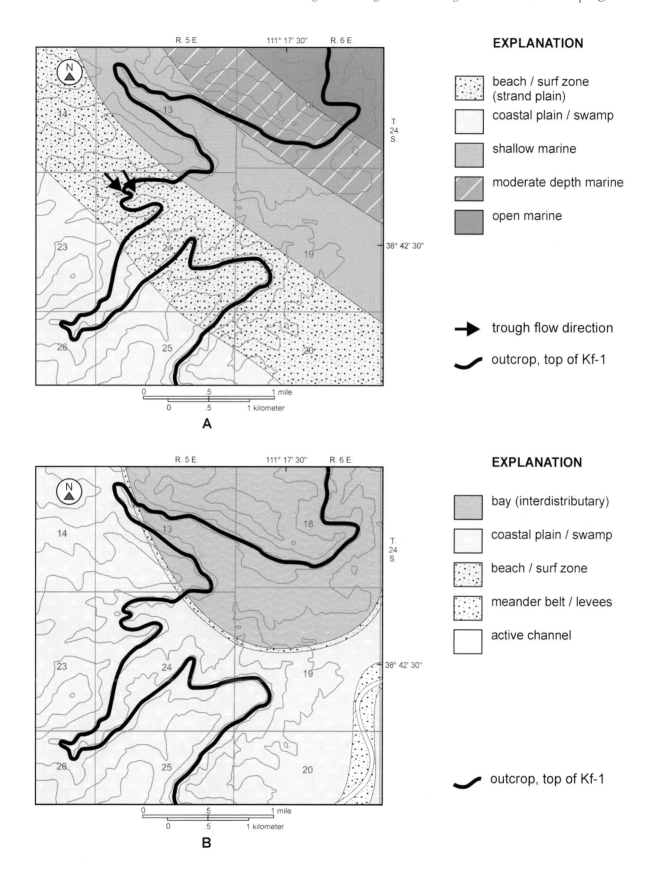

Figure 19. Paleogeographic maps of the study area during deposition of parasequence Kf-1-IC-d. (A) shows strand plain deposits alongshore from distributary mouths during Stage 1. (B) shows the interdistributary bay that formed as a result of prograding distributary channels during Stage 2. Both sets of deposits occur within depositional unit D.

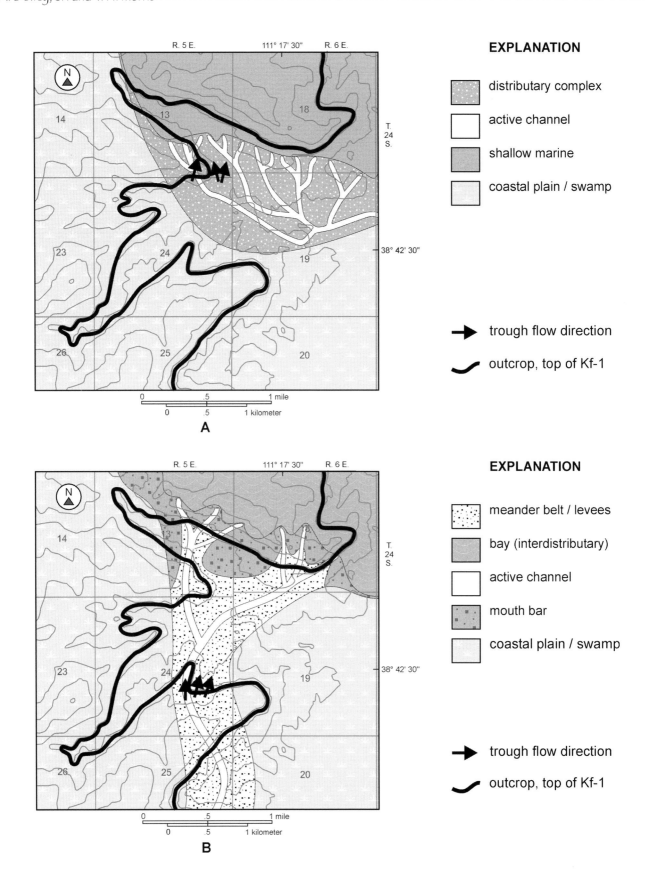

Figure 20. Paleogeographic maps of the study area during Stage 3 (A) and Stage 4 (B) of parasequence Kf-1-IC-d. (A) shows northwestward progradation of a crevasse splay system into an interdistributary bay. During Stage 4, a distributary avulsed into the immediate study area. (B) shows the multilateral channel bifurcating into several smaller multistory channels.

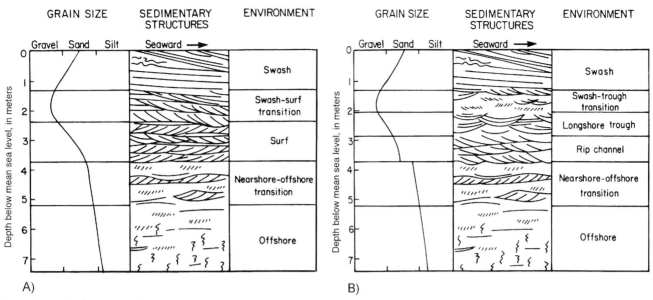

Figure 21. Vertical sequence of sedimentary structures observed within (A) non-barred and (B) barred shorelines of the Oregon coast. The orientation of sedimentary structures within "the Wall" eliminates the non-barred shoreline model. The orientation of sedimentary structures in "The Wall" roughly coincides with orientations observed for barred shorelines. The abrupt increase in grain size and the evidence for a significant scoured horizon are not observed. After Hunter et al. (1979).

DISCUSSION

Variations in Depositional Models

The precise nature of the depositional mechanism that formed the thick-bedded, trough and tabular cross-stratified lithofacies (Lithofacies 4) in depositional unit C is not definitely known. Previously, this lithofacies has been interpreted to form within an upper shoreface environment. The close association of this facies with facies that have been interpreted as lower and middle shoreface and foreshore seems to support this interpretation. However, two other important observations need to be addressed. First, paleocurrent data taken from within this lithofacies on the southeast side of Indian Canyon suggest that, with some minor variation, sediment transport was generally in an offshore direction (Figure 18A). Second, the presence of a tidal inlet in the immediate vicinity suggests that, at least during Stage 1, tides may have affected deposition within this facies. Comparisons of these deposits with models of barred and non-barred shorelines and tidal deposits give some insight into possible variations of the depositional environmental significance of this lithofacies.

Barred and Non-Barred Shoreline Models

Work by Clifton et al. (1971) along the Oregon coast illustrates the types of sedimentary structures, and perhaps more importantly, the orientation of sedimentary structures along high-energy shorelines. They examined non-barred, high-energy shorelines. The vertical se-

quence of sedimentary structures they described is shown in Figure 21A. They divided the high-energy nearshore into five zones. From base to top these zones are: (1) offshore, (2) nearshore-offshore transition, (3) surf zone, (4) swash-surf transition, and (5) swash zone. The surf zone represents the zone of shoaling waves. It is characterized by lunate megaripples and resulting trough cross-stratification. This trough cross-stratification is oriented onshore. This differs from the orientation of the bedforms studied within the trough cross-stratified facies of "The Wall" in Indian Canyon.

Hunter et al. (1979) examined the deposits within barred, high-energy systems and found a variety of possible bedform orientations. Bedforms deposited within longshore bars should be oriented onshore or obliquely onshore. Deposits within longshore troughs show longshore orientations. The rip-channel facies contains bedforms oriented offshore or obliquely offshore. Rip-channel mouth bars also show seaward-oriented bedforms. Hunter et al. (1979) also predicted the vertical sequence of structures that would be produced as a barred shoreline progrades (Figure 21B). The vertical facies sequence would show a progression from offshore and nearshore-offshore transition facies to rip-channel, longshore trough, and finally, a transition to the swash facies. In general, the grain size would increase upward in the section through the longshore trough facies with an abrupt increase across the base of the rip-channel facies. This abrupt increase in grain size is the result of scour into finer-grained sediments by rip currents and the deposition of a basal lag.

Davidson-Arnott and Greenwood (1976) studied deposits of shore-parallel bars. These deposits vary somewhat from those of oblique bars. Shore-parallel bars contain three lithofacies. The seaward side of the bar contains ripple lamination and seaward-dipping parallel laminations. The bar crest contains horizontal bedding and trough cross-stratification. The landward facies consists primarily of landward-dipping planar lamination. Successive bars are separated by troughs with current ripples oriented perpendicular to the shoreline. Rip-channel deposits are characterized by scoured bases overlain by abrupt increases of grain size and seaward-oriented cross-stratification.

The seaward-oriented cross-stratification within Lithofacies 4 is consistent with the deposits of rip-channels or rip-channel mouth bars. However, neither the abrupt increase in grain size nor the type of scour that would result from longshore migration of rip-channels have been observed in the study area. Furthermore, the landward-dipping, planar lamination associated with the landward side of shore-parallel bars has not been observed. The absence of prominent grain size variations is common within the study area. With the exception of fluvial channels, the grain sizes within the Ferron deltaic system generally range from very fine to upper fine sandstone and some medium sandstone. However, the lack of any evidence of landward-oriented or landward-migrating structures within the lower two-thirds of Lithofacies 4 is not explained.

Ebb-Tidal Delta Model

An alternative to the barred and non-barred shoreline models is the ebb-tidal delta model. Deposition within an ebb-tidal delta can explain the seaward-oriented cross-stratification within "The Wall." The landward and variably oriented section of cross-stratification overlying this could represent deposits within swash bars flanking and overlying the ebb-tidal delta (M.O. Hayes, verbal communication, 1995).

Ebb-tidal deltas develop best under mesotidal conditions; however, they can form under high microtidal conditions as well (M.O. Hayes, verbal communication, 1995). Davis and Hayes (1984) found that the critical condition to be met is not absolute tidal range, but the balance between tidal range and wave energy. If tidal energy and wave energy are balanced, a localized increase in tidal range resulting from local variations in shoreline morphology could result in an increase in the dominance of tidal deposition.

The Ferron Sandstone is generally thought of as having been formed under wave-dominated, microtidal conditions (T. A. Ryer, verbal communication, 1995). The importance of waves is certainly evident in the thick, sand-rich deposits forming "The Wall" of Indian Canyon. The existence of ebb-tidal delta deposits would imply tidal ranges that are higher than previously thought for the Ferron Sandstone. Several pieces of evidence suggest that an ebb-tidal delta could have existed in this area. First is the observation, already made above, that ebb-tidal deltas can form under microtidal conditions. Second, along embayments such as the modern Georgia Bight, tidal range increases away from the headlands and towards the apex of the concavity of the bight. With this increase in tidal range, the dominance of tidal deposits increases as well (M.O. Hayes, verbal communication, 1995). Such an embayment may have existed at this time. Third, the identification of tidal-inlet deposits within Indian Canyon suggests that an ebb-tidal delta is a reasonable association of facies. Finally, though many sedimentologists believed that ebb-tidal deltas are usually reworked and redeposited by some subsequent depositional mechanism (such as longshore drift), some authors suggest that ebb-tidal deposits could be preserved under prograding shoreline conditions, (Heward, 1981). It is certain that this shoreline prograded, and the upward- and seaward-stepping parasequence architecture suggests that at least on longer scales, sedimentation was overwhelming sea-level rise.

Realistically, a hybrid involving both the ebb-tidal delta model and a barred shoreline model may have produced the observed features within Indian Canyon. Slight differences in the depositional models may affect hydrocarbon reservoir potential, however, conclusive evidence to eliminate either model may not exist.

Sequence Stratigraphic Implications

Sequence stratigraphy has become a powerful tool in the interpretation of ancient rock successions. The use of these concepts has brought new insight into areas of stratigraphy previously thought to be well understood. In light of this, it is appropriate to consider the consequences of the previous discussion on the sequence stratigraphy of the Ferron Sandstone in the study area.

As discussed earlier, the four depositional units are divided into three allocyclic, marine-dominated parasequences (Kf-1-IC-a, -b, and -c) and one autocyclic, fluvial-dominated parasequence (Kf-1-IC-d). This interpretation is based on the identification of marine-flooding surfaces from vertical offsets of landward pinchouts of marine sandstone bodies and detailed facies analysis. The identification of different parasequence types may seem to be an insignificant detail, but it has numerous implications.

Let us consider the implications of this interpretation. According to the paleogeographic interpretation stated above, a river system avulsed into the area during or soon after deposition of parasequence Kf-1-IC-c. Accordingly, deposition within the study area changed from wave-dominated deposition to fluvial-dominated deposition. However, there should have existed at the same time some portion of the shoreline that was directly affected by wave energy (eastward). This might have been an area that was completely unaffected by distributary systems or a portion of a deltaic system that was unprotected from approaching waves. Wave-dominated deposits in this area are within the same parasequence as the fluvial-dominated deposits in Indian Canyon. They are, however, spatially distinct.

One application of sequence stratigraphic interpretation and parasequence identification is to identify single, communicable, reservoir-quality facies in different wells. Sequence stratigraphic interpretations have the potential to reflect more accurately relationships between sandstone bodies and this greater understanding can be applied to production enhancement (Van Wagoner et al., 1990).

The benefit of correctly identifying allocyclic and autocyclic parasequences can be seen when attempting to predict facies distributions. Taking the example of parasequence Kf-1-IC-d, we will first predict facies distributions with an "incorrect" interpretation. The coarsening-upwards sequence observable in the interdistributary bay and crevasse splay deposits of Kf-1-IC-d superficially resembles the coarsening-upwards sequence observed in the transition from prodelta, lower shoreface, and middle shoreface facies of Kf-1-IC-a, -b, -c. Assuming the (erroneous) interpretation that these were deposits of another allocyclically controlled, wave-dominated shoreface, we should expect to find a thick, reservoir-quality sandstone body landward of the location where we collected our data from Kf-1-IC-d and at a position stratigraphically higher than (or vertically offset from) the reservoir-quality shoreface succession within Kf-1-IC-c. This, however, is not the case. What we do see in the stratigraphically higher and more landward location is a succession of fine-grained delta plain carbonaceous shales, siltstones and tight, thin, very fine-grained sandstones.

Erroneously interpretating the controls over parasequence formation within the study area could also result in miscalculations of rates of sea-level rise (or the rate of subsidence) occurring during a given interval of time or confusion between scoured channel sandstone bases and incised valleys. Subsequent errors in sequence boundary identification and lithofacies prediction would, in turn, result.

It should be noted that a correct parasequence interpretation is dependent upon correct lithofacies and depositional environment interpretations. In an oil field, the actual amount of data available, regardless of how mature the field, is small in comparison to the data available from outcrop. The ability of a geologist to make detailed interpretations of depositional environments is limited by the amount of available core, the resolution of any seismic data, and the degree to which wireline log responses to lithofacies and depositional environments are understood. This could also limit the ability of the geologist to make a correct sequence stratigraphic interpretation. Hopefully, data such as those presented here will provide an analog that will allow the geologist to formulate more working hypotheses. This may also allow the geologist to correctly identify and understand subsurface reservoir rocks.

Reservoir Quality

Analysis of the facies and depositional environments within the Willow Springs Wash study area allows us to predict which facies would serve as better hydrocarbon reservoirs. This is done by identifying those facies which have the thickest accumulations of clean sandstone, the coarsest grain size, the largest areal extent, and the fewest barriers to fluid flow.

Wave-dominated deposits in parasequences Kf-1-IC-a and -c appear to meet the criteria of greater thickness, larger areal extent, and abundance of coarser grains. The fluvial-deltaic deposits of Kf-1-IC-d are generally finer grained, thinner, less extensive areally, and typically have more abundant heterolithic breaks (such as the interdistributary bay-fill and crevasse splay/minor distributary mouth bar deposits). The deposits of medium-grained sandstone are generally thin (1–2 ft [0.3–0.6 m]) and not very extensive. Based upon these criteria, we can conclude that the wave-dominated deposits create better-quality reservoir facies.

In reality, however, the geologist and the reservoir engineer can not choose the type of deposits that hold the hydrocarbon. Indeed, the problem is that the hydrocarbons are found within a variety of fluvial-deltaic facies. Because of this circumstance, the geologist and the reservoir engineer need to correctly identify the facies characteristics through the use of detailed facies identification, sequence stratigraphic analysis, and the use of outcrop analogs in an effort to develop optimal production strategies.

CONCLUSIONS

Six important conclusions can be drawn from this case study of the Ferron Sandstone in the Willow Springs Wash area:

1. Eleven distinct depositional facies can be identified in the Kf-1 parasequence set in Indian Canyon of the Willow Springs Wash study area. These are: (1) prodelta-inner shelf, (2) lower shoreface, (3) middle shoreface, (4) upper shoreface-ebb tidal delta, (5) foreshore, (6) tidal inlet, (7) interdistributary bay-fill, (8) crevasse-splay, (9) multistory channel, (10) multilateral channel, and (11) strand plain/beach ridge facies. These facies can be grouped into two depositional systems: wave-dominated shoreline and fluvial-dominated delta.

2. Based on lateral and vertical facies associations, the rocks can be divided into four mappable depositional units (parasequences). In ascending order they are Kf-1-IC-a, -b, -c, and -d. Marine-flooding surfaces, which are identified by vertical offset in landward pinchouts as well as the juxtaposition of more "distal" deposits over more "proximal," separate the -a, -b, and -c depositional units. In contrast, there is no marine flooding surface separating the -c and -d units and therefore no vertical offset at their landward pinchouts. Hence, the Kf-1-IC-c and -d parasequences are in communication in an updip direction even though they are vertically separated in a downdip direction. Subsequently, we divide all of the depositional units into two categories of parasequences: allocyclically controlled parasequences (Kf-1-IC-a, -b, -c) and autocyclically controlled parasequences (Kf-1-IC-d).

3. During the deposition of parasequences Kf-1-IC-a, -b, and the early part of Kf-1-IC-c, the study area was characterized by northeastward progradation of a northwest-southeast striking shoreline. Parasequence Kf-1-IC-a and the early part of parasequence Kf-1-IC-c represent longer pulses of progradation while parasequence Kf-1-IC-b represents a shorter pulse. During the later part of parasequence Kf-1-IC-c, river avulsion created an autocyclic change in deposition that ultimately resulted in parasequence (Kf-1-IC-d). During this time, sedimentation within the study area was characterized by deposition of a crevasse-splay within an interdistributary bay.

4. These interpretations suggest that parasequence formation may be independent of changes in relative sea level. Misinterpretation of these autocyclic parasequences as marine-dominated allocyclic parasequences may cause the geologist to incorrectly visualize lithofacies relationships and compartmentalization (or the lack of it) within a reservoir. Because of this, it is important to accurately identify lithofacies and depositional environments and then use these conclusions to modify sequence stratigraphic interpretations where possible.

5. Depositional environments may greatly affect the quality of a sandstone body as a hydrocarbon reservoir. The presence of fine-grained breaks and heterolithic facies will affect the way that fluids flow through a rock body. These characteristics of sandstone bodies are largely a result of depositional processes. For this reason, detailed facies analysis is a useful tool in oil field development. However, limitations on the available knowledge of rock characteristics from logs (e.g. sedimentary structures, internal architecture, etc.) may prohibit a complete and accurate interpretation. This illustrates the need for outcrop analogs.

6. The complex interplay between environments within a fluvial-deltaic system coupled with the effects of relative sea-level change can make the identification of depositional environments and parasequences a difficult task. Log, core, and high resolution seismic data may be integrated with realistic outcrop analogs to reduce error and risk when creating field development strategies. Accurate interpretations can lead to more detailed conclusions about the type and scale of heterogeneity within a reservoir and the locations of prospective targets.

ACKNOWLEDGMENTS

Special thanks to Tom Ryer whose expertise on the Ferron pointed the way to this most enlightening field study area. Photomosaic images were originally acquired by Tom Ryer and Paul Anderson. This research was funded by the Utah Geological Survey (contract number 94-2067), the U.S. Department of Energy (contract number DE-AC22-93BC14896), and in part by the Department of Geology at Brigham Young University. We thank Douglas A. Sprinkel (Utah Geological Survey) and Kevin Hae Hae (Anadarko Petroleum Corp.) for their constructive technical reviews.

REFERENCES CITED

Anderson, P. B., K. McClure, T. C. Chidsey, Jr., T. A. Ryer, T. H. Morris, J. A. Dewey, Jr., and R. D. Adams, 2003, Interpreted regional photomosaics and cross section, Cretaceous Ferron Sandstone, east-central Utah: Utah Geological Survey Open-File Report 412, compact disk.

Clifton, H. E., R. E. Hunter, and R. L. Phillips, 1971, Depositional structures in the non-barred high-energy nearshore: Journal of Sedimentary Petrology, v. 41, no. 3, p. 651–670.

Coleman, J. M., 1976, Deltas — processes of deposition and models for exploration: Champaign, Continuing Education Publication Company, Inc., 102 p.

Davidson-Arnott, R. D. G., and B. Greenwood, 1976, Facies relationships on a barred coast, Kouchibouguac Bay, New Brunswick, Canada, in R. A. Davis, Jr., and R. L. Ethington, eds., Beach and nearshore sedimentation: Society for Sedimentary Geology (SEPM) Special Publication 24, p. 149–168.

Davis, R. A., and M. O. Hayes, 1984, What is a wave-dominated coast?: Marine Geology, v. 60, p. 313–329.

Dott, R. H., Jr., and J. Bourgeois, 1982, Hummocky stratification — significance of its variable bedding sequences: Geological Society of America Bulletin, v. 93, no. 8, p. 663–680.

Emery, D. and K. L. Myers (eds.), 1996, Sequence Stratigraphy: Malden, Blackwell Science, 297 p.

Heward, A. P., 1981, A review of wave-dominated clastic shoreline deposits: Earth Science Reviews, v. 17, p. 223–276.

Hunter, R. E., H. E. Clifton, and R. L. Phillips, 1979, Depositional processes, sedimentary structures, and predicted vertical sequences in barred nearshore systems, Southern Oregon Coast, Journal of Sedimentary Petrology, v. 49, no. 3, p. 711–726.

Miall, A. D., 1985, Architectural element analysis — a new method of facies analysis applied to fluvial deposits: Earth Science Reviews, v. 22, p. 261–308.

Ryer, T. A., 1981, Deltaic coals of Ferron Sandstone Member of Mancos Shale — predictive model for Cretaceous coal-bearing strata of Western Interior: AAPG Bulletin, v. 65, no. 11, p. 2323–2340.

—1991, Stratigraphy, facies, and depositional history of the Ferron Sandstone near Emery, Utah, in T. C. Chidsey, Jr., ed., Geology of east-central Utah: Utah Geological Association Publication 19, p. 45–54.

Ryer, T. A., J. A. Dewey, Jr., and T. H. Morris, 1995, Distinguishing allocyclic and autocyclic causes of parasequence-level cyclicity — lessons from deltaic strata of the Upper Cretaceous Ferron Sandstone, central Utah (abs.): AAPG Bulletin, v. 79, no. 6, p. 924.

Van Wagoner, J. C., R. M. Mitchum, K. M. Campion, and V. D. Rahmanian, 1990, Siliciclastic sequence stratigraphy in well logs, cores and outcrops: AAPG Methods in Exploration, no. 7, 55 p.

"Indian picture writing," near Emery, Utah, circa 1910. Photograph courtesy of the family of C. T. Lupton.

Analog for Fluvial-Deltaic Reservoir Modeling: Ferron Sandstone of Utah
AAPG Studies in Geology 50
T. C. Chidsey, Jr., R. D. Adams, and T. H. Morris, editors

Geologic Framework, Facies, Paleogeography, and Reservoir Analogs of the Ferron Sandstone in the Ivie Creek Area, East-Central Utah

Paul B. Anderson[1], Thomas C. Chidsey, Jr.[2],
Thomas A. Ryer[3], Roy D. Adams[4], and Kevin McClure[2]

ABSTRACT

The Ferron Sandstone of east-central Utah has world-class outcrops of dominantly fluvial-deltaic, Turonian-Coniacian-aged strata deposited along the margins of the rapidly subsiding Cretaceous foreland basin. The Ferron consists of a series of stacked, transgressive-regressive cycles which form an eastward-thinning wedge. The Ivie Creek area contains abrupt facies changes in two of these cycles referred to as Kf-1 and Kf-2.

Kf-1 consists of unusual river-dominated delta deposits that prograde southeast to northwest across the Ivie Creek area. Progradation is parallel or onshore to the regional shoreline trend. Distinctive, steeply inclined bedsets or clinoforms, defined by bounding surfaces are classified into four facies: proximal, medial, distal, and cap. Clinoform facies are based on grain size, sedimentary structures, bedding thickness, inclination angle, and stratigraphic position. These deposits accumulated on an arcuate delta lobe which was prograding into a deeper water, fully marine bay. The main delta, which we interpret to have been located to the east and northeast, created a protected embayment in the northwest part of the Ivie Creek area. The Kf-1 clinoforms represent deposition into the embayment fed by river channels from the southeast.

Kf-2 is represented by wave-modified deltaic deposits that generally coarsen east to west, and consist of shoreface and distributary complex facies. These relatively clean, sand-rich deposits accumulated along a local north-south shoreline trend defined by a landward pinchout of marine shoreface facies, as opposed to the more common regional northwest-southeast shoreline trend recognized in other Ferron cycles above and below Kf-2. In the western part of the Ivie Creek area, east- to northeast-flowing distributary channels deposited large amounts of sand in north-south-trending distributary-mouth bars. Shallow- to moderate-depth marine conditions existed in the eastern part of the area. An uncommon transition from shoreface, to bay, to coastal plain/swamp occurred during the late stage of Kf-2 deposition.

As a reservoir analog, the Ferron Sandstone in the Ivie Creek area displays variations in sedimentary structures and lithofacies that influence both its compartmentalization and permeability. Bounding layers like those observed on outcrop were identified from core and geophysical well-log data. These features can be incorporated into reservoir models and simulations for oil field development and secondary or enhanced oil recovery programs.

[1]Consulting Geologist, Salt Lake City, Utah; [2]Utah Geological Survey, Salt Lake City, Utah;
[3]The ARIES Group, Inc., Katy, Texas; [4]A&R Resources, LLC., Salt Lake City, Utah

INTRODUCTION

The Ivie Creek area (Figure 1) contains abrupt facies changes in the delta-front sandstones or parasequence sets of the Turonian-Coniacian-aged Ferron Sandstone, east-central Utah. These deposits represent the primary lithofacies typically found in a fluvial-dominated deltaic reservoir. Sedimentary structures, lithofacies, bounding surfaces, and permeabilities measured along closely spaced traverses were combined with drill-hole data to develop the paleogeographic picture and a three-dimensional view of the reservoir analogs within the area.

The Ferron Sandstone in the Ivie Creek area consists of two regional-scale parasequence sets as described by Anderson and Ryer (this volume), the Kf-1 and Kf-2 (Figure 2). However, in the Ivie Creek area the Kf-1 parasequence set consists of only one parasequence, Kf-1-Ivie Creek (Iv), with two bedsets referred to here as Kf-1-Ivie Creek[a] and Kf-1-Ivie Creek[c]. The older Kf-1-Iv[a] bedset has been interpreted as a fluvial-dominated delta deposit which changes from proximal to distal from east to west. The Kf-2 parasequence set represents wave-modified deposits. Each type of deltaic system produces differing geometries that can be characterized by geologic parameters for petroleum exploration or reservoir modeling efforts.

The objectives of this study were, therefore, to determine: (1) how variations in sedimentary structures and lithofacies influence both reservoir compartmentalization and permeability, (2) how bounding surfaces can be identified from core and geophysical well-log data, and (3) ultimately, how such features should be considered in oil field development and secondary or enhanced oil recovery programs.

Location

The Ivie Creek area is located in the central part of the Ferron outcrop belt (Figure 1), about 6 mi (9.7 km) south of the town of Emery, Utah, in the center of T23S, R6E, Salt Lake Base Line (SLBL), Emery County. The Ferron Sandstone outcrops straddle Interstate 70 (I-70). A small stream, Ivie Creek, flows just north of the interstate highway, hence the name Ivie Creek area.

Access to the area is excellent because of proximity to I-70. The "Ivie Creek amphitheater" is an informal name applied to a broad, curving area of cliffs north of I-70 where the most detailed work was conducted (Figure 3).

Methods

Within the study area, outcrops along Ivie Creek, Quitchupah Creek, and the I-70 road-cut, along with Utah Geological Survey (UGS) drill holes provide information in three dimensions on rock bodies of the Kf-1 and Kf-2 parasequence sets. Surface and subsurface

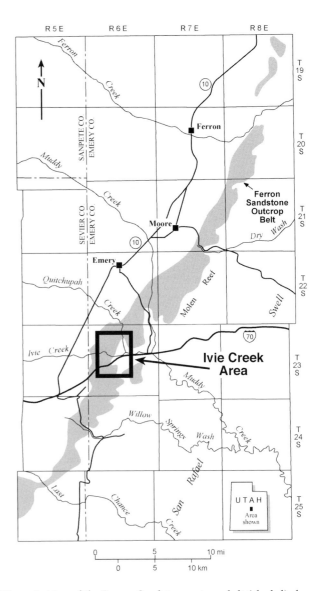

Figure 1. Map of the Ferron Sandstone outcrop belt (shaded) showing location of the Ivie Creek area.

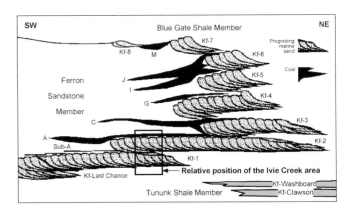

Figure 2. Diagrammatic cross section showing relative positions of Ferron parasequence sets with their associated coal zones (black), and the relative stratigraphic position of the Ivie Creek area within the Ferron Sandstone (from Ryer, 1991; Anderson et al., 1997). The diagram has no scale.

data (Figure 3) were collected to: (1) map variations in lithofacies, sedimentary structures (size, geometry, orientation/dip, and percent of sedimentary structures), and bedding types, (2) characterize bounding surfaces both between and within sandstone bodies, and (3) merge with permeability transect data. These data consist of measured stratigraphic sections (23), paleocurrent measurements (about 400), scaled interpreted photomosaics (21), drill-hole core logs (four), and geophysical logs (five sets) (Anderson et al., 2003). Measured sections were correlated to interpret stratigraphy, lithofacies, depositional environment, and define reservoir components in the statistical and reservoir-modeling parts of a U.S. Department of Energy-funded project entitled *Geological and Petrophysical Characterization of the Ferron Sandstone for 3-D Simulation of a Fluvial-Deltaic Reservoir* (see Jarrard et al., this volume; Mattson et al., this volume; Forster et al., this volume).

The drill holes in the Ivie Creek area were located downdip 200–1200 ft (60–265 m) from the Ferron outcrop to provide data for a three-dimensional (3-D) morphologic interpretation of individual lithofacies and to capture the various reservoir changes in the Kf-1 and Kf-2 parasequence sets (Figure 3). Where borehole conditions allowed, drill holes were logged with sonic, density, neutron, focused resistivity, spectral gamma ray, and dipmeter tools. Cuttings were described and tied to existing wells in the area to pick tops and core points. A total of 586 ft (179 m) of core was recovered and described from the Kf-1 and Kf-2 parasequence sets.

STRATIGRAPHY

The Ferron Sandstone in the Ivie Creek area consists of two, regional-scale parasequence sets, the Kf-1 and Kf-2 (Figures 2 and 4 [see detailed cross sections in Anderson et al., 2002) and photomosaics in Plates 1 through 11 on the CD-ROM, and in Anderson et al., 2003]). The Kf-1 represents fluvial-dominated delta deposits that change from proximal to distal. In the Ivie Creek area, two bedsets (see Anderson and Ryer, this volume, for a definition of bedset) are present within the Kf-1-Iv parasequence: the Kf-1-Iv[a] and Kf-1-Iv[c] (Figures 4 and 5). Kf-1 prograded from the south-southeast to the north-northwest as fluvial-dominated delta lobes represented by two bedsets. Most of the sediments in Kf-1 cycles accumulated as delta lobes, sourced from a point. Locally, the pods of sediment in Kf-1 cycles indicate deposition in a protected environment such as a bay. Kf-1 lies conformably upon the Tununk Shale Member of the Mancos Shale. Kf-1 is actually the second parasequence set from the base of the "upper" Ferron (the Kf-Washboard and Kf-Clawson lie within the

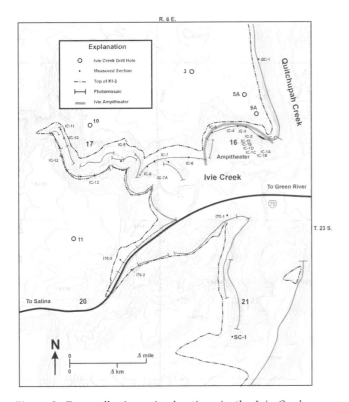

Figure 3. Data collection point locations in the Ivie Creek area. Selected photomosaics, Plates 1 through 11, are found on the CD-ROM.

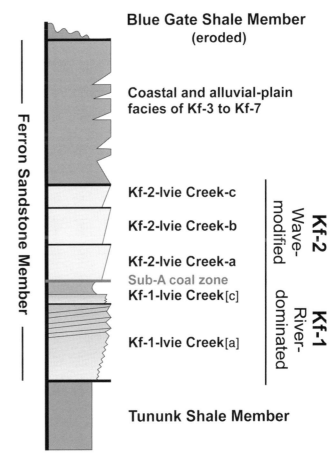

Figure 4. Ivie Creek area Ferron stratigraphy including parasequence sets, parasequences, and named bedsets.

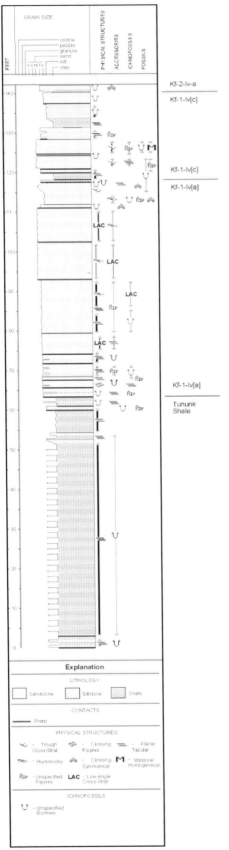

Figure 5. Stratigraphic section from the Ivie Creek amphitheater of Kf-1 (originally at a scale of 1 inch = 10 ft [2.54 cm = 3 m]) showing lithology, nature of contacts, sedimentary structures, ichnofossils, and bedset designations.

"lower" Ferron). The first parasequence set feathers into the Tununk Shale before reaching the Ivie Creek area.

The Ferron sediment source switched from the east-southeast to the west going from Kf-1 to Kf-2 deposition. Kf-2 contains more and cleaner sand, indicating a more wave-influenced environment of deposition. The Kf-2 parasequence set contains three parasequences: the Kf-2-Iv-a, Kf-2-Iv-b, and Kf-2-Iv-c (Figure 6). These parasequences show less lateral variation in lithofacies than the Kf-1 bedsets, due to greater wave influence. Within the study area, there is also little lateral variation in thickness of sand-rich Kf-2 lithofacies, even when lateral change occurs from one depositional subfacies to another. Kf-2 cycles accumulated in sheet-like bodies that pin-

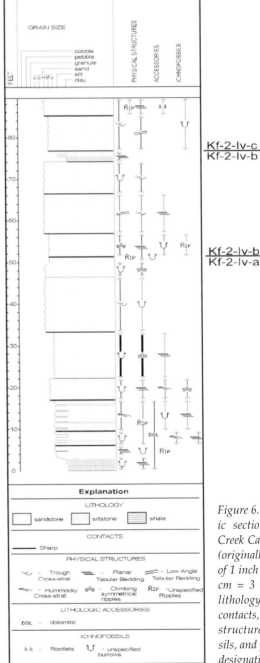

Figure 6. Stratigraphic section from Ivie Creek Canyon of Kf-2 (originally at a scale of 1 inch = 10 ft [2.54 cm = 3 m]) showing lithology, nature of contacts, sedimentary structures, ichnofossils, and parasequence designations.

334

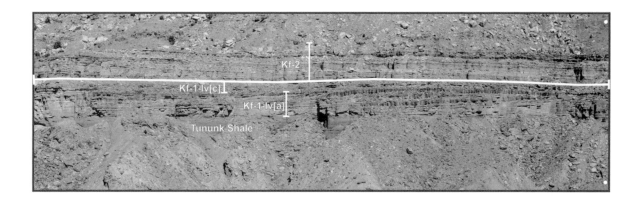

Figure 7. Photomosaic of Ferron parasequence sets, view to the north, of the Ivie Creek area displaying contrasting delta-front architectural styles (top is left panel, bottom is right panel). The heavy, white, horizontal line is a parasequence-set boundary separating Kf-1 and Kf-2 and drawn on top the Sub-A coal zone. Kf-1-Iv[a] has steeply inclined (10–15°) clinoforms representing fluvial-dominated deposition. Kf-2 has gently inclined (< 3°) clinoforms representing wave-modified deposition. Outcrop in photomosaic located in SW1/4NE1/4 Sec. 16, T23S, R6E, Salt Lake Base Line, Emery County, Utah.

chout laterally, forming a wedge due to wave-action along the delta front. The contact between Kf-1 and Kf-2 is generally drawn at the top of the Sub-A coal zone in the Ivie Creek area (Figure 7). The top of Kf-2 lies near the top of the C coal, and includes all of the A coal zone and delta-plain strata which separate the A and C coal zones (see Figure 2). Above the Kf-2 parasequence set are coastal-plain and alluvial-plain facies which, although not divided here, represent the landward equivalents of the marine portions of Kf-3 through Kf-7 parasequence sets.

Kf-1-Iv[a] Bedset

The Kf-1-Iv[a] bedset can be traced on outcrop photomosaics to its landward-most position in mid-Blue Trail Canyon south of I-70 to its seaward extent in mid-Quitchupah Canyon. The distance from landward to seaward pinchout along a depositional dip line is about 1.5 mi (2.4 km). Although the photo-documentation of the landward pinchout is poor, the unit clearly does not exhibit offset into the non-marine lithofacies and hence, by our definition, has been designated a bedset instead

of a parasequence. This bedset thins rapidly to the west as well as to the more typical seaward direction of north. This is anomalous for Ferron delta lobes.

The Kf-1-Iv[a] bedset represents fluvial-dominated deltaic lobes prograding and dipping to the north-north-west. This bedset is a fan-like deposit that thickens to the east and was deposited into an area having minimal wave influence; therefore, the primary bed forms are preserved (Figure 8).

Delta-front sandstones of a modified Gilbert delta make up these sand-rich deposits which change from proximal to distal (where the sandstone pinches out) as they pass from east to west across the Ivie Creek area. Smaller sedimentary packages than the bedset are recognized within Kf-1-Iv[a]. Figure 9 illustrates these smaller units. In descending order they are: (1) subcycle, (2) clinoform, and (3) bounding layer.

Subcycle Definition

A **subcycle** divides the bedset into smaller, recognizable, coarsening-upward, bed-thickening-upward units. The mappable extent of the subcycle is distinctively smaller than the bedset. It is possible that the unit's for-

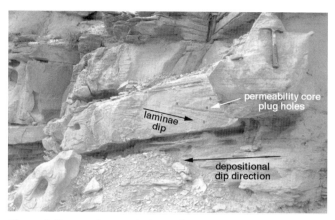

Figure 8. Low-angle cross-beds are the primary sedimentary structure dominant in the proximal portions of the Kf-1-Iv[a] bedset. The general direction of the depositional dip is from right to left. Note the gentle up-depositional-dip inclinations of many laminae in the bed below the hammer. Holes in the rock are from outcrop permeability plugs taken in the unit. Location is near the top of transect 3 (see Figure 3).

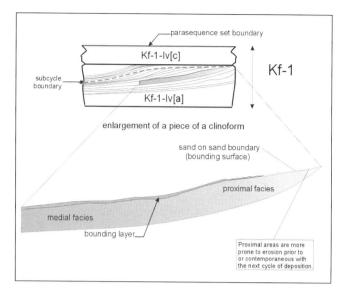

Figure 9. Stratigraphic nomenclature and hierarchical units of the Kf-1-Iv[a] bedset, Ivie Creek area.

mation is in response to a localized relative sea-level change (perhaps compaction related), or to a change in sediment supply.

Clinoform Definition

The Kf-1-Iv[a] bedset was the focus of much of the geologic investigation which fed into the development of a geologic model used in flow simulation (see Mattson et al., this volume; Forster et al., this volume). Using photomosaics, the architectural elements of the Kf-1-Iv[a] were quantified as off-lapping packages of beds referred to here as clinoforms. **Clinoform** is used to identify a group of beds which are inclined seaward in an *en echelon* pattern and generally separated from one another by a distinctive bounding surface observable on outcrop. The clinoforms are visually defined on photomosaics of the outcrop (Figures 7 and 10). The clinoform shape is defined on the photomosaics based on a slightly more recessive break in the cliff face, chiefly along the

bounding surfaces which define the clinoform. Four general lithofacies categories, described in detail later, were assigned to clinoforms (polygons) on the scaled photomosaic panels: clinoform proximal, clinoform medial, clinoform distal, and clinoform cap.

Bounding Layer Definition

The term **bounding surface** was originally used to describe the line on the photomosaics that separates one clinoform from another. Upon closer examination, it was clear that the lines drawn as the boundaries of clinoforms represent units that generally have some thickness on outcrop. In order to signify this thickness feature, the term **bounding layer** is used to describe the thin rock layer that creates the erosive contrast on the cliff face and is appropriately called a bounding surface at the scale of the subcycle or bedset. The bounding layer is defined by the base of the overlying clinoform and by the top of the underlying main sand body of the genetically associated

Figure 10. Inclined bedforms (clinoforms) in the Kf-1-Iv[a] bedset at the Ivie Creek amphitheater, Ivie Creek case-study area, defined by the inclined gray lines. These lines represent the bounding elements of the clinoform — bounding layers. Vertical lines are permeability transects.

clinoform. The bounding layer is an important element of the flow characteristics of a clinoform-type reservoir because it generally contains beds and laminae which are finer grained, mud and carbonaceous-detritus rich, and probably have distinctly lower permeability than the main sand body of the clinoform.

The measured section data, photomosaics, and overlaying line work (which defines the clinoforms and lithofacies), and permeability transect data were used to target specific clinoforms for study. Bounding layers were described with the following information: location (plotted on the photomosaic), type of bounding surface (between what lithofacies), thickness of the bounding layer, slope profile, detailed description of lithology, a general description of the rocks above and below the bounding layer, photographs of the layer and the overlying and underlying rock, and geologists' in-field interpretation of the cause of the recessive break in the cliff face.

Descriptive field data were gathered to better define the characteristics potentially affecting flow across clinoform-to-clinoform boundaries. The emphasis was on bounding layers in portions of the clinoforms that are designated proximal or medial lithofacies. It was assumed, based on field observations along the outcrop, that the distal lithofacies in the clinoforms are all similar in permeability and essentially act as strong baffles or barriers to flow.

Clinoform Characteristics

The basal Kf-1-Iv[a] bedset of the Kf-1 is characterized by clinoform geometries that dip basinward. Clinoforms in the delta front dip 10–15°, and shale out laterally within a mile (1.6 km) down depositional dip. The clinoforms change characteristics from the shallower water, more proximal locations, to the deeper water, more distal locations (Figure 11A). Clinoform facies designations were based on sedimentary structures (Figure 11B), grain size (Figure 11C), bedding thickness, inclination angle, and stratigraphic position. The artificially abrupt end of the medial facies in Figure 9 (end of shading) is representative of most of the contacts portrayed between proximal, medial, and distal clinoform facies in this study because polygonal packages of rock are necessary for reservoir modeling and simulation. In reality, the transition from medial to distal (as well as medial to proximal) is gradational.

As the bounding layer is followed up the inclined surface of the clinoform exposure, the last visible evidence of the bounding layer is often one or several laminae of carbonaceous material forming a very slight recess in the outcrop. On rare occasions, the bounding layer does not reach the upper termination of the clinoform, resulting in a true sand-on-sand contact between clinoforms.

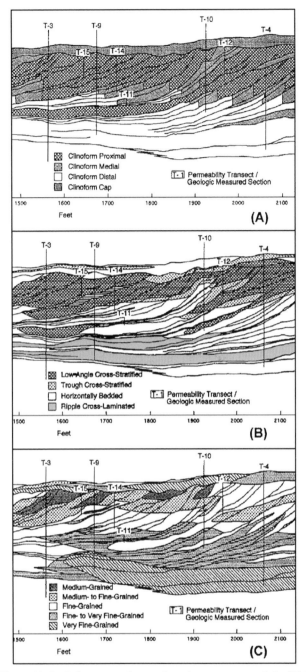

Figure 11. Scaled cross sections, oriented west to east, of the Kf-1-Iv[a] bedset from the Ivie Creek amphitheater based on a portion of the interpreted photomosaic (modified from Mattson, 1997). Vertical lines represent permeability transect locations. (A) Cross section showing clinoform lithofacies assigned to the Kf-1-Iv[a] bedset. (B) Dominant sedimentary structure distribution within clinoform facies. (C) Average grain size distribution within clinoform facies.

The generally abrupt nature of the contact between the clinoform below, and the bounding layer above raises the question of which clinoform the bounding layer is most closely related to. The bottoms of most clinoforms (top of the underlying bounding layer) exhibit a smooth, sharp, concave-up contact. Occasionally, erosional truncation of the underlying bounding layer (Figure 12)

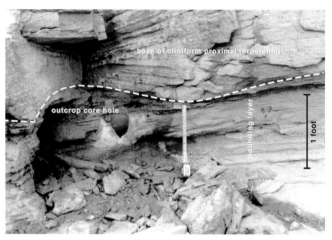

Figure 12. Low-angle cross-beds preserved in the proximal portion of Kf-1-Iv[a] cut into the underlying bounding layer at location B-7 (see Figure 13 for measurement locations).

and/or the underlying clinoform is present, particularly in the more proximal part of the clinoforms, hence, we conclude that each bounding layer is more closely related to the depositional episode of the underlying clinoform. Sand-on-sand contacts represent either an area of no deposition of the finer-grained sediment during the waning stage of deposition of the underlying clinoform, or an area of erosion prior to, and associated with, the deposition of the next clinoform.

The abrupt end of sand-dominated deposition in a clinoform, marked by a bounding layer, is interpreted to represent the temporary change in sand distribution at the delta. During the temporary cessation of rapid sedimentation, the small amount of wave energy at the delta front continues to move and rework some bed-load sediments. Suspended load (and associated carbonaceous debris) "rained down" on the delta front almost continuously. Mud, silt, and carbonaceous debris dominated deposition in the deeper water portions of Kf-1-Iv[a], indicating the ability of wave-energy to winnow the "fines" to deeper and lower energy environments. The presence of the bounding layer/clinoform couplet is indicative of high frequency depositional cyclicity during deposition of the subcycle.

Characteristics of Clinoform Bounding Layers

The lower surface of the bounding layer is generally sharp, but occasionally is gradational over a thin interval. Most of the bounding layers examined contained two common elements: (1) they are finer grained, less cemented, and less resistant than the overlying and underlying clinoforms, and (2) they contain laminations of carbonaceous material, which are consistently poorly cemented and more easily eroded, hence, are expressed as recesses on outcrop. The bounding layers range in thickness from less than an inch to a few feet thick. They generally decrease in thickness and grain size in a down-depositional-dip direction. The bounding layers generally range from fine-grained sandstone to mudstone, and commonly contain bedding planes with laminae rich in carbonaceous debris. The bounding layers are chiefly horizontal to slightly irregular bedded with minor oscillation ripples and flaser bedding. Bed thickness is generally thin to laminated. Occasionally, some portion of the bounding layers contains gypsum veinlets that are probably related to weathering, since gypsum has not been observed in bounding layers examined in core. The lithologic nature of the bounding layers is generally very similar to the clinoform distal facies.

Bounding layer thickness was measured at 23 locations (Figure 13) and is summarized in Table 1. The clinoform facies above and below each measurement point is also listed in Table 1. The bounding layer is believed to be genetically related to the underlying clinoform. Using this assumption, the mean thickness and range for the bounding layer associated with clinoform proximal facies is 0.358 ft (0.109 m) (0.05–1.05 ft [0.02–0.32 m]), while the associated thickness for the clinoform medial facies is 0.600 ft (0.183 m) (0.11–1.63 ft [0.03–0.5 m]). Only one measurement was taken at the clinoform distal facies location (B-1). Table 2 is a summary of the thickness changes on a single bounding layer. With the limited number of data points common to the same bounding layer, there is high variability in the thickness to length ratio.

The focus of the outcrop investigations was on the bounding layers associated with clinoform proximal and

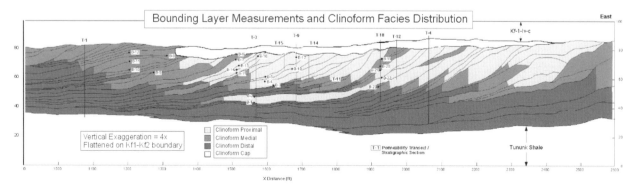

Figure 13. Bounding layer measurement locations and clinoform facies distribution, Ivie Creek amphitheater (modified from Mattson, 1997).

Table 1. Bounding layer measurements and characteristics.

Measurement ID	Thickness (in ft)	Clinoform Facies	
		Above the layer	Below the layer
B-1	0.63	Proximal	Distal
B-2	0.40	Proximal	Proximal
B-3	0.05	Proximal	Proximal
B-4	0.17	Internal-proximal*	Internal-proximal
B-5	0.35	Proximal	Proximal
B-6	0.26	Cap	Medial
B-7	0.86	Medial	Proximal
B-8	1.25	Medial	Medial
B-9	0.70	Medial	Medial
B-10	0.26	Medial	Medial
B-11	0.38	Medial	Medial
B-12	0.49	Medial	Medial
B-13	0.13	Proximal	Proximal
B-14	0.54	Proximal	Proximal
B-15	1.63	Proximal	Medial
B-16	0.19	Proximal	Proximal
B-17	0.08	Internal-proximal	Internal-proximal
B-18	1.05	Proximal	Proximal
B-19	0.11	Proximal	Medial
B-20	0.50	Medial	Medial
B-21	0.07	Medial	Proximal
B-22	0.40	Proximal	Proximal
B-23	0.40	Medial	Medial

* Internal-proximal is "bounding layer" character with a clinoform

Table 2. Tabulation of thickness changes along specific bounding layers.

From ID to ID	Change in thickness (ft)	Distance between points	Clinoform facies from pt. to pt.	Ratio Thickness/length
B-6 to B-8	0.99	200 ft	medial-medial	0.0046
B-16 to B-14	0.35	72 ft	prox.-prox.	0.0049
B-22 to B-2	0	375 ft	prox.-prox.	0
B-23 to B-1*	0.23	356 ft	medial-distal	0.0007

*jump of one clinoform up at B-1

medial facies. These two facies have the best reservoir characteristics. As the bounding layer is tracked seaward, or down the clinoform, it becomes essentially indistinguishable from the underlying and overlying clinoform distal facies. Another way to state this is the bounding layer is a tongue of clinoform distal facies which separates clinoform proximal-proximal facies, cli-noform proximal-medial facies, and clinoform medial-medial facies.

The mapping of the bounding layer thickness is based on an estimate of thickness from a "best guess" examination of the detailed outcrop photomosaics and bounding layer measurements from the field (Figure 14). The bounding layer thickness is broken into three thick-

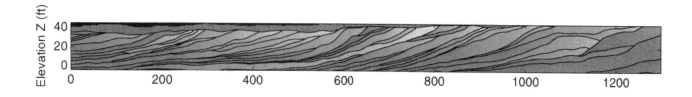

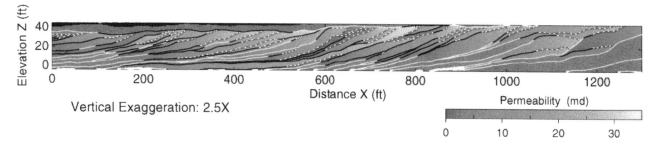

Vertical Exaggeration: 2.5X

Upper diagram shows each bounding layer in black and modeled as uniform in character.
Lower diagram shows bounding layers indicating the layer thickness.
Solid White: thickness greater than 0.75 feet
Black: thickness between 0.25 and 0.75 feet
Dotted White: thickness less than 0.25 feet

Figure 14. Ivie Creek, 40-ac (98.8-ha) spacing model estimating bounding layer thicknesses.

ness categories: (1) < 0.25 ft (0.08 m), (2) ≥ 0.25 ft (0.08 m) and ≤ 0.75 ft (0.23 m), and (3) > 0.75 ft (0.23 m).

Close inspection of the photography and associated line work, and outcrop observations (Figure 10) reveals the presence of internal bounding layers within the designated clinoforms. Generally these internal layers are less continuous than the external bounding layers that define the clinoform. These internal bounding layers are similar in nature, and should have similar permeability characteristics, to those found external to the clinoform.

Recognition of a bounding layer in the subsurface is essential in transferring knowledge from the outcrop analog to producing reservoirs. Using the detailed outcrop descriptions of these layers, cores from the drill holes near the outcrop at Ivie Creek (Figure 3) were examined to identify bounding layers. Unfortunately, no proximal facies were encountered in any of the drill holes. The greatest thickness of high-energy facies is found in cores from drill hole Ivie Creek no. 9A and are medial facies. Several bounding layers were identified in core from this drill hole.

The identification of bounding layers in the core is interpretive. It is impossible to correlate individual bounding layers from outcrop into the subsurface, even with drill holes in close proximity to the outcrop (Figure 3), but features similar to those observed on outcrop can be identified in core.

Kf-1-Iv[c] Bedset

The Kf-1-Iv[c] bedset is sand rich and varies in thickness within the Ivie Creek area. In contrast to the Kf-1-Iv[a], delta-front, subtle clinoforms of the lower part of the bedset dip less than 5°. The sand-rich lithofacies thin in both up-depositional-dip and down-depositional-dip directions due to lateral lithofacies changes. The lower section of the Kf-1-Iv[c] laps onto the more distal parts of the Kf-1-Iv[a] in the western part of the area and represents the distal portion of another delta lobe, probably sourced from the southwest. This lower section may be completely absent in some locations as a result of erosion and/or non-deposition. It also includes a channel sandstone body, lenticular in cross section, deposited by a northwesterly flowing stream (Figure 15).

The uppermost part of the Kf-1-Iv[c] is continuous across the entire Ivie Creek area and indicates wave and fluvial influences (Figures 7 and 15). It is capped by unidirectional, trough cross-bedded sandstone. The Kf-1-Iv[c] contains slump or rotated block features near the mouth of Ivie Creek Canyon, and thickens to the north as the Kf-1-Iv[a] pinches out. In the amphitheater area, above the basal cross-bedded sandstone, are 10–15 ft (3–6 m) of bay-fill deposits. These deposits consist of carbonaceous mudstone; thin, rippled-to-bioturbated sandstone and siltstone; fossiliferous mudstone to sandstone; oyster coquina; and ash-rich coal. Although no contacts were mapped in the immediate Ivie Creek area, the upper portion of this bay interval is related to a younger marine progradation found north and east of the Ivie Creek area. The uppermost, carbonaceous mudstone or ash-rich coal is the Sub-A coal zone. The coal zone shows considerable variation in thickness (0–1 ft [0–0.3 m] in thickness) and intertongues with underlying bay deposits. A flooding surface has been identified

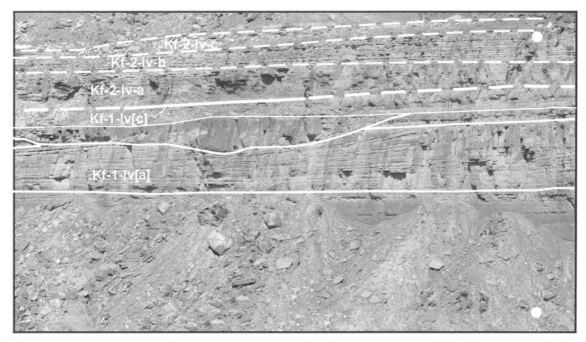

Figure 15. Photomosaic, view to the north, of Ferron parasequences in the Ivie Creek area displaying Kf-1-Iv[a] clinoforms and a Kf-1-Iv[c] distributary channel. The Kf-1-Iv[a] contains more clinoform distal facies than observed to east on Figure 7. Outcrop in photomosaic located in SW1/4NE1/4SW1/4 Sec. 16, T23S, R6E, Salt Lake Base Line, Emery County, Utah.

at the top of the Sub-A coal zone. The boundary with the overlying Kf-2 parasequence set is drawn at this flooding surface.

The Kf-1-Iv[c] thickens rapidly in the western portion of the Ivie Creek area. On the west, the Kf-1-Iv[c] has onlapped the Kf-1-Iv[a]. It represents a slightly younger episode of progradation, filling space that the Kf-1-Iv[a] delta left unfilled, likely due to avulsion of the sediment source. At the mouth of Ivie Creek Canyon, it is anomalously thick due to large slump features or rotated blocks (see Bhattacharya and Davies, this volume). Failure of the rotated blocks is consistently toward the north to northwest, the direction that the delta lobe appears to have prograded. The unusual abundance of rotated blocks in the Ivie Creek area and Muddy Creek area to the northeast (see Bhattacharya and Davies, this volume) may be related to a zone of flexure in the foreland basin, the axis of which is to the northwest. More rapid subsidence toward the northwest may have encouraged failure of the delta-front in that direction.

Kf-2-Iv-a Parasequence

The oldest parasequence of the Kf-2 parasequence set in the Ivie Creek area is the Kf-2-Iv-a. It is separated from the overlying Kf-2-v-b parasequence by a surface that was initially interpreted to be a "marine-flooding surface" because it appeared to separate two vaguely upward-coarsening depositional sequences. Truncation of channel deposits in the lower part of Ivie Creek

Canyon has subsequently been recognized at the top of Kf-2-Iv-a, confirming the initial interpretation of a flooding surface and parasequence status.

The base of the Kf-2-Iv-a parasequence consists of interbedded sand and minor shale deposited in prodelta to lower-shoreface environments. These deposits are thin, typically less than 10 ft (3 m) thick. In some places along the basal contact of the parasequence, a thin (1 ft [0.3 m]) bed of transgressive deposits is present just above the Sub-A coal zone. The prodelta and lower-shoreface deposits are overlain by a 0.5–1-ft-(0.2–0.3-m-) thick zone of highly carbonaceous to coaly sandstone which grades into 20–30 ft (6–9 m) of very fine-grained, silty, and slightly carbonaceous sandstone representing a middle shoreface environment. The middle-shoreface unit is intensely bioturbated. The flooding surface and parasequence boundary at the top of the Kf-2-Iv-a is difficult to recognize because there is frequently no offset in facies with the overlying Kf-2-Iv-b. Kf-2-Iv-a becomes thin and downlaps onto the parasequence set boundary a short distance to the east and within the amphitheater area.

The Kf-2-Iv-a parasequence thickens (to 50 ft [15 m]) westward across the area. Near the mouth of Ivie Creek Canyon, it consists of a thin sequence of lower shoreface heterolithics overlain by about 28 ft (8.5 m) of middle-shoreface deposits. Farther west up the canyon, a distributary channel deposit cuts into the upper half of the shoreface deposits. The coarser (medium-grained) channel is easy to distinguish from the darker-colored, shoreface deposits. Possible foreshore deposits are found near

Figure 16. Photomosaic, view to the southwest, of Ferron parasequences in the Ivie Creek area displaying the mud-rich, distal delta-front deposits of the Kf-1 and distributary-mouth-bar deposits of the Kf-2. The Kf-1 has now changed laterally from the sand-rich, proximal delta-front deposit. The distributary-mouth-bar deposits of the Kf-2-Iv-b parasequence are broadly channelized. Outcrop in photomosaic located in SE1/4SE1/4 Sec. 17, T23S, R6E, Salt Lake Base Line, Emery County, Utah.

the last outcrop of the Kf-2-Iv-a in the bottom of Ivie Creek.

Kf-2-Iv-b Parasequence

In the Ivie Creek amphitheater, the Kf-2-Iv-b parasequence was deposited in a middle- shoreface environment. Kf-2-Iv-b exhibits gently seaward-inclined beds that are very conspicuous when viewed from west to east along the outcrop. The parasequence has at least two sub-cycles of upward grain size coarsening and bed thickening, as clearly exposed on the south side of I-70. The Kf-2-Iv-b parasequence consists of horizontally bedded, silty sandstone at the base (middle shoreface) and unidirectional, trough cross-bedded sandstone at the top (distributary channel to mouth-bar deposits). These units are generally intensely bioturbated.

Along Ivie Creek Canyon, Kf-2-Iv-b is dominantly unidirectional, trough cross-bedded sandstone of a 24-ft-thick (7.3 m) mouth-bar complex (Figure 16) which continues westward up the canyon and is present in the subsurface at UGS drill hole Ivie Creek no. 11, 0.5 mi (0.8 km) to the southwest. The unit thickens slightly from west to east, in contrast to the underlying parasequence.

Kf-2-Iv-c Parasequence

The Kf-2-Iv-c parasequence is separated from the underlying Kf-2-Iv-b parasequence by a siltstone to shale interval that varies in thickness across the east portion of the Ivie Creek area. Generally, the entire parasequence fines from west to east. In the Ivie Creek amphitheater, Kf-2-Iv-c is interpreted as a bay-fill deposit (although it is devoid of body fossils). At the top of this sequence is a thin, medium-grained carbonaceous sandstone, which may represent the migration of

a low-energy beach (foreshore deposits) across the bay fill prior to capping by coastal-plain deposits and deposition of the overlying A-coal zone (which is locally burned).

The type area of Kf-2-Iv-c is the mouth of Ivie Creek Canyon (Figure 16). This unit undoubtedly warrants designation as a parasequence inasmuch as the associated transgressive surface is clearly recognizable both in Ivie Creek Canyon and to the south in the I-70 road cut. The landward pinchout of the marine facies of Kf-2-Iv-c trends just slightly east of south toward I-70. The shoreline sandstone unit displays some interesting and unusual changes at the mouth of Ivie Creek Canyon, changing over a distance of about 300 ft (90 m) from a strongly wave-modified shoreface unit to a much lower energy unit that contains mud interbeds and finer sand, and that has a silvery-gray color in outcrop. This change suggests a change from a coast directly facing the sea to one that was sheltered from wave energy. The environment of this shoreline unit is a wave-modified coast, probably shoreface facies in the proximal part, transforming laterally to low-wave-energy bay facies. There is evidence for bay-head deltas and tidal channels feeding the bay to the northeast, in Quitchupah Canyon.

In Ivie Creek Canyon, Kf-2-Iv-c forms a 10-ft (3.1-m) cliff. Excellent upper shoreface facies are exposed. The top of the unit is rooted by the overlying coastal-plain vegetation. Just a short distance up Ivie Creek Canyon, there is a thin, but well developed, carbonaceous shale at the top of Kf-2-Iv-b, with the flooding surface for Kf-2-Iv-c immediately above.

An interesting feature of the Kf-2-Iv-c parasequence is a structural high mapped by Mattson (1997) in the southern portion of the Ivie Creek area. This closed high is the result of differential compaction around the sides

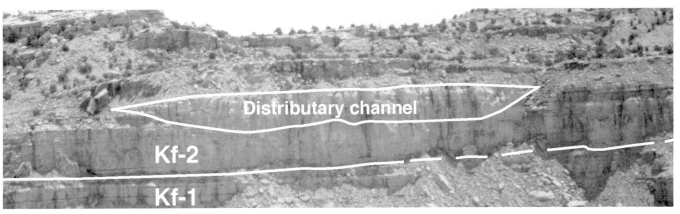

Figure 17. Photomosaic near I-70, view to the south, showing both Kf-1 and Kf-2. A major distributary channel (center of photo) associated with the Kf-2-Iv-c parasequence has cut down to near the base of the Kf-2-Iv-a parasequence. Outcrop in photomosaic located in NW1/4NE1/4NW1/4 Sec. 21, T23S, R6E, Salt Lake Base Line, Emery County, Utah.

of a large distributary channel, previously described by Barton (1994) (Figure 17).

FACIES

A general description of the facies used in the Ferron regional work is provided in Ryer and Anderson, this volume. The following descriptions offer additional detail and were used in the description and analysis of the Ivie Creek area.

Distributary Complex

The distributary complex facies is characterized by a predominance of sandstone and trough cross-stratification (parasequence Kf-2-Iv-a) which is typically unidirectional (Figures 18 and 19). In places, this facies can be subdivided into distributary channel, distributary mouth-bar, and clinoform facies. This facies is often characterized by the complex geometry of cross-bedding and large-scale bounding surfaces that are related to cut and fill processes in contrast to the flat to very gently inclined surfaces of the lower delta-front. Figure 20 shows that the distributary complex facies has a higher gamma count than expected. Clay rip-ups and a poorer sorting are probably the cause of the high gamma count.

Distributary Channel

The distributary channel facies is common in river-dominated delta-front deposits and is also found within the wave-modified shoreline deposits of the Ivie Creek area (Figures 15 and 20). It is characterized by channels with high height-to-width ratios, and unidirectional trough cross-stratified and current ripple cross-laminated deposits. Channel fills are sandstone dominated, but heterolithic channel fills are also found. In sand-filled channels, the grain size is often coarser than the surrounding delta-front or shoreface and fines upward. Troughs in the channel base generally contain mud rip-

Utah Geological Survey
Ivie Creek No. 3
NWNW Sec. 16, T. 23 S., R. 06 E., SLBL

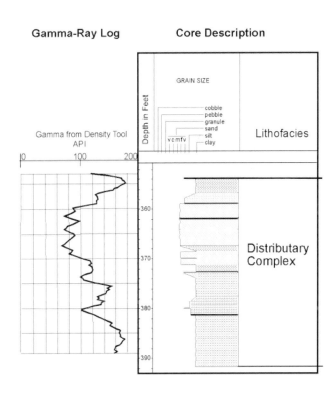

Figure 18. Example of the Ferron Sandstone reservoir gamma-ray type log and core description for the distributary complex facies.

up clasts, woody fragments, and rare shark teeth. This facies grades seaward into the distributary-mouth bar facies.

Distributary-Mouth Bar

The distributary-mouth bar facies is found in the

Figure 19. Distributary complex facies cut into the shoreface. The lighter colored distributary complex facies, below the geologist, is dominated by unidirectional trough cross-stratification. The underlying middle shoreface is dark brown. Note the burrowing which is iron-stained, just above the scoured base and contact with the middle shoreface.

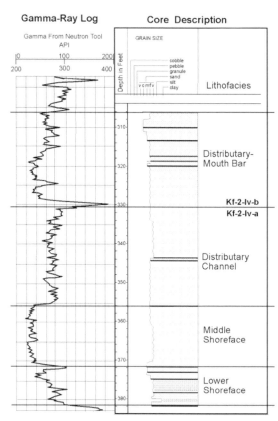

Utah Geological Survey
Ivie Creek No. 11
NENW Sec. 20, T. 23 S., R. 06 E., SLBL

Figure 20. Example of the Ferron Sandstone reservoir gamma-ray type log and core description for the distributary-mouth bar, distributary channel, middle shoreface, and lower shoreface facies.

upper parts of delta-front sequences of Kf-2, and is associated with distributary channel deposits (Figures 16 and 20). This facies is characterized by fine-grained, or coarser, trough-cross-stratified sandstone, with moderate, to intense burrowing developed in areas that had lower flow velocities and decreased sedimentation rate; some intervals between trough sets are completely bioturbated. *Ophiomorpha* is common; escape burrows are less common. Paleoflow directions show a strong offshore component with the amount of scatter increasing with increased wave influence and distance from the distributary channel. Trough heights are generally less than 1.5 ft (0.5 m). Major bed boundaries are inclined in a seaward direction (parasequence Kf-2-Iv-b). Traced laterally and seaward, the facies commonly grades into middle shoreface or lower delta-front facies (fluvial-dominated shoreline).

Clinoform Facies

The clinoforms change characteristics from the shallower water, more proximal (near the shoreline) loca-

tions to the deeper water, more distal locations (Figures 5 and 21). Four clinoform facies are designated within the Kf-1-Iv[a] bedset: clinoform proximal (cp), clinoform medial (cm), clinoform distal (cd), and clinoform cap (cc). Clinoform facies were based on grain size, sedimentary structures, bedding thickness, inclination angle, and stratigraphic position. Facies cp, cm, and cd are assigned to clinoforms only, and facies cc is a bounding facies above the clinoforms (Anderson et al., 1997; Chidsey, 1997).

Clinoform proximal (cp): Facies cp is sandstone, mostly fine to medium grained. The chief sedimentary structure is low-angle cross-stratification with minor horizontal and trough cross-stratification and rare hummocky bedding. The facies is dominantly thick- to medium-bedded, well to moderately indurated, with permeabilities ranging from 2 to 600 millidarcies (md) and a mean of about 10 md. The inclination of bed boundaries is generally greater than 10°. This facies is interpreted to be the highest energy, and most proximal

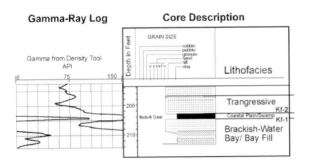

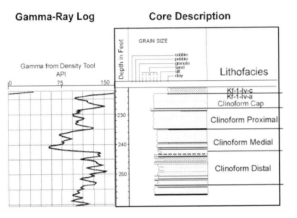

Figure 21. *Example of the Ferron Sandstone reservoir gamma-ray type log and core description for the clinoform distal, medial, proximal, and cap facies; and the brackish-water bay/bay fill, coastal plain/swamp, and transgressive facies.*

Figure 22. *Clinoform medial facies exposed just east of permeability transect 1 (T1) at the Ivie Creek amphitheater. The facies is dominated by horizontal bedding as found in the thicker bedded units. The recessive parts of the outcrop are typically ripple laminated and finer grained than the resistant sandstone beds. Hammer for scale.*

Figure 23. *Clinoform distal facies in the Ivie Creek amphitheater. The base of the Ferron Sandstone is exposed at the base of the heterolithic sequence of interbedded sandstone, siltstone, and mudstone. Ripple and planar bedding are common.*

to the sediment supply. The steep inclinations are interpreted to represent deposition into a relatively localized deep portion of an open bay environment. The dominance of low-angle cross-stratification, along with inclinations within the bed or clinoforms in an up-depositional dip direction, indicates the influence of on-shore wave energy (Figure 8).

Clinoform medial (cm): Facies cm is dominantly sandstone with about 5% shale. The sandstone is primarily fine grained with slightly more fine- to very fine-sized grains than fine- to medium-sized grains. Horizontal beds dominate with some rippled, trough, and low-angle cross-stratified beds (Figure 22). Bed thickness ranges from laminated to very thick, but most beds are medium. The beds are generally well to moderately indurated, but are occasionally noted as friable. The permeability values range from non-detectable to 100 md with the mean about 3 md. Inclination on the clinoform boundaries is between 2° and 10°. This facies is generally transitional between end members clinoform proxi-

mal and distal. Clinoform medial is occasionally present at the erosional truncation, or off-lapping boundary, of the clinoforms, with no visible connection to facies clinoform proximal.

Clinoform distal (cd): Facies cd is sandstone (sometimes silty) and about 10% shale. The sandstone grain size is dominantly fine to very fine grained, with considerable variation (Figure 23). Sedimentary structures in this facies are chiefly horizontal laminations and ripples in medium to thin beds. Induration of the beds ranges from well cemented to friable. Average facies permeability is just at the detection limit of 2 md, but ranges up to 80 md. This facies is gradational with facies cm and represents the deepest water and lowest energy deposits within the clinoform. It can be traced distally into prodelta to offshore facies.

Clinoform cap (cc): Facies cc is sandstone, generally

Figure 24. Clinoform cap facies with trough cross-bedding exposed at the top of the photo. The hammer sandwiched between two hard reddish-beds, is within the underlying clinoform proximal facies. Kf-1-Iv[a] bedset in the Ivie Creek amphitheater.

very fine to fine grained. The beds are horizontal, with some trough and low-angle cross-stratification in thick to medium beds (Figure 24). Burrowing is rare. The sandstone is mostly well indurated, with permeabilities ranging from non-detectable to 100 md and a mean of about 2 md. Clinoform cap facies are present stratigraphically above the subcycle line (truncated clinoforms), where the line is near the top of the parasequence and bed boundaries show little to no inclination. This facies is interpreted to represent an eroded and reworked delta top.

Brackish-Water Bay/Bay Fill

The brackish-water/bay-fill facies consists of heterolithic mudstone and sandstone, coarsening and bed-thickening upward, with wave ripple to horizontal laminations which are sparsely to intensely burrowed. The presence of a brackish-water fauna (*Crassostrea, Corbula securis, Lucinid, Caryo corbulais,* aff. *Varia,* and *Serrthid*), is the distinguishing characteristic of this facies. This fauna is present within "dirty" sandstone, siltstone, mudstone, and carbonaceous mudstone. Oyster coquinas commonly are found within this facies (Figure 25). Typically the rocks are rich in carbonaceous debris. The facies is often underlain and/or overlain by coastal plain facies (the top is often rooted), and may grade into shoreface deposits below (Figure 21).

Coastal Plain/Swamp

The coastal plain/swamp facies is represented by a sequence of non-marine rocks dominated by mudstone, carbonaceous mudstone, and siltstone with minor sandstone. Coal is commonly interstratified with the other rock types (Figures 21, 26, and 27). These rocks are interpreted as inter-fluvial environments along a low-gradient coast.

Figure 25. Oyster coquina containing brackish-water fauna within carbonaceous mudstone. Core from UGS drill hole Ivie Creek no. 11.

Figure 26. Thin transgressive unit (Kf-2) above the thin carbonaceous bed that is part of the Sub-A coal zone (Kf-1) and set boundary between Kf-1 and Kf-2 in the Ivie Creek area. This is near the seaward downlap of the Kf-2-Iv-a parasequence in the Ivie Creek amphitheater. Hammer for scale.

Figure 28. Middle shoreface facies usually form a vertical cliff and consist of very fine to fine-grained sandstone that is horizontally bedded and often moderately burrowed to bioturbated. Note the horizontal bedding indicated by the light horizontal lineations in the photograph. Parasequence Kf-2-Iv-a near permeability transect 20 (T20), Ivie Creek area.

Figure 27. Lower shoreface facies found above the Kf-1/Kf-2 parasequence set boundary in the Ivie Creek area. Typically the facies consists of interbedded very fine-grained sandstone, siltstone, and shale, and coarsens and bed thickens upward. This is stratigraphically the lowest facies in the wave-modified delta deposits in the area.

Transgressive

The transgressive facies consists of siltstone and very fine to fine grained sandstone with an erosional base. Lag deposits are common and the facies is bioturbated. Transgressive units are 8 ft (2.4 m) or less in thickness (Figure 21). This facies represents a significant marine-flooding event (onset of Kf-2) with the base usually being a parasequence set boundary (Figures 26 and 27).

Shoreface

The shoreface facies is the relatively steeply dipping zone from the subaerial beach to a poorly defined point where the slope flattens on the sea floor. Wave energy is sufficient to move sand-sized grains in this zone. At the seaward end of the shoreface is the prodelta facies. This mud-dominated facies represents the area just seaward

of the dominant influence of wave-energy and typically interfingers with the lower shoreface defined below.

Lower Shoreface

The lower shoreface facies consists of thinly interbedded shale to siltstone and very fine to fine-grained sandstone. Wave ripples to horizontal laminations dominate this facies (Figure 27). Burrowing is generally found on the top of thin sandstone beds and the shale is often bioturbated. This facies is very similar to the transition facies, but sandstone is more abundant than shale and siltstone (Figure 20). Hummocky stratification is common.

Middle Shoreface

The middle shoreface facies is very fine to fine grained sandstone composed of hummocky, swaley, and planar laminations with minor ripple laminations (Figure 28). This facies is generally thick, representing 50% or more of the shoreface sequence (Figure 20). The most common burrow types are *Thalassinoides* and *Ophiomorpha* and the amount of burrowing varies from moderately to intensely bioturbated.

Upper Shoreface

The upper shoreface facies is characterized by fine to medium grained, multidirectional to bimodal, cross-stratified sandstone (Figure 29), in sets that are occasionally separated by planar laminations, and generally

Figure 29. Upper shoreface and foreshore facies exposed near perme-ability transect 17 (T17) in Ivie Creek Canyon. The hammer marks the boundary between the two facies. Note the cross-beds in the base of the upper shoreface, typical of this facies. Ophiomorpha is common in both facies. The foreshore bedding is horizontal to sometimes slightly inclined. Note the vertical rootlets penetrating the facies and originating from the overlying A coal.

moderate to well sorted. This facies is about 10 ft (3 m) thick (Figure 30), but greater in the vicinities of the land-ward pinchouts of the parasequences, where the upper shoreface may reach 20 ft (6.1 m) in thickness. The facies is slightly to moderately burrowed, with *Ophiomorpha* as the most common trace fossil. This facies is one of the most permeable in the Ferron.

Foreshore

The foreshore facies consists of fine- to medium-grained sandstone with planar to inclined bedding, slightly to intensely burrowed, and sometimes rooted (Figure 29). This facies is not always present at the top of the shoreface sequence, but when present ranges up to a few feet in thickness (Figure 30).

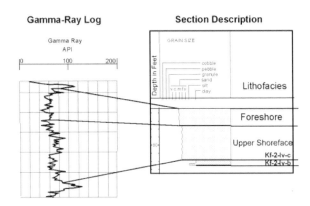

Figure 30. Example of the Ferron Sandstone reservoir gamma-ray type log and core description for the upper shoreface and foreshore facies.

PALEOGEOGRAPHY AND DEPOSITIONAL PROCESSES

Kf-1-Iv[a] Bedset

The Kf-1-Iv[a] bedset in the Ivie Creek area records an episode of delta-lobe progradation into a deeper water, fully marine bay. However, the main delta was probably located to the east and northeast (Anderson et al., 1997; Chidsey, 1997). That delta allowed a protected embayment to develop in the northwest part of the area. The direction of delta-lobe progradation into the pro-tected bay, west-northwest, is somewhat unique for the Ferron Sandstone. The direction of progradation is well constrained by control from outcrops in the Ivie Creek area (see Figure 3). Inclination of these outcropping cli-noforms is away from a point located in space at about the intersection of I-70 and the trend of Quitchupah Canyon. The clinoform outcrops together form an arcu-ate feature (convex surface to the northwest), which is interpreted as a fluvial-delta complex.

Figure 31 A-E is a series of paleogeographic maps of the Ivie Creek area at about the time the delta lobe pro-graded into the Ivie Creek amphitheater. The distribu-tary complexes/delta front, shallow marine, and deep marine environments produced the clinoform proximal, medial, and distal facies, respectively. Based on the geometry of the clinoform deposits, the source for the delta lobe is believed to be to the southeast. Outcrops to the southeast of the Ivie Creek area, along the south side of I-70, contain numerous distributary-channel deposits

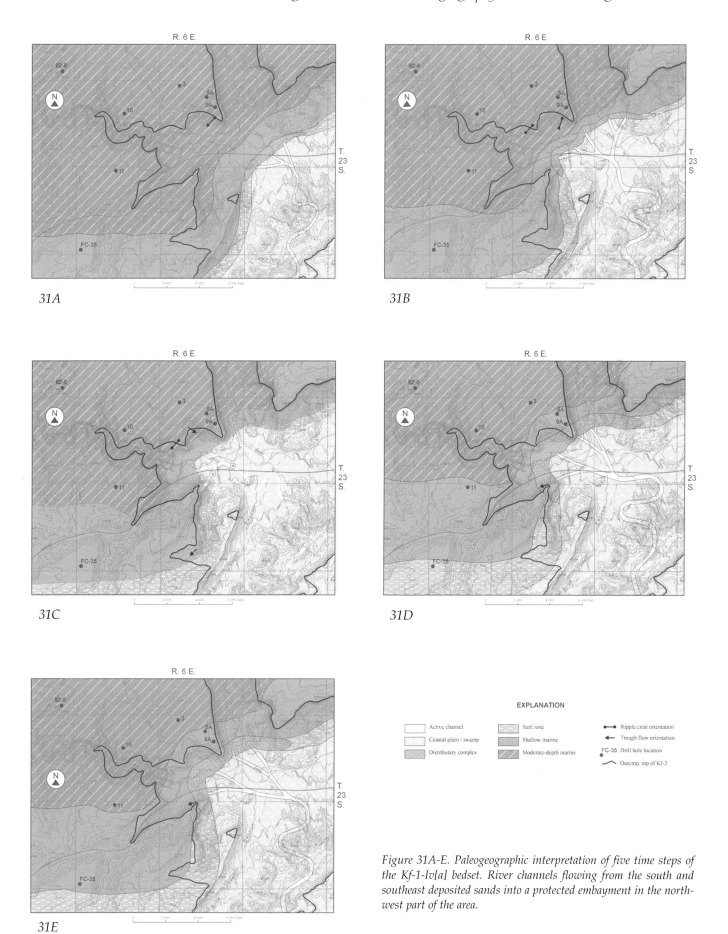

31A

31B

31C

31D

31E

EXPLANATION

Active channel		Surf zone	⊢■⊣ Ripple crest orientation
Coastal plain / swamp		Shallow marine	← Trough flow orientation
Distributary complex		Moderate-depth marine	● FC-35 Drill hole location
			⌒ Outcrop, top of Kf-2

Figure 31A-E. Paleogeographic interpretation of five time steps of the Kf-1-Iv[a] bedset. River channels flowing from the south and southeast deposited sands into a protected embayment in the northwest part of the area.

Time one - post river avulsion and oversteepening of shallow shoreface by clinoform deposition

Typical inclination of beds

Time two - minor erosion from either wave energy or currents associated with the deposition of the next clinoform. A thin layer of mixed bedload/suspended load is deposited down-slope from the eroding area as a bounding layer.

eroding

Time three - deposition of two clinoforms with bounding layers - little erosion.

Time four - erosion (gray lens) by storm energy.

Time five - major clinoform deposition followed by minor proximal erosion (gray lens).

Time six - three episodes of clinoform deposition with only minor hiatuses between. Bounding layers form with each clinoform.

Time seven - erosion of the tops of the previously deposited clinoforms, related to a major storm event or slight rise in relative sea level.

Time eight - post-erosional event, deposition of cap facies and related clinoform. Defines the bottom of a "subcycle" and a change in sediment supply.

cap facies

Figure 32. Diagrammatic cross sectional view of clinoform depositional time steps in the Ivie Creek area.

350

at the top of Kf-1. The Kf-1-Iv[a] bedset thins westward from a 60-ft- (18-m-) thick, sand-dominated, deltaic deposit to less than 10 ft (3 m) of shale and siltstone as it is traced from the mouth of Quitchupah Creek to the entrance into the narrow canyon of Ivie Creek, near the west line of section 16 (Figure 3).

Figure 32 contains a series of cartoons illustrating an interpretation of the depositional history of a portion of the lower subcycle of the Kf-1-Iv[a] bedset at the Ivie Creek amphitheater area. These diagrammatic cross sections step through time. Figure 33 is a companion to Figure 32 and shows the map view of each time step. Time Step 1 begins during active progradation of delta deposits and after the establishment of a major delta-front deposit prograding into the area. At the base of this illustration on Figure 32 is a line that represents more typical inclinations of delta-front deposits in the Ferron Sandstone. Note the steeper angle on the delta profile of the Kf-1-Iv[a] bedset. Minor erosion of the uppermost portion of the delta-front (clinoform proximal facies) is portrayed in Time Step 2. Time Steps 3 to 6 illustrate the addition of several clinoforms with some minor erosion. At the scale of the illustrations in Figure 32, the bounding layers were not specifically depicted, but are

deposited with each clinoform.

At Time Step 7, a significant change in deposition occurs. Either a relative rise of sea level at this location (most likely caused by minor tectonics or compaction and subsidence of the delta lobe), or a change in the wave-energy regime of the bay resulted in reworking of the previous upper portion of the delta and deposition of the cap facies during Time Step 8. It is likely that this change, marked by the clinoform cap facies, was related to a change in the volume, or point of discharge, of sediments as well. The recognition of an overall fining of the Kf-1-Iv[a] bedset at this stratigraphic level points to some change in the depositional regime. This event marks the base of the next subcycle.

Distributary channel switching is the most probable mechanism for clinoform cyclicity as shown in Figure 33; however, no distributary channel deposits have been found juxtaposed to the clinoforms. This association shown on Figures 32 and 33 is conjectural. Either these channels were located just landward of the clinoforms found on outcrop and have been removed by erosion, or they were present within the clinoforms but were located in areas presently eroded away. Either way, wave energy within the bay must have played an important

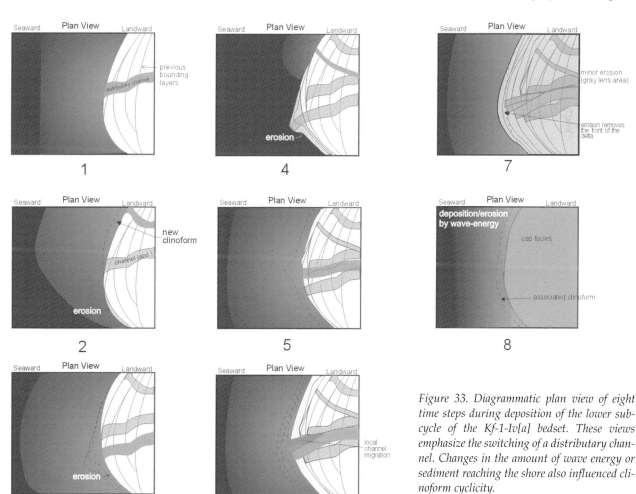

Figure 33. Diagrammatic plan view of eight time steps during deposition of the lower subcycle of the Kf-1-Iv[a] bedset. These views emphasize the switching of a distributary channel. Changes in the amount of wave energy or sediment reaching the shore also influenced clinoform cyclicity.

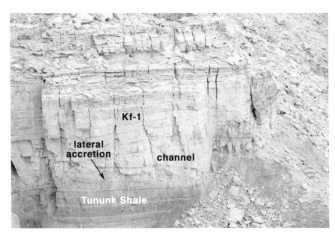

Figure 34. Ferron-Tununk contact in upper Quitchupah Canyon. The first regressive unit is the Kf-1-Iv[a] bedset. Note the inclined heterolithics which are stratigraphically equivalent to the channel-shaped sandstone body to the right. This is an indication of the shallow water depths that existed during the earliest deltaic deposition of the Ferron in the Quitchupah Canyon area. This is very near the seaward pinchout of the Kf-1 parasequence set.

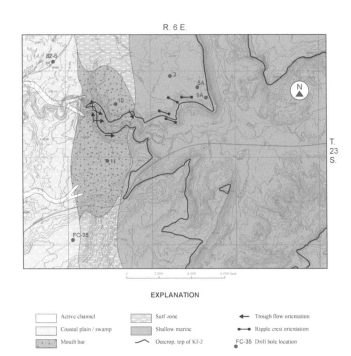

Figure 35. Paleogeographic interpretation of the Kf-2-Iv-a parasequence.

role in distributing the sediment along the delta front. Low-angle cross-stratification common in the clinoform proximal facies typically shows inclinations of cross-laminae in an up-depositional dip direction, an unlikely direction for fluvial-generated bed forms. Some of these structures have characteristics similar to hummocky bedding, indicating the influence of oscillating wave energy during deposition.

Kf-1-Iv[c] Bedset

In the Ivie Creek amphitheater, this unit is separated from the underlying Kf-1-Iv[a] bedset by a distinctive break in slope. Interbedded shale and siltstone are responsible for the slope change. Several channel sandstone deposits are present in the area, and one occurs very near the contact with the unit below. These were small distributary channels feeding the prograding shoreline of Kf-1-Iv[c] seaward of the amphitheater. The typical fluvial-deltaic deposits outcrop to the northeast of the amphitheater on the west side of Quitchupah Canyon. Here, the deposits can be traced northward into distal deltaic facies.

Across Quitchupah Canyon the tie to the section in the Ivie Creek amphitheater is more difficult. The unit has been identified directly across the canyon from the amphitheater but poor outcrops farther east and north into Quitchupah Canyon made mapping difficult, and the unit is not defined on photomosaics in these areas. Fluvial-deltaic marine rocks that lie below the Kf-1-CC parasequence (see Anderson et al., this volume) in upper Quitchupah Canyon may be the Kf-1-Iv[c] bedset. These rocks exhibit steep inclinations similar to the Kf-1-Iv[a] bedset in the eastern part of the Ivie Creek amphitheater.

Channel deposits at the base of the Ferron Sandstone, and probably in the Kf-1-Iv[c] bedset, or slightly younger, indicate that the water depth during the first regressive deposition was on the order of 15–20 ft (5–6 m) deep in this area of Quitchupah Canyon (Figure 34). The paleogeography of these earliest progradational deposits was a series of small deltas prograding into shallow seas. The geometry and architecture are often characterized by steeper inclinations in the delta-front. The combination of shallow water, rapid deposition, and relatively low wave energy created a unique environment which preserved steeper clinoforms in the delta-front deposits of the first regression into the Tununk sea.

Kf-2-Iv-a Parasequence

The shoreline orientation of the Kf-2-Iv-a parasequence rotated to the west to become primarily north-south (Figure 35). This unit downlaps onto the parasequence set boundary in the Ivie Creek area. A similar stratigraphic relationship to the underlying unit is exposed south of I-70. Perhaps the downlapping of this unit is related to growth of a peripheral bulge at this location in the foreland basin. The westernmost exposure of Kf-2-Iv-a is in Ivie Creek Canyon where the facies is distributary-mouth bar. This indicates that the shoreline was probably not too far to the west. Drill hole Ivie Creek no. 11 contained distributary-mouth bar facies of the unit, but interpretation of drill holes farther west indicate the shoreline has been crossed and only nonmarine deposits exist there.

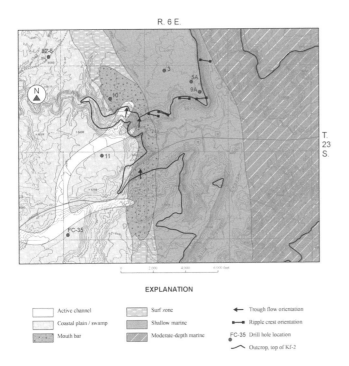

Figure 37. Soft-sediment deformation from the lower Kf-2-Iv-b on the east side of Quitchupah Creek. The "rib cage" like feature is caused by the more cohesive rippled beds which are rolled during an episode of soft-sediment deformation, probably related to seismic events in the thrust belt to the west. Similar deformation is common at this stratigraphic interval in the surrounding outcrops.

EXPLANATION

☐ Active channel	☐ Surf zone	← Trough flow orientation
☐ Coastal plain / swamp	☐ Shallow marine	•—• Ripple crest orientation
☐ Mouth bar	☐ Moderate-depth marine	FC-35 Drill hole location
		⌒ Outcrop, top of Kf-2

Figure 36. Paleogeographic interpretation of the Kf-2-Iv-b parasequence.

Kf-2-Iv-b Parasequence

Like the preceding unit, the shoreline orientation during deposition of the Kf-2-Iv-b parasequence was generally north-south. In the landward exposures at Ivie Creek Canyon, the unit is dominated by distributary-mouth bar and distributary channels associated with wave-modified delta deposits (Figure 36). Further seaward, the distributary-mouth bar deposits give way to middle-shoreface deposits. Seaward (east) of the Ivie Creek area, a well-developed lower shoreface is present at the base of the unit. Several subcycles of upward coarsening, middle shoreface or distal, mouth-bar deposits can be identified in the eastern Ivie Creek area, probably related to the proximity of the sediment distribution point and wave energy through the history of progradation. Several wave-ripple crest measurements were taken in the area and have a consistent east-west orientation (Figure 36), perhaps indicating a shoreline-parallel longshore drift. Paleoflow of several of the distributary channels indicates a general northerly trend in the local area.

At the I-70 road cut, a new highway drainage channel was cut through solid rock to accommodate storm runoff. This cut provides superb views of Kf-2-Iv-b mouth-bar deposits. From the downstream end of the drainage channel, cross-bedded, mouth-bar deposits can be visually followed into distal bar or middle shoreface deposits to the east.

Kf-2-Iv-b has abundant soft-sediment deformation features that occur throughout the area in specific stratigraphic intervals. These features are particularly evident near the mouth of the canyon south of I-70, and in Quitchupah Canyon (Figure 37). Proximity to a seismically active thrust belt, and the occurrence of these features in specific stratigraphic intervals, indicates their cause may be disturbance by earthquakes shortly after deposition.

Kf-2-Iv-c Parasequence

The shoreline of this unit is well exposed in the western portion of the Ivie Creek area, both at the top of the I-70 road cut, and in Ivie Creek Canyon (Figure 38A). Connecting these two points creates a line bearing of just west of north. Wave-dominated delta deposits with a well-developed foreshore and upper shoreface are found east of the landward pinchout throughout the narrow portion of Ivie Creek Canyon. This wave-dominated coast prograded to a point just beyond where Ivie Creek breaks through the Coal Cliffs today. This point is indicated on Figure 38A by the change from surf zone to shallow marine depths. It is here that the character of the rocks changes dramatically within about 100 ft (33 m) horizontally.

Typical, shallow, shoreface deposits intertongue with thin- to medium-bedded, horizontal to rippled, very fine to fine grained sands, which grossly coarsen upward. We interpret these deposits as a bay-fill sequence. Some event must have shut off the main wave energy of the coast and partially restricted the flow along the coast. Figure 38B illustrates our vision of the paleogeography of the area. The figure places a surf

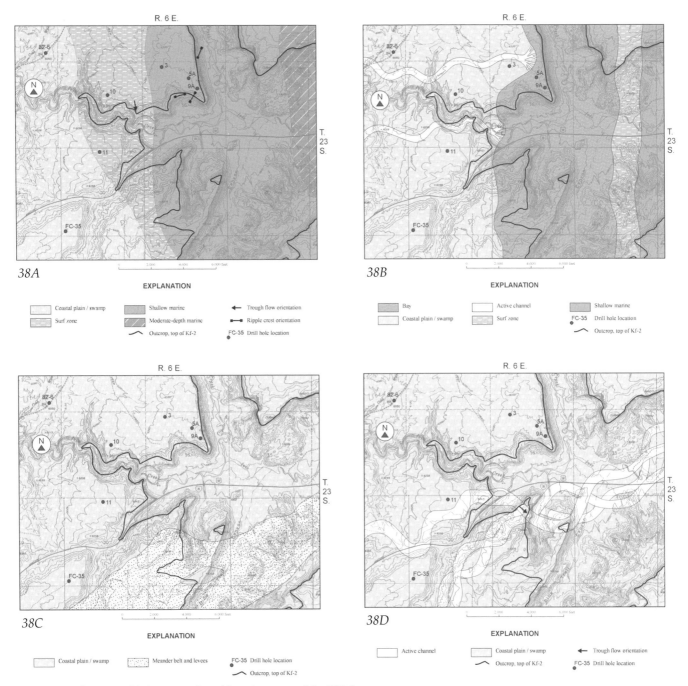

R. 6 E.

38A

EXPLANATION

Coastal plain / swamp   Shallow marine   ← Trough flow orientation

Surf zone   Moderate-depth marine   ⊷ Ripple crest orientation

Outcrop, top of Kf-2   • FC-35 Drill hole location

R. 6 E.

38B

EXPLANATION

Bay   Active channel   Shallow marine

Coastal plain / swamp   Surf zone   • FC-35 Drill hole location

Outcrop, top of Kf-2

R. 6 E.

38C

EXPLANATION

Coastal plain / swamp   Meander belt and levees   • FC-35 Drill hole location

Outcrop, top of Kf-2

R. 6 E.

38D

EXPLANATION

Active channel   Coastal plain / swamp   ← Trough flow orientation

Outcrop, top of Kf-2   • FC-35 Drill hole location

Figure 38. Paleogeographic interpretation of four time steps of the Kf-2-Iv-c parasequence.

zone on the east side of Quitchupah Canyon. This is supported by a well-developed shoreface exposed to the southeast along cliffs facing I-70 at the same stratigraphic interval. The transition from bay-fill to shoreface on the east side of the bay is either poorly exposed or eroded so this part of the interpretation is somewhat tentative. Some distributary channels are present in outcrops in Ivie Creek Canyon (Figure 38B), and on the west side of Quitchupah Canyon, a bay-head delta is present in the Kf-2-Iv-c parasequence.

A large meanderbelt channel system cuts into the top of the Kf-2-Iv-c parasequence just south of the immediate Ivie Creek area (Figure 38C). The channel system

flowed to the northeast, but much of this Ferron channel system has been removed by erosion. A late-stage episode of lateral channel migration across the top of the Kf-2-Iv-c delta-plain deposits is recorded by scours into the meanderbelt deposits (Figure 38D). The best example of this channel system is exposed on the south side of I-70, just east of the road cut (see Figure 17). The Texas Bureau of Economic Geology has taken numerous permeability measurements from this channel (Barton, 1994; White and Barton, 1998, 1999; Knox and Barton, 1999). This same channel is exposed in the road cut on the north side of I-70, but the channel orientation and the nature of its exposure in the cut face do not present the classic channel shape exhibited in the southern exposure. The meanderbelt and younger channel systems feed the continued progradation of Kf-2-Iv-c and stratigraphic equivalents to the east and northeast.

SUMMARY

The Upper Cretaceous Ferron Sandstone was deposited predominantly in a fluvial-deltaic setting along the western margin of a rapidly subsiding foreland basin. The Ferron consists of a series of stacked, transgressive-regressive cycles that form an eastward-thinning wedge. Exposures and drill-hole data in the Ivie Creek area, located in the central part of the Ferron outcrop belt, show abrupt facies changes in the delta-front sandstones of two of these cycles or parasequence sets: Kf-1 and Kf-2. These deposits also represent the primary lithofacies typically found in fluvial-dominated deltaic reservoirs. Each type of deltaic system produces differing geometries that are characterized by geologic parameters for exploration or reservoir modeling efforts. Sedimentary structures, lithofacies, bounding surfaces, and permeabilities measured along closely spaced traverses combined with drill-hole data produced a paleogeographic picture and a three-dimensional view of the reservoir analogs within the area.

In the Ivie Creek area, the Kf-1 parasequence set consists of one parasequence, the Kf-1-Iv, with two bedsets, the Kf-1-Iv[a] and Kf-1-Iv[c], composed of unusual river-dominated delta deposits that prograde southeast to northwest across the Ivie Creek area. Progradation is parallel or onshore to the regional shoreline trend. Distinctive, steeply inclined bedsets or clinoforms, defined and generally separated from one another by bounding surfaces or layers, were deposited seaward in an *en echelon* pattern. The bounding surface or layer observable on outcrop would create a barrier or baffle to fluid flow in a reservoir because it generally contains beds and laminae which are finer grained, mud and carbonaceous-detritus rich, and probably has distinctly lower permeability than the main sand body of the clinoform. Thus, the clinoforms are delineated as polygon-shaped packages of rock which are necessary for reservoir modeling and simulation.

The clinoforms in the Ivie Creek area are classified into four facies: proximal, medial, distal, and cap. Clinoform facies are based on grain size, sedimentary structures, bedding thickness, inclination angle, and stratigraphic position. These deposits accumulated on an arcuate, prograding lobe of a delta into a deeper water, fully marine bay. The main delta, which was located to the east and northeast, allowed a protected embayment to develop in the northwest part of the Ivie Creek area. The clinoforms represent deposition into the embayment fed by river channels from the southeast.

Kf-2 is represented by wave-modified, deltaic deposits that generally coarsen east to west, and consist of shoreface (lower, middle, upper, and foreshore) and distributary complex facies (distributary channel and distributary mouth-bar). These relatively clean, sand-rich deposits accumulated along a north-south shoreline trend defined by a landward pinchout of marine facies, as opposed to the more regional northwest-southeast trend recognized in other Ferron cycles. In the western part of the Ivie Creek area, east- to northeast-flowing distributary channels deposited large amounts of sand in north-south-trending distributary-mouth bars. Shallow marine to moderate depth marine conditions existed in the eastern part of the area. An uncommon transition from shoreface to bay to coastal plain/swamp occurred during the late stage of Kf-2 deposition.

As a reservoir analog, the Ferron Sandstone in the Ivie Creek area displays variations in sedimentary structures and lithofacies that influence both its compartmentalization and permeability. Bounding layers, such as those observed on outcrop, and identified from core and geophysical well-log data, are an important element that must not be overlooked in reservoir characterization. These features provide the important ingredients for reservoir modeling and simulation studies for oil field development and secondary or enhanced oil recovery programs of fluvial-dominated deltaic reservoirs.

ACKNOWLEDGMENTS

This paper was supported by the Utah Geological Survey (UGS), contract numbers 94-2488 and 94-2341, as part of a project entitled *Geological and Petrophysical Characterization of the Ferron Sandstone for 3-D Simulation of a Fluvial-Deltaic Reservoir*, M. Lee Allison and Thomas C. Chidsey, Jr., Principal Investigators. The project was funded by the U.S. Department of Energy (DOE) under the Geoscience/Engineering Reservoir Characterization Program of the DOE National Petroleum Technology Office, Tulsa, Oklahoma, contract number DE-AC22-93BC14896. The Contracting Office Representative was Robert Lemmon.

We thank the Mobil Exploration/Producing Technical Center and Amoco Production Company for their technical and financial support of the project which was conducted between 1993 and 1998. We also thank the following individuals and organizations for their technical contributions at the time this work was performed: R. L. Bon, B. P. Hucka, D. A. Sprinkel, D. E. Tabet, Kevin McClure, K. A. Waite, S. N. Sommer, and M. D. Laine of the UGS, Salt Lake City, Utah; M. A. Chan, C. B. Forster, Richard Jarrard, Ann Mattson, and S. H. Snelgrove of the University of Utah, Salt Lake City, Utah; J. V. Koebbe of Utah State University, Logan, Utah; Don Best, Jim Garrison, Bruce Welton, F. M. Wright III of Mobil Exploration/Producing Technical Center, Dallas, Texas (now

Exxon/Mobil); and R. L. Chambers, Chandra Rai, and Carl Sondergeld of Amoco Production Company (now BP), Tulsa, Oklahoma.

Jim Parker (UGS) drafted figures and Cheryl Gustin (UGS) prepared the manuscript. Finally, we thank David E. Tabet and Mike Hylland, UGS, Robert Lemmon, U.S. Department of Energy, National Petroleum Technology Office, and Viola Rawn-Schatzinger, RMC Consultants, Inc., for their careful reviews and constructive criticism of the manuscript.

REFERENCES CITED

Anderson, P. B., T. C. Chidsey, Jr., and T. A. Ryer, 1997, Fluvial-deltaic sedimentation and stratigraphy of the Ferron Sandstone, *in* P. K. Link and B. J. Kowallis, eds., Mesozoic to Recent geology of Utah: Brigham Young University Geology Studies, v. 42, pt. 2, p. 135–154.

Anderson, P. B., T. C. Chidsey, Jr., K. McClure, A. Mattson, and S. H. Snelgrove, 2002, Ferron Sandstone stratigraphic cross sections, Ivie Creek area, Emery County, Utah: Utah Geological Survey Open-File Report 390DF, 10 p., 7 pl.

Anderson, P. B., K. McClure, T. C. Chidsey, Jr., T. A. Ryer, T. H. Morris, J. A. Dewey, Jr., and R. D. Adams, 2003, Interpreted regional photomosaics and cross section, Cretaceous Ferron Sandstone, east-central Utah: Utah Geological Survey Open-File Report 412, compact disk.

Barton, M. D., 1994, Outcrop characterization of architecture and permeability structure in fluvial-deltaic sandstones, Cretaceous Ferron Sandstone, Utah: Ph.D. dissertation, University of Texas, Austin, 255 p.

Chidsey, T. C., Jr., compiler, 1997, Geological and petrophysical characterization of the Ferron Sandstone for 3-D simulation of a fluvial-deltaic reservoir — annual report for the period October 1, 1995 to September 30, 1996: U.S. Department of Energy, DOE/BC/14896-15, 57 p.

Knox, P. R., and M. D. Barton, 1999, Predicting interwell heterogeneity in fluvial-deltaic reservoirs B effects of progressive architecture variation through a depositional cycle from outcrop and subsurface observations, *in* R. A. Schatzinger and J. F. Jordan, eds., Reservoir characterization — recent advances: AAPG Memoir 71, p. 57–72.

Mattson, A., 1997, Characterization, facies relationships, and architectural framework in a fluvial-deltaic sandstone — Cretaceous Ferron Sandstone, central Utah: Master's thesis, University of Utah, Salt Lake City, 174 p.

Ryer, T. A., 1991, Stratigraphy, facies, and depositional history of the Ferron Sandstone in the canyon of Muddy Creek, east-central Utah, *in* T. C. Chidsey, Jr., ed., Geology of east-central Utah: Utah Geological Association Publication 19, p. 45–54.

White, C. D., and M. D. Barton, 1998, Comparison of the recovery behavior of contrasting reservoir analogs in the Ferron Sandstone using outcrop studies and numerical simulation: The University of Texas at Austin, Bureau of Economic Geology Report of Investigations No. 249, 46 p.

—1999, Translating outcrop data to flow models, with applications to the Ferron Sandstone: Society of Petroleum Engineers, Reservoir Evaluation & Engineering (SPE 57482), v. 2, no. 4, p. 341–350.

Reservoir Permeability, Modeling, and Simulation Studies

Meal time at C. T. Lupton's U. S. Geological Survey field camp, circa 1910.
Photograph courtesy of the family of C. T. Lupton.

Analog for Fluvial-Deltaic Reservoir Modeling: Ferron Sandstone of Utah
AAPG Studies in Geology 50
T. C. Chidsey, Jr., R. D. Adams, and T. H. Morris, editors

Modeling Permeability Structure and Simulating Fluid Flow in a Reservoir Analog: Ferron Sandstone, Ivie Creek Area, East-Central Utah

Craig B. Forster[1], Stephen H. Snelgrove[2], and Joseph V. Koebbe[3]

ABSTRACT

Numerical waterfloods are simulated within two- and three-dimensional (2-D and 3-D) permeability structures developed for the Ivie Creek area along the Ferron Sandstone (Cretaceous) outcrop belt. Permeability structures are constructed by combining outcrop facies architecture data with permeability measurements made on both outcrop and drill-core samples. Simulated waterfloods are used to explore: (1) how detailed, fluvial-deltaic facies-derived permeability structures might influence oil production, and (2) the ability of different permeability upscaling approaches to capture the impact of the detailed structures on oil production. Permeability values incorporated in the models range from 0.1 to 50 millidarcies. The role of facies architecture is preserved in the 2-D gridded models (grid blocks are 2.5 ft long and 0.5 ft high) by assigning facies-related permeability trends within amalgamations of distinct clinoform bodies separated by thin shaley bounding layers. Results of a series of 2-D numerical waterflood simulations illustrate the sensitivity of total oil production and the timing of water breakthrough to the nature of the thin, interclinoform, shaley bounding layers. Permeability upscaling experiments indicate that common averaging approaches (computing arithmetic, harmonic, or geometric means) are inadequate to upscaling permeability in this fluvial-deltaic setting. An upscaling technique based on perturbation analysis yields 2-D simulation results similar to those obtained with detailed permeability models. Detailed permeability structures are upscaled and assigned in more coarsely gridded 3-D models (grid blocks are 20 ft by 20 ft in plan and 4 ft thick) by defining permeability facies that encompass portions of adjacent clinoform bodies. Results of a series of 3-D numerical waterflood simulations with 5-spot and 9-spot production well patterns illustrate the significant impact that the upscaled permeability facies geometry exerts on oil production. Comparing 2-D and 3-D simulation results confirms that it can be misleading to use 2-D simulation results to predict oil recovery and water cut in a reservoir with the internal 3-D geometry inferred at the Ivie Creek site.

[1]Department of Geology & Geophysics, University of Utah, Salt Lake City, Utah
[2]Department of Civil & Environmental Engineering, University of Utah, Salt Lake City, Utah
[3]Mathematics and Statistics Department, Utah State University, Logan, Utah

INTRODUCTION

Detailed, two- and three-dimensional petrophysical and fluid-flow modeling of the Cretaceous Ferron Sandstone in the Ivie Creek area of east-central Utah (Figure 1), is based on data from two sources: (1) the stratal architecture on outcrop, derived mainly from cliff-face photomosaic panels; and (2) permeability measurements made on core plugs extracted from the outcrop and from slabbed borehole cores (Forster and Snelgrove, 2002). Descriptions of the procedures used to establish the outcrop architecture and to measure permeability, and an analysis of the permeability measurements, are contained in Mattson (1997) and Mattson and Chan (this volume). These topics are reviewed only briefly here. The key objective of the flow-simulation studies is to evaluate the impact of outcrop-based geology on numerical simulations of reservoir performance.

The stratal architecture exposed at Ivie Creek consists of the juxtaposition of many variants of a single architectural element, the clinoform, and indicates prograding, shallow-marine deposition of a fluvial-dominated delta-front system (Figure 2; see Anderson et al., this volume). The sandstone beds within each sigmoid-shaped clinoform body record an episode of deposition, followed by partial erosion. Depositional episodes were likely related to variations in the sediment load carried by the distributary channel(s) feeding the delta front, and by upstream channel switching. Clinoform bodies are inclined seaward, arranged *en echelon*, and are usually separated from each other by a thin (less than 1 ft thick [< 0.3 m]) mudstone and siltstone bounding layer. The resulting architecture provides the spatial framework for distributing permeability and porosity within numerical models of this delta-front reservoir analog.

Underlying this approach is the assumption that the similar geometry shared by individual clinoforms indicates that they were deposited under the influence of similar hydrodynamic and geological processes. The interplay between these processes results in a characteristic spatial distribution within each clinoform of sediment grain sizes, sedimentary structures, and sediment mineralogy. These factors, in turn, are key controls on primary porosity and permeability. Those parts of a clinoform deposited in high energy hydrodynamic conditions with an ample supply of coarse, angular quartz grains would be expected to produce comparatively high porosity and high permeability rocks. Lower energy conditions that favor the deposition of smaller grains and tabular clay particles yield lower permeability deposits. In short, since individual clinoforms were originally deposited under conditions similar to those in which every other clinoform was deposited, all clinoforms exhibit similar spatial patterns of primary porosity and permeability (Weber, 1978).

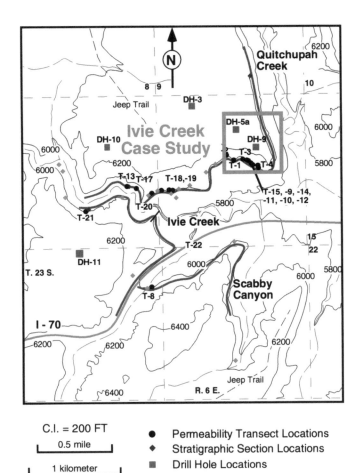

C.I. = 200 FT

0.5 mile

1 kilometer

● Permeability Transect Locations
◆ Stratigraphic Section Locations
■ Drill Hole Locations
— Photomosaic Coverage

Figure 1. Location of Ivie Creek study area.

Trends in permeability and porosity are difficult to establish at the Ivie Creek area because of the strategy used to collect core plug samples from the outcrop (discussed below). However, most clinoforms at Ivie Creek show a systematic decrease in grain size and a clear succession of sedimentary structures in the direction of depositional dip. This trend indicates a transition from a high energy depositional setting proximal to the sediment source to a low energy depositional setting distal to the source. A corresponding, but weak, progression from high permeability and porosity to lower permeability and porosity exists within clinoforms, even though much of the primary porosity has been modified by diagenesis.

A gridded, two-dimensional (2-D) digital model of the stratal architecture was generated by interpreting a photomosaic of the exposed cliff-face in the Ivie Creek amphitheater (Figure 1). Individual clinoforms are preserved in the digital model, and are the basis for developing two- and three-dimensional permeability models of the case study. Each clinoform is assumed to result from a similar set of depositional conditions, and thus to exhibit similar trends in permeability. The validity of this approach is based on the discussion above.

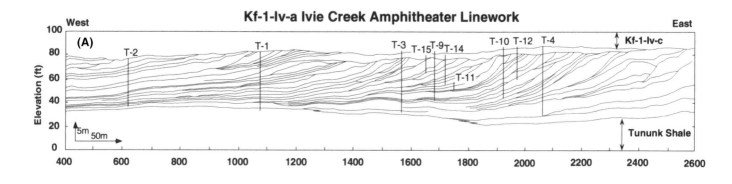

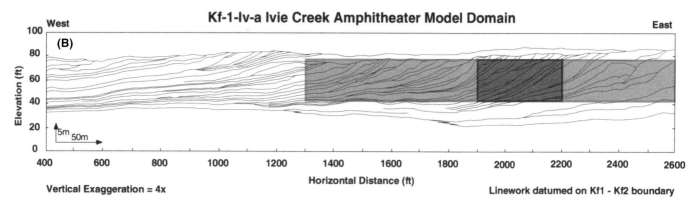

Figure 2. (A) Linework interpretation of Ivie Creek clinoform facies architecture (from Chidsey and Allison, 1996; Mattson, 1997). The labeled vertical lines are locations of permeability transects. (B) The shaded area indicates the extent of gridded modeling domain used in fluid-flow simulations. The sub-domain shaded in darker gray was used in permeability upscaling experiments.

Permeability distributions within a clinoform are assigned in the models by first subdividing the clinoform into facies. The mean permeability of each facies, derived from permeability measurements made on plugs extracted from the outcrop, is then used to assign permeability throughout each clinoform. An empirical relationship that relates permeability and porosity at the Ivie Creek area is used to compute porosity values from permeability.

Several petrophysical models are constructed, each a different representation of the degree to which individual clinoforms are surrounded by thin, siltstone and mudstone bounding layers. Numerical simulations of a waterflood in a two-phase (oil and water) system illustrate the impact the bounding layers might have on reservoir performance. The fraction of the original in-place oil that is ultimately recovered is only slightly affected by the various bounding layer models. The timing of production, however, is closely related to the spatial configuration of the bounding layers. A permeability upscaling study performed using a sub-region of one the permeability models indicates that none of the simple averaging methods (arithmetic, harmonic, and geometric) are appropriate for deposits of this type. An upscaled permeability model obtained using a perturbation technique, however, yields simulation results similar to those obtained using the detailed permeability model.

Three-dimensional (3-D) petrophysical models are created in several steps as described by Chidsey and Allison (1996), Chidsey (1997), Mattson (1997), and briefly described here. Eight detailed vertical cross sections containing clinoform geometry and facies are derived from outcrop and drill hole control (Figure 1). Two cross sections correspond to the south and east facing outcrops (including the Ivie Creek amphitheater cliff-face, used in the 2-D modeling studies), four define the sides of the modeling domain, and two (tied to drill holes) cut diagonally across the block. Detailed features contained in the gridded cross sections are incorporated in the computational grid of the 3-D flow-simulation model by coarsening the grid to match the computational grid. The dominant clinoform facies type found within each computational block is assigned in the coarsened model as an attribute for the whole block. Facies data from these coarsened vertical sections are transferred to horizontal slices created throughout the modeling domain and the facies are interpolated between control points. The permeability of a grid block is defined as the arithmetic mean of the permeability values obtained for the specific clinoform facies assigned to the block. The resulting model (Reference Case) is consistent from layer to layer, and agrees with our hypothesis that sediment deposition created arcuate lobes sourced from the east-southeast. An isotropic permeability tensor is assumed for each grid block in the Reference Case.

Three other 3-D petrophysical models are developed to examine whether any significant improvement might be made by including outcrop-based information in numerical waterflood simulations. The first model (Anisotropic Case) is obtained by estimating 3-D permeability anisotropy from the 2-D permeability modeling and applying the result to the facies-based model described above. The second model (Layered Case) is a simple, isotropic, layered model derived by re-sampling the gridded facies architecture of the Reference Case at the four corners of the model domain. Using only these data represents a conventional, and simplified, approach to building a reservoir model given borehole data obtained at a 40-ac (16.2-ha) spacing. The third, and simplest model (Homogeneous Case) is constructed by assigning a single value of isotropic permeability throughout the model domain.

Numerical simulations are performed using the TETRAD-3D finite-difference reservoir simulation software, provided for use during the time of the study courtesy of Mobil Exploration and Producing Company. This package allows the implicit pressure/explicit saturation (IMPES) solution technique to be used to minimize computational requirements. However, when simulation runs involving large changes in oil saturation through time cannot be treated explicitly, a fully implicit solution method is used. All simulations involve two fluids, oil and water. The bulk of the 2-D simulations were completed before 3-D modeling began. Thus, differences in the model parameterization between the 2-D and 3-D modeling efforts reflect changes in thinking that occurred as the study evolved.

TWO-DIMENSIONAL MODELING

Developing a Gridded Architecture Model

The architectural framework of the Kf-Iv-1[a] bedset was established by Anderson and Ryer (this volume) and Anderson et al. (this volume). The framework was built using a cliff-face photomosaic of the study area, augmented by careful field checking. The interpretation is based on the thin (typically less than 1 ft [<0.3 m]), mudstone and siltstone bounding layers that often separate individual clinoforms. These layers are generally less resistant to weathering than the sandstone clinoforms, producing recessive breaks in the cliff-face. The breaks are clearly seen on the photomosaic, highlighting the prograding clinoform architecture of the cliff face. The architecture is sometimes difficult to interpret in the proximal portion of the outcrop, where the bounding layers separating adjacent clinoforms are often absent. Interpretation is also challenging in the distal region, where clinoforms thin and become increasingly finer-grained,

approaching the character of a bounding layer. Indeed, clinoforms and bounding layers are indistinguishable in the most distal reaches of the outcrop. The linework representation of the architecture is shown in Figure 2A (Mattson, 1997). Figure 2B indicates the portion of the study area eventually used in 2-D flow simulations, denoted the "model domain" in subsequent discussions.

The linework was digitized and rectified as described by Mattson (1997). Rectification corrects for distortion in the linework interpretation caused by projecting the cliff-face, which has spatial depth, onto the 2-D image of the photomosaic.

As discussed above, the digital petrophysical model is constructed by treating each clinoform individually. The underlying assumption is that the spatial configuration of petrophysical properties within a particular clinoform is similar to that in all other clinoforms. Thus, a single set of rules can be used to assign properties within any particular clinoform. This approach requires that each clinoform be represented as a closed polygon, analogous to the piece of a puzzle, and that all such polygons completely fill the model domain.

Several steps are needed to transform the rectified, digitized linework into a structure suitable for developing a numerical model. First, gaps in the linework due to minor digitizing errors and interpretation uncertainty are repaired (closed), enabling each clinoform to be represented as a polygonal body. Next, individual clinoform boundaries are established by cutting the appropriate line segments from the digitized linework and joining them to form a closed polygon. Finally, the entire model domain is gridded so that petrophysical properties can be assigned to rectangular grid blocks within each clinoform.

Figure 3 shows a small portion of the digitized linework and illustrates how gaps in the linework are repaired. The gaps must be eliminated so that clinoform boundaries can be represented as closed polygons. The arrows in Figure 3A indicate breaks in the linework due to digitizing errors. These errors are manually fixed by extending the lines along trend, as shown in Figure 3B, so that they terminate against another line. The circles indicate gaps caused by interpretation uncertainty, where the clinoform boundary could not be established either from the photomosaic or from field checking. These particular examples are from the thin, fine-grained, distal portions of the affected clinoforms, where it is difficult to distinguish a clinoform from a bounding layer. A similar problem often occurs in the proximal reaches of a clinoform, where the bounding layer is thin or nonexistent. These gaps are repaired either by terminating a line against another line (as in the two upper circles), or by extending the line to the domain boundary. The decision to either terminate or extend a line is dictated mainly by geological principles, although at times modeling convenience is also a consideration.

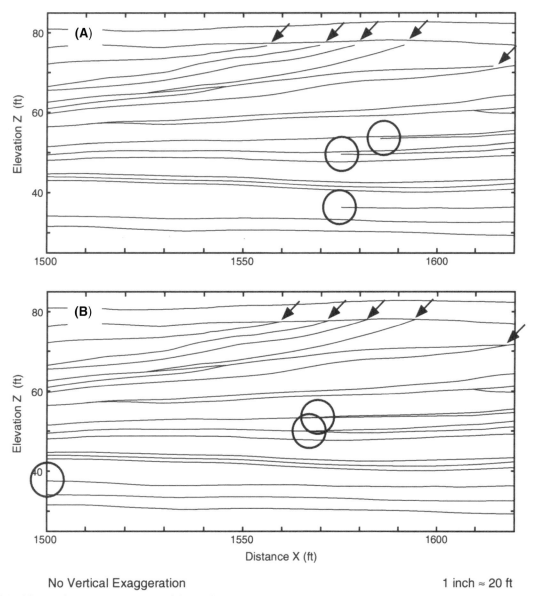

Figure 3. Digitized linework repair in a portion of the modeling domain. In (A) some lines do not terminate against other lines or extend to a boundary. In (B) the linework is repaired. Arrows indicate simple digitizing errors; circles indicate ambiguity in the interpretation.

The linework derived from the photomosaic was digitized without regard to individual clinoforms. Instead, the longest continuous line segments possible were digitized in order to expedite the digitizing process. As a consequence, any single digitized line usually contains portions of the boundary of many different clinoforms. In order to treat clinoforms individually, the boundary of each clinoform in the modeling domain is constructed by cutting the appropriate line segments from the digitized linework and joining them to form a closed polygon. The concept behind a computer program written to accomplish this task is depicted using the sub-domain shown in Figure 4. Four digitized lines (labeled 1a35, 1a36, 1a37, and 1a50) contribute to the boundary of the gray clinoform in Figure 4A). The lines also contribute to the boundaries of other clinoforms, in particular the clinoform to the right and below the gray

clinoform (other lines, and hence clinoforms, in the sub-domain are omitted for the sake of clarity). Also note that part of the boundary is shared by the two clinoforms. The closed circles in Figure 4B show where each of the lines in Figure 4B is cut and joined to form the boundary of the gray clinoform.

The closed polygon defining the boundary of a clinoform simply specifies the shape of the clinoform. It does not provide a way to associate an attribute, such as permeability, to regions on the interior of the clinoform. Gridding imposes a structure on the modeling domain that can be used to assign attributes to specific regions within every clinoform in the domain. It is a crucial step, allowing the linework representation of the Ivie Creek facies architecture to be used in conjunction with fluid-flow modeling. The clinoforms in Figure 5A are shown in their gridded form in Figure 5B. The shading in Fig-

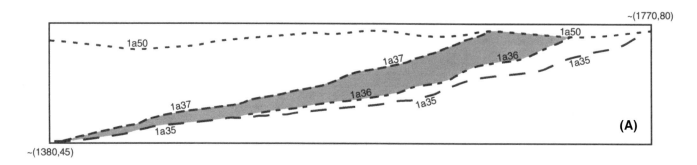

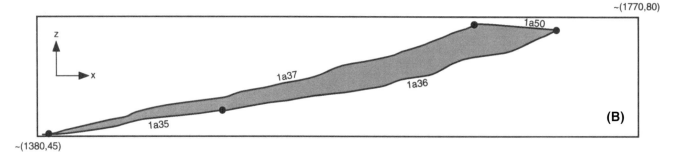

2.5X Vertical Exaggeration

Horizontal Scale: 1 in = 45 ft

Figure 4. Extracting the boundary of an individual clinoform from the digitized linework. (A) Shows that four digitized lines contribute to the boundary of the clinoform indicated by the gray shading. (B) Shows the portions of each of the lines that were cut and joined to form the boundary of the clinoform.

ure 5B indicates varying values of permeability, although any attribute can be represented in this way. A computer program designed to allow different gridding increments in the x and z directions, was written to accomplish this step. The grid increments used to construct Figure 5, x = 2.5 ft (0.8 m) and z = 0.5 ft (0.2 m), are the same as those used in fluid-flow simulations. Simple rectangular gridding is used because subsequent flow modeling is performed using finite-difference flow simulation software.

Each clinoform is gridded with respect to the modeling domain shown in Figure 2. The domain is 1302.5 ft (397 m) long in the x-direction and 40 ft (12 m) high in z. The gridding increments are x = 2.5 ft (0.8 m) and z = 0.5 ft (0.2 m), resulting in nx = 521 grid blocks in x and nz = 80 blocks in z, for a total of 41,680 grid blocks. The domain is sized to capture the transition from highest to lowest values of permeability at the interwell scale. The grid-block dimensions are sized to allow a reasonably detailed description of the bounding layers, while at the same time keeping the number of blocks small enough that the computational resources available to the project would not be overwhelmed.

Distributing Permeability Within Clinoforms

The original intent of the study was to construct a permeability model of the Ivie Creek area by merging the facies architecture with outcrop-based permeability measurements, and then applying geostatistical meth-

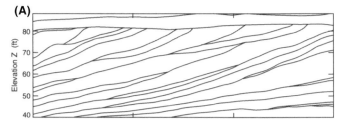

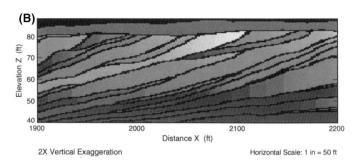

2X Vertical Exaggeration

Horizontal Scale: 1 in = 50 ft

Figure 5. Sub-domain showing: (A) closed polygons enclosing individual clinoforms and (B) gridded clinoforms. The polygons in (A) are simply lines in the plane. The grid in (B) allows petrophysical properties, represented here in black and as shades of gray, to be assigned to throughout the modeling domain. This sub-domain was used in subsequent permeability upscaling experiments.

ods to distribute permeability throughout the model domain. Our ability to follow this approach was inhibited by the steep, broken terrain of the outcrop, which made the collection of core plugs very difficult. Most of the outcrop was inaccessible except by rope, so the loca-

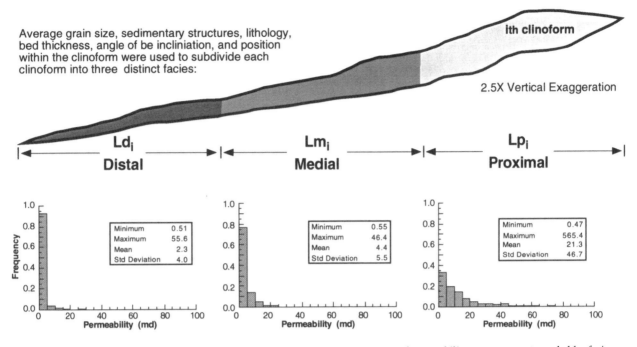

Figure 6. Method used to subdivide each clinoform into distinct facies, and histograms of permeability measurements pooled by facies.

tions of permeability profiles were dictated by safety concerns. Plugs were therefore taken along widely spaced vertical profiles, as shown in Figure 2 (the vertical sample spacing was 0.3 ft [9.1 cm]). As a result, parts of many clinoforms were sampled, but no single clinoform contained enough well distributed measurements to adequately characterize the spatial structure of permeability within it. Plausible estimates of permeability trends and correlation lengths were not feasible. With the benefit of retrospect, the ideal approach would be to characterize several clinoforms with well-distributed measurements. The results of such a test could be used to build the case study permeability model in a more rigorous fashion.

Relationships between absolute permeability and clinoform facies described by Chidsey and Allison (1996), Chidsey (1997), and Mattson (1997) are used to surmount the problem of incomplete permeability characterization within a clinoform. Each clinoform in the gridded model domain is populated with permeability using the simple approach illustrated in Figures 6 and 7. Each clinoform is subdivided into three facies (distal, medial, and proximal) based on average grain size, sedimentary structure, lithology, bed thickness, angle of bed inclination, and position within the clinoform (Chidsey and Allison, 1996; Chidsey, 1997; Mattson, 1997; Anderson et al., this volume). An example of this procedure, together with permeability histograms of measurements grouped by facies, is shown in Figure 6. The arithmetic mean of permeability of a particular facies is assigned to the grid cell at the mean x-coordinate of that facies (Figure 7). Absolute permeability in any other grid cell is calculated by linear interpolation or extrapolation, based

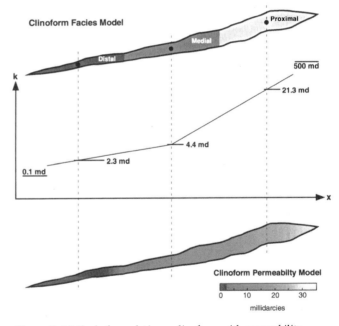

Figure 7. Method of populating a clinoform with permeability.

on the x-coordinate of the cell. A lower limit of 0.1 millidarcies (md) is imposed on the extrapolation in the distal part of a clinoform, while an upper limit of 500 md is imposed on the extrapolation in the proximal part of a clinoform (however, the maximum calculated permeability is only 33.9 md). These limits roughly correspond to the minimum and maximum permeability measurements in the respective facies. Note that this scheme imparts only a horizontal permeability trend within a clinoform, since the interpolation or extrapolation is based solely on x coordinates. The data do not support

Table 1. Facies preservation in model domain clinoforms.

Facies Preserved*	Number of Clinoforms
Complete	18
Proximal	8
Medial and Proximal	5
Medial	13
Distal and Medial	19
Distal	3

* Does not include cap facies, which is present in only one clinoform.

defining vertical permeability trends at specified x coordinate locations, thus this possibility is not included in the models.

The approach described above is possible only when a clinoform is "complete" in the sense that it contains each of the three clinoform facies types: proximal, medial, and distal (Mattson, 1997). However, only 18 of the 66 clinoforms in the model domain are complete; the remaining 48 clinoforms are "incomplete," containing only one or two of the clinoform facies. Incomplete clinoforms are the result of erosion or non-deposition of a particular facies. Table 1 lists the possible combinations of facies that can be preserved in a clinoform, and the corresponding number of clinoforms of each combination type incorporated in the model domain.

Absolute permeability is distributed within incomplete clinoforms using the average length ratios R_{mp} and R_{md} computed from the complete clinoforms. These ratios are defined as:

$$R_{mp} = \frac{1}{N} \sum_{i=1}^{N} \frac{Lm_i}{Ld_i} \qquad (1)$$

and

$$R_{md} = \frac{1}{N} \sum_{i=1}^{N} \frac{Lm_i}{Lp_i} \qquad (2)$$

where N = 18, and Lp_i, Lm_i, and Ld_i are as shown in Figure 6. R_{mp} = 0.823 is the average ratio of x-distance spanned by the medial facies to the x-distance spanned by the proximal facies. R_{md} = 0.383 is the average ratio of x-distance spanned by the medial facies to the x-distance spanned by the distal facies. The ratios are used to compute the length of the "missing" facies so that the linear interpolation/extrapolation method outlined above for populating complete clinoforms with permeability can be applied to incomplete clinoforms. For example, consider the kth incomplete clinoform in which the distal and medial facies are preserved but the proximal facies is absent. The estimated x-distance Lp_k spanned by the missing proximal facies is computed in:

$$Lp_k = \frac{Lm_k}{R_{mp}} \qquad (3)$$

With this computed length Lp_k, interpolation of perme-

ability between the medial and proximal facies can be performed as discussed earlier. A similar approach is used for incomplete clinoforms in which a facies other than proximal is missing.

Petrophysical Models Used for Flow Simulation

Plate 1 shows the four permeability models built for subsequent use in flow simulation studies (the black region in the top left of each model represents null grid blocks that are inactive during numerical flow simulations) (Plates 1 through 19 are on the CD-ROM). Each is a different portrayal of the extent to which bounding layers surround individual clinoforms. At one extreme, no bounding layers separate clinoforms (Plate 1a): all contacts between adjacent clinoforms are sandstone-on-sandstone. At the other extreme, every clinoform in the model domain is completely encased by bounding layers (Plate 1d). Between these extremes, and presumably closer to reality, are two models in which bounding layers are absent over a portion of the modeling domain. The first of these models is based on the detailed interpretation of bounding layers described in Anderson et al. (this volume) and is shown in Plate 1b. The second (Plate 1c) is built by assuming that a sandstone-on-sandstone contact exists between two neighboring clinoforms wherever the proximal facies of the two clinoforms are adjacent. The permeability tensor is taken to be isotropic in all models. A single relative permeability curve, from Odeh (1981), is used in the flow simulations.

Boundary layers are assigned a constant permeability of 0.01 md. This is lower than any permeability measured in the study, and an order of magnitude lower than the smallest permeability permitted within a clinoform. Selecting this plausible value provides enough contrast between boundary layer permeability and clinoform permeability to compare the impact that various boundary layer configurations have on reservoir performance.

No thickening of bounding layers is considered, although field observations indicate that boundary layers thicken in the depositional dip direction. The thick-

Table 2. Key reservoir and fluid properties.

Parameter	Description	Value
P_i	Initial reservoir pressure	5000 psia
Oil gravity	API oil gravity	45° API
S_{wc}	Connate water saturation	0.30
S_{or}	Residual oil saturation	0.15
c_f	Formation compressibility	2×10^{-6} psi^{-1}
c_o	Oil compressibility	1×10^{-4} psi^{-1}
c_w	Water compressibility	5×10^{-5} psi^{-1}
μ_o	Oil viscosity	2.5 cp
μ_w	Water viscosity	1.0 cp

ness of the modeled bounding layers is nominally 0.5 ft (0.2 m), the z-dimension of the model grid blocks. Greater thicknesses occur where bounding layer grid blocks are "stacked" as a result of the conversion from polygonal to gridded clinoforms, and where the bounding layers of adjacent clinoforms intersect.

Porosity Φ is computed from absolute permeability k using the empirical relationship obtained from permeability and porosity measurements of outcrop core plugs (Jarrard et al., this volume):

$$\Phi = (\log_{10} k + 2.64)/22.72 \qquad (4)$$

where Φ is fractional porosity and k is absolute permeability. Porosity is computed from the permeability model in which bounding layers are absent (Plate 1a). This porosity model is used in all numerical flow simulations, regardless of the permeability model used, in order to facilitate comparisons between the flow simulation results of the four permeability models. Thus, the resulting distribution of both Φ and k reflect the distribution of the various clinoform facies. Note that different porosity models, derived from and used with each permeability model, would yield different pore volumes. The different pore volumes, in turn, would produce different estimates of original oil in place (OOIP) and also affect the timing of water breakthrough in waterflood simulations.

Two-Dimensional Flow Simulation Studies

The effect of permeability layering imparted to the model reservoir by juxtaposing clinoform facies, and the impact of bounding layer configuration on reservoir performance is studied by simulating a waterflood in each of the petrophysical models illustrated in Plate 1 and discussed in the previous section. The flow simulations involve two fluid phases, water and oil. Water injection begins at the start of the simulations, without any period of primary production. Injection is accomplished by a vertical well completed in all active grid cells on the left (west) side of the model. The water injection rate is 10 stock tank barrels (STB)/day (1.6 m^3) (approximately 0.007 pore volumes/day). Production is from a vertical well completed in all active grid blocks on the right (east) of the model, subject to a 3500 pounds per square inch absolute (psia) (24,133 Kpa) bottom-hole pressure constraint. The time period simulated is three years (1095 days). Simulation experiments indicate that reversing the positions of the injector and producer has only a minor impact, and does not materially affect the main results of the study.

The total pore space in each model is 8145 cubic ft (231 m^3), yielding an average fractional porosity for the entire model domain of about 0.16. Initial oil saturation in all models is 0.70, with an initial reservoir pressure of 5000 psia (34,475 Kpa). OOIP is 1003 STB (160 m^3). Other key reservoir and fluid properties are listed in Table 2. These properties, the pressure-volume-temperature (PVT) properties of the reservoir fluids (Appendix, on the CD-ROM), as well as the relative permeability curve and capillary pressure curve used are taken from Odeh (1981).

The modeling results are presented as a set of figures showing the spatial distribution of oil saturation in each petrophysical model, from model **a** to **d** (for example, Figures 8A to 8D), at a series of time steps. Plate 2 shows oil saturation at 180 days after the start of water injection. Plate 3 shows oil saturation at 365 days, Plate 4 at 540 days, and Plate 5 at 730 days. The arrangement of the oil saturation plots in each figure is consistent with the arrangement of the permeability models, **a** through **d**, displayed in Plate 1. Production curves presented in Figure 8 summarize the production performance of each petrophysical model.

A common, though hardly surprising, feature of the simulation results obtained from each of the four petrophysical models is the tendency for the injected water to preferentially invade the upper, higher permeability part of the domain. This tendency is enhanced by the absence of bounding layers surrounding either all or part of the clinoforms comprising the domain architecture. Model **a** contains no bounding layers to impede flow; the preferred flow in the upper part of the domain simply reflects the higher permeabilities due to the juxtaposition of proximal facies. In models **b** and **c**, the water invasion is enhanced by the absence of bounding layers in the proximal parts of clinoforms, combined with the presence of bounding layers surrounding the medial

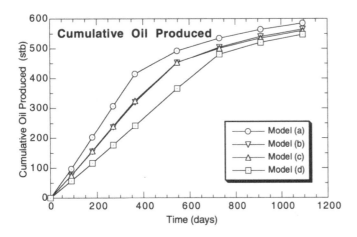

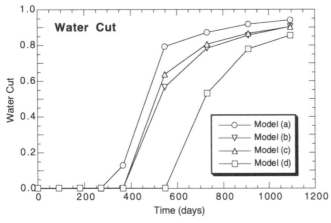

Figure 8. Cumulative production and water cut curves for the four petrophysical models. OOIP is 1003 STB for all models.

and distal parts of clinoforms. Water flow is impeded in the lower part of the domain by the low permeability bounding layers between adjacent clinoforms, while in the upper part of the domain flow is enhanced because bounding layers are nearly absent.

The partial bounding layer models enhance the permeability stratification imparted to the model by the *en echelon* juxtaposition of clinoforms and the resulting arrangement of clinoform facies. A finger of high water saturation extends into the upper domain, so that near and behind the water front only about half the reservoir thickness is involved in the flood. The tendency for water to override the lower part of the domain is diminished in model **d**, where all clinoforms are completely enclosed by bounding layers. In this case, bounding layers surround and retard flow through the proximal, higher permeability facies concentrated in the upper part of the domain. The finger of high water saturation in the upper part of the model is much shorter in **d** that in **a** through **c** so that just behind the water front a greater vertical proportion of the reservoir is involved in the flood.

The facies-based bounding layer model in Plate 1c is used to estimate effective horizontal and vertical

absolute permeability, and thus permeability anisotropy, for the model domain. The computed permeability anisotropy is used in the 3-D modeling work discussed more fully in a later section.

Total oil recovery from each model after three years of production is similar (Figure 8). Cumulative production ranges from a high of 585 STB (93 m^3) (0.58 of OOIP) from model **a** to a low of 548 STB (87 m^3) (0.55 of OOIP) from model **d**; models **b** and **c** yield cumulative production, respectively, of 565 STB (90 m^3) (0.56 of OOIP) and 560 STB (89 m^3) (0.558 of OOIP). At any fixed time after the start of production, model **a** has the highest cumulative production and model **d** the lowest. These results are expected, since the more realistic permeability configurations of models **b** and **c** are intermediate between the extremes represented by models **a** and **d**. The shapes of the production curves in Figure 8 suggest that the differences in cumulative production between the models would persist even for considerably longer production times. The curves also suggest that the incremental additions to cumulative production would be small, and possible only under the penalty of high water cut conditions, as discussed below.

Although cumulative production from each of the four models is similar after three years, it is clear from Figure 8 that bounding layer configuration has a large impact on the timing of oil production and water breakthrough. Water breakthrough occurs first in model **a**, at about 365 days. Water cut ([water produced]/[total fluids produced]) climbs rapidly, to nearly 0.80 at 547 days. In contrast, breakthrough in model **d** occurs at about 547, and a water cut of 0.80 is reached after approximately 1000 days. Models **b** and **c** exhibit similar water breakthrough curves, with breakthrough between 365 days and 547 days; a water cut of 0.80 is reached at roughly 800 days.

These modeled production results have clear economic consequences. For example, model **a** yields a pessimistic estimate of the time to water breakthrough, and an optimistic estimate of cumulative production at a fixed time after the start of water injection. Note that at 365 days cumulative production from model **a** is 415 STB (66 m^3), nearly 30% higher than the 320 STB (51 m^3) produced from model **b**. Economic calculations based on model **a** would result in high cash flow early in the operation of the field, and a short lifetime of operation if the decision to abandon the field is based on water cut. Conversely, predictions based on model **d** would indicate a prolonged period of low water cut production at low production rates.

There are only minor differences in the production curves for models **b** and **c**, suggesting that the facies-based approach to modeling the extent of bounding layers is a good approximation to the model based on field observations. If this method is to be used, it is crucial that good estimates of clinoform dimensions and the relative lengths of the facies comprising clinoforms be obtained by comparing downhole information to similar facies mapped in outcrop analogs.

Several issues that would likely have an impact on the performance of the modeled reservoir were not explored in the study. These include:

- Bounding layer thickness typically increases in the depositional dip direction. Including this observation in the permeability models, combined with the absence of bounding layers between the proximal facies of adjacent clinoforms would increase the apparent stratification of flow exhibited by models **b** and **c**. Water invasion in the lower part of the model domain would be diminished, enhancing the finger of high water saturation advancing in the upper part of the domain.

- Flow between adjacent clinoforms is likely retarded by permeability anisotropy in bounding layers. Modeling bounding layer anisotropy is difficult, however, because the principal directions of the permeability tensor are probably aligned parallel (maximum permeability) and perpendicular (minimum permeability) to bedding rather than coincident with coordinate axes. Moreover, the bedding orientation of a particular bounding layer varies in space. Bounding layer bedding at Ivie Creek is near horizontal in the distal part of a clinoform, but may reach up to 10° proximally. The effect of anisotropy is likely to enhance the stratification of flow, as in the case of increasing boundary layer thickness discussed above.

- Different relative permeability and capillary pressure curves for the clinoform interiors and bounding layers are not used. Differences in capillary pressure between the two regions may result in the flow of oil from the bounding layers to the clinoform interiors, even though the bounding layers retard the flow of oil induced by hydrodynamic forces.

Permeability Upscaling (Homogenization)

Given the computational resources available to the project team, reservoir simulations were not possible using the grid block size (2.5 ft x 0.5 ft [0.8 x 0.2 m], yielding 41,680 grid blocks) used in the 2-D Ivie Creek modeling. A 3-D model of the same thickness (40.5 ft [12.3 m]) and of areal extent equal to the squared length (13,002 ft^2 [1208 m^2], slightly less than 40 ac [16.2 ha]) of the 2-D model would require nearly 22,000,000 grid blocks. Permeability upscaling, or homogenization, addresses this issue by attempting to find the equivalent permeabilities of a group of large grid blocks which will yield the same simulation results as those obtained using many smaller grid blocks. Coarser gridding can then be used to discretize the problem domain.

Two-dimensional permeability upscaling experiments are conducted using a computer program developed by Koebbe (2002). The program is a flow-based averaging scheme which yields homogenized absolute permeability using a perturbation method applied to the equations for flow in a porous medium. The method provides a homogenized, full tensor, absolute permeability. It does not address the issue of upscaling relative permeabilities, and thus is not a "pseudo-function" approach (for example, Kossack et al., 1990).

The portion of the 2-D modeling domain highlighted in Figure 2 (including clinoform facies and bounding layers) and shown detail in Figure 5 is used in the upscaling experiments. Fluid and reservoir properties are the same as those used in the previous 2-D modeling, except that the injection and production rates are adjusted to reflect the smaller domain size. In addition, the porosity is fixed at 50%.

Four grid block sizes are used in the upscaling experiments: the original, fine-scale 2.5 ft x 0.5 ft (0.8 x 0.2 m) blocks, denoted 1 x 1; blocks scaled by a factor of 2 in both x and z (5.0 ft x 1.0 ft [1.5 x 0.3 m]), denoted 2 x 2; blocks scaled by a factor of 4, denoted 4 x 4; and blocks scaled by a factor of 8, denoted 8 x 8. An upscaled, absolute permeability tensor is obtained by supplying the homogenization program with the scaling factor and the detailed, 1 x 1 permeability model. Water injection simulations were then run with the TETRAD-3D simulator on a model domain discretized at the larger grid spacing and populated with the upscaled permeabilities. Off-diagonal terms of the upscaled permeability tensor were ignored because TETRAD-3D assumes that the grid axes are parallel to the principal directions of the permeability tensor. Upscaled permeabilities are also computed by taking the harmonic and geometric means of the permeabilities within the larger grid blocks. Water injection simulations are also run using the permeabilities obtained by these simple averaging techniques.

Cumulative production and water cut curves obtained from the upscaling experiments are shown in Figure 9. Also shown for reference are the results of water injection into a homogeneous domain with a permeability of 8 md, the arithmetic mean of the permeabilities in the 1 x 1 model. Results are presented for harmonic averaging, geometric averaging, and perturbation method homogenization.

Harmonic averaging tends to emphasize lower permeabilities. It is the appropriate averaging technique for computing the effective permeability for a stratified medium when flow of a fully saturating, single-fluid phase is perpendicular to the stratification. When used to upscale the permeabilities in the small Ivie Creek sub-

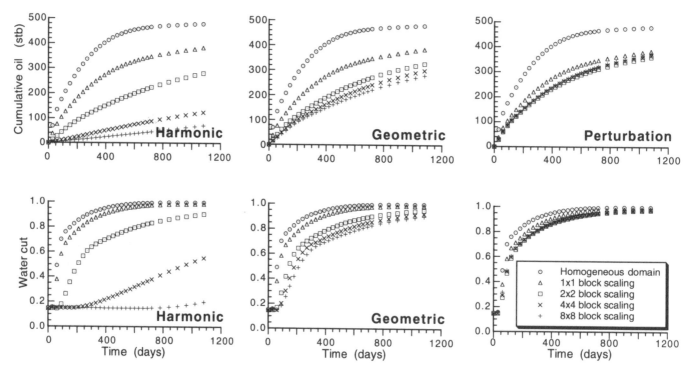

Figure 9. Results of permeability upsealing experiments. Top row: cumulative oil production plots obtained using harmonic, geometric, and perturbation method averaging at various block sizes. Bottom row: corresponding water cut plots.

domain, both cumulative production and water cut are underestimated. The production estimates are very sensitive to grid block size, and degrade dramatically as the size of the grid block increases: in the 8 x 8 case, cumulative production after three years of simulated waterflooding is only 70.2 STB (11.2 m^3), compared with 383.2 STB (60.9 m^3) in the 1x1 case; water cut at three years is 0.20, compared with 0.98 in the 1 x 1 case. This occurs because a grid block is more likely to contain low permeability values as the size of the grid block increases.

Geometric averaging yields a number intermediate between arithmetic and harmonic averaging. It is considered a quick and dirty upscaling method after the work of Warren and Price (1961), who found it to be the best estimator of bulk permeability for a medium in which permeabilities are distributed randomly and without spatial correlation. Cumulative production and water cut are both underestimated when geometric averaging is used to upscale permeabilities, as was the case with harmonic averaging. However, production estimates are less sensitive to grid block size than with harmonic averaging. Cumulative production in the 8 x 8 case is 278.1 STB (44.2 m^3) and water cut is 0.90 after three years of simulated waterflooding.

Permeability upscaling by the perturbation method also leads to lower simulated values of cumulative production and water cut. However, the simulated results are better than those obtained using either harmonic or geometric permeability. Cumulative production is 372.0 STB (59.1 m^3) and water cut is 0.97 after three years of

simulated waterflooding. In addition, the computed results are far less sensitive to grid block size than they are when harmonic or geometric permeability averaging is used. Indeed, production estimates obtained in the 8 x 8 case are actually better than those obtained from the 2 x 2 or the 4 x 4 case. This surprising result warrants further attention, which we are unable to provide in this study.

Not shown are the results obtained using simple arithmetic averaging as an upscaling method. Experiments performed using a model in which clinoform interiors have a constant permeability indicate that arithmetic averaging leads to simulation results that overestimate production estimates. Production estimates tend towards the results obtained from the homogeneous model as the size of the grid block increases.

Summary and Conclusions

Two-dimensional petrophysical and fluid-flow modeling at the Ivie Creek area is based on the outcrop architecture along with permeability measurements made on core plugs extracted from the outcrop and on slabbed core from drill holes.

The clinoform is the main architectural element observed at the site. Each sigmoidal clinoform consists of sandstone bedsets that record an episode of delta-front deposition followed by partial erosion. Clinoforms are inclined seaward, arranged *en echelon*, and are usually separated from each other by a thin (less than 1 ft [<

0.3 m]) mudstone and siltstone-bounding layer. The resulting architecture provides the spatial framework for modeling the configuration of petrophysical properties, and the subsequent numerical simulation of two-phase (oil and water) fluid flow.

A gridded, digital model of the site architecture preserves individual clinoforms, and is the basis for distributing absolute permeability throughout the modeling domain. Permeability modeling is accomplished by subdividing each clinoform into three facies (distal, medial, and proximal). The arithmetic means of permeabilities grouped by facies are used to linearly interpolate or extrapolate permeability throughout a particular clinoform. A method based on ratios of mean lengths of the three facies was developed to handle those clinoforms in which all three facies are not present. Resulting clinoform permeabilities in the modeling domain range from 0.1–33.9 md. Bounding layer permeability is a constant 0.01 md.

Four permeability models are built for the flow-simulation studies. Each portrays a different extent to which constant thickness bounding layers surround individual clinoforms. At one extreme, no bounding layers separate clinoforms and all contacts between adjacent clinoforms are sandstone-on-sandstone. At the other extreme, every clinoform in the model domain is completely encased by bounding layers. Between these extremes, and presumably closer to reality, are two models in which bounding layers are absent over a portion of the modeling domain. One of the intermediate models is based on field observations. The other is obtained by assuming that the bounding layer between clinoforms is thin or absent where the proximal facies of two neighboring clinoforms are adjacent. The same porosity model, obtained using an empirical relationship between permeability and porosity for the study, was used together with each of the four permeability models in subsequent flow simulations.

The production performance of each reservoir model was studied by simulating a waterflood in each of the reservoirs. Water injection began at the start of the simulations, without a period of primary production. A single relative permeability curve and a single capillary pressure curve were used in all simulations. In addition, absolute permeability tensor was considered to be isotropic. The simulations run for a time period of three years (1095 days).

All models exhibit a common, unsurprising feature: preferred flow of the waterflood in the upper, higher permeability portion of the models. This is caused by the juxtaposition of proximal facies, which imparts a stratified permeability structure to the models. This effect is most pronounced in the two models in which bounding layers are absent between proximal clinoform facies.

The timing of oil production and water breakthrough is sensitive to the nature of the bounding layers.

Earliest breakthrough occurs, at roughly 365 days, in the model without bounding layers. This model also yields the highest cumulative production at any fixed time after the start of water injection: after three years 0.583 of the OOIP is produced. The longest time to breakthrough occurs in the model in which bounding layers completely enclose clinoforms, at between 547 and 730 days. This model also yields the lowest cumulative production at any fixed time after the start of water injection: after three years 0.546 of the OOIP is produced. The two models with partial bounding layers are similar to one another, and yield simulation results between those of other two models. In these intermediate cases water breakthrough occurs between 365 days and 547 days, while after three years of production about 0.561 of the OOIP is produced. Economic calculations based on cumulative oil production and time to water breakthrough will clearly be affected by the type of bounding layer model used.

Permeability upscaling experiments indicate that common averages (arithmetic, harmonic, and geometric) are inadequate approaches to upscaling absolute permeability in this depositional environment. An upscaling technique based on perturbation analysis yields simulation results similar to those obtained with detailed permeability models.

THREE-DIMENSIONAL MODELING

Model Domain

The 3-D model grid is 2000 ft by 2000 ft (610 x 610 m) in plan view and 60 ft (18 m) thick (Figure 10). The domain contains 150,000 rectilinear grid blocks each 20 ft by 20 ft (6 x 6 m) in plan view and 4 ft (1.2 m) thick. Thus, there are 15 layers each with a grid of 100 by 100 blocks. The southeast corner of the model lies just outside the point where the cliff that bounds the north side of Ivie Creek meets the cliff that bounds the west side of Quitchupah Canyon (Figure 11).

Production Strategies and Wellbore Configurations

A simple 5-spot pattern (Figure 10) containing a central, vertical water injection well and four vertical oil production wells, is used to model a waterflood scenario for most simulations. In these cases a maximum injection rate of 1516 STB (241 m^3) of water are injected per day. In one additional case, a 9-spot pattern containing a central injection well (maximum injection rate of 1516 STB water per day) and eight production wells (Figure 10) is simulated. In all cases the wells are assumed completed

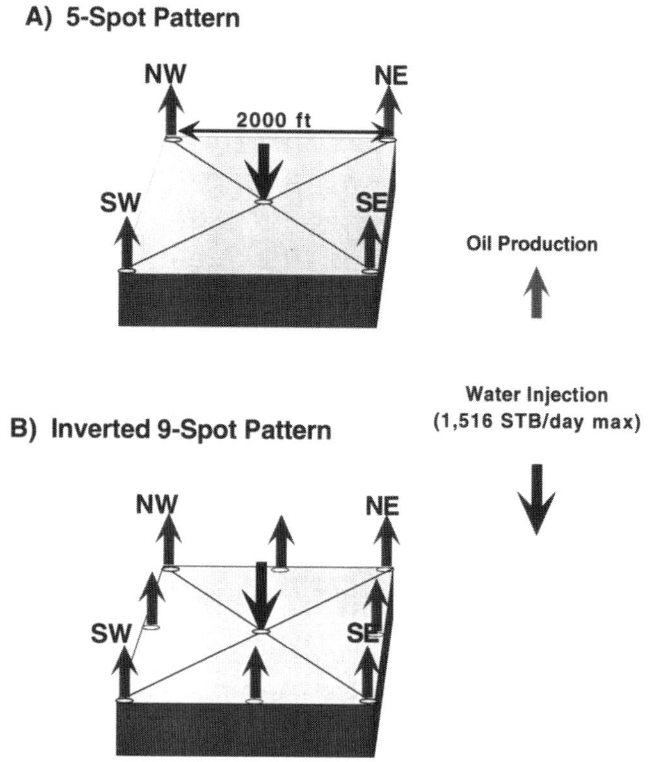

A) 5-Spot Pattern

B) Inverted 9-Spot Pattern

Oil Production

Water Injection
(1,516 STB/day max)

Figure 10. Schematic diagram showing well patterns used in simulated reservoir production.

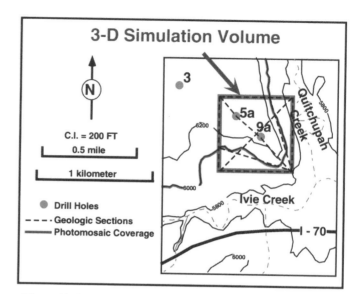

Figure 11. Location map showing the relationship between the 3-D simulation volume, cliff-face outcrops, drill holes, and geological cross sections.

across the entire thickness of the reservoir. A wellbore radius of 0.25 ft (7.62 cm) is specified for both injection and production wells. Wellbore skin effects are assumed negligible.

Physical Properties

Initial Conditions

A hypothetical, but plausible, set of initial reservoir conditions are selected for the 3-D fluid-flow modeling. Each model run started at the same initial oil and water saturations (0.5 for each fluid), rather than modeling primary production prior to performing a waterflood. This approach provides a straightforward basis for comparing the relative differences in reservoir performance. Note that an initial oil saturation of 0.7 is assigned in the 2-D simulations described in the previous section. A uniform initial fluid pressure of 5000 psi (34,475 Kpa) and an isothermal temperature of 166° F (60° C) are also assumed throughout the model domain. A minimum bottom-hole pressure of 2685 psi (18,513 Kpa) (corresponding to the bubble-point pressure of black oil) is specified at the water injection well. This constraint prevents the reservoir pressure from falling below the bubble-point pressure, ensuring that gas is not released from solution and restricting the simulation to only two phases (oil and water).

Absolute Permeability

The distribution of clinoform facies described in previous sections is used to construct the different permeability structures simulated within the model domain. The reference permeability structure is derived from the 3-D clinoform facies structure (proximal, medial, and distal facies). Each facies is assigned an absolute permeability equal to the arithmetic mean of the permeability measurements made in that facies. These mean permeability values, listed in Table 3 are the same as those used to construct the 2-D, trending permeability models discussed in the previous section (Figure 7). The proximal facies is assigned a permeability of 21 md, the medial facies is assigned 4 md, and the distal facies is assigned 2 md. The fourth facies (clinoform cap) and the fifth facies (overlying Kf-1-Iv[c] bedset) are assigned permeabilities of 3 and 0.01 md, respectively. The base of the model is assumed to mark the upper boundary of the low-permeability Tununk Shale. Four different permeability structures (Reference Case, Homogeneous Case, Layered Case, and Anisotropic Case) are constructed, representing four different ways that drill-hole data might be used to interpolate absolute permeabilities within the model domain (Figure 12).

The Reference Case (Figure 12) is defined as the most detailed permeability structure constructed for the model domain. Because a unique permeability and a unique porosity is associated with each clinoform facies type, these parameters are also shown in the facies distributions illustrated in the vertical and horizontal sections of Plate 6. The maps and sections shown in Plate 6 are taken directly from the detailed, 3-D facies structure. The horizontal sections (labeled with reference to their height, in ft, above the base of the model) show three of

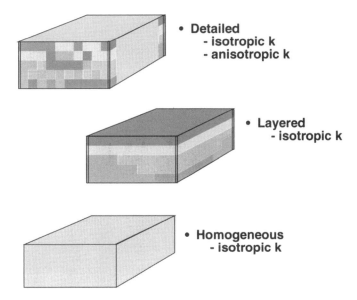

Figure 12. Schematic diagram showing the four permeability structures considered in the 3-D reservoir simulations, Reference case (detailed isotropic and anistropic).

the 15 layers involved in the fluid-flow simulator. The vertical sections are constructed along the two diagonals that extend between opposite corners of the model domain and intersect at the central water injection well (Figure 10). Only isotropic permeability values are assigned in the Reference Case. Thus, although facies-based permeability variations are included in the permeability structure, the bulk anisotropy imparted by the shaley bounding layers discussed under 2-D modeling is not included.

The Anisotropic Case includes the same level of detail as the Reference Case, however, anisotropic permeabilities are assigned at each gridblock in an effort to incorporate the possible impact of the shaley bounding layers discussed and modeled in the previous section on 2-D modeling. The 2-D modeling results indicate that these features can play an important role in controlling oil production. The bulk effect of the shaley bounding layers is crudely approximated in a four-step process of building an anisotropic permeability structure outlined in Appendix and shown in Plate 7. Note that a radial character, representing the inference that the rocks included within the model volume were deposited in a fan-like lobe, is imparted to the anisotropic permeability

structure that is, in turn, superimposed on the heterogeneous facies structure incorporated in the Reference Case (Plate 6). In addition, it is important to recognize that the rules used to create the Anisotropic Case yield a wider range of permeability values than are included in the corresponding Reference Case. One consequence of our approach to constructing the Anisotropic Case is that both north-south permeability (k_{NS}) and west-east (k_{WE}) respectively, are less than the isotropic k values assigned at the south and east boundaries of the model domain (Plate 8). Elsewhere, using a radial anisotropy ratio of k_{arc}/k_r (permeability k in annular direction [k_{arc}] perpendicular to radial permeability [k_r]) equal to 10 can yield either smaller or larger values of both k_{NS} and k_{WE}. It is clear, however, that important elements of the original permeability heterogeneity are retained.

The Layered Case is a simplified permeability structure constructed by sampling the vertical distribution of facies found at each of the four corners of the Reference Case. This sampling mimics the facies data that might be collected by drilling wells at the four locations. With this information in hand, a simple geological model is constructed by assuming that each facies has a tabular character. The resulting permeability structure (Plate 9) has a layered appearance that mimics some of the gross features of the Reference Case (Plate 6). The simplest permeability structure is provided by the Homogeneous Case. In this case, the harmonic mean of all permeabilities assigned in the Reference Case (permeability of 3.8 md and porosity of 14%) is assigned throughout the model domain.

Porosity

Porosity is defined throughout each model domain using a functional relationship between porosity Φ and absolute permeability k. Because each facies type is assigned a single permeability value, a single porosity value is also assigned to each facies. Permeabilities and porosities measured on representative core plugs collected from the Ivie Creek area suggest a reasonable correlation between porosity and permeability (Jarrard et al., this volume). The empirically derived relationship was given in equation 4. Because we assume negligible reservoir consolidation by matrix subsidence, a coefficient of reservoir matrix compressibility is not required. Permeability and porosity as a function of facies type are shown in Table 3.

Table 3. Permeability and porosity for each clinoform facies.

Facies	Permeability (md)	Porosity (%)
Proximal	21.3	17
Medial	4.4	14
Distal	2.3	13
Clinoform Cap	3.0	14
Kf-1-Iv[c]	0.01	3

Relative Permeability and Capillary Pressure

Computed reservoir performance can depend strongly on the character of relative permeability relationships used to represent variations in permeability as a function of oil saturation (S_o) or water saturation (S_w). Although absolute permeabilities derived from outcrop provide good insight regarding the permeability structure of the Ivie Creek area, we concluded that measuring the relative permeability characteristics of outcrop-derived samples would yield only limited new insight regarding relative permeabilities at reservoir conditions. As a consequence, plausible relative permeability relationships are derived from previous measurements of relative permeabilities for the Ferron Sandstone (Barton and Tyler, 1991, 1995; Fisher et al., 1993a, 1993b; Barton, 1994) and Berea Sandstone (Honarpour et al., 1986).

Developing oil-water relative permeability relationships for the Ferron Sandstone involves a two-step process. First, brine-air relative permeability relationships available for both Berea and Ferron rocks are compared to develop a transformation that relates the characteristics of the Berea Sandstone to those of the more poorly sorted Ferron Sandstone. Once the transformation relationship is developed, oil-water relative permeability data for the Berea Sandstone is transformed into an oil-water relative permeability relation for Ferron Sandstone using the Brooks-Corey relations. Ultimately, relative permeability relationships for oil (k_{ro}) and water (k_{rw}) are derived for each facies included in the model domain. The approach used to derive the relative permeability relationships is outlined in the Appendix.

Water capillary pressure P_c (Appendix), or the difference in fluid pressure across the oil-water interface, is obtained from the standard water-wet relation provided by Honarpour et al. (1986). In a water-wet reservoir a decrease in water saturation causes a corresponding increase in capillary pressure.

Fluid Properties

Fluid properties are defined to be consistent with an initial reservoir fluid pressure of 5000 psia (34,475 Kpa). Wherever possible fluid properties (Appendix) were specified to be the same as those used to conduct the black oil simulations performed for the Society of Petroleum Engineers Comparative Solution Project (Odeh, 1981). A two-phase oil/water system is assumed.

Three-Dimensional Flow Simulation Results

Two-phase fluid-flow simulations are used to explore how the four different permeability structures might influence oil production. In each case production is initiated with a waterflood rather than starting with an initial period of primary production.

Oil saturations obtained for a 5-spot production pattern in the Reference Case at times of 1, 3, and 10 years after the start of water injection are shown in three horizons (Plate 10) and two vertical cross sections (Plate 11). The detailed heterogeneities included in the Reference Case clearly influence the progress of the waterflood. Note that the low permeability clinoform cap facies (Kf-1-Iv[c] bedset) that dips down into the northwest quadrant of the model causes a localized increase in oil saturation to 70% from the initial value of 50%. The distribution of oil saturation at early times is clearly influenced by the details of the permeability structure. Higher permeabilities assigned in the proximal facies cause the waterflood to break through first at the southeast corner production well, followed soon after by breakthrough at the southwest corner.

A 9-spot pattern of wells (Figure 10) is simulated to evaluate the relative merits of a more closely spaced pattern of production wells. Oil saturations obtained at times of 1, 3, and 10 years after the start of water injection are shown in Plates 12 and 13. Although the simulation results clearly indicate an increased sweep efficiency relative to the Reference Case, localized areas with increased oil saturation indicate regions where oil is likely to remain in the reservoir after production is halted. If the radial geometry assumed when building the permeability structure can be identified through analysis of bedding orientations logged in the initial 5-spot pattern, then a more efficient pattern of wells might be devised.

A 5-spot pattern simulated in the simple Layered Case is used to illustrate how an incomplete understanding of the facies architecture might contribute to uncertainty in assessing reservoir performance. The layered permeability structure has a much reduced level of heterogeneity than that of the Reference Case (Plates 6, 8, and 9). As a consequence, a more uniform sweep of the reservoir is achieved in the Layered Case (Plates 14 and 15). The low permeability clinoform cap facies encountered in the northwest corner of the model domain, however, produces a region of enhanced oil saturation similar to that found in the Reference Case. Although the graphical results suggest a qualitative similarity in production characteristics for both the Reference and Layered Cases, a later section uses plots of cumulative oil production and water cut to more quantitatively illustrate the differences between the simulation results.

Results obtained by simulating the 5-spot pattern in the Anisotropic Case are shown in Plates 16 and 17. Recall that the higher permeabilities built into this case reflect the relatively high radial anisotropy ratio (k_{arc}/k_r) of 10 used in the model. We suspect that this choice yields an upper limit for the possible effect of the bulk anisotropy imparted to each facies type by the lower permeability bounding layers found within each facies. Additional numerical experiments with a variety of anisotropy ratios are required, however, to more fully evaluate this hypoth-

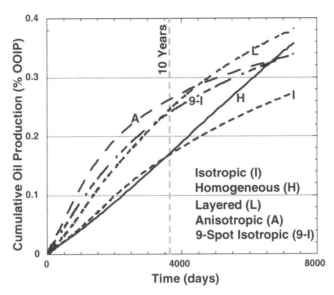

Model	Mean k (md)			Mean ϕ (%)	OOIP (MSTB)	10 Yr Recovery (% OOIP)
	WE	NS	Z			
Isotropic		4		18	3.0	17
Homogeneous		4		14	3.0	17
Layered		5		18	3.2	25
Anisotropic	5	5	0.5	18	3.0	26
9-Spot Isotropic		4		18	3.0	25

Figure 13. Plot of cumulative oil production against time for each of the permeability models used in the 3-D simulations.

esis. Given the anisotropic permeability structure shown in Plate 7 the modeling results show enhanced and more uniform sweep when compared to results obtained in simulating the isotropic Reference Case (Plates 10 and 11). Note that the zones of increased oil saturation computed in the Reference Case are not found in the Anisotropic Case. This result likely reflects the overall enhanced permeability of the Anisotropic Case combined with the heterogeneous structure imparted to permeabilities found within each facies type. Although the permeabilities assigned within each facies of the Anisotropic Case attain relatively high values, the internal heterogeneity seems more realistic than the homogeneous permeabilities of each facies modeled in the Reference Case. Developing a radially anisotropic permeability structure might be motivated by two factors. First, dipping shaley drapes that might form low permeability bounding layers are identified in drilling logs. Second, analysis of dips associated with the shaley drapes suggest a radially symmetric character with a likely position for the point of origin of the sediments.

Discussion

Quantitative Comparison of 3-D Simulation Results

Simulation results presented in the previous section are quantitatively summarized in plots of cumulative oil

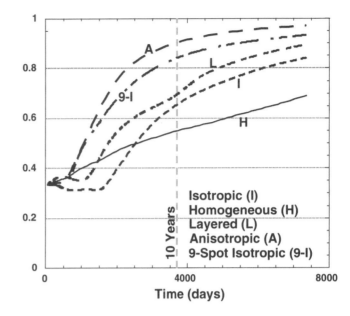

Water Cut = $W_{Prod}/(Oil_{Prod} + W_{Prod})$

Figure 14. Plot of water cut against time for each of the permeability models used in the 3-D simulations.

production (Figure 13) and water cut (Figure 14). Although not discussed in the previous section, results obtained for the Homogenous Case are also shown in (Figures 13 and 14). Table 4 summarizes the cumulative oil production obtained for each case after 10 years of waterflooding. Water cut shown in Figure 14 is computed as the volume of water produced divided by the total volume of oil plus water produced.

The cumulative oil production from both the Reference and Homogeneous Cases is remarkably similar (17% of OOIP) for the first 10 years of waterflooding (Figure 13). After 10 years of production, however, the homogeneous case yields an increasingly large deviation from the oil recovery obtained for the heterogeneous Reference Case. Although computed cumulative oil production is similar during the first 10 years for both the Reference and Homogeneous Cases, the plot of cumulative water cut (Figure 14) reveals markedly different water production (65% and 55% water cut, respectively). The inherent heterogeneity imparted by the facies-based petrophysical properties of the Reference Case yields less water cut in the first four years of production while much greater water cut is produced at later times. This result suggests that even a relatively coarsely defined heterogeneity in petrophysical properties can have a significant impact on oil production economics. For example, the northwest quadrant of the Homogeneous Case has higher permeability than that of the Kf-1-Iv[c] bedset facies included in this region of the Reference Case (Plate 6). Increased sweep efficiency in this region of the Homogeneous Case likely accounts for the primary differences between the two simulation results. Early

Table 4. Cumulative oil production after 10 years of waterflooding.

Model	Mean k (md)			Mean N (%)	OOIP (MSTB)*	10 Yr Recovery (% OOIP)	10 Yr Water Cut
	k_{WE}	k_{NS}	k_z				
Reference		4		18	3.0	17	.65
Homogeneous		4		14	3.0	17	.55
Layered		5		18	3.2	25	.68
Anisotropic	5	5	5	18	3.0	26	.90
9-Spot - Reference		4		18	3.0	26	.85

* MSTB = thousand stock tanks barrels.

breakthrough at the southeast corner yields a marked increase in water cut for the Reference Case at the five-year mark.

Let us assume that the Anisotropic Case represents an upper limit for realistically distributing petrophysical properties within the 3-D model domain. Thus, depending upon the radial anisotropy ratio (k_{arc}/k_r) assigned to each facies type, it appears that both the Homogeneous and Reference Cases could yield moderate to severe underestimates of the volume of oil that might be produced from fluvial-deltaic reservoirs with architectures similar to that exposed in the Ivie Creek outcrops. This result strongly suggests that reasonable efforts should be made to incorporate the impact of the shaley bounding layers in reservoir simulators. Developing appropriate empirical rules for defining values of bulk anisotropy associated with different architectural styles might provide a workable approach. It is important to note that the underestimate of total oil produced in the Reference and Homogeneous Cases is derived at reduced cost in terms of water cut. Results obtained for the Anisotropic Case consistently yield much higher values of cumulative water cut.

Attempting to represent clinoformic sandy bodies and shaley bounding layers as layered facies (for example, Layered Case) provides little improvement over the other approaches used to represent the 3-D structure of petrophysical properties. The model results suggest that after 10 years of waterflooding, estimates of cumulative oil production could range between 17 to 26% depending on the approach used to represent the reservoir architecture. Breakthrough of the waterflood is much earlier in the Layered Case, in part, due to the more uniform distribution of petrophysical properties in the Layered Case. In the absence of a comprehensive understanding of the producing reservoir, it seems reasonable to expect a range of uncertainty in estimating cumulative oil production on the order of ±40%. A similar range of uncertainty is indicated for estimates of cumulative water cut. The Layered Case provides much higher oil production than that of both the Reference and Homogeneous Cases.

Comparing saturation plots of Plates 10 through 13 indicates that the 9-spot pattern yields improved sweep efficiency. This result appears to be obtained with increased costs expressed as larger values of water cut.

Comparing 2-D and 3-D Simulation Results

It is valuable to compare the results of the very detailed 2-D simulation results to those obtained in the 3-D simulations. Although the comparison is complicated by the fact that somewhat different initial conditions and relative permeability properties were used in each case, a qualitative comparison can be made between the 3-D Reference Case and the 2-D simulations.

Plate 18 shows the analogous permeability structures defined for the 2-D and 3-D model domains. Section A-A' extracted from the 3-D model corresponds exactly with the 2-D section constructed along the Ivie Creek amphitheater outcrop. Note that the bounding layers fully encase the sandy bodies in this version of the 2-D model. Although the facies-based permeabilities used in the 3-D case do not incorporate the detailed internal permeabilities of sandy bodies and their shaley bounding layers, the overall permeability structure of the 2-D case is preserved in the 3-D structure of section A-A' (Plate 18). Recognizing the general radial character of the 3-D permeability structure section C-SE is extracted to obtain a reasonable analog for the fluid-pressure gradients imposed between the injection and production wells of the 2-D case (Plate 18). Comparing the permeability structure of the 2-D, A-A', and C-SE sections shows that similar permeability structures are obtained in each case. The corresponding C-SE permeability structures obtained for the Layered and Anisotropic cases are shown in the vertical sections of Plates 9 and 7, respectively.

Oil saturations computed in the 2-D case and the corresponding 3-D Reference, Layered, and Anisotropic Cases are shown in Plate 19. Although somewhat different time steps are available for presenting the results (two years in the 2-D case and three years in the 3-D case), the overall appearance of the simulation results is similar for each case. Closer inspection of Plate 19, however, suggests that results obtained for the 3-D Anisotropic Case are most similar to the 2-D simulation results. The primary difference between the 2-D results and those obtained for the Reference and Layered Cases is the lack of a zone of enhanced saturation near the top of the reservoir as the production well is approached. The upper region of the 2-D case is likely better swept than the corresponding 3-D Reference and Layered Cases because the proximal facies contains permeability values higher than the average value used in the 3-D cases (Plate 19). Apparently the reduced permeability provided by the shaley bounding layers does little to moderate the effect of the enhanced proximal facies permeabilities. Thus, the simulation results suggest that the impact of the lower permeability shaley bounding layers might reasonably be represented by assigning characteristic, but variable, values of bulk anisotropy for each facies. Additional numerical experiments are required, however, to more fully explore the implications of this result.

Conclusions

Oil production by waterflooding a 2000 ft by 2000 ft by 60 ft (610 m x 610 m x 18 m) analog reservoir volume with a 5-spot pattern of wells is simulated for four, 3-D permeability/porosity structures based on outcrop studies in the Ivie Creek area. Each permeability/porosity structure represents different approaches taken to preserve the geologic details that might influence oil production. A relatively fine 3-D grid (150,000 nodes with 20 ft [6 m] spacing in both horizontal directions and 4 ft [1 m] spacing in the vertical direction) is developed to preserve the features observed in outcrop. The Reference Case represents a geologist's best estimate of the 3-D geometry of clinoform-related facies mapped on two intersecting cliff-face exposures and encountered in behind-the-outcrop drill holes. Although the overall geometry of each of four facies types is preserved, the internal variations in stratigraphy and petrophysical properties explored in the 2-D simulations are not retained. It is unlikely that such a detailed model could be developed with confidence based on subsurface data alone. The Layered Case is a simplified permeability structure constructed by sampling the vertical distribution of facies found at each of the four corners of the Reference Case without any benefit of the detailed stratigraphic insight that might be provided by outcrop analog studies. This case represents a structure that might reasonably be inferred by a reservoir geologist in the absence of information regarding the environment of deposition that created the clinoforms observed in outcrop. The simplest model contains a homogeneous, isotropic permeability structure numerically equivalent to the bulk average permeability of the Reference Case. The most complicated model is an idealized anisotropic permeability structure intended to capture the effect of the dipping shale drapes identified within each facies but not incorporated in the isotropic Reference Case.

Although simulated distributions of oil and water saturation appear fairly similar at each point in time, plots of cumulative oil recovery and water cut reveal significant differences in the production response of each analog reservoir model. For example, although the isotropic Reference Case and the Homogeneous Case each produce about 17% of OOIP after 10 years of production, the Homogeneous Case severely underestimates the water cut predicted using the Reference Case (0.55 for the Homogeneous Case and 0.65 for the Reference Case). Meanwhile, after 10 years of production the Layered Case severely over predicts oil recovery (25% of OOIP for the Layered Case and 17% for the Reference Case) with a relatively small increase in water cut (0.68 for the Layered Case and 0.65 for the Reference Case). These differences in production response (primarily inferred from computed water cut) suggest that every effort should be made to develop permeability models that preserve the effect of the clinoform-related facies observed in outcrop. Given the lack of detail typically available from subsurface data and current computational constraints, however, it is likely that these effects will usually be approximated within larger simulation grid blocks than used in this study.

Simulations performed to produce oil from the Reference Case with an inverted 9-spot pattern suggest only a possible advantage over the 5-spot pattern. Although 10-year oil recovery is much improved with the 9-spot pattern (25% OOIP for the 9-spot pattern and 17% for the 5-spot pattern) the resulting water cut is much less attractive (0.85 for the 9-spot pattern and 0.65 for the 5-spot pattern). The reservoir-specific economics of drilling additional producer wells, and the desired timing of production must be known before the relative economic advantages of either approach can be fully identified.

Even without attempting to incorporate the effect of shale drapes within each clinoform-related facies, it is clear that the overall facies geometry exerts a significant impact on oil production. Analogous fluid-flow simulations performed with the more finely gridded 2-D permeability models indicate that relatively thin shale drapes within each clinoform-related facies can have a significant impact on oil production and water cut. The possible impact of the shale drapes is crudely approximated in the 3-D permeability model by computing anisotropic permeability tensors for each node that reflect the characteristic radial geometry associated with the inferred depositional environment. The results of this effort are somewhat ambiguous because we are restricted to using a 5-point differencing scheme that cannot represent the off-diagonal terms of permeability tensors oriented in directions other than those of the simulator coordinate axes. We are convinced, however, that reasonable estimates of vertical and horizontal anisotropy can be made for the Ivie Creek area using the 2-D permeability structures. Additional simulations with higher-order differencing schemes or finite element models are required to properly evaluate the advantages of adopting this approach.

Recommendations

The following are recommendations for additional 3-D modeling and reservoir fluid-flow simulations:

- Additional numerical experiments should be performed using the 3-D permeability/porosity models. Digital files containing the three heterogeneous permeability models developed as input to the reservoir simulator are available from the Utah Geological Survey (Appendix). The detailed, isotropic, heterogeneous Reference Case model provides a foundation for assessing different approaches for 3-D upscaling. In addition, the clinoforms mapped in the Ivie Creek area have slopes much steeper than typically observed at other locations in the same geologic environment. Possible differences in computed oil production and water cut, as a function of clinoform slope, can be explored by stretching the x and y coordinates of the permeability models while maintaining a constant vertical scale and well spacing.

- Ground-water hydrology testing methods should be used to perform wellbore-to-wellbore (interwell) scale, shallow *in situ* tests of permeability anisotropy in well understood, unfractured, and water-saturated sandstone analogs. Such data sets, combined with detailed numerical fluid-flow simulations, would help substantially to develop and evaluate methods for assigning permeability anisotropy at simulator grid-block scales. Unfortunately, the few outcrop analog studies that include behind-the-outcrop drill holes have been carried out where the drill holes fail to reach the water table and fracturing is common. One viable strategy could integrate detailed 3-D ground penetrating radar (GPR), high resolution seismic, detailed drill-hole logging, and shallow interwell testing in closely spaced holes drilled in modern unconsolidated analogs near mappable cliff-face exposures.

- Commercial software available for building stochastic 3-D permeability models (for example, Heresim3D™) should be used in an effort to mimic the detailed, isotropic, heterogeneous Reference Case. Such an exercise would provide a sound basis for evaluating how the commercial software results might best be tuned to work with clinoform-related facies. Fluid-flow simulations performed using the stochastic model would provide insight into the

level of uncertainty that might be associated with extrapolating from sparse wellbore data in producing reservoirs. Although Heresim3D™ has been used to create detailed 3-D clinoform structures similar to those of the Ivie Creek area (Joseph et al., 1993), the results of any fluid-flow simulations are not reported.

- Additional effort should be expended to quantify how permeability anisotropy associated with shale drapes found within the clinoform-related facies might be incorporated at appropriate scales in numerical fluid-flow simulations. The formal homogenization method applied in the 2-D modeling should yield improved estimates of upscaled bulk anisotropy if applied to 3-D volumes on the order of 200 by 200 ft (61 x 61 m) in plan and 60 ft (18 m) thick. Higher order differencing and finite element methods are needed, however, in fluid-flow simulations used to test different methods for computing effective 3-D permeability anisotropy at grid-block scales.

- Simulations performed in this 3-D study have not fully accounted for the possible variations in relative permeability and capillary pressure curves associated with different, clinoform-related facies types. Additional fluid-flow simulations would aid in establishing the relative importance of these effects as compared to the obvious impact of facies geometry illustrated in this study.

IMPLICATIONS OF RESERVOIR ANALOG MODELING RESULTS FOR OIL RESERVOIR EXPLORATION AND DEVELOPMENT

Geehan (1993) notes that development planning can be strongly influenced by heterogeneities that cannot be correlated between wells or mapped with seismic data. Yet, such heterogeneities are difficult to include in the performance prediction models needed to establish optimal well spacing, the economic viability of horizontal wells, and estimate hydrocarbon recovery factors. Outcrop analog studies help to define which heterogeneities should be preserved in performance prediction models, and provide a foundation for estimating the range of uncertainty implicit in predicted oil recoveries. Outcrop analog studies help to address two issues of concern: (1) the geometry of architectural elements that might be distinguished with high-resolution sequence stratigraphy, and (2) the internal permeability structure of different architecture types. Both issues can be addressed using outcrop-to-simulation studies like the Ivie Creek study. Outcrop studies typically provide only a 2-D view of architectural elements and permeability structure. Even with two, nearly orthogonal cliff faces and several behind-the-outcrop drill holes, it is still difficult to unequivocally resolve the 3-D facies architecture in the Ivie Creek area. Significant 3-D insight is derived, however, in more recent outcrop-based studies where we use high-resolution, 3-D GPR to establish the 3-D geometry of architectural elements observed in adjacent cliff faces (Thurmond et al., 1997; Snelgrove et al., 1998). The resulting data sets provide a sound basis for constructing geologically plausible permeability models using wellbore and seismic data from producing reservoirs. The Ivie Creek study provides an outcrop-to-simulation case study that adds to the growing suite of reservoir analogs available to constrain geological and petrophysical models used in reservoir simulation studies.

Spacing of Vertical Wells

The 3-D simulation results show that creating a more closely spaced 9-spot pattern of wells within our analog reservoir significantly boosts oil recovery at the expense of increased water cut. If shale drapes could be explicitly included in the 3-D model, as they are in the 2-D models, we would expect to see a different response because interwell connectivity would likely be reduced. Unfortunately, we lack the outcrop-based information required to estimate the geometry of sandy bodies as viewed in a vertical section that traces an annular transect between the two cliff-face outcrops. If the spacing between shale drapes is short in the annular transect, then a significant increase in compartmentalization should be expected that, in turn, may cause the 9-spot pattern to perform better than the 5-spot pattern.

Viability of Horizontal Wells

Economides et al. (1989) and Geehan (1993) suggest that the economic viability of horizontal wells depends upon the ratio of horizontal permeability k_h to vertical k_v permeability within the volume swept by an individual well. For example, horizontal wells are most likely to be effective when the vertical permeability approaches that of the horizontal permeability. In attempting to estimate 3-D anisotropy we used our detailed 2-D permeability models to compute ratios of k_h/k_v in a radial direction that corresponds to the orientation of the cliff-face exposures. Using this approach we found that permeability ratios vary as a function of facies type and position relative to the sediment source. Unfortunately, we lack a sound basis for attempting to compute permeability ratios in the annular direction. It is clear, however, that steeply inclined clinoforms are more likely to provide the permeability ratios needed to justify horizontal wells because the steeply dipping shale drapes tend to reduce bulk permeability in the horizontal direc-

tion. Creating 3-D permeability models with different clinoform slopes by stretching the x and y coordinates provides a basis for computing and plotting the corresponding variations in k_h/k_v that could be used to assess the viability of horizontal production wells.

Oil Recovery Estimates

The 2-D simulation results show similar magnitudes of ultimate oil production for each permeability model, but different water breakthrough times. The 3-D simulation results, however, show different 10-year and ultimate oil recoveries as well as different water cut patterns. This confirms that it can be misleading to use 2-D simulation results to predict oil recovery and water cut in a reservoir with the internal 3-D geometry inferred at the Ivie Creek site. It is clear, however, that a more complete assessment of oil production from the Ivie Creek analog reservoir would require explicit inclusion of the shale drapes in the 3-D permeability model in the same way that they are included in the 2-D model. We lack both the data and computational capacity to perform such an analysis.

We suggest above that the 2-D simulation results should not be used to predict oil recovery for the corresponding 3-D case. The 2-D results, however, are valuable in that they clearly show that the timing of oil production and water breakthrough can be affected by both facies geometry and internal shale drapes. Furthermore, the comparison of permeability averaging methods performed using the 2-D permeability models suggests that the clinoform geometry observed in outcrop is poorly approximated by traditional averaging approaches. A formal, flow-based homogenization method (Koebbe, 2002) performs much better in preserving the impact of the dipping shale drapes. Because we are restricted to using a 5-point differencing scheme, however, we are unable to fully exploit the inherent strength of the homogenization approach as it should be applied in a 3-D context.

Application to Reservoir Studies

In their description of an integrated study of the Meren E-01/MR05 sands in the Niger Delta, Cook et al. (1999) provide a good, real world foundation for explaining how outcrop-to-simulation work similar to our Ivie Creek study might aid reservoir studies. Located in offshore Nigeria, the reservoir comprises interstratified sandstones and shales deposited in shoreface to shelf environments. Cook et al. (1999) combined geological attribute information obtained from well logs with depositional environment information inferred from sequence stratigraphic analysis. Using a sequence stratigraphic approach with 1-ft (0.3-m) vertical resolution provides an opportunity to match outcrop analogs to the reservoir by comparing not only the geometries and distributions of depositional features, but also sediment supply and accommodation rates (as expressed by sandy body geometries and fine scale bedding statistics [Geehan, 1993]).

Although Cook et al. (1999) did not incorporate an outcrop analog model in their analysis, their subsurface studies provide the information needed to evaluate the possible match to an analog reservoir and to exploit a suitable analog. Cook et al. (1999) constructed variograms to characterize permeability structure in terms of sand quality and sea/baffle quality using wellbore data obtained from each parasequence. Increased confidence in the variograms could be derived if a matching outcrop analog were available to provide a quantitative, observation-based assessment of both characteristics. In addition, outcrop-based observations would aid in assessing the likelihood that sandy bodies with inclined shaley drapes similar to those observed in the Ivie Creek area might be found between sub-horizontal flooding surfaces. Thus, although shale volumes might be computed from gamma-ray logs, dipmeter logs, an outcrop analog would aid in identifying the possibility of dipping beds.

Cook et al. (1999) developed a 14.5-million cell geological model that contains too many nodes to be used in practical numerical fluid-flow simulations. As a consequence, Cook et al. (1999) used an in-house homogenization software package that incorporates the approach of Durlofsky et al. (1995) to upscale the geological model to a 250,000 node simulation model that explicitly retained the flooding surfaces defined by high-resolution sequence stratigraphic mapping. The homogenization approach adopted by Cook et al. (1999) is similar to the flow-based homogenization approach implemented by Koebbe (2002) and used in our study.

Outcrop-Based Reservoir Analog Studies

Outcrop-to-simulation studies performed for the Ivie Creek area provide valuable insights that we have been incorporating in subsequent outcrop analog studies in the Ferron Sandstone and elsewhere. For example, we now target formations where the top surface of the analog facies is available for detailed mapping, GPR surveys, and comprehensive continuous core drilling. The drill core enables us to quantify the possible effects of weathering on the outcrop permeability measurements, and provides a base for closely spaced mini-permeability measurements throughout the section of interest. The drill core provides a complete record of all facies in the section, including those that are weathered out of the nearby cliff faces. The GPR surveys provide detailed 3-D geometries that cannot be fully characterized even with two orthogonal cliff-face exposures.

Although we used a deterministic approach to building the 3-D permeability model for the Ivie Creek area, stochastic approaches should be used to estimate the uncertainty that results from our inability to fully map the detailed 3-D geometry of outcrop analogs. Joseph et al. (1993) illustrate how a stochastic modeling technique might be used to create geologically plausible clinoform architectural elements similar to those observed in the Ivie Creek case-study area. Before constructing the stochastic model, Joseph et al. (1993) built a deterministic model using Heresim3D™ and included inferences regarding environments of deposition and all available outcrop data from exposures of the Roda deltaic complex in Spain. The deterministic model provides a reference for evaluating the success of the stochastic model built with a random genetic simulation approach. In attempting to mimic typical reservoir contexts, only a subset of the data incorporated in the deterministic model are used in building the stochastic model. Because insufficient data are available to apply a completely geostatistical approach, a genetic model is used to simulate the accumulation of sedimentary bodies that produce the clinoform architecture observed in outcrop. Sufficient data are available for the Ivie Creek area, and additional Ferron outcrop sites to perform similar comparative studies.

ACKNOWLEDGMENTS

This paper was supported by the Utah Geological Survey (UGS), contract numbers 94-2079 and 94-2141, as part of a project entitled *Geological and Petrophysical Characterization of the Ferron Sandstone for 3-D Simulation of a Fluvial-Deltaic Reservoir*, M. Lee Allison and Thomas C. Chidsey, Jr., Principal Investigators. The project was funded by the U.S. Department of Energy (DOE) under the Geoscience/Engineering Reservoir Characterization Program of the DOE National Petroleum Technology Office, Tulsa, Oklahoma, contract number DE-AC22-93BC14896. The Contracting Office Representative was Robert Lemmon.

We thank the Mobil Exploration/Producing Technical Center for its technical and financial support of the project that was conducted between 1993 and 1998. We also thank the following individuals and organizations for their technical contributions at the time this work was performed: Kevin McClure and S. N. Sommer of the UGS, Salt Lake City, Utah; M. A. Chan, Susan Colarullo, Hongmei Huang, and Ann Mattson of the University of Utah, Salt Lake City, Utah; and Don Best, Jim Garrison, Bruce Welton, F. M. Wright III of Mobil Exploration/Producing Technical Center, Dallas, Texas.

Cheryl Gustin (UGS) prepared the manuscript. Finally, we thank David E. Tabet, UGS, Wil E. Culham, REGA Inc, Dallas, Texas, and Milind D. Deo, Chemical and Fuels Engineering Department, University of Utah, Salt Lake City, Utah, for their careful reviews and constructive criticism of the manuscript.

REFERENCES CITED

Barton, M. D., 1994, Outcrop characterization of architecture and permeability structure in fluvial-deltaic sandstones, Cretaceous Ferron Sandstone, Utah: Ph.D. dissertation, University of Texas, Austin, 255 p.

Barton, M. D., and Noel Tyler, 1991, Quantification of permeability structure in distributary channel deposits Ferron Sandstone, Utah, in T. C. Chidsey, Jr., ed., Geology of east-central Utah: Utah Geological Association Publication 19, p. 273–282.

—1995, Sequence stratigraphy, facies architecture, and permeability structure of fluvial-deltaic reservoir analogs — Cretaceous Ferron Sandstone, central Utah: University of Texas, Bureau of Economic Geology, Field Trip Guidebook, August 21–23, 1995, 80 p.

Chidsey, T. C., Jr., compiler, 1997, Geological and petrophysical characterization of the Ferron Sandstone for 3-D simulation of a fluvial-deltaic reservoir — annual report for the period October 1, 1995 to September 30, 1996: U.S. Department of Energy, DOE/BC/14896-15, 57 p.

Chidsey, T. C., Jr., and M. L. Allison, compilers, 1996, Geological and petrophysical characterization of the Ferron Sandstone for 3-D Simulation of a fluvial-deltaic reservoir — annual report for the period October 1, 1994 to September 30, 1995: U.S. Department of Energy, DOE/BC/14896-13, 104 p.

Cook, G., L. D. Chawathé, H. Legarre, and E. Ajayi, 1999, Incorporating sequence stratigraphy in reservoir simulation — an integrated study of the Meren E-01/MR-05 sands of the Niger Delta: Proceedings of Society of Petroleum Engineers 15th Symposium on Reservoir Engineering, SPE 51892, p. 147–159.

Durlofsky, L. J., R. A. Behrens, R. C. Jones, and A. Bernath, 1995, Scale up of heterogeneous three dimensional reservoir descriptions: Proceedings of 1995 Society of Petroleum Engineers Annual Technical Conference and Exhibition, SPE 30702, p. 147–159.

Economides, M. J., J. D. McLennan, E. Brown, and J. Roegiers, 1989, Performance and stimulation of horizontal wells: World Oil, v. 208, no. 6, p. 41–77.

Fisher, R. S., M. D. Barton, and Noel Tyler, 1993a, Quantifying reservoir heterogeneity through outcrop characterization — architecture, lithology, and permeability distribution of a seaward-stepping fluvial-

deltaic sequence, Ferron Sandstone (Cretaceous), central Utah: Gas Research Institute Report 93-0022, 83 p.

—1993b, Quantifying reservoir heterogeneity through outcrop characterization — architecture, lithology, and permeability distribution of a landward-stepping fluvial-deltaic sequence, Ferron Sandstone (Cretaceous), central Utah: Gas Research Institute Report 93-0023, 132 p.

Forster, C. B., and S. H. Snelgrove, 2002, Ferron Sandstone permeability database, Ivie Creek area, Emery County, Utah: Utah Geological Survey Open-file Report 389, 97 p., 16 plates, compact disc.

Geehan, G., 1993, The use of outcrop data and heterogeneity modeling in development planning, in R. Eschard, and B. Doligez, eds., Subsurface reservoir characterization from outcrop observations: Paris, Editions Technip, p. 53–64.

Honarpour, M., L. Koederitz, and H. A. Harvey, 1986, Relative permeability of petroleum reservoirs: Boca Raton, Florida, C.R.C. Press.

Joseph, P., L. Y. Hu, O. Dubrule, D. Claude, P. Crumeyrolle, J. L. Lesueur, and H. J. Soudet, 1993, The Roda deltaic complex (Spain) — sedimentology to reservoir stochastic modelling, in R. Eschard, and B. Doligez, eds., Subsurface reservoir characterization from outcrop observations: Paris, Editions Technip, p. 97–109.

Koebbe, J. V., 2002, Documentation and installation guide for HOMCODE — a code for scaling up permeabilities using homogenization: Utah Geological Survey Open-file Report 392, 416 p.

Kossack, C. A., J. O. Aasen, and S. T. Opdal, 1990, Scaling up heterogeneities with pseudofunctions: Society of Petroleum Engineers Formation Evaluation, v. 5, no. 9, p. 226–232.

Mattson, Ann, 1997, Characterization, facies relationships, and architectural framework in a fluvial-deltaic sandstone — Cretaceous Ferron Sandstone, central Utah: Master's thesis, University of Utah, Salt Lake City, 174 p.

Odeh, A. S., 1981, Comparison of solutions to a three-dimensional black-oil reservoir simulation problem: Journal of Petroleum Technology, v. 33, no. 1, p. 13–25.

Snelgrove, S. H., G. McMechan, R. B. Szerbiak, D. Wang, R. Corbeanu, K. Soegaard, J. Thurmond, C. B. Forster, L. Crossey, and K. Roche, 1998, Integrated 3-D characterization and flow simulation studies in a fluvial channel, Ferron Sandstone, east-central Utah (abs.): AAPG Annual Convention Extended Abstracts, v. 2, p. A617.

Thurmond, J., G. McMechan, K. Soegaard, R. Szerbiak, G. Gaynor, C. B. Forster, and S. H. Snelgrove, 1997, Integration of 3-D ground-penetrating radar, outcrop, and borehole data applied to reservoir characterization and flow simulation (abs.): AAPG Annual Convention Official Program, v. 6, p. A116.

Warren, J. E., and H. S. Price, 1961, Flow in heterogeneous porous media: Society of Petroleum Engineering Journal, p. 153–169.

Weber, K. J., 1978, Influence of common structures on fluid flow in reservoir models: Journal of Petroleum Technology, v. 34, no. 3, p. 665–672.

Analog for Fluvial-Deltaic Reservoir Modeling: Ferron Sandstone of Utah
AAPG Studies in Geology 50
T. C. Chidsey, Jr., R. D. Adams, and T. H. Morris, editors

Facies Architecture and Permeability Structure of Valley-Fill Sandstone Bodies, Cretaceous Ferron Sandstone, Utah

Mark D. Barton[1], Noel Tyler[2], and Edward S. Angle[3]

ABSTRACT

Exposures of the Ferron Sandstone Member of the Cretaceous Mancos Shale formation, east-central Utah provide large-scale cross-sectional views of valley-fill sandstone bodies within a well-constrained sequence stratigraphic framework. Quantitative data collected from these outcrops help constrain the modeling of interwell volumes of analogous valley-fill reservoirs and permit a better evaluation of reservoir potential. The study has implications for how heterogeneities should be modeled in analogous reservoirs. Differences in net-to-gross, connectivity, and petrophysical property structure of valley-fill deposits can be related to stratigraphic stacking pattern (progradational versus aggradational) and position along depositional profile (proximal versus distal).

Valley-fill deposits display a general decrease in net-to-gross and sand body connectivity when stacking patterns of delta-front sandstone bodies change from strongly progradational to aggradational or from proximal to distal positions within a valley fill. Valley-fill deposits associated with strongly progradational delta-front sandstone bodies are preserved as relatively homogenous, sand-rich bodies that have a narrow shoestring-like geometry. Sandstone beds are highly amalgamated and mudstone interbeds are generally rare. The principal heterogeneity within these deposits are caused by relatively low-permeability sandstones containing abundant clay clasts that occur along the base and margins of individual channel-form bodies. In contrast, valley fills associated with aggradational delta-front units are preserved as relatively heterogeneous, mud-rich bodies that have a broad ribbon like geometry. Internal heterogeneities result from mudstone drapes that line the base and margin of individual channel-form bodies and from the interbedding of sandstone dominated channel-form bodies with mudstone-dominated channel fills. In going from a proximal to distal position within a valley-fill deposit the frequency and continuity of mudstone drapes and frequency of mudstone-dominated channel fills increases. Vertical and horizontal variograms indicate permeability is correlated over a distance of about 4–8 m (13–26 ft) vertically and approximately 40–60 m (130–200 ft) laterally. The vertical and lateral correlation ranges are similar to the average channel-form width and thickness.

[1]Shell International Exploration and Production Inc., Houston, Texas;
[2]ARC Group, LLC, Leander, Texas; [3]Texas Water Development Board, Austin, Texas

INTRODUCTION

Sandstone reservoirs typically display a complex internal architecture that fundamentally controls flow paths, recovery efficiency, and ultimately the volume of oil or gas left in a reservoir at abandonment. Reservoir models provide geoscientists with insights to understanding and managing hydrocarbon reservoirs. However, the limited well data and complex sedimentary architecture associated with many hydrocarbon reservoirs make the construction of a detailed model difficult. This has lead to an increased use of analog outcrop data to guide or constrain reservoir description and model construction.

This study investigates the sedimentary architecture and permeability structure of valley-fill deposits from superb exposures of the Cretaceous Ferron Sandstone, east-central Utah. The Ferron Sandstone outcrop study combines regional stratigraphic relationships with a detailed reservoir to interwell scale view of bedding geometry, facies distribution, and permeability structure. Patterns of facies architecture and permeability structure within valley-fill deposits are documented from (1) valley-fill deposits associated with progradational and aggradational delta-front units, and (2) from relatively proximal to distal positions within a valley-fill sandstone body. Outcrop data collected in this study were used to construct reservoir models and investigate the impact of sedimentary architecture on reservoir behavior (White and Barton, 1999).

GEOLOGIC SETTING

The Ferron Sandstone is a lithostratigraphically defined member of the Mancos Shale formation exposed in east-central Utah (Figure 1) (Lupton, 1916). It is interpreted as a fluvial-deltaic system deposited during a widespread regression of the Western Interior Seaway as thrust-belt sediments were shed eastward and accumulated along the margin of a rapidly evolving foreland basin during Late Cretaceous (Turonian) time (Armstrong, 1968; Ryer and McPhillips, 1983). The Ferron Sandstone is composed of two distinct clastic wedges (Hale, 1972; Ryer and McPhillips, 1983), an early wedge derived from the northwest (the Clawson and Washboard sandstones) and a later wedge derived from the southwest (the Ferron clastic wedge).

Marine shales divide the Ferron clastic wedges into a series of sandstone-rich tongues, each comprising a delta-front sandstone body (Ryer, 1981). The tongues are up to 30 m (100 ft) thick and extend basinward 10–40 km (6–25 mi). Early tongues step seaward whereas later tongues stack vertically or step landward. Each delta-front sandstone body contains many upward-coarsening and upward-shoaling shallow-marine successions (Gardner, 1993; Barton, 1994) or parasequences (Van

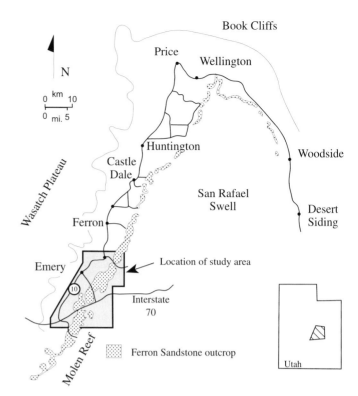

Figure 1. Location map of the Ferron Sandstone outcrop and field study area, east-central Utah.

Wagoner et al., 1989). Individual parasequences are 5–15 m (16–50 ft) thick and extend basinward 1–10 km (0.6–6 mi). Within each delta-front sandstone body the parasequences stack in progradational, aggradational, and retrogradation patterns (Barton, 1994; Barton et al., this volume) (Figure 2). A widespread erosion surface overlies the initial progradational parasequence set and is onlapped by coastal-plain and marginal-marine deposits associated with a set of aggradational to retrogradational parasequences. The erosion surface is characterized by deep incision into underlying strata and an abrupt basinward shift in facies. The erosion surface at the top of the shallow-marine strata is interpreted as an unconformity associated with a fall in relative sea level. The large, lens-shaped bodies that are deeply incised into the shallow marine deposits are interpreted as valley fills (Barton et al., this volume).

The valley-fill deposits display narrow to broad ribbon-like geometry that are up to 25 m (80 ft) thick and from 0.2–4 km (0.1–2.5 mi) across. Internally, they are composed of many channel-form bodies that vary in thickness from 2–8 m (7–26 ft) and in width from 20–200 m (70–660 ft). The channel-form bodies are similar in scale and structure to channel storeys defined by Allen (1983) or macroforms defined by Miall (1985). They are interpreted as deposits of fluvial channels and bars that accumulated within the erosional valleys during a subsequent rise in relative sea level. Based on internal bedding architecture, four types of channel-form bodies

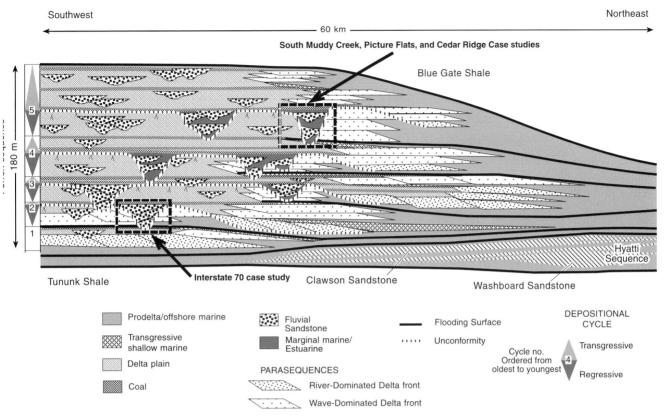

Figure 2. Schematic cross section depicting stratigraphic relationships of the Ferron Sandstone in east-central Utah, based on the study. Interstate 70 case study is located within unit 2. South Muddy Creek, Picture Flats, and Cedar Ridge case studies are located within unit 5.

were recognized. They include massive or amalgamated, convex-upward, concave-upward, and lateral (Figure 3). The channel-form bodies displaying the massive and concave-upward bedding geometry are interpreted as passive channel fills. The channel-form bodies displaying the inclined bedding geometry are interpreted to have formed by the lateral accretion of a side-attached or point bar. The channel-form bodies displaying the convex-upward bedding geometry are interpreted to have formed by the downstream accretion of a mid-stream bar.

OUTCROP STUDY APPROACH

Data collection was designed to quantify geologic and petrophysical variability within representative portions of the valley-fill sandstone bodies. A valley-fill sandstone body from a seaward-stepping sandstone tongue (see Figure 2) was examined at a site referred to as Interstate 70 (Figure 4). From the retrogradational (landward-stepping) portion of the Ferron Sandstone (see Figure 2), proximal to distal changes are documented in the South Muddy Creek, Picture Flats, and Cedar Ridge localities, respectively (Figure 4).

Geometric, lithologic, and petrophysical data were collected from laterally continuous outcrop exposures over distances of several thousand meters. Each site was

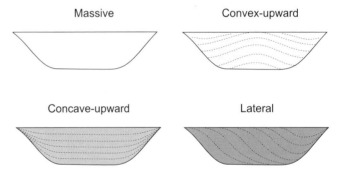

Figure 3. Schematic representations of channel-form infill or internal bedding architecture. Types recognized include massive, convex-upward, concave-upward, and lateral.

characterized by describing and mapping facies, tracing bedding surfaces, and measuring lithologic and petrophysical properties along closely spaced vertical transects. A probe permeameter was used extensively (Barton, 1994). The data sets discussed in this paper contain hundreds of sandstone and shale layers and thousands of sedimentologic and petrophysical measurements. Each outcrop is further described in terms of the channel-form types present, their maximum width and thickness, and their overall grain size.

Several studies have demonstrated the portability of outcrop observations to the subsurface (Stalkup and

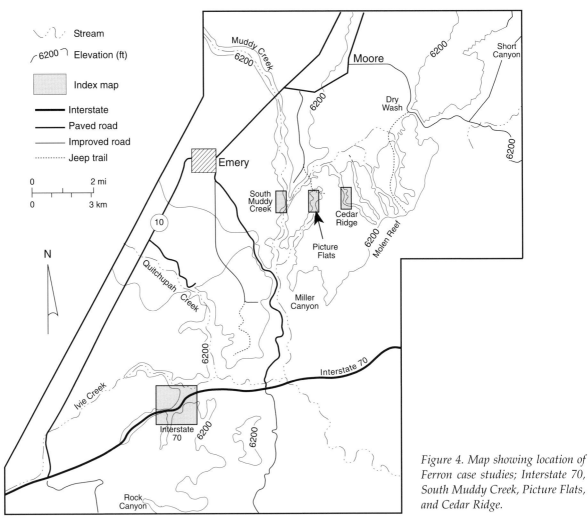

Figure 4. Map showing location of Ferron case studies; Interstate 70, South Muddy Creek, Picture Flats, and Cedar Ridge.

Ebanks, 1986; Goggin, 1988; Kittridge, 1988; Barton, 1994). In general, the mean of the permeability measurements is not portable. However, distribution type, coefficient of variation, and correlation measures are portable between outcrop and subsurface. Two methods were used to examine the permeability structure; direct correlation of permeability profiles and statistical semivariograms (Stalkup and Ebanks, 1986). Both methods measure the lateral and vertical continuity of zones with similar permeability.

SEAWARD-STEPPING: INTERSTATE 70 LOCALITY

The facies architecture and permeability structure within a seaward-stepping tongue (No. 2) were examined at a site on the south side of Interstate 70 and Ivie Creek, Emery County, Utah (Figure 5a). The exposure provides a view parallel to the northeasterly direction of shoreline progradation. The interval is continuously exposed for nearly 1 km (0.6 mi). Ten vertical logs, spaced 20 m (70 ft) apart, were measured and described along the exposure.

Geologic Description

The interval examined consists of a shallow-marine sandstone body that is up to 20 m (70 ft) thick and extends the length of the outcrop (Figure 6a). It is bounded at the base by a marine shale and at the top by a coal. Marine-flooding surfaces and their correlative transgressive surfaces of erosion divide the shallow-marine sandstone body into three upward-coarsening successions or parasequences. The parasequences are up to 10 m (30 ft) thick and dip gently to the northeast where they gradually thin and become more fine grained. The parasequences consist of well sorted, fine-grained, hummocky cross-stratified sandstones overlain by well sorted, medium-grained, low angle, multi-directional trough cross-stratified sandstones. Mudstone drapes are rare and burrowing is common through out the succession.

Near the northeastern end of the exposure, the shallow-marine strata have been completely eroded and replaced by a large lens-shaped body that is up to 25 m (80 ft) thick and 200 m (660 ft) wide. The body is a composite of smaller channel-form sandstone bodies (Figure 6b) that vary from 2–8 m (7–26 ft) in thickness and

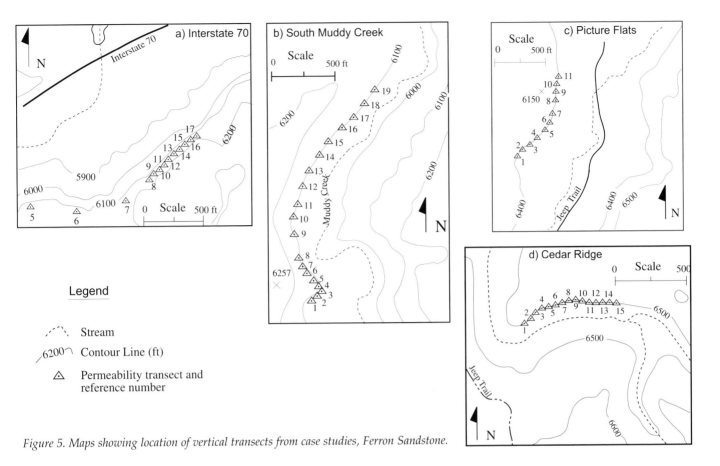

Legend

- ‑‑‑ Stream
- 6200 Contour Line (ft)
- △ Permeability transect and reference number

Figure 5. Maps showing location of vertical transects from case studies, Ferron Sandstone.

a) Facies architecture and permeability profiles

b) Channel-form architecture

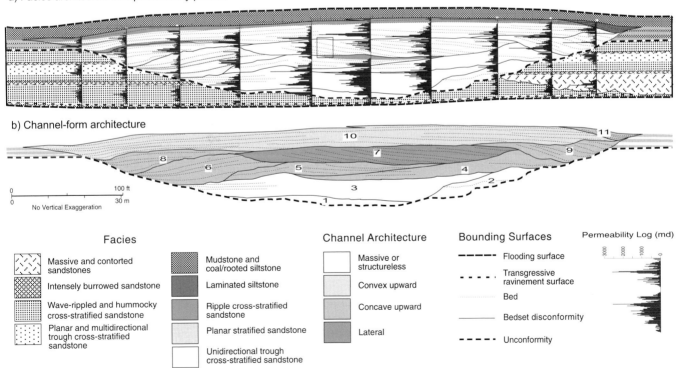

0   100 ft
0   30 m
No Vertical Exaggeration

Facies

- Massive and contorted sandstones
- Intensely burrowed sandstone
- Wave-rippled and hummocky cross-stratified sandstone
- Planar and multidirectional trough cross-stratified sandstone
- Mudstone and coal/rooted siltstone
- Laminated siltstone
- Ripple cross-stratified sandstone
- Planar stratified sandstone
- Unidirectional trough cross-stratified sandstone

Channel Architecture

- Massive or structureless
- Convex upward
- Concave upward
- Lateral

Bounding Surfaces

- ‑ ‑ ‑ Flooding surface
- ‑ ‑ ‑ Transgressive ravinement surface
- ‑‑‑‑‑ Bed
- ——— Bedset disconformity
- ‑ ‑ ‑ Unconformity

Permeability Log (md)

Figure 6. (A) Facies architecture and permeability profiles (upper cross section) from the Interstate 70 case study (see Figure 5). Location of sample grid indicated by rectangle. Channel-form architecture illustrated in lower cross section (B). Channel elements numbered in ascending stratigraphic order.

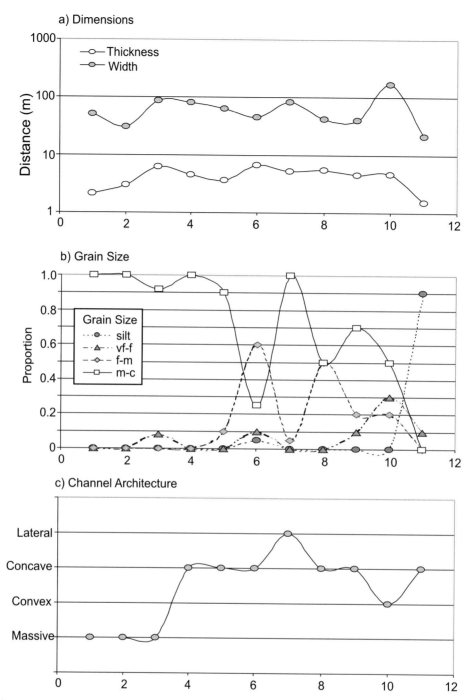

Figure 7. Channel-form dimensions, grain size, and internal architecture from Interstate 70 case-study site. Channel elements numbered from oldest to youngest in ascending stratigraphic order.

20–200 m (70–660 ft) in width (Figure 7a). A total of 11 channel-form bodies are visible in the exposure. Massive and concave-upward channel-forms are more common then convex-upward or lateral channel-form elements. The massive channel-form bodies (nos. 1, 2, and 3) are more frequent within the older part of the valley fill whereas a mixture of concave-upward, convex-upward, and inclined channel-form bodies occur in the younger portion (Figure 7c). Lithologically, all the channel-form bodies are composed of sandstone, with the exception of

the final channel-form body (no. 11), which is composed of a mixture of mudstone and siltstone. Mudstone inter-beds are uncommon but occur in a discontinuous fashion within channel-form element no. 6 (see Figures 6a and 6b). Grain sizes of individual channel-form bodies decrease in an erratic fashion from medium- to coarse-grained sandstone within the older portion of the valley fill to fine- to medium-grained sandstone within the younger portion of the valley fill (Figure 7b). The massive channel-form bodies consist of a basal erosional lag

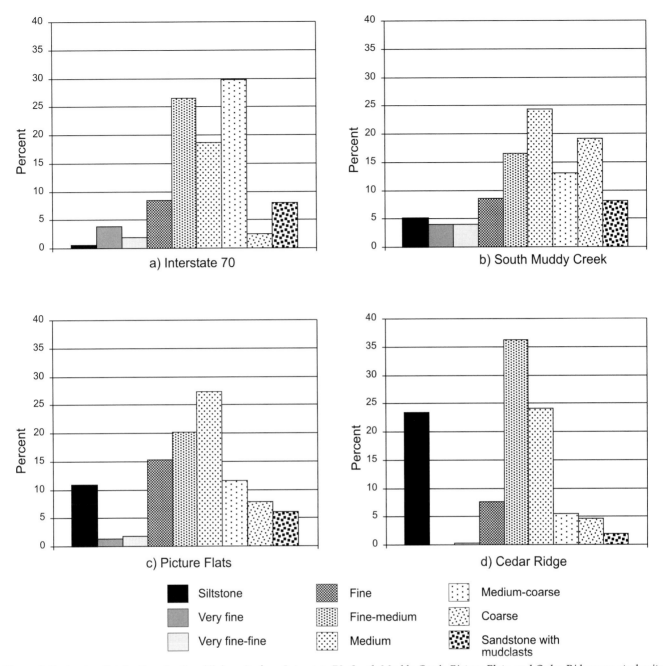

Figure 8. Grain-size distribution of valley-fill deposits form Interstate 70, South Muddy Creek, Picture Flats, and Cedar Ridge case-study sites, Ferron Sandstone, Utah.

of siderite nodules and clay clasts overlain by poorly sorted, medium- to coarse-grained sandstones that have a massive appearance or display unidirectional trough cross-stratification. The convex-upward and inclined channel-form bodies display an upward-fining succession of medium- to coarse-grained trough cross-stratified sandstones that in some cases pass upward into fine- to medium-grained ripple and parallel-laminated sandstones. Overall, 45% of the sandstones vary from medium-fine to medium-coarse, 15% from very fine to fine, and 33% medium-coarse to coarse (Figure 8). Another 8% of the sandstones are coarse grained but

contain abundant clay clasts. Mudstone and siltstone make up 1% of the overall composition.

Permeability Description

Sandstone permeabilities vary from 0.1 md to 1020 md, average 166 md, and have a coefficient of variation of 0.99 (Figure 9). With the exception of sandstones containing abundant clay clasts permeability shows a good correlation with grain size (Figure 10a). Coarse-grained sandstones average 314 md, medium-grained sandstones average 110 md, fine-grained sandstones average

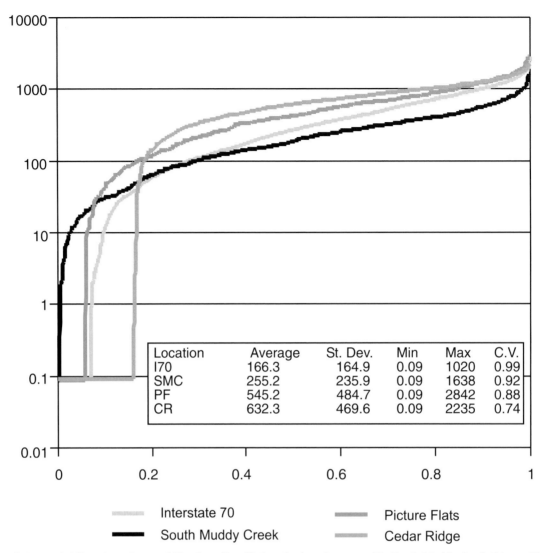

Location	Average	St. Dev.	Min	Max	C.V.
I70	166.3	164.9	0.09	1020	0.99
SMC	255.2	235.9	0.09	1638	0.92
PF	545.2	484.7	0.09	2842	0.88
CR	632.3	469.6	0.09	2235	0.74

Interstate 70 Picture Flats
South Muddy Creek Cedar Ridge

Figure 9. Cumulative probability plots of permeability for valley-fill deposits from Interstate 70, South Muddy Creek, Picture Flats, and Cedar Ridge case-study sites, Ferron Sandstone, Utah.

33 md, and very fine grained sandstones average 1 md. Coarse-grained sandstones containing abundant clay clasts average 40 md.

Permeability patterns are highly variable with erratic, uniform, upward-increasing and upward-decreasing permeability trends (see Figure 6a). The trends roughly match the arrangement of channel-form with individual trends often being correlatable from one profile to the next within a channel-form body but not between channel-form bodies. Zones of high permeability, on the order of several hundred to a thousand millidarcies, correspond to the distribution of medium- to coarse-grained sandstones. The zones extend several meters to ten's of meters vertically and several tens of meters to several hundreds of meters laterally. The zones of high permeability are separated by zones of reduced permeability that often occur near the tops and bases of the channel-form bodies. The reduced permeabilities near the top of the channel-form bodies are associated with fine-grained ripple laminated sandstones. Reduced per-

meabilities near the base of the channel-form bodies are associated with coarse-grained sandstones containing an abundance of intraformational clay clasts. These deposits are several centimeters to several meters thick and extend laterally in a discontinuous fashion along the base and margins of the channel-from bodies. A small 6 by 6 m (20 by 20 ft) sample grid constructed overlapping an inclined channel-form body shows how permeability trends follow the bedding geometry (Figure 11). Highest permeabilities also tend to occur in the center of the fill and decrease near the margins. Vertical and horizontal variograms indicate permeability is correlated over a distance of about 4–5 m (13–16 ft) vertically and approximately 40–50 m (130–160 ft) laterally (Figure 12a). Inspection of the variograms indicates that more than half the variability occurs at distances of less than a meter (0.6–0.9 m [2–3 ft]) vertically and 10 m (30 ft) (6–9 m [20–30 ft]) horizontally. The vertical and lateral correlation ranges are similar to the average channel-form width and thickness.

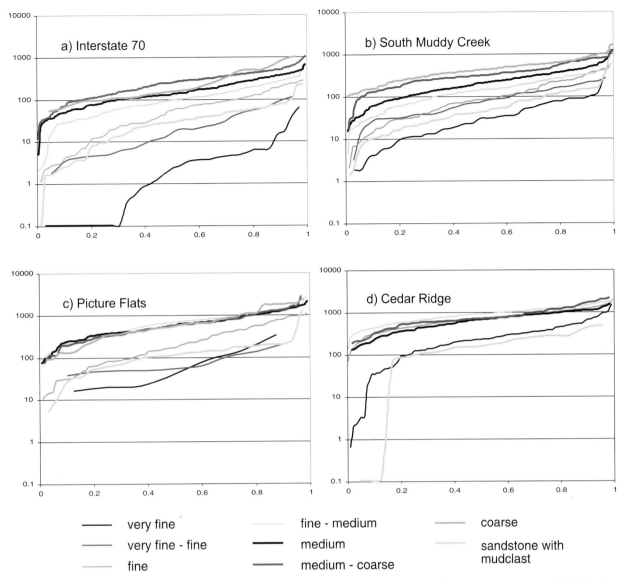

Figure 10. Cumulative probability plots of permeability for grain-size classes of valley-fill deposits from Interstate 70, South Muddy Creek, Picture Flats, and Cedar Ridge case-study sites, Ferron Sandstone, Utah.

LANDWARD-STEPPING: SOUTH MUDDY CREEK LOCALITY

The South Muddy Creek site is located on the west side of Muddy Creek Canyon. The exposure is approximately 1 km (0.6 mi) in length and aligned roughly perpendicular to the axis of the valley-fill deposits. Stratigraphically, the exposure rests near the southern margin of the valley fill near the landward pinch-out of the shallow-marine facies tract of sandstone tongue No. 5 (Figure 13). Lithologic and permeability data were collected from 19 vertical logs, spaced 30–50 m (100–160 ft) apart (Figure 5b).

Geologic Description

At the South Muddy Creek site shallow-marine strata have been extensively eroded and replaced by valley-fill deposits (Figure 14). In areas where near-shorezone strata have been preserved they are up to 12 m (39 ft) thick and composed of an upward-coarsening and bed-thickening succession of hummocky and swaley cross-stratified sandstones. Burrowing is common throughout the succession and mudstone interbeds rare. A meter thick, intensely burrowed sandstone lies at the base of the succession and marks a marine transgression.

Erosional surfaces subdivide the valley-fill strata into a complex of channel-form bodies that stack vertically and laterally to form a body that is up to 20 m (70 ft) thick and 1 km (0.6 mi) wide (Figure 15). Individual channel-form bodies range from 2–8 m (7–26 ft) in thickness and from 15–160 m (50–520 ft) in width (Figures 15 and 16a). Convex-upward channel-form bodies dominate the older portion of the valley fill while a mixture of convex-upward and lateral channel-form elements dominate the younger portion of the valley fill (Figure 16c).

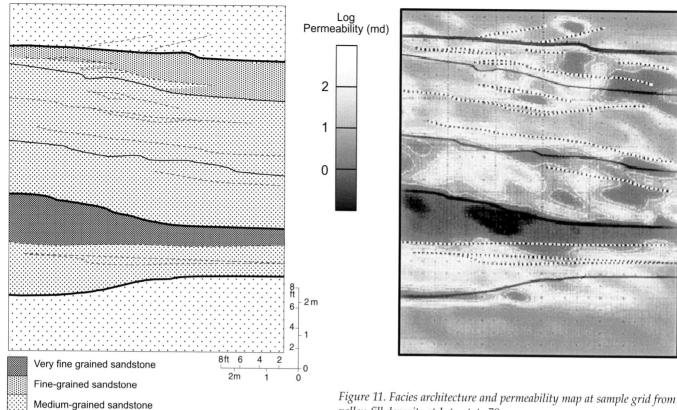

Log
Permeability (md)

Very fine grained sandstone

Fine-grained sandstone

Medium-grained sandstone

Coarse-grained sandstone

Figure 11. Facies architecture and permeability map at sample grid from valley-fill deposits at Interstate 70.

Overall, 33% of the sandstones are medium-coarse to coarse, 42% of the sandstones range from medium-fine to medium, and 17% are fine to very fine (see Figure 8b). About 8% of the sandstones contain abundant clay clasts. Grain size of individual channel-form bodies decrease in an erratic fashion from medium- to coarse-grained sandstone within the older portion of the valley fill to fine- to medium-grained sandstone within the younger portion of the valley fill (Figure 16b). Channel-form element nos. 1 and 19 consist almost entirely of mudstone and siltstone. Channel-form element nos. 10 and 13 contain sub-equal amounts of mudstone and sandstone. The concave-upward and massive channel-form bodies consist of a basal erosional lag of siderite nodules and clay clasts overlain by poorly sorted, medium- to coarse-grained sandstones that have a massive appearance or display unidirectional trough cross-stratification. The concave-upward channel-form bodies tend to fine upward but vary in composition from medium- to coarse-grained sandstones to muddy siltstones in some cases. The lateral channel-form bodies are composed from base to top of a mud-clast-rich sandstone lag, overlain by an upward-fining succession of trough cross-stratified, parallel-laminated, and ripple-laminated sandstones. Mudstone and siltstone make up 5% of the overall composition.

Permeability Description

Sandstone permeabilities vary from 0.1 md to 1638 md, average 255 md, and have a coefficient of variation of 0.92 (see Figure 9). With the exception of sandstones containing abundant clay clasts permeability shows a good correlation with grain size (see Figure 10b). The highest permeabilities are observed within the medium- to coarse-grained sandstones that have an average permeability of 708 md. Medium- to fine-grained sandstones average 330 md. Fine-grained, ripple-laminated sandstones average 55 md. Coarse-grained sandstones containing abundant mudclasts have the lowest permeabilities averaging 44 md.

In general, permeabilities have an erratic upward decreasing trend that mirrors the vertical grain layering (see Figure 14). Zones of relatively high permeability are separated vertically and laterally by low-permeability sandstones that are rich in intraformational mudclasts. Individual trends range from 1–8 m (3–26 ft) vertically and 10–100 m (30–330 ft) laterally. Zones of low permeability occur along the base and margin of the channel-form bodies and vary in thickness from a few decimeters to a few meters. Laterally, they may extend a few meters to tens of meters but only rarely do they form completely continuous zones. The majority of the inclined and

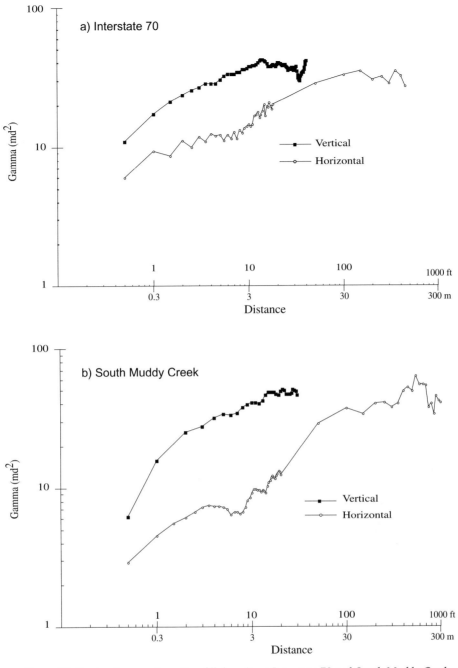

Figure 12. Semivariograms for valley-fill deposits at Interstate 70 and South Muddy Creek.

concave-upward channel-form bodies display a persistent upward decreasing permeability profile. The massive channel-form bodies tend to show no persistent permeability trend. The vertical variogram indicates a correlation length of 4–6 m (13–20 ft) whereas the horizontal variogram indicates a correlation length of 100–150 m (330–490 ft) (Figure 12b).

PICTURE FLATS LOCALITY

The Picture Flats site is located about 1.6 km (1 mi) to the northeast of the South Muddy Creek site. Stratigraphically it is positioned about a 1.5 km (0.9 mi) sea-

ward of the South Muddy Creek locality but within the same stratigraphic unit. The exposure is about 0.5 km (0.3 mi) in length and aligned perpendicular to the axis of the valley-fill deposits (see Figure 13). Lithologic and permeability data were measured at 0.15 m (0.5 ft) intervals from 11 vertical sections spaced about 30 m (100 ft) apart (Figure 5c).

Geologic Description

Near-shorezone strata are poorly preserved due to erosion by the overlying valley-fill deposits (Figure 17). Thickness rarely exceeds 5 m (15 ft) and in many places

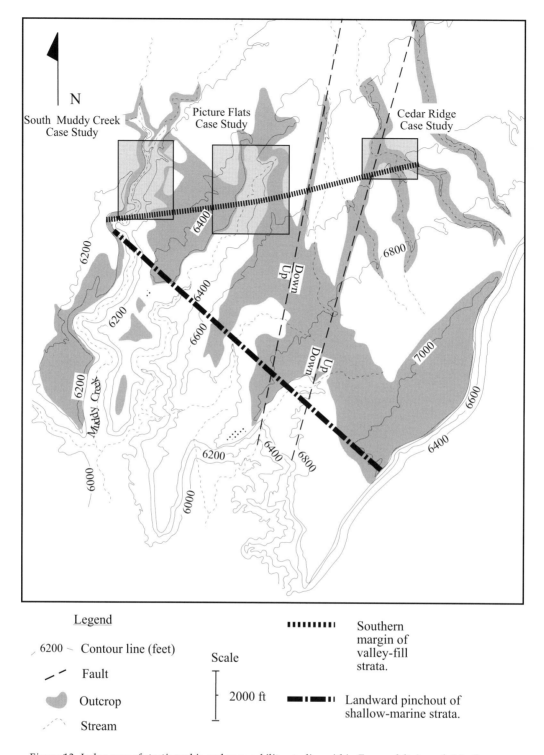

Figure 13. Index map of stratigraphic and permeability studies within Ferron deltaic cycle No. 2.

the strata have been completely removed. Where present the near-shorezone strata consist of a meter-thick, intensely burrowed sandstone that is gradationally overlain by an upward-coarsening and bed-thickening succession of wave-rippled and hummocky cross-stratified sandstones interstratified with burrowed to laminated mudstones. The upward increase in bed thickness reflects increased amalgamation of hummocky cross-stratified sandstones.

Overlying valley-fill deposits vary in thickness from 13–20 m (43–70 ft). Erosional surfaces subdivide the deposits into a stack of lenticular channel-form bodies that range from 2–7 m (7–23 ft) in thickness and from 20–130 m (70–430 ft) in width (Figure 18 and Figure 19a).

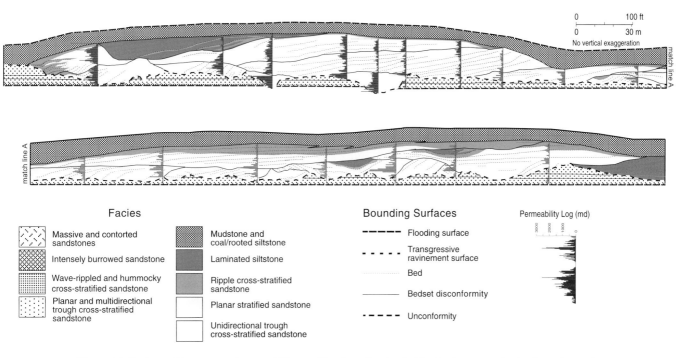

Figure 14. *Facies architecture and permeability profiles from South Muddy Creek case-study site (see Figure 5).*

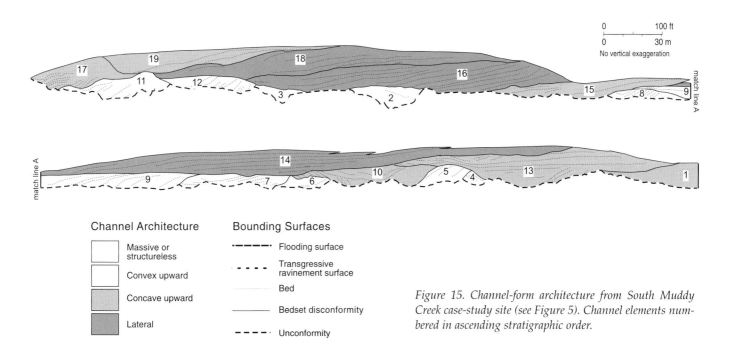

Figure 15. *Channel-form architecture from South Muddy Creek case-study site (see Figure 5). Channel elements numbered in ascending stratigraphic order.*

Early portions of the valley fill are dominated by a mixture of concave-upward and convex-upward channel-form types (Figure 19c). Most of the concave-upward and convex-upward channel-form bodies consist of a basal erosional lag of clay clasts overlain by poorly sorted, medium- to coarse-grained sandstones that have a massive appearance or display unidirectional trough cross-stratification. The upper portion of the valley fill is dominated by a mixture of concave-upward and lateral channel-form bodies that generally display a well-developed, upward-fining grain size trend. Unlike Interstate 70 and South Muddy Creek, mudstone interbeds were relatively common often occurring as 0.3–3 m (1–3 ft) drapes that lined the margins and base of individual channel-form bodies (nos. 2, 3, 4, and 6). Channel-form element nos. 5, 12, 16, 22, and 23 consist of a heterolithic mix of mudstone, siltstone, and sandstone. Overall, 20% of the sandstones are medium-coarse to coarse, 49% of

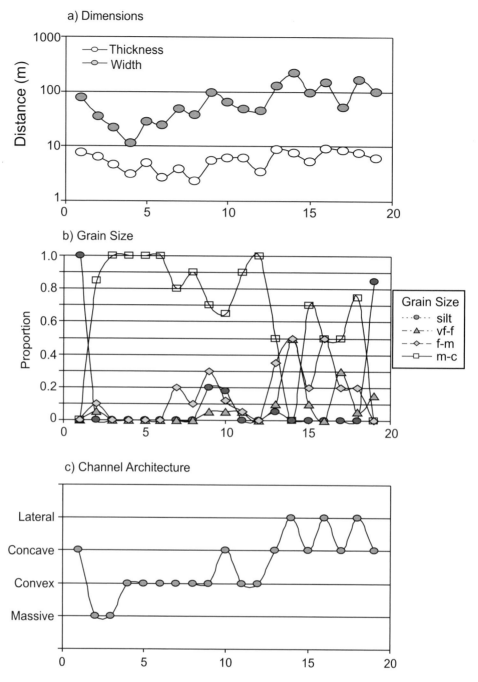

Figure 16. Channel-form dimensions, grain size, and internal architecture from South Muddy Creek case-study site. Channel elements numbered from oldest to youngest in ascending stratigraphic order.

the sandstones range from medium-fine to medium, and 20% are fine to very fine (see Figure 8c). About 8% of the sandstones contain abundant clay clasts. Mudstone and siltstone interbeds make up 6% of the composition. Mudstone and siltstone make up 11% of the overall composition.

Permeability Description

Sandstone permeabilities vary from 0.1 md to 2842 md, average 550 md, and have a coefficient of variation

of 0.88 (Figure 9c). Unlike Interstate 70 and South Muddy Creek, the permeabilities of fine-medium through very coarse grained sandstones are not significantly different with all grain sizes groups averaging between 700 to 800 md (Figure 10c). Coarse-grain sandstones containing mudclasts have significantly lower permeabilities averaging 102 md.

Unlike Muddy Creek, where zones of high permeability are concentrated near the base of the valley, at Picture Flats they are distributed irregularly throughout. The permeability patterns tend to match the vertical and

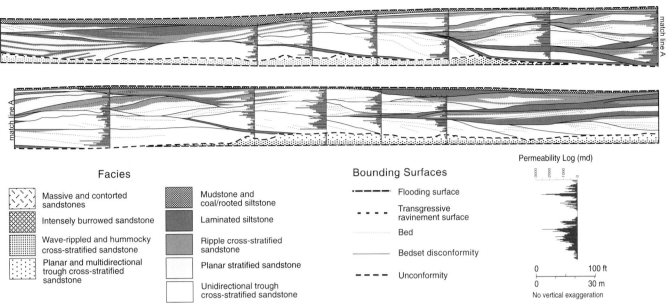

Facies

Massive and contorted sandstones

Intensely burrowed sandstone

Wave-rippled and hummocky cross-stratified sandstone

Planar and multidirectional trough cross-stratified sandstone

Mudstone and coal/rooted siltstone

Laminated siltstone

Ripple cross-stratified sandstone

Planar stratified sandstone

Unidirectional trough cross-stratified sandstone

Bounding Surfaces

Flooding surface

Transgressive ravinement surface

Bed

Bedset disconformity

Unconformity

Permeability Log (md)

3000 2000 1000 0

0   100 ft
0   30 m
No vertical exaggeration

Figure 17. Facies architecture and permeability profiles from Picture Flats case-study site (see Figure 5).

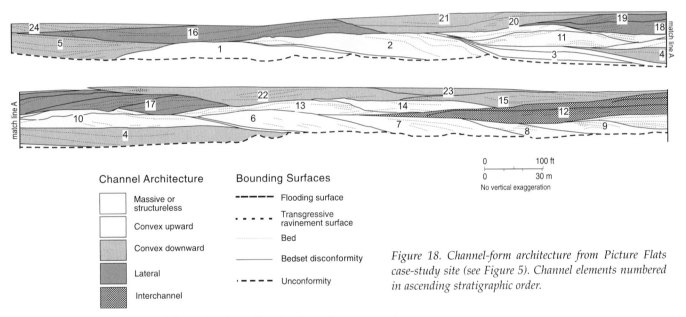

Channel Architecture

Massive or structureless

Convex upward

Convex downward

Lateral

Interchannel

Bounding Surfaces

Flooding surface

Transgressive ravinement surface

Bed

Bedset disconformity

Unconformity

0   100 ft
0   30 m
No vertical exaggeration

Figure 18. Channel-form architecture from Picture Flats case-study site (see Figure 5). Channel elements numbered in ascending stratigraphic order.

lateral stacking of channel-form bodies. Continuity of high-permeability zones between channel-form bodies is severely disrupted by mudstone drapes, mudstone-dominated channel fills, and mud-clast-rich sandstones bounding the base of individual channels.

CEDAR RIDGE LOCALITY

The Cedar Ridge site is within the same stratigraphic interval as the South Muddy Creek and Picture Flats sites (Figure 13). It is located near the distal end of the valley-fill facies tract in cycle 5, approximately 1.5 km (1 mi) to the northeast (seaward) of the Picture Flats site. Stratigraphic and petrophysical data were collected at 0.15 m (0.5 ft) intervals from 15 vertical logs located on the north side of the canyon (Figure 5d).

Geologic Description

Near-shorezone strata are up to 15 m (50 ft) thick, and composed of an upward-coarsening and bed-thickening succession of interbedded mudstones and hummocky cross-stratified sandstones that grade upward into swaley cross-stratified sandstones. The shallow-marine strata are incised and replaced by valley-fill deposits that are up to 22 m (72 ft) in thickness (Figure 20a). Erosional surfaces divide the valley-fill deposits into a complex of channel-form bodies that vary in thickness from 2–7 m (7–23 ft) and from 20–140 m (70–460 ft) in width (Figure 20b and Figure 21a). The fill consists of sub-equal amounts convex-upward channel-form elements (nos. 2, 3, 5, and 13), concave-upward channel-form elements (nos. 1, 4, 6, 7, 8, and 9) and lateral chan-

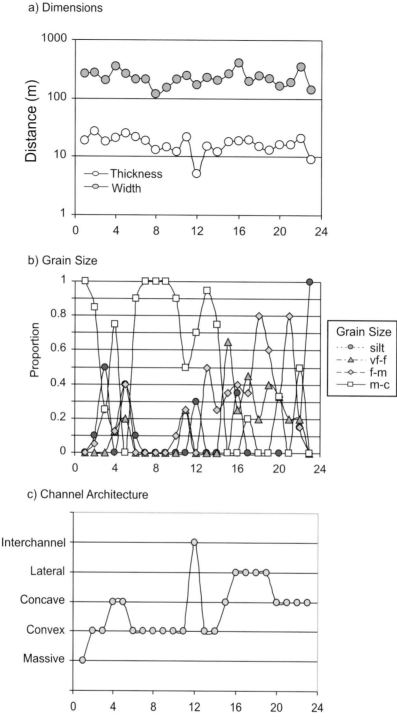

Figure 19. *Channel-form dimensions, grain size, and internal architecture from Picture Flats case-study site. Channel elements numbered from oldest to youngest in ascending stratigraphic order.*

nel-form elements (nos. 12, 14, and 15). Channel-form element nos. 1, 4, and 9 are composed entirely of decimeter thick beds of mudstone, siltstone, and shell debris/hash. Bioturbation is common throughout the mudstone-dominated successions. The convex-upward channel-form bodies are composed largely of medium-grained, trough cross-stratified sandstone with a poorly developed upward-fining grain-size trend. Overall, 63% of the sandstones range from medium-fine to medium,

11% are medium-coarse to coarse, and 8% are very fine to fine (Figure 8d). About 2% of the sandstones contain abundant clay clasts. Mudstone and siltstone make up 23% of the overall composition.

Permeability Description

Sandstone permeabilities vary from 0.1 md to 2235 md, average 632 md, and have a coefficient of variation

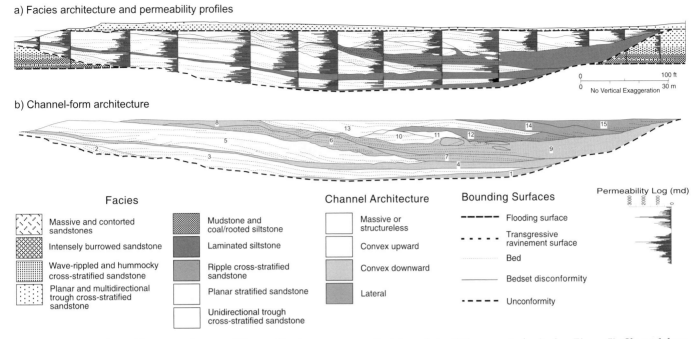

Figure 20. (A) Facies architecture and permeability profiles (upper cross section) form Cedar Ridge case-study site (see Figure 5). Channel-form architecture illustrated in lower cross section (B). Channel elements numbered in ascending stratigraphic order.

of 0.74 (Figure 9d). Similar to Picture Flats, grain sizes from fine-medium to coarse do not have significantly different permeabilities and average between 800–900 md (Figure 10d).

The valley fill consists of discrete zones of high permeability (200 to 2500 md) that are vertically and laterally isolated by low permeability mudstone drapes and mudstone-dominated channel fills (see Figure 20a). Permeabilities within the sandstone-dominated channel forms are uniformly high, on the order of 500–2000 md, and display few internal heterogeneities. Lateral trends indicate permeabilities decreasing slightly toward the margins of individual channel-form bodies.

DISCUSSION

Within seaward-stepping sandstone tongues of the Ferron Sandstone the valley-fill deposits are preserved as relatively homogenous, sand-rich bodies that have a narrow shoestring-like geometry. Channel-form bodies are highly amalgamated and dominated by a mixture of massive and concave-upward channel-form types (82% of total) (Figure 22). The lack of convex-up, lateral, and mudstone-dominated channel-form types suggest the incised valley was occupied by low-sinuosity channels that were passively filled with sand during abandonment and avulsion. This type of channel abandonment may reflect the gradual loss of gradient associated with rapid delta lobe progradation and abandonment; a scenario that seems consistent with well-developed, river-dominated delta-front deposits that characterize the seaward-stepping portion of the Ferron Sandstone. Low

gradients due to an extended coastal plain may have also inhibited the development of significant channel sinuosity and lateral channel migration. The lack of significant lateral channel migration and sinuosity would prevent the development of midstream and side-attached bars (convex-upward and lateral channel-form types) as well as meander loop cutoffs that might result in mudstone-dominated channel fills. The controlling processes on the facies architecture of the valley-fill deposits could be high rates of sediment supply, low rates of accommodation, or marine processes that were insufficient to rework incoming sediment.

In the landward-stepping portion of the Ferron Sandstone, valley-fill deposits have a broad, ribbon-like geometry. Convex and lateral channel-form types compose on average 60% of the fill at South Muddy Creek, Picture Flats, and Cedar Ridge (see Figure 22). Mudstone-dominated channel fills comprise 10% of the total at South Muddy Creek, 20% at Picture Flats, and 20% at Cedar Ridge. The common presence of convex-upward and lateral channel-form suggests the valleys were occupied by meandering channels with well-developed midstream and side-attached bars. The occurrence of mudstone-dominated channel fills suggests that the channels were occasionally abandoned due to neck or meander-loop cut-offs. The high degree of burrowing and abundant shell hash within the mudstone-rich channel fills suggests a marine influence within the channels due to possible storm or tidal currents. The development of meandering channels with midstream and side-attached barforms suggests the channels were relatively stable and long-lived features. This type of channel morpholo-

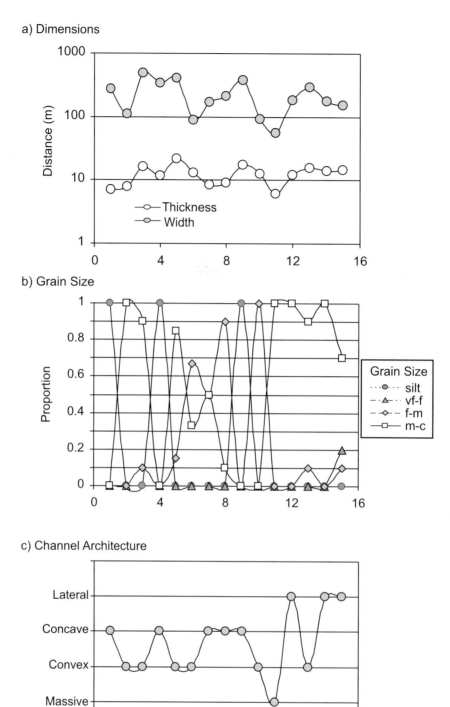

Figure 21. Channel-form dimensions, grain size, and internal architecture from Cedar Ridge case-study site. Channel elements numbered from oldest to youngest in ascending stratigraphic order.

gy may reflect the relatively high coastal-plain gradients associated with relatively slow delta-front progradation. A scenario that seems consistent with wave-dominated delta-front deposits that characterize the landward-stepping portion of the Ferron Sandstone.

Some of the downstream changes in overall grain size distribution of the valley-fill deposits associated with the landward-stepping portion of the Ferron Sandstone are predictable and others surprising (Figure 8b, c, and d). Predictably, there is a reduction in the medium-coarse and coarse-grain sizes from 33% at South Muddy Creek, to 20% at Picture Flats, to 11% at Cedar Ridge. There is also an increase in mud content from 1% at South Muddy Creek, to 6% at Picture Flats, to 16% at

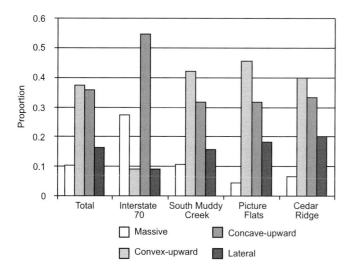

Figure 22. Channel-form distribution at each case study.

Cedar Ridge. However, somewhat surprising is the reduction in the very fine to fine-grain fraction from 9% at South Muddy Creek, to 4% at Picture Flats, to 1% at Cedar Ridge. Petrographic analysis by Fisher et al. (1993a, 1993b) on sandstones at the three localities indicates there is also a downstream increase in sorting, from moderately sorted to well sorted, and an increase in mineralogical maturity, from lithic arenites to quartz arenites. The limited grain-size range, mineralogical maturity, and well-sorted nature of the sandstones near the distal end of the valley fill suggests a significant portion of the sands may have been reworked in a near-shorezone environment prior to being redeposited within channels of the valley fill. Storm or tidal currents may have been responsible for transporting the sands up-valley from the near-shorezone environment.

At Interstate 70 and South Muddy Creek permeability is closely correlated with grain size. The only exception is the coarse-grained sandstones containing abundant clay clasts which display roughly an order of magnitude (10X) reduction in permeability compared to similar sandstones containing no mudclasts. Petrographic analysis by Fisher et al. (1993a, 1993b) on the sandstones indicates the unusually low permeabilities are due to the plugging of pore space by pseudo-matrix formed from the deformation of clay clasts during compaction. At Picture Flats and Cedar Ridge there is no significant difference in permeability for grain sizes from fine-medium to coarse.

In a downstream direction there is approximately a four-fold (4x) increase in permeability and a 25% reduction in the variability within similar grain-size classes. Petrographic analysis of the sandstones by Fisher et al. (1993a, 1993b) suggests that the increase is due to improved sorting and a decrease in the presence of ductile rock fragments. This trend may reflect progressive reworking of the sandstones in a downstream direction

resulting in the removal of ductile rock fragments as well as the incorporation of well-sorted, wave-reworked sands eroded from underlying shoreline deposits and carried upstream by storm or tidal currents.

Implications for reservoir modeling are that predictable styles of valley-fill architecture can be related to styles of delta-front progradation. The strongly progradational or seaward-stepping portion of the Ferron Sandstone is characterized by valley fills that have a narrow shoestring geometry. Internally, these deposits are relatively sand-rich and homogeneous. In contrast, the landward-stepping portion of the Ferron Sandstone is characterized broad by valley fills that have a broad ribbon-like geometry. Internally, these deposits contain significant amounts of mudstones in the form of mudstone-dominated channel fills and channel base drapes. The proportion of mudstone-dominated channels fills increases in a downstream direction. Five to 6 km (3–4 mi) from the seaward extent of the valley fill the proportion of mudstone-dominated channels fills is 5% of the total (South Muddy Creek). Three to 4 km (2–2.5 mi) from the seaward extent of the valley fill the proportion of mudstone-dominated channels fills is 11% of the total (Picture Flats). One to 2 km (0.6–1.2 mi) from the seaward extent of the valley fill the proportion of mudstone-dominated channels fills is 23% of the total (Cedar Ridge). Permeability correlation lengths within the valley-fill sandstones were 4–6 m (13–20 ft) vertically and 40–60 m (130–200 ft) laterally. The correlation lengths closely match the average thickness and width of the channel-form elements and suggest that in analogous reservoir permeability correlation lengths should be scaled to the dimensions of the channel-form elements. The principal heterogeneity within the sandstones is due to the presence of relatively low permeability mudclast-rich sandstones that occur in a discontinuous fashion along the base and margins of individual channel-form bodies.

CONCLUSION

An outcrop analysis of the Ferron Sandstone indicates that valley-fill architecture can be related to the stacking pattern of delta-front parasequences and to its position along the depositional profile. Valley fills associated with strongly progradational delta-front sandstone bodies are preserved as shoestring like bodies consisting largely of sand-rich fluvial deposits. The dominance of massive and concave-upward channel-form elements within the valley fill suggest the valley was occupied by low-sinuosity channels that were passively filled with sand. The principal heterogeneity within these deposits are related to relatively low permeability sandstones containing abundant clay clasts that occur along the base and margins of individual channel-form bodies. In contrast, valley fills associated with aggradational delta-front units are preserved as broad ribbon-like bodies consisting of heterogeneous mixture of fluvial sandstones, mud-rich abandoned channel fills, and marginal-marine mudstones. The dominance of convex-upward and lateral channel-form elements within the valley fill suggest the valley was occupied by meandering channels with well-developed midstream and side-attached barforms. Internal heterogeneities result from mudstone drapes that line the base and margin of individual channel-form bodies and from the interbedding of sandstone-dominated channel-form bodies with mudstone-dominated channel fills. The presence of mudstone within the valley fill increases from 5–25% over a distance of 4 km (2.5 mi) from the most proximal position studied to the seaward extent of the valley fill.

Sandstone permeability was closely correlated with grain size. The only exception is the coarse-grained sandstones containing abundant clay clasts which display roughly an order of magnitude (10X) reduction in permeability compared to similar sandstones containing no mudclasts. Sandstone permeability displayed a fourfold (4X) increase over a distance of 4 km (2.5 mi) from the most proximal position studied to the seaward extent of the valley fill. Vertical and horizontal variograms indicate permeability is correlated over a distance of about 4–6 m (13–20 ft) vertically and approximately 40–60 m (130–200 ft) laterally. The vertical and lateral correlation ranges are similar to the average channel-form width and thickness.

ACKNOWLEDGMENTS

This paper represents part of a Ph.D. study conducted at the University of Texas at Austin under the supervision of Dr. William Fisher and Noel Tyler. Financial support for this project was provided by the Gas Research Institute under contract no. 5089-260-1902. The project was conducted at the Bureau of Economic Geology at the University of Texas in Austin under the supervision of Noel Tyler and Steve Fisher. Ideas on the Ferron Sandstone benefited from discussions with Brian Willis, Janok Bhattacharya, Michael Gardner, and Paul Knox. Reviewers Peter McCabe and Tony Gorody are thanked for the help in shaping the final paper.

REFERENCES CITED

Allen, J. R. L., 1983, Studies in fluviatile sedimentation: bars, bar complexes and sandstone sheets (low sinuosity braided streams) in the Brownstones (l. Devonian), Welsh Borders: Sedimentary Geology v. 33, p. 237–293.

Armstrong, R. L., 1968, Sevier orogenic belt in Nevada and Utah: Geological Society of America Bulletin, v. 79, p. 429–458.

Barton, M. D., 1994, Outcrop characterization of architecture and permeability structure in fluvial-deltaic sandstones, Cretaceous Ferron Sandstone, Utah: Ph.D. dissertation, University of Texas at Austin, 259 p.

Fisher R. S., N. Tyler, and M. D. Barton, 1993a, Quantifying reservoir heterogeneity through outcrop characterization — architecture, lithology, and permeability distribution of a seaward-stepping fluvial-deltaic sequence, Ferron Sandstone (Cretaceous), central Utah: Gas Research Institute, Topical Report 93-0022, 83 p.

—1993b, Quantifying reservoir heterogeneity through outcrop characterization — architecture, lithology, and permeability distribution of a landward-stepping fluvial-deltaic sequence, Ferron Sandstone (Cretaceous), central Utah: Gas Research Institute, Topical Report 93-0023, 132 p.

Gardner, M. H., 1993, Sequence stratigraphy and facies architecture of the Upper Cretaceous Ferron Sandstone Member of the Mancos Shale, east-central Utah: Ph.D. dissertation, Colorado School of Mines, Golden, 528 p.

Goggin, D. J., 1988, Geologically sensible modeling of the spatial distribution of permeability in eolian deposits: Page Sandstone (Jurassic), northern Arizona: Ph.D. dissertation, University of Texas at Austin, 731 p.

Hale, L. A., 1972, Depositional history of the Ferron Formation, central Utah, in J. L. Baer and E. Callaghan, eds., Plateau-basin and range transition zone: Utah Geological Association Publication 2, p. 115–138.

Kittridge, M. G., 1988, Analysis of permeability variation-San Andres Formation (Guadalupian) — Algerita Escarpment, Otero County, New Mexico: M.S. thesis, University of Texas at Austin, 361 p.

Lupton, C. T., 1916, Geology and coal resources of Castle Valley in Carbon, Emery, and Sevier Counties, Utah: U.S. Geological Survey Bulletin 628, 88 p.

Miall, A. D., 1985, Architectural element analysis — a new method of facies analysis applied to fluvial deposits: Earth Science Reviews, v. 22, p. 261–308.

Ryer, T. A., 1981, Deltaic coals of the Ferron Sandstone Member of the Mancos Shale — predictive model for Cretaceous coal-bearing strata of the Western Interior: AAPG Bulletin, v. 65, no. 11, p. 2323–2340.

Ryer, T. A., and M. McPhillips, 1983, Early Late Cretaceous paleogeography of east-central Utah, in M. W. Reynolds and E. D. Dolly, eds., Mesozoic paleogeography of west-central United States: Society of Economic Paleontologists and Mineralogists, Rocky Mountain Section Paleogeography Symposium, p. 253–272.

Stalkup, F. I., and W. J. Ebanks, 1986, Permeability variation in a sandstone barrier-tidal channel-tidal delta complex, Ferron Sandstone (Lower Cretaceous), central Utah: Society of Petroleum Engineers, Paper 15532, 13 p.

Van Wagoner, J. C., H. W. Posamentier, R. M. Mitchum, P. R. Vail, J. F. Sarg, T. S. Loutit, and J. Hardenbol, 1989, An overview of sequence stratigraphy and key definitions, *in* C. W. Wilgus, B. S. Hastings, C. G. St. C. Kendall, H. W. Posamentier, C. A. Ross, and J. C. Van Wagoner, eds., Sea level changes — an integrated approach: Society of Economic Paleontologists and Mineralogists Special Publication 42, p. 39–45.

White C. D., and M. D. Barton, 1999, Translating outcrop data to flow models, with applications to the Ferron Sandstone: Society of Petroleum Engineers, Paper No. 38741, p. 233–248.

Price-Emery Stage, Castle Valley, Utah, circa 1910. Photograph courtesy of the family of C. T. Lupton.

Analog for Fluvial-Deltaic Reservoir Modeling: Ferron Sandstone of Utah
AAPG Studies in Geology 50
T. C. Chidsey, Jr., R. D. Adams, and T. H. Morris, editors

3-D Fluid-Flow Simulation in a Clastic Reservoir Analog: Based on 3-D Ground-Penetrating Radar and Outcrop Data from the Ferron Sandstone, Utah

Craig B. Forster[1], Stephen H. Snelgrove[1], Siang Joo Lim[1],
Rucsandra M. Corbeanu[2], George A. McMechan[3], Kristian Soegaard[4],
Robert B. Szerbiak[3], Laura Crossey[5], and Karen Roche[5]

ABSTRACT

An integrated data set containing detailed three-dimensional ground-penetrating radar surveys, outcrop mapping, and both *in situ* and lab-based petrophysical measurements provides an unusually detailed foundation for simulating fluid flow through the three-dimensional permeability structure of a channel sandstone reservoir analog within a study volume 40 m x 20 m (130 ft x 60 ft) (in plan) and 15 m (45 ft) deep. Permeability (*k*) data (more than 1100 data points ranging from 0.5–300 md) are obtained from probe permeameter measurements made on core plugs collected from five vertical cliff-face transects and on slabbed drill-hole core from four, 14-m-deep (42-ft) holes drilled 10–20 m (30–60 ft) behind the outcrop. Analytical petrography and comparison of outcrop *k* values with those from drill-hole core show that weathering has caused a factor of 2 to 5 increase in cliff-face permeability with a significantly different univariate *k* distribution from that of the behind-the-outcrop drill-hole core. The permeability differences are accounted for by "correcting" the weathered cliff-face values to mimic the distribution of *k* values obtained behind the outcrop using an inverse transformation method. The field-based *k* data are integrated into two classes of three-dimensional, geostatistically generated, stochastic permeability models. One model type uses ground-penetrating radar data solely to define important bounding surfaces with sandstone permeability extrapolated directly from the *k* data. The other model type uses the ground-penetrating radar data to explicitly constrain a sandstone permeability structure and to define the geometry, thickness, and distribution of shale/mudstone units.

A series of two-phase fluid-flow simulations within 15 m x 15 m (45 ft x 45 ft) (in plan) by 10.8 m (35 ft) deep model domains, using grid blocks 1 m^2 (9 ft^2) (in plan) and 0.2 m (0.7 ft) thick, yield several conclusions. First, replacing the entire volume with a single *k* value computed with arithmetic, geometric, or harmonic mean of all *k* values in the volume yields a poor approximation of oil produc-

[1]*Department of Geology & Geophysics, University of Utah, Salt Lake City, Utah;* [2]*Agip, The Hague, The Netherlands;*
[3]*Geosciences Department, University of Texas at Dallas, Richardson, Texas;* [4]*E&P Research Center, Norsk Hydro ASA, Bergen, Norway;*
[5]*Department of Earth and Planetary Sciences, University of New Mexico, Albuquerque, New Mexico*

tion computed using the original detailed, heterogeneous k structure. Second, in most cases upscaling using a vertically averaged geometric mean (as opposed to using arithmetic or harmonic means) provides a reasonable match to the oil production computed using the original detailed, heterogeneous k structure. Third, it is important to distinguish between fully continuous versus discontinuous shale unit geometries, and to be able to estimate angle of shale unit dip, when selecting methods for upscaling outcrop permeability values for use in reservoir simulators. Integrating three-dimensional ground-penetrating radar surveys with outcrop-based sedimentological and petrophysical data can provide the estimates of size, dip, continuity, and distribution of shale units needed at the scale of individual simulator grid blocks to help improve our approaches for deriving the upscaled k values used as input to production-scale reservoir simulators.

INTRODUCTION

It is increasingly apparent that detailed, three-dimensional (3-D) characterization of sedimentary reservoir rocks is needed to develop improved predictions of petroleum reservoir performance. There are two issues of concern. First, we need to improve our ability to map and quantify the 3-D geometry and distribution of shaley low-permeability units that impede fluid flow through clastic sedimentary rocks. Second, we need to establish appropriate methods for upscaling the influence of the low-permeability units, and permeability variations within higher permeability sandstone, for the relatively large grid-block volumes used in reservoir simulation. Over the past decade, increasing use of outcrop analogs has provided important insight regarding detailed facies architecture and permeability structure (Eschard and Doligez, 1993; Fisher et al., 1993a, 1993b; Barton, 1994; Begg et al., 1994; Doyle and Sweet, 1995; Liu et al., 1996; Robinson and McCabe, 1997; Eschard et al., 1998; Dutton and Willis, 1998). Unfortunately, outcrops can only provide a series of two-dimensional (2-D) views from cliff faces or surface pavements. What is needed is a way to dissect, or image, the internal character of sedimentary rocks found behind cliff-face exposures.

A new technology for characterizing sedimentary rocks in 3-D is now emerging through the use of ground-penetrating radar (GPR) (Baker and Monash, 1991; Gawthorpe et al., 1993; McMechan et al., 1997). Ground-penetrating radar is a high-resolution geophysical technique that can provide indirect information on the lithologic and petrophysical properties of shallow subsurface rock units. The vertical resolution of GPR is on the order of a few decimeters, and the depth of penetration is in the range of meters to tens of meters (Davis and Annan, 1989). As a consequence, the GPR method is well suited to exploring the detailed facies architecture of shallow outcrop analogs of important reservoir facies. The results of GPR surveys can be used to map the geometry and distribution of shaley units while also providing the insight needed to develop geologically plausible permeability structures.

At the Coyote Basin field site in east-central Utah (Corbeanu et al., 2001, 2002, this volume; Szerbiak et al.,

2001) we have investigated the petrophysical structure of fluvial channel-belt deposits in the Cretaceous Ferron Sandstone. Field, laboratory, and numerical studies have been performed within a rock volume (20 m x 40 m [60 ft x 120 ft] in plan, 15 m [45 ft] deep) that approximates the size of a simulator grid block that might be found in a full-field reservoir simulation. A unique aspect of this study is that 3-D GPR survey results obtained from behind the outcrop are closely tied to petrophysical and stratigraphic data collected from an adjacent cliff face and from four drill holes penetrating the rock volume. Details of the project are outlined in a series of project reports submitted to the U.S. Department of Energy (McMechan et al., 1998, 1999, 2001).

This paper builds on the work described by Szerbiak et al. (2001) and Corbeanu et al. (2001, 2002, this volume) by using fluid-flow simulations to illustrate how their GPR-based studies at Coyote Basin might be used as a foundation for developing improved approaches for upscaling petrophysical properties in fluvial channel-belt, and other, sedimentary rocks. In this study, the thickness and distribution of sub-horizontal shale horizons are quantified using two different approaches and are then incorporated into detailed fluid-flow simulation models. In the first approach, the GPR results are used to map the topography of the shale-sand and sand-sand contacts while permeability values for the sandy bodies are extrapolated using data derived from cliff-face outcrops and drill-hole core using a geostatistical approach (Szerbiak et al., 2001). In a second approach, the GPR results are used to infer the thickness and distribution of the shale units and to explicitly constrain the permeability values extrapolated from the field-based data (Corbeanu et al., 2001). A series of two-phase fluid-flow simulations performed using fine grid blocks (0.2 m [0.7 ft] thick by 1 m² [9 ft²] in plan) are combined with simulations performed where upscaled, uniform k values are assigned to a progressively greater number of grid blocks. The simulation results help to explore how 3-D GPR data integrated with outcrop- and drill hole-derived petrophysical and sedimentological data aid in assessing alternative methods for assigning geologically plausible, upscaled permeability structures within appropriately sized simulator grid blocks.

PROJECT SETTING

The Coyote Basin field site (Corbeanu et al., 2001, 2002; Szerbiak et al., 2001) is located in east-central Utah, at the top of the upper Turonian SC3 Member of the Cretaceous Ferron Sandstone (Gardner, 1995; Garrison et al., 1997). The Ferron Sandstone outcrops along the southwestern flank of the San Rafael Swell (Figure 1) and contains a series of fluvial-deltaic complexes that extend to the northeast. Excellent exposures occur along vertical cliffs oriented both parallel, and perpendicular, to the direction of deposition.

A variety of high-resolution sedimentologic and stratigraphic data acquired at the site provide outcrop control for constructing a detailed 3-D representation of the distribution of lithotypes within the 3-D GPR data volume (15 m [45 ft] thick and 40 m x 16.5 m [120 ft x 54 ft] in plan). The top surface of the site clearly shows cross-bed sets that are characteristic of a fluvial channel and indicate flow in a southeasterly direction. Photomosaic panels of the 12-m-high (36-ft) cliff face adjacent to the GPR grid (Figure 1) and five detailed, measured stratigraphic sections on the cliff face indicate an upper channel complex that contains at least five sandy units separated by discontinuous muddy to silty layers (Plate 1 on compact disc). Detailed logging of continuous core from drill holes A, B, C, and D (Figure 1) provides additional subsurface control.

The upper channel complex (approximately 12 m [36 ft] thick) overlies floodplain mudstone, coal, and crevasse-splay deposits. The cliff-face facies section on Plate 1 (all plates are on the CD-ROM) concentrates exclusively on the upper fluvial sandstone. Several architectural elements are mapped and referred to, in ascending stratigraphic order, as *Units 1* through *5*. *Units 1* through *4* in the lowest 7 m (21 ft) of the channel complex, consist of fine-grained lenticular sandstone bodies that pinch out over distances of several tens of meters parallel to the cliff face (in the direction of the channel axis). These elements are low-angle, parallel-laminated sandstones that scour into the underlying parallel-laminated sandstone. The bases of *Units 1* through *4* are erosional and commonly have mudstone intraclast conglomerate along the basal scour. The upper parts of *Units 1* through *4* are capped by a 5–10-cm (2–4-in.) thick mudstone layer which is laterally discontinuous because of truncation by the overlying unit. *Units 1* through *4* are interpreted as scour and fill elements deposited during flood events within a fluvial channel.

Unit 5, the uppermost 4.5–5.5 m (14.8–18.1 ft) of the channel sandstone (Plate 1), consists exclusively of medium- to large-scale, trough cross-bedded, medium-grained sandstone. In the lower half of *Unit 5* a greater proportion of trough cross-beds are preserved (between 10–30 cm [4–12 in.] in thickness) whereas trough cross-

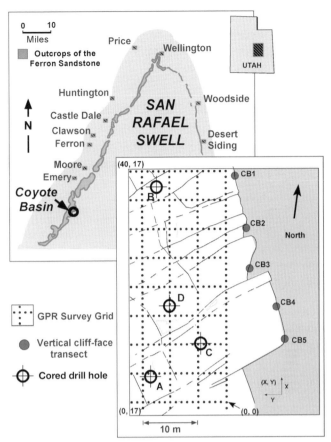

Figure 1. Site map showing GPR grid and locations of drill holes and cliff-face transects at the Coyote Basin study site established in the Ferron Sandstone on the western flank of the San Rafael Swell in east-central Utah.

beds tend to be less than 10 cm (4 in.) thick in the upper part of *Unit 5* because of scouring by overlying cross-beds. Paleocurrent measurements made in *Unit 5* clearly indicate that flow was to the southeast during deposition. The long axis of the GPR grid (Figure 1) is aligned approximately parallel to the direction of inferred flow.

Drilling activities included collecting core plugs from outcrop, continuous coring behind the cliff face, and collecting core plugs from the continuous drill-hole core. Outcrop-derived 2.4 cm (0.9 in.) in diameter, 4–10-cm-long (2–4 in.) core plugs were collected from the east-facing cliff face adjacent to the GPR survey volume (Figure 1) on transects coincident with the measured stratigraphic sections (Plate 1). A total of 485 plugs were collected at 10 cm (4 in.) spacing along the vertical transects.

Drill-hole core 6.35 cm (2.5 in.) in diameter was collected from the Coyote Basin drill holes A, B, C, and D (Figure 1) using a truck-mounted, rotary triple-tube coring rig. Each hole was drilled to a depth of approximately 14 m (42 ft) and bottomed in thick, gray shale. Core recovery was greater than 97% with most core loss occurring in shale zones or where two pieces of sandstone core happened to spin and wear against each other in the core barrel. A total of 93 core plugs were cut from the slabbed core as a foundation for permeability/porosity testing, petrologic/diagenesis studies, and electrical property measurements.

GROUND-PENETRATING RADAR SURVEYS

Corbeanu (2001), Corbeanu et al. (2001, 2002, this volume), and Szerbiak et al. (2001) outline the details of the Coyote Basin GPR surveys, data processing, and interpretation. A 100 MHz common offset 3-D GPR survey was performed behind the outcrop on a rectangular grid of 34 north-south oriented lines at a spacing of 0.5 m (1.6 ft) between adjacent lines (Figure 1). Each GPR line contains 81 traces, equally spaced at 0.5 m (1.6 ft). The GPR transmitter voltage was 1000 V. Dipole antennas are oriented parallel to each other and perpendicular to the in-line direction. The separation between antennae was 3 m (9 ft) and depth of penetration was about 15 m (45 ft).

Prior to interpreting the survey results, processing and migration of the GPR data is required to focus the GPR image and reduce artifacts unrelated to primary lithology. Processing and migration methods are adapted from those commonly used to process seismic survey data. Because 3-D GPR data volumes have a format similar to that of 3-D seismic data, 3-D seismic interpretation software provides a flexible and efficient means for display, attribute computation, and analysis of the GPR data (Corbeanu et al., 2002).

Principal lithologic units and bounding surfaces mapped in outcrop and drill-hole core are also mapped in the migrated GPR data volume, thus, a sequence of radar facies is identified that matches the sedimentary facies. Ground-penetrating radar reflections in the lower approximate 7 m (21 ft) of the data volume correlate well with bounding surfaces between sandstone layers and mudstone or mudstone intraclast conglomerate layers as these correspond to a significant change in electrical properties (Corbeanu et al., 2001, 2002). In the uppermost 5 m (15 ft), the GPR reflections are interpreted as correlating with bounding surfaces of the trough cross-bed co-sets, where these co-sets are comparable in thickness with the vertical GPR resolution (about 0.5 m [1.5 ft]) and present a significant change in grain size at the facies boundary. There are no mudstone layers visible in the upper 5 m (15 ft) of the GPR data volume.

SANDSTONE PERMEABILITY AND POROSITY DATA

Probe permeameter, and Hassler cell permeability and porosity tests were performed on samples collected from the study volume. The probe permeameter was used to test: (1) one end (the sawcut end opposite the outcrop exposure) of the core plugs collected from outcrop, (2) both sawcut ends of the 93 core plugs cut from the continuous core, and (3) closely spaced points along the entire length of slabbed, non-shale drill-hole core collected from behind the outcrop (at drill holes A, B, C, and D in Figure 1). Porosity and Hassler cell permeability tests were performed by Terra Tek Inc. of Salt Lake City, Utah, on 33 of the core plugs cut from the continuous drill-hole core.

Probe permeameter measurements (Hurst and Goggin, 1995; Meyer and Krause, 2001) were made using a computer-controlled, stage-mounted, electronic probe permeameter at the University of Utah. A steady state method is used to test approximately 1 cm^3 (with a 3 mm diameter tip) volume of rock using nitrogen gas as the permeant. Two sample spacings were employed: 5 cm (2 in.) spacing along the entire length of each continuous drill-hole core, and 1 cm (0.4 in.) spacing over selected intervals containing fine-grained layers separating the sandstone units.

Probe permeameter data obtained from outcrop and drill-hole core range from 0.5–326 md. Vertical variations in k, as measured on outcrop-derived core plugs, are shown in Plate 1 for the five vertical transects. Low permeability (≤ 0.5 md) in mudstone and siltstone units is excluded from the drill-hole data histogram because the cliff-face data include very few measurements from these highly erodable layers. Although the overall range of permeability is similar for both outcrop and cores, it is important to note that k values obtained from outcrop-derived core plugs have a larger mean (90 versus 32 md), and a greater standard deviation (61 versus 30 md), and are skewed to the higher permeability end of the data range (Figure 2). This difference between outcrop- and drill-hole core-derived k data is also seen when the data are grouped by individual facies units (Figure 3).

Steady state permeability tests performed in the Hassler cell on 33 of the core plugs cut from the drill-hole core provide an average value for each entire core plug. The Hassler cell test results correspond well, over a permeability range of 1–150 md, to the averaged results of probe permeameter tests performed on each end of the same core plug.

Porosity values determined with a Boyles's Law approach were obtained for the 33 core plugs cut from drill-hole core. Porosity values range from 0.10–0.23 with a mean of 0.18. A cross-plot of porosity against $\log_{10}$ permeability for the 26 sandy core plugs (Figure 4) suggests that sandy plugs from *Unit 5* have a higher permeability for given porosity when compared to sandy plugs from the other units. Recall that *Unit 5* contains trough cross-bedded, medium-grained sandstone deposited within the main fluvial channel while *Units 1* through *4* are mainly parallel laminated, fine-grained

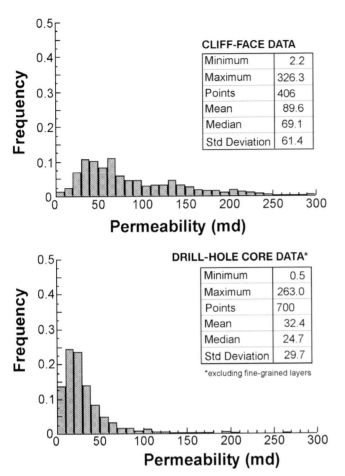

Figure 2. Histograms of cliff-face and drill-hole core permeability data derived from probe permeameter tests performed on core plugs collected from both cliff-face and slabbed drill-hole core.

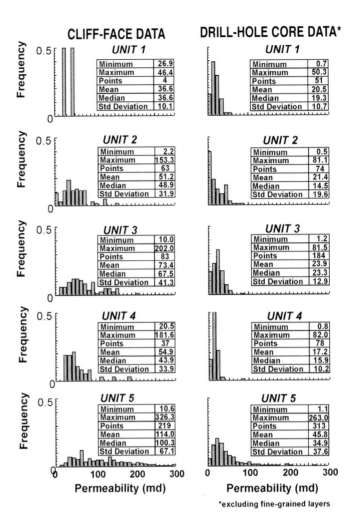

Figure 3. Histograms and statistics of drill-hole and cliff-face permeability data shown in Figure 2, but separated by unit.

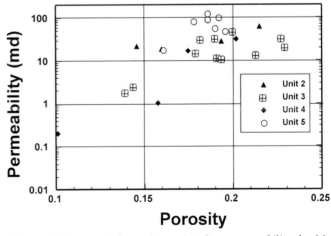

Figure 4. Cross-plot of porosity against $\log_{10}$ permeability for 26 sandy cores cut from slabbed drill-hole core.

sandstone bodies (commonly with a basal intraclast conglomerate) associated with flooding events. The *Unit 5* porosity values span a relatively narrow range of permeability when compared to porosity values obtained for the other units. The cross-plot (Figure 4) suggests that a single least squares line of the form $\log_{10} k = f(\phi)$ is not appropriate for representing the porosity characteristics of all units.

Analytical petrography performed on a suite of 204 samples helps to assess relationships between petrophysical measurements and sandstone petrology, and provides a foundation for interpolating and extrapolating results obtained at the Coyote Basin reservoir analog. Point counting using standard petrographic techniques enables us to characterize and quantify sandstone framework composition, authigenic mineralogy, porosity/microporosity, and texture (Roche, 1999). The samples are classified as ranging across the lithic arkose and feldsarenite compositional fields. Samples obtained from outcrop are consistently lower in lithic content and generally higher in quartz than samples obtained from drill holes located behind the outcrop. Both lithic fragments and feldspar appear to be destabilized in the

weathering process and tend to occur as irresolvable matrix in thin sections (Roche, 1999). The weathering-enhanced macroporosity observed in thin section provides a logical explanation for the enhanced permeability values obtained from the cliff face when compared to those obtained from drill-hole core behind the outcrop.

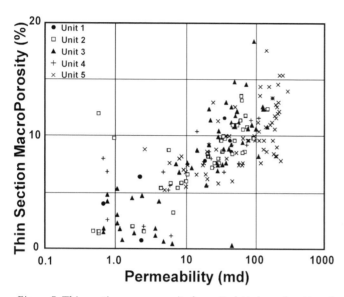

Figure 5. Thin-section macroporosity by unit plotted as a function of probe permeameter k obtained from adjacent locations in the drill-hole core.

Figure 5 shows the apparent correlation between increasing permeability as determined using the probe permeameter and increasing macroporosity as determined using analytical petrography.

PERMEABILITY MODELS

Several permeability models are constructed using the integrated 3-D GPR, petrophysical and sedimentological data sets. Two principal differences in model type reflect two different approaches for exploiting the information contained in the GPR data. In the first, stochastic permeability model type (Type-A), the GPR data are used to infer the topography of several sub-horizontal bounding surfaces identified as the tops of relatively thin, low-permeability shale units or as well-defined sand-on-sand contacts. Several permeability structures are created, some with and some without the thin shale units, by distributing values of sandstone permeability within the study volume using sequential Gaussian simulation (Deutsch and Journel, 1992; Goovaerts, 1997). In the second, geophysics-derived permeability model type (GPR), the GPR data are used to infer the thickness and distribution of the thin, low-permeability shale units while also explicitly constraining the inferred distribution of sandstone permeability within the volume. Ultimately, the detailed permeability structures are upscaled to assess how the results of fluid-flow simulations are distorted by simple upscaling procedures.

The coordinate axes of the permeability model are parallel to the axes of the 3-D GPR survey grid (Figure 1). The model volume is 39 m x 15 m (117 ft x 45 ft) in plan and 10.6 m (35 ft) thick with a uniform grid-block size of 1 m x 1 m (3 ft x 3 ft) and 0.2 m (0.7 ft) thick. The model grid contains a total of 34,560 grid blocks.

Type-A Permeability Models: GPR-Defined Surface Topography on Shaley Units

Type-A models are developed by using the 3-D GPR data to identify the location and topography of a series of four bounding surfaces corresponding to sand on sand contacts and the upper sand-shale contact of low-permeability shaley units mapped in both outcrop and drill-hole core. The details of the GPR data collection, processing, and interpretation are outlined by McMechan and Soegaard (1998), McMechan et al. (1999, 2001), Szerbiak et al. (2001), and Corbeanu et al. (2001, 2002).

Merging Outcrop and Drill-Hole Core Permeability Data Sets

In developing permeability models within the study volume it is inappropriate to directly combine the raw cliff-face data together with the differently distributed drill-hole permeability data. Thus, the cliff-face data are transformed, on a unit-by-unit basis, to have the same statistical distribution as the drill-hole data. An underlying assumption is that the drill-hole data are "correctly" distributed (recognizing that both data sets likely suffer to varying degrees from the effects of weathering, but that the degree of weathering is less severe in the drill-hole rocks).

We adopt the inverse transformation method (Bain and Engelhardt, 1992), shown in Figure 6 for *Unit 5*, to correct to outcrop k data. Figure 6a shows the empirical cumulative distribution functions $F(x)$ (Bain and Engelhardt, 1992) for both drill-hole core permeability and cliff-face permeability, regarding each as a random variable. $F(x)$ is the probability that some drill-hole core (cliff face) k is less than or equal to x. The objective is to transform the random variable corresponding to cliff-face permeability into the random variable corresponding to drill-hole core permeability. This is done by observing that, for a fixed $F(x)$, the two random variables (drill-hole core permeability and cliff-face permeability) have different values. Figure 6b shows how this shared value of $F(x)$ provides the link for mapping a cliff-face permeability of 155.5 md to an adjusted permeability of 55.2 md. Note that the transformation is not simply a constant static shift applied to all cliff-face permeability measurements; the shift depends on the magnitude of the permeability to be transformed.

Figure 7 shows the effect of the transformation on the permeability data of cliff-face transect CB5. Figure 7a is a plot of the original, raw permeability data for CB5. Figure 7b shows the transformed permeability data. For reference, the permeability profile for core C (the drill hole nearest to CB5, Figure 1) is plotted in Figure 7c. The overall character of the transformed CB5 data is consistent with the character of the drill-hole core data, in con-

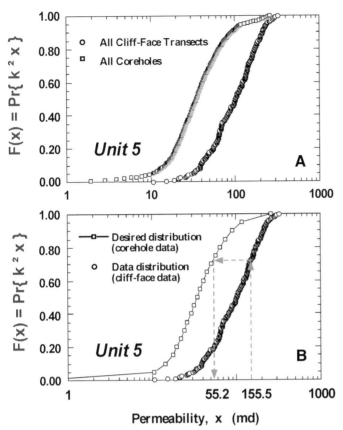

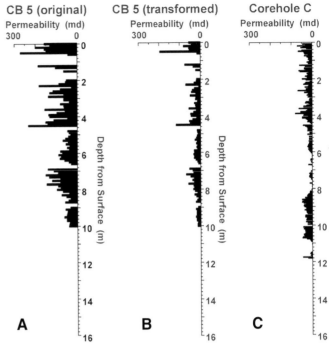

Figure 6. Method of transforming raw cliff-face permeability data to have the same univariate statistical distribution as the drill-hole core data. A — empirical cumulative distribution functions F(x) for both drill-hole core permeability and cliff-face permeability, regarding each as a random variable. B — shared value of F(x) providing the link for mapping a cliff-face permeability to an adjusted permeability.

Figure 7. Comparison of CB5 permeability data before and after transformation, with nearby drill hole C (Figure 1) permeability data for reference. A — plot of the original, raw permeability data for CB5. B — the transformed permeability data. C — the permeability profile for core C.

trast to the large difference in character between the original CB5 data and the data from drill hole C. The thicker bars on the CB5 plots reflect the fact that the sampling interval of the cliff-face permeability data is twice that of the drill-hole core data.

All Units Type-A Background Permeability Structure

A background Type-A permeability structure (without mudstone or shale) is created with sequential Gaussian simulation (Deutsch and Journel, 1992; Goovaerts, 1997) using the merged permeability data set and ignoring the underlying facies architecture. The spatial correlation structure of the permeability model is estimated by first using the normal scores transform to convert the k distribution into data that are distributed as a standard random variable (i.e., a Gaussian random variable with a mean of zero and a variance of one) then constructing vertical variograms using the merged permeability data from the nine vertical profiles. The principal axes of the 3-D variogram ellipse are assumed to be parallel to the grid coordinate axes because the direction of paleoflow

during deposition is approximately parallel to the long axis of the study volume. A clear correlation structure is evident in the normal scores vertical variogram of Figure 8. The experimental data are well fitted by a model variogram comprising three structures (nugget, exponential, and spherical with: nugget of 0.25; exponential range 0.75 m [2.46 ft] and contribution 0.40; and spherical range 3.5 m [11.5 ft] and contribution 0.35). Variogram model parameters are summarized in Table 1.

A horizontal normal scores model variogram is shown in Figure 8b. It was computed using data from two short (3.4 m and 5.3 m [11.2 ft and 17.4 ft]) cliff-face transects in *Unit 3*. These transects are shorter than a plausible correlation length that might be inferred from purely geological arguments. The experimental variogram of Figure 8b suggests that the horizontal correlation length is at least 3 m (9 ft). Beyond lags of 3 m (9 ft) the variogram values become erratic because of the low number of data pairs available for averaging. The horizontal model variogram in Figure 8b includes a nugget of 0.35 and an exponential structure with range of 10.0 m (30 ft) and contribution of 0.65. An isotropic variogram is assumed in the horizontal (x, y) plane.

A single vertical slice cut from one of five realizations of the Type-A (all units), 3-D background permeability structure is shown in Plate 2a. The tendency for permeability stratification in this layered sedimentary environment is evident.

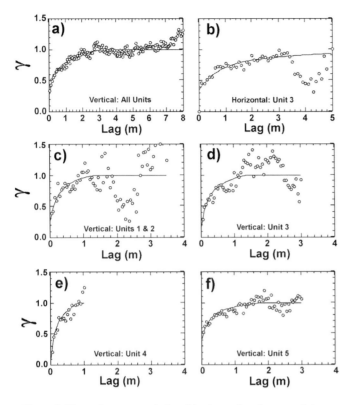

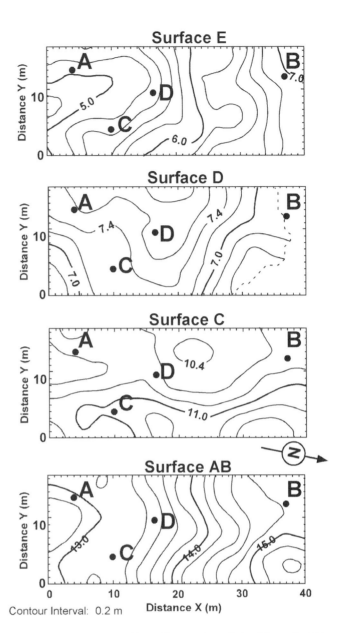

Figure 8. Normal scores vertical and horizontal variograms: (a) vertical, all units, (b) horizontal Unit 3, (c) vertical, Units 1 and 2 combined, (d) vertical, Unit 3, (e) vertical, Unit 4, and (f) vertical, Unit 5. The solid line is the model variogram with parameters given in Table 1.

Defining and Superimposing Type-A Low-Permeability Shale Units

The reference permeability model used as a basis for comparing fluid-flow simulation results (Plate 2b) is created by superimposing low-permeability (0.01 md) shale units on the background Type-A permeability model (Plate 2a). The topography of the top bounding surface of each shale unit is inferred from the results of the 3-D GPR surveys as constrained by stratigraphic data collected from drill-hole core and the adjacent cliff face. Contour maps of the bounding surfaces are shown in Figure 9. Surfaces AB, C, D, and E correspond to surfaces having the same designation as the bounding surfaces mapped on the cliff face (Plate 1). Surface AB is a composite of surfaces A and B identified on the outcrop. Details of the approaches used to collect, process, and interpret the GPR data are outlined by Corbeanu et al. (2001, 2002) and Szerbiak et al. (2001).

The thickness of the shale units superimposed on the Type-A (all units) background permeability model is set to ensure that all shale layers are fully continuous across the model domain. Thus, the shale units are everywhere one or two grid blocks thick to provide thickness values ranging from 0.2–0.4 m (0.7–1.3 ft). The resulting, stratified permeability structure within a sin-

Figure 9. Contour maps of bounding surfaces derived from the 3-D GPR survey. Note that surface D pinches out against surface E. Contours are in depth in meters below GPR datum.

gle vertical slice cut from the model volume is shown in Plate 2b.

Type-A (Unit-by-Unit) Permeability Structure

We recognize that it may be important to build permeability structures on a facies, or unit-by-unit, basis rather than relying on geostatistical modeling within the entire model domain to produce geologically plausible k distributions. The Type-A (unit-by-unit) background permeability model (Plate 2c) is constructed following an approach similar to that outlined above. The principal difference is that merged data for each unit are used to compute the spatial correlation structures and provide a foundation for the sequential Gaussian simula-

Table 1. Type-A model variogram structures.

	Nugget	Exponential Model		Spherical Model	
		Range (m)	Contribution	Range (m)	Contribution
All Units	0.25	0.75	0.40	3.5	0.35
Units 1 & 2	0.25	0.60	0.50	1.5	0.25
Unit 3	0.10	0.45	0.50	1.5	0.40
Unit 4	0.20	0.60	0.50	1.5	0.30
Unit 5	0.30	0.36	0.40	2.0	0.30
Horizontal*	0.35	10.00	0.65	NA	NA

* Horizontal variogram computed from data in *Unit 3*. NA = not available

tions needed to distribute *k* values within combined *Units 1* and *2*, and within each individual *Unit 3, 4,* and *5* (Plate 1). The model variogram structures that correspond to the experimental variograms shown in Figure 8 are summarized in Table 1. Permeability models constructed for each unit within the model domain are trimmed to fit between the appropriate bounding surfaces inferred from the 3-D GPR survey results. As before, the low-permeability shale units (0.01 md) defined in the previous section are superimposed on the Type-A (unit-by-unit) background permeability structure (Plate 2a) to produce the unit-by-unit stratified permeability structure shown in a vertical slice cut from the model volume (Plate 2c).

Type-B Permeability Models: GPR-Constrained Shale Characteristics and Permeability Distribution within Sandstone

The Type-B (unit-by-unit) permeability models (Plates 2d and 2e) are created by using the 3-D GPR data set to explicitly constrain inferred 3-D shale/mudstone distribution and thickness, and also the distribution of sandy body permeability. The principal assumption underlying this approach is that a link between lithology and electrical properties leads to a relationship between radar reflection attributes and lithology (Corbeanu et al., 2002). Given that permeability and porosity variations often reflect variations in lithology, the radar survey results provide insight regarding permeability variations within the study volume. Instantaneous amplitude (*IA*) and instantaneous frequency (*IF*) are used to estimate permeability and shale distributions from the GPR data set (Taner et al., 1979). Both GPR attributes contain information on electrical properties of the subsurface that are, in turn, related to permeability, porosity, and clay content. Because the physical relationship between GPR attributes and rock properties is unclear for consolidated rocks, an empirical relationship is derived for the Coyote Basin GPR data following the example of Schultz et al. (1994).

A linear regression is used to relate permeability values measured (for each sedimentologic unit, or radar facies) at specified points in the four drill holes and vertical section CB1 (described in previous sections, Figure 1) to the GPR instantaneous amplitude and frequency found at the same locations. The resulting function has the form:

$$\ln[P(x,y,z)] = c_0 + c_1 * IA + c_2 * IF$$

where, c_0 to c_2 are empirically estimated weighting coefficients, $\ln(P)$ is the natural logarithm of permeability, and *IA* and *IF* are the instantaneous radar amplitude and frequency, respectively. Permeability throughout the study volume (on a unit-by-unit basis) is predicted from the instantaneous amplitude and frequency at each point in the 3-D GPR volume by weighting the regression coefficients according to the relative distance of the points from each borehole. Predicted permeability values compare well with the measurements made at specific points (Corbeanu et al., 2002). About 98% of the predicted permeability values fall within the range of measured values (0.1–290 md). The remaining 2% of the predicted values appear to be a consequence of poor data support and are replaced with the average of adjacent values. The resulting GPR (unit-by-unit) background permeability structure is shown in the vertical section of Plate 2d.

Three-dimensional distribution of shale units is evaluated, within each radar facies, using cluster analysis to identify locations where specific combinations of radar attributes associated with shale units mapped in the boreholes are identified elsewhere in the study volume (Corbeanu et al., 2002). Shale/mudstone units are found to have characteristic GPR responses at the drill holes that can be used as a foundation for identifying shale units away from the control points. The size and shape of the shale/mudstone units are quantified using indicator geostatistics (Deutsch and Journel, 1998). Overall, *Unit 2* through *Unit 4* contain about 8% mudstone and mudstone intraclast conglomerate by volume. *Unit 5* contains no mudstone. Mudstone units are typically small bodies with lengths of 2.3–3.5 m (7.5–11.5 ft), combined with a few units longer — 25–30 m (75–90 ft) bodies. The latter have ribbon-like shapes rather than tabular or sheet-like layers. Computed shale distributions are superimposed on the Type-B (unit-by-unit) background permeability model of Plate 2d to produce

a)

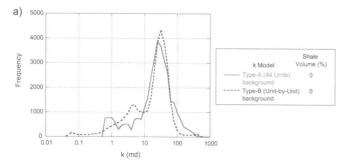

b)

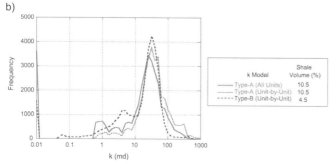

c)

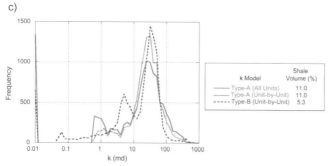

Figure 10. Frequency distribution plots of permeability values assigned within each permeability model: (a) Full Domain background Type-A (all units) and Type-B (unit-by-unit) models without shale units, (b) Full Domain Type-A (all units) and Type-B (unit-by-unit) models with shale units superimposed, and (c) Block 1 Type-A (all units) and Type-B (unit-by-unit) models with shale units superimposed.

the Type-B (unit-by-unit) permeability structure with shale/mudstone units shown in the vertical section of Plate 2e.

Comparison of Permeability Models

Several factors illustrate differences and similarities between the permeability models illustrated in Plate 2 — visual appearance, shale volume, and univariate permeability distributions. All permeability models have a stratified appearance that is accentuated when low-permeability shale units are incorporated into the models (Plates 2b, 2c, and 2e). It is important to note, however, that the shale units in the Type-B model are discontinuous and vary in thickness. As a consequence, the Type-B shale units form incomplete barriers to vertical fluid flow and eliminate the sandy body pinchout (where two shale units intersect, Plates 2b and 2c) found in the Type-

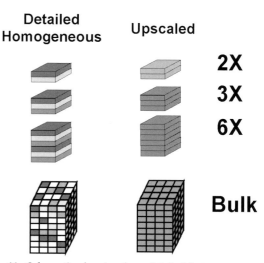

Figure 11. Schematic showing how detailed heterogeneous permeability structures are upscaled for different levels of upscaling. Note that grid-block size is unchanged in the upscaling process.

A models. Variations in shale volume, defined as a percentage of the total volume within the permeability model, reflect the assumed difference in shale continuity within each model. As a consequence, the Type-A models (with continuous shale units) have a shale volume of 10.5% while the Type-B model (with discontinuous shale units) has a shale volume of 4.5%.

Plots illustrating the univariate distribution of k within each permeability model are shown in Figure 10. First, consider the background Type-A (all units) and Type-B (unit-by-unit) permeability models shown in Figure 10a. The overall form of both distributions is similar with most frequent values of k falling between 12–50 md and maximum k values of about 300 md. The shoulders of each distribution differ because of the different approaches used to interpolate k values between the data points. Important changes to the distributions are noted when low-permeability shale units are superimposed on the background k models. These differences result because grid blocks formerly filled with sand-related k values become filled with low-permeability shale values.

Permeability Upscaling

Simple permeability upscaling approaches are used to explore how the detailed insight extracted from the 3-D GPR data might aid in upscaling and assigning permeability values in sedimentary rocks. Each upscaling exercise involves computing the arithmetic, geometric, and harmonic mean permeability of groups of grid blocks drawn from the model volume. The mean permeability values are computed for vertically distributed groups of grid blocks (2, 3, and 6 blocks) found only in the same column (Figure 11) to complete a simple vertical upscaling. Although permeability values are lumped, the original grid-block sizes (1 m x 1 m [3 ft x 3 ft] in plan

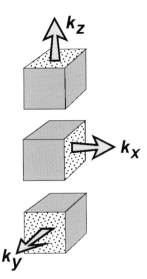

Figure 12. Numerical Darcy experiments involve computing flow in a specified coordinate direction between two opposing faces, combined with the known pressure gradient, to compute the equivalent bulk permeability in that coordinate direction. A series of three numerical experiments are required to compute an apparent permeability tensor for the model domain.

and 0.2 m [0.7 ft] thick) are retained to facilitate comparison of the numerical results. The bulk upscaling method involves computing the arithmetic, geometric, and harmonic mean permeability of all grid blocks to produce a single, isotropic k value for each model (Figure 11). Differences in computed oil and water production caused by differences between the original, detailed permeability structure and the upscaled equivalent are explored using the simulated waterflood exercises outlined in the next section.

FLUID-FLOW SIMULATION APPROACH AND SETUP

The permeability models outlined above, and their upscaled equivalents, form the foundation for a series of numerical waterfloods. Both single- and two-phase simulations are used to gain insight regarding differences and similarities in the way that each permeability model might perform if incorporated within a full-scale reservoir simulator.

Single-Phase Fluid-Flow Simulation Approach

Single-phase fluid-flow simulations are performed using a standard groundwater flow model (MODFLOW) embedded within the user-friendly Groundwater Modeling System (GMS) interface developed by Environmental Modeling Systems, Inc. MODFLOW is a transient, 3-D finite difference model developed by the U.S. Geological Survey (McDonald and Harbaugh, 1988) to compute rates and directions of groundwater flow through saturated porous media subject to a range of boundary conditions. Steady-state single-phase flow simulations are performed to test our permeability models prior to embarking on the two-phase simulations and to derive bulk, apparent permeability tensors for each permeability model.

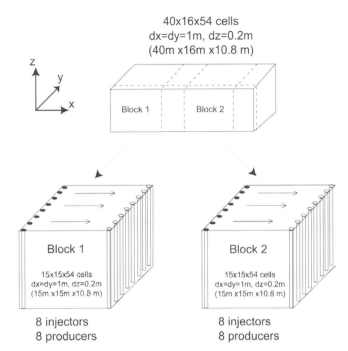

Figure 13. The Full Domain models are cut into two, equally sized sub-domains to reduce computational time and effort. Most simulations are performed using Block 1 models. Flow regimes modeled in two-phase fluid-flow simulations in single directional flow configuration. The Full Domain simulations involve 40 x 16 cells in plan and 54 cells high within a model volume of 40 m x 16 m (120 ft x 48 ft) in plan and 10.8 m (35.4 ft) thick. The Block 1 and Block 2 model domains contain 15 x 15 cells in plan with 54 cells high for a model volume of 15 m x 15 m (45 ft x 45 ft) in plan and 10.8 m (35.4 ft) thick. In all cases individual grid blocks (cells) are 1 m x 1 m (3 ft x 3 ft) in plan and 0.2 m (0.7 ft) thick. In the models the upper (x-y), lower (x-y), front (x-z), and back (x-z) boundary planes are impermeable.

Apparent permeability tensors are computed by performing a set of three, independent and mutually orthogonal, numerical Darcy experiments on each permeability model (Figure 12). Assigning a pressure gradient between two opposing faces of each model volume enables us to compute the bulk volumetric flow through the volume and the corresponding bulk equivalent permeability in each coordinate direction (k_x, k_y, k_z). With the results of three simulations, one performed in each coordinate direction, we produce apparent permeability tensors for each model volume (Figure 12). In addition to modeling flow through the full model domains described above, we also work with consistently sized Block 1 and Block 2 model sub-domains (Figure 13) to reduce the computational effort involved in subsequent two-phase flow simulations. The form of the k distributions for Block 1 (Figure 10c) and Block 2 (not shown) sub-domains is similar to those of the full domain models (Figures 10a and 10b). The shale volume of the Block 1 models (Type-A = 11%, Type-B = 5.3%) is slightly greater than the average values computed for the full domains (Type-A = 10.5%, Type-B = 4.5%).

Table 2. Relative permeability data (Louisiana State University, 1997).

Oil Saturation (%)	Relative k - Oil	Relative k - Water
0	0.00	1.00
2	0.00	0.71
12	0.00	0.51
20	0.00	0.27
30	0.00	0.18
40	0.03	0.12
50	0.09	0.07
60	0.17	0.04
70	0.30	0.02
80	0.50	0.00
90	0.75	0.00
100	1.00	0.00

Two-Phase Fluid-Flow Simulation Approach

Two-phase, 3-D finite difference flow simulations are performed to model the progress of a waterflood through each model domain using the Black Oil Application Simulation Tool (BOAST98) developed for the U.S. Department of Energy (Fanchi et al, 1982; Louisiana State University, 1997; Heemstra, 1998). BOAST98 simulates transient, 3-D, isothermal, three-phase fluid flow (oil, gas, and water phases) through porous media. In this study we consider the flow of only water and dead oil; the gas phase is excluded by setting the oil expansivity and volume of solution gas both to zero. In addition, capillary pressure is set to zero, irreducible water saturation is set to 20%, and both fluid and rock compressibility values are constant. Oil/water relative permeability values as a function of oil saturation (Table 2) are those used in testing BOAST (Louisiana State University, 1997).

The two-phase model domain geometry is similar to that outlined above for the single-phase fluid-flow simulations. An approximately planar source of injected water (eight fully penetrating injection wells) at one boundary causes flow across the model domain with oil and water produced from eight fully penetrating production wells at the opposite boundary (Figure 13). The upper (x-y), lower (x-y), front (x-z), and back (x-z) boundary planes are impermeable. Simulations were run to evaluate differences between flow induced in either the x- or y- coordinate directions. The water injection rate is 16 stock tank barrels per day. Bottom-hole pressures cannot fall below 4020 psi (27,720 kPa) in the production wells. Average initial reservoir pressure is 4400 psi (30,300 kPa); initial oil saturation is 80%. A uniform, homogeneous porosity of 0.20 is assigned in all simulations so that the effects of porosity variations cannot influence our ability to discern the possible effects of permeability variations on oil production

SIMULATION RESULTS

Single-Phase Simulation Results

Computed apparent permeability tensors (Figure 14) provide upscaled, equivalent homogeneous anisotropic permeability values for each model domain. The stratified character of the background permeability models produces an inherent vertical anisotropy of approximately 1:3 for the Type-A (all units) model (9 to about 27 md) and for the Type-B (unit-by-unit) model (7 to about 20 md) for vertical versus horizontal permeability. In both Type-A and Type-B models, horizontal permeability is approximately isotropic ($k_x \cong k_y \cong$ 19–27 md). Apparent permeability tensors computed for model sub-domains (Blocks 1 and 2) show that bulk k values for the sub-domains differ little from those of the corresponding full domain (Figure 14).

Superimposing low-permeability, sub-horizontal shale units (0.01 md) on the Type-A (all units) background permeability model causes a significant, two order of magnitude, reduction in k_z (from about 8–0.1 md) and small reductions in both k_x and k_y (Figure 14). The presence of a sandy body pinchout in the Type-A model volume causes k_x to be reduced somewhat more than k_y (Figure 14). The discontinuous shale units superimposed on the Type-B model yields a value of $k_z = 7$ md; only slightly smaller than that of the original background model ($k_z = 9$ md). Values of k_x and k_y are little affected by the superposition of shale units in the Type-B model. Apparent permeability tensors computed for model sub-domains (Blocks 1 and 2) show that bulk k values in each coordinate direction are similar for the corresponding full domain and partial sub-domains.

The apparent Type-A (all units) permeability tensor values are similar to the bulk k values computed for the Block 1 and Block 2 sub-model domains (Table 3). For example, the Type-A (all units) background permeability model has k_z values (about 10 md) that lie between the

Type-A (All Units)   Type-A (Unit-by-Unit)   Type-B (Unit-by-Unit)

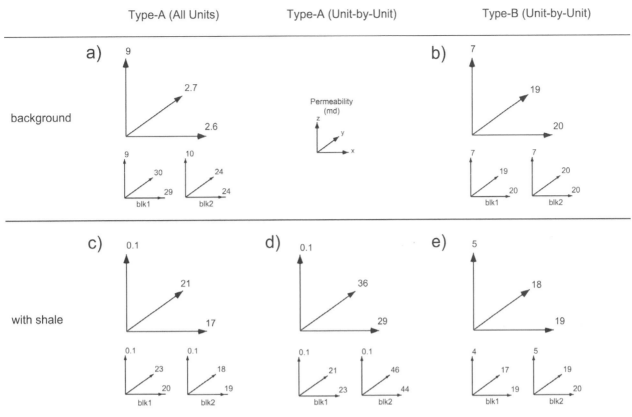

Figure 14. Apparent permeability tensors (k in md) computed for each model type using single-phase flow simulations with flow in the x- y-and z-coordinate directions: (a) Type-A (all units) background k without shale, (b) Type-B (unit-by-unit) background k without shale, (c) Type-A (all units) with shale, (d) Type-A (unit-by-unit) with shale, and (e) Type-B (unit-by-unit) with shale.

Table 3. Computed bulk mean permeability values (md) for Type-A and Type-B, Block 1 and Block 2 models.

	Type-A (All Units) Background		Type-A (All Units) With Shale	
	Block 1	Block 2	Block 1	Block 2
Arithmetic Mean	40	30	33	27
Geometric Mean	17	17	7	7
Harmonic Mean	4.8	5.9	0.09	0.09

			Type-A (Unit-by-Unit) With Shale	
			Block 1	Block 2
Arithmetic Mean			30	61
Geometric Mean			11	21
Harmonic Mean			0.08	0.09

	Type-B (Unit-by-Unit) Background		Type-B (Unit-by-Unit) With Shale	
	Block 1	Block 2	Block 1	Block 2
Arithmetic Mean	26	25	25	24
Geometric Mean	12	13	8	10
Harmonic Mean	1.5	3.6	0.17	0.21

computed geometric and harmonic means (Table 3) while k_x and k_y values (24–30 md) lie between the arithmetic and geometric means (Table 3). When continuous sub-horizontal shale units are superimposed, k_z is approximated by the harmonic mean while k_x and k_y values remain between the arithmetic and geometric

means. Similar results are obtained for the Type-A (unit-by-unit) and Type-B (unit-by-unit) models (Table 3).

The apparent k tensor for the Type-A (unit-by-unit), layered model of Figure 14d differs from that of the Type-A (all units) model with shale (Figure 14c) because the higher k values of the uppermost *Unit 5* (Plate 1) are better preserved. This leads to greater vertical and horizontal bulk anisotropy in the Type-A (unit-by-unit) model. In addition, greater differences are found between apparent k tensors computed for Block 1 and Block 2 in the Type-A (unit-by-unit) model (Figure 14d).

Two-Phase Simulations for Detailed, Heterogeneous Permeability Structures

Plots of cumulative oil and water production in Figure 15 illustrate the similarity in waterflood progress computed within the full domain for both Type-A (all units) and Type-B (unit-by-unit) permeability models. Figure 15 provides cumulative plots of the percentage of original oil in place (OOIP) produced by injecting a specified number of pore volumes of water (PV). The OOIP is calculated by multiplying the uniform porosity of 20% times the initial oil saturation of 80% times the total volume of the model domain. One pore volume of injected water is similarly calculated by multiplying the uniform porosity of 20% times the total volume of the model domain. Water production is computed as the cumulative total stock tank barrels of water produced during the waterflood simulation. Cumulative oil production rises rapidly to 40% of OOIP at 0.4 PV then more slowly approaches an ultimate sweep of about 50% to 60% of OOIP (Figure 15a). Conversely, water production rises slowly in early stages of the waterflood (at PV less than 0.4) before rapidly increasing in later stages of the waterflood (Figure 15b).

The permeability distributions assigned in the Full Domain models (both background k distributions and with shale units superimposed) are shown in Plates 2A and 2B. Despite small differences in the details of each distribution, the computed progress of the waterflood through the Type-A and Type-B full domain models is almost exactly the same (Figures 16a and 16b).

The bulk of our simulations are performed on the Block 1 sub-domain to reduce computational cost and time. Figures 16c through 16f show simulation results for Block 1 and 2 sub-domains when two permeability models were simulated in the x-coordinate flow direction; one without shale units (background k only) and one with shale units superimposed. It is readily apparent that adding the shale units causes a significant, but small, reduction in ultimate oil production (from 2–4% at PV about 0.7 in Table 4); even though the waterflood is migrating parallel to the sub-horizontal shale units. This result reflects the fact that oil is retained in and near the shale units where 20% porosity and 0.01 md permeability are assigned.

Comparing results obtained for Blocks 1 and 2 (Figures 15c through 15f) also show small, but significant, differences in the ability of the waterflood to sweep oil from the model volume. Superimposing the shale units

causes about 2% less production at PV approximately 0.7. Because differences between the Block 1 and 2 results are small, it seems reasonable to perform the remaining simulations using the Block 1 model subdomains of Figure 13, rather than the Full Domain models (Figure 13).

Plate 3 provides a 2-D view of a vertical section cut along the x-coordinate direction of oil saturation at three times (30, 90, and 180 days corresponding to PV of 0.16, 0.47, and 0.94) during the simulated x-direction waterflood through Block 1 of both the Type-A (all units) and Type-B (unit-by-unit) models. The corresponding permeability models (with shale units superimposed) are also shown in Plate 3. Apparent differences in the distribution of unswept oil found in the Type-A and Type-B models reflect the fundamentally different approaches used to define the distribution and geometry of shale units. Yet, there is little difference in computed oil and water production (Figure 15). The plots of cumulative oil production shown in Figure 17 indicate that the Type-A unit-by-unit permeability model is more readily swept than the Type-A (all units) and Type-B (unit-by-unit) models.

Two-Phase Simulations for Upscaled Permeability Structures

Four levels of upscaling are performed on the Type-A (all units) and Type-B (unit-by-unit) permeability models. Figure 11 shows how two (2X), three (3X), and six (6X) vertically connected blocks are used to produce equivalent, isotropic bulk permeability values for the group of blocks that are the arithmetic, geometric, and harmonic mean permeability values for each group. Note that the original grid-block size is preserved to facilitate the comparison of numerical results. The average, isotropic k values of the equivalent homogeneous Blocks 1 (labeled *Bulk* in Figure 11 and Table 4) are obtained by computing the three means using all the grid-block permeability values assigned in the detailed heterogeneous case. Figure 16 shows the permeability distributions plotted for the detailed, heterogeneous Block 1 Type-A (all units) and Type-B (unit-by-unit) models as compared to those plotted for the 2X, 3X, and 6X levels of upscaling. As the level of upscaling increases for both permeability models a greater number of low permeability grid blocks are included in the models. This suggests a degradation in the ability of the upscaled permeability structure to preserve the influence of the detailed heterogeneous structure on waterflood progress. Conversely, upscaling in the Type-B models has relatively little impact on the k distribution (Figure 16).

Figure 18 provides a summary of the effects of upscaling on computed oil production for x-direction flow. In each case, the simple bulk mean permeability

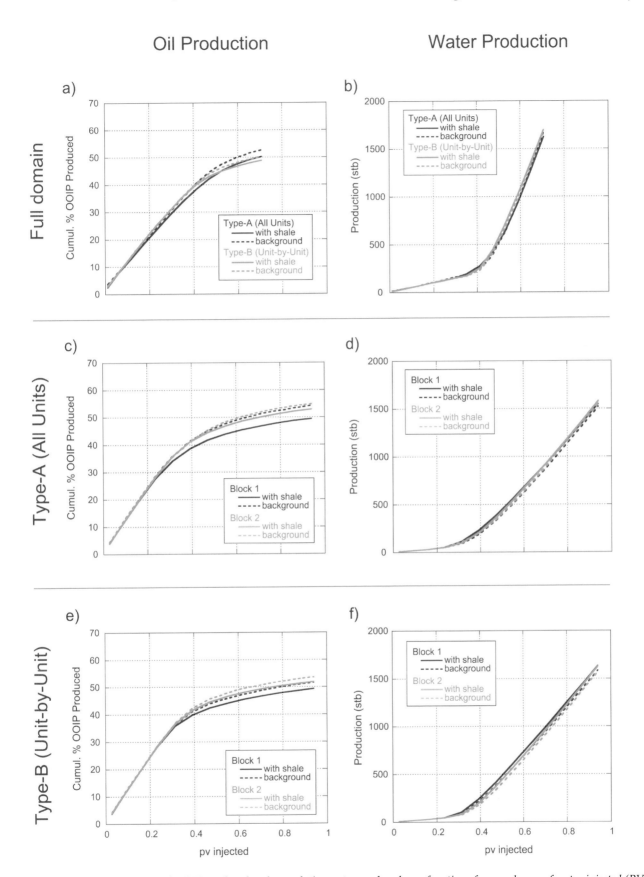

Figure 15. Plots of cumulative % of OOIP produced and cumulative water produced as a function of pore volumes of water injected (PV): (a) and (b) Full Domain model with Type-A and Type-B (all units) — oil production and water production, respectively; (c) and (e) Block 1 and Block 2 models for Type-A and Type-B (all units) — oil production; (d) and (f) Block 1 and Block 2 models for Type-A and Type-B (all units) — water production.

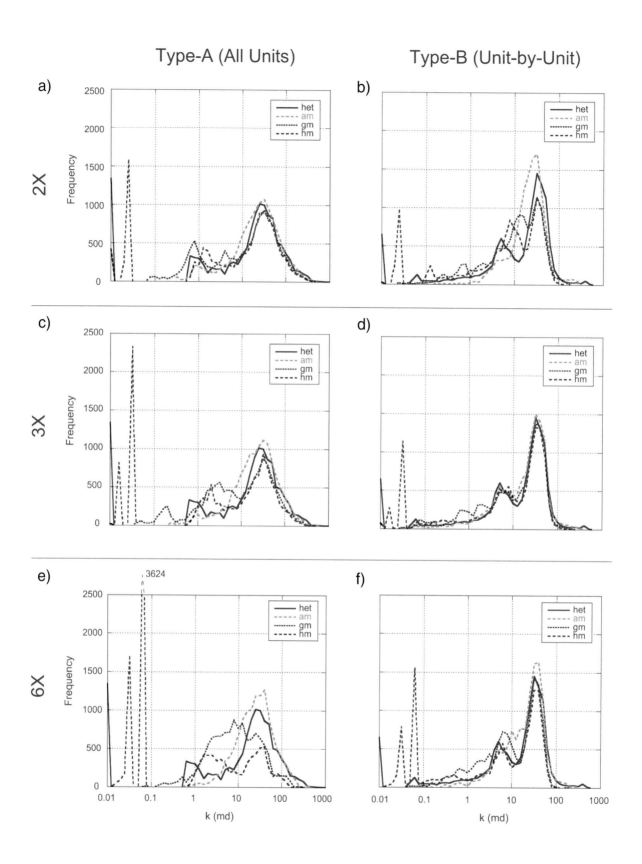

Figure 16. Permeability distributions assigned for each vertically computed mean (arithmetic, geometric, and harmonic) for each level of upscaling (2X, 3X, and 6X) and for the detailed heterogeneous case (het) for Block 1 Type-A (all units) and Type-B (unit-by-unit) models.

Table 4. Cumulative % of OOIP produced at PV about equal to 0.7 for Block 1 models.

	Type-A (All Units)			Type-B (Unit-by-Unit)		
Background	51			49		
With Shale	47			47		
Upscale With Shale	AM	GM	HM	AM	GM	HM
2X	49	49	45	54	52	47
3X	52	50	43	50	47	43
6X	53	51	39	52	47	40
Bulk	55	57	57	55	57	57

computed using the arithmetic, geometric, or harmonic means fails to provide a reasonable approximation of the flow properties of the two permeability structures. All three approaches used to compute a single bulk k value for the model provide similarly poor approximations to the computed oil production (Figure 18).

As the level of upscaling with arithmetic or harmonic mean is increased from 2X to 3X to 6X, computed oil production is less well approximated (Figure 18 and Table 4). The arithmetic mean method produces increasing overestimates (up to 112%) of production when compared to the detailed heterogeneous case; while the harmonic mean method produces increasing underestimates (down to 83%) of production (Figure 18 and Table 4). With the exception of the 2X upscaling for the Type-B model, the geometric mean yields the best approximation of oil production. For the Type-A model, however, the geometric mean consistently overestimates oil production computed using the detailed heterogeneous model. On the other hand, the geometric mean applied at the 3X and 6X levels provide oil production results that essentially match those of the heterogeneous model. Thus, the effect of discontinuous shale distributions, as estimated on the basis of the Type-B surveys, are better approximated in this simple upscaling approach than those assumed to be laterally continuous throughout the model domain.

DISCUSSION

Simulation results presented in the preceding sections illustrate the potential value in using 3-D GPR surveys of outcrop analogs to improve our understanding of the detailed distribution and thickness of shale units contained in sedimentary reservoir rocks. In particular, the ability of 3-D GPR surveys to help assess the degree of continuity of thin, sub-horizontal shale horizons can provide important clues regarding the degree of upscaling that might be appropriate when attempting to compute average grid-block permeability values for input to numerical reservoir simulators. For example, if the thin shale units found at Coyote Basin are laterally continuous (Type-A models) then it seems inappropriate to be using a 3X or greater upscaling approach. Thus, relative-

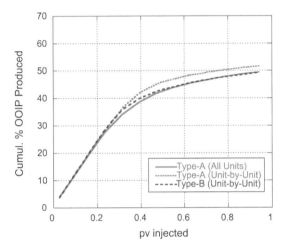

Figure 17. Plots of cumulative % of OOIP produced as a function of pore volumes of water injected (PV) of Block 1 with Type-A (all units) and Type-B (unit-by-unit).

ly thin grid blocks may be required (less than 0.5 m [1.5 ft]). On the other hand, if the shale units are as discontinuous as is inferred from the GPR data and embodied in the Type-B permeability model, even a 6X or greater upscaling approach may be appropriate and enable us to use grid blocks with thickness greater than 1 m (3 ft).

The results presented to this point involve only the sub-horizontal shale units we studied at Coyote Basin. Extensions of this work at other Ferron Sandstone outcrops at nearby Corbula Gulch (Li and White, 2003), however, provide an opportunity to evaluate the benefits of using the 3-D GPR method to assess the characteristics of dipping shale units in a consolidated point-bar deposit (Corbeanu, 2001; McMechan et al., 2001). Although we have yet to incorporate the results of the Corbula Gulch GPR surveys in our flow simulations, we suspect that the GPR method should be of even greater value in helping to assess how best to upscale detailed permeability structures in such rocks. For example, if the drape model is divided into two halves, each would provide similar, better oil sweep results because none of the dipping shale units completely cross the model. As a consequence, improved GPR-derived understanding of the degree of both shale continuity and shale dip are expected to provide valuable insight regarding which

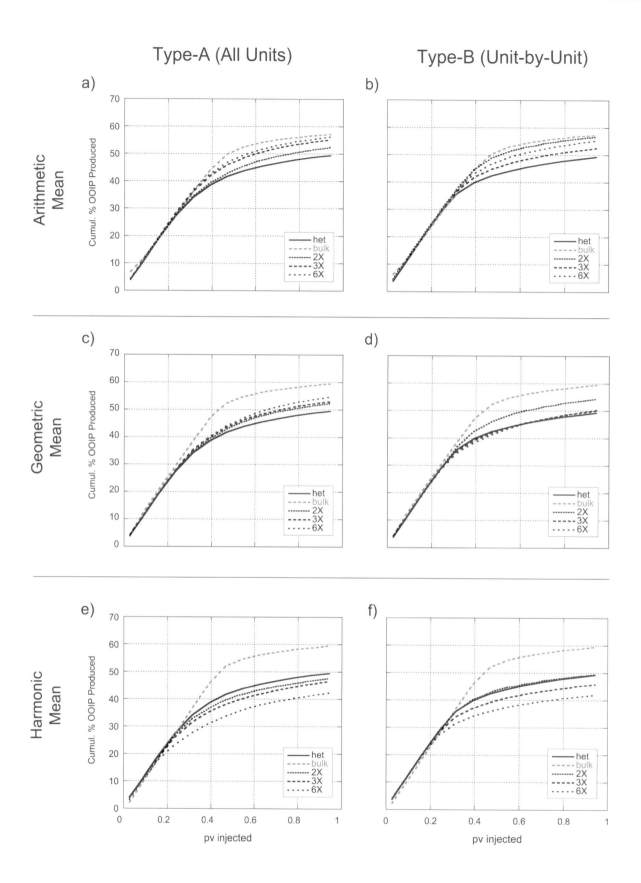

Figure 18. Plots of cumulative % of OOIP produced as a function of pore volumes of water injected (PV) into the Block 1 models Type-A (all units) and Type-B (unit-by-unit): (a) and (b) arithmetic mean k structure; (c) and (d) geometric mean k structure; (e) and (f) harmonic mean k structure.

approaches should be used to define upscaled permeability structures in similar reservoir rocks. Li and White (2003) report the results of detailed, 3-D, two-phase waterflood simulations for shale drapes found in marine-influenced distributary channel deposits of the Ferron Sandstone. Their results indicate that the shale features have the greatest impact on vertical components of flow by decreasing breakthrough time, sweep efficiency, and vertical permeability.

CONCLUSIONS

Comparing permeability values obtained from outcrop and values from continuous core collected behind the outcrop shows that weathering of the lithic arkose sandstone at Coyote Basin has caused a factor of 2 to 5 increase in cliff-face permeability with a significantly different univariate k distribution from that of the rock drilled at least 10 m (30 ft) behind the outcrop. This permeability difference is noted, to varying degrees, for each of the five sandstone facies studied at Coyote Basin. We account for these permeability differences in our permeability models by "correcting" the cliff-face values using an inverse transformation method.

Two approaches are employed for developing permeability models using 3-D GPR data integrated with outcrop and drill-hole core studies. One approach uses GPR-derived bounding surfaces as a foundation for defining the distribution, thickness, and geometry of continuous, low-permeability shale units. The other approach involves explicitly constraining models of the shale distribution/thickness and sandstone permeability structure using the GPR data. The principal difference between these approaches is that geological insight alone is used in the first approach to define the thickness and continuity of the shale units while GPR-derived insight is used in the second.

Single-phase flow simulations provide insight regarding the bulk permeability anisotropy of each permeability model. In the absence of mudstone or shale, the stratified character of the background permeability models produces an inherent vertical anisotropy of approximately 1:3 (vertical:horizontal k) in the Type-A and Type-B models. In both Type-A and Type-B models, horizontal permeability is approximately isotropic. Superimposing continuous shale units ($k = 0.01$ md) on the Type-A model produces a much greater vertical anisotropy of about 1:200. Superimposing shale units ($k = 0.01$) on the Type-B model causes a small increase in vertical anisotropy (from 1:3 to 1:4) because the discontinuous shale units act as partial barriers to vertical fluid flow.

Two-phase fluid-flow simulations indicate that it is important to be able to distinguish between fully continuous versus discontinuous shale unit geometries when selecting methods for upscaling outcrop permeability values for use in reservoir simulators. Simula-

tions performed using a hypothetical permeability model with dipping shale units suggest that it is even more important to be able to assess the angle of dip of thin shale drapes when selecting an upscaling method. Integrating 3-D GPR surveys with outcrop-based sedimentological and petrophysical data can provide the estimates of size, dip, continuity, and distribution of shale units needed at the scale of individual simulator grid blocks to help improve our approaches for deriving the upscaled k values used as input to production-scale reservoir simulators.

The detailed permeability structures derived from integrated data collected at the Coyote Basin outcrop analog provide an unusual opportunity to visualize the distribution of unswept oil in a fine-scale, geologically plausible permeability structure associated with a fluvial-channel complex. The permeability models, and corresponding fluid-flow simulation results, form an important basis for assessing different methods that might be used to mimic the impact of permeability structures found in similar reservoir rocks when constructing full-scale reservoir simulators.

Applying simple upscaling approaches (vertical averaging with arithmetic, geometric, and harmonic means) at different levels of upscaling suggests that the relatively simple permeability structure found in the Coyote Basin study volume (10.6 m [34.8 ft] thick and 39 m x 15 m [117 ft x 45 ft] in plan) should not be approximated by a single permeability value. For flow configurations with continuous, sub-horizontal shale units it may be necessary to maintain a grid-block thickness less than 0.5 m (1.6 ft). Grid blocks used in this study are 1 m x 1 m (3 ft x 3 ft) in plan. Additional work would help to evaluate if larger x- and y-coordinate dimensions might still provide reasonable representation of the detailed permeability structure in upscaled models. Although we have insufficient information to suggest what grid-block sizes might be appropriate when upscaling systems with dipping shale units, it is clear that we will require different approaches than the simple averaging methods outlined here.

Upscaling based on computing vertically averaged geometric means (compared to arithmetic and harmonic means) provides the best approximation of the impact of the detailed permeability structures on progress of a simulated waterflood through the fluvial-channel complex studies at Coyote Basin.

ACKNOWLEDGMENTS

The research leading to this paper was funded primarily by the U.S. Department of Energy under Contract DE-FG03-96ER14596 to the University of Texas at Dallas through a sub-contract between the University of Utah and the University of Texas at Dallas. Additional financial support was provided by the University of Texas at

Dallas GPR Consortium and a consortium of industry sponsors working through the University of Utah. The computer-controlled, stage-mounted, electronic probe permeameter was generously donated to the University of Utah by the Mobil Exploration and Producing Technical Center in 1996. We wish to thank Ari Menitove and John Thurmond who helped us collect field data. This paper is contribution No. 1010 from the Department of Geosciences at the University of Texas at Dallas. We thank Anthony Gorody for his careful review of the manuscript.

REFERENCES CITED

Bain, L. J., and M. Englehardt, 1992, Introduction to probability and mathematical statistics: Boston, PWS-Kent, 644 p.

Baker, P. L., and U. Monash, 1991, Fluid, lithology, geometry, and permeability information from ground-penetrating radar for some petroleum industry applications: Society of Petroleum Engineers, paper SPE 22976, p. 277–286.

Barton, M. D., 1994, Outcrop characterization of architecture and permeability structure in fluvial-deltaic sandstones, Cretaceous Ferron Sandstone, Utah: Ph.D. dissertation, University of Texas at Austin, Austin, 255 p.

Begg, S. H., A. Kay, E. R. Gustason, and P. F. Angert, 1994, Characterization of complex fluvial deltaic reservoir for simulation: Society of Petroleum Engineers, paper 28398, p. 375–364.

Corbeanu, R. M., 2001, Detailed internal architecture of ancient distributory channel reservoirs using ground-penetrating radar, outcrop, and borehole data — case study of Cretaceous Ferron Sandstone, Utah: Ph.D. dissertation, University of Texas at Dallas, Dallas, 122 p.

Corbeanu, R. M., K. Soegaard, R. B. Szerbiak, J. B. Thurmond, G. A. McMechan, D. Wang, S. H. Snelgrove, C. B. Forster, and A. Menitove, 2001, Detailed internal architecture of a fluvial channel sandstone determined from outcrop, cores and 3-D ground-penetrating radar — example from the Mid-Cretaceous Ferron Sandstone, east-central Utah: AAPG Bulletin, v. 85, no. 9, p. 1583–1608.

Corbeanu, R. M., G. A. McMechan, R. B. Szerbiak, and K. Soegaard, 2002, Permeability and mudstone prediction from GPR attribute analysis — example from the Cretaceous Ferron Sandstone Member, east-central Utah: Geophysics, v. 67, p. 1495–1504.

Davis, J. L., and A. P. Annan, 1989, Ground penetrating radar for high resolution mapping of soil and rock stratigraphy: Geophysical Prospecting, v. 37, p. 531–551.

Deutsch, C. V., and A. G. Journel, 1992, GSLIB — Geostatistical software library and user's guide: New York, Oxford University Press, 340 p.

Doyle, J. D., and M. L. Sweet, 1995, Three-dimensional distribution of lithofacies, bounding surfaces, porosity, and permeability in a fluvial sandstone — Gypsy Sandstone of Northern Oklahoma: AAPG Bulletin, v. 79, p. 70–96.

Dutton, S. P., and B. J. Willis, 1998, Comparison of outcrop and subsurface sandstone permeability distribution, lower Cretaceous Fall River Formation, South Dakota and Wyoming: Journal of Sedimentary Research, v. 68, no. 5, p. 890–900.

Eschard, R., and B. Doligez (eds.), 1993, Subsurface reservoir characterization from outcrop observations: Proceedings 7th IFP Exploration and Production Research Conference, Scarborough, April 12–17, 1992, Editions Technip, 198 p.

Eschard, R., P. Lemouzy, C. Bacchiana, G. De'saubliaux, J. Parpant, and B. Smart, 1998, Combining sequence stratigraphy, geostatistical simulations, and production data for modeling a fluvial reservoir in the Chaunoy field (Triassic, France): AAPG Bulletin, v. 82, p. 545–568.

Fanchi, J. R., Harpole, K. J., and S. W. Bujnowski, 1982: BOAST — a three-dimensional, three-phase black oil applied simulation tool (version 1.1), technical description and FORTRAN code: U.S. Department of Energy, DOE/BC/10033-3 (v. 1).

Fisher, R. S., M. D. Barton, and N. Tyler, 1993a, Quantifying reservoir heterogeneity through outcrop characterization — architecture, lithology, and permeability distribution of a landward-stepping fluvial-deltaic sequence, Ferron Sandstone (Cretaceous), central Utah: Gas Research Institute Report 93-0022, 132 p.

—1993b, Quantifying reservoir heterogeneity through outcrop characterization — architecture, lithology, and permeability distribution of a seaward-stepping fluvial-deltaic sequence, Ferron Sandstone (Cretaceous), central Utah: Gas Research Institute Report 93-0023, 83 p.

Gardner, M. H., 1995, Tectonic and eustatic controls on the stratal architecture of mid-Cretaceous stratigraphic sequences, central western interior foreland basin of North America, in S. L. Dorobek and B. M. Ross, eds., Stratigraphic evolution of foreland basins: Society for Sedimentary Geology (SEPM) Special Publication 52, p. 243–281.

Garrison, J.R., Jr., T. C. V. van den Bergh, C. E. Barker, and D. E. Tabet, 1997, Depositional sequence stratigraphy and architecture of the Cretaceous Ferron Sandstone: implications for coal and coalbed methane resources — a field excursion, *in* P. K. Link and B. J. Kowallis, eds., Mesozoic to Recent geology of Utah: Provo, Brigham Young University Geology Studies, v. 42, pt. 2, p. 155–202.

Gawthorpe, R. L., R. E. L. Collier, J. Alexander, J. S. Bridge, and M. R. Leeder, 1993, Ground penetrating radar — application to sandbody geometry and heterogeneity studies, *in* C. P. North and D. J. Prosser, eds., Characterization of fluvial and aeolian reservoirs: Geological Society Special Publication No. 73, p. 421–432.

Goovaerts, P., 1997, Geostatistics for natural resource evaluation: Oxford, Oxford University Press, 483 p.

Heemstra, R., 1998, Boast98 and EdBoast — user's guide and documentation manual: Tulsa, BDM-Federal, Inc.

Hurst, A., and D. Goggin, 1995, Probe permeametry — an overview and bibliography: AAPG Bulletin, v. 79, p 463–473.

Li, Hongmei, and C. D. White, 2003, Geostatistical models for shales in distributary channel point bars (Ferron Sandstone, Utah) — from ground-penetrating radar to three-dimensional flow modeling: AAPG Bulletin, v. 87, no. 12. p. 1851–1868.

Liu, K., P. Boult, S. Painter, and L. Paterson, 1996, Outcrop analog for sandy braided stream reservoirs — permeability patterns in the Triassic Hawkesbury Sandstone, Sydney Basin, Australia: AAPG Bulletin, v. 80, p. 1850–1866.

Louisiana State University, 1997, BOAST 3: User's Guide and Documentation Manual: U.S. Department of Energy, DOE/BC/14831-18.

McDonald, M. G., and A. W. Harbaugh, 1988, A modular three-dimensional finite-difference ground-water flow model: U.S. Geological Survey Techniques of Water Resources Investigations 06-A1.

McMechan, G. A., G. C. Gaynor, and R. B. Szerbiak, 1997, Use of ground-penetrating radar for 3-D sedimentological characterization of clastic reservoir analogs: Geophysics, v. 62, p. 786–796.

McMechan, G. A., M. C. Wizevich, C. Aiken, R. M. Corbeanu, X. Zeng, M. Balde, X. Xue, C. B. Forster and S. H. Snelgrove, 1998, Integrated 3-D ground-penetrating radar, outcrop, and borehole data applied to reservoir characterization and flow simulation: U.S. Department of Energy Annual Report, Contract DE-FG03-96ER14596.

McMechan, G. A., R. M. Corbeanu, X. Xu, X. Zeng, C. Aiken, J. Bhattacharya, S. Hammon, C. B. Forster and S. H. Snelgrove, 1999, Integrated 3-D ground-penetrating radar, outcrop, and borehole data applied to reservoir characterization and flow simulation: U.S. Department of Energy Annual Report, Contract DE-FG03-96ER14596.

McMechan, G. A., R. M. Corbeanu, C. B. Forster, K. Soegaard, X. Zeng, C. Aiken, R. B. Szerbiak, J. Bhattacharya, M. C. Wizevich, X. Xu, S. H. Snelgrove, K. Roche, S. J. Lim, D. Novakovic, C. D. White, L. Crossey, D. Wang, J. R. Thurmond, W. S. Hammon III, M. Balde and A. Menitove, 2001, Integrated 3-D ground-penetrating radar, outcrop, and borehole data applied to reservoir characterization and flow simulation: U.S. Department of Energy Final Report, Contract DE-FG03-96ER14596.

Meyer, R., and F. F. Krause, 2001, A comparison of plug-derived and probe-derived permeability in cross-bedded sandstones of the Virgelle Member, Alberta, Canada — the influence of flow directions on probe permeametry: AAPG Bulletin, v. 85, no. 3, p. 477–489.

Robinson, J. W., and P. J. McCabe, 1997, Sandstone-body and shale-body dimensions in a braided fluvial system — Salt Wash Sandstone Member (Morrison Formation), Garfield County, Utah: AAPG Bulletin, v. 81, p. 1267–1291.

Roche, K. N., 1999, A high-resolution petrographic study of the Cretaceous Ferron Sandstone, Coyote Basin, Utah — integrating petrology and petrophysics: M.S. thesis, University of New Mexico, Albuquerque, 110 p.

Schultz, P. S., R. Shutz, M. Hattori, and C. Corbett, 1994, Seismic-guided estimation of log properties, Part 1 — a data driven interpretation methodology: The Leading Edge, v. 13, p. 305–311.

Szerbiak, R. B., G. A. McMechan, R. M. Corbeanu, C. B. Forster, and S. H. Snelgrove, 2001, 3-D characterization of a clastic reservoir analog; from 3-D GPR data to a fluid permeability model: Geophysics, v. 66, p. 1026–1037.

Taner, M. T., F. Koehler, and R. E. Sheriff, 1979, Complex seismic trace analysis: Geophysics, v. 44, p. 1041–1063.

Williams coal mine in the Ferron Sandstone, Muddy Creek, 3 miles east of Emery, Utah, 1911.
Photograph courtesy of U. S. Geological Survey.

Analog for Fluvial-Deltaic Reservoir Modeling: Ferron Sandstone of Utah
AAPG Studies in Geology 50
T. C. Chidsey, Jr., R. D. Adams, and T. H. Morris, editors

Three-Dimensional Architecture of Ancient Lower Delta-Plain Point Bars Using Ground-Penetrating Radar, Cretaceous Ferron Sandstone, Utah

Rucsandra M. Corbeanu[1], Michael C. Wizevich[2], Janok P. Bhattacharya[3], Xiaoxiang Zeng[3], and George A. McMechan[3]

ABSTRACT

Accurate three-dimensional description of reservoir architecture using outcrop analogs is hampered by limited exposure of essentially two-dimensional outcrops. This study contains the first fully three-dimensional description of ancient marine-influenced point bar sandstones of lower delta-plain distributary channels and is based on the integration of detailed outcrop and drill-hole data, and two- and three-dimensional ground-penetrating radar data. The studied outcrops are in the Cretaceous Ferron Sandstone of east-central Utah.

Point bars deposited in marine-influenced, lower delta-plain channels show complex facies and geometries that resemble both fluvial point bars (upward-fining grain-size distribution and laterally stacked inclined bedsets), and tidally influenced point bars (extensive mud drapes on the inclined bedset surfaces and upstream migration of inclined bedsets).

The bankfull width and mean bankfull depth were estimated at 225–150 m (738–492 ft) and at 3.9–5.2 m (12.8–17.1 ft), respectively. The heterogeneities in these point-bar deposits include mudstone drapes on the upper bounding surfaces of the inclined bedsets, and mudstone intraclast conglomerates lying on basal erosional scours of inclined bedsets. The spatial distribution of these heterogeneities is determined by direct mapping in outcrop in conjunction with modeling ground-penetrating radar amplitudes by geostatistical techniques. Mudstone layers are generally 5 m (16 ft) in length in the direction parallel to flow with a small percentage of mudstone layers 15 m (49 ft) in length, and 10 m (33 ft) perpendicular to flow, downdip along the inclined beds. The detailed distribution of heterogeneities inside reservoirs potentially affects flow behaviour.

[1]Agip, The Hague, The Netherlands; [2]Department of Earth and Atmospheric Sciences, Cornell University, Ithaca, New York;
[3]Geosciences Department, University of Texas at Dallas, Richardson, Texas

INTRODUCTION

The three-dimensional (3-D) geometry of ancient point-bar deposits, either in alluvial or coastal channels is not known directly, but is usually inferred from two-dimensional (2-D) outcrops (Puigdefabregas and van Vliet, 1978; Flint and Bryant, 1993) combined with drill-hole data behind the outcrops (Van Wagoner et al., 1990; Falkner and Fielding, 1993). Rarely, two or more out-crops are oriented in orthogonal directions, yielding a pseudo 3-D view of the deposits. Three-dimensional geometry of point-bar deposits is better described in modern rivers deposits than in ancient deposits, but modern river deposits cannot be mapped below the water table, and lack complete documentation of the deposition over large time and space scales typical of outcrop examples. Studies of modern rivers also lack information about what is ultimately preserved in ancient deposits (Jackson, 1976; Bridge, 1985, 1993; Bridge et al., 1995). Theoretical models are used to pre-dict variation in 3-D geometry of point bars under dif-ferent conditions of channel migration (Bridge, 1984; Willis, 1989, 1993). However, these models are common-ly simplified and cannot account for all parameters involved in the development of a point bar (e.g. the vari-able flow parameters, styles of migration, sediment load) and do not consider marine-influenced channels. New methods are needed to test inferred or theoretical point-bar models with real 3-D examples.

In this study we use 3-D ground-penetrating radar (GPR), to describe the 3-D internal structure of an ancient point-bar sandstone. Two-dimensional GPR sur-veys have been used in both modern unconsolidated sediments (Gawthorpe et al., 1993; Alexander et al., 1994; Bridge et al., 1995, 1998; Bristow et al., 2000) and ancient consolidated sedimentary rocks (Baker and Monash, 1991; Stephens, 1994; Bristow, 1995) to investigate the internal structure of sedimentary deposits, but true 3-D GPR studies are few (Beres et al., 1995; McMechan et al., 1997; Hornung and Aigner, 2000; Corbeanu et al., 2001).

This study presents a detailed 3-D description of marine-influenced point-bar deposits of lower delta-plain distributary channels from the Ferron Sandstone Member of the Upper Cretaceous Mancos Shale in east-central Utah, by integrating detailed outcrop and drill-hole sedimentologic data with 100 MHz 2-D and 3-D GPR data. Published examples of point-bar deposits in deltaic distributary channels are all 2-D studies and refer especially to upper delta-plain channels (Elliott, 1976; Cherven, 1978; Hobday, 1978; Plint, 1983; Fielding, 1984; Hopkins, 1985).

Generally it is thought that lower delta-plain dis-tributary channels are rather stable, do not migrate, and do not form point-bar or meander-belt deposits (Cole-man and Prior, 1982, p. 155). Most fluvial point-bar models assume an upward overall decrease in mean

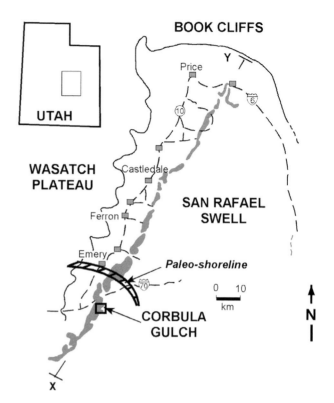

Figure 1. Location of the Corbula Gulch site in the Ferron Sandstone outcrop belt (the shaded area) along the southwestern flank of the San Rafael Swell in east-central Utah. X-Y represents the location of stratigraphic cross section in Figure 2. Paleo-shoreline during depo-sition of SC7 cycle is represented by hatched area.

grain size (from lag gravel and intraclast conglomerate, to sand and mud) and of the scale of the sedimentary structures, and a downstream decrease in grain size (Allen, 1964, 1970). In reality, fluvial point-bar deposits show more complexity than most idealized models (Jackson, 1976, 1978). Also, in relatively low-energy streams with significant suspended sediment load (Jack-son, 1981), or in streams with minor tidal influence, muddy layers may drape point-bar surfaces, especially in the upper part of a point bar (Smith, 1987; Jordan and Pryor, 1992). In this paper we show that point bars deposited in marine-influenced, lower delta-plain chan-nels show complex facies and geometries that share fea-tures of both fluvial- and tidal-point bars.

REGIONAL SETTING

Ferron Sandstone Stratigraphic Setting

The study area is located at Corbula Gulch along the western flank of the San Rafael Swell in east-central Utah. The outcrop is in the upper portion of the Ferron Sandstone known as the "Last Chance Delta" (Hale, 1972; Garrison et al., 1997) (Figure 1). The Ferron Sand-stone Member is one of several clastic wedges that pro-graded from the uplifted Sevier orogenic thrust belt

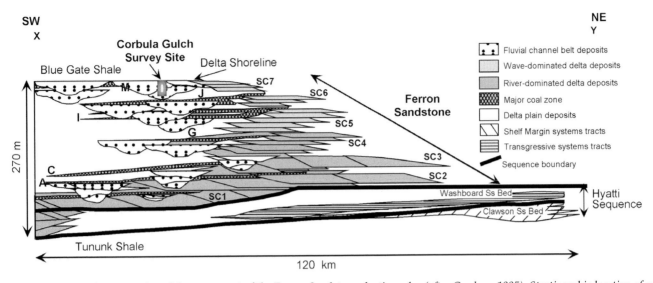

Figure 2. Generalized cross section of the upper part of the Ferron Sandstone clastic wedge (after Gardner, 1995). Stratigraphic location of survey site at Corbula Gulch and the paleo-shoreline are illustrated. See Figure 1 for location of cross section. Letters A to M identify major coal zones; SC1 to SC7 are short-term stratigraphic cycles.

northeast into the Mancos Sea during Turonian time (Hale, 1972; Cotter, 1975; Ryer, 1981; Gardner, 1992). The upper portion of the Ferron Sandstone is a deltaic complex that Ryer (1981, 1991) subdivided into seven deltaic cycles. Gardner (1992, 1995) defined the Ferron delta complex as a Type-2, third-order depositional sequence and also divided it into seven stratigraphic cycles, SC1 through SC7 (Figure 2). These cycles were also defined as parasequence sets by Ryer and Anderson (1995) and Garrison et al. (1997).

Gardner (1995) described facies tracts in successive short-term cycles, SC1 to SC3, as progradational, formed during an intermediate-term relative sea-level fall. Facies tracts of cycles SC4 and SC5 are aggradational and formed during a slow relative sea-level rise. Facies tracts of short-term stratigraphic cycles SC6 and SC7 are landward stepping and interpreted to be formed during a rapid relative sea-level rise. Each short-term cycle is capped by a major coal zone (Figure 2).

Corbula Gulch Stratigraphic Setting

The channel complex studied at Corbula Gulch is 12–15 m (39–49 ft) thick and consists of four erosively based, stacked channel bars within cycle SC7 (Figure 2). During cycle SC7 the coastline lay 24 km (15 mi) northeast of Corbula Gulch and became more embayed and influenced by marine conditions than the progradational cycles (SC1 to SC3) (cf. Ryer, 1991; Ryer and Anderson, 1995; Garrison et al., 1997) (Figure 1). The base of cycle SC7 rests on the J major coal zone that caps cycle SC6 (Garrison et al., 1997) and contains fine deposits composed of carbonaceous mudstones, fissile shales, dark-gray mudstones, coals, and small-scale cross-stratified, fine-grained sandstones with root traces (Figures 3 and 4A). In some locations, the uppermost coal bed contains

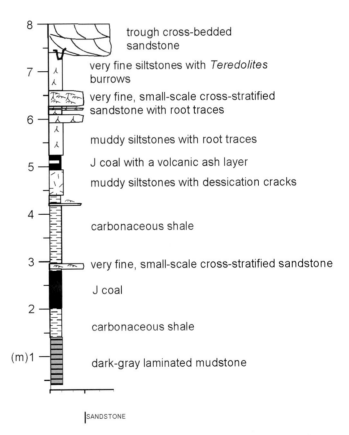

Figure 3. Generalized vertical measured section through the lower delta-plain deposits that underline the sandstone deposits. Vertical scale is in meters.

a 15-cm-thick (6-in.) volcanic ash layer (Figure 4A). The muddy siltstone deposits contain desiccation cracks, abundant root traces (Figure 4B), and *Teredolites* burrows (Figure 4C). This combination of elements indicates immature paleosols and suggests a humid depositional environment.

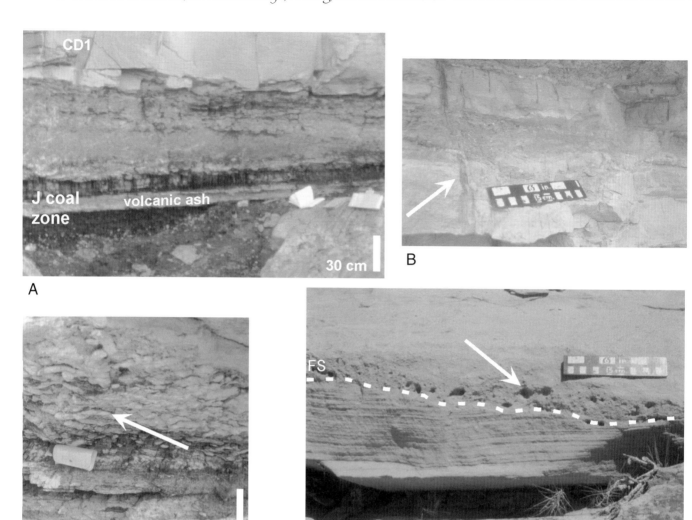

Figure 4. (A) Example of paleosols and coal zone with volcanic ash layer that underlie the channel deposits CD1 sandstone; (B) example of root-ed (arrow), very fine, ripple cross-laminated sandstone; (C) example of <u>Teredolites</u> burrows (arrow) within coalified logs; (D) transgressive lag deposits containing <u>Ophiomorpha</u> (arrow) that represents the flooding surface (dashed white line FS) that caps SC7. Color version on CD-ROM.

The mapped location of the paleo-shoreline at only 24 km (15 mi) north (Garrison et al., 1997) and the presence of marine, wood-boring trace fossils such as *Teredolites*, which are common in the underlying floodplain at Corbula Gulch, show that the delta plain experienced periodic marine inundation, followed by subaerial exposure. The channel deposits at Corbula Gulch are thus interpreted to be deposited on a marine-influenced lower delta plain.

The upper boundary of SC7 is a flooding surface (Garrison et al., 1997) represented locally by a transgressive-lag deposit containing *Ophiomorpha* burrows (Figure 4D). Most of this surface is absent from Corbula Gulch due to erosion.

METHODS AND DATABASE

The site at Corbula Gulch contains two cliff faces oriented approximately east-west and north-south (Figure

5). The cliff faces present horizontal and vertical exposures of about 1000 m (3300 ft) and 15 m (50 ft), respectively.

In order to reference the data in space to a unique datum, the study area was mapped in absolute coordinates using a combination of global positioning system (GPS) and a laser range-finder system (Xu, 2000). An accurate terrain model was generated (Xu, 2000).

The Ferron Sandstone is tilted a few degrees toward the northwest in the Corbula Gulch area. To represent the true depositional dip of the bedset surfaces we removed the regional dip of the Ferron outcrop. To compute this regional dip of the Ferron at the Corbula Gulch site, the J coal horizon was assumed to have been horizontal initially. The coal horizon was mapped at different outcrop locations with decimeter accuracy over a 1 km² (0.6 mi²) area around Corbula Gulch using a laser range finder, referenced to absolute GPS positions (Xu, 2000). A plane is fitted through the range-finder points

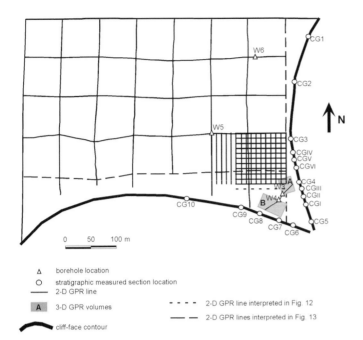

Figure 5. Map of the GPR survey site at Corbula Gulch showing the location of the 3-D GPR surveys and 2-D GPR lines on the flat mesa top related to the cliff-face outcrop positions. CG1 to CG10 are locations of measured stratigraphic sections at the cliff face. CGI to CGVI are location of short measured sections at the cliff face. W3 through W6 are locations of drill holes from which cores were extracted.

with known (x, y, z) coordinates yielding a strike of 038° and a dip of 2.5° northwest (Xu, 2000). The regional dip, computed at Corbula Gulch, is very similar in magnitude and orientation with the regional dip of the Ferron documented by previous workers (Ryer, 1981; Barton, 1994; Garrison et al., 1997).

Cliff-Face and Drill-Hole Data

The geologic data collected at Corbula Gulch include detailed facies maps both the east-west and north-south-oriented outcrops (Figures 5 and 6), ten stratigraphic sections (CG1 through CG10) evenly spaced along both outcrops (Figure 6), and four 16-m-long (55-ft), 6.3-cm (2.5 in.) cores obtained from drill holes behind the outcrops (W3 through W6). The bounding surfaces identified in the outcrop were also correlated with drill-hole cores. Paleoflow measurements from cross-strata are also recorded (Figure 6).

In order to describe inaccessible exposures we used a telescope with 60X magnification. Where possible we accessed the outcrop directly to document in detail the mudstone drapes and mudstone-intraclast conglomerate layers inside the channel deposits. The mudstone and mudstone-intraclast conglomerate layers are interpreted as the most important potential fluid-flow barriers within channel-sandstone reservoirs and were mapped in detail.

Permeability measurements were performed on core plugs extracted from the outcrop along each stratigraphic section at a sample spacing of 0.1 m (0.3 ft). A probe permeameter was used to test one end of each core plug. The permeability measurements on drill-hole cores were made at sample spacing of 0.05 m (0.16 ft) using a computer-controlled, stage-mounted, electronic probe permeameter.

Ground-Penetrating Radar Data

Ground-Penetrating Radar Overview

Ground-penetrating radar is a high-resolution geophysical technique that is based on recording the energy from an electromagnetic pulse that is reflected and diffracted at natural boundaries that have high contrast in electromagnetic properties (Davis and Annan, 1989). For ancient sedimentary rocks of mixed sandstone and mudstone lithology, the depth of penetration and vertical resolution can be on the order of 10–15 m (33–49 ft) and 0.25–0.5 m (0.82–1.6 ft), respectively, depending on the recording frequency and the electrical properties of the rocks (Szerbiak et al., 2001). The maximum depth of penetration depends on attenuation of the GPR signal. The attenuation decreases as the effective electrical resistivity increases and as the signal frequency decreases. The vertical resolution depends on the propagation velocity of electromagnetic waves and the signal bandwidth, which in turn depends mainly on the complex dielectric permitivities of the materials encountered and the dominant frequency in the source, respectively (Davis and Annan, 1989). There is always a trade off between maximum depths of penetration, obtained at low frequencies (e.g., 25 or 50 MHz) and the minimum dimension of features that are resolved at high frequencies (e.g., 200 MHz or more).

Flat, barren mesas, like those found at the top of the Ferron Sandstone outcrops, represent optimal environments for application of GPR technology. A shallow ground-water table will strongly attenuate the GPR signal in the saturated zone below the water table. In well-drained semi-arid environments, such as the site discussed in this paper, the water table is sufficiently deep that it is not the limiting factor for optimizing the depth of penetration.

Ground-Penetrating Radar Acquisition and Processing

The GPR survey covers three different orders of areal extent frequently used in flow-simulation studies, and consists of 2-D and 3-D GPR data sets (Figure 5). The dimension of one large grid unit in a reservoir model, equivalent to an inter-well distance, is covered by the large 2-D GPR grid in Figure 5. The dimension of a typ-

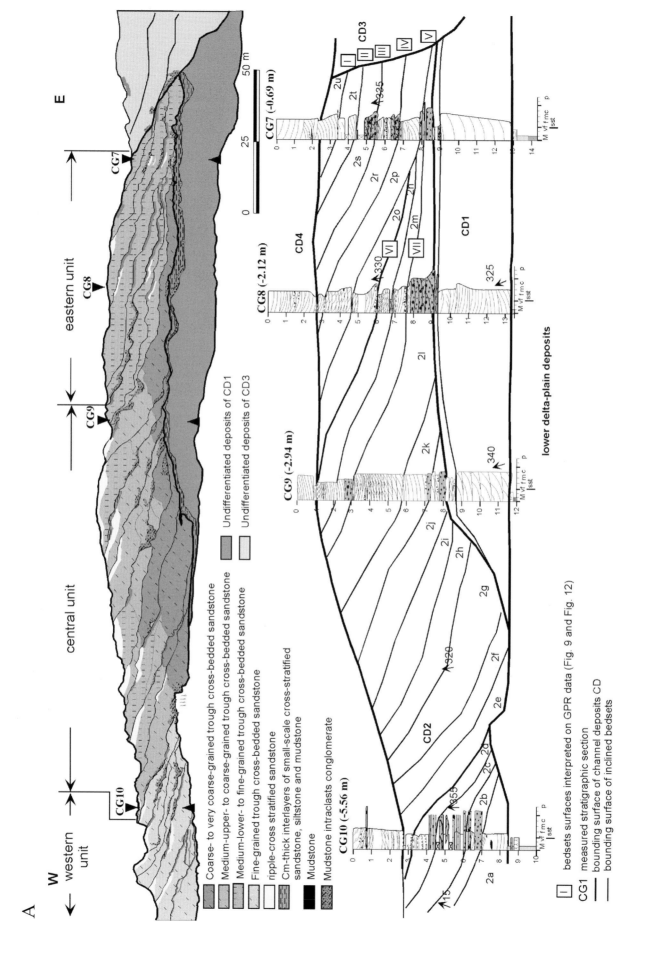

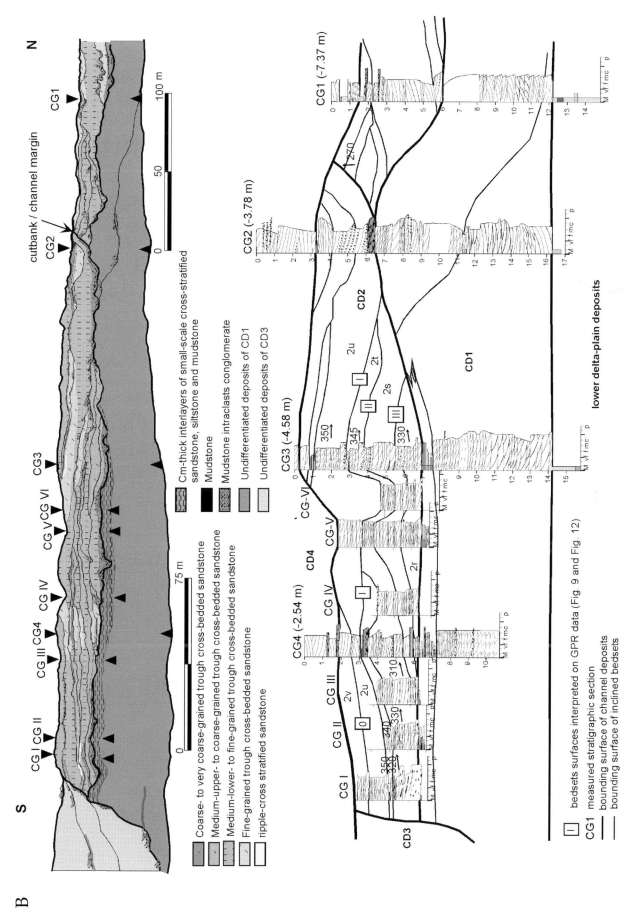

Figure 6. Sedimentary facies map (upper) of the cliff faces at Corbula Gulch. The position of the cliff faces at the (A) west-east cliff face, and (B) south-north cliff face relative to the GPR survey is shown in Figure 5. The measured stratigraphic sections (lower) at the cliff face are shown with details of sedimentary structures and grain size. The depth of each measured section from the reference datum of the GPR data is shown in brackets. Color version on CD-ROM.

ical voxel (volume pixel) in a reservoir flow simulator (Dreyer, 1993; Lowry and Jacobsen, 1993) is represented by the small 3-D GPR volumes in Figure 5. To link the two sets of GPR data, a third survey of 2-D GPR lines of intermediate scale was acquired.

We recorded two 3-D common-offset digital GPR data sets (50 m x 28 m [164 ft x 92 ft] and 31 m x 22 m [102 ft x 72 ft], respectively) with inter-line and inter-trace spacing of 0.5 m (1.6 ft). A survey consisting of 100 m x 100 m (328 ft x 328 ft) grid of 2-D common-offset GPR lines oriented north-south and east-west at 10 m (33 ft) spacing between lines, and 0.25 m (0.82 ft) spacing between traces on each line was also acquired (Figure 5). Finally, we recorded a 375 m x 550 m (1230 ft x 1806 ft) grid of 2-D common-offset GPR lines oriented north-south and east-west with 75 m (246 ft) spacing between lines, and 0.5 m (1.6 ft) spacing between traces on each line. A PulseEKKO IV GPR system with a transmitter voltage of 1000 V was used to collect all GPR data sets. The half-wave dipole antennas were oriented parallel to each other and perpendicular to the survey lines at an offset of 3 m (10 ft) and a central recording frequency of 100 MHz (resulting in a vertical resolution of about 0.5 m [1.6 ft]) in all surveys. Common-midpoint data were collected at 50, 100, and 200 MHz to determine optimal recording parameters and to estimate the propagation velocity. Drill-hole GPR data sets were recorded at four drill holes (Figure 5) to allow high-accuracy estimation of the vertical velocity distribution within the 3-D GPR volumes (Hammon et al., 2002).

All GPR data sets were pre-processed and the two 3-D GPR cubes and intermediate grid of 2-D lines were depth migrated before interpretation. Pre-processing included trace editing, direct wave removal (to reduce near-surface interference), time zero correction, band-pass filtering (to discriminate high-frequency events associated with subtle sedimentary features from high-amplitude energy near the median signal frequency), topographic corrections, and gain analysis (to compensate for the rapid attenuation of the signal). The pre-processing techniques used in GPR surveys are fully described by Szerbiak et al. (2001). Velocity analysis included estimation of one-dimensional vertical velocity profiles at wells and measured sections, and interpolation between these profiles to obtain a 3-D velocity model. Finally, a 3-D pre-stack Kirchhoff depth migration was applied to provide direct and accurate 3-D correlation of the geologic features and the GPR reflections. The vertical sampling of the depth migrated data is 0.1 m (0.3 ft). Due to signal attenuation in the mudstone layers in the first few meters below the surface, the 100 MHz GPR data had a maximum penetration depth of about 10 m (33 ft), so we are able to investigate with the GPR only the inclined heterolithic beds of the channel deposits.

General Ground-Penetrating Radar Interpretation

The similarities between GPR data and seismic data allow the procedures developed for interpreting 3-D seismic data (Brown, 1996) to be successfully utilized in 3-D GPR investigations (Beres et al., 1995; Hornung and Aigner, 2000; Corbeanu et al., 2001). Both 3-D GPR volumes at Corbula Gulch were examined repetitively on in-line and cross-line profiles and on horizontal slices to spatially map the inclined beds observed in outcrop. Unlike 2-D lines that may or may not be recorded parallel to the dip direction of the beds, a 3-D cube may be sliced arbitrarily according to different azimuths suggested by the sedimentologic interpretation. Also, detailed interpretation of the 2-D GPR intermediate-scale data set was possible by interactive correlation with the 3-D GPR interpretation using seismic interpretation software that permitted a direct link between different 2-D and 3-D data sets. Ground-penetrating radar interpretation is built on the principles used in seismic stratigraphic interpretation (Vail, 1987; Gawthorpe et al., 1993; Corbeanu et al., 2001; and many others). For example, a radar facies is similar to a corresponding seismic facies; both are units mapped in 3-D with distinct shape and composed of groups of reflections with specific configuration, continuity, amplitude, frequency, and interval velocity, recognizably different from adjacent radar/seismic facies (Baker and Monash, 1991; Gawthorpe et al., 1993). A radar sequence is used in this paper to describe a relatively conformable succession of GPR reflections that are discordant with surrounding reflections. The discordances match the erosional surfaces associated with different channel elements. Ground-penetrating-radar reflections can be generated at bedding planes, fracture planes, or any other boundary separating rock types with different electrical properties. Electrical properties of a rock correlate mainly to lithologic composition (sand/clay ratio, grain size, sorting, etc.) and water saturation, which is generally a measure of permeability and porosity of rocks (Knight and Nur, 1987; Annan et al., 1991; Rea and Knight, 1998). The main geologic surfaces that produce GPR reflections at Corbula Gulch are layers with high clay content as mudstone drapes and mudstone-intraclast conglomerate layers. We also made use of various GPR attributes such as instantaneous amplitude, frequency, and phase to recognizing complex interfaces of layers with small thickness (Tanner et al., 1979; Robertson and Nogami, 1984).

SEDIMENTARY FACIES

Four channel deposits (CD) were identified within the sandstone body and are referred to as channel deposits CD1 to CD4, in ascending stratigraphic order (Figures 7A and 7B). These channel deposits are composed of a set of more-or-less conformable, large-scale

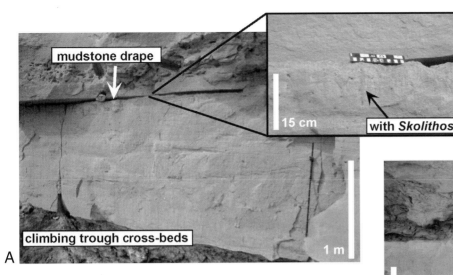

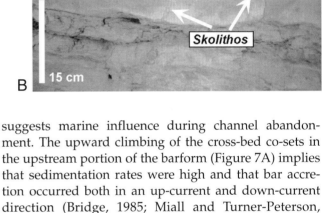

Figure 7. Outcrop examples of channel deposit CD1. (A) Channel bar composed of upstream climbing trough cross-beds (paleoflow is toward right) and draped by a thin mudstone layer with rare Skolithos burrows. (B) centimeter-thick alternation of fine-grained, ripple cross-stratified sandstone, siltstone, and mudstone with rare burrows. Color version on CD-ROM.

inclined bedsets and have erosional basal surfaces. Each channel deposit is characterized by complex internal features including lithofacies, bedding features, and erosional features, and distinctive external shapes. Channel deposit CD2 is characterized also by internal GPR features of specific radar facies, and GPR reflection termination at the bounding surfaces.

Channel Deposit 1

The lowermost channel deposit in the succession, CD1, varies in thickness from 0–7 m (0–23 ft) and has a basal erosion surface (Figure 6). Channel deposit CD1 is tabular to lenticular and is truncated by basal erosional surface of CD4 and CD3 (Figure 6). The paleoflow direction measured from trough cross-beds and basal scours is toward the north-northwest (azimuth 310° to 350°).

Channel deposit 1 consists mainly of trough cross-bedded sandstone and planar cross-bedded sandstone that in places form 1–2 m (3–6 ft) thick, coarsening- or fining-upward co-sets. Trough cross-beds in CD1 typically climb in an upstream direction (Figure 7A) and the co-sets are draped locally by centimeter-thick mudstone layers containing *Skolithos* burrows (Figure 7A). These mudstone drapes extend for 15–20 m (49–66 ft) in the paleoflow direction (Figure 7B). The upper part of CD1 is a 0.25–0.5-m (0.8–1.6-ft) thick alternation of thin, fine-grained, ripple cross-stratified sandstone and/or siltstone, and mudstone. In the fine sandy and silty layers are rare *Skolithos*, *Planolites*, and *Thalassinoides* burrows.

The mudstone-draped co-sets in the lower part of CD1 are interpreted as abandoned fluvial bars (Bridge, 1985). The presence of some brackish-water burrows

suggests marine influence during channel abandonment. The upward climbing of the cross-bed co-sets in the upstream portion of the barform (Figure 7A) implies that sedimentation rates were high and that bar accretion occurred both in an up-current and down-current direction (Bridge, 1985; Miall and Turner-Peterson, 1989). The uniform nature of trough cross-bedded sandstone of CD1 and the low variance of the paleocurrent data (Figure 6) suggest that CD1 was deposited in a low sinuosity distributary channel. The fine-grained deposits in the upper part represent the abandonment stage of the distributary channel (Figure 7B).

Channel Deposit 2

Sedimentologic Description

Channel deposit 2 consists of a series of low-angle, inclined bedsets dipping apparently toward the east (Figure 6A) that are similar to the inclined heterolithic stratification (IHS) and inclined stratification (IS) of Thomas et al. (1987). Channel deposit 2 has a generally erosive base with moderate local relief, and cuts into the underlying deposits of CD1. Channel deposit 2 cuts directly into underlying muddy delta-plain deposits in places where it removes CD1 (Figure 6A). Channel deposit 2 has a tabular shape and a mean thickness of about 7 m (23 ft), and is truncated toward the east and south by CD3 and on top by CD4 (Figure 6). Toward the north, CD2 cuts into another unit composed of inclined bedsets dipping south (Figure 6B).

Each inclined bedset cuts erosionally into the underlying inclined bedset and forms an accreting unit. Com-

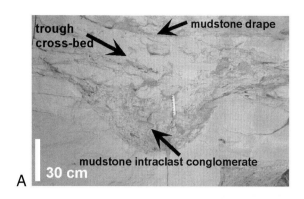

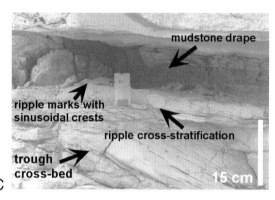

Figure 8. Facies and trace fossils characteristic of channel deposit CD2. (A) Mudstone intraclast conglomerate along the erosional basal bounding surfaces of an inclined bed; (B) Teredolites burrows common within the fossilized log fragments within the basal conglomerate; (C) trough cross-bedded sandstone, ripple cross-laminated, fine-grained sandstone/siltstone and mudstone facies succession typical of inclined beds within CD2; and (D) Skolithos and Arenicolites burrows in the upper parts of the inclined bedsets. Color version on CD-ROM.

monly, the basal bounding surface of the inclined beds contains mudstone intraclast conglomerates along scours (Figure 8A). The mudstone intraclast conglomerate beds can locally reach more than 1 m (3 ft) thickness within deep scours and can contain mudstone clasts and woody debris up to 0.5 m (1.6 ft) long. The coalified woody debris commonly contains extensive *Teredolites* burrows (Figure 8B). Laterally, mudstone intraclast conglomerate beds have small extent up to 4–5 m (13–16 ft) in outcrop. Inclined bedsets fine upward and generally contain trough cross-bedded to ripple cross-laminated sandstone and decimeter-thick alternation of siltstone/mudstone beds and mudstone drapes (Figure 8C). The siltstone/mudstone beds and mudstone drapes are 0.1–0.2 m (0.3–0.7 ft) thick, but extend downdip, parallel to the inclined surfaces of the beds, up to 7–8 m (23–26 ft) and occasionally more than 10 m (33 ft) in outcrop. Rare *Skolithos*, *Arenicolites*, (Figure 8D) *Planolites*, and *Thalassinoides* burrows occur in the upper, fine-grained parts of these inclined bedsets.

In the west-east outcrop we identified three different units of inclined bedsets characterized by specific grain-size, thickness of beds, and facies (Figure 6A).

The western-most unit of inclined bedsets (Figure 6A) is about 6 m (20 ft) thick and consists of bedsets dipping at 5° east. The thickness of each inclined bedset is

between 1 and 2 m (3 and 6 ft). Internally, these inclined beds consist mostly of low-angle to parallel-laminated and rarely trough cross-bedded sandstone, and ripple cross-laminated, fine-grained sandstone. In the most western part of this unit, the inclined bedsets are composed entirely of fine-grained, ripple cross-laminated sandstone. Each inclined bedset has an erosional basal bounding surface with mudstone intraclast conglomerate along scours and is capped by a decimeter-thick mudstone layer. In the eastern part of the unit, the mudstone intraclast conglomerate is more common along the erosional scours than in the western most part. Overall, within the bedset unit, there is a decrease in mudstone content laterally, in the direction of bed dip. Paleoflow direction measured from trough cross-beds is north (azimuth of 355° to 015°).

The central unit of inclined bedsets (Figure 6A) is about 7 m (56 ft) thick and consists of bedsets dipping more steeply at 8–10° eastward. The thickness of each inclined bedset is between 1.5 m and 3 m (4.9 ft and 10 ft), and the mean thickness of the trough cross-bed sets is about 0.14 m (0.46 ft) with an overall, upward thinning of the cross-bed sets (CG9 in Figure 6A). The inclined bedsets consist of very coarse- to medium-grained, trough cross-bedded sandstone to fine-grained, ripple cross-laminated sandstone. Locally, decimeter-thick mudstone drapes separate the inclined beds. The lower part of the unit of bedsets is mainly composed of coarse- to very coarse grained trough cross-bedded sandstone with mudstone intraclast conglomerates scattered along foresets. Locally, the upper part of the central unit of bedsets consists of fine-grained, ripple cross-laminated sandstone. Overall, within the unit of bedsets, there is a decrease in grain size upward and in the direction of bedset dip. The paleoflow measured from trough cross-beds is toward the northwest (azimuth 320°).

The eastern-most unit of inclined bedsets (between CG8 and CG7 in Figure 6A) is about 6 m (20 ft) thick and consists of bedsets dipping less steeply than the central bedset, at 5° toward the east. Thickness of inclined bedsets ranges from 1–1.5 m (3–4.9 ft), and the mean thickness of the trough cross-bed sets is 0.13 m (0.43 ft). Each inclined bedset consists of medium-grained, trough cross-bedded sandstone to very fine grained ripple, cross-laminated sandstone to mudstone. Within each inclined bedset there is a distinct upward-fining trend perpendicular to the inclined bounding surface of the bedset. The basal bounding surface of each bedset is erosional and is locally overlain by mudstone intraclast conglomerate. Mudstone drapes are parallel with the inclined bounding surface of the bedsets and is locally discontinuous due to erosion. Overall there is more mudstone in the central and eastern part of this unit of bedsets. Paleoflow measured from trough cross-beds is northwest (azimuth 330° to 335°).

Part of the eastern-most unit of inclined bedsets crops out in the north-south cliff face and the inclined beds are nearly horizontal, slightly convex upward surfaces. Each inclined bed consists of medium-grained, trough cross-bedded sandstone and in the southern part of the outcrop of coarse- to medium-grained, trough cross-bedded sandstone to fine-grained, ripple cross-laminated sandstone to mudstone. In places, especially toward the central and northern part of the outcrop, the inclined beds consist only of medium-grained to fine-grained sandstone. The mudstone drapes are discontinuous and present mostly in the southern and northern part of the outcrop. Paleoflow measured from trough cross-beds is generally northwest and varies vertically from azimuth 310°, 320°, and 330° in the lower part of the bedset to 345° and 350° in the upper part.

Ground-Penetrating Radar Interpretation

The GPR interpretation is focused on CD2. The bounding surface between CD2 and CD4 on the GPR record is not a continuous reflection, but a surface that envelops the truncations and toplap terminations of the oblique reflections below (Figures 9 and 10). The bounding surface between CD2 and underlying CD1 is a composite reflection produced by the downlapping terminations of the inclined reflections with long tangential toes.

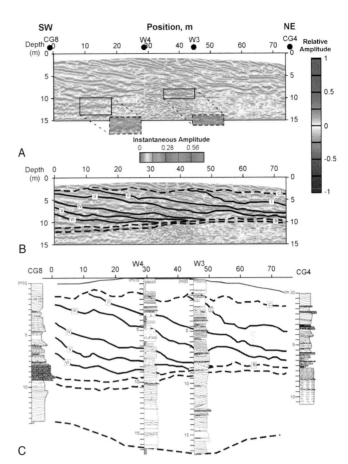

Figure 9. Ground-penetrating radar traverse line from CG8 to CG4 through both 3-D GPR volumes (Figure 5). The middle portion, between W4 and W3, is a 2-D GPR line that links the interpretation from the two data cubes. (A) Uninterpreted and (B) interpreted GPR reflections as inclined surfaces (0 to V) within CD2. Surfaces B and C are respectively the basal and top bounding surfaces of CD2. Notice the inclined, continuous, high-amplitude reflection that characterizes the main radar facies within CD2 (also in Table 1). Rectangular insets show examples of composite reflections resulting from constructive interference from thin layers. These composite reflections are resolved in the GPR instantaneous amplitude display. (C) Cross section showing the correlation between the 3-D GPR interpretation and the outcrop and drill-hole lithologic columns. CG8 and CG4 are projected from 6 m (20 ft) and 10 m (33 ft), respectively, onto the GPR profile. Color version on CD-ROM.

Mudstone drapes on the top surface of inclined bedsets produce good contrast in electrical properties and give rise to readily identifiable GPR reflections. The mudstone intraclast conglomerate beds at the bases of the inclined bedsets can also produce GPR reflections. The surfaces separating cross-stratification sets can produce GPR reflections if there is significant variation in grain size, but the electrical contrast is very small, and the reflections have low amplitudes and are discontinuous (Gawthorpe et al., 1993; Bridge et al., 1995; Corbeanu et al., 2001). In many cases these boundaries do not have the same orientation and dip as the inclined bedsets, and interfere constructively or destructively with the stronger reflections given by the inclined bedsets producing composite reflections.

Ground-penetrating radar interpretation of the three-dimensional data sets: Channel deposit 2 represents a radar sequence composed of high-amplitude, continuous, oblique GPR reflections dipping at an angle of 6–8° northeast (Figures 9A and 9B). Each inclined bedset was identified on outcrop (Figure 9C) and then interpreted throughout both 3-D GPR volumes. Where the inclined beds are thinner than the vertical resolution of 0.5 m (1.6 ft), composite reflections can cause interference of high positive or negative amplitudes due to tuning effects. These composite reflections were interpreted by analyzing other GPR attributes like instantaneous amplitude (Figure 9B).

A horizontal slice through both volumes at a depth of about 5 m (16 ft) shows the general structure of dipping beds toward the northeast, consistent with the sedimentologic interpretation (Figure 11). There is a slight change in the strike of the inclined reflections from Grid B to Grid A, from azimuth 340° to azimuth 320°, respectively. The radar facies interpreted to compose the radar sequence CD2 in the 3-D GPR volumes are presented and described in Table 1.

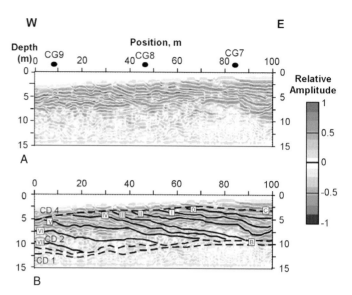

Figure 10. (A) Uninterpreted and (B) interpreted GPR line from the intermediate (100 m x 100 m [328 ft x 328 ft]) 2-D GPR grid (Figure 5). The interpretation shows a series of inclined surfaces dipping east and correlating with the 3-D GPR cubes (Figure 9) and outcrop (Figure 6). Recognized GPR facies are described in Table 1. CG7 to CG9 are the projections of the measured sections at the cliff face onto the GPR lines. Surfaces B, C, and 0 to VII are the same as in Figures 6 and 9. Color version on CD-ROM.

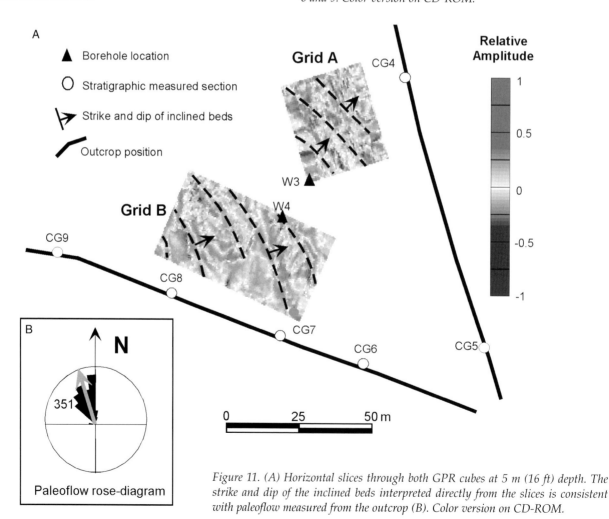

Figure 11. (A) Horizontal slices through both GPR cubes at 5 m (16 ft) depth. The strike and dip of the inclined beds interpreted directly from the slices is consistent with paleoflow measured from the outcrop (B). Color version on CD-ROM.

438

Table 1. Radar vs. sedimentologic facies of point-bar deposits. Color version on CD-ROM.

GPR Reflection Configuration	Radar Pattern	Sedimentologic description
High amplitude, oblique, parallel reflections		Lateral accreting surfaces of point bar Inclined Heterolithic Stratification (view perpendicular on depositional dip)
High amplitude, oblique, continuous, concave upward reflections		Lateral accreting surfaces of point bar Inclined Heterolithic Stratification (view perpendicular to depositional dip)
High and low amplitude, mounded reflections		Lenticular sandstone developed on the down-stream accreting part of the point bar (view parallel to depositional dip)
High and low amplitude, hummocky reflections		Accretionary topography on the point bar surfaces
		Mudstone intraclast conglomerate bodies accumulated at the base of scours
Low amplitude, chaotic reflections		Massive and/or cross-bedded sandstones without significant variation in grain size

Geostatistical analysis of three-dimensional ground-penetrating radar amplitudes: To quantify the lateral extent of mudstone and conglomerate layers inside CD2, assumed to mimic the continuity of the corresponding GPR reflections, experimental variograms are computed from the GPR relative amplitude data. Variogram modeling has been successfully used by Rea and Knight (1998) and Corbeanu et al. (2001) to characterize heterogeneities of the subsurface in 2-D and 3-D, computing the correlation lengths of radar reflections along maximum and minimum correlation directions. The assumption is that there exists a link between the lithology of layers and their electrical properties and thus a relationship between lithology and the correlation structure of GPR reflections. This spatial relationship can be expressed through standard variograms, a measure of the spatial autocorrelation of a regionalized variable (Rea and Knight, 1998; Corbeanu et al., 2001). The experimental variograms are computed from GPR relative amplitudes within the 3-D GPR facies of inclined reflections of CD2, using the equation:

$$\gamma_{(h)} = (1/2N(h))\Sigma(x_i - y_i)^2 \qquad (1)$$

where γ is the computed variance of GPR amplitudes, h is the separation distance between two data points (the lag), $N(h)$ is the number of pairs of data points separated by h, x_i is the data value at one of the points of the i-th pair, and y_i is the corresponding data value at the second point (Deutsch and Journel, 1998).

The maximum correlation direction of the 3-D GPR amplitudes was inferred from the dip and strike of the inclined beds measured in outcrop, which average 8° and 340°, respectively. The minimum correlation direction is perpendicular to the maximum correlation direction. The experimental variogram along the maximum correlation direction is modeled with a simple exponential structure with the parameters: sill = 1.13, nugget = 0.2, and range = 10 m (33 ft) (Figure 12A). Ground-penetrating radar reflections are a direct response of the existence of a contrast between the mud drapes (or mudstone intraclast conglomerate beds) and the surrounding sandstone, and so the correlation of the GPR amplitude correspond, generally, to the continuity of the mudstone or mudstone intraclast conglomerate layers. The range of 10 m (33 ft), derived from modeling the variogram of GPR amplitudes, is comparable with the lengths of the mudstone drapes and mudstone intraclast conglomerate layers measured in outcrop in the dip direction. The experimental variogram along the minimum correlation direction is modeled with a nested structure composed of an exponential model with sill = 0.95, nugget = 0.0, and range = 5.3 m (17.4 ft) combined with a gaussian model with sill = 0.15, nugget = 0.0, and range = 18 m (59 ft) (Figure 12B).

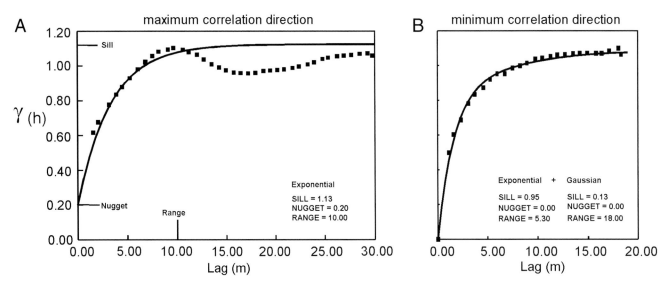

Figure 12. Experimental variograms (squares) and fitted models (continuous line) derived from the GPR amplitudes and shown along (A) maximum and (B) minimum correlation directions within CD2; the results of the variogram analysis are also presented. The nugget is the nonzero y-intercept of the experimental variogram. The sill is a constant at which the variance no longer increase and flattens. The range is the maximum distance at which the data are no longer correlated and the sill is obtained. For definitions of symbol used in the figures see equation 1 in the text.

Ground-penetrating radar interpretation of the two-dimension intermediate scale data set: The main radar sequence interpreted on the intermediate scale GPR grid corresponds also to CD2 in outcrop. The radar facies consist mainly of inclined, parallel reflections and inclined, concave upward reflections in the direction parallel to the dip of inclined beds (Figure 10). Other radar facies containing hummocky, distorted, chaotic, and less common, mounded reflection configurations were identified on the direction parallel to the paleoflow (Table 1). Each radar facies characterizes different parts of the inclined beds. The hummocky, distorted, and chaotic radar facies are characteristic of the basal downlapping end of the inclined beds. The geometry of the inclined surfaces (Figure 10) show beds dipping between 3° and 7° east, perpendicular to the direction of the flow.

Ground-penetrating radar interpretation of the two-dimensional large scale data set: Only large scale features (GPR sequences) were interpreted in the largest 2-D GPR grid (Figure 5). This large grid was not migrated in depth like the previous GPR data sets due to lack of velocity information and well ties in the northwestern part of the survey (Figure 5). A simple time-to-depth conversion was applied using an average velocity of 0.1 m/ns.

Four different radar sequences were identified in the largest GPR grid. The radar sequence, corresponding to CD2 in outcrop, is composed mainly of inclined, parallel to concave upward reflections identified in the eastern and southern part of the grid (Figure 13). To the north, the CD2 radar sequence pinches out against a radar sequence composed of inclined, parallel reflections apparently dipping south. The radar sequence of CD1 is composed mainly of hummocky reflections (Figure 14A). To the west, the CD2 radar sequence is truncated also by a radar sequence containing exclusively hummocky reflections, probably representing the infilling of a younger channel (Figure 13B). Only the nearest 2-D GPR lines from the outcrops were interpreted on the large scale GPR grid, because the rest of the grid was not suitable for detailed interpretation due to coarse sampling and lack of geologic control.

Integrated Sedimentologic and Ground-Penetrating Radar Interpretation

Correlating outcrop, drill-hole, and GPR data allows the relationship between vertical facies successions of the inclined beds in CD2 and their lateral geometry to be established. Two bounding surfaces from the central inclined bedset (surfaces VI and VII in Figures 6 and 14) and six bounding surfaces from the eastern-most bedset (surface 0 to V in Figures 6 and 14) were interpreted in outcrop, drill-hole, and GPR data.

The interpreted inclined surfaces migrate laterally from west to east, and also upstream (Figures 14 and 15). The strike of the inclined beds is essentially parallel with the paleoflow and rotates from a northwest-southeast direction within the central bedset (inclined surfaces VI and VII) to a north-south direction within the eastern-most bedset (inclined surfaces 0 to V). The inclined beds extend more than 250 m (820 ft) in the direction of flow based on information from the 2-D GPR large grid and north-south outcrop. This suggests that CD2 is located near the outer bank of the channel (Figures 6B and 13A).

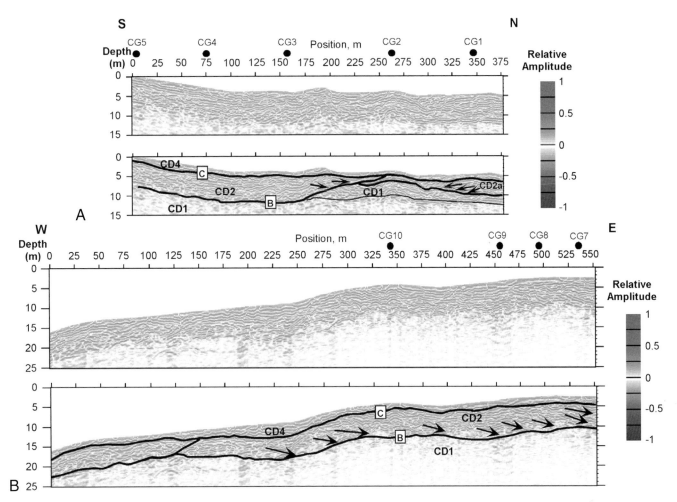

Figure 13. Two-dimensional GPR lines from the largest GPR grid parallel with the (A) south-north and (B) east-west cliff face (Figure 5). The uninterpreted and interpreted versions are displayed for correlation. CD2 deposits pinch out north against the channel cut-bank (see also Figure 6B). The westward truncation of CD2 is interpreted only from GPR data. Black arrows show the inclined beds within CD2. CG1 to CG10 are the projections of the measured sections at the cliff face onto the GPR lines. Color version on CD-ROM.

The inclined beds of CD2 were most likely deposited by lateral and upstream accretion of a point bar within a highly sinuous distributary channel with paleoflow essentially toward the north-northwest. However, upstream migration of point bars is mostly described in tidal creeks (Thomas et al., 1987), although examples from braided rivers have also been described (Bristow, 1993). The paleocurrent directions measured in CD2 are essentially unidirectional, unlike the tidal channels that typically show bimodal paleoflow directions (Thomas et al., 1987). The marine incursions observed in CD2 are thus interpreted as seasonal, representing times of river flooding, rather than tidal. The muds drape bar-scale features rather than occurring at the scale of individual cross-strata such as the double mud drapes and sigmoidal cross-stratification diagnostic of tidal stratification (e.g. Nio and Yang, 1991). No tidal bundles, tidal rhythmites, nor double mud laminae were observed. Extensive bioturbation by *Teredolites* on large logs deposited on the basal, erosional scours of the inclined beds and within the underlying delta plain suggests marine influ-

ence (Bromley et al., 1984). The location of the coastline 24 km (15 mi) from Corbula Gulch made possible a deposition on the lower delta plain affected by repetitive incursion of marine waters, probably related to times of seasonally low discharge. Bioturbation like *Planolites*, *Thalassinoides*, *Skolithos*, and *Arenicolites* in the siltstone/mudstone layers draping the inclined beds are also indicative of brackish water influence suggesting mixing of river and marine waters (Pemberton et al., 1992).

Paleogeometry of the Point-Bar Deposits

The most important parameters for estimating the dimensions of a meandering paleochannel are the maximum horizontal width of one inclined bed measured in direction perpendicular to the flow and the vertical depth from the toe to the top of the inclined beds (Ethridge and Schumm, 1978). These widths and depths are the expression of the bankfull channel width and the maximum bankfull channel depth, respectively (Ethridge and Schumm, 1978). The horizontal width (W) of one inclined bed is up to 150 m (492 ft) within the central

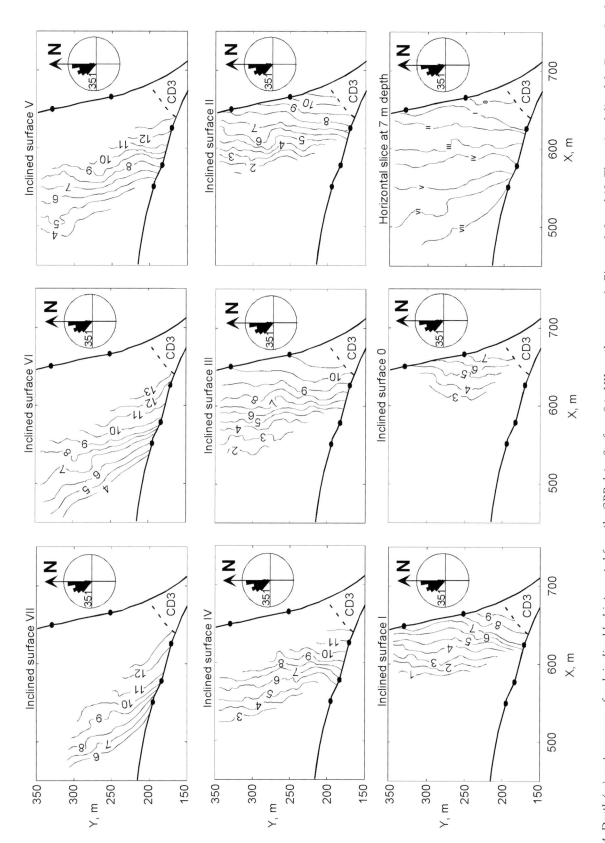

Figure 14. Depth (m) contour maps of each inclined bed interpreted from the GPR data. Surfaces 0 to VII are the same as in Figures 6, 9, and 12. The regional dip of the Ferron Sandstone outcrop (38° strike and 2.5° northwest dip) was removed to accurately represent the true depositional dip of the point-bar surfaces. The strike of the inclined beds is consistent with the paleoflow as indicated by paleocurrent measurements in the outcrop. The dashed line represents the interpreted CD3 channel-margin. The map in the lower-right corner represents a horizontal slice through the interpreted inclined beds at 7 m (23 ft) depth. The inclined beds are generally parallel to slightly divergent toward north.

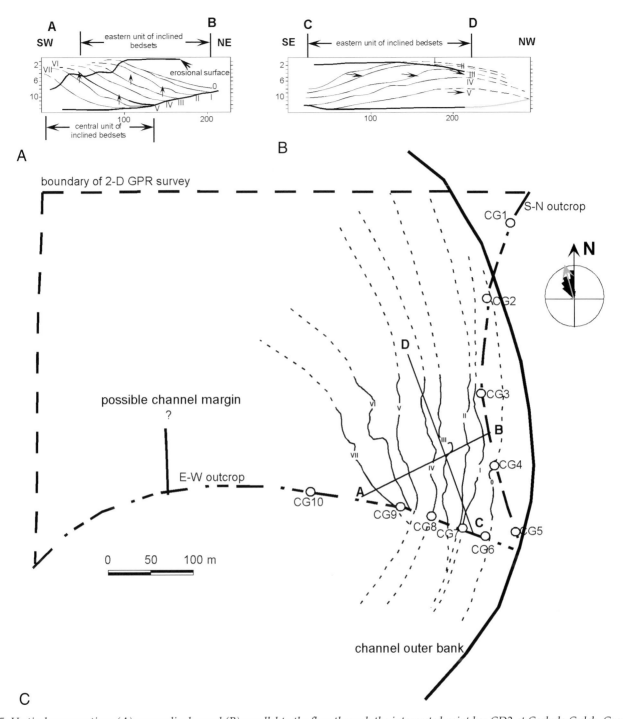

Figure 15. Vertical cross sections (A) perpendicular and (B) parallel to the flow through the interpreted point bar CD2 at Corbula Gulch. Grey lines are estimated inclined surfaces before erosion; (C) reconstruction of channel margin and inclined surfaces (VII to 0) in horizontal section at 7-m (23-ft) depth. Continuous lines are interpretation from the GPR data, dashed lines are inferred based on the local trend of each mapped inclined surface. A-B and C-D show the location of the cross sections in (A) and (B) panels.

inclined bedset unit and up to 100 m (330 ft) within the eastern inclined bedset unit (Figures 6A and 14). The vertical depth (D) of an inclined bedset varies from 6 m (20 ft) within the central inclined bedset to 8 m (26 ft) within the eastern inclined bedset (Figures 6A and 14). Applying formulas used by Ethridge and Schumm (1978) to estimate the bankfull width (W*1.5) and mean bankfull depth (D*0.585/0.9), resulted in a value of

225–150 m (738–492 ft) for the bankfull width of the channel and a value of 3.9–5.2 m (12.8–17.1 ft) for the mean bankfull depth. However, these values should be considered as minimum estimates because the upper part of the point bar may be completely removed by erosion associated with CD4.

The central inclined bedset shows the typical fining-upward and thinning-upward succession of a fluvial point bar as described by Allen (1970). However, it has been demonstrated that this fining- and thinning-upward succession does not cover all possible vertical successions in meandering channels, and is characteristic only of certain models of channel migration (Bluck, 1971; Bridge and Jarvis, 1976; Jackson, 1976, 1978). Jackson (1976, 1978) showed that necessary conditions of velocity magnitude, intensity of spiral motion, and depth of channel required to produce the "standard" point bar depositional model could be reached only in the downstream parts of point bars and at moderate curved bends. Other positions along the bend can produce very different facies successions. Thus, the central bedset unit is more likely to be deposited in the downstream half of the bend, while the eastern-most bedset unit may represent deposition in a more upstream position. The eastern-most unit of inclined bedsets was deposited in a more marine-influenced environment, characterized by thinner inclined beds and more extensive mud drapes covering parts of the dip-length of the bed surfaces. The distinct fining-upward succession within each inclined bed (e.g. CG7 and CG8, Figure 6A) is interpreted to be indicative of repetitive floods characterized by flashy discharge with rapid dumping of the sediments (presence of well-developed mudstone intraclast conglomerate at the base of each inclined bed and the normal grading inside each inclined bed). Each inclined bed is interpreted as the deposit of one flood event (Thomas et al., 1987). The mudstone drape represents fines settling out from suspension at much-reduced velocity, probably a consequence of marine influence. At times of lower discharge, the salt wedge at the downstream end of the distributary channel can migrate tens of kilometers upstream (e.g. Nelson, 1970).

The different units of inclined bedsets described in outcrop and identified in GPR appear to represent successive, major phases in the growth history of a single point bar affected by seasonal variations in discharge with marine influence (Figure 15).

Channel Deposit 3

Channel deposit 3 cuts deeply into the underlying CD1 and CD2 deposits and is confined to the southwest edge of the outcrop. Channel deposit 3 has a highly erosive base with a maximum relief of about 10 m (33 ft) over a lateral distance of 50 m (164 ft) (Figure 6) and contains thick basal mudstone intraclast conglomerates along the scours. Internally, CD3 consists mostly of bedsets of 3–4.5 (10–15 ft) thickness at their maximum in the outcrop. The bedsets are composed of structureless sandstone with small amounts of large-scale trough cross-bedded, tangential and sigmoidal cross-bedded sandstone and contains mudstone intraclast conglomerate along basal erosional scours. Locally, mud clasts are up to 0.5 m (1.6 ft) long. In places, the upper part of these

internal units comprises centimeter-scale alternations of sandstone/siltstone and mudstone layers. The paleocurrent measurements range from northeast (azimuth 43°) to southeast (azimuth 109°) direction suggesting that the channel flowed broadly eastward, rather than CD1, which flowed broadly to the north.

The structureless nature, large clasts, and internal scour surfaces suggests that channel deposit CD3 was deposited during intervals of extremely high-flow conditions (Wizevich, 1992) within a distributary channel. A lack of burrowing, compared to CD1, suggests that channel discharge was too high to allow any significant incursion of marine water. Channel deposits of CD3 are not recognized anywhere on the 2-D intermediate GPR grid. The outcrop thus intersects the western margin of a large, northeastward flowing channel deposit.

Channel Deposit 4

Channel deposit 4 is the uppermost unit and also lies at the top of SC7. Channel deposit 4 has an erosive basal bounding surface with moderate local relief and overlies all previously described channel deposits (Figure 7). Overall, CD4 is tabular, with significant increase in thickness toward the west (about 4.5 m [15 ft] interpreted from the GPR data). Element CD4 consists of a complex mixture of medium- and large-scale, trough cross-bedded sandstone, planar-tabular cross-bedded sandstone with some mudstone drapes on foresets, and sandstone containing convolute stratification. Grain size within CD4 is highly variable ranging from medium- to coarse-grained sandstone in fining- and coarsening-upward successions.

Paleocurrent directions reflect the depositional complexity of this uppermost element and ranges from a northeast (azimuth 60°) paleoflow on planar cross-beds to a southeast (azimuth 140°) paleoflow on trough cross-beds. These alternating current directions suggest tidal influence, probably related to the overall transgression that caps SC7 at the top of the Ferron clastic wedge. Poor exposure of CD4 in outcrops (due to erosion) made it difficult to determine the detailed bedding architecture and further interpretation has not been attempted.

DISCUSSION

The inclined bedsets deposited at Corbula Gulch are an excellent example of marine influenced point bars deposited within lower delta-plain distributary channels, and are quite different from the non-migratory, straight channels that are inferred to be typical of lower delta-plain distributary channels (e.g. Colemen and Prior, 1982). These channels were affected by seasonal and possibly longer-term changes in channel discharge, but show little evidence of tidal processes. The characteristics of this point bar deposited in a lower delta plain differs from classical fluvial point bars, but also does not match well with standard models for distributary channels. The 3-D geometry of the inclined surfaces within the point bar indicates that they were formed by lateral and upstream migration (Figures 15A and B). The laterally accreting bedset surfaces show, in vertical section, perpendicular to flow, more linear to concave-upward shapes, with long asymptotic ends toward the channel axis (Figure 15A). This tendency of the inclined beds to follow the shape of the channel was related by Hopkins (1985) to the existence of "passive" point bars that plug the channel before abandonment. Probably the easternmost inclined bedset represents the initial phase of channel infilling and abandonment. Also, these concave-upward shapes are more linear and divergent in plan view (Figure 15C) compared to the arcuate pattern of the swell and ridge topography of other fluvial point bars (cf. Puigdefabregas and van Vliet, 1978; Gibling and Rust, 1993).

Compared to simple theoretical models of fluvial point bars (Bridge, 1984; Willis, 1989) the inclined bedsets at Corbula Gulch show a much more complex grain-size distribution than the theoretical fining-upward and downstream trend. Also, the inclined bed surfaces are concave upward and migrate both laterally and upstream, while theoretical model generally predict convex upward inclined surfaces, migrating laterally and downstream. Upstream migration of the point bars is not usually depicted in theoretical models of fluvial point bars (Willis, 1989). Other factors, such as marine influence, decrease in discharge and slope within distributary channels, may be required to generate more representative theoretical models of point-bar deposits associated with distributary channels.

The lithologic heterogeneities at Corbula Gulch are represented by mudstone drapes along inclined accretion surfaces, and by mudstone intraclast conglomerate along minor erosional surfaces separating the inclined beds. Quantification of 3-D distribution of the mudstone drapes is performed through geostatistical analysis of 3-D GPR amplitudes. These mudstones are not extensive parallel to the flow and are never longer than 15 m (49 ft). Initially longer mudstones are typically fragmented to smaller dimensions of 5 m (16 ft) as a consequence of erosional scour by overlying units. Along the downdip direction, the mudstones average 10 m (33 ft) long. Combining the 3-D architecture of point-bar deposits and the 3-D distribution of heterogeneities constitutes important input to reservoir flow simulation (e.g. Novakovic et al., 2002).

CONCLUSION

Point-bar deposits within lower delta-plain, high-sinuosity distributary channels are described and inter-

preted at Corbula Gulch, in the fluvio-deltaic Cretaceous Ferron Sandstone Member of the Mancos Shale. The paleoshoreline is located at 24 km (15 mi) north, and marine (*Teredolites*) and brackish water (*Skolithos, Arenicolites, Thalassinoides*) burrows are common within the delta-plain deposits and the mudstone/siltstone drapes at the tops of the inclined beds.

The grain-size distribution is more complex than the upward- and downstream-fining prediction from theoretical point-bar models.

The inclined bedset surfaces migrated laterally and upstream and show concave-upward shapes paralleling the channel. In horizontal section these inclined surfaces are linear and divergent, parallel to the channel cutbank. Present fluvial theoretical models do not predict these geometries and these geometries may be more characteristic of point bars formed within marine-influenced distributary channels (Bhattacharya et al., 2001).

The main radar facies characterizing the point-bar deposits in 3-D and 2-D GPR surveys are inclined, parallel or concave upward, continuous, high-amplitude reflections that truncate upward against the erosional surface of the overlying deposits and downlap downward onto the base of the channel.

The estimated minimum paleochannel bankfull width and mean depth are 150 m (492 ft) and 4 m (13 ft), respectively, but the upper part of the inclined beds is probably completely removed by erosion and thus we may be underestimating the dimensions of the inclined beds.

The main heterogeneities within the point-bar deposits at Corbula Gulch are shale drapes on the inclined accretion surfaces, and mudstone intraclast conglomerate along basal erosional surfaces of the inclined beds. These high clay content layers were originally longer parallel to the flow (15 m [49 ft]) but are now discontinuous (< 5 m [16 ft]) because of deep erosion by the overlying units. Perpendicular to the flow (downdip the inclined beds) the mudstone and mudstone intraclast conglomerate are more continuous, reaching about 10 m (33 ft) in length. This detailed distribution of the heterogeneities within a reservoir analog interpreted from GPR data is important to predict flow behaviour.

ACKNOWLEDGMENTS

The research leading to this paper was funded primarily by the U.S. Department of Energy under Contract DE-FG03-96ER14596 with auxiliary support from the University of Texas at Dallas GPR Consortium. The interpretation of the migrated data was done using the PC-based seismic interpretation software WinPics of Kernek Technologies Ltd. The geostatistical analysis was done with the Geostatistical Software Library (GSLIB) programs. The permeability measurements were collected and analyzed by Steve Snelgrove and Craig Forster from the University of Utah. An earlier version of this paper benefited from reviews by John Bridge. This paper is contribution number 996 from the Geosciences Department at the University of Texas at Dallas.

REFERENCES CITED

Alexander, J., J. S. Bridge, M. R. Leeder, R. E. L. Collier, and R. L. Gawthorpe, 1994, Holocene meander-belt evolution in an active extensional basin, southwestern Montana: Journal of Sedimentologic Research, v. B64, p. 542–559.

Allen, J. R. L., 1964, Studies in fluviatile sedimentation: six cyclothems from the Lower Old Red Sandstone, Anglo-Welsh basin: Sedimentology, v. 3, p. 163–198.

—1970, Studies in fluviatile sedimentation: a comparison of finning upwards cyclothems with special reference to coarse-member composition and interpretation: Journal of Sedimentary Petrology, v. 40, p. 298–323.

Annan, A. P., S. W. Cosway, and J. D. Redman, 1991, Water table detection with ground-penetrating radar (abs.): SEG Annual International Meeting and Exposition, Program with Abstracts, E/G1.4, p. 494–497.

Baker, P. L., and U. Monash, 1991, Fluid, lithology, geometry, and permeability information from ground-penetrating radar for some petroleum industry applications: Society of Petroleum Engineer, paper SPE 22976, p. 277–286.

Barton, M. D., 1994, Outcrop characterization of architecture and permeability structure in fluvial-deltaic sandstones, Cretaceous Ferron sandstone, Utah: Ph.D. dissertation, University of Texas, Austin, Texas, 260 p.

Beres, M., A. Green, P. Huggenberger, and H. Horstmeyer, 1995, Mapping the architecture of glaciofluvial sediments with three dimensional georadar: Geology, v. 23, p. 1087–1090.

Bhattacharya, J. P., R. M. Corbeanu, and C. Olariu, 2001, Downstream causes of avulsion in distributary channels, as seen in Cretaceous sandstones of Central Utah (abs): 7th International Conference on Fluvial Sedimentology, program and abstracts, Open-File Report 60, Conservation and Survey Division, Institute of Agriculture and Natural Resources, University of Nebraska-Lincoln, p.59.

Bluck, B. G., 1971, Sedimentation in the meandering River Endrick: Scottish Journal of Geology, v. 7, p. 93–138.

Bridge, J. S., 1984, Flow and sedimentary processes in river bends: comparison of field observation and theory, in M. Elliot, ed., River meandering: Proceedings Rivers '83, American Society of Civil Engineers, p. 857–872.

—1985, Paleochannel patterns inferred from alluvial deposits: a critical review: Journal of Sedimentary Petrology, v. 55, p. 579–589.

—1993, Description and interpretation of fluvial deposits: a critical perspective: Sedimentology, v. 42, p. 801–810.

Bridge, J. S., and J. Jarvis, 1976, Flow and sedimentary processes in the meandering River South Esk, Glen Clova, Scotland: Earth Surface Processes, v. 2, p. 401–416.

Bridge, J. S., J. Alexander, R. E. L. Collier, R. L. Gawthorpe, and J. Jarvis, 1995, Ground-penetrating radar and coring used to study the large-scale structure of point-bar deposits in three dimensions: Sedimentology, v. 42, p. 839–852.

Bridge, J. S., R. E. L. Collier, and J. Alexander, 1998, Large-scale structures of Calamus River deposits (Nebraska,U.S.A.) revealed using ground-penetrating radar: Sedimentology, v. 45, p. 977–986.

Bristow, C. S., 1993, Sedimentary structures exposed in bar tops in the Brahmaputra River, Bangladesh, in J. L. Best and C. S. Bristow, eds., Braided rivers: Geological Society of London Special Publication 75, p. 277–289.

—1995, Internal geometry of ancient tidal bedforms revealed using ground penetrating radar, in B. W. Flemming and A. Bartholoma, eds., Tidal signatures in modern and ancient sediments: International Association of Sedimentologists Special Publications, v. 24, p. 313–328.

Bristow, C. S., P. N. Chroston, and S. D. Bailey, 2000, The structure and development of foredunes on a locally prograding coast: insights from ground-penetrating radar surveys, Norfolk, UK: Sedimentology, v. 47, p. 923–944.

Bromley, R. G., G. Pemberton, and R. A. Rahmani, 1984, A Cretaceous woodground: the teredolites ichnofacies: Journal of Paleontology, v. 58, p. 488–498.

Brown, A. R., 1996, Interpretation of three-dimensional seismic data: AAPG Memoir 42, 4th ed., 444 p.

Cherven, V. B., 1978, Fluvial and deltaic facies in the Sentinel Butte Formation, central Eilliston basin: Journal Sedimentary Petrology, v. 48, p. 159–170.

Coleman, J. M., and D. B. Prior, 1982, Deltaic environments of deposition, in P.A. Scholle and D. Spearing, eds., Sandstone depositional environments: AAPG Memoir 31, p. 139–178.

Corbeanu, R. M., K. Soegaard, R. B. Szerbiak, J. B. Thurmond, G. A. McMechan, D. Wang, S. H. Snelgrove, C. B. Forster, and A. Menitove, 2001, Detailed internal architecture of a fluvial channel sandstone determined from outcrop, cores and 3-D ground-penetrating radar: example from the Mid-Cretaceous Ferron Sandstone, east-central Utah: AAPG Bulletin, v. 85, p. 1583–1608.

Cotter, E., 1975, Late Cretaceous sedimentation in a low energy coastal zone: the Ferron Sandstone of Utah: Journal of Sedimentary Petrology, v. 45, p. 669–685.

Davis, J. L., and A. P. Annan, 1989, Ground-penetrating radar for high resolution mapping of soil and rock stratigraphy: Geophysical Prospecting, v. 37, p. 531–551.

Deutsch, C. V., and A. G. Journel, 1998, GSLIB geostatistical software library and user's guide: Oxford, Oxford University Press, 340 p.

Dreyer, T., 1993, Geometry and facies of large-scale flow units in fluvial-dominated fan-delta-front sequences, in M. Ashton, ed., Advances in reservoir geology: Geological Society Special Publication 69, p. 135–174.

Elliott, T., 1976, The morphology, magnitude and regime of a Carboniferous fluvial-distributary channel: Journal of Sedimentary Petrology, v. 46, p. 70–76.

Ethridge, F. G., and S. A. Schumm, 1978, Reconstructing paleochannel morphologic and flow characteristics: methodology, limitations, and assessment, in A. D. Miall, ed., Fluvial sedimentology: Canadian Society of Petroleum Geology Memoir 5, p. 703–721.

Falkner, A., and C. R. Fielding, 1993, Quantitative facies analysis of coal-bearing sequences in the Bowen Basin, Australia: applications to reservoir description, in S. S. Flint and I. D. Bryant, eds., The geological modeling of hydrocarbon reservoirs and outcrop analogs: International Association of Sedimentologists Special Publication 15, p. 81–89.

Fielding, C. R., 1984, Upper delta plain lacustrine and fluviolacustrine facies from the Westphalian of the Durham coalfield, NE England: Sedimentology, v. 31, p. 547–567.

Flint, S. S., and D. Bryant, eds., 1993, The geological modeling of hydrocarbon reservoirs and outcrop analogs: International Association of Sedimentologists Special Publication 15, 269 p.

Gardner, M. H., 1992, Sequence stratigraphy of the Ferron Sandstone, east-central Utah, in N. Tyler, M. D. Barton, and R. S. Fisher, eds., Architecture and permeability structure of fluvial-deltaic sandstones: a field guide to selected outcrops of the Ferron Sandstone, east-central Utah: Austin, University of Texas, Bureau of Economic Geology Guidebook, p. 1–12.

—1995, Tectonic and eustatic controls on the stratal architecture of mid-Cretaceous stratigraphic sequences, central western interior foreland basin of North America, in S. L. Dorobek and G. M. Ross, eds., Stratigraphic evolution of foreland basins: Society for Sedimentary Geology (SEPM) Special Publication 52, p. 243–281.

Garrison, J. R., Jr., T. C. V. van den Bergh, C. E. F. Barker, and D. E. Tabet, 1997, Depositional sequence stratigraphy and architecture of the Cretaceous Ferron Sandstone: implications for coal and coalbed methane resources — a field excursion, in P. K. Link and B. J. Kowallis, eds., Mesozoic to recent geology of Utah: Provo, Brigham Young University Geology Studies v. 42, pt. 2, p. 155–202.

Gawthorpe, R. L., R. E. L. Collier, J. Alexander, J. S. Bridge, and M. R. Leeder, 1993, Ground- penetrating radar: application to sandbody geometry and heterogeneity studies, in C. P. North and D. J. Prosser, eds., Characterization of fluvial and aeolian reservoirs: Geological Society Special Publication 73, p. 421–432.

Gibling, M. R., and B. R. Rust, 1993, Alluvial ridge-and swale topography: a case study from the Morien Group of Atlantic Canada, in M. Marzo and C. Puygdefabregas, eds., Alluvial sedimentation: International Association of Sedimentologists Special Publications 17, p. 133–150.

Hale, L. A., 1972, Depositional history of the Ferron Formation, central Utah, in J. L. Baer, ed., Plateau-basin and range transition zone: Utah Geological Association Publication 2, p. 115–138.

Hammon, W. S., III, X. Zeng, R. M Corbeanu, and G. A. McMechan, 2002, Estimation of the spatial distribution of fluid permeability from surface and tomographic GPR data and core, with a 2-D example from the Ferron Sandstone, Utah: Geophysics, v. 67, p. 1505–1515.

Hobday, D. K., 1978, Fluvial deposits of the Ecca and Beaufort Groups in the eastern Karoo Basin, Southern Africa, in A. D. Miall, ed., Fluvial sedimentology: Canadian Society of Petroleum Geology Memoir 5, p. 413–429.

Hopkins, J. C., 1985, Channel-fill deposits formed by aggradation in deeply scoured, superimposed distributaries of the Lower Kootenai Formation (Cretaceous): Journal of Sedimentary Petrology, v. 55, p. 42–52.

Hornung, J., and T. Aigner, 2000, Integration of outcrop sedimentology, petrophysics and georadar for enhanced 3-D reservoir-characterization of fluvial sandstones (Triassic, Germany): Geological Society of America International Meeting, Abstracts with Program, A-80.

Jackson, R. G., II, 1976, Depositional model of point bars in the lower Wabash River: Journal of Sedimentary Petrology, v. 46, p. 579–594.

—1978, Preliminary evaluation of lithofacies models for meandering alluvial streams, *in* A. D. Miall, ed., Fluvial sedimentology: Canadian Society of Petroleum Geologists Memoir 5, p. 543–576.

—1981, Sedimentology of muddy fine-grained channel deposits in meandering streams of the American Middle West: Journal of Sedimentary Petrology, v. 51, p. 1169–1192.

Jordan, D. W., and W. A. Pryor, 1992, Hierarchical levels of heterogeneity in a Mississippi River meander belt and application to reservoir systems: AAPG Bulletin, v. 76, p. 1601–1624.

Knight, R. J., and A. Nur, 1987, The dielectric constant of sandstones, 60kHz to 4MHz: Geophysics, v. 52, p. 644–654.

Lowry, P., and T. Jacobsen, 1993, Sedimentological and reservoir characteristics of a fluvial-dominated delta-front sequence: Ferron Sandstone Member (Turonian), east-central Utah, U.S.A., *in* M. Ashton, ed., Advances in reservoir geology: Geological Society Special Publication 69, p. 81–103.

Miall, A. D., and C. E. Turner-Peterson, 1989, Variations in fluvial style in the Westwater Canyon Member, Morrison Formation (Jurassic), San Juan Basin, Colorado Plateau: Sedimentary Geology, v. 63, p. 21–60.

McMechan, G. A., G. C. Gaynor, and R. B. Szerbiak, 1997, Use of ground-penetrating radar for 3-D sedimentological characterization of clastic reservoir analogs: Geophysics, v. 62, p. 786–796.

Nelson, B. W., 1970, Hydrography, sediment dispersal and recent historical development of the Po River delta, Italy, *in* J. P. Morgan and R. H. Shaver, eds., Deltaic sedimentation, modern and ancient: Society for Sedimentary Geology (SEPM) Special Publication 15, p. 152–184.

Nio, S. D., and C. S. Yang, 1991, Diagnostic attributes of clastic tidal deposits: a review, *in* D. G. Smith, G. E. Reinson, B. A. Zaitlin, and R. A. Rahmani, eds., Clastic tidal sedimentology: Canadian Society of Petroleum Geologists Memoir 16, p. 3–28.

Novakovic, D., C. D. White, R. M. Corbeanu, W. S. Hammon III, J. P. Bhattacharya, and G. A. McMechan, 2002, Effects of shales in fluvial-deltaic deposits: ground-penetrating radar, outcrop observations, geostatistics and three-dimensional flow modeling for the Ferron Sandstone, Utah: Mathematical Geology, v. 34, p. 857–893.

Pemberton, S. G., J. A. MacEachern, and R. W. Frey, 1992, Trace fossil facies models: environmental and allostratigraphic significance, *in* R. G. Walker and N. P. James, eds., Facies models response to sea level change: Geological Association of Canada Geotext 1, p. 47–72.

Plint, A. G., 1983, Facies, environments and sedimentary cycles in the Middle Eocene, Bracklesham Formation of the Hampshire Basin: evidence for global sea-level changes: Sedimentology, v. 30, p. 625–653.

Puigdefabregas, C., and A. van Vliet, 1978, Meandering stream deposits from the Tertiary of the Southern Pyrenees, *in* A. D. Miall, ed., Fluvial sedimentology: Canadian Society of Petroleum Geologists Memoir 5, p. 469–485.

Rea, J., and R. Knight, 1998, Geostatistical analysis of ground-penetrating radar data: a means of describing spatial variation in the subsurface: Water Resources Research, v. 34, p. 329–339.

Robertson, J. D., and H. H. Nogami, 1984, Complex seismic trace analysis of thin beds: Geophysics, v. 49, p. 344–352.

Ryer, T. A., 1981, Deltaic coals of Ferron Sandstone Member of Mancos Shale: predictive model for Cretaceous coal-bearing strata of Western Interior: AAPG Bulletin, v. 65, p. 2323–2340.

—1991, Stratigraphy, facies and depositional history of the Ferron Sandstone in the Canyon of Muddy Creek, east-central Utah, *in* T. C. Chidsey, Jr., ed., Geology of east-central Utah: Utah Geological Association Publication 19, p. 45–54.

Ryer, T. A., and P. B. Anderson, 1995, Parasequence sets, parasequences, facies distributions, and depositional history of the Upper Cretaceous Ferron deltaic clastic wedge, central Utah (abs.): AAPG Bulletin, v. 79, p. 924.

Smith, D. G., 1987, Meandering river lithofacies models: modern and ancient examples compared, in Recent developments in fluvial sedimentology: Society for Sedimentary Geology (SEPM) Special Publication, Proceedings of 3rd International Fluvial Sedimentology Conference, Fort Collins, Colorado, August 1985.

Stephens, M., 1994, Architectural element analysis within the Kayenta Formation (Lower Jurassic) using ground-probing radar and sedimentological profiling, southwestern Colorado: Sedimentary Geology, v. 90, p.179–211.

Szerbiak, R. B., G. A. McMechan, R. M. Corbeanu, C. B. Forster, and S. H. Snelgrove, 2001, 3-D characterization of a clastic reservoir analog: from 3-D GPR to a 3-D fluid permeability model: Geophysics, v. 66, p. 1026–1037.

Tanner, M. T., F. Koehler, and R. E. Sheriff, 1979, Complex seismic trace analysis: Geophysics, v. 44, p. 1041–1063.

Thomas, R. G., D. G. Smith, J. M. Wood, J. Visser, E. A. Calverley-Range, and E. H. Koster, 1987, Inclined heterolithic stratification — terminology, description, interpretation and significance: Sedimentary Geology, v. 53, p. 123–179.

Vail, P. R., 1987, Seismic stratigraphic interpretation procedure, *in* A. W. Bally, ed., Atlas of seismic stratigraphy: AAPG Studies in Geology 27, p. 1–10.

Van Wagoner, J. C., R. M. Mitchum, K. M. Campion, and V. D. Rahmanian, 1990, Siliciclastic sequence stratigraphy in well logs, cores and outcrops: concepts for high-resolution correlation of time and facies: AAPG Methods in Exploration 7, 55 p.

Willis, B J., 1989, Paleochannel reconstructions from point bar deposits: a three-dimensional perspective: Sedimentology, v. 36, p. 757–766.

—1993, Interpretation of bedding geometry within ancient point-bar deposits, *in* M. Marzo and C. Puigdefabregas, eds., Alluvial sedimentation: International Association of Sedimentologists Special Publication 17, p. 101–114.

Wizevich, M. C., 1992, Pennsylvanian quartzose sandstones of the Lee Formation: fluvial interpretation based on lateral profile analysis: Sedimentary Geology, v. 78, p. 1–47.

Xu, X., 2000, Three-dimensional virtual geology: photorealistic outcrops, and their acquisition, visualization and analysis: Ph.D. dissertation, University of Texas at Dallas, Dallas, 400 p.

Colorado Plateau wildcat, Emery County, Utah, circa 1920.

Analog for Fluvial-Deltaic Reservoir Modeling: Ferron Sandstone of Utah
AAPG Studies in Geology 50
T. C. Chidsey, Jr., R. D. Adams, and T. H. Morris, editors

The Geometry, Architecture, and Sedimentology of Fluvial and Deltaic Sandstones Within the Upper Ferron Sandstone Last Chance Delta: Implications for Reservoir Modeling

T. C. V. van den Bergh[1,2] and James R. Garrison, Jr.[1,3]

ABSTRACT

The Turonian-Coniacian Upper Ferron Sandstone Last Chance Delta was deposited along the western margin of the Western Interior Seaway as a wave-modified, river-dominated deltaic system. The Last Chance Delta was deposited during a slow relative sea-level rise whose rate of rise decreased with time. The sedimentation rate progressively decreased throughout the deposition of the Last Chance Delta.

Architectural and sedimentological data for deltaic near-marine sandstones indicate that primary deltaic depositional style is directly correlated with degree of wave-modification, which is controlled by the ratio of sedimentation rate to the rate of relative change in sea level. The progradational parasequence sets have a mean sandstone dip length/thickness aspect ratio of 788. The aggradational parasequence sets are shorter with a mean length/thickness of 520. The retrogradational parasequence sets are shorter and thinner with a mean length/thickness of 397. River-dominated progradational parasequences have a mean length/thickness of 611, a mean width/thickness of 212, and a mean length/width of 1.9. River-dominated, wave-modified progradational parasequences have longer dip lengths and a higher length/thickness of 845. The aggradational parasequences have similar lengths as the wave-modified parasequences, with a mean length/thickness of 606. The retrogradational parasequences are short and thin, with a mean length/thickness of 793.

Stream-mouth bar, reworked stream-mouth bar, and upper shoreface deposits show trends of length/thickness changing systematically with degree of wave-reworking, from a mean length/thickness of 479 (width/thickness = 256; length/width = 1.9) in river-dominated parasequences to 546 and 595 in reworked stream-mouth bar and upper shoreface deposits, respectively. Retrogradational parasequences have higher upper shoreface mean length/thickness aspect ratios of 649. Proximal delta-front, reworked proximal delta-front, and middle shoreface deposits show similar trends. River-dominated parasequences have mean proximal delta-front length/thickness of 425 (width/thickness = 472; length/width = 1.8) and reworked proximal delta-front and middle shoreface deposits have a mean length/thickness of 827 and 912, respectively. Retrogradational parasequences have a mean

[1]The Ferron Group Consultants, Emery, Utah
[2]Present address: SGS Minerals Services, Huntington, Utah
[3]Present address: Colorado Plateau Field Institute, Price, Utah

middle shoreface length/thickness of 807. Distal delta-front, reworked distal delta-front, and lower shoreface deposits also show similar trends. River-dominated parasequences have mean distal delta-front length/thickness ratios of 518 and reworked distal delta-front and lower shoreface deposits have mean length/thickness ratios of 819 and 2469, respectively. Retrogradational parasequences have a mean lower shoreface length/thickness of 981.

Architectural and sedimentological data for fluvial channel-belt sandstones indicate that over-all geometry, internal architecture, and preserved sedimentary structures are directly correlated with sedimentation rate and rate of relative change in sea level. Internal channel belt architecture is controlled by the response of the river equilibrium profile to changes in relative sea level and shoreline position. Channel belts, from progradational parasequence sets, deposited during times of high sedimentation rate and moderate relative sea-level rise, are laterally restricted and multi-storied with channel-fill elements stacked vertically within the channel-belt boundaries. Fluvial channel belts in the upper delta plain have average width/thickness aspect ratios of 28.8; distributary channel belts located near the paleoshoreline have average aspect ratios of 19.0. Fluvial channel belts from aggradational parasequence sets deposited during times when sedimentation rate was approximately equal to the rate of relative sea-level rise are laterally extensive and multi-storied with channel-fill elements stacked laterally en-echelon. Fluvial channel belts in the upper delta plain have average width/thickness aspect ratios of 59.2; distributary channel belts, located near the paleoshoreline have a mean aspect ratio of 12.1. Channel belts from retrogradational parasequence sets deposited during times when sedimentation rate was less than the rate of relative sea-level rise are laterally extensive and sheet-like with average aspect ratios of 100.0. Their channel-fill elements generally stacked vertically within the channel-belt boundaries. Amalgamated, braided fluvial deposits occur within small high-gradient incised valleys developed during periods of 4th- and 5th-order relative falls in sea level. The preserved incised-valley fluvial deposits, within the Last Chance Delta, range in width from 1.3–8.8 km (0.8–5.5 mi) and in thickness from 9–32 m (27–96 ft); the average width/thickness aspect ratio is 169.4 near the valley mouths and 644.1 at 10–17 km (6–11 mi) inland from the mouth.

INTRODUCTION

Both the external and internal architecture of fluvial and deltaic sandstone reservoirs determines, not only the paths of fluid migration during oil and gas production, but the sizes and shapes of reservoirs, the internal heterogeneity within these reservoirs, and their production characteristics. With a better understanding of the architectural elements of these fluvial-deltaic reservoir systems, better reservoir models, both deterministic and stochastic, can be developed resulting in an improved prediction of a reservoir's internal heterogeneity, recovery factor, and production/performance characteristics. These predictions can be made with more certainty and, more efficient drilling and production strategies can be subsequently developed.

This paper describes the geometry, architecture, and sedimentology of fluvial channel-belt and near-marine sandstones within the informally named Last Chance Delta of the Upper Ferron Sandstone and briefly discusses the implications of the Upper Ferron Sandstone data for geologic modeling of reservoir systems. In this study, particular attention is given to the architectural and sedimentological complexities and the evolution of the channel belts and near-marine sandstones that are a complex function of relative changes in sea level and changes in sedimentation rate.

All available quantitative and semi-quantitative architectural data from the Upper Ferron Sandstone channel-belt sandstones and near-marine sandstones have been compiled from previous works by Lowry and Jacobsen (1993) and (Barton, 1994) and combined with more recently acquired quantitative data of van den Bergh and Garrison (1996) and Garrison and van den Bergh (1997, this volume) to provide the most complete and comprehensive analysis of the sedimentological and architectural systematics of the sandstone facies of the Upper Ferron Sandstone that is possible at this time.

Definitions, Terminology, and Nomenclature

Fluvial-Deltaic Systems

In this study, the term "fluvial-deltaic system" is broadly used to describe a depositional system in which there are genetically related coastal non-marine fluvial-dominated depositional facies and near-marine shoreline depositional facies, both originally deposited by the same coastal plain fluvial systems. This broad definition does not require that there be a shoreline protuberance (i.e., as in the strict definition of a delta as defined in Elliott, 1986), since the extent to which a shoreline protuberance develops, especially in tectonically active areas, is very sensitive to sedimentation rate, amount of wave energy, and rate of relative rise in sea level.

Fluvial-deltaic systems consist of a non-marine "delta plain facies association" and a "near-marine facies association." The non-marine delta-plain association is composed of fluvial channel belts and their associated, and laterally correlative, flood-plain facies (e.g., levee deposits, crevasse-splay deposits, overbank mudstones, and lateral swamp and mire facies) and interdistributary bay and lagoonal facies. Near-marine deposits are sand-rich delta-front deposits that have been subjected to various degrees of wave and tidal destruction and reworking (i.e., these deposits include not only delta-front sandstones and mudstones, but also shoreface deposits produced by wave reworking of delta-front deposits). Marine facies are those deposits formed under deeper water marine conditions, below wave base, where shelf and basin processes dominate (e.g., prodelta and offshore deposits).

Nomenclature

In this study, both fluvial and deltaic deposits are qualitatively classified according to the degree of wave and tidal modification of the overall deltaic system to which they are genetically related. Although this classification is a continuum (Galloway, 1975), in this study, three broad classes of Ferron deltas are recognized: river-dominated, wave-reworked, and wave-dominated (Figure 1). This classification is based on the sedimentological and ichnological characteristics of the near-marine sandstones (see Table 1 in Garrison and van den Bergh, this volume). In addition, these fluvial-deltaic deposits are also classified according to their position within the parasequence stacking pattern as outlined by Garrison and van den Bergh (this volume).

Near-marine sandstone deposits: The near-marine sandstones of river-dominated and wave-reworked fluvial-deltaic systems are separated into stream-mouth-bar facies (SMB), proximal delta-front facies (pDF), distal delta-front facies (dDF), and prodelta facies (PD). Near-marine sandstones of wave-dominated deltaic systems are divided into upper shoreface facies (USF), middle shoreface facies (MSF), lower shoreface facies (LSF), and offshore facies (OS). The general criteria for these divisions are shown in Table 1 of Garrison and van den Bergh (this volume).

Fluvial channel and channel-belt sandstone deposits: In this study, the Upper Ferron Sandstone fluvial channels and channel belts are additionally classified according to the type of fluvial system in which they occur (i.e., alluvial, distributary, or incised-valley), and to the approximate position of the channel or channel belt cross section with respect to the inferred paleoshoreline position.

Channels and channel belts in the delta-/alluvial-plain facies association that generally bifurcate as they approach the paleoshoreline, are defined as distributary channels and channel belts, although realizing that locally minor tributary drainage may also have occurred. Those channel belts that are dominantly part of tributary systems are designated as alluvial channel belts. To apply this classification, to fluvial channels and channel belts in the ancient rock record, requires that the channels and channel belts be, at least partially, exposed in 3-D along the outcrop belts. Outcrops in the study area are generally not suitable for mapping the areal extent of channels and channel belts. Only one channel-belt complex is exposed sufficiently and has been mapped in enough detail to demonstrate its branching systematics. The proximity of the channels and channel belts to the near-marine deposits and their association with laterally extensive coal zones (Garrison and van den Bergh, 1997, this volume) suggest that, in general, the Upper Ferron Sandstone fluvial channels and channel belts are probably the preserved parts of distributary systems along the Last Chance Delta coastal plain. For the purposes of this study, they will all be considered distributary systems.

Preserved distributary channels are defined, in this study, as single storied channel bodies occurring within the delta-plain facies association. They may be sandstone filled and/or mud filled. Distributary channels range from low to moderate sinuosity (i.e., meandering) and consequently exhibit barforms and sedimentary structures common to the range of channel sinuosities. Generally, distributary channel sandstone systems are associated with crevasse-splay and overbank facies and laterally extensive, although not necessarily continuous, coal beds. When preserved distributary channel sandstone bodies exhibit, internally, multiple episodes of erosion and sediment infilling, they are designated as distributary channel belts, reflecting the fact that they record a history of avulsion, abandonment, reoccupation, and/or changes in discharge rate. Smaller distributary channel bodies have a tendency to not exhibit (i.e., record) multiple erosional and depositional episodes, although channels with a single channel-fill event are very rare, in the Last Chance Delta. All of the fluvial bodies encountered in this study are classified as channel belts.

In order to qualitatively examine architectural changes in the Ferron fluvial channel belts as a function of proximity to the paleoshorelines, an arbitrary spatial criterion is established. Since outcrops generally do not allow one to trace a channel belt, from an outcrop cross section at some arbitrary point within the delta-plain facies association to its mouth-bar deposits at the paleoshoreline, we simply classify the channel belts as proximal or distal distributaries. The reference point for this classification is the landward margin of the delta-plain

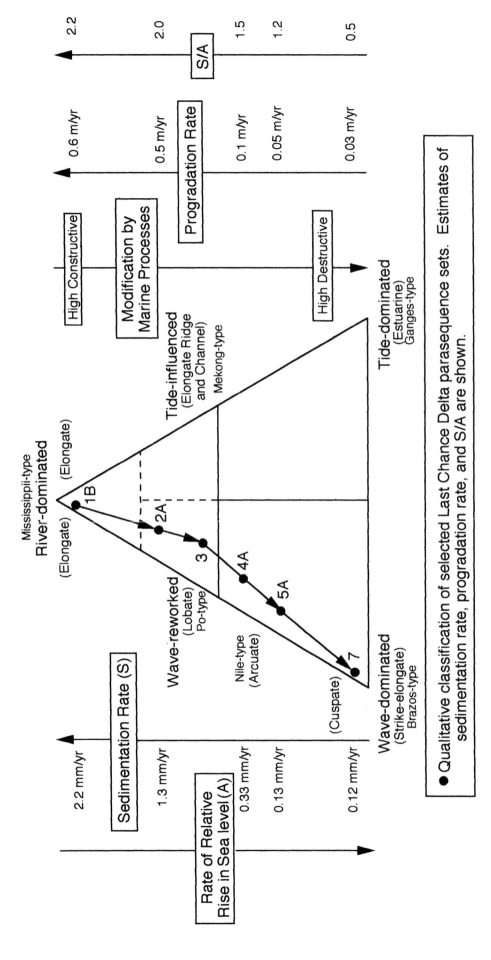

Figure 1. Ternary diagram showing a classification of deltas showing the relative effects of sedimentation rate, relative rise in sea level, progradation rate on delta morphology and degree of marine reworking. Modified from Galloway (1975). The approximate locations of various phases of Last Chance Delta deposits are shown along with the estimates of sedimentation rate (S), progradation rate, rate of relative rise in sea level (A), and the ratio S/A.

facies association. Distal distributary channel belt outcrop cross sections are defined as those less than 10 km (6 mi) from the paleoshoreline of associated near-marine parasequence. Proximal distributary channel-belt outcrop cross sections are defined as those located greater than 10 km (6 mi) landward from the position of the paleoshoreline of the associated near-marine parasequence, as defined in the cross sections presented by Garrison and van den Bergh (this volume). It should be noted that, in a progradational fluvial-deltaic setting, the sandstone preserved in distributary channel-belt outcrop cross sections, at any given location, represent material deposited over a specific time period and at progressively increasing distances from the paleoshoreline. Therefore, a channel-belt cross section, represents materials deposited at a range of distances from the paleoshoreline. In this study, the distances of channel-belt cross sections, from inferred paleoshorelines, are given as a range.

The amalgamated braided channel-belt deposits within the Upper Ferron Sandstone Last Chance Delta are considered to belong to an incised-valley fill facies association based on their (1) stratigraphic position above sub-regional erosional unconformities, (2) level of fluvial incision, (3) association with estuarine and near-marine incised-valley fill deposits, and (4) complete enclosure within the Upper Ferron Sandstone incised valleys. For a detailed description of the non-fluvial incised-valley deposits, the reader is referred to Garrison and van den Bergh (this volume).

Depositional Sequence Stratigraphy

The depositional sequence stratigraphic interpretations, presented in this paper, utilize the terminology and concepts presented by Posamentier et al. (1988), Posamentier and Vail (1988), Van Wagoner et al. (1990) and Van Wagoner (1995). For detailed discussions of the terminology and the application of depositional sequence stratigraphic concepts, the reader is referred to is referred to those definitive publications. In this study, the depositional sequence represents a relatively conformable succession of genetically related strata bounded by unconformities or their correlative conformities. Parasequences and parasequence sets are the building blocks of the depositional sequence. A parasequence is a relatively conformable succession of related bed and bedsets bounded by marine-flooding surfaces or their correlative surfaces. A parasequence set is a set of genetically related parasequences forming a distinctive stacking pattern and bounded by marine-flooding surfaces or their correlative surfaces. Within the depositional sequence, systems tracts are defined based on stratal properties (e.g., see Van Wagoner, 1995).

The stacking pattern of parasequences within a parasequence set is either progradational, aggradational, or retrogradational. These parasequence stacking patterns reflect the general relationship of the rate of sedimentation to the rate of relative sea-level rise (Garrison and van den Bergh, this volume). In a progradational stacking pattern, each successively younger parasequence has the landward pinchout of its near-marine sandstone displaced seaward relative to the landward pinchout of the near-marine sandstone of the underlying parasequence (i.e., the paleoshorelines associated with successive parasequences exhibit significant seaward migration). In an aggradational stacking pattern, the landward pinchout of the near-marine sandstone each successively younger parasequence is located generally in the same areal position relative to the landward pinchout of the near-marine sandstone of the underlying parasequence (i.e., to a 1st-order approximation, the paleoshorelines associated with successive parasequences do not exhibit significant landward or seaward migration). In a retrogradational stacking pattern, each successively younger parasequence has the landward pinchout of its near-marine sandstone displaced landward relative to the landward pinchout of the near-marine sandstone of the underlying parasequence (i.e., the paleoshorelines associated with successive parasequences exhibit significant landward migration).

Since the parasequence sets within the Upper Ferron Sandstone Last Chance Delta themselves exhibit a stacking pattern, we have defined an "external stacking pattern" nomenclature to describe qualitatively the general 1st-order relative stacking of parasequence sets. The external parasequence set stacking patterns can be either progradational, aggradational, or retrogradational. These parasequence set stacking patterns also reflect the general relationship of the rate of sedimentation to the rate of relative sea-level rise (Garrison and van den Bergh, this volume). If a parasequence set is assigned a progradational external stacking pattern, the landward pinchout of its oldest near-marine sandstone is displaced seaward relative to the landward pinchout of the oldest near-marine sandstone of the underlying parasequence set. These parasequence sets general have internally progradational stacking patterns. If a parasequence set is assigned an aggradational external stacking pattern, the landward pinchout of its oldest near-marine sandstone is located generally in the same areal position relative to the landward pinchout of the oldest near-marine sandstone of the underlying parasequence. These parasequence sets generally have internally aggradational stacking patterns. If a parasequence set is assigned a retrogradational external stacking pattern the landward pinchout of its youngest near-marine sandstone is displaced landward relative to the landward pinchout of the youngest near-marine sandstone of the underlying parasequence set. These parasequence sets generally have an internally retrogradational stacking pattern.

General Geology and Stratigraphy of the Upper Ferron Sandstone Last Chance Delta

The Ferron Sandstone Last Chance Delta, which crops out in Castle Valley of east-central Utah (Figure 2), can be divided into two distinct clastic wedges. The lower portion of the Ferron Sandstone is referred to as the Lower Ferron Sandstone (Ryer and McPhillips, 1983). The upper portion of the Ferron Sandstone, named the Upper Ferron Sandstone (Ryer and McPhillips, 1983), is a younger, thicker wave-modified river-dominated deltaic system, informally called the Last Chance Delta by Hale (1972). Along the outcrop belt, the Blue Gate Shale overlies the Upper Ferron Sandstone, which in turn overlies the Tununk Shale containing the distal facies of the Lower Ferron Sandstone.

Garrison and van den Bergh (this volume) have recognized at least 42 distinct depositional parasequences forming 14 parasequence sets (Figure 3 [also see Table 2 in Garrison and van den Bergh, this volume]). Parasequence Sets 1A and 1B form a sand-poor, river-dominated deltaic system composed of small delta lobes. These lobes consist of well-defined near-marine and associated delta-plain facies associations. Parasequence Sets 2A, 2B, 2C, 2D, and 3 consist of river-dominated, wave-reworked near-marine and associated delta-plain facies associations. Parasequence Sets 4A, 4B, 5A, 5B, 6, 7, and 8 consist of wave-dominated near-marine and associated delta-plain facies associations. Parasequence Sets 1A through 3 have a progradational external stacking pattern. Parasequence Sets 4A through 5B appear to be stacked into an aggradational external stacking pattern. Parasequence Sets 6 through 8 appear to be stacked in a retrogradational external stacking pattern (Garrison and van den Bergh, this volume). In the non-marine to transitional near-marine facies associations, the upper boundaries of parasequence sets are coal zones, when not coincident with sequence boundaries or transgressive ravinement surfaces.

Within the Last Chance Delta, these 14 parasequence sets form four 4th-order depositional sequences (denoted FS1-FS4) (see Table 2 in Garrison and van den Bergh, this volume). Laterally extensive unconformities, with 20–30 m (60–90 ft) of erosional relief, locally mark the upper boundaries of sequences FS1, FS2, and FS3. These unconformities, type-1 sequence boundaries (denoted SB2, SB3, and SB4) produced by major fluvial events, record basinward shifts of paleoshorelines by up to 3–7 km (2–4 mi) and the development of very small incised-valley systems (Garrison and van den Bergh, this volume). Based on the molluscan fauna, the depositional sequence boundaries SB2, SB3, and SB4 are assigned ages of 90.0 Ma, 89.7 Ma, and 89.2 Ma, respectively (Garrison and van den Bergh, this volume).

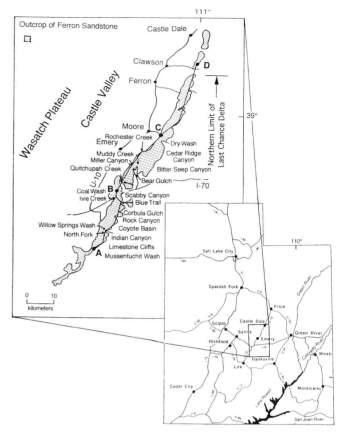

Figure 2. Location map showing the outcrop belt of the Ferron Sandstone in the southern Castle Valley (after Cotter, 1975).

The deposition of the Last Chance Delta occurred during a slow sea-level rise, interrupted only by three minor 4th-order sea-level falls at 90.0 Ma (Garrison and van den Bergh, this volume). Stratigraphic, geometric, and architectural data, for both near-marine and non-marine sandstone facies indicate that Last Chance Delta architecture was controlled by the ratio of sedimentation rate to relative changes in sea level. The systems tract style and parasequence set stacking patterns within 4th-order sequences FS1 through FS4 also reflect the changes in sediment supply (Garrison and van den Bergh, this volume).

Methodology

The depositional sequence stratigraphic study of the Upper Ferron Sandstone Last Chance Delta of Castle Valley represents a detailed field sedimentological and stratigraphic analysis of the entire Ferron Sandstone outcrop belt from just north of Last Chance Creek to Ferron Creek (Garrison and van den Bergh, this volume). The data sets include detailed sedimentological measured sections, geometric sections, core descriptions, and interpretations of photomosaics that represent complete photomosaic coverage of the outcrop belt. The regional cross sections were generated from these data sets. A 63.5-km-long (39-mi) cross section general-

456

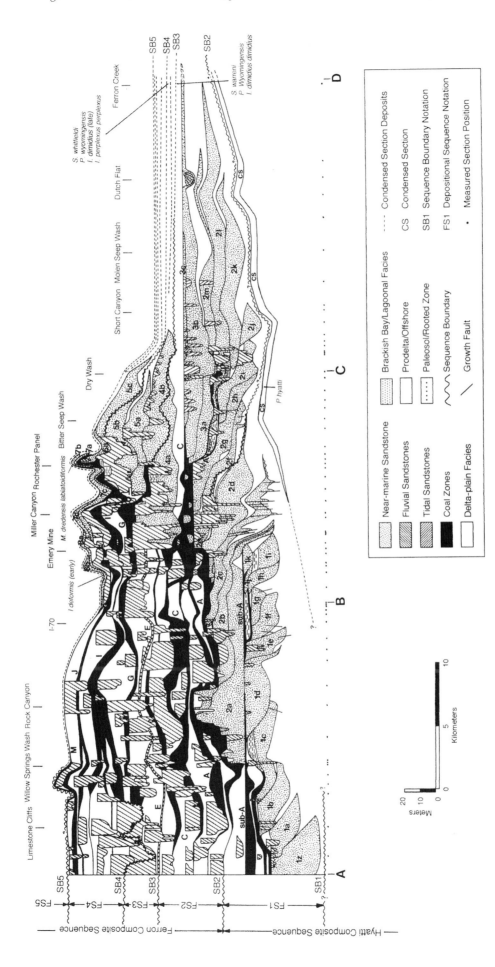

Figure 3. Cross section of the Upper Ferron Sandstone Last Chance Delta clastic wedge generalized from regional cross sections (see Plates II through IV of Garrison and van den Bergh [this volume] on CD-ROM).

ly parallel to the depositional dip of the Last Chance Delta and an intersecting 3 km (2 mi) strike cross section in Willow Springs Wash were generated (Garrison and van den Bergh, this volume).

These regional cross sections show not only the geometry and architecture of the stratal elements within the depositional sequences, but detailed depositional facies, as well. The dimensions of architectural elements are calculated from these cross sections. The mean dimensions of Upper Ferron Sandstone architectural elements are summarized in Table 1. Table 1 includes some data taken from Upper Ferron Sandstone studies of Lowry and Jacobsen (1993) and Barton (1994).

Near-Marine Sandstones

For the sedimentological and architectural analysis of near-marine sandstones, the regional dip cross sections of Garrison and van den Bergh (this volume) are assumed to generally cut all parasequences and parasequence sets in a depositional dip direction. The exceptions are Parasequence Set 3, which is cut in an oblique-dip cross section and Parasequences 1e, 1f, 1g, and 1i of Parasequence Set 1B, which are cut in a strike cross section. Errors introduced by this general assumption will be generally less than ±5%. Errors associated with graphical representations of the geometries displayed on the cross sections of Garrison and van den Bergh (this volume) are approximately ±1%. Since the cross sections cut the sandstones of Parasequence Set 3 in a depositional dip-oblique direction, data from this parasequence set are not included in the dip cross-section analysis. Parasequence Set 2B is also not included in the dip cross-section analysis because of the severe truncation of its near-marine facies by sequence boundary SB2 and the overlying incised-valley fluvial facies of Parasequence 2C. The calculated dip cross-section geometries for the near-marine sandstones and their depositional facies are presented in Tables A-1 through A-5 on the CD-ROM.

Channel-Belt Sandstones

A favorable strike outcrop section that exposes almost the complete delta-plain facies association of the Ferron Sandstone stratigraphic interval is located in Willow Springs Wash, Willow Springs Quadrangle, Emery and Sevier Counties (Figure 2). Outcrop studies to quantify the sizes and shapes of Upper Ferron Sandstone distributary channel belts were conducted in Willow Springs Wash, along both the north and south canyon walls (van den Bergh, 1995; van den Bergh and Garrison, 1996; Garrison and van den Bergh, 1997; this volume). These data represent the majority of the data for the following discussion and subsequent statistical analysis. Lowry and Jacobsen (1993) and Barton (1994) acquired data for some fluvial and distributary channel belts

north of Willow Springs Wash. For completeness, this data is also included in the following analysis.

The outcrop studies, to quantify the sizes and shapes of Ferron Sandstone fluvial channel belts in Willow Springs Wash, consisted of (1) the mapping, on photomosaics, of the architectural components of the Upper Ferron Sandstone channel belts exposed in both the north and south canyon walls of Willow Springs Wash and (2) laser surveying of the boundaries of the channel-belt bodies. The boundaries of the channel belts are located very accurately with the combination of photo mapping and laser surveying, resulting in channel dimensions accurate to within 0.05 m (0.16 ft). The errors associated with the dimensional data presented in the studies of Lowry and Jacobsen (1993) and Barton (1994) is probably about 0.5 m (1.6 ft). This data is presented in Table A-6 on the CD-ROM.

Amalgamated fluvial incised-valley deposits are exposed in Miller and Muddy Creek Canyons, Mesa Butte and Emery East Quadrangles, Emery County (Figure 2). The external and internal architecture and sedimentology of these fluvial deposits were quantified by (1) the mapping, on photomosaics, of the architectural components of the incised-valley fill deposits in Miller and Muddy Creek Canyons and (2) the construction of detailed sedimentological measured sections through the fluvial deposits. The data for these fluvial deposits is also included in Tables A-1 and A-6.

UPPER FERRON SANDSTONE NEAR-MARINE SANDSTONES

Sedimentological Character

River-Dominated Near-Marine Sandstone Deposits

The river-dominated deltaic sandstones of the Upper Ferron Sandstone Last Chance Delta are exemplified by the near-marine facies of Parasequence Set 1B (Figures 1 and 3). Ferron Parasequence Set 1B contains 12 river-dominated, fluvial-deltaic parasequences (denoted 1z, 1a through 1k), and exhibits a progradational parasequence stacking pattern. This river-dominated parasequence set consists of a system of very small, river-dominated sub-delta lobes. The outcrop belt cuts these parasequences in both dip and strike-oblique directions. These sub-delta lobes are small, frequently less than 6 km (4 mi) in length and 3 km (2 mi) in width (Figure 4).

The river-dominated near-marine facies of the parasequences in Ferron Parasequence Set 1B exhibit both vertical and lateral facies changes from (1) stream-mouth bar, to (2) proximal delta front, to (3) distal delta

Table 1. Mean dimensions of Upper Last Chance Delta architectural elements.

Element or Facies*	Parasequence Stacking	Total Length (km)	Maximum Thickness (m)	Total Width (m)	Length/Thickness Aspect Ratio	Width/Thickness Aspect Ratio	Length/Width Aspect Ratio	Delta Class
PSS	Progradational	22.6	28.5	4310	788	-	4.1	River-dominated
PSS	Aggradational	11.4	22.2	-	520	-	-	Wave-reworked
PSS	Retrogradational	5.3	8.9	-	611	-	-	Wave-dominated
PS	Progradational	6	17.2	3080	397	212	1.9	River-dominated
PS	Progradational	11.4	16.1	-	845	-	-	Wave-reworked
PS	Aggradational	11.1	18.5	-	606	-	-	Wave-dominated
PS	Retrogradational	3	5	-	793	-	-	Wave-dominated
DC**	Progradational	-	12.1	198	-	19	-	River-dominated
DC**	Aggradational	-	21.1	255	-	12.1	-	Wave-dominated
DC***	Progradational	-	9.1	250	-	28.8	-	River-dominated
DC***	Aggradational	-	11.9	660	-	59.2	-	Wave-reworked
DC***	Retrogradational	-	5.2	457	-	100	-	Wave-dominated
SMB	Progradational	2.6	5.9	1100	479	256	1.9	River-dominated
SMB	Progradational	-	11.5	1100	-	97.1	-	Wave-reworked
SMB[1]	Progradational	4	-	7900	-	-	2.1	Wave-reworked
SMB[2]	Progradational	4.4	9.3	673	545	155	3.5	Wave-dominated
USF	Aggradational	6.5	13.9	-	595	-	-	Wave-dominated
USF	Retrogradational	1.7	4.1	-	649	-	-	Wave-dominated
pDF	Progradational	3.3	8.3	2000	425	243	1.8	River-dominated
pDF	Progradational	5.1	7.1	4410	827	472	1.8	Wave-reworked
MSF	Aggradational	8	9.8	-	912	-	-	Wave-dominated
MSF	Retrogradational	1.6	2.7	-	807	-	-	Wave-dominated
dDF	Progradational	3	5.6	3130	518	256	2	River-dominated
dDF	Progradational	4.8	5.7	4900	819	1342	0.6	Wave-dominated
LSF	Aggradational	7.8	3.4	-	2469	-	-	Wave-dominated
LSF	Retrogradational	1.5	1.5	-	981	-	-	Wave-dominated
IVF**	Progradational	-	13.9	2217	-	188	-	Wave-reworked
IVF***	Progradational	-	13.8	8889	-	644	-	Wave-reworked
IVF**	Aggradational	-	29.2	4292	-	151	-	Wave-dominated

* PSS = parasequence set; PS = parasequence; DC = distributary channel; FC = fluvial channel; SMB = stream mouth bar; pDF = proximal delta front
dDF = distal delta front; USF = upper shoreface; MSF = middle shoreface; LSF = lower shoreface; IVF = incised-valley fluvial deposit
** Distal distributary channel or incised-valley fill fluvial deposit
*** Proximal distributary channel or incised-valley fluvial deposit
[1]Barton (1994); [2]Lowry & Jacobsen (1993); All other measurements were compiled in this study

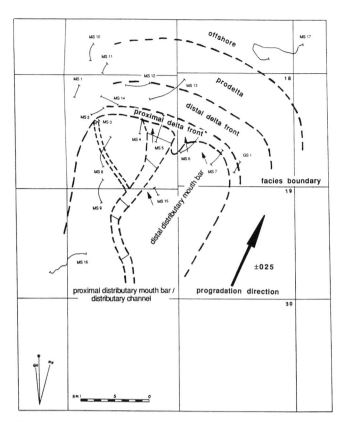

Figure 4. Map showing the general shape and size of river-dominated deltaic Ferron Parasequence 1c in Willow Springs Wash. Measured section locations and nomenclature is that of van den Bergh (1995).

front, to (4) prodelta. Figure 5 shows a measured section through Ferron Parasequence 1c in Willow Springs Wash displaying the vertical facies change through the river-dominated near-marine deposits. The delta-front deposits are shale-rich (van den Bergh, 1995). Based on the observations of van den Bergh (1995), in Willow Springs Wash, Parasequence Set 1B near-marine facies is estimated to consist of approximately 68% sandstone and 32% silty mudstone. In general, Parasequence Set 1B near-marine deposits exhibit little to mild evidence of wave influence. The near-marine deposits locally may exhibit poorly developed hummocky cross-stratification, and only rarely herringbone cross-stratification and bi-directional ripple cross-stratification. These small delta lobes frequently have well-developed clinoforms, particularly on what is interpreted to be the side opposite the prevailing swell direction (e.g., Scabby Canyon to Ivie Creek to South Quitchupah Creek) (Garrison and van den Bergh, this volume).

The stream-mouth-bar deposits are typically very fine to fine-grained, and occasionally medium-grained, sandstones that locally can contain up to 10% interbedded to massive to laminated mudstones and siltstones, in the abandonment phase at their tops (Figure 5) (van den Bergh, 1995). In general, the bedform profile ranges from trough cross-stratification, in the lower portion, to

trough cross-beds containing climbing ripples to ripple cross-stratification near the top of the deposit. Trough bed thickness generally ranges from 20–30 cm (8–12 in.) and the ripple bedset thickness is generally about 10 cm (4 in.). Commonly the base of the stream-mouth bar is a scour surface. The upper surface of the stream-mouth bar is a marine-flooding surface that commonly exhibits burrowing by marine organisms (*Arenicolites*). Delta lobe abandonment was typically accompanied by mild to moderate wave-reworking, such that the uppermost parts of the stream-mouth-bar facies may contain herringbone cross-stratification, *Skolithos* and *Ophiomorpha* burrows, and low-angle cross-stratification.

The proximal delta-front deposits are typically very fine grained, and occasionally fine-grained, sandstones with up to 30% mudstone and silty mudstone interlayers (van den Bergh, 1995). The sandstone bedsets generally thicken upward through the facies. The sandstone bedsets range in thickness from 0.1–1.0 m (0.3–3.3 ft). The cross-stratification is dominantly planar to ripple cross-stratification, although sporadic hummocky cross-stratification also occurs. The sandstone bedsets may either fine upward from very fine grained sandstone to siltstone or silty mudstone or coarsen upward from silty mudstone or siltstone to very fine grained sandstone. Commonly the sandstone beds are moderately burrowed, containing mud-filled *Teichichnus*, *Arenicolites*, *Planolites*, *Skolithos*, *Cylindrichnus*, and *Diplocraterion* burrows. In areas where the proximal delta-front deposits exhibit mild wave-reworking, the sandstones may be burrowed by *Ophiomorpha*, *Thalassinoides*, and *Teichichnus*.

The distal delta-front deposits are typically very fine grained sandstones with up to 50% interlayers of massive to laminated mudstones and silty mudstones (van den Bergh, 1995). The sandstone bedsets generally thicken upward through the facies. The cross-stratification is dominantly planar to ripple cross-stratification; hummocky cross-stratification is infrequent. The sandstone bedsets range in thickness from 10–60 cm (4–24 in.). These sandstone bedsets generally thicken upward and also fine upward in grain size. Distal delta-front sandstones may contain *Skolithos*, *Planolites*, and *Arenicolites* burrows.

Wave-Reworked Near-Marine Sandstone Deposits

Wave-reworked, river-dominated deltaic deposits of the Upper Ferron Sandstone Last Chance Delta are exemplified by the near-marine facies of Parasequence Set 2A (Figures 1 and 3). Parasequence Set 2A contains four wave-modified, river-dominated parasequences (denoted 2a through 2d), and exhibits an internal progradational parasequence stacking pattern. Parasequence Set 2A exhibits an external progradational stacking pattern, relative to the underlying and overlying parasequence sets.

The near-marine parasequences in Ferron Parasequence Set 2A exhibit both vertical and lateral facies changes from (1) reworked stream-mouth bar, frequently preserved as upper shoreface deposits, to (2) reworked delta front, frequently preserved as middle and lower shoreface deposits, to (3) prodelta. Figure 6 shows a measured section through Ferron Parasequence 2d in Muddy Creek amphitheater illustrating the facies succession. These reworked near-marine deposits typically contain less than 5% mudstone and siltstone. In general, reworked stream-mouth bar and distal bar (i.e., pDF and dDF) deposits of Parasequence Set 2A are very sand-rich and exhibit evidence of moderate wave influence. Minor tidal influence is evident as a small tidal-channel complex in Parasequence 2a in Coyote Basin (Garrison and van den Bergh, this volume).

The wave-reworked stream-mouth-bar sandstones are typically very fine to medium-grained sandstones. The bedform profile ranges from trough cross-stratification, in the lower portion to ripple cross-stratification near the top of the deposit; occasionally herringbone cross-stratification occurs in the middle portions. Trough bed thickness generally ranges from 10–70 cm (4–28 in.), with bedsets ranging from 0.2–1.5 m (0.7–5.9 ft); herringbone cross-stratified beds range in thickness from 10–30 cm (4 in.–1 ft). Planar stratified bedsets range in thickness from 0.2–1.0 m (0.7–3.3 ft); the ripple bedset thickness ranges from 10 cm–1.0 m (4 in.–3.3 ft). Wave-reworked stream-mouth bar sandstones are frequently burrowed by *Ophiomorpha, Planolites, Teichichnus, Skolithos,* and *Thalassinoides* (ichnofacies *Skolithos* and *Cruziana*).

The wave-reworked proximal delta-front sandstones are typically very fine to fine-grained sandstones, that locally can contain up to 5% mudstone and siltstone lamina. Beds generally thicken upward and grain size coarsens upward. These sandstones range from hummocky cross-stratified to horizontal stratified to ripple cross-stratified. Hummocky cross-stratified beds range in thickness from 10–60 cm (4 in.–2 ft), with bedset thickness ranging from 0.2–2.0 m (0.7–6.6 ft), and the ripple bedset thickness ranging from 10–70 cm (4–28 in.). Wave-reworked proximal delta-front sandstones frequently contain *Ophiomorpha, Arenicolites, Planolites, Teichichnus, Palaeophycus, Skolithos,* and *Thalassinoides* burrows (ichnofacies *Skolithos* and *Cruziana*).

The wave-reworked distal delta-front sandstones are typically very fine to fine-grained sandstones, that locally can contain up to 20% mudstone and siltstone lamina. The beds generally thicken and coarsen upward. These sandstones range from hummocky cross-stratified to ripple cross-stratified; occasionally horizontal stratification is present. Hummocky cross-stratified beds range

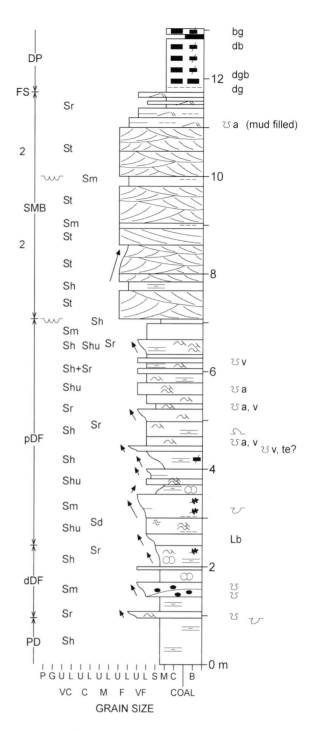

Figure 5. Measured section through Ferron Parasequence 1c in Willow Springs Wash showing the vertical facies change through the river-dominated, near-marine facies (after van den Bergh, 1995).

in thickness from 5–20 cm (2–8 in.), with bedset thickness ranging from 25–50 cm (10–20 in.). Ripples range in size from 1–10 cm (0.4–4 in.), with the ripple bedset thickness ranging from 5–10 cm (2–4 in.). Wave-reworked distal delta-front sandstones frequently contain *Ophiomorpha, Arenicolites, Planolites, Teichichnus, Diplocraterion, Skolithos,* and *Thalassinoides* burrows (ichnofacies *Skolithos* and *Cruziana*).

Sequence Stratigraphy and Depositional Facies

SB	Depositional sequence boundary
PSS	Parasequence set
PS	Parasequence
MFS	Maximum flooding surface
FS	Flooding surface
TL	Transgressive lag
TRS	Transgressive ravinement surface
SMB	Stream-mouth bar
DP	Delta-plain facies
FC	Fluvial channel
DC	Distributary channel
PD	Prodelta
pDF	Proximal delta front
dDF	Distal delta front
BB	Brackish-bay facies
USF	Upper shoreface
MSF	Middle shoreface
LSF	Lower shoreface
OS	Offshore

General Rock Description

sst	Sandstone
sist	Siltstone
▨	Altered volcanic ash layer
▬	Coal
➡ ➡	Almost coal
mst ---	Mudstone
▬ ▬	Carbonaceous shale
✶	Carbonaceous plant fragments
➡	Coal fragment
C	Clay grain size
M	Mudstone grain size
S	Silt grain size
VFL	Very fine lower grain size
VFU	Very fine upper grain size
FL	Fine lower grain size
FU	Fine upper grain size
ML	Medium lower grain size
MU	Medium upper grain size
CL	Coarse lower grain size
CU	Coarse upper grain size
G	Granule grain size
P	Pebble grain size
B	Altered volcanic ash deposit (including tonsteins and bentonites)
↖	Coarsening-upward grain size
↗	Fining-upward grain size

b	Brown	y	Yellow
bl	Black	gr	Green
g	Gray	L	Light color
r	Red	d	Dark color

Surface Hierarchy

1	Bed surface	5	Base of major channel system within channel belt
2	Bedset surface		
3	Base of channel element	6	Base of channel belt
4	Base of barform	7	Depositional sequence boundary

Sedimentary Structures and Detailed Rock Descriptions

Sh ---	Horizontal lamination in mudstone
Sh ═	Horizontal stratification in sandstone
Sr ∿	Wave ripples
Sr ⌒	Current ripples
St	Trough cross-stratification
St$_{La}$	Large-scale trough cross-stratification
St$_S$	Small-scale trough cross-stratification
St = 090	Paleocurrent direction
Shu	Hummocky trough cross-stratification
Ssw	Swaley trough cross-stratification
Sm	Massive bedding
Sher	Herringbone trough cross-stratification
\\\\\\\	Wedge-tabular cross-stratification
↑ML-FU	Fining-upward on lamina scale
≈	Contorted bedding
∞	Bumpy weathering
⤚	Mudstone rip-up clasts
⌶	Scour surface
ᴜ	Flutes and grooves
∩	Flame structure
◉	Septarian nodules
ooo	Extrabasinal pebbles
●●●	Intrabasinal pebbles

Ichnofossils and Other Biostratigraphic Information

➤	Oysters
•)	Brackish-bay fauna
ʊ	Burrow (unidentified)
ʊ H	Horizontal burrow
ʊ esp	Escape burrow
ʊ sk	*Skolithos*
ʊ te	*Teichichnus*
ʊ th	*Thalassinoides*
ʊ oph	*Ophiomorpha*
ʊ pl	*Planolites*
ʊ ch	*Chondrites*
ʊ con	?
⋏	Roots
✐	Logs and trees
✐+t	Log with *Teredolites*
⬭	Bivalve

Figure 5 continued.

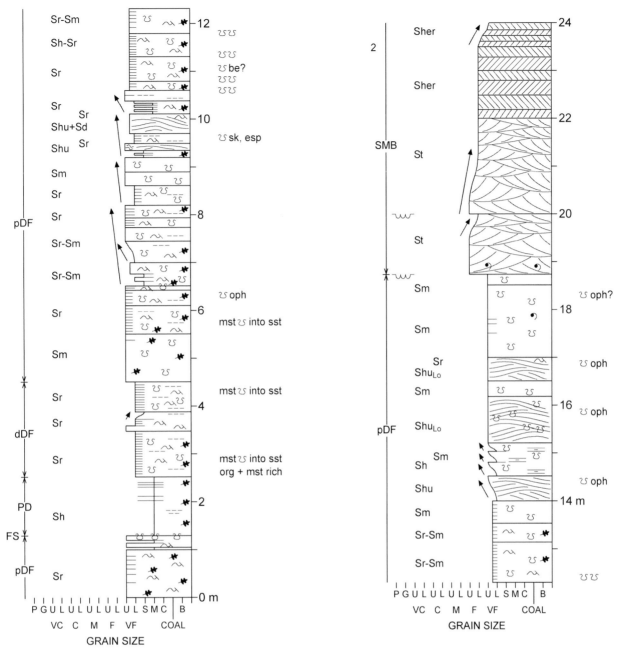

Figure 6. A portion of measured section GV16 in Muddy Creek Amphitheater, in Muddy Creek Canyon showing the vertical facies succession through Ferron Parasequence 2d. See Figure 5 for explanation of symbols and notation.

Wave-Dominated Near-Marine Sandstone Deposits

Wave-dominated deltaic deposits of the Upper Ferron Sandstone Last Chance Delta are exemplified by the near-marine facies of Parasequence Set 4A (Figures 1 and 3). Parasequence Set 4A contains only a single parasequence (denoted 4a), but is given hierarchical equivalence to a parasequence set (Garrison and van den Bergh, this volume). Parasequence Set 4A exhibits and external aggradational stacking pattern, relative to the underlying and overlying parasequence sets.

The near-marine facies of the single parasequence in Parasequence Set 4A represent wave-dominated stream-mouth-bar and delta-front deposits, generally preserved as upper, middle, and lower shoreface deposits. The near-marine facies exhibits vertical and lateral facies changes from (1) reworked stream-mouth-bar, preserved as upper shoreface deposits (USF), to (2) reworked proximal delta-front deposits (pDF), preserved as middle shoreface (MSF) deposits, to (3) reworked distal delta-front deposits (dDF), preserved as lower shoreface (LSF) deposits, to (4) prodelta. Figure 7 shows measured sections through Ferron Parasequence 4a, in Miller Canyon,

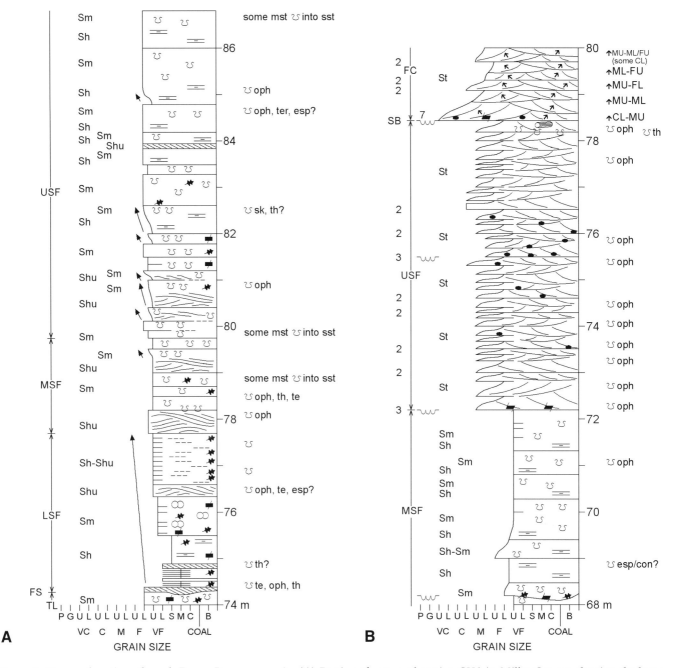

Figure 7. Measured sections through Ferron Parasequence 4a. (A) Portion of measured section GV6 in Miller Canyon showing the lower shoreface facies of Parasequence 4a. (B) Portion of measured section GV8 in Bear Gulch showing the middle and upper shoreface facies of Parasequence 4a. See Figure 5 for explanation of symbols and notation.

displaying the lower shoreface facies and, in Bear Gulch, displaying the middle and upper shoreface facies. These near-marine deposits are preserved as a wave-dominated shoreface profile. The near-marine deposits of Parasequence Set 4A typically contain less than 1% interlayered siltstone and mudstone.

Parasequence Set 4A upper shoreface deposits are fine- to medium-grained sandstones. These sandstones are trough cross-stratified with beds ranging in thickness from 10–30 cm (4–12 in.) and trough bedsets ranging from 0.3–2.0 m (10–6.6 ft) in thickness. Troughs frequent-

ly fine upward from medium- to fine-grained sandstone. These sandstones frequently contain the ichnofossil *Ophiomorpha* (vertical) and less commonly *Thalassinoides*.

The middle shoreface deposits are very fine to fine-grained sandstones containing less than 1% mudstone and siltstone interlayers. Individual bedsets may coarsen upward from very fine grained to fine-grained sandstones. Occasionally bedsets may have a fine-grained base and then rapidly fine upward into a uniform very fine grained sandstone. The middle shoreface deposits contain hummocky to planar cross-stratification and are frequently intensely burrowed by *Ophiomorpha* (horizontal), *Diplocraterion*, *Teichichnus*, *Planolites*, *Skolithos*, and *Thalassinoides* (ichnofacies *Skolithos* and *Cruziana*), such that no bedding is preserved. When preserved, the hummocky cross-stratification is 10–20 cm (4–8 in.) thick, with bedsets ranging from 30–70 cm (12–28 in.) in thickness.

Parasequence Set 4A lower shoreface deposits are very fine to fine-grained sandstones containing up to 25% sandy siltstone interlayers. The facies generally coarsens upward from burrowed sandy siltstone, at the base, to sandstone bedsets that fine upward from very fine grained sandstone to siltstone, in middle portions of the facies, to sandstone bedsets at the top that coarsen upward from very fine grained sandstone to fine-grained sandstone, at the top. The lower shoreface deposits are hummocky cross-stratified to horizontal stratified to ripple cross-stratified, and are commonly intensely burrowed, such that no bedding is preserved. Bedsets range from 10–70 cm (4–28 in.) in thickness.

Architectural Systematics

Parasequences and Parasequence Sets

A plot of dip length versus thickness for Upper Ferron Sandstone Parasequence Sets and Parasequences is shown in Figure 8. Table 1 summarizes the mean dimensions of architectural elements of the Upper Ferron Sandstone Last Chance Delta. Tables A-1 and A-2 present detailed quantitative architectural data for the Ferron parasequence sets and parasequences, respectively. A detailed examination of the dimensions of parasequence sets and parasequences (Figure 8, and Tables A-1 and A-2), indicates some clear trends based on analysis of delta class, parasequence stacking patterns, and the systematics of sediment supply and available accommodation space.

Parasequence sets: When classified according to their general parasequence set stacking and delta class, the near-marine sandstones of the parasequence sets occupy distinct fields on the plot of dip length versus thickness. The progradational Parasequence Sets 1B, and 2A, 2C, and 2D have mean dip lengths of 23 km (14 mi) with mean thicknesses of 28 m (69 ft); the mean aspect ratio (i.e., length/thickness) is 788. The aggradational Parasequence Sets 4A, 4B, 5A, and 5B have mean dip lengths of 11 km (7 mi) with mean thicknesses of 22 m (66 ft); the mean aspect ratio is 520. The retrogradational Parasequence Sets 6, 7, and 8 have mean dip lengths of 5.3 km (3 mi) with mean thicknesses of 9 m (27 ft); the mean aspect ratio is 611.

Examining the parasequences of Parasequence Set 1B (i.e., Parasequences 1e, 1f, and 1g) that exhibit strike sections along the cross section (see Plate II of Garrison

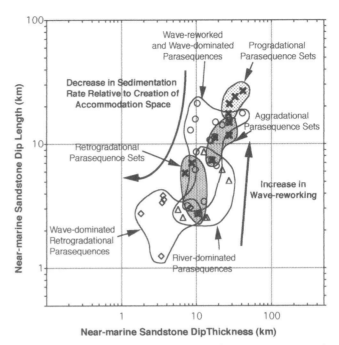

Figure 8. Plot of dip length versus thickness for Upper Ferron Sandstone parasequence sets and parasequences.

and van den Bergh, this volume, on the CD-ROM), it is possible to obtain some limited information on the areal aspect ratio (i.e., dip length/strike width) of river-dominated Parasequence Set 1B. These parasequences have a total strike width of about 6 km (4 mi), which suggests an areal aspect ratio of about 4.1.

Parasequences: Examination of the dip length and thickness systematics, at the parasequence scale, reveals groupings based on the degree of wave-reworking (Figure 8). In particular, in Parasequence Sets 1B, 2A, 2C, and 2D, that exhibit a progradational stacking pattern, the effects of degree of wave-reworking is very evident. In the very river-dominated Parasequence Set 1B, the average near-marine facies of the Parasequences 1a, 1b, 1c, 1d, and 1k is 17 m (56 ft) thick and 6.0 km (3.7 mi) long, with a mean aspect ratio of 397. The strike dimensions of Parasequences 1e, 1f, and 1g suggest that these deltas had a mean strike width/thickness aspect ratio of 212 and mean dip length/strike width aspect ratios near 1.9. The river-dominated, but wave-reworked near-marine facies of parasequences of Parasequence Sets 2A, 2B, and 2D are similar in thickness, with a mean thickness of 16 m (48 ft), but have substantially longer dip lengths, averaging 11.0 km (6.8 mi), and higher aspect ratios (e.g., mean aspect ratio is 845). The length and thickness systematics of the parasequences belonging to the externally aggradational parasequence sets are similar to the wave-reworked, externally progradational parasequence sets (i.e. mean thickness is 18 m [59 ft] and mean dip length is 11.1 km [6.9 mi]); the mean aspect ratio is 606. This decrease in progradational length reflects the overprint of a declining sediment supply, as discussed above. The parasequences that belong to parasequence sets exhibiting a retrogradational stacking pattern (i.e., Parasequence Sets 6, 7, and 8) are shorter in dip length, with mean lengths of only 3.0 km (1.9 mi), and are thin, with mean thicknesses of only 5.0 m (16.4 ft), resulting in a mean aspect ratio of 793.

Sandstone Depositional Facies

A plot of dip length versus thickness for stream-mouth bar, proximal delta-front, and distal delta-front depositional facies from Ferron parasequence sets are shown in Figures 9, 10, and 11. Wave-reworked facies are also included in these plots. Although the number of measurements is relatively small, examining the dimensions of individual depositional facies within the Ferron Last Chance Delta parasequences and parasequence sets (Tables 1, A-3, A-4, and A-5), clear trends can be discerned.

Stream-mouth-bar facies and reworked equivalent facies: A plot of dip length versus thickness for stream-mouth-bar depositional facies and their reworked equivalents, including upper shoreface deposits, from Upper Ferron Sandstone parasequence sets are shown in Figure 9. The dimensions of stream-mouth-bar and equivalent deposits, when examined as a function of position within the Upper Ferron Sandstone Last Chance Delta, indicate clear trends.

The stream-mouth-bar deposits from the river-dominated parasequences of progradational Parasequence Set 1B have mean dip lengths of approximately 2.6 km (1.6 mi) and a mean thickness of about 6 m (18 ft), with

a mean length/thickness aspect ratio of 479. The mean strike length of Parasequence Set 1B stream-mouth-bar deposits, exhibiting a strike cross section, is 1.1 km (0.7 mi). This suggests a mean strike width/thickness aspect ratio of 256 and a mean dip length/strike width aspect ratio of about 1.9.

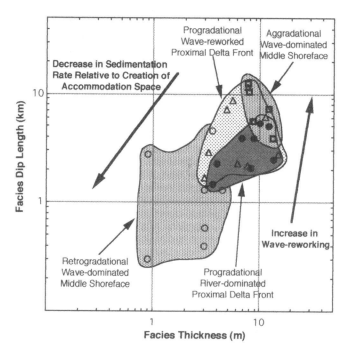

Figure 10. Plot of dip length versus thickness for proximal delta-front depositional facies and their reworked equivalents, including middle shoreface deposits, from Ferron parasequence sets.

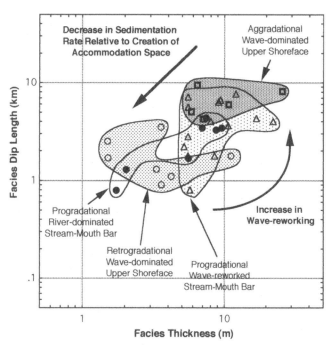

Figure 9. Plot of dip length versus thickness for stream-mouth-bar depositional facies and their reworked equivalents, including upper shoreface deposits, from the Upper Ferron Sandstone parasequence sets.

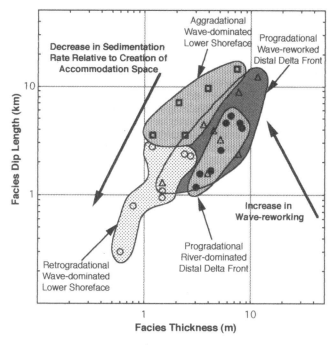

Figure 11. Plot of dip length versus thickness for distal delta-front depositional facies and their reworked equivalents, including lower shoreface deposits, from Ferron parasequence sets.

466

The wave-reworked stream-mouth-bar deposits, from externally progradational Parasequence Sets 2A, 2B, and 2C, have a mean dip length of approximately 4.4 km (2.7 mi) and a mean thickness of about 9 m (27 ft), with a mean length/thickness aspect ratio of 545. The mean strike length of these wave-reworked stream-mouth-bar deposits, exhibiting a strike cross section, is 0.7 km (0.4 mi). This suggests a mean strike width/thickness aspect ratio of 155. Barton (1994) reported an even smaller width/thickness aspect ratio of 97. We calculated a mean dip length/strike width aspect ratio of about 3.5. This is somewhat larger than the 2.1 reported by Lowry and Jacobsen (1993).

The upper shoreface deposits from externally aggradational Parasequence Sets 4A, 4B, 5A, and 5B, representing extensively reworked stream-mouth-bar deposits, are much thicker and have longer dip lengths than the stream-mouth-bar and reworked stream-mouth-bar deposits of progradational Parasequence Sets 1, 2A, 2C, and 2D. These deposits have a mean dip length of 6.5 km (4.0 mi) and a mean thickness of about 14 m (42 ft). Although these deposits are longer and thicker than the externally progradational parasequence sets, they have a similar mean dip length/thickness aspect ratio of 595. No data exists to speculate on the dip length/strike width aspect ratio of these upper shoreface deposits.

The upper shoreface deposits from retrogradational Parasequence Sets 6, 7, and 8, representing extensively wave-modified stream-mouth-bar deposits, are much thinner and have shorter dip lengths than the stream-mouth-bar and reworked stream-mouth-bar deposits of progradational Parasequence Sets 1, 2A, 2C, and 2D, and the aggradational Parasequence Sets 4A, 4B, 5A, and 5B. These deposits have a mean dip length of about 1.7 km (1.1 mi) and a mean thickness of about 4 m (16 ft). These deposits have a mean dip length/thickness aspect ratio of 649. No data exists to speculate on the dip length/strike width aspect ratio of these upper shoreface deposits.

Proximal delta-front facies and reworked equivalent facies: A plot of dip length versus thickness for proximal delta-front depositional facies and their reworked equivalents, including middle shoreface deposits, from Ferron parasequence sets are shown in Figure 10. The dimensions of proximal delta-front and equivalent deposits, when examined as a function of position within the Upper Ferron Sandstone Last Chance Delta, also exhibit clear trends.

The proximal delta-front deposits from the river-dominated parasequences of progradational Parasequence Set 1B have a mean dip length of approximately 3.3 km (2.1 mi) and a mean thickness of about 8 m (24 ft), with a mean length/thickness aspect ratio of 425. The mean strike length of Parasequence Set 1B proximal

delta-front deposits is 2.0 km (1.2 mi). This suggests a mean strike width/thickness aspect ratio of 243 and a mean dip length/strike width aspect ratio of about 1.8.

When compared to the river-dominated proximal delta-front deposits, the wave-reworked proximal delta-front deposits, from externally progradational Parasequence Sets 2A, 2B, and 2C, have a slightly longer mean dip length of about 5.1 km (3.2 mi), but a similar mean thickness of about 7 m (21 ft). They have a mean length/thickness aspect ratio of 827. The mean strike length of these reworked stream-mouth-bar deposits, exhibiting a strike cross section, is about 4.4 km (2.7 mi). This suggests a mean strike width/thickness aspect ratio of 472 and a mean dip length/strike width aspect ratio of 1.8.

The middle shoreface deposits from externally aggradational Parasequence Sets 4A, 4B, 5A, and 5B, representing extensively reworked stream-mouth-bar deposits, are slightly thicker and have longer dip lengths than the proximal delta-front and reworked proximal delta-front deposits of progradational Parasequence Sets 1, 2A, 2C, and 2D. These deposits have a mean dip length of 8.0 km (5.0 mi), a mean thickness of about 10 m (30 ft), and a mean dip length/thickness aspect ratio of 912. No data exists to speculate on the dip length/strike width aspect ratio of these middle shoreface deposits.

The middle shoreface deposits from retrogradational Parasequence Sets 6, 7, and 8, representing extensively reworked stream-mouth bar deposits, are much thinner and have shorter dip lengths than the upper shoreface deposits of progradational Parasequence Sets 1, 2A, 2C, and 2D, and the aggradational Parasequence Sets 4A, 4B, 5A, and 5B. These deposits have a mean dip length of about 1.6 km (1.0 mi) and a mean thickness of about 3 m (9 ft). These deposits have a mean dip length/thickness aspect ratio of 807. No data exists to speculate on the dip length/strike width aspect ratio of these middle shoreface deposits.

Distal delta-front facies and reworked equivalent facies: A plot of dip length versus thickness for distal delta-front depositional facies and their reworked equivalents, including lower shoreface deposits, from Ferron parasequence sets are shown in Figure 11.

The distal delta-front deposits from the river-dominated parasequences of progradational Parasequence Set 1B have a mean dip length of approximately 3.0 km (1.9 mi) and a mean thickness of about 6 m (18 ft), with a mean length/thickness aspect ratio of 518. The mean strike length of Parasequence Set 1B proximal delta-front deposits is 3.1 km (1.9 mi). This suggests a mean strike width/thickness aspect ratio of 256 and a mean dip length/strike width aspect ratio of about 2.0.

When compared to the river-dominated distal delta-front deposits, the wave-reworked distal delta-front deposits, from externally progradational Parasequence

Sets 2A, 2B, and 2C, have a slightly longer mean dip length of approximately 4.8 km (7.7 mi), but a similar mean thickness of about 6 m (18 ft). They have a mean length/thickness aspect ratio of 819. Only one measurement of a strike cross section was obtained. Although there is substantial error in using this single measurement, the strike measurement suggests a mean strike length of about 4.9 km (3.0 mi), which indicates in a mean strike width/thickness aspect ratio of 1342 and a mean dip length/strike width aspect ratio of 0.6.

The lower shoreface deposits from externally aggradational Parasequence Sets 4A, 4B, 5A, and 5B, representing extensively reworked distal delta-front deposits, are slightly thinner and have longer dip lengths than the distal delta-front and reworked distal delta-front deposits of progradational Parasequence Sets 1, 2A, 2C, and 2D. These deposits have a mean dip length of 7.8 km (4.8 mi) and a mean thickness of about 3 m (9 ft). These deposits have a mean dip length/thickness aspect ratio of 2469. No data exists to speculate on the dip length/strike width aspect ratio of these lower shoreface deposits.

The lower shoreface deposits from retrogradational Parasequence Sets 6, 7, and 8, representing extensively reworked distal delta-front deposits, are much thinner and have shorter dip lengths than the lower shoreface deposits of progradational Parasequence Sets 1, 2A, 2C, and 2D, and the aggradational Parasequence Sets 4A, 4B, 5A, and 5B. These deposits have a mean dip length of about 1.5 km (0.9 mi) and a mean thickness of about 1.5 m (4.5 ft). These deposits have a mean dip length/thickness aspect ratio of 981. No data exists to speculate on the dip length/strike width aspect ratio of these lower shoreface deposits.

Near-Marine Sandstone Sedimentology and Architecture as a Function of Varying Sediment Supply and Parasequence Set Stacking

Sedimentology

The sandstones from the river-dominated, wave-reworked, and wave-dominated near-marine parasequences exhibit clearly distinct sedimentological characteristics that reflect the changes in sediment supply, progradation rate, and relative sea level. The ratio of sedimentation rate to rate of relative change in sea level and progradation rate are strongly correlative with the degree of delta modification by marine processes, which is reflected in the delta class (Figure 1).

River-dominated near-marine parasequences: Garrison and van den Bergh (this volume) calculated the sedimentation rate and average relative rise in sea level dur-

ing the deposition of river-dominated Parasequence Set 1B to be 2.2 mm/year and 1.0 mm/year, respectively; the average progradation rate was calculated to be 0.6 m/year (2 ft/yr). In the river-dominated near-marine parasequences, distal prodelta/shelf slumping, contorted bedding, flame structures, escape burrows, and the cannibalization of distal facies of the stream-mouth bars by proximal facies of the stream-mouth bars indicate the sedimentological response to rapid sedimentation and progradation (van den Bergh, 1995).

Wave-reworked near-marine parasequences: Garrison and van den Bergh (this volume) calculated the average sedimentation rate during the deposition of wave-reworked Parasequence Set 2A to be about 1.3 mm/year and the average relative rise in sea level to be about 0.9 mm/year. This decrease in sedimentation rate, relative to the sedimentation rate during the deposition of Parasequence 1B, slowed progradation to about 0.5 m/year (1.6 ft/yr). The decrease sedimentation and progradation rates allowed wave processes to become more effective in reworking the delta-front deposits (Figure 1). The sedimentological character of Parasequence 2A reflects these changes. The reworked stream-mouth-bar deposits are frequently moderately burrowed (ichnofacies *Skolithos* and *Cruziana*) and may exhibit well-developed trough and, less frequently, herringbone stratification, suggesting the initiation of the development of a shoreface profile. The reworked delta-front deposits are frequently moderately burrowed (ichnofacies *Skolithos* and *Cruziana*), and exhibit well-developed hummocky and planar stratification and bi-directional ripple-stratification (van den Bergh, 1995).

Wave-dominated near-marine parasequences: The sedimentation rate during the deposition of Parasequence Set 4A has been estimated to average about 0.3 mm/year and the rise in relative sea level is estimated to be 0.2 mm/year (Garrison and van den Bergh, this volume). This decrease in sedimentation rate and in the rate of relative rise in sea level during Parasequence Set 4A deposition, relative to that during Parasequence Set 2A deposition, slowed progradation to 0.1 m/year (0.3 ft/yr), promoted aggradation, and allowed wave processes to rework the delta-front deposits into a shoreface profile.

Architecture

The architectural systematics of parasequence sets suggest that the general trend in parasequence set evolution within the Last Chance Delta clastic wedge is one of first progradations of thick, long parasequence sets, followed by the aggradation of thick, short parasequence sets, and finally followed by the retrogradation of thin, short parasequence sets. When considered in the context of the slow 3rd-order relative rise in sea level and sediment supply systematics presented by Garrison

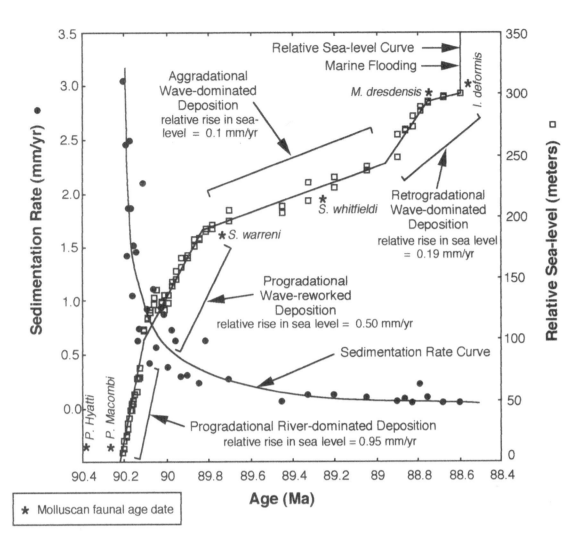

Figure 12. Plot showing interplay of sedimentation rate and relative change in sea level as a function of time for the Upper Ferron Sandstone Last Chance Delta.

and van den Bergh (this volume) (Figure 12), this overall trend can be explained by a progressive decrease in sediment supply throughout the evolution of the delta.

The architectural systematics of the near-marine facies of the river-dominated progradational parasequences of Parasequence Set 1B are consistent with the typical morphology and avulsion frequency exhibited by river-dominated deltas. The progressive decrease in sediment supply, during deposition of the wave-reworked parasequences of Parasequence Sets 2A-3 and the wave-dominated aggradational parasequences of Parasequence Sets 4A through 5B, dramatically slowed the seaward progradation of these deltaic events and allowed sufficient time for wave reworking. The dimensions of these parasequences reflect both the effects of low sedimentation rate and increased wave-reworking. The parasequences that belong to retrogradational parasequence sets have lengths and thicknesses that appear to be mainly controlled by the continued, but

dramatic, decrease in sediment supply, as well as moderate wave-reworking.

It is clear from the dimensional data (Figures 1 and 8) that the sedimentation rate progressively and systematically decreased and the degree of wave-reworking increased throughout the deposition of the Last Chance Delta clastic wedge. Even though this analysis is made by examining the dimensions of parasequences and parasequence sets, instead of the genetic sequences defined by Gardner (1992, 1993), this study is in full agreement with the conclusions of the sediment volume calculations of Gardner (1993).

No consistent geometrical or architectural relationships can be found that correlate with the positions of parasequences or parasequence sets within the systems tracts of the higher frequency 4th-order depositional sequences (see Table 2 in Garrison and van den Bergh, this volume). Therefore, it can be assumed that the effects of the 4th-order sea-level fluctuations were overshadowed by the overwhelming effects of the sediment supply from 2.2 mm/year to 0.1 mm/year, during a period of continual slow decreasing rate of sea-level rise (i.e., from 1.0 mm/year to 0.1 mm./year) over a period of 1.7 m.y. (Garrison and van den Bergh, this volume).

Progradational parasequence sets: In the progradational parasequence sets, sediment supply is large relative to available accommodation space, resulting in substantial progradation of successive delta events. In the beginning stages of the development of the Last Chance Delta (i.e., during the deposition of Parasequence Set 1B) when sedimentation rates were very high (i.e. 2.2 mm/year with a rise in relative sea level of 1.0 mm/year), the deltas were extremely river-dominated, with little wave-reworking (Figure 1). This allowed rapid seaward progradation of the delta. The geometry of the near-marine sandstones of this river-dominated system was determined by frequent local avulsions, high sedimentation rates, and very rapid seaward progradation rates.

During the deposition of Parasequence Sets 2A, 2C, and 3, there was still a moderate sedimentation rate (0.9 mm/year), although the rate of rise in relative sea level slowed (0.5 mm/year). This change still allowed for major seaward progradation and an increase in the degree of wave-reworking (Figure 1).

The study of Heller and Paola (1996) has suggested that frequency of local avulsions is strongly coupled to sedimentation rate. High sedimentation rates lead to rapid shoreline progradation which, in turn, forces rivers to rapidly readjust their equilibrium profiles. This rapid adjustment of the stream equilibrium profile is accompanied by upstream vertical aggradation and frequent local avulsion. In a back-tilted basin with high sedimentation rates, there will be a high frequency of local avulsions. This high avulsion frequency would have the effect of preventing a delta lobe from prograding too far, before avulsion, abandonment, and subsequent re-establishment of deltaic deposition in another adjacent location. This factor is probably also manifest in the differences in near-marine sandstone architecture between the river-dominated progradational parasequences of Parasequence Set 1B and the wave-reworked progradational parasequences of Parasequence Sets 2A, 2C, and 2D. The river-dominated parasequences are shorter and thinner than the wave-reworked parasequences. The dimensions of the stream-mouth-bar, proximal delta-front, and distal delta-front deposits are, consequently, shorter and thinner, as well.

Aggradational parasequence sets: Midway through the development of the Last Chance Delta, during the deposition of the externally aggradational Parasequence Sets 4A, 4B, 5A, and 5B, the sedimentation rate declined somewhat (0.2 mm/year) and the rate of rise in relative sea level slowed (0.1 mm/year) (Garrison and van den Bergh, this volume). This also slowed the seaward pro-gradation, allowing vertical aggradation of the deltaic events and substantial wave-reworking.

Within parasequence sets that exhibit an aggradational stacking pattern, the near-marine sandstones of the individual parasequences and the individual depositional sandstone facies within the near-marine sandstones, are generally thicker and longer than those belonging to the parasequence sets that exhibit either a progradational or retrogradational stacking pattern. These parasequence sets were deposited during a period where the sedimentation rate was more balanced with the production of accommodation space. Following the arguments above, it is likely that the decrease in sedimentation rate resulted in a decrease in avulsion frequency. The morphological effect of this balance in sedimentation rate and accommodation space production was that individual delta lobes remained stable for longer periods of time, which results in a more continuous seaward progradation (i.e., longer near-marine sandstone profiles). This stability also allows for more intensive wave-reworking such that the delta lobes become, in fact, wave-dominated, and the leading edges develop prograding shoreface profiles. The balance in sedimentation rate and accommodation space production furthermore produces vertical aggradation, resulting in thicker parasequences. These systematics are also manifest in thicker and longer upper, middle, and lower shoreface facies, within each parasequence.

Retrogradational parasequence sets: In the final phase of the development of the Last Chance Delta, during the deposition of retrogradational Parasequence Sets 6, 7, and 8, the sedimentation rate decreased further (0.1 mm/year) and the rate of rise in relative sea level increased slightly (0.2 mm/year). This resulted in the development of small retrogradational parasequence sets and a retreat of the paleoshorelines.

Within parasequence sets that exhibit a retrogradational stacking pattern, the near-marine sandstones of the individual parasequences and the individual depositional sandstone facies within the near-marine sandstones, are the shortest and thinnest of all Last Chance Delta parasequences. These near-marine sandstones were deposited during a period of Last Chance Delta deposition when sedimentation rates were at their lowest levels and the rate of relative sea-level rise slightly increased. These low sedimentation rates, coupled with an increase in relative sea-level rise, slowed the seaward progradation of individual delta lobes and increased the degree of wave-reworking, resulting in short dip length delta lobes. Although well-developed shoreface profiles evolved, the rising relative sea level prohibited vertical aggradation, resulting in very thin near-marine sandstone profiles. These parasequences also have thin, short upper, middle, and lower shoreface facies, within each parasequence.

Implications of Near-Marine Sandstone Geometry and Architecture for Estimating Lateral and Vertical Continuity and Lithologic Heterogeneity

The dimensional information for near-marine sandstone parasequences and depositional facies, discussed above, provide key insight into the potential for interwell connectivity, reservoir continuity, and reservoir properties.

The grain size and sorting characteristics of the stream-mouth bar, proximal delta-front, and distal delta-front depositional facies (and their reworked equivalents), respectively, correspond roughly to units of decreasing reservoir quality. Outcrop permeability studies by Lowry and Jacobsen (1993), Barton (1994), and Mattson (1997) suggest that the air permeability varies by as much as a factor of 3–10 between each of these facies. The air permeability also varies between the river-dominated, the wave-reworked, and the wave-dominated near-marine deposits, as well. The mean air permeability of the river-dominated stream-mouth bar, wave-reworked stream-mouth bar, and wave-dominated upper shoreface sandstones are 40 millidarcies (md), 325 md, and 246 md, respectively. The mean air permeability of the river-dominated proximal delta-front, wave-reworked proximal delta-front, and wave-dominated middle shoreface sandstones are 3 md, 55 md, and 132 md, respectively. The mean air permeability of the river-dominated distal delta-front, wave-reworked distal delta-front, and wave-dominated lower shoreface sandstones are 1.5 md, 7 md, and 29 md, respectively. These variations are a complex function of grain size, sorting, and bedform and lamina style. Therefore, an understanding of the size, shape, and proportion of each of these facies in near-marine parasequences is essential for a better understanding of reservoir properties.

The variations in sedimentation rate and accommodation space (i.e., the rate of relative change in sea level) during the deposition of the Last Chance Delta clastic wedge, resulted in distinct near-marine sandstone geometries that affect the lateral continuity and lateral lithologic heterogeneity. As discussed above, the lateral extent of the near-marine sandstones of the aggradational and wave-reworked progradational parasequence sets are quite large (i.e., averaging 11.1 km [6.9 mi] and 11.4 km [7.1 mi], respectively). The river-dominated progradational parasequences are somewhat more restricted in lateral continuity (i.e., averaging 6.0 km [3.7 mi] in dip length). The retrogradational parasequence sets have near-marine sandstones of the most restricted lateral extent (i.e., averaging only 3.0 km [1.9 mi] in dip length).

Examining the distribution of depositional facies within each of the delta lobe types discussed above (i.e.,

those belonging to the aggradational, wave-reworked progradational, river-dominated progradational, and retrogradational parasequence sets), it is possible to speculate on the distribution of reservoir quality types within each, as well as, speculate on lateral continuity. The deltaic deposits of the wave-dominated aggradational parasequence sets have the largest proportion of good reservoir quality facies. Their total sandstone facies is dominated by the high reservoir quality upper shoreface facies, which average 46% of the total sandstone facies. The lowest reservoir quality sandstone is the lower shoreface facies, which averages only 14% of the total sandstone facies. Although similar in lateral continuity, the sandstone facies of the deltaic deposits belonging to wave-reworked progradational parasequence sets contain a lower proportion of high reservoir quality sandstone facies (i.e., re-worked stream-mouth-bar deposits average only 39% of the total sandstone facies). The lowest reservoir quality facies (i.e., reworked distal delta-front deposits) make up an average of 26% of the total sandstone facies. The near-marine sandstone facies of the river-dominated progradational parasequences contain only an average of 26% of high reservoir quality sandstone facies (i.e., stream-mouth-bar deposits). The sandstones are dominated by the moderate reservoir quality proximal delta-front deposits (i.e., averaging 45% of the total sandstone facies). Although the retrogradational parasequences contain a high proportion of good reservoir quality upper shoreface sandstones (i.e., averaging 51% of the total sandstone facies), the restricted length (i.e., averaging 3.0 km [1.9 mi]) and lower thicknesses (i.e., averaging 5 m [15 ft]) makes these retrogradational sandstones less attractive as potential reservoirs.

Based on the above discussions and considerations of lateral continuity and distributions of reservoir quality, it is concluded that the parasequence sets within the Upper Ferron Last Chance Delta clastic wedge can be ranked in terms of their overall reservoir potential. This ranking from best to worst is (1) aggradational parasequence sets, (2) wave-reworked progradational parasequence sets, (3) river-dominated progradational parasequence sets, and (4) retrogradational parasequence sets.

Implications of Near-Marine Sandstone Geometry and Architecture for Geologic Modeling

In the development of any stochastic modeling (i.e., geo-object model) (e.g., Williams et al., 1993) or deterministic geologic modeling (e.g., Johnson and Krol, 1989) of a sandstone reservoir, it is necessary to know the external size and shape distributions of the individual sandstone bodies to be included in the model. The reservoir can be modeled at a variety of scales, from the

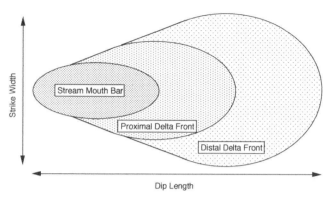

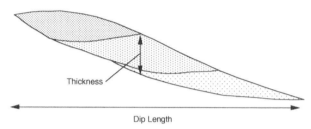

Figure 13. Diagram of the simple geo-object proposed for the modeling of river-dominated near-marine sandstones. The object is internally divided into three (shaded) sub-volumes representing the stream-mouth bar, the proximal delta-front, and the distal delta-front depositional facies.

scale of individual beds and bedsets to the scale of a fluvial channel belt to the scale of a complete delta lobe or shoreline sand body.

Practical experience has shown that the near-marine parasequence is the most useful of the possible architectural elements that can be chosen for building geologic models for near-marine sandstones. The parasequence, the fundamental building block of depositional sequence stratigraphy (Van Wagoner et al., 1990), has dimensions on the order of the same scale as giant oil fields (Reynolds, 1999). In the deltaic depositional environment, the parasequence is generally formed by the abandonment or marine flooding of a delta lobe. The near-marine portion of a fluvial-deltaic parasequence is generally sandstone bounded or separated from overlying and underlying strata by easily recognizable marine-flooding surfaces or erosional unconformities (i.e., a depositional sequence boundary). Although internally a heterogeneous reservoir element (i.e., composed of distinct depositional facies of differing reservoir properties), the parasequence provides an appropriately sized and an easily definable stratal element for making correlations and building geological models of deltaic reservoirs. The architectural and sedimentological data for the near-marine sandstones of the Upper Ferron Sand-

stone Last Chance Delta provide information for utilizing the near parasequence as a building block in geologic modeling.

The 3-D shapes of ancient delta-front sandstones are not well understood. Size and shape information from modern deltaic systems do not provide enough information on the size of shape of the final preserved sandstone system, but furnish only a snapshot of the sand being deposited during a short instant in geologic time. Outcrops are our best source of information on the size and shapes of these sandstone units. Based on an extensive literature analysis of sandstone dimensions, Reynolds (1999) suggested that river-dominated distributary-mouth-bar sandstone bodies, including their distal-bar facies, have a 2-D elliptical plane view emanating from a point source river mouth; the vertical cross-sectional shape was not addressed. Reynolds (1999) also described wave-dominated shoreline sandstones as sheet-like objects, thicker in the near-shore direction.

Based on the outcrop cross-sectional analysis of Garrison and van den Bergh (this volume), it is possible to enhance these two simple deltaic/shoreline sandstone shapes described by Reynolds (1999). The result is the generation of two simple objects whose shapes honor the geologic parameters determined from outcrop studies and whose internal volumes are divided into sub-volumes representing depositional facies (i.e., reservoir quality zonations) (Figures 13 and 14).

THE UPPER FERRON SANDSTONE DISTRIBUTARY CHANNEL AND CHANNEL-BELT SANDSTONES

Each of the 14 Upper Ferron Sandstone parasequence sets have distributary channel-belt sandstones that represent the preserved portions of channel systems that originally supplied sediments to the prograding near-marine facies (Figure 3). These distributary channel belts are located within the delta-plain facies association landward of the landward pinchouts of the near-marine facies. These distributary channel belts are the reservoir building blocks for constructing either deterministic or stochastic geologic models in the alluvial-plain and/or delta-plain environment.

Three of the Upper Ferron Sandstone parasequence sets have well-developed incised-valley systems containing high-gradient, amalgamated braided fluvial systems that originally supplied sediments to the near-marine facies formed at the mouths of the incised-valley systems (Garrison and van den Bergh, this volume). These incised-valley fluvial systems cut down into delta-plain facies associations and into proximal near-marine sandstones. These incised-valley systems are excellent large-scale reservoirs that are frequently encased in non-reservoir rocks.

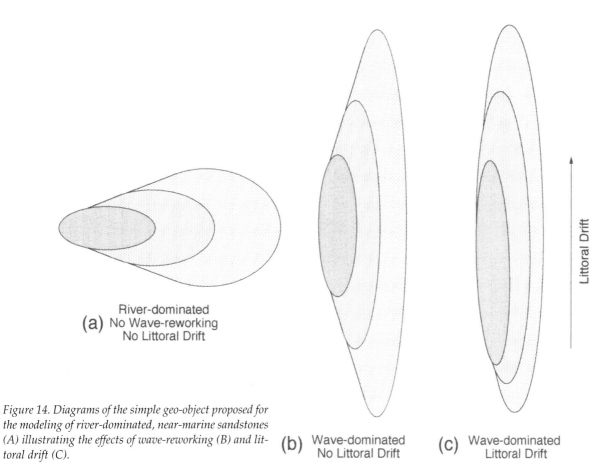

Figure 14. Diagrams of the simple geo-object proposed for the modeling of river-dominated, near-marine sandstones (A) illustrating the effects of wave-reworking (B) and littoral drift (C).

External Geometry of Non-Marine Channel-Belt Sandstones within the Ferron Last Chance Delta

Distributary Channel Belts

A portion of the outcrop section that exposes almost the complete delta-plain facies association of the Ferron Sandstone stratigraphic interval is located in Willow Springs Wash, Willow Springs Quadrangle, Emery and Sevier Counties (Figures 2 and 3 [Plate V of Garrison and van den Bergh, this volume, on the CD-ROM]). Examination of the channel belts in Willow Springs Wash reveals clear differences in channel-belt geometry and internal architecture as a function of delta class and channel position within the parasequence and parasequence set stacking arrangement, in the Upper Ferron Sandstone Last Chance Delta clastic wedge (Figure 15).

Progradational distributary channel belts: The progradational proximal distributary channel belts, in Willow Springs Wash, have a bimodal thickness distribution ranging from 2.5–14.5 m (8.2–47.6 ft) with thicknesses between 3–5 m (9–15 ft) and 12–15 m (36–45 ft) being most common. Their widths range from 79–580 m (240–1900 ft), with widths near 70–140 m (230–460 ft) and 180–220 m (590–1500 ft) being most common. The width/thickness aspect ratios range from 10.2–52.9, with

an average of 28.8 (Figure 16). The progradational distal distributary channel belts are generally narrower and thicker than the proximal channel belts with widths ranging from 43–510 m (140–170 ft), averaging 198 m (650 ft), and thicknesses ranging from 3.4–22.9 m (11.2–75.1 ft), with thickness of 3–7 m (9–21 ft) and 14–19 m (42–57 ft) being most common. The aspect ratios of the distal distributary channel belts range from 7.8 to 45.5, averaging 19.0.

The distal distributary channel belts are generally narrower that the proximal channel belts, although their thicknesses are comparable. The aspect ratios of the distal distributary channel belts are much less than those of the upper delta plain fluvial channel belts. The change in size and shape between the proximal and distal distributary channel belts is consistent with the change in channel morphology resulting from river bifurcation as they approach the paleoshoreline (van den Bergh and Garrison, 1996; Garrison and van den Bergh, 1997).

Aggradational distributary channel belts: In Willow Springs Wash, the aggradational proximal distributary channel belts have a bimodal thickness distribution ranging from 4.5–18.3 m (14.8–60.0 ft), with thicknesses between 8–11 m (24–33 ft) and 17–18 m (51–54 ft) being most common. Their widths range from 380–1448 m (1140–4344 ft), with widths near 380–570 m (1140–1710 ft) being most common. The width/thickness aspect

Figure 15. The North Canyon wall of Willow Springs Wash showing the strike cross sections of distributary channels.

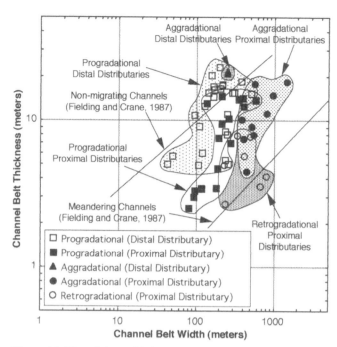

Figure 16. Plot of channel-belt thickness versus channel-belt strike width within the Upper Ferron Sandstone Last Chance Delta.

ratios range from 32.0–97.4, with an average of 59.2 (Figure 16). The aggradational distal distributary channel belts, reported by Barton (1994), are generally narrower and thicker than the distributary channel belts of the upper delta plain, with a mean width of 255 m (770 ft) and a mean thickness of 21 m (63 ft). The mean aspect ratio of 12.1, for these distal distributary channel belts, is also much less than those of the distributary channel belts in Willow Springs Wash.

The proximal aggradational distributary channel belts in Willow Springs Wash are generally much wider than the progradational distributary channel belts, while the thicknesses are generally slightly greater; aspect ratios of the aggradational fluvial channel belts are a factor of two higher that those of the proximal distributary channel belts. This change in size and shape between the proximal and distal distributary channel belts is also consistent with the change in channel morphology resulting from river bifurcation as they reach the paleoshoreline (van den Bergh and Garrison, 1996; Garrison and van den Bergh, 1997).

Retrogradational distributary channel belts: In Willow Springs Wash, the distributary channel belts from the retrogradational parasequence sets are all proximal in position and have a thickness distribution ranging from 2.7–7.8 m (8.9–25.6 ft), averaging 5.2 m (17.1 ft) thick. Their widths are uniformly distributed from 228–808 m (684–2420 ft). The width/thickness aspect ratios range from 43.8 to 195.0, with a mean aspect ratio of 100.0 (Figure 16). The retrogradational distributary channel belts in Willow Springs Wash are similar in width to the proximal aggradational distributary channel belts, but generally much wider than the progradational distributary channel belts; these channel belts are commonly thinner than either the aggradational or progradational distributary channel belts; aspect ratios of the retrogradational distributary channel belts are usually higher that those of the aggradational and progradational distributary channel belts.

Channel-fill elements: The aspect ratio distributions for the individual channel-fill elements within the distributary channel belts, are similar, regardless of their position within the parasequence set external or internal stacking pattern, suggesting that only the channel-fill element stacking patterns vary between the distributary channel belts within the Upper Ferron Sandstone. Data indicates that the proximal, distributary channel belt, channel-fill elements are generally thicker and wider than the distal, distributary channel belt fill elements. This change in size of the distal distributary channel belt macroforms is also consistent with the change in channel morphology resulting from river bifurcation.

Incised-Valley Fluvial Systems

Progradational parasequence set: Parasequence Set 2C contains a well-developed incised-valley with amalgamated fluvial deposits. Parasequence Set 2C is part of the 4th-order lowstand systems tract of Ferron Sequence FS2; the valley-fill successions represent 5th-order depositional sequences within Parasequence Set 2C (Garrison and van den Bergh, this volume). The best exposed valley-fill sequence in Parasequence Set 2C occurs between Anderson Ranch, along the Coal Cliffs, to the amphitheater of Muddy Creek Canyon (see Plate I of Garrison and van den Bergh, this volume, on the CD-ROM). This incised-valley fill sequence is referred to in this paper as the South Muddy Creek Incised Valley (Figure 17). An additional, stratigraphically higher, 5th-order depositional sequence and very small incised-valley system occurs farther seaward, in the area between the North Goosenecks of Muddy Creek and the Muddy Creek

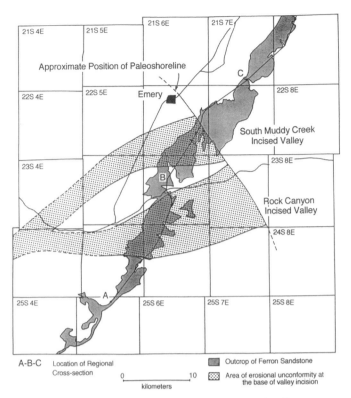

Figure 17. Map showing the location of the incised-valley system belonging to Upper Ferron Sandstone Parasequence Set 2C.

Aggradational parasequence sets: Parasequence Set 4B contains a well-developed incised-valley with amalgamated fluvial deposits. These fluvial deposits are considered representative of incised-valley fill deposits in aggradational parasequence sets. Parasequence Set 4B is part of the 4th-order lowstand systems tract of Ferron Sequence FS3; the valley-fill succession represents 4th-order systems tracts within Ferron Sequence FS3. The valley-fill sequence in Parasequence Set 4B occurs between Anderson Ranch, along the Coal Cliffs, to Rochester Creek (Garrison and van den Bergh, this volume). This incised-valley fill sequence is referred to in this paper as the Miller Canyon Incised Valley. The incised-valley system appears to bifurcate in the area southwest of Quitchupah Creek, north of I-70, such that in the Miller Canyon/Muddy Creek area, the incised-valley system is composed of two distinct, parallel incised-valley fill sequences (denoted here as the North Miller Canyon Incised Valley and the South Miller Canyon Incised Valley). Both incised-valley segments are up to 32 m (96 ft) in depth. The incised-valley system has a floor with substantial topographic relief. The incised-valley system is oriented approximately 60°. Near the mouth of the valleys (i.e., segment 1 of Zaitlin et al., 1994), the North and South Miller Canyon Incised Valleys are approximately 2.6 km (1.6 mi) and 5.2 km (3.2 mi) wide, respectively.

Internal Architecture and Sedimentological Character of Upper Ferron Sandstone Channel-Belt Sandstones

Examination of the distributary channel belts in Willow Springs Wash reveals clear differences in channel belt internal architecture and sedimentology as a function of delta class and channel-belt position within the parasequence set stacking arrangement of the Upper Ferron Sandstone Last Chance Delta clastic wedge. Although these distributary channel belt cross sections represent sediment deposited over a substantial period of time (i.e., greater than 5000–10,000 years), and at a range of distances from the paleoshoreline, clear patterns in architecture and sedimentological style exist. Within a class of channel belts, channel belt internal architecture and sedimentological style do not change as a function of distance to the paleoshoreline. The distributary channel belts of river-dominated and wave-reworked deltaic progradational parasequence sets are generally laterally restricted and multi-storied with channel-filling elements (i.e., macroforms and/or barforms) generally stacked vertically within the channel-belt boundaries. The distributary channel belts of wave-dominated deltaic aggradational parasequence sets are generally quite laterally extensive and multi-storied with channel-filling elements generally stacked en eche-

Graveyard. This is referred to in this paper as the North Muddy Creek Incised Valley. It incises into the highstand deposits of the South Muddy Creek Incised Valley. Farther south, along the outcrop belt, a second valley-fill sequence, the same age as the South Muddy Creek Incised-Valley Fill, occurs between the north side of Willow Springs Wash and Scabby Canyon, south of Interstate 70 (I-70). We informally refer this as the Rock Canyon Incised Valley (Figure 17).

The South Muddy Creek Incised Valley and the Rock Canyon Incised Valley form a large incised-valley complex (Figure 17). The incised-valley system appears to bifurcate in the area south of Oak Spring, north of I-70. The South Muddy Creek Incised Valley is an asymmetric valley. It ranges in depth from 8.6 m (28.2 ft) on its southern margin, south of Bear Gulch, to depths of 25–30 m (75–90 ft) along the northern part, south of the South Goosenecks of Muddy Creek. The incised valley is oriented approximately 70°. Near the mouth of the valley (i.e., segment 1 notation of Zaitlin et al., 1994), it is approximately 6.1 km (3.8 mi) wide. The Rock Canyon Incised Valley has a floor with substantial topographic relief; maximum depths reach 30 m (90 ft) in the areas near Scabby Canyon and Coyote Basin. This incised valley is oriented 60°. At Rock Canyon, it has an inner valley (i.e., Segment 3) width of about 7 km (4 mi) and a width at the mouth of about 10 km (6 mi).

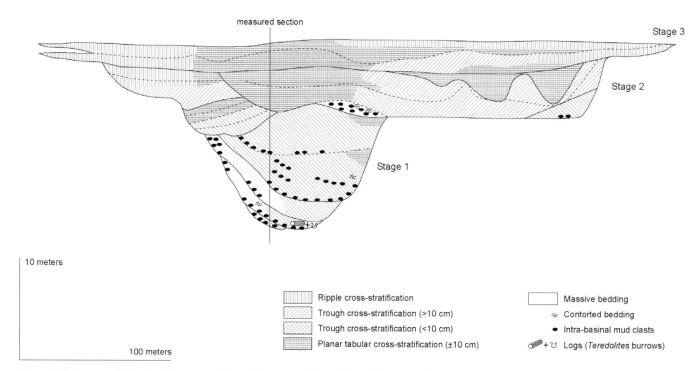

measured section

Stage 3

Stage 2

Stage 1

10 meters

100 meters

	Ripple cross-stratification
	Trough cross-stratification (>10 cm)
	Trough cross-stratification (<10 cm)
	Planar tabular cross-stratification (±10 cm)

	Massive bedding
≈	Contorted bedding
•	Intra-basinal mud clasts
	Logs (*Teredolites* burrows)

Figure 18. Exaggerated schematic cross section of the County Line Channel showing the internal architecture and the distribution of sedimentary structures (bedforms). Bounding surfaces that are less well defined and are internal to major bounding surfaces are denoted as dashed lines. The vertical line is the line of measured section shown in Figure 19.

lon laterally within the channel-belt boundaries. The distributary channel belts of wave-dominated deltaic retrogradational parasequence sets are laterally extensive and sheet-like with channel-filling elements generally stacked vertically within the channel-belt boundaries.

The County Line Channel — An Analog for Distributary Channel Belts from Progradational Parasequence Sets

Examples of channel belts from river-dominated to waver-reworked, progradational parasequence sets are the informally named County Line Channel and the related Coal Miner's Channel in Willow Springs Wash which occur in Parasequence 1d of Parasequence Set 1B (Figures 3 and 15). Figure 18 shows the internal character of the County Line Channel and Figure 19 is a measured section through the County Line Channel. Figure 20 shows a map view of the County Line Channel and the related Coal Miner's Channel reconstructed from outcrops in Willow Springs Wash, Indian Canyon, North Fork Canyon, and the Limestone Cliffs (Garrison and van den Bergh, this volume).

The County Line Channel is a low sinuosity distributary channel belt occurring within the progradational phase of Last Chance Delta deposition. The County Line Channel generally trends 40°. It has a meander wavelength of about 5 km (3 mi) and a meander amplitude of only 1 km (0.6 mi) indicating a low sinuosity channel pattern (Figure 20). Channel sinuosity, defined as the

ratio of the thalweg length along two successive identical meanders disposed symmetrically about a straight meander axis to the total axis length of the meanders, is calculated as 1.09. The County Line Channel bifurcates just south of Indian Canyon, resulting in a smaller channel belt, denoted here as the Coal Miner's Channel, generally trending 360° north (Figure 20). The Coal Miner's Channel has a meander wavelength of about 3.3 km (2.1 mi) and a meander amplitude of only 0.3 km (0.2 mi) indicating a low sinuosity channel pattern, similar to the County Line Channel.

The main branch of this distributary complex, denoted here as the County Line Channel, feeds a mouth-bar complex near the mouths of Coyote Basin and Rock Canyon (Figure 20). The regional work of Garrison and van den Bergh (this volume) suggests that the delta-front deposits, near the mouths of Coyote Basin and Rock Canyon, prograded seaward about 3.5 km (2.2 mi).

The County Line Channel is a 379 m (1140 ft) wide and 18.3 m (60 ft) thick distributary channel belt. It incises through the Sub-A1 coal zone into Parasequence 1c (Figure 3). The channel belt aspect ratio is 20.7. The Coal Miner's Channel is 107.9 m (354 ft) wide and 4.9 m (16 ft) thick, resulting in an aspect ratio of 22.0. The smaller Coal Miner's Channel branch does not cut through the Sub-A_1 coal zone. This suggests bifurcation occurred fairly late in the history of the County Line Channel. In this paper, only the County Line Channel will be discussed in more detail, architecturally and sedimentologically.

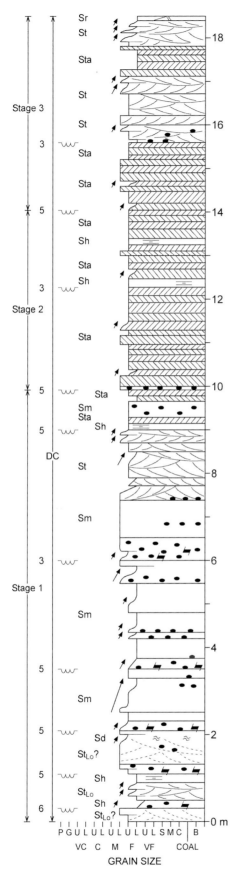

Figure 19. *Measured section through the central portion of the County Line Channel showing the internal sedimentology. See Figure 5 for explanation of symbols and notation.*

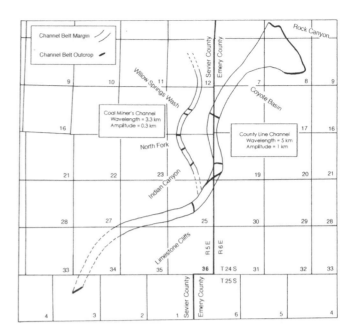

Figure 20. *Map view of the County Line Channel and the Coal Miner's Channel, in Willow Springs Wash, Willow Springs Quadrangle, Emery and Sevier Counties, Utah, showing the nature of the meandering and branching of the distributary channel belt.*

The County Line Channel is composed of 14 channel-fill elements. Figure 18 is an exaggerated schematic cross section of the County Line Channel showing the distribution of sedimentary structures (i.e., bedform types) within the internal channel-belt architecture. The major, well-defined, bounding surfaces are denoted by solid lines in Figure 18; surfaces that are less well-defined and are internal to the elements bound by the major surfaces are denoted with dashed lines.

The external geometry of the County Line Channel suggests that there are at least three stages (denoted Stage 1, Stage 2, and Stage 3) in the development of the distributary channel-belt morphology. Each stage preserves from 4–10 m (12–30 ft) of sand and each stage becomes progressively wider and less confined than the previous stage. The lower five elements represent scour and fill elements deposited within a narrow 101-m (330-ft) wide, confined channel. The preserved thickness of this first stage of the channel belt is approximately 10 m (30 ft), resulting in a width/thickness aspect ratio of 10.0. The second stage of channel-belt development is recorded by the next seven higher channel-fill elements, reflecting both cut and fill characteristics and lateral accretion structures. These channel-fill elements were deposited in a much wider, yet confined channel 341-m-wide (1120-ft) channel. The preserved thickness of this second stage of channel-belt development is approximately 6.7 m (21 ft), resulting in a width/thickness aspect ratio of 50.9. The final stage of the channel-belt development is dominated by lateral accretionary bedforms and bedsets, which interfinger with the laterally equivalent delta-plain facies associations. This channel-fill event has well-developed levees and overbank facies. The preserved width and thickness of this phase are 379 m (1240 ft) and 4.1 m (13 ft), respectively, resulting in of a width/thickness aspect ratio of 91.4.

Internal channel-belt architecture and sedimentology:
The lower portion of the channel-belt complex, representing Stage 1, is a thick and narrow convex down, confined, symmetric channel shape that incises into the underlying delta-plain and near-marine rocks of Parasequences 1c and 1d (Figure 3). The contact of the County Line Channel with underlying rocks may be sandstone on sandstone or sandstone on shale drape. Wood and clay pebble gravel lags are common near the base of the Stage 1 channel-fill elements.

The lower five channel fill-elements are associated with the narrow, confined Stage 1 development of the County Line Channel belt. These five channel-fill elements represent major scour and fill events (Figure 18). The bounding surfaces of these channel-fill elements are frequently delineated by thick, clay pebble lag deposits, although clay pebble lag zones, composed of clay pebbles floating in a sand matrix, are also common. These channel-fill elements are generally medium-grained sandstones that fine upward from medium to fine-grained sandstones. They generally have large trough cross-stratified beds, greater than 15 cm (6 in.) thick, near their bases, which change upward into faint large troughs and massive sandstones. The trough cross-stratification decreases in size upward, to less than 10 cm (4 in.) thick. Occasionally, trough cross-beds near bounding surfaces are contorted. Grain size and bedform size also decrease toward the channel margins. Trough cross-stratification and load casts near the channel base suggest an average paleocurrent direction of 28°±10° for the County Line Channel outcrop on the north cliff face of Willow Springs Wash.

Stage 2 of channel-belt development appears as a much thinner and wider channel-belt complex. This second stage has a flat base that appears to lie conformably on the underlying delta-plain shales, the Sub-A$_1$ coal zone, and brackish-water bay deposits of Parasequence 1d. Near the outer 5–10 m (15–30 ft) of the Stage 2 channel fill, the channel belt base becomes curved upward indicating at least 7 m (21 ft) of incision into the underlying rocks. Local overbank deposits and small crevasse-splay deposits occur on the eastern margin of Stage 2 of the channel belt.

There are seven channel-fill elements in Stage 2 of the development of the County Line Channel. These channel-fill elements are dominated by lateral accretion sedimentary surfaces, structures, and bedforms. The bounding surfaces between these channel-fill elements are frequently defined by bedding surfaces between major bedform domains.

The channel-fill elements in Stage 2 are generally medium to fine in grain size and exhibit an overall fining upward grain-size trend. The lower channel-fill elements, near the central portion of the channel, exhibits faint large trough cross-beds to massive, structureless sandstones that are transitional into smaller trough cross-stratified beds and planar and wedge tabular cross-stratified beds that suggest lateral migration. Some channel-fill elements are dominantly planar and wedge tabular cross-bedded suggesting unrestricted flow with a westward lateral migration direction (Figure 18). Paleocurrent measurements in the planar and wedge tabular cross-bedded sandstones suggest flow directions of 357°± 7° for these laterally migrating bedforms. The channel-fill elements of Stage 2 become finer grained and bedforms become smaller in size towards the channel margins. On the western side, at the base of the Stage 2 channel-belt fill, a channel-fill element has well-defined sets of westward dipping lateral accretion surfaces with planar tabular cross-stratification migrating up the surfaces (Figure 18). A small convex upward clay pebble barform occurs at the base of Stage 2 on the eastern side of the channel belt (Figure 18).

The final stage (i.e., Stage 3) of channel-belt development is a much thinner and wider, poorly confined channel. These channel-fill elements appear to be conformable on top of the Stage 2 elements and interfinger with the lateral delta-plain facies associations.

The upper two channel-fill elements are the final phase of the development and preservation of the County Line Channel. These channel-fill elements interfinger with the lateral delta-plain facies associations. These channel-fill elements are generally medium- to fine-grained sandstones. They exhibit large scale planar tabular cross-beds near the central portion of the channel (i.e., suggesting lateral migration based on paleocurrent orientations), but display small scale planar tabular cross-beds, ripple cross-stratification, and climbing ripple cross-stratification near the top and laterally towards the channel margins.

Architectural evolution of the County Line Channel:
Garrison and van den Bergh (this volume) calculated that during the development of the County Line Channel, the sedimentation rate was approximately 2.4 mm/year and the relative rise in sea level was 1.3 mm/year. The delta progradation rate was calculated to be about 0.5 m/year (1.6 ft/yr). The high sedimentation rate and high progradation rates resulted in a delta having a "birds foot" Mississippi style morphology, with low sinuosity, confined, vertically aggrading distributary channels dominated by fluvial processes.

The County Line Channel evolved through three distinct stages. These three distinct stages of distributary channel belt development are related to normal river-dominated deltaic progradation, sedimentation rate, and changes in relative sea level. The width/thickness aspect ratios of the preserved sandstones of each Stages 1, Stage 2, and Stage 3 are 10.0, 50.9, and 91.4, respectively. The overall channel belt width/thickness aspect ratio of the County Line Channel changes from 10.0 to 24.3 to 20.7 at the end of Stage 1, Stage 2, and Stage 3, respectively. These systematic changes are interpreted to be a consequence of normal channel-belt evolution as a result of river-dominated delta progradation and delta-plain aggradation due to a relative rise in sea level (van den Bergh and Garrison, 1996; Garrison and van den Bergh, 1997).

The change in aspect ratio during distributary channel-belt development is a direction consequence of delta progradation. The cross section of the County Line Channel, exposed on the north wall of Willow Springs Wash, is only about 1.7 km (1.2 mi) from the initial paleoshoreline. The effects introduced by distributary channel evolution during progradation are accentuated, but the overall geometry of the Country Line Channel is not unique to channel belts occurring very near the paleoshoreline. In Willow Springs Wash, a progradational Parasequence Set 3 channel belt, located almost 25 km (16 mi) from the paleoshoreline, also exhibits this overall three-stage evolution and geometry (e.g., see Plate V of Garrison and van den Bergh, this volume).

The narrow, confined Stage 1 channel-fill event represents the channel belt's cross section at an initial position 1.7 km (1.1 mi) from the paleoshoreline. At this period, the channel belt is a very low sinuosity, confined segment, with an aspect ratio of 10.0. This aspect ratio is consistent with the lower average value for distal-channel belts occurring in the lower delta plain, as presented in Figure 18. The narrow, confined County Line Channel's Stage 1 channel belt cross section abruptly widens during Stage 2, while still remaining confined. As discussed earlier, channel belt cross sections generally become narrower as they approach the paleoshoreline — a consequence of distributary bifurcation. The wider Stage 2 event represents the channel belt cross section, at a later time, after the paleoshoreline had migrated seaward an additional 1–2 km (0.6–1.2 mi) and there had been at least 7 m (21 ft) of delta-plain aggradation. Stage 2 development represents deposition in a wider upstream channel-belt segment. This superposition of a wider channel-belt segment upon a narrower downstream segment, would suggest that (1) the channel-belt position remained stable for a long period of time, and that (2) that local avulsions resulted in a wider young channel-belt segment reoccupying the position of the older original channel-belt site. At the end of Stage 2, the Stage 2 channel fill had an aspect ratio of 50.9, although the overall channel belt aspect ratio was only 24.3. As evolution proceeded, at the end of Stage 3, the Stage 3 channel-fill event had an aspect ratio of 91.4, although

the overall channel belt aspect ratio was only 20.7. These changes in the overall geometry of the County Line Channel are consistent with the relationships illustrated in Figure 18.

The final channel fill and abandonment, recorded by the channel-fill elements of Stage 3, are not significantly wider than Stage 2, but are, comparatively, unconfined and show interfingering with the laterally equivalent delta-plain facies associations. This suggests that Stage 3 is simply a normal abandonment phase of Stage 2.

Vertical sediment aggradation within the Country Line Channel is a direct consequence of stream equilibrium profile adjustment in response to rapid deltaic progradation. The high sedimentation rate relative to a slower relative rise in sea level resulted in a progressively seaward-migrating paleoshoreline. The original stream equilibrium profile at the initiation of equilibrium fluvial deposition is shown in Figure 21a. This stream equilibrium profile is graded to sea level. During rapid seaward progradation, the point to which the stream equilibrium profile is graded progressively moves seaward at a relatively rapid rate (Figure 21b). This migration of the paleoshoreline results in the production of increased accommodation space upstream. Vertical stream aggradation occurs in an attempt to grade the stream to the new equilibrium profile and new paleoshoreline position.

An increase in the upstream sedimentation rate and the resulting vertical upstream stream aggradation frequently results in local avulsions (Heller and Paola, 1996). Local avulsions are common in the river-dominated Mississippi-type deltas class. High avulsion rates and the rapid evolution of channel belts would tend to prevent stabilization of channel belts and minimizes extensive lateral migration, hence the common vertical aggradation profiles in the Ferron progradational distributary channel belts. This observation would also suggest that vertical aggradation and local avulsions are also common in the wave-reworked progradational delta class as well (Figure 1).

The Kokopelli Channel — An Analog for Distributary Channel Belts from Aggradational Parasequence Sets

An example of a channel belt from an aggradational parasequence set is informally named the Kokopelli Channel, in Willow Springs Wash, which occurs in Parasequence Set 4B (Figures 3 and 15). Figure 22 shows the internal character of the Kokopelli Channel. Figure 23 shows a measured section through the eastern end of the Kokopelli Channel.

The Kokopelli Channel, an aggradational distributary channel belt within Parasequence Set 4B, is exposed in the north canyon wall in Willow Springs Wash. The

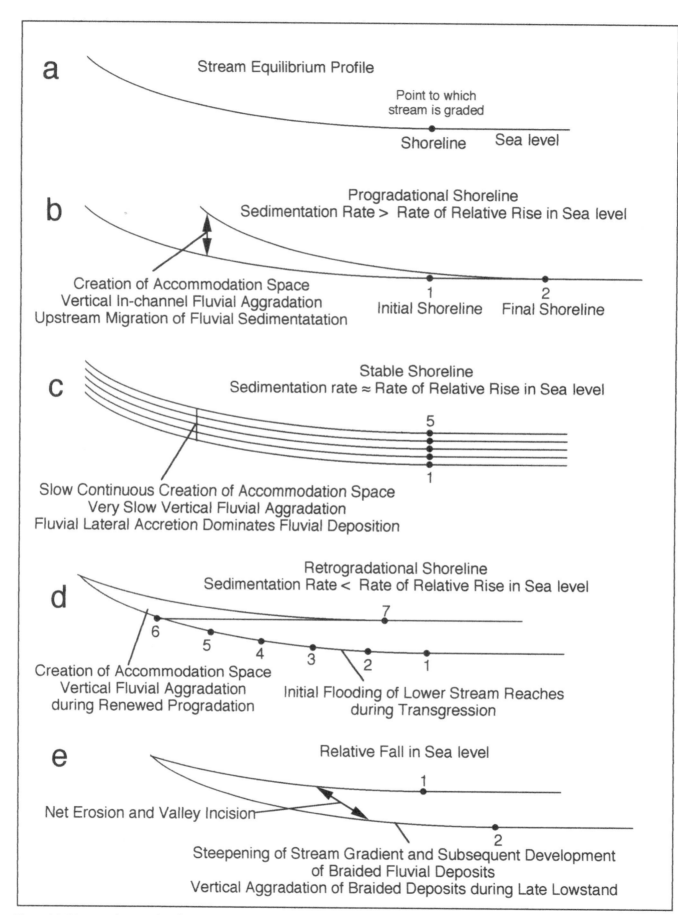

Figure 21. Diagram showing the effects of sedimentation rate and rate of relative change in sea level on stream equilibrium profiles.

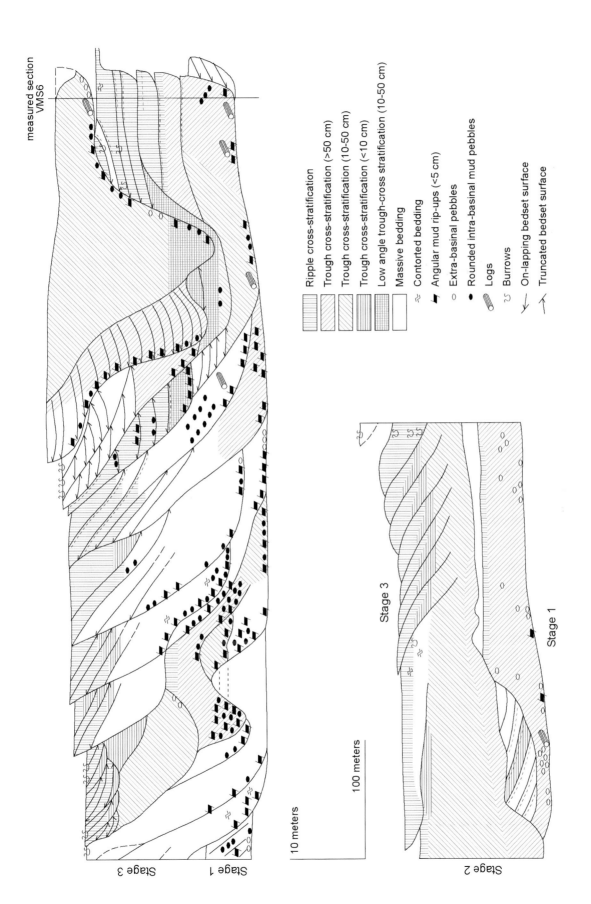

Figure 22. Exaggerated schematic cross section of the Kokopelli Channel showing the internal architecture and the distribution of sedimentary structures (bedforms). Bounding surfaces that are less well defined and are internal to major bounding surfaces are denoted as dashed lines. The vertical line is the line of measured section VMS6 shown in Figure 23.

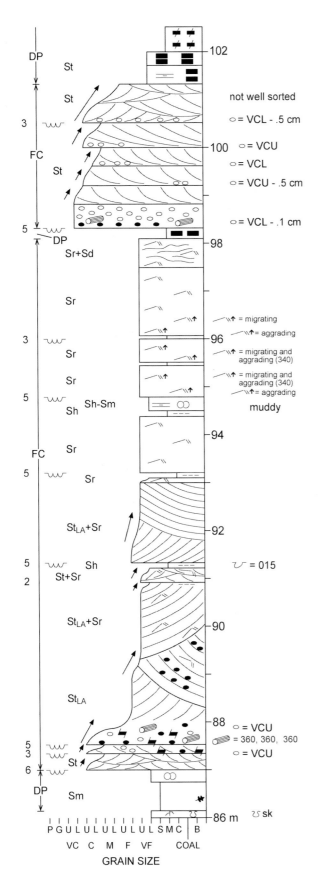

Figure 23. A portion of measured section VMS6 through the eastern part of the Kokopelli Channel showing the internal sedimentology. See Figure 5 for explanation of symbols and notation.

eastern end is located almost directly above the County Line Channel (Figure 15). The channel belt is 1448 m (4750 ft) wide and 18.3 m (60 ft) thick, with an aspect ratio of 79.2. It is not clear whether the entire extent of this channel belt is exposed in the study area, since the limit of the Willow Springs Wash outcrop along the county road into Willow Springs Wash may have truncated the belt's western end (see Plate V of Garrison and van den Bergh, this volume). This 1448–m (4750–ft) width may represent a minimum width for the channel-belt complex. The Kokopelli Channel generally trends 55°.

The Kokopelli Channel is multi-storied with channel-fill elements, dominated by lateral accretion structures, and generally stacked en echelon laterally within the channel-belt boundaries. In general, the channel-fill elements have a complete sedimentary profile from initial scour to abandonment phase, although locally truncation of the abandonment phase is common. The Kokopelli Channel cross section in Willow Springs Wash is at least 22 km (14 mi) from the Ferron Parasequence Set 4B near-marine sandstones, marking the paleoshoreline.

The Kokopelli Channel is a composite channel belt that consists of three sub-channel belts, denoted, in this paper, as sub-channel belt Stages 1, 2, and 3. Figure 22 is a schematic cross section of the eastern half of the Kokopelli Channel complex, showing sub-channel belt Stage 1, the eastern part of Stage 2, and Stage 3, with the internal architecture of the channel-fill elements and the internal sedimentary structures. The exact width of sub-channel belt Stages 1 and 2 cannot be determined because of local truncation. The Stage 1 sub-channel belt phase is deeply incised by the Stage 2 and Stage 3 phases, such that only a small 258-m-wide (774-ft), 5.8-m-thick (19-ft) segment remains near the base of the channel-belt complex. The Stage 2 phase, which is incised on its eastern margin by the Stage 3 sub-channel belt phase, appears to have been at least 908 m (2720 ft) wide and has a preserved thickness of 9.5 m (31 ft). The aspect ratio of this preserved segment is 95.4. The Stage 3 sub-channel belt phase is 653 m (1960 ft) wide and 18.3 m (54.9 ft) thick, with an aspect ratio of 35.7. Only the youngest channel element of sub-channel belt Stage 3 has been completely preserved. It has a width of 159 m (477 ft) and a thickness of 15 m (45 ft). The aspect ratio of this preserved segment is 10.6.

Internal channel-belt architecture and sedimentology: The Stage 1 sub-channel belt phase of the Kokopelli Channel complex is a sub-channel belt element which is truncated towards the western-end by Stage 2 and on its eastern end by Stage 3. The Stage 1 sub-channel belt consists of two channel-fill elements and several indistinct truncation surfaces. The Stage 1 sub-channel belt appears to locally cut completely through the informally named E Rooted Zone of Garrison and van den Bergh (this volume) that caps the underlying Parasequence Set 4A.

The lower-most Stage 1 channel-fill element is, in general, characterized by trough cross-stratification; trough beds are 50 cm (20 in.) or more in thickness. Ripple-stratification is well-developed near the top. The grain size varies from coarse to medium at the base, and fine to very fine at the top. Extra-basinal, subangular to well-rounded pebbles, composed of chert, quartz, sandstone, and volcanic clasts, occur as floating pebbles, within the sandstone matrix, as lags along the base of the channel, and as lags at the bases of trough cross-stratified beds. The pebbles range up to 2.5 cm x 1.5 cm (1 in. x 0.6 in.) in diameter and the average pebble size is 13 cm x 0.75 cm (5 in. x 0.3 in.). Angular mud rip-up clasts and well-rounded to sub-angular, intra-basinal mud clasts occur along the base of the major channel and along reactivation surfaces. Occasionally, laminae exhibit fining-upward grain-size profiles from coarse to fine.

The upper channel-fill element of the Stage 1 sub-channel belt is composed of accretionary barforms and bedsets (Figure 22). The accretionary surfaces dip at an angle of 14° to the east. Each successively younger accretionary barform shows a progressive decrease in grain size; from coarse to medium to very fine. Each successively younger accretionary barform also shows a progressive change in sedimentary structure; from massive, to trough cross-stratified (trough bedset thickness up to 10 cm [4 in.]), to ripple-stratified. The accretionary barforms are separated by gray-green colored mudstone laminae.

The Stage 2 sub-channel belt is composed of two channel-fill elements (Figure 22). The lower-most and volumetrically largest channel-fill element has a preserved width of 834 m (2500 ft) and thickness of 9.5 m (31 ft). The aspect ratio of this preserved segment is 87.7. The mean paleocurrent orientation is 55°±2°. This large channel-fill element is incised by a smaller, younger channel-fill element, along its western margin. This smaller channel is a single convex downward channel with a width of 113 m (339 ft) and a thickness of 8.1 m (27 ft), resulting in a width/thickness aspect ratio of 13.9. The mean paleocurrent orientation, for this smaller channel, is 88°±4°. The Stage 2 sub-channel belt generally rests on top of the E Rooted Zone. Sporadic extra-basinal pebble lags occur along the base of the sub-channel belt. These pebbles are chert, quartz, feldspar, sandstone fragments, and mudstone fragments, and may reach a maximum diameter of 5 mm. Locally large tree logs occur along the base. Angular mud rip-ups and well-rounded, subangular, intra-basinal mud clasts are not uncommon. The Stage 2 sub-channel belt occasionally contains burrows of *Planolites* and *Skolithos* at the top.

The vertical sedimentary bedform profile, within the main body of the Stage 2 sub-channel belt, in general, is trough cross-stratification, with ripple-stratification at the very top. Trough bedset thickness ranges from greater than 50 cm (20 in.) at the bottom to 10–50 cm (4–20 in.) in the middle to less than 10 cm (4 in.) near the top. Less frequently, low-angle trough-stratification may occur at the base. The trough cross-stratified bedsets occur along accretion surfaces that have an easterly dipping sigmoid shape. These surfaces, bounding the trough cross-stratified bedsets, are generally horizontal in the northwestern part of the channel and are inclined to the east, towards the southeastern part of the sub-channel belt (van den Bergh, 1995). The ripple-stratified bedsets also dip laterally, towards the east. In general, the vertical grain-size profile of the Stage 2 sub-channel belt is medium and occasionally coarse at the base, to very fine at the top. Occasionally, on the lamina-scale, there is a fining-upward profile developed that ranges from coarse to fine at the base to fine to very fine at the top.

The Stage 3 sub-channel belt consists of at least 18 channel-fill elements which are, generally, laterally stacked from west to east. Figure 22 shows the distribution of sedimentary structures (i.e., bedform types) within the internal channel-fill elements. Figure 23 is a measured section through the Stage 3 sub-channel belt. Most of the Stage 3 sub-channel belt elements incise through the E Rooted Zone.

In general, the western margin of each successive laterally stacked, channel-fill element is located progressively farther east than the previous element, suggesting a progressive lateral migration of each successive element towards the east. This progressive scouring of older channel-fill elements and migration eastward has, in general, left only the western margins of the channel-fill elements preserved (Figure 22). The youngest channel-fill element is a completely preserved convex downward channel. This channel element has a width of 158 m (474 ft) and thickness of 15 m (45 ft), resulting in a width/thickness ratio of 10.6. This channel has incised almost in the center of underlying channel elements, giving the appearance of a vertical stacking pattern, but the western channel margin of the youngest channel-fill element has actually moved laterally to the east of the underlying channel-fill elements. The mean paleocurrent orientation of these laterally stacked, partially scoured channel-fill elements is 55°±5°. The mean paleocurrent direction of the youngest channel-fill element, in the southeastern part of the sub-channel belt, is 45° near the base. Near the tops of the youngest two channel-fill elements, trough cross-stratified beds have a mean paleocurrent direction of 351°±4°. This suggests laterally migrating bedforms, at some angle to the channel orientation of 45°.

The lower parts of the oldest ten laterally stacked channel-fill elements in Stage 3 sub-channel belt are either trough cross-stratified or massive. Trough bedset thickness ranges from about 50 cm (20 in.) at the base to

10–50 cm (4–20 in.) in the middle of the channel-fill elements. The bedsets of the upper parts of these channel-fill elements are trough cross-stratified, with trough bed thickness less than 10 cm (4 in.) and ripple-stratified at the very top. The bedsets are bound by lateral accretion surfaces that generally dip towards the east. Bedsets generally thin upward. Contorted bedding is not uncommon. Thick, angular mud rip-up clasts and well-rounded to angular intra-basinal mud clast lags, up to several meters thick, are common at the bases of channel-fill elements. Individual rip-ups may be up to 40 cm x 10 cm (16 in. x 4 in.), although the average size is 7.5 cm x 3.3 cm (3 in. x 1.3 in.). The mud rip-up clasts and intra-basinal mud clasts are sporadically iron stained. Locally, the tops of channel-fill elements are burrowed (i.e., *Skolithos* and *Planolites*) and rooted. Sporadically, extra-basinal pebble lags occur along the base of the channel-fill elements. These pebbles can range up to 5.0 mm in diameter, but average about 2 mm. Logs can also be found along the bases of these channel-fill elements (van den Bergh, 1995). The vertical grain-size profile of these laterally stacked channel-fill elements is medium (locally coarse) at the base to very fine at the top. The average grain size is medium to fine.

The youngest seven channel-fill elements generally have trough cross-stratification (trough-set thickness varies from 10–40 cm [4–16 in.]) and large-scale, low-angle trough-stratification (trough set thicknesses up to 40 cm [16 in.]) at their bases. Towards the top and edges of the channel-fill elements, the stratification changes into ripple cross-stratification, climbing-ripple, and migrating-ripple cross-stratification. Ripples migrate in a 340° direction. The bedsets of the youngest of these seven channel elements are slightly concave-upward, slightly dipping towards the west, and are stacked vertically, suggesting that the bedsets are probably part of sigmoid shaped, easterly dipping accretionary barforms.

Several of the channel-fill elements, in the Stage 3 sub-channel belt, are separated by massive mudstones. The vertical grain-size profile is medium to coarse at the base, to fine and very fine at the top and edges of the channels. The average grain size is fine. Sporadically, the grain size fines-upward on a laminae-scale, from coarse to fine. Locally, the tops of channel-fill elements are very intensely burrowed (i.e., *Planolites*, *Skolithos*, and *Scoyenia*). Well-rounded to subangular extra-basinal clasts, consisting of chert, quartz, volcanic rock fragments, sandstone, and mudstone occur along the base of channels, beds, and troughs. These clasts can be up to 1.5 cm x 0.8 cm (0.6 in. x 0.3 in.), averaging 8 mm by 4 mm. Angular mud rip-ups are up to 1.8 cm x 0.8 cm (0.7 in. x 0.6 in.), and the average is 1.5 cm x 1.8 cm (0.6 in. x 0.7 in.), along the base of the channels and sporadically along bedset surfaces.

The final channel-fill element of the Kokopelli Channel is a single channel with a convex downward shape, of which the abandonment phase has not been completely preserved. This channel has a width of 158 m (474 ft), and a thickness of 15.0 m (49.2 ft). The aspect ratio of this channel is 10.6. The general sedimentary structure profile is low-angle large-scale trough stratification at the base to trough cross-stratification towards the top. Trough set thickness ranges from 10–50 cm (4–20 in.). Bedsets generally thin upward. The grain size varies from very coarse at the base to fine at the top, averaging coarse to medium. Angular mud rip-up clasts, well-rounded to subangular intra-basinal mud clasts, and well-rounded to subangular extra-basinal clasts are common along the base of the channel. Extra-basinal clasts are also common along bedset surfaces. Large trough cross-stratified beds at the base of the channel suggest a paleocurrent direction of 45°. The mean paleocurrent orientation of trough cross-stratified beds within the upper part of the channel is 352°±5°, suggesting lateral migration at an angle to the mean channel direction.

Architectural evolution of the Kokopelli Channel complex: Garrison and van den Bergh (this volume) calculated that during the development of Kokopelli Channel, the sedimentation rate was approximately 0.1 mm/year and there was a relative rise in sea level of about 0.05 mm/year. Deltaic shoreline progradation is calculated as 0.04 m/year (0.13 ft/yr). This modest sedimentation rate and relative rate of sea-level rise resulted in a wave-dominated delta class (Figure 1).

The Kokopelli Channel complex is a 1448-m-wide (4340 ft), 18.3-m-thick (60 ft) channel belt that generally trends 55°. The Kokopelli Channel complex is multi-storied with at least 22 channel-fill elements deposited in three stages (denoted as Stages 1, 2, and 3). These channel-fill elements are dominated by lateral accretion structures, and are generally stacked en echelon laterally within the channel belt boundaries.

The Kokopelli Channel complex cross section, in Willow Springs Wash, is at least 22 km (14 mi) from the paleoshoreline and that the maximum seaward progradation of the Ferron Parasequence Set 4B delta front was 9.5 km (5.9 mi). This suggests that the three sub-channel belt cross sections of the Kokopelli Channel complex all formed from 22–32 km (14–20 mi) from the paleoshoreline. Based on the Kokopelli Channel's stratigraphic position (i.e., incising into the E Rooted Zone of Parasequence Set 4A and capped by the G coal zone at the top of Parasequence Set 4B) suggests that the Stage 1 cross section occurred about 22 km (14 mi) from the paleoshoreline and that the final channel-fill element of the Stage 3 cross section represents deposition at about 32 km (20 mi) from the paleoshoreline (Garrison and van den Bergh, this volume).

The preserved channel belt thickness increases progressively from 5.8 to 9.5 to 18.3 m (10-31-60 ft), for Stage 1, Stage 2, and Stage 3, respectively. The Stage 1 sub-channel belt is mostly removed by erosion, therefore little can be inferred about its original width or aspect ratio. The sub-channel belt width (908 m [2720 ft]) and aspect ratio (95.4) for Stage 2 is larger than that of the younger Stage 3 (i.e., 653 m [1960 ft] and 35.7 m [117 ft], respectively). This suggests that the channel-belt complex became progressively thicker and narrower during each stage of development (i.e., more confined or restricted).

Early Stage 1 and Stage 2 development of wide, thin sub-channel belts dominated by large, poorly defined, lateral accretion surfaces and inclined trough and ripple cross-stratification bedsets, suggest unconfined, progressive lateral migration to the east. Lateral migration of the Stage 2 sub-channel-belt element was at least 760 m (2280 ft). Little evidence exists for major truncations, incisions, or cut-and-fill events during this lateral migration.

Near the beginning of Stage 3, the individual channel-fill elements become more confined with deposition restricted to relatively narrow channel profiles, and widths on the order of 150–190 m (450–570 ft). The Stage 3 sub-channel belt cuts down through the E Rooted Zone, about 1.5 m (4.9 ft) lower than the base of Stage 2. The first channel-fill element of Stage 3 has a top about 1.5 m (4.9 ft) higher than the top of Stage 2 and a base almost 4 m (12 ft) higher. At each pulse-wise lateral step of channel-belt development, signified by the development of each new channel incision and filling, lateral migration was only on the order of 35–40 m (105–120 ft) east. The third oldest channel-fill element of Stage 3 cuts down 5.5 m (18 ft) lower than the oldest two Stage 3 channel-fill elements. From the deposition of the first channel-fill element to the final channel-fill element, there was a total stratigraphic climb (i.e., vertical aggradation) of channel tops of about 5 m (15 ft). The change in channel development style, size, and degree of lateral migration, near the beginning of Stage 3 sub-channel-belt development, suggests that there was first a downcutting and then substantial aggradation (i.e., stratigraphic climb). Lateral migration slowed progressively, and stratigraphic climb (i.e., vertical channel aggradation) increased, during the evolution of the channel complex as the channel belt's development continued until near the end of the development of Parasequence Set 4B.

The youngest Stage 3 channel-fill element is somewhat coarser grained than any of the previous channel-fill elements in the Kokopelli Channel. There is evidence that this final fill element was somewhat narrower than the previous elements (Figure 22). Therefore, it can be suggested that the final channel-fill event may also record a change in bed load properties, as well as channel geometry.

The Kokopelli Channel Belt cross section represents a proximal distributary at a distance greater than 22 km (14 mi) from the paleoshoreline throughout its develop-

ment. The variations in external geometry and internal architecture are directly related to the very slow 0.04 m/year (0.13 ft/yr) progradation rate, although during the channel-belt development there was a 9.5 km (5.9 mi) paleoshoreline seaward migration. The development of the channel belt is dominated by extensive lateral accretion sedimentation, although vertical aggradation is clearly present and increases progressively throughout the development of the channel belt.

Lateral accretion and modest aggradation within the Kokopelli Channel is a direct consequence of stream equilibrium profile adjustment in response to very slow deltaic progradation (i.e., 0.04 m/year [0.13 ft/yr]). The low sedimentation rate and the comparable slow relative rise in sea level resulted in a stable, slightly seaward-migrating paleoshoreline. The delta-plain and shoreface sediments underwent an extensive period of vertical accretion during the slow relative rise in sea level. The original stream equilibrium profile at the initiation of equilibrium fluvial deposition is shown in Figure 21a. During the period of shoreline stability, the point to which the stream equilibrium profile is graded progressively moves upward at a relatively slow rate as the delta-plain and shoreline sediments aggrade (Figure 21c). This vertical migration of the paleoshoreline results in the production of only a minimal amount of new accommodation space upstream. This, in turn, results in a very slow vertical sediment aggradation upstream in the fluvial system. The development of the channel belt is therefore dominated by lateral stream aggradation and migration.

A low rate of upstream sedimentation and the resulting lateral upstream stream accretion frequently results in rare local avulsions (Heller and Paola, 1996). Local avulsions are rare in the wave-dominated Nile-type delta class (Figure 1). Low avulsion rates and the slow evolution of channel belts would tend to stabilize the channel belts and lead to extensive lateral migration, hence the common lateral accretion profiles in the Ferron aggradational distributary channel belts.

The Caprock Channel — An Analog for Channel Belts from Retrogradational Parasequence Sets

The Caprock Channel in Willow Springs Wash is an example of a channel belt occurring within an externally retrogradational parasequence set (e.g., Parasequence Set 7) (Figure 3). Figure 24 shows a schematic cross section illustrating the internal character of the Caprock Channel. Figure 25 is a measured section through the channel belt.

The Caprock Channel is a channel belt within Ferron Parasequence Set 7, and is exposed along the North Canyon wall of Willow Springs Wash (Figure 3). This channel belt has a sheet-like geometry with a width of

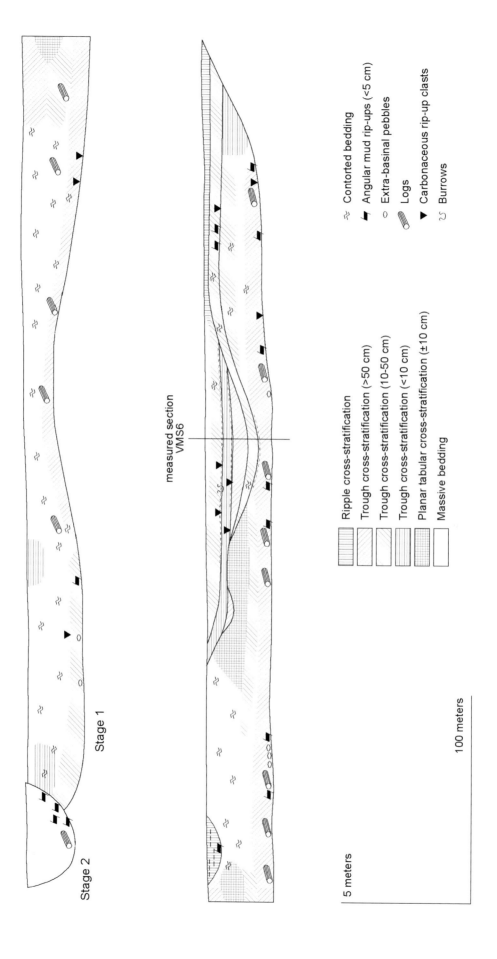

Figure 24. Exaggerated schematic cross section of the Caprock Channel showing the internal architecture and the distribution of sedimentary structures (bedforms). Bounding surfaces that are less well defined and are internal to major bounding surfaces are denoted as dashed lines. The vertical line is the line of measured section VMS6 shown in Figure 25.

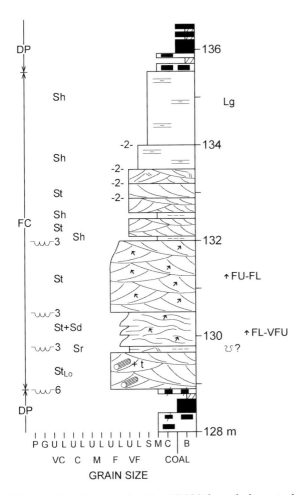

Figure 25. A portion of measured section VMS6 through the central part of the Caprock Channel showing the internal sedimentology. See Figure 5 for explanation of symbols and notation.

808 m (2420 ft) and a thickness of 4.1 m (13.5 ft). The aspect ratio of this channel belt is 195.0. It is not clear whether the entire extent of this channel belt is exposed in the study area, since the limit of the Willow Springs Wash outcrop along the county road into Willow Springs Wash may have truncated the belt's western end (see Plate V of Garrison and van den Bergh, this volume, on the CD-ROM). This 809-m (2420-ft) width may, as a result, represent a minimum width for the channel belt complex.

The Caprock Channel consists of two sub-channel belt stages, although the older channel belt phase composes over 95% of the preserved channel belt cross section. The second phase of channel-belt development is represented simply by the incision of a small, convex downward channel at the western end of the complex. The channel belt trends 28°. In general, the Caprock Channel sets on top of the J_2 coal zone of Garrison and van den Bergh (this volume).

Internal channel-belt architecture and sedimentology: Stage 1 consists of three, vertically stacked, channel-fill

elements. The mean paleocurrent orientation of the sub-channel belt is 28°±9°. The oldest and largest channel-fill element in the channel belt has a preserved thickness of 4.1 m (13.5 ft) and a preserved width of 769 m (2300 ft). The aspect ratio of this channel element is at least 187.6. The younger two channel-fill elements are much smaller. The older of these two channel-fill elements is partially truncated by the youngest channel-fill element. It has a preserved width of 238 m (714 ft) and a preserved thickness of only 1.1 m (3.6 ft), giving an apparent aspect ratio of 216.4. The youngest channel fill is completely preserved and has a width of 312 m (936 ft) and a thickness of 3.0 m (9.8 ft). The aspect ratio of this channel element is 104.0.

The vertical sedimentary structure profile in the oldest channel-fill element in Stage 1 is trough cross-stratification at the base, to locally small-scale planar tabular cross-stratification, and to ripple stratification at the top (Figure 24). Trough bedset thickness changes laterally from greater than 50 cm (20 in.), to 10–50 cm (4–20 in.), and vertically to less than 10 cm (4 in.). Small-scale planar tabular cross-bed thickness is about 10 cm (4 in.). Contorted bedding is very common in the upper and middle parts of the channel fill of the western and central part of the channel belt (Figure 24). There are abundant tree logs, and locally, abundant angular mud rip-ups and subangular to well-rounded intra-basinal mud clasts along the base of the channel. Mud clasts have a maximum length of 5 cm (2 in.) and average length is 1 cm (0.4 in.). Locally, extra-basinal pebbles occur as matrix-supported lags along the base of the channel. The maximum pebble diameter is 5 mm, and the matrix is coarse-grained sandstone. Other features that occur along the base of the channel are locally abundant plant fragments, flutes, grooves, and load casts (van den Bergh, 1995). The vertical grain-size profile is in general medium to coarse at the base to very fine at the top. Individual planar-stratified beds fine upward on a laminae scale, from coarse to medium grain size.

Sedimentary structures in the younger two channel-fill elements of Stage 1 are trough cross-stratified bed-sets. The bedsets of the youngest channel-fill element are interbedded with muddy siltstones (van den Bergh, 1995). The trough bedset thickness ranges from greater than 50 cm (20 in.) to less than 10 cm (4 in.). In general, bedsets thin upward from 1.5–0.5 m (4.9–1.6 ft). Locally, the top of the youngest channel-fill element has ripple-stratified bedsets, and is burrowed (ichnofossil *Plano-lites*).

Stage 2 of the Caprock Channel is composed of the incision of a relatively simple, single channel body, 54 m (160 ft) wide and 6.7 m (22 ft) thick (i.e., aspect ratio of 8.1), into the Stage 1 sub-channel belt complex. Sedimentary structures, within the Stage 2 channel, are not well developed, and consequently have not been documented. Grain size is uniform (medium-lower) within this small channel, and angular mud rip-ups and logs are common along the base of the channel.

Architectural evolution of the of the Caprock Channel: Garrison and van den Bergh (this volume) calculated that during the development of the Caprock Channel, the sedimentation rate was approximately 0.1 mm/year and relative sea level was rising at a rate of about 0.3 mm/year. The progradation rate was calculated as 0.03 m/year (0.1 ft/yr). The Caprock Channel is a relatively simple channel-belt complex, with a thin, 4.1-m-thick (13.5 ft), sheet-like geometry. It has a width of 808 m (2650 ft). It consists dominantly of one sub-channel belt stage, incised by a small channel at its western end (i.e., Stage 2).

The Caprock Channel cross section, in Willow Springs Wash, represents sandstones deposited in a channel-belt segment located from 18–24 km (11–15 mi) from the paleoshoreline of retrogradational Ferron Parasequence Set 7. In general, the Caprock Channel appears to have formed as a single, very wide confined channel system, with only minor vertically stacked scour and fill, channel-fill events along the eastern margin of the channel belt. The two final channel-fill events probably represent the final abandonment stage of the Caprock Channel. The lack of accretionary structures within the wide Stage 1 channel-fill element, suggests that lateral migration was minimal. It is possible that the extensive development of contorted bedding may have prevented these structures from being preserved, if they were developed. It is also possible that the final incision of the confined, sheet-like Caprock Channel by the small confined channel at the western end, may represent a much later event and not genetically related to the Caprock Channel. The deposition of the Caprock Channel occurred during a period when the sedimentation rate far outpaced by the rate of relative rise in sea level. The potential for delta-plain aggradation was high. Delta-plain aggradation leads to a channel confinement and stabilization, and vertical accretion.

The extensive development of contorted bedding indicates a high water table during the development of this channel belt. This is a common feature of channel belts within the retrogradational Upper Ferron Sandstone Parasequence Sets 6 and 7 (van den Bergh, 1995). Since Parasequence Set 7 is associated with the final retrogressive phase of the Last Chance Delta, the rise of the local water table may be expected to be concurrent with the modest 0.3 mm/year relative rise in sea level (Garrison and van den Bergh, this volume). The occurrence of contorted bedding, interpreted by van den Bergh (1995) to be associated with a rise in the local water table, and the abundance of shale laminae and drapes in the Caprock Channel, are similar to observations described by

Shanley and McCabe (1995) for transgressive systems tract fluvial channels exhibiting tidal influences; as much as 80–100 km (50–60 mi) inland from paleoshorelines. Shanley and McCabe (1995) have described rising water tables and increases in clay drapes and heterolithic stratification in fluvial strata, which they attributed to an increase in tidal influence in fluvial channels associated with transgressions and marine floodings near the top of depositional sequences. Tidally influenced sedimentary structures *sensu stricto* have not been recognized in the Caprock Channel, but the occurrence of tree logs with marine borings (i.e., *Teredolites*) within channel belts associated with Ferron Parasequence Sets 6 and 7 (van den Bergh, 1995), suggests that there may be some marine (i.e., tidal) influence. This is not inconsistent with the Caprock Channel's position within the transgressive systems tract of the Ferron Composite Sequence (Garrison and van den Bergh, 1997; this volume).

Vertical aggradation and extensive development of contorted bedding, indicating a high water table within the Caprock Channel, is a direct consequence of stream equilibrium profile adjustment in response to slow progradation (0.03 m/year [0.1 ft/yr]), slow sedimentation (0.1 mm/year), and steady paleoshoreline transgression (i.e., a relative rise in sea level of 0.3 mm/year). The low sedimentation rate and the moderate relative rise in sea level resulted in a transgressive paleoshoreline. The delta-plain and shoreface sediments underwent an extensive period of flooding during the slow relative rise in sea level. The original stream equilibrium profile at the initiation of equilibrium fluvial deposition is shown in Figure 21a. During the period of shoreline transgression, the point to which the stream equilibrium profile is graded progressively moved upward and landward at a relatively slow rate, as the delta-plain and shoreline sediments were flooded (Figure 21d). This slow vertical and landward migration of the paleoshoreline results in the production of only new accommodation space upstream. This, in turn, results in a slow vertical sediment aggradation upstream in the fluvial system. The development of the channel belt is therefore dominated by slow vertical stream aggradation and minimal lateral migration. In an attempt to adjust to the new higher point to which the stream is graded and to reach equilibrium with the increased local discharge, the channel belts become wider without deepening. The lower reaches of the stream channel are flooded and the sediment becomes water saturated and sediment deformation occurs.

A low rate of upstream sedimentation results in rare local avulsions (Heller and Paola, 1996). Local avulsions are rare in the transgressive, extremely wave-dominated Brazos-type delta class (Figure 1). Low avulsion rates and the slow evolution would tend to stabilize the channel belts.

Distributary Stream Channel Size

There is not much evidence to suggest that the thickness of preserved channel-fill elements change significantly as a function of distance to the paleoshoreline, but limited, scattered data do suggest that preserved channel-fill elements do become thinner downstream. The widths and aspect ratios of single channel fills, such as Stage 1 of the County Line Channel (i.e., a width of 101 m [300 ft] and an aspect ratio of 10.0), the last phase of Kokopelli Channel Stage 3 (i.e., a width of 159 m [477 ft] and an aspect ratio 10.6), and the final element of the Caprock Channel (i.e., a width of 59 m [177 ft] and an aspect ratio 8.1), suggest that the rivers existing during the evolution of the Last Chance Delta had widths that remained relatively constant, although the geometry of the overall channel belts changed throughout the history of the Last Chance Delta.

Distributary Channel Belt Stacking Density

Channel belt stacking density can be qualitatively ranked by examining the channel belt to non-channel delta-plain facies ratio (i.e., reflecting number of channel belts) and the mean channel belt aspect ratios (i.e., reflecting interconnectedness). In Willow Springs Wash, the ratio of channel belt to non-channel delta-plain facies (i.e., sand-body proportion) in vertical sections, ranges from 0.43 to 1.18 to 0.43 for the progradational, aggradational, and retrogradational parasequence set intervals, respectively, and mean proximal distributary channel belt width/thickness aspect ratios range from 28.8 to 59.2 to 100.0. This suggests that channel belt stacking density increases with time as parasequence set stacking changes from progradational to aggradational; channel belt stacking density decreases somewhat between aggradational phase and the retrogradational phase, although it still remains higher than in the progradational phase.

Figure 26 shows a plot of sedimentation rate versus the rate of change of relative sea level (i.e., change in accommodation space) for the Last Chance Delta (Garrison and van den Bergh, this volume). Garrison and van den Bergh (this volume) have argued that relative sea level changes are dominantly controlled by local tectonic subsidence. Therefore, it can be concluded that sedimentation rate is correlated with relative change in sea level and subsidence. This implies that channel-belt size, interconnectedness, and stacking density within the Last Chance Delta are controlled, in part, by tectonic subsidence. Tectonic subsidence directly affects the point to which streams are ultimately graded. It is also concluded that in the Last Chance Delta channel belt width/thickness (interconnectedness) and channel belt stacking density are inversely proportional to sedimentation rate. This conclusion is in agreement with general

arguments regarding alluvial sedimentation in foreland basins (e.g., see Heller and Paola, 1996).

Regional Avulsion and Basinal Controls on Distributary Channel-Belt Architecture

Recently proposed elaborate 3-D models for alluvial architecture (e.g., Mackey and Bridge, 1995; Heller and Paola, 1996) have attempted to explain fluvial architecture as a function of sedimentation rate, avulsion rate, subsidence rate, and basin geometry. These models for alluvial architecture (e.g., Mackey and Bridge, 1995; Heller and Paola, 1996) have established some relationships that are useful in interpreting the architecture of alluvial systems. From studies of these models, it has been concluded that (1) alluvial architecture records relative changes in basin subsidence rate, (2) basin geometry controls downstream alluvial stacking patterns, (3) sedimentation rate is inversely correlated with stacking density and interconnectedness, (4) avulsion frequency is related to sedimentation rate and avulsion frequency can either increase faster or slower that sedimentation rate, and (5) avulsion style controls alluvial stacking patterns. Evaluating the data from the Upper Ferron Sandstone, in light of these models, has clarified some of the controls exhibited by these factors on Upper Ferron Sandstone channel belt architecture and has confirmed some of the relationships put forth in these studies.

The Last Chance deltaic events responded mainly to changes in sedimentation rate. Evaluating the stacking density and architectural changes in the Upper Ferron Sandstone distributary channel belts, some generalizations can be made regarding some of the controls on the architecture and sedimentology that result from these parameters. The Ferron Sandstone was deposited in a back-tilted foreland basin along the western margin of the Cretaceous Interior Seaway. In a back-tilted basin, subsidence rate and sedimentation rate decrease outward into the basin. If the avulsion rate is constant, this implies that channel belt stacking density and interconnectedness should increase downstream. The outcrops of the Upper Ferron Sandstone do not provide enough strike cross sections to allow a detailed evaluation of changes in downstream channel-belt architecture. However, it is clear from the discussions above that in a single strike cross section, channel belt stacking density and interconnectedness (i.e., sand-body proportion, channel-belt size, and aspect ratio) increase with time and with the decreasing sedimentation rate.

Heller and Paola (1996) concluded that, in such a back-tilted basin, regional avulsions are the dominant processes that control channel-belt architecture (i.e., stacking density and patterns). The relationship between avulsion frequency and sedimentation rate is not clear, in most cases. Heller and Paola (1996) have proposed that avulsion frequency may either increase faster or

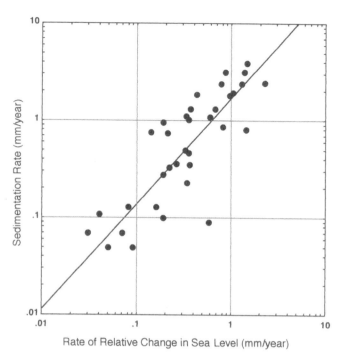

Figure 26. Plot of calculated sedimentation rate versus the calculated rate of change of relative sea level for the Upper Ferron Sandstone Last Chance Delta.

slower than sedimentation rate. In their models, a decrease in channel belt stacking density and interconnectedness with increasing sedimentation rate, such as described for the Upper Ferron Sandstone, occurs only if avulsion frequency increases at a slower rate than sedimentation rate.

In light of the modeling of Heller and Paola (1996), some generalizations can be made regarding the origin of the internal architecture of channel belts. In the progradational parasequence sets (formed when sedimentation, progradation, and subsidence rates were relative high), the channel belts were confined and aggraded vertically. Under these conditions, it can be concluded that avulsion rates were also relatively high. The sand-body frequency (i.e., channel belts per unit time) can also be considered high. High avulsion rates and the rapid evolution of channel belts would tend to prevent stabilization of channel belts and minimize extensive lateral migration. The channel belts, occurring in aggradational parasequence sets, are unconfined and laterally aggraded, resulting in good interconnectivity and en echelon stacking. These channel belts formed at relatively low sedimentation and subsidence rates. The sand-body frequency (i.e., channel belts per unit time) can also be considered low. At these lower sedimentation rates, avulsion was infrequent. Low avulsion rate and slower rate of channel evolution (i.e., low sand-body frequency) allowed more time for lateral migration and aggradation. The Upper Ferron Sandstone channel belts, from retrogradational parasequence sets, are generally thin

and vertically aggraded and, although very wide, are relatively poorly connected and isolated from each other. During the formation of these channel belts, sedimentation was low, subsidence rate was moderate, and sand-body frequency is considered to be very low. The conditions during the deposition of these channel belts were quite similar to those existing during the deposition of the aggradational parasequence sets, although the subsidence rate was slightly higher and sand-body frequency was somewhat lower. Therefore, it appears that the change from thick, wide, laterally migrating channel belts to thin, wide, confined channel belts can be attributed to the increase in subsidence rate (transgression) and the subsequent slow rate of delta-plain aggradation.

Based on the above analysis, it appears that the driving forces behind the development of different channel belt architectural styles as a function of delta class and parasequence set stacking pattern are changes in rate of sedimentation, relative changes in sea level, progradation rate, changes in delta-plain aggradation rate (i.e., sand-body frequency), and regional avulsion frequency, as these factors affect the stream equilibrium profiles.

INCISED-VALLEY FLUVIAL SYSTEMS

Internal Architecture and Sedimentology

Progradation Parasequence Set

The fluvial deposits of the inner segment 3 of the South Muddy Creek Incised-Valley Fill in Parasequence Set 2C appear to have formed as amalgamated braided stream deposits (Garrison and van den Bergh, this volume). These deposits exhibit multi-storied, internally scoured vertical stacking that prevents them from being reliably sub-divided into systems tracts. These deposits are generally exposed in vertical cliffs along Miller and Muddy Creek Canyons, where they obtain a thickness of 30 m (90 ft). Locally, these amalgamated deposits contain heterolithic channel fills.

These heterolithic channel-fill deposits consist of very fine to fine-grained, trough cross-stratified to rippled sandstones interlayered with laminated mudstone containing very thin lamina of very fine grained sandstone. The sandstone layers range in thickness from 10–60 cm (4–24 in.) and the mudstone layers range from 5–35 cm (2–14 in.) in thickness. Trough cross-beds commonly contain climbing ripples on their surfaces. The sandstones frequently contain burrows (ichnofossils *Scoyena* and *Skolithos*) and fragments of coaly material.

The sandstone-rich braided stream deposits consist of amalgamated braided channel belts, 6–8 m (18–24 ft) thick, composed of trough cross-stratified, medium-grained to fine-grained sandstone. Individual channel belts may contain a thin abandonment phase consisting of ripple cross-stratified, very fine grained sandstone channel barforms. These channel-belt barforms have scour bases and range in thickness from 1.5–2 m (4.9–6.6 ft). The barforms have a sigmoidal shape in the dip direction and are channel-form in cross section. These barforms consist of trough bedsets that range in thickness form 20–60 cm (8–24 in.) and fine upward from medium-grained to fine-grained sandstone. In some barforms, ripples are found climbing up the trough cross-stratification. Trough cross-stratification fines upward on the lamina scale from medium-grained to fine-grained sandstone. Clay rip-up pebbles up to 3 cm (1 in.) in diameter are found along bedset boundaries and also floating within the sandstone bedsets. Locally, fine carbonaceous and coaly material can be found along lamina set boundaries.

Aggradational Parasequence Sets

The fluvial deposits of the inner segment 3 of the Miller Canyon Incised-Valley fill appear to have formed as amalgamated braided stream deposits (Garrison and van den Bergh, this volume). These deposits exhibit a multi-storied, internally scoured vertical stacking. It is not possible to reliably subdivide these multi-storied amalgamated fluvial deposits into systems tracts. These deposits are generally exposed in vertical cliffs along Miller Canyon, where they obtain a thickness of over 30 m (90 ft). Locally, these amalgamated deposits contain heterolithic channel-fill deposits.

These heterolithic channel-fill deposits consist of very fine to fine-grained, rippled sandstones interlayered with laminated carbonaceous mudstone and mudstone containing very thin lamina of siltstone. The sandstone layers range in thickness from 2 cm (0.8 in.) to 2.7 m (8.9 ft) and the mudstone layers range from 25 cm (10 in.) to 2.6 m (8.5 ft) in thickness. Ripple bedsets range in thickness from 2 cm (0.8 in.) to 2 m (6 ft). The sandstones frequently contain burrows, fragments of coaly material, and clay clasts. Rare lags of pelecypod fragments occur in the sandstone layers.

The sandstone-rich braided stream deposits consist of amalgamated braided channel belts, 4–8 m (12–24 ft) thick, composed of trough cross-stratified, coarse-grained to fine-grained sandstone barforms (Figure 27). These channel belt barforms have scour bases and range in thickness from 2.3–5 m (7.5–15 ft). The barforms have a sigmoidal shape in the dip direction and are channel-form in cross section. These barforms consist of trough cross-beds, 15–30 cm (6–12 in.) in thickness and bedsets, 60 cm to 2.2 m (2–7.2 ft) in thickness, and fine upward from medium-grained to fine-grained sandstone and from coarse-grained to medium-grained sandstone. In some barforms, ripples are found climbing up the trough cross-stratification. Trough cross-stratification fines upward on the lamina scale from medium-grained to fine-grained sandstone. Extrabasinal material ranging

in size from very coarse-grained sand to pebbles up to 0.5 cm (0.2 in.) in diameter are found along cross-bed boundaries, and also floating within the sandstone bedsets. Locally, fine carbonaceous and coaly material can be found along lamina set and bedset boundaries.

Effects of delta progradation, sedimentation rate, and relative rise in sea level on fluvial system development: There appears to be very little change in the fluvial style, internal architecture, or sedimentology of the incised-valley amalgamated braided fluvial systems with delta class or the rate of progradation, sedimentation, and relative rise in sea level. These factors are more important in producing variations in the overall incised-valley fill facies association (Garrison and van den Bergh, this volume) that produce variability in the fluvial incised-valley fill deposits.

The incised-valley fill facies association that developed in wave-reworked, progradational Parasequence 2C was filled during a time when the sedimentation rate was high and exceeded the rate of relative rise in sea level. Moderate sediment by-pass allowed this incised-valley fill to develop elements of 5th-order lowstand, transgressive, and highstand systems tracts (Garrison and van den Bergh, this volume). The inner segment of the incised-valley fill contains fluvial-fill facies representing lowstand, amalgamated braided stream fluvial facies overlain by transgressive and highstand fluvial deposits; transgressive and highstand fluvial deposits were not clearly distinguishable from the lowstand deposits. The middle segment consists of amalgamated lowstand distributary channel and interdistributary bay facies overlain by brackish bay and bay-head delta transgressive deposits, which are in turn overlain by highstand distributary and interdistributary bay facies. The outer segment of the incised-valley fill consists of small lowstand wave-reworked delta-front sandbodies overlain by transgressive estuarine bay facies; locally the underlying lowstand facies are truncated by wave ravinement. The highstand systems tract consists of wave-reworked stream-mouth-bar facies.

The incised-valley fill facies association that developed in wave-dominated aggradational Parasequence 4B was filled during a period when sedimentation rate was balanced with the rate of relative rise in sea level. Fluvial and shoreline aggradation occurring during this period allowed the incised valley to be almost completely filled with lowstand, amalgamated braided fluvial valley-fill facies (Garrison and van den Bergh, this volume). The transgressive systems tract is represented by thin brackish water estuarine facies occurring locally at the top of the valley fill. The highstand systems tract is represented by wave-modified shoreface sandstones that "over-filled" across the top of the older incised-valley fill deposits. The incised-valley fill facies association that developed in wave-dominated aggradational Parasequence 5B was

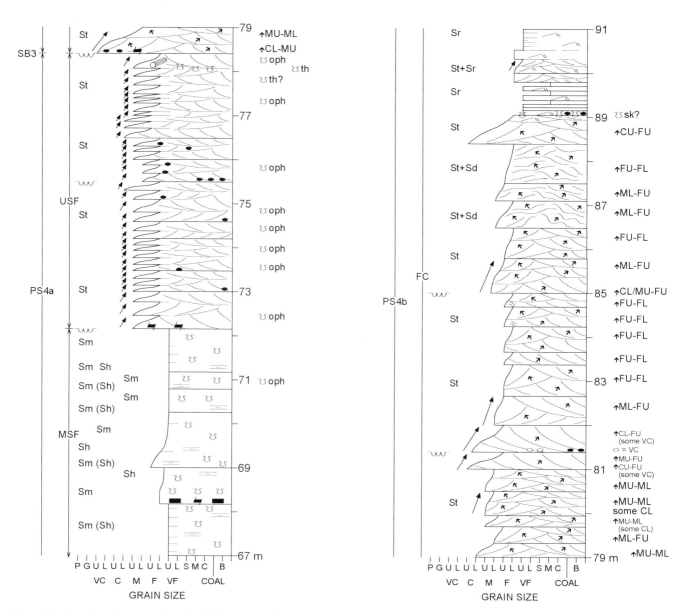

Figure 27. A portion of measured section GV8 showing the erosional unconformity at the base of Upper Ferron Sandstone Parasequence Set 4B fluvial valley fill and the incision into the near-marine deposits of Parasequence Set 4A. See Figure 5 for explanation of symbols and notation.

filled during a period when sedimentation rate was slightly less than the rate of relative sea-level rise. Fluvial and shoreline aggradation occurring during this period allowed the incised valley to be filled completely with 4th-order, lowstand fluvial fill facies (Garrison and van den Bergh, this volume). The 4th-order early-transgressive systems tract is represented by thin brackish water estuarine facies occurring locally at the top of the valley fill. The sedimentation rate was outpaced by the rise in relative sea level and the 4th-order late-transgressive systems tract, represented by thin retrogradational wave-modified shoreface parasequences, "over-filled" across the top of the older incised-valley fill deposits.

Effects of the adjustment of stream equilibrium profile on fluvial system development: The contrast in fluvial

style between the distributary channel belts, formed during periods of progradation, aggradation, or transgression of the paleoshoreline, and the incised-valley fluvial facies, formed following a period of relative fall in sea level, is a direct function of the development of a new stream equilibrium profile following movement of the paleoshoreline. The trajectory of the point at the paleoshoreline to which the streams are graded during periods of progradation, aggradation, or transgression is positive. During a relative fall in sea level the trajectory is negative.

During a relative fall in sea level, the point to which the stream profile is graded is shifts seaward and stratigraphically downward (Figure 21e). This trajectory steepens the stream gradient which forces the original low gradient meandering fluvial systems to become braided. The original equilibrium fluvial deposition gives way to fluvial net erosion and the development of an incised valley. The erosion rate decreases upstream to the knick-point. Once the paleoshoreline has been established at the lower sea-level position, the steepened gradient persists until the braided fluvial system begins to aggrade. Eventually the stream becomes graded again and meandering fluvial systems are established. The aggradation of the braided system produces the amalgamated braided deposits that completely fill the inner segment of the incised valley.

IMPLICATIONS OF DISTRIBUTARY CHANNEL BELT GEOMETRY AND ARCHITECTURE ON LATERAL AND VERTICAL LITHOLOGIC HETEROGENEITY

The ratio of net sand thickness to gross stratigraphic thickness (net/gross), calculated for intervals within the Willow Spring Wash section, also vary as a function of the 3rd-order, external parasequence set stacking pattern. The overall Ferron Sandstone net/gross ratio in Willow Springs Wash ranges from 0.22 to 0.47, averaging 0.31. The net/gross ratios calculated for the externally progradational, aggradational, and retrogradational parasequence set intervals are 0.22±0.08 (range = 0.09–0.33), 0.43±0.18 (range = 0.14–0.65), and 0.32±0.9 (range = 0.23–0.45), respectively. The aggradational interval has the largest net/gross ratios, reflecting the aggradation and lateral stacking of the channel belts. The externally progradational parasequence set interval has the lowest net/gross reflecting the wide spacing of confined channel belts.

The probability of interwell connectivity (i.e., the probability that two wells, at a specified well spacing, will penetrate the same lithologic unit), at typical 40-ac (16-ha) (430 m [4300 ft]) and 80-ac (32-ha) (860 m [2600 ft]) well spacings is also a strong function of position within the overall external stacking pattern of the 3rd-order Last Chance deltaic parasequence sets. In the externally progradational parasequence sets, the average width is 223±126 m (669±378 ft). The probability of a channel belt extending between two wells spaced 40 ac (16 ha) apart is only about 10%, and there is very little probability at an 80-ac (32-ha) spacing. In the externally aggradational parasequence sets, the average width is 623±336 m (1870±1000 ft). The probability of a channel belt extending between two wells spaced 40 ac (16 ha) apart is about 75%, and the probability at an 80-ac (32-ha) spacing is only about 22%. In the externally retrogradational parasequence sets, the average width is 457±217 m (1370±651 ft). The probability of a channel belt extending between two wells spaced 40 ac (16 ha)

apart is about 50%, and the probability at an 80-ac (32-ha) spacing is less than 10%.

Since, in general, the outcrops at Willow Springs Wash represents a series of channel belt cross sections within the upper delta plain, these general statements about net/gross ratios and interwell connectivity apply only to the delta-plain facies association. Overall, within the delta-plain facies association, there appears to be little vertical incision of channel belts from one parasequence set into those of another, hence the low net/gross ratios and consequently low vertical connectivity. In Willow Springs Wash, only locally do fluvial channel belts of a parasequence set incise into the fluvial systems of another resulting in vertical connectivity and continuity (e.g., the incision of fluvial systems of Parasequence Set 5A into those of Parasequence Set 4B (see Plates II and V of Garrison and van den Bergh, this volume), but this is the exception rather than the rule. Farther south in the Limestone Cliffs, in an alluvial-plain setting, the channel belts appear to frequently incise into underlying fluvial systems. This produces a quite different set of systematics.

MODELING OF DISTRIBUTARY CHANNEL SANDSTONE BODIES

Deterministic models (i.e., graphic interpolations) and correlations of fluvial channel belts in the subsurface, are generally difficult and contain substantial errors due to the ribbon-like plan-view geometry, the restricted cross sectional dimensions, and the lack of information about channel belt sinuosity and quantitative interconnectedness. Given channel belt thickness determined from cores and well logs using bar thickness and relationships of trough cross-stratification thickness to channel depth (e.g., Bridge and Tye, 2000), channel belt cross sectional widths can be estimated. These estimates use a variety of approaches from empirical relationships of channel belt width to channel belt thickness (e.g., Fielding and Crane, 1987; Bridge and Mackey, 1993; and Bridge and Tye, 2000) or derived from mathematical samplings of channel belt thickness, and aspect ratio distributions (Figure 16). Application of cross sectional widths to deterministic models requires knowledge of paleoflow direction relative to cross section orientation. Knowledge of channel belt body information in the downstream direction is much more difficult to determine from cores and well logs. Ghosh (2000) has presented a relationship that allows the calculation of channel-belt sinuosity from paleocurrent information in outcrops and modern systems. This approach has not been tested in cores, although it is promising. Given these limitations, it is clear that deterministic graphic interpolations are less than ideal. Bridge and Tye (2000) used a sophisticated fluvial process model to generate cross sections of a fluvial reservoir in an attempt to apply deterministic data in 3-D. This approach requires information about channel avulsion rate, channel-belt aggradation style and rate, and channel-belt connectedness. These parameters are extremely difficult to determine in outcrops and cores and impossible from well logs. Given these limitations, geoscientists have sought more mathematical modeling methods to create 3-D cross sections of fluvial reservoir systems (e.g., Williams et al., 1993).

Geo-object modeling or Boolean modeling methods are mathematically based in that they create, in an unbiased fashion, a geologic model that honors only the well data (i.e., deterministically) and the statistical distributions and rules governing the architecture of the reservoir bodies. This is accomplished by distributing the geometrical objects in the background volume according to well-defined sets of rules that specify overlap (i.e., isolated or overlapping). These models are conditioned to available well control and reservoir net sand/gross thickness ratios. Reservoir sand bodies are modeled as simple geometric objects (e.g., half cylinders, disks, complex polygons, half ellipsoids, etc.) embedded in a background of shale (i.e., non-reservoir) (e.g., see Haldorsen and MacDonald, 1987; Begg and Williams, 1991; Jones et al., 1993; Williams et al., 1993).

The challenge is to first create objects that accurately represent the geology and architecture of reservoir sandstone bodies. Secondly, it is necessary to determine the sinuosity, thickness, width, length, or aspect ratio (i.e., width/thickness, length/thickness, or length/width ratios) probability distributions that allow stochastic modeling algorithms to size these reservoir objects. These probability distributions must describe the preserved channel-belt sandstones. This information is generally obtained from 2-D and 3-D quantitative architectural studies of outcrop analogs. Two-dimensional and 3-D quantitative architectural studies of outcrop analogs can also provide data for conditioning more traditional deterministic models. For example, having a knowledge of sinuosity, thickness, width, length, or aspect ratio probability distributions for reservoir sandstone bodies can significantly increase ones ability to make intelligent correlations into the interwell space.

Models for Distributary Channels and Channel Belts

Distributary channel sand bodies are, historically, the most documented and stochastically modeled reservoir sand objects. The most common approach used to model these objects is to use a half cylinder as a channel segment. These half cylinder objects are placed end to end to create a fluvial or distributary-channel system, conditioned by available well control.

The quantitative parameters needed to describe a fluvial or distributary channel or channel-belt architecture are summarized in Figure 28. The most obvious parame-

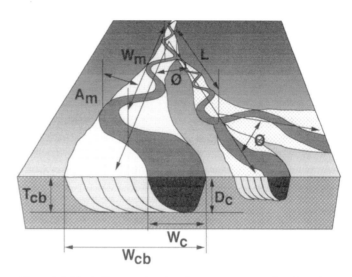

Figure 28. Block diagram showing the parameters used to describe a meandering distributary channel and channel-belt system (from van den Bergh and Garrison, 1996).

ters are the channel width (W_c) and depth (D_c) (i.e., thickness). It should be noted that for highly meandering river systems, the depth and width of the river channel are not simply related to the width (W_{cb}) and thickness (T_{cb}) of the preserved channel-belt sand. The width (W_{cb}) and thickness (T_{cb}) distributions are easily obtained from outcrops such as the Ferron Sandstone (Figure 16). Since rivers meander and are rarely straight, the most obvious additional parameter necessary to describe a fluvial or distributary channel belt system's architecture is the sinuosity (S) of the river system (Ghosh, 2000), and the amplitude (A_m) and wavelength (W_m) of the meanders. As implied above, except for low sinuosity river systems, the sinuosity of the river and the sinuosity of the preserved channel belts are not simply related. Modern rivers can give some indications of river sinuosity characteristics; they cannot clearly provide the architectural information necessary to stochastically model ancient river systems in outcrops or in the subsurface. Again, outcrop analogs, such as the Ferron Sandstone, are the *solution du jour* — they provide the 2-D and 3-D quantitative architecture of preserved ancient river systems. In the Ferron Sandstone, although much data exists on channel belt widths (W_{cb}) and thicknesses (T_{cb}), only the County Line and the Coal Miner's Channels in Willow Springs Wash are exposed sufficiently to quantify the channel belts system's sinuosity; A_m = 1 km (0.6 mi) and W_m = 4.8 km (3.0 mi), and A_m = 0.3 km (0.2 mi) and W_m = 3.2 km (2.0 mi) for the County Line Channel and Coal Miner's Channel, respectively (Figure 20). The calculated sinuosity of the County Line Channel is 1.09.

A natural feature of river systems is that, while in their alluvial plain, major rivers have tributaries that join them. When a river enters its delta plain, it will begin to bifurcate or branch on their way to the coastline. River bifurcation into distributaries is the inverse of the joining of tributaries to a river, therefore the architectural systematics are the inverse of each other. To quantify branching or river bifurcation, it is essential to know (1) the maximum length of a river segment prior to branching, (2) the angle formed between the two resultant distributaries, and (3) the decrease in a river's width upon branching. Garrison et al. (1991) suggested that, for a distributary system, the total number of distributaries and the lengths of each subsequent distributary segment after each subsequent river bifurcation can be described as a power law function of the form

$$\log(N(L)) = C_1 - D_L'\log(L) \qquad (1)$$

where the number of distributaries $N(L)$ of length L, C_1 is a coefficient and D_L' is the apparent effective length fractal dimension (i.e., scaling factor) of the branching distributary system. It follows that the one can calculate a constant branch-scaling ratio

$$R_b = L_{n-1}/L_{1n} \qquad (2)$$

where L_n is the length of a distributary before bifurcation and L_{n-1} is the length of the resulting segments after bifurcation. The only remaining parameter needed to describe the branching architecture is the branching angle between the two resultant distributaries Ø. Modern rivers can give substantial information about a river's branching characteristics, but again they cannot clearly provide the architectural information necessary to stochastically model ancient river systems in outcrops or in the subsurface. Outcrop analogs can provide the 2-D and 3-D quantitative architecture of preserved ancient river systems. Rarely are fluvial or distributary channel belts exposed sufficiently to allow quantification of branching. In the Ferron Sandstone, only the County Line Channel system has been mapped in sufficient detail to show the nature of the branching; Ø = 40° (see Figure 20).

It follows from the above discussions, and the data from Figure 4, that upon branching, the resulting distributaries become smaller in width, probably without substantial increases in channel depth. The reduction in width of each successive distributary segment can be described by a power law relationship similar to equation 1:

$$\log(N(W_{cb})) = C_1 - D_w'\log(W_{cb}) \qquad (3)$$

where the number of distributaries $N(W_{cb})$ of width W_{cb}, C_1 is a coefficient and D_w' is the apparent effective width fractal dimension (i.e., scaling factor) of the branching distributary system. It should be pointed out that the above discussions also apply to tributary coalescence in fluvial systems as well.

Since it is clear that outcrops with sufficient exposure to document the architectural ramifications of distributary branching and tributary coalescence do not

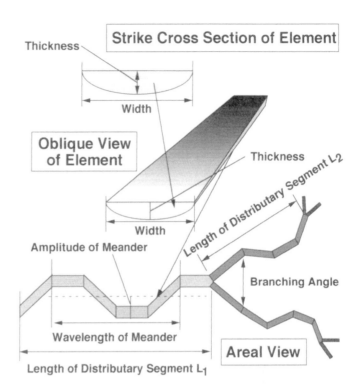

Figure 29. Diagram showing the proposed object for modeling sinuous, branching distributary channel and channel-belt systems (from van den Bergh and Garrison, 1996).

exist, the quantitative documentation of the relationships in equations 1, 2, and 3 must come from a combination of the analysis of modern rivers and the detailed architectural mapping of outcrop analogs.

Figure 29 is a proposed enhancement of the linearly connected, half cylinder object designed for use in stochastic geologic object modeling of channel and channel-belt sandstones (e.g., Williams et al., 1993). As with the traditional approach, these half cylinder objects are placed end to end to create a fluvial or distributary channel system, conditioned by available well control, with the notable exception that the branching characteristics of distributaries is allowed. The distributions describing the lengths of a subsequent distributary can be obtained from equation 1. Branching ratios (R_b) and branching angles (Ø) can be specified, instead of a length distribution relationship. One key issue is that during branching, the aspect ratio of the channel segments will change according to equation 3.

Enhancements and the Implications for Reservoir Volumetrics

An important consequence of this branching model, outlined above, is that it will facilitate the connection of fluvial or distributary channels of different widths and thicknesses within the constraints of equations 1, 2, and

3. This allows much more flexibility, in connectivity and reservoir volumetrics, than traditional approaches. Volumetrics and reservoir connectivity will, therefore, obviously become more realistic and reservoir performance more easily modeled.

A further enhancement of the simple half cylinder channel object is to allow a more realistic convex downward lower part of the channel object by permitting the shape to change in response to the style of incision into the country rock. The different types of complex convex downward shapes are well illustrated in the schematic diagrams of the County Line Channel (Figure 18) and the Kokopelli Channel (Figure 22). The simplest way to accomplish this is to specify aspect ratios at various points within a channel belt cross section. For example, if a channel belt cross section is divided into five equal-width segments, an overall width/local thickness aspect ratio could be specified for each of the boundaries between the five segments. For example, to model the shape of the County Line Channel (Figure 18), these aspect ratios would be 82.7, 20.7, 50.3, and 50.3, from west to east, respectively. These values may even be allowed to vary as a function of channel sinuosity. The obvious ramification of this more realistic shape is improved channel-belt volumetrics. This concept warrants further investigation, but is beyond the scope of this paper.

In addition, when the thicknesses of stochastically modeled channel cross sections are conditioned by available sand thicknesses interpreted from well data, errors can also be introduced, both into the model and subsequently into the volumetric calculations, by allowing the wellbore penetration of the modeled channel to be at any random point in a channel cross section. Some of this error can be reduced, if we allow the conditioning of the thicknesses of the penetrated channel cross sections to be made more realistic by (1) choosing the penetration points of the wellbore into the channel, and (2) conditioning to the vertical bedform and bedset profiles. The sedimentological data presented above indicates that vertical sedimentary profiles vary as a function of position along a channel belt cross section and as a function of parasequence set stacking pattern. This information and the analog models for seaward-stepping, aggradational, and landward-stepping channel belts, outlined above, will allow penetration points to be chosen more realistically by using these models to estimate the position of the wellbore into a channel belt cross section.

CONCLUSIONS

Near-Marine Sandstones

Architectural data, for near-marine sandstone facies, indicate a pattern of overall deltaic deposition that is directly correlated with parasequence set stacking patterns and controlled by sediment supply systematics and degree of wave modification.

The sediment supply (i.e., sedimentation rate) and accommodation space (i.e., the rate of change in relative sea level) systematics operating during the deposition of the Last Chance Delta clastic wedge, resulted in distinct near-marine sandstone geometries that affect the lateral continuity and lateral lithologic heterogeneity.

It is concluded that the parasequence sets within the Upper Ferron Last Chance Delta clastic wedge can be ranked in terms of their overall reservoir potential. This ranking from best to worst is (1) aggradational parasequence sets, (2) wave-reworked progradational parasequence sets, (3) river-dominated progradational parasequence sets, and (4) retrogradational parasequence sets.

Fluvial Channel-Belt Sandstones

The distributary channel belts of progradational parasequence sets generally have laterally restricted geometries and are multi-storied with channel-filling elements generally stacked vertically within the channel-belt boundaries; aspect ratios average near 29. The abandonment phase of each channel-fill element is truncated by successively younger elements. The distributary channel belts of aggradational parasequence sets are generally quite laterally extensive and multi-storied with channel-fill elements usually stacked en echelon laterally within the channel belt boundaries; aspect ratios average near 59. In general, the channel-fill elements have a complete sedimentary profile from initial scour to abandonment. The distributary channel belts of retrogradational parasequence sets are laterally extensive and sheet-like with channel-fill elements usually stacked vertically within the channel-belt boundaries; aspect ratios average near 100. The channel-fill elements have in poorly developed sedimentary profiles, commonly having the abandonment phase of each channel-fill element truncated by successively younger elements. A reduction in the size and aspect ratios of channel belts downstream are a manifestation of river bifurcation and branching. The distributary channel belts of progradational and aggradational parasequence sets have distal distributary aspect ratios of 19 and 12, respectively.

Changes in architectural style between distributary channel belts of progradational, aggradational, and retrogradational parasequence sets can be explained in terms of simple changes in the sedimentation rate, rate of relative change in sea level, and avulsion frequency as these parameters affect the trajectory of the point to which the streams are graded. The adjustment of the streams to a new paleoshoreline position dictates the style of aggradation and accretion and subsequently the internal architecture.

Distributary channel belts forming at high sedimentation rates have low connectivity and stacking density;

high avulsion rates prevent extensive lateral migration. Distributary channel belts forming at low sedimentation rates have higher connectivity and stacking density; lower avulsion rates stabilize channel belts and promote lateral accretion. Distributary channel belts forming during transgression exhibit moderate vertical stream aggradation and minimal lateral migration; the lower reaches of the stream channels become water saturated and sediment deformation is common.

Fluvial systems occurring within incised-valley systems form in response to a steepening of the stream gradient due to a relative fall in sea level. The steepened the stream gradient forces the original low gradient meandering fluvial systems to become braided as compensation. The original equilibrium fluvial deposition gives way to fluvial net erosion and the development of an incised valley. The erosion rate decreases upstream to the knick-point. Once the paleoshoreline has been established at the lower sea-level position, the steepened gradient persists until the braided fluvial system begins to aggrade. Eventually the stream becomes graded again and meandering fluvial systems are established.

ACKNOWLEDGMENTS

The Ferron Sandstone depositional sequence stratigraphy research of the Ferron Group Consultants, L.L.C., was supported, in part, by grants from Arco Exploration and Production Technology, Anadarko Petroleum, Amoco Production Company, British Petroleum Exploration (Alaska), Chevron Overseas Petroleum, Live Earth Products, Norsk-Hydro ASA, Phillips Petroleum Company, Shell E&P Technology, Texaco, U.S.A., and Union Pacific Resources in 1997, 1998, and 1999. Field discussions of the depositional sequence stratigraphy of the Ferron Sandstone with J. P. Bhattacharya, J. M. Boyles, K. Soegaard, C. Jenkins, M. Scheihing, C. J. O'Byrne, and K. W. Shanley proved invaluable. Discussions with J. Thomas of Amoco always provided timely wisdom. J. M. Boyles of Union Pacific Resources and K. W. Shanley of Amoco provided graphics and reprographics support during the compilation phase of this project. Past discussions of geo-object modeling with N. Saad and K. Soegaard have proved invaluable. We thank Jim Parker, Utah Geological Survey, for drafting many of the figures.

REFERENCES CITED

Barton, M. D., 1994, Outcrop characterization of architecture and permeability structure in fluvial-deltaic sandstones, Cretaceous Ferron Sandstone, Utah: Ph.D. Dissertation, University of Texas, Austin, 259 p.

Begg, S. H., and J. K. Williams, 1991, Algorithms for generating and analyzing sandbody distributions, in L. W. Lake, ed., Reservoir characterization II: London, Academic Press, 726 p.

Bridge, J. S., and S. D. Mackey, 1993, A theoretical study of fluvial sandstone body dimensions, in S. S. Flint and I. D. Bryant, eds., Geological modelling of hydrocarbon reservoirs: International Association of Sedimentologists Special Publication 15, p. 213–236.

Bridge, J. S., and R. S. Tye, 2000, Interpreting the dimensions of ancient fluvial channel bars, channels, and channel belts from wireline-logs and cores: AAPG Bulletin, v. 84, p. 1205–1228.

Cotter, E., 1975, Late Cretaceous sedimentation in a low energy coastal zone — the Ferron Sandstone of Utah: Journal of Sedimentary Petrology, v. 45, p. 669–685.

Elliott, T., 1986, Deltas, in H. G. Reading, ed., Sedimentary environments and facies: Oxford, Blackwell Scientific Publications, p. 113–154.

Fielding, C. R., and R. C. Crane, 1987, An application of statistical modelling to the prediction of hydrocarbon recovery factors in fluvial reservoir sequences, in Recent Developments in Fluvial Sedimentology: Society for Sedimentary Geology (SEPM) Special Publication 39, p. 321–327.

Galloway, W. E., 1975, Process framework for describing the morphologic and stratigraphic evolution of deltaic depositional systems, in M. L. S. Broussard, ed., Deltas, models for exploration: Houston Geological Society, p. 87–98.

Gardner, M. H., 1992, Sequence stratigraphy of the Ferron Sandstone, east-central Utah, in N. Tyler, M. D. Barton, and R. S. Fisher, eds., Architecture and permeability structure of fluvial-deltaic sandstones: a field guide to selected outcrops of the Ferron Sandstone, east-central Utah: University of Texas, Austin, Bureau of Economic Geology Guidebook, p. 1–12.

—1993, Sequence stratigraphy of the Ferron Sandstone (Turonian) of east-central Utah: Ph.D. Dissertation, Colorado School of Mines, Golden, 406 p.

Garrison, J. R., Jr., and T. C. V. van den Bergh, 1997, Coal zone and high-resolution depositional sequence stratigraphy of the Upper Ferron Sandstone, in P. K. Link and B. J. Kowallis, eds., Mesozoic to Recent geology of Utah: Brigham Young University Geology Studies, v. 42, pt. II, p. 160–178.

Garrison, J. R., Jr., W. C. Pearn, and D. U. von Rosenberg, 1991, The fractal nature of geological data sets — power law processes everywhere!: Society of Petroleum Engineers, paper 22842, p. 261–272.

Ghosh, P., 2000, Estimation of channel sinuosity from paleocurrent data: a method using fractal geometry: Journal of Sedimentary Research, v. 70, p. 449–455.

Haldorsen, H. H., and C. J. MacDonald, 1987, Stochastic modeling of underground reservoir facies (SMURF): Society of Petroleum Engineers, paper 16751.

Hale, L. A., 1972, Depositional history of the Ferron Formation, central Utah, *in* J. L. Baer and E. Callaghan, eds., Plateau-Basin and Range transition zone, central Utah: Utah Geological Association Publication 2, p. 29–40.

Heller, P. L., and C. Paola, 1996, Downstream changes in alluvial architecture — an exploration of controls on channel stacking patterns: Journal of Sedimentary Research, v. 66, p. 297–306.

Johnson, H. D., and D. E. Krol, 1989, Geological modeling of a heterogeneous sandstone reservoir, Lower Jurassic Statfjord Formation, Brent Field, *in* L. W. Lake, ed., Reservoir characterization — 1: Society of Petroleum Engineers Reprint Series 27, p. 46–58.

Jones, A., J. Doyle, T. Jacobsen, and D. Kjonsvik, 1993, Which sub-seismic heterogeneities influence waterflood performance? A case study of a low net-gross fluvial reservoir: 7th European IOR Symposium, Moscow, p. 35–46.

Lowry, P., and T. Jacobsen, 1993, Sedimentological and reservoir characteristics of a fluvial-dominated delta-front sequence — Ferron Sandstone Member (Turonian), east-central Utah, *in* M. Ashton, ed., Advances in reservoir geology: Geological Society Special Publication 69, p. 81–103.

Mackey, S. D., and J. S. Bridge, 1995, Three-dimensional model of alluvial stratigraphy — theory and application: Journal of Sedimentary Research, v. B65, p. 7–31.

Mattson, A. 1997, Characterization, facies relationships, and architectural framework in a fluvial-deltaic sandstone — Cretaceous Ferron Sandstone, central Utah: M.S. thesis, University of Utah, Salt Lake City, 174 p.

Posamentier, H. W., and P. R. Vail, 1988, Eustatic controls on clastic deposition II — Sequence and systems tract models, *in* C. K Wilson, B. S. Hastings, C. G. St. C. Kendall, H. W. Posamentier, C. A. Ross, and J. C. Van Wagoner, eds., Sea-level changes — an integrated approach: Society for Sedimentary Geology (SEPM) Special Publication 42, p. 125–154.

Posamentier, H. W., M. T. Jervey, and P. R. Vail, 1988, Eustatic controls on clastic deposition I — Conceptual framework, *in* C. K Wilson, B. S. Hastings, C. G. St. C. Kendall, H. W. Posamentier, C. A. Ross, and J. C. Van Wagoner, eds., Sea-level changes — an integrated approach: Society for Sedimentary Geology (SEPM) Special Publication, v. 42, p. 109–124.

Reynolds, A. D., 1999, Dimensions of paralic sandstone bodies: AAPG Bulletin, v. 83, p. 211–229.

Ryer, T. A., and M. McPhillips, 1983, Early Late Cretaceous paleogeography of east-central Utah, *in* M. W. Reynolds and E. D. Dolly, eds., Mesozoic paleogeography of west-central United States: Society for Sedimentary Geology (SEPM), Rocky Mountain Section Paleogeographic Symposium 2, p. 253–272.

Shanley, K. W., and P. J. McCabe, 1995, Sequence stratigraphy of Turonian-Santonian strata, Kaiparowits Plateau, southern Utah, U.S.A. — implications for regional correlation and foreland basin evolution, *in* J. C. Van Wagoner and G. T. Bertram, eds., Sequence stratigraphy of foreland basin deposits: AAPG Memoir 64, p. 103–136.

van den Bergh, T. C. V., 1995, Facies architecture and sedimentology of the Ferron Sandstone Member of the Mancos Shale, Willow Springs Wash, east-central Utah: M.S. thesis, University of Wisconsin, Madison, 255 p.

van den Bergh, T. C. V., and J. R. Garrison, Jr., 1996, Channel belt architecture and geometry — a function of depositional parasequence set stacking pattern, Ferron Sandstone, east-central Utah: AAPG Rocky Mountain Section Meeting, Rocky Mountain Section Meeting, expanded abstracts volume, Montana Geological Society, p. 37–42.

Van Wagoner, J. C., 1995, Overview of sequence stratigraphy of foreland basin deposits — terminology, summary of papers, and glossary of sequence stratigraphy, *in* J. C. Van Wagoner and G. T. Bertram, eds., Sequence stratigraphy of foreland basin deposits: AAPG Memoir 64, p. ix-xxi.

Van Wagoner, J. C., R. M. Mitchum, K. M. Campion, and V. D. Rahmanian, 1990, Siliciclastic sequence stratigraphy in well-logs, cores, and outcrop: AAPG Methods in Exploration Series, v. 7, 55 p.

Williams, J. K., A. J. Pearce, and G. W. Geehan, 1993, Modeling complex channel environments — assessing connectivity and recovery: 7th European IOR Symposium, Moscow, p. 47–57.

Zaitlin, B. A., R. W. Dalrymple, and R. Boyd, 1994, The stratigraphic organization of incised-valley systems associated with relative sea-level change, *in* R. W. Dalrymple, R. Boyd, and B. A. Zaitlin, eds., Incised-valley systems — origin and sedimentary sequences: Society for Sedimentary Geology (SEPM) Special Publication 51, p. 45–60.

The Ferron Coalbed Methane Play

South Main Street, Huntington, Utah, circa 1900.
Photograph used by permission, Utah State Historical Society, all rights reserved.

Analog for Fluvial-Deltaic Reservoir Modeling: Ferron Sandstone of Utah
AAPG Studies in Geology 50
T. C. Chidsey, Jr., R. D. Adams, and T. H. Morris, editors

Coalbed Gas in the Ferron Sandstone Member of the Mancos Shale: A Major Upper Cretaceous Play in Central Utah

Scott L. Montgomery[1], David E. Tabet[2], and Charles E. Barker[3]

ABSTRACT

Drilling for coalbed gas in the Upper Cretaceous Ferron Sandstone of central Utah during the 1990s has resulted in one of the most successful plays of this kind. Development through the year 2003 has resulted in three fields, as well as a potential fairway 6–10 mi (10–16 km) wide and 20–60 mi (32–96 km) long, corresponding to shallow coal occurrence at depths from 1100–3500 ft (330–1060 m) in the Ferron, a sequence of interbedded fluvial-deltaic sandstone, shale, and coal in the lower part of the Cretaceous Mancos Shale. The major reservoirs in this interval consist of thin to moderately thick (3–30 ft [1–10 m]) coalbeds of relatively low rank (high-volatile B bituminous) and variable gas content, ranging from 100 scf/ton (3.0 cm^3/g) or less in the south to as high as 500 scf/ton (15.6 cm^3/g) in the north. Other lithologies also contain gas and contribute a minor portion of the produced gas. Productive wells have averaged over 500 mcf/day and, after several years of production, continue to typically show increases in gas production. In the major productive area, Drunkards Wash unit, the first 33 producers averaged 974 mcf and 85 bbl of water per day after five years of continuous production. Estimated ultimate recoverable reserves for individual wells in this unit average about 1.9 bcf, with one standard deviation about that mean of ±1.5 bcf.

Based on several criteria, including gas content, thermal maturity, and chronostratigraphy, the play is divided into northern and southern parts. The northern part is characterized by coals that have the following characteristics: (1) high gas contents; (2) moderate thermal maturity (e.g. vitrinite reflectance [R$_o$] values of 0.6–0.8%); (3) good permeabilities (4–20 md); (4) lack of exposure; and (5) overpressuring, due to artesian conditions. Southern coals have much lower gas contents (<100 scf/ton [3.0 cm^3/g]), lower thermal maturity (R$_o$ = 0.4–0.6%), and they are partially exposed along an extensive, 35-mi (56-km) outcrop belt that may have allowed a degree of gas flushing. These coals, however, are slightly thicker and more extensive than those to the north and thus may retain some potential. Northern coals appear to contain a mixture of gas from three sources: in-situ thermogenic methane, migrated thermogenic methane from more mature sources, and late-stage biogenic gas. Current development is focused on the northern portion of the stated fairway, where well control and an existing infrastructure are present. Indications are that coalbed methane development in the Ferron will increase by at least one hundred in the near future from the 754 wells producing as of the end of 2003.

[1]*Petroleum Consultant, Seattle, Washington;* [2]*Utah Geological Survey, Salt Lake City, Utah;*
[3]*U.S. Geological Survey, Denver, Colorado*

501

INTRODUCTION

Within the coterminous United States, the gas resource in coalbeds has been estimated at 675 tcf, distributed among more than two-dozen basins (Scott, 1993; Tyler et al., 1995). No less than 79% of this total is identified in basin of the Rocky Mountain region, where extensive deposits of Late Cretaceous and early Tertiary coals are found at attractive drilling depths of 500–5000 ft (150–1500 m). Beginning in the 1990s, several new plays in Rocky Mountain basins have exhibited considerable success, owing in large part to advances in the ability to characterize and produce the relevant coalbed methane (CBM) reservoirs. Among these plays are the Fort Union Formation (Paleocene) of the Powder River Basin, the Vermejo Formation (Late Cretaceous) of the Raton Basin, and the Ferron Sandstone Member of the Mancos Shale (Late Cretaceous) of the Uinta Basin. In particular, the Ferron play is impressive in terms of high gas content and production and may well represent one of the most successful CBM plays in North America.

The Ferron Sandstone play extends off the southwest corner of the Uinta Basin along the western margin of the San Rafael Swell, a prominent Laramide (Late Cretaceous-Eocene) uplift in central Utah (Figure 1). The play is estimated to be 6–10 mi (10–16 km) wide and 30–60 mi (48–96 km) long, corresponding to shallow coal occurrence in the Ferron Sandstone, a sequence of interbedded fluvial-deltaic sandstone, shale, and coal in the lower portion of the Cretaceous Mancos Shale (Figure 2). Drilling activity has been concentrated along the northeastern portion of the fairway, where three fields have been opened (see Figure 1). The largest of these is Drunkards Wash field, consisting of over 535 wells on 160-acre (64 ha) spacing. Coal reservoirs in this field average 24 ft (7.2 m) in net thickness and occur at depths of 1100–3500 ft (330–1060 m). Productive coalbeds consist of high-volatile B bituminous coal that exhibit uncommonly high gas contents. Gas production from Drunkards Wash field began in 1992 and has increased steadily. Average daily flow rates have increased from less than 300 mcf to over 620 mcf, in part as a result of increased dewatering with a rapid growth in the number of producers. Initial negative declines have typified gas production from most wells along with decreases in water production. By December of 2003, cumulative production for the unit exceeded 438 bcf from a total of 549 wells. Analysis of well performance suggest a productive life of 17 yr or more for individual wells, with ultimate recoverable reserves averaging 1.9 bcf. This translates into a minimum of 1.0 tcf recoverable for the field itself and as much as 2 tcf for the entire fairway.

Owing, in part, to their unique qualities and high potential, Ferron coal reservoirs have been the subject of growing investigation (Tabet et al., 1995; Burns and Lamarre, 1997; Conway et al., 1997; Lamarre and Burns, 1997; Barker and Dallegge, 1998; Tabet, 1998; Buck et al., 2001; Lamarre et al., 2001). This work has included study in the following areas: (1) reservoir properties; (2) gas content and coal geochemistry; (3) coal thickness and quality; (4) hydrologic setting; and (5) well performance and completion effectiveness. In addition, a separate series of detailed studies on the sequence stratigraphy, depositional history, and sandstone reservoir character of the Ferron have been performed during the past decade, in part under the sponsorship of the Gas Research Institute, the U.S. Department of Energy, and other organizations (see, for example, Gardner et al., 1992; Fisher et al., 1993; Gardner and Cross, 1994; Anderson et al., 1997; Mattson, 1997; Garrison et al., 1997; White and Barton, 1998). A significant amount of geologic information consequently exists regarding the Ferron interval. Yet, with regard to CBM potential in particular, data remain preliminary in a number of areas, and important questions, such as the origin and distribution of the gas itself, are still unresolved.

This paper summarizes available information on the Ferron CBM play and discusses its implications for future activity. Data have been derived from a wide variety of published and unpublished sources to provide an overview of major topics and questions related to the geology of the play. We hope this summary may serve as a basis for further discussions on the significance of this CBM deposit to resource developments elsewhere.

REGIONAL GEOLOGY

Structure and Tectonics

The Ferron Sandstone CBM play extends from the southwest corner of the Uinta Basin along the northwest flank of the San Rafael Swell, a large, basement-cored anticline with a steep eastern margin and gentle (5–15°) western margin. Northward, the prospective fairway terminates where the Ferron coals pass under the Book Cliffs, which form the resistant southern border of the Uinta Basin proper. Westward, the fairway margins are defined by increasing depths of burial, and thinning of the Ferron coals beneath the central Wasatch Plateau. In the southeast portion of the fairway, the Ferron Sandstone crops out along the western margin of the San Rafael structure, providing excellent exposures that have been the subject of detailed study (e.g., Cotter, 1975; Ryer, 1981, 1991; Gardner et al., 1992; Anderson et al., 1997; Mattson, 1997; Garrison et al., 1997). This work focuses on the sedimentological and depositional character of the Ferron and has confirmed that it consists of stacked, fluvial-deltaic deposits included within a foreland sequence of alternating marine, marginal marine, and non-marine sediments shed eastward from the Cre-

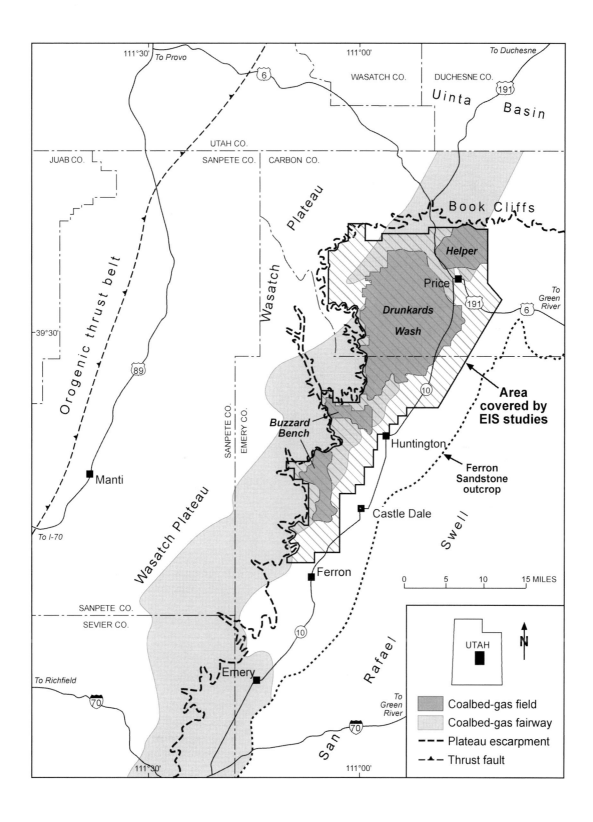

Figure 1. Regional setting and extent of Ferron coalbed methane (CBM) fairway, central Utah.

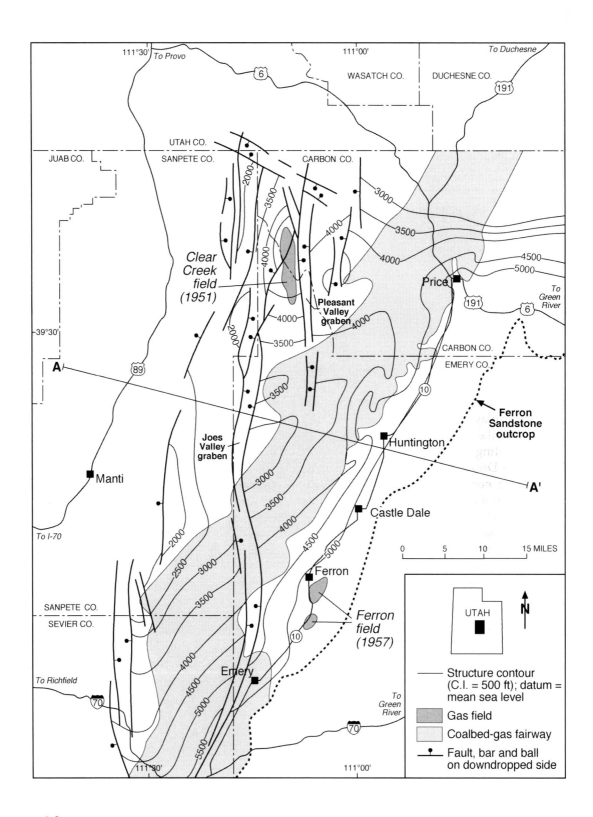

2A

Figure 2. (A) Regional structure contour map, top of Ferron Sandstone (modified from Sprinkel, 1993), and (B) diagrammatic cross section, eastern Wasatch Plateau, showing structural styles in the vicinity of the current Ferron CBM fairway (modified from Neuhauser, 1988).

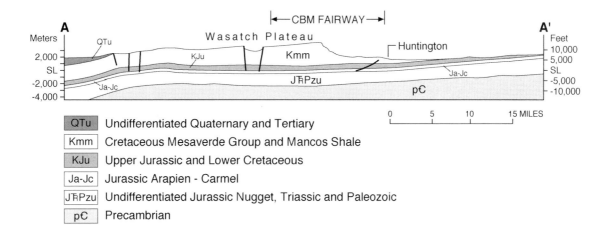

2B

taceous-Tertiary Sevier orogenic belt (Gardner and Cross, 1994).

Structurally, the area encompassed by the current fairway lies in a transition zone between the San Rafael Swell and the Wasatch Plateau. Along the eastern side of the fairway, the Ferron dips at about 200 ft/mi (40 m/km) northwestward off the San Rafael Swell, and local folding and reverse faulting are also evident (Tripp, 1989; Burns and Lamarre, 1997). At Drunkards Wash field, reverse faults display up to 150 ft (45 m) of vertical offset and generally strike parallel to the axis of the San Rafael feature, suggesting a genetic relationship. West and southwest of the Drunkards Wash area, the structural style changes to normal faulting and associated folding, both evident at the Ferron Sandstone level (Figure 2). Normal faults in this area appear associated with Basin and Range extensional deformation and are thus considered Miocene and later in age. These structures have vertical displacements up to 450 ft (136 m) (Doelling, 1972) and form several prominent north-south grabens (e.g., Pleasant Valley, Joes Valley) that extend over 45 mi (70 km) to the vicinity of Emery (Figure 2). These features attracted early hydrocarbon exploration along their margins. Larger volumes of gas, however, were found in structural-stratigraphic traps associated with anticlines at the Clear Creek and Ferron fields.

The fairway area, particularly the northern end, was intruded during the Oligocene and Miocene by a series of mafic, alkaline igneous dikes (Tingey et al., 1991). The east-west orientation of the older dikes (25–18 Ma) differs from the north-south orientation of the youngest dikes (8–7 Ma) and reflects the change in tectonic regime from east-west compression to east-west extension.

Burial history curves indicate that maximum overburden for Cretaceous deposits was achieved in the Eocene (Barker and Dallegge, 1998; Dallegge and Barker, 2000). Subsequent uplift and exhumation associated with Basin and Range faulting have caused removal of over 5000 ft (1500 m) of material across much of the Wasatch Plateau and San Rafael Swell.

Stratigraphy and Lithology

The Ferron Sandstone Member is Turonian-Coniacian in age (Cotter, 1975; Gardner et al., 1992; Schwans and Campion, 1997; Barker et al., 1999) and is placed within the lower part of the Mancos Shale. The Mancos Shale comprises a thick sequence of marine shale, which interfingers with several eastward-prograding sandstone tongues that reflect episodes of deltaic sedimentation along the western margin of the Cretaceous Western Interior Seaway. Throughout the Cretaceous, the western margin of this seaway in central Utah was a foreland basin, accommodating material shed from exposed Paleozoic and basement rocks along the active Sevier orogenic belt to the west (Ryer, 1981, 1991) (see Figure 3). To the west, proximal foreland deposits equivalent in age to the more distal Mancos Shale include conglomerate and sandstone of the Indianola Group. Underlying the Mancos Shale are conglomerate, sandstone, shale, and coal of the Dakota Sandstone. Overlying deposits include sandstone, shale, and coal of the Mesaverde Group. Within the foreland basin, the Mancos Shale consists of a series of marine shale tongues that interfinger with continental and nearshore clastic wedges shed from the overthrust belt to the west. These Mancos members in ascending order are the Tununk Shale, the Ferron Sandstone, the lower Blue Gate Shale, the Emery Sandstone, and the upper Blue Gate Shale. The Ferron overlies the Tununk with apparent disconformity, and is overlain conformably by the Blue Gate (Schwans and Campion, 1997).

Previous studies of the Ferron interval have tended to focus on either its northern or southern parts. The presence of world-class exposures along the southeast margin of the current fairway, east and south of the town

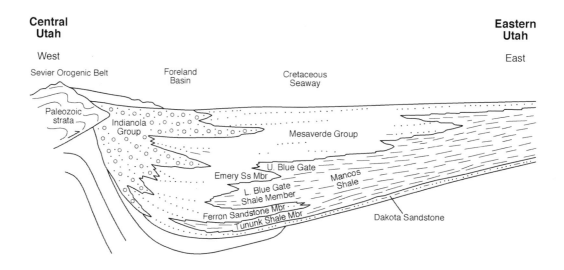

Figure 3. Generalized regional cross section of Upper Cretaceous strata, central and eastern Utah. The Ferron Sandstone occurs in the lower portion of the thick marine shales of the Mancos interval. Cross section modified from Armstrong (1968).

of Emery, has encouraged a particular concentration of outcrop investigations and stratigraphic-depositional interpretations (see, for example, Ryer, 1991; Gardner et al., 1992; Fisher et al., 1993; Gardner and Cross, 1994; Anderson et al., 1997; Garrison et al., 1997; Mattson, 1997; White and Barton, 1998). In this area, a dip-oriented section of nearly the entire Ferron Sandstone Member (referred to in this paper as Ferron Sandstone), including the major coal-bearing horizons, is exposed for up to 30–35 mi (48–54 km). In the northern part of the fairway, however, where the great majority of CBM production has occurred, the coal-bearing section is buried and outcrops consist of thin, shallow-marine tongues of the Ferron. Comprehensive study of well data from Drunkards Wash, Buzzard Bench, and Helper fields has been conducted by the companies involved, but many results of those studies are only recently being published (see Buck et al., 2001; Lamarre, 2001, and Lamarre, this volume; Klein et al., this volume). A limited study of the northern area (Bunnell and Holberg, 1991) indicates the stratigraphic-depositional models developed for the southern area may also apply to the northern end of the fairway. However, what follows is a summary based on the more extensive published work from the outcrop in the south.

The Ferron Sandstone is divided informally into lower and upper parts on the basis of extensive outcrop study. The lower part consists of two, seaward-stepping sandstone intervals, known as the Clawson and Washboard sandstone units. Together, these units are up to 100 ft (30 m) thick and consist of fine- to very fine-grained, shelf-margin sandstones that have a northerly source. These units exist only along the eastern part of the Ferron clastic wedge; to the south and west, they thin and pinch out into marine or delta-plain shale (Ryer and

McPhillips, 1983). The upper part of the Ferron comprises the major part of the formation and includes a series of stacked, transgressive-regressive deposits indicative of fluvial-deltaic deposition. Lithologically, these deposits are comprised of coarse to very fine-grained sandstone of quartz-rich and lithic arkose composition, interbedded with sandy shale, siltstone, shale, and coal. Where the Clawson and Washboard sandstone units are absent, upper Ferron deposits compose the entire member and thicken westward from about 180 ft (55 m) near Emery to as much as 750 ft (230 m) beneath the western Wasatch Plateau (Tripp, 1989). Ryer (1981) identified seven major sandstone tongues within the upper Ferron, ranging in thickness from a few tens of feet (10 m) up to 110 ft (33 m) and extending west-to-east from 3–30 mi (5–50 km). Each sandstone unit consists of multistoried and multilateral tabular bodies that exhibit complex internal architecture and facies relationships.

Coal stratigraphy within the Ferron is based on the delineation of eight coal zones, consisting of specific coalbeds and laterally equivalent, associated lithologies. Lupton (1916) gave these coal zones letter designations that have been retained by more recent studies, as shown in Figure 4. Individual zones commonly divide into two or more "splits," within which the coal content decreases westward. At any single location, up to 13 coalbeds may be present, having single beds ranging from 1–30 ft (0.3–10 m) thick. In nearly all cases, each coalbed contains one or more laterally continuous kaolinitic-bentonitic partings known as "tonsteins." Tonsteins represent altered volcanic ash deposits and serve as excellent chronostratigraphic and log markers.

Data from outcrop, core, and well log study reveal that the coals thin and pinch out both to the east and west. Eastward, they give way to shoreline sandstones

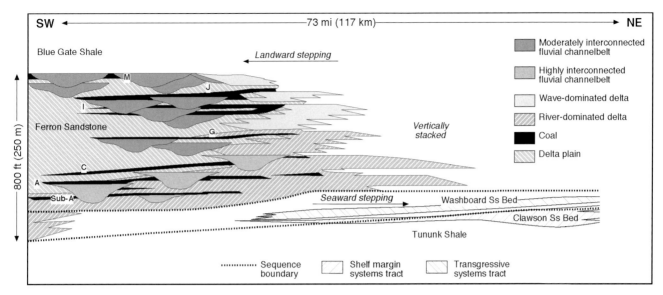

Figure 4. Schematic cross section, showing interpreted stratigraphic and facies relationships and depositional cycles of the Ferron Sandstone. Cross section modified from Fisher et al. (1993).

and delta front/margin deposits, which are in turn succeeded by marine shales and siltstones (Ryer, 1981; Gardner et al., 1992). To the west, coals pass into fluvial sandstones and floodplain/upper delta plain shales (Ryer, 1981). Detailed studies indicate that individual coal zones can show several thickness maxima, possibly due to the effects of differential compaction and original swamp/mire topography (Garrison et al., 1997).

Facies and Depositional History

Extensive study of outcrops and available core data has lead to the interpretation of two major deltaic complexes in the Ferron (Hale and Van DeGraaff, 1964; Hale, 1972; Cotter, 1975; Ryer, 1981; Gardner and Cross, 1994; Garrison et al., 1997). These two proposed complexes, known as the Vernal and Last Chance deltaic complexes (Figure 5A), correspond to two distinct clastic wedges that appear to exhibit significant differences in thickness, provenance, and possibly facies development (Ryer, 1981; Ryer and Lovekin, 1986; Gardner and Cross, 1994). The Vernal delta is interpreted as a relatively thin (180–300 ft [55–90 m]), storm- and wave-dominated shoreline/deltaic sequence that has a northern source area (Ryer, 1981). This complex may be older than its southern counterpart, but this is poorly known since only a few distal, seaward tongues of the Vernal delta are exposed. The Last Chance complex to the south is a thicker (300–600 ft [90–180 m]), wave-modified, river-dominated deltaic system, displaying evidence of a south-southwest sediment source (Ryer, 1981). Recent workers have interpreted an evolution for this complex consisting of three separate deltaic depocenters having major regional avulsion and marine flooding events that cause shifts in the principal loci of deposition (Garrison

and van den Bergh, 1999). Related deposits are exposed along a 35-mi (5-km) outcrop belt that reveals an almost continuous dip section. Because of this, the great majority of published studies on the Ferron have focused on its southern part.

Subsurface work by Tripp (1989) offers an alternate view of the Ferron depositional setting with no gap, or embayment, between the Vernal and Last Chance deltas. Her isopach map of the total Ferron interval (Figure 5B) shows an intervening thick sediment lobe extending eastward from the orogenic belt in the area between the Vernal and Last Chance deltas.

Outcrop studies of the Last Chance delta complex have identified a wide range of specific facies and depositional environments. Identified facies range from fluvial valley fill and flood-plain deposits to delta-plain, distributary channel, mouth-bar, estuarine/bay, shoreline, tidal, and shallow marine deposits (Ryer, 1981, 1991; Gardner et al., 1992; Anderson et al., 1997; Garrison et al., 1997). A combination of sedimentological features, sandstone petrography, sandstone architecture, and stacking patterns has been used to support interpretation of these facies and to divide the major portion of the Ferron member into a series of fluvial-deltaic pulses. Sequence stratigraphic analysis suggest that the southern Ferron corresponds to fourth- and fifth-order sequences and parasequences deposited during a slow third-order sea level rise interrupted by several minor sea level falls (Garrison et al., 1997; Garrison and van den Bergh, 1999). As shown in the schematic cross section of Figure 4, a progression from seaward-stepping (progradational), to vertically stacked (aggradational), to landward-stepping contributions is interpreted for these deposits. Sediment supply is inferred to have controlled the depositional pattern (Ryer, 1981, 1991). Thus,

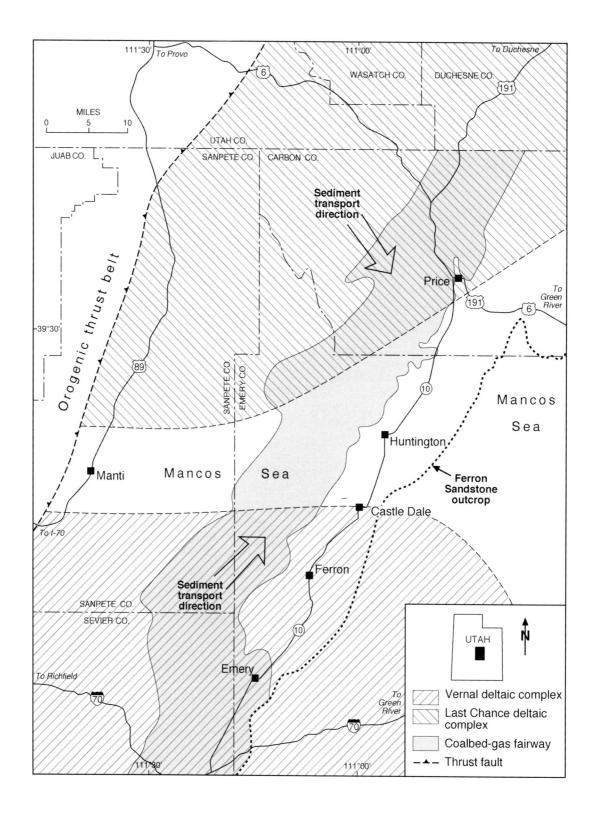

5A

Figure 5. (A) Interpreted regional extent of the Vernal and Last Chance delta systems within the Ferron Sandstone interval, Utah (modified from Ryer, 1991), and (B) Ferron Sandstone interval isopach map showing three delta lobes (modified from Tripp, 1989).

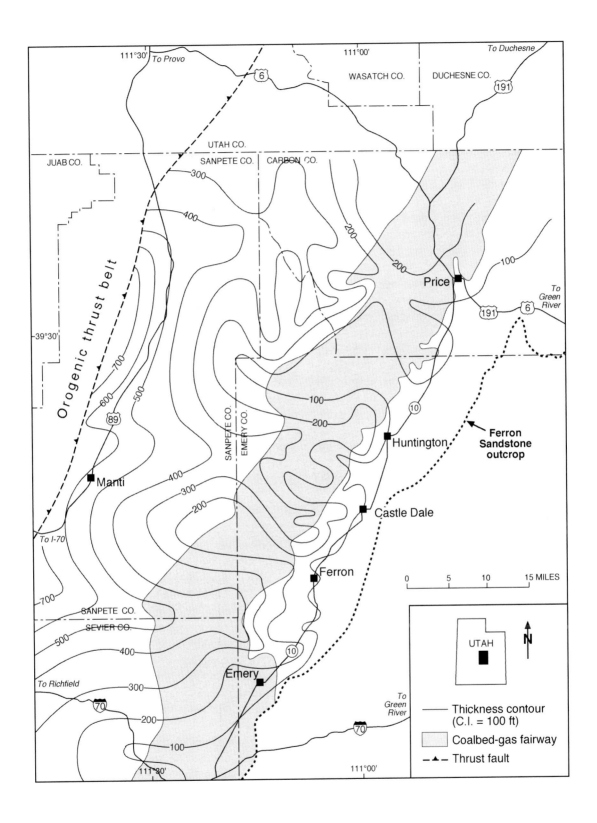

5B

the interpreted progression reflects a high initial rate of sediment supply relative to creation of accommodation space, followed by a period of balance between influx and accommodation, and finally a decrease in relative influx. Coals are thickest and most continuous in aggradational deposits, where they most commonly occur immediately below the flooding surface that caps each deltaic cycle (Garrison et al., 1997). They occupy a position landward of delta front and shoreline sandstones, suggesting deposition in delta plain or transitional (flood plain-delta plain) settings.

It is not clear to what degree this interpretative scheme, developed for the Last Chance deltaic complex to the south, is applicable to the unexposed Vernal deltaic complex to the north. Data from the Drunkards Wash field area suggests that the two areas are quite similar in terms of depositional cycles and the number of coalbeds (Bunnell and Holberg, 1991, Burns and Lamarre, 1997, Buck et al., 2001). It seems notable that isopach data of the overall Ferron interval and the total net coal between Price and Emery show no significant gap, but instead show continuity of the coal and sediments between the two interpreted deltas (Ryer and Lovekin, 1986; Tripp, 1989; Tabet et al., 1995). Future study is required to fully evaluate these relations and their importance to exploration within the Ferron CBM fairway.

GENERAL CHARACTER OF FERRON COALS

Coalbeds in the Ferron Sandstone differ from those commonly targeted for CBM exploration in that although thin and of relatively low rank, they are very high in gas content. Few coalbeds exceed 10 ft (3 m) in thickness, and all coals are high-volatile B or C bituminous in rank. For coals of this rank in Rocky Mountain basins, gas content is commonly in the range of 50–150 scf/ton (1.6–4.6 cm^3/g) (Tyler et al., 1995), however, for Ferron coals in Drunkards Wash field, the gas content is as high as 500 scf/ton (15.6 cm^3/g) (Burns and Lamarre, 1997; Barker and Dallegge, 1998; Dallegge and Barker, 2000). Gas content of this magnitude normally occurs in coals of high-volatile A and medium-volatile bituminous rank, corresponding to maximum thermal generation of methane and to Ro values of 0.78–1.5%. The elevated gas content of the Ferron coals near Drunkards Wash may be the result of a number of factors: (1) the effects of coal composition; (2) the presence of any biogenic gas; (3) the possibility of migrated gas; and (4) suppressed vitrinite reflectance. All of these factors may have contributed to the resulting gas content of Ferron coals.

Thickness Patterns

Regional thickness patterns for total net coal in the potential Ferron CBM fairway are shown in Figure 6.

Data are given for all coalbeds at least 1 ft (0.3 m) thick. Wells in Drunkards Wash field commonly show three to six coalbeds over a stratigraphic interval of 150–200 ft (45–60 m) (Burns and Lamarre, 1997). A few wells contain up to 10 or more beds, however, not all of which may be productive. Isopach trends are oriented largely northeast-southwest (Tabet et al., 1995). This orientation is orthogonal to paleo-shorelines in the Last Chance delta, and parallel to subparallel to those of the Vernal delta (Ryer and McPhillips, 1983; Tabet et al., 1995). General continuity of coal occurrence is interpreted for the entire fairway trend; however, there are significant gaps in the available data, for example west of Ferron. Nonetheless, several important observations can be made from Figure 6.

First, within the overall trend, at least three areas of increased coal thickness exist, and isopach maxima are present in three of these areas: west of Price in the Drunkards Wash field, west of Castle Dale in the Buzzard Bench field, and southwest of Emery. At Drunkards Wash (Burns and Lamarre, 1997), well data indicate the net coal thickness ranges from 4–48 ft (1.1–14.5 m); in the Buzzard Bench field (see Lamarre's Buzzard Bench paper in this volume) the net thickness ranges from 4–40 ft (1.1–11.0 m); and to the south, near Emery (Tabet et al., 1995), exposures of the Ferron reveal the net thickness ranges from 4–55 ft (1.1–17 m). Second, distinct thinning along narrow west-east- and northwest-southeast-oriented zones is evident at intervals of approximately 8–10 mi (13–19 km) along trend. These zones appear to correlate with areas of increased net sandstone thickness in the Ferron and have been interpreted as marking distributary channel systems (Tripp, 1989). This interpretation, however, appears to conflict with traditional definitions of the Last Chance and Vernal deltaic complexes, which propose an intervening embayment. Third, Figure 6 shows that significant coal development extends at least 12–15 mi (19–24 km) farther eastward in the southern portion of the trend, compared with the northern portion, reflecting a southeasterly shift in the southern shoreline orientation. A fairly abrupt boundary is evident in the vicinity of Emery, where the coals have been truncated by erosion along the outcrop. Northeast-trending pods of increased net coal in this area may correspond to lobes of the Last Chance delta complex (Tabet et al., 1995).

While past studies have focused solely on the coalbeds themselves as reservoirs, Lamarre et al. (2001) indicate other carbonaceous lithologies associated with the coalbeds are also gas bearing and should be considered in reservoir modeling and evaluation of the play area. These authors found that by including lithologies with log-derived densities ranging from 1.75–2.55 g/cm^3 the potential reservoir thickness can be tripled, and the in-place gas resource doubled, for a well in the play area.

510

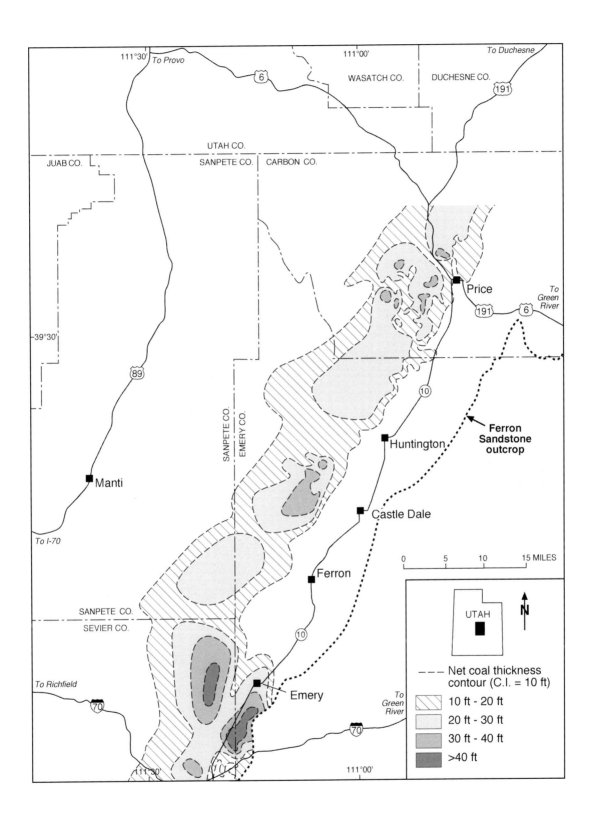

Figure 6. Isopach map of total net coal in the Ferron Sandstone, including all coalbeds greater than or equal to 1 ft in thickness. Contour interval is 10 ft; map modified from Tabet et al. (1995).

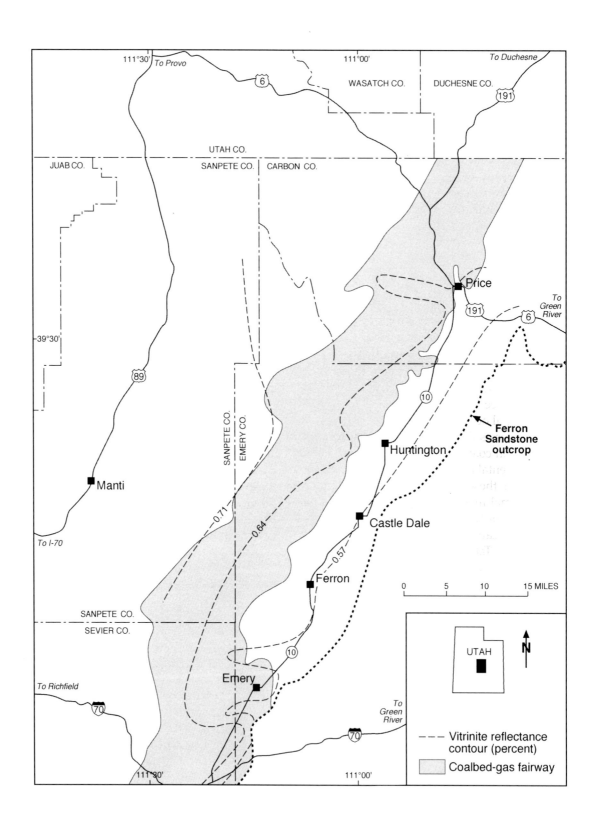

Figure 7. Contour map of vitrinite reflectance values determined for Ferron Sandstone coals in relation to the proposed coalbed methane fairway trend. Map modified from Tabet et al. (1995).

Vitrinite Reflectance and Burial History

Vitrinite reflectance (mean random value) distribution within the potential fairway is given in Figure 7. The data shown are generalized. According to Burns and Lamarre (1997), average R_o in Drunkards Wash field is 0.7%, corresponding to high-volatile B bituminous rank. Values in the Emery area are somewhat lower, averaging about 0.5% (high-volatile C bituminous), which suggests reduced thermal maturity (Dallegge and Barker, 2000). Cross plots of R_o versus sample depth imply that significant gas generation in Ferron coals required burial depths of at least 9300 ft (2820 m) (Tabet, 1998). These coals are presently exhumed to depths of less than 5500 ft (1670 m) within the CBM fairway.

Burial history studies also suggest significant differences between northern and southern coals in the Ferron. To determine possible maximum burial depth patterns, Tabet (1998) combined Cretaceous and Tertiary formation thickness data from Hintze (1988) to derive the map of Figure 8. This map argues for up to 4500–5000 ft (1360–1500 m) of increased burial in the vicinity of Drunkards Wash compared with the southern, Emery area. Burial history reconstructions, such as those discussed by Dallegge and Barker (2000), indicate a comparable difference in overburden at 50 to 40 Ma, when maximum burial occurred.

Other work suggests that R_o values are suppressed in the northern Ferron coalbeds (Quick and Tabet, 1999, 2003). Relevant elemental analysis and study of thermoplastic properties in these coals appear to indicate a higher level of thermal maturity than that indicated by vitrinite reflectance alone. In this respect, it is important to note that R_o values can vary significantly for different beds in a single well (Tabet et al., 1995): in Drunkards Wash field, for example, it is typical for R_o to range from 0.62–0.68% for samples within a 100 ft (30 m) interval, whereas in the Emery area the range is commonly 0.53–0.61%. Such variance provides strong evidence that individual coals responded differently to the thermal maturation process. One explanation for variable R_o, and for R_o suppression, is coal composition. Another cause of the Ro suppression may have been rapid burial and the ensuing overpressured conditions resulting from hydrocarbon generation and effective stratigraphic seals around the coals (Quick and Tabet, 2003).

Cleating

Hucka (1991) investigated fracture patterns and coal cleat at numerous mines in the overlying Blackhawk Formation coals, and one mine in the Ferron Sandstone. She noted several lines of evidence supporting the interpretation that cleating is either a result of, or heavily influenced by, regional tectonic events. Such evidence includes: (1) regular spacing and continuity of cleat orientation over large areas; (2) frequent parallelism between face cleat orientation and fault strike; and (3) the presence of slickensided surfaces, pulverized coal, and open-space mineralization along many cleat planes. This information suggests that many fracture planes in the coal of the Blackhawk and underlying Ferron experienced shearing.

Figure 9 presents relevant data for cleat orientations measured in underground locations from the overlying Blackhawk Formation (Mesaverde Group), along with one measurement from a mine in the Ferron coals. These locations lie along the western portion of the projected CBM fairway, that is, within the Wasatch Plateau province, and across the Book Cliffs province to the north. In broad terms, two patterns of cleating are discernible. Around the flanks and nose of the San Rafael Swell, face cleats exhibit a consistent west-northwest orientation that has a secondary northeast trend. Within the Wasatch Plateau, in contrast, northeast trends are dominant and largely parallel major normal fault trends. These patterns support the more recent fracture study of Condon (1999) in the Ferron. In several mines located in this area, Hucka (1991) noted rotation in the orientation of face cleats and surface joints toward parallelism with fault strike.

Such patterns suggest at least two major phases of fracture development in coals of the region. An initial phase of cleating may have occurred in the Eocene, coinciding with maximum burial of the Ferron and late-stage movement on the San Rafael uplift. A later phase, related to major normal faulting in the Wasatch Plateau, may have occurred in the Miocene-Pliocene as a result of Basin and Range extension. Mapping of a similar pattern of igneous dikes for the northeastern CBM fairway area (e.g., between Helper and Castle Dale) indicates a dual pattern of fractures also developed in this area: an early set oriented west-northwest and a later set oriented more northerly (Tingey, 1989; Tingey et al., 1991).

SPECIFIC CONSIDERATIONS: NORTHERN AND SOUTHERN COALS

Information presented previously in this paper suggests that important differences distinguish the northern and southern areas in terms of the depositional setting, burial history, and gas potential of Ferron coals. These differences are noted in several lines of evidence.

Coal Composition

Compositionally, northern and southern coals exhibit apparent differences that have a direct bearing on CBM-generating capacity (Barker et al., 1999). At a given maturity level, coals with a higher vitrinite content and lower inertinite and ash/mineral matter contents have both higher generating and sorption (ability to retain)

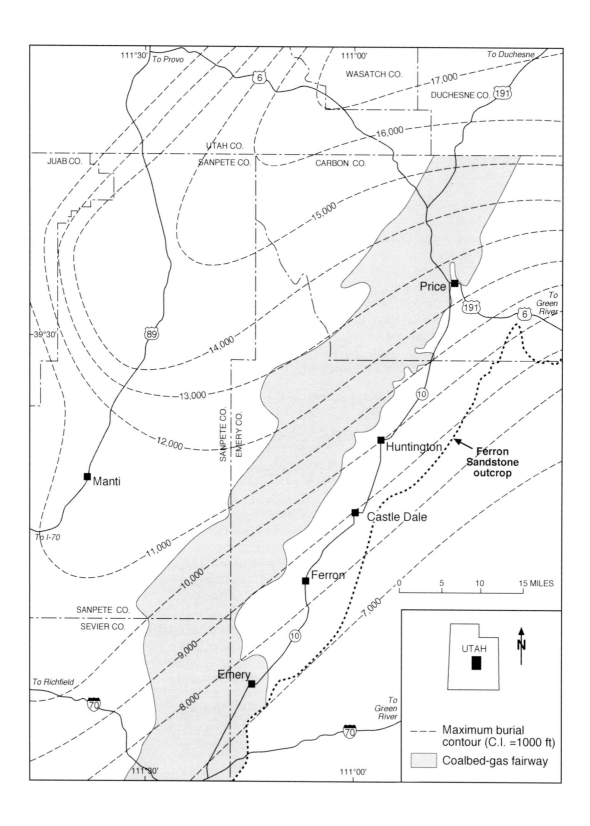

Figure 8. Maximum depth of burial isopach map for the Ferron Sandstone based on stratigraphic data from Hintze (1988). Map modified from Tabet (1998).

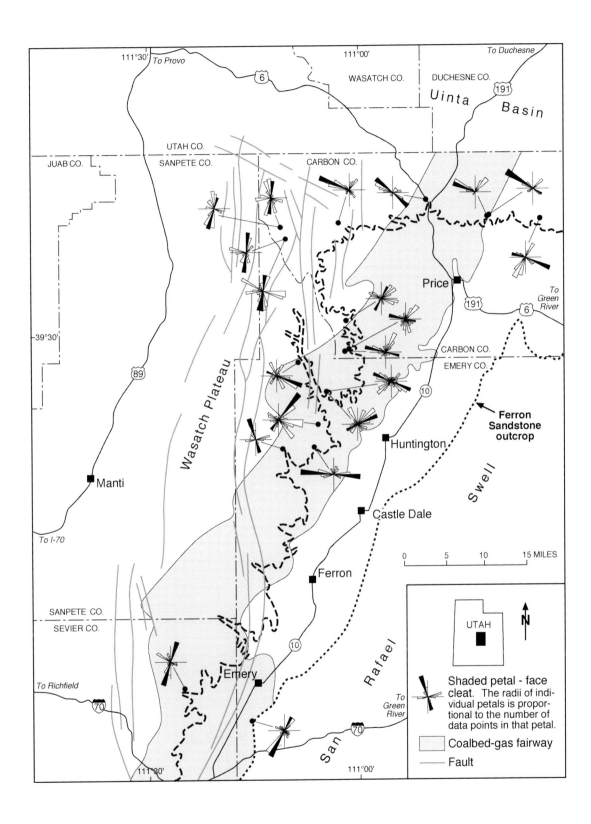

Figure 9. Map of cleat orientations measured in underground mines in the coals of the Upper Cretaceous Blackhawk Formation, and one mine in the Ferron Sandstone, central Utah. Map shows how the regional, northwest-trending face cleat orientation is modified under the Wasatch Plateau by north-south normal faulting. Map modified from Hucka (1991).

Table 1. Comparison of average coal quality for the northern and southern Ferron coals.

Sample Area	Analysis Basis*	Volatile Matter	Fixed Carbon	Ash	Sulfur	Btu/lb
North**	dry	36.0	46.6	17.4	1.9	12230
	d.a.f.	43.6	56.4	----	2.3	14810
South+	dry	39.8	47.9	12.3	1.8	12589
	d.a.f.	45.4	54.6	----	2.1	14355

*d.a.f. = dry, ash-free; values in weight percent except for heat content
** U.S. Steel, unpublished data from 366 coal-core analyses in the Drunkards Wash area; Btu/lb calculated from ultimate analyses
+ Consolidation Coal Company, unpublished data from 32 coal-core analyses near Emery, Utah

Table 2. Comparison of measured gas contents to the theoretical amount generated thermogenically and the experimentally measured adsorption capacity of the Ferron coals.

Area	Direct Measurements	Estimated Thermogenic Gas	Measured Adsorption Capacity
South	0 to 150 scf/ton*	3 scf/ton**	160 scf/ton**
	(0 to 4.7 cm³/g)	(0.1 cm³/g)	(5 cm³/g)
North	200 to 500 scf/ton+	256 scf/ton**	300 to 600 scf/ton#
	(6.4 to 15.6 cm³/g)	(8 cm³/g)	(9.5 to 18.8 cm³/g)

* Smith, 1986 (eight samples)
** Barker and Dallege, 1998
+ Burns and Lamarre, 1997
# Anadarko, unpublished data

capacities compared with coals having less vitrinite and more inertinite and ash/mineral matter.

Table 1 compares the compositional analysis of coal cores taken from both ends of the CBM trend. Samples from the northern part of the trend average slightly more ash, but on a dry, ash-free (d.a.f.) basis show a slightly higher rank as indicated by a higher fixed carbon content and a higher estimated heat content. These data therefore confirm the slightly higher rank for the northern coals indicated by the vitrinite reflectance measurements and burial history.

Maceral composition studies by Barker and Pierce (1999) indicate that the northern coals may have less inertinite and mineral matter and more liptinite than do southern coals. Samples from 36 southern coalbeds have mean vitrinite contents that range from 70–87%, mean liptinite contents that range from 3–6%, and mean inertinite contents that range from 10–24% (Hucka et al., 1997). Lamarre (this volume) reports maceral compositions from two wells in the Buzzard Bench field of the central part of the fairway that fall within the same compositional ranges as given for the southern Ferron coals.

For the better-studied southern coals, Hucka et al. (1997) data reveal that, on a mineral-free basis, there is considerable variation in maceral composition, both between and within beds. Since so few samples of Fer-

ron coals have been analyzed for maceral composition, more study is needed to determine if the preliminary petrographic differences between northern and southern coals are statistically significant.

Gas Content and Gas Generation

During early CBM development, only a limited number of direct gas content measurements were available for Ferron coals. Smith (1986) reported eight gas desorption measurements for coals in the Emery area (Table 2) that range from no gas to 150 scf/ton (4.7 cm³/g). For the Drunkards Wash area near Price, Lamarre and Burns (1997) indicate gas contents ranging from 200–500 scf/ton (6.4–15.6 cm³/g). Recently, Lamarre (2001) has shown that gas contents of the Ferron coals decrease in a regular and dramatic fashion from north to south, decreasing from over 400 scf/ton (12.8 cm³/g) in the north near Price, Utah down to less than 50 scf/ton (1.6 cm³/g) south of Emery, Utah (Figure 10).

In addition to direct gas content measurements, Dallegge and Barker (2000) used burial history and coal rank data to derive theoretical estimates of how much thermogenic gas Ferron coals may have generated (Table 2). The relevant data suggest that southern coals, near Emery, should have generated 3 scf/ton (0.1 cm3/g),

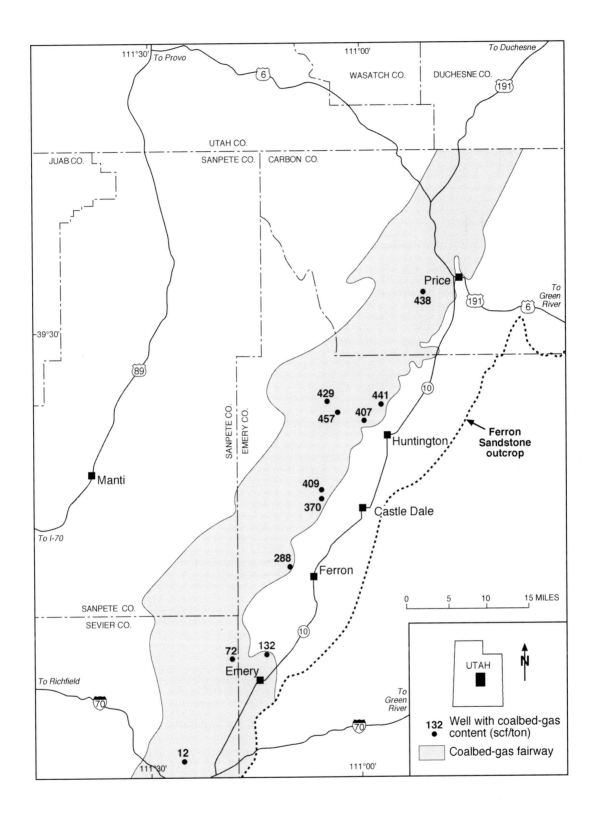

Figure 10. Map showing gas contents decreasing from north to south along the Ferron CBM fairway. Map modified from Lamarre (2001).

whereas northern coals should have generated 256 scf/ton (8 cm3/g). Because these estimates are generally much less than the maximum measured amounts of gas in both southern and northern coals, it is assumed that other sources of gas must exist. Gas composition data (Lamarre and Burns, 1997; Barker and Dallegge, 1998; Collett, 1999; Dallegge and Barker, 2000) show the coalbed gas contains 94–98% methane, 3–5% nitrogen, 1–2.5% carbon dioxide, and 0.1–0.3% ethane. Analysis of the carbon isotopes from the methane and carbon dioxide indicate the additional gas in the Ferron coals has come from biogenic and migrated thermogenic sources.

Adsorption Capacity

Laboratory studies of adsorption capacity, or reservoir capacity, have been performed on Ferron coals to determine the maximum volume of gas they could hold under varying pressures and temperatures. Here, too, significant differences were found between northern and southern Ferron coals (Table 2). In particular, adsorption isotherms have been constructed for two wells in the northern coal area and for six samples from different coal exposures in the southern area. The two wells in the north are the Phillips Petroleum Utah Federal 25-9-1, in the Drunkards Wash field, and the Anadarko A-2 Helper Federal, located in the Helper field approximately 6 mi (10 km) to the north-northeast. The relevant data are shown in Figure 11.

In Figure 11A, isotherms for three separate samples from the Anadarko A-2 Helper Federal are shown. All three agree quite closely and indicate a gas content of 550–600 scf/ton (17.1–18.8 cm³/g; on a dry, ash-free basis) for a reservoir pressure of 1300 psia (91.4 kg/cm², measured initial bottom-hole pressure). Figure 11B, however, displays two isotherms meant to compare data for northern and southern Ferron coals. The "N. Ferron Coal" represents data plotted for Drunkards Wash field. These data indicate a sorptive capacity of 438 scf/ton (13.6 cm³/g), corresponding to a measured reservoir pressure of 765 psi (53.8 kg/cm²). Comparison with Figure 11A shows that this isotherm agrees closely with those determined for Helper field. Lamarre (this volume) provides a field-average isotherm for the Buzzard Bench field in the central portion of the Ferron fairway, which shows a dry, ash-free gas content of the coals there at 474 scf/ton (14.8 cm³/g), for a reservoir pressure of 1300 psia (91.4 kg/cm²) and a temperature of 92°F (33°C).

For southern Ferron coals, an average isotherm was derived using six, weathered outcrop samples from one of the stratigraphically lowest coals in the Ferron (A2 split of the A coal), analyzed at 77°F (25°C). Figure 11B clearly suggests a considerably lower sorptive capacity for these southern coals compared with those in Drunkards Wash. Again, this contrast must be qualified to

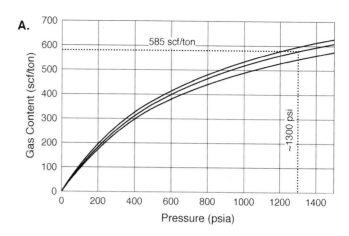

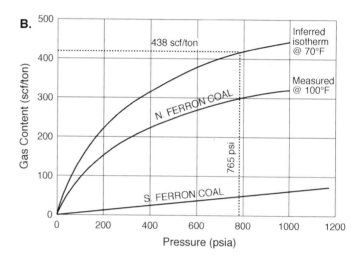

Figure 11. Adsorption isotherms for coals in the Ferron CBM play, east-central Utah. (A) Isotherms for three different samples from the Anadarko A-2 Helper Federal well, Helper field (Sec. 2, T14S, R10E). Initial bottom hole pressure at this well was approximately 1300 psi, corresponding to a gas content of 550–600 scf/ton. Temperature is approximately 100°F (38°C). (B) Isotherms for both northern and southern coals. Upper plots are for the Phillips Petroleum 25-9-1 Utah Federal (Sec. 25, T14S, R9E), Drunkards Wash field at approximately 70°F (21°C) and 100°F (38°C). Lower isotherm represents an average for six samples taken in the Emery area. Graph modified from Barker and Dallegge (1998).

some degree by the fact that unweathered and weathered coals are being compared. Yet, the noted differences in coal composition, vitrinite reflectance, and gas-generating capacity all support Lamarre's data (2001) showing a progressive decrease in gas content and carrying capacity from the northern to the southern end of the Ferron trend.

Gas Isotope Data

Isotope studies of coalbed gases from Drunkards Wash field were performed to help determine gas origin (Burns and Lamarre, 1997; Collett, 1999). Resulting data suggest that a significant component of this gas has a biogenic source. Coalbed gases of biogenic origin are common in coals from various basins of the Rocky Mountain region (Scott, 1993; Scott et al., 1994). Such gases result from degradation of the relevant coals and wet gases by anaerobic bacteria, which are introduced through groundwater influx commonly at a late stage in the uplift and erosional history of the basin.

Relative to thermogenic gasses, biogenic coalbed gasses in Rocky Mountain basins are characterized by isotopically light methane ($\delta^{13}C$ > -55°/oo) and carbon dioxide ($\delta^{13}C$ > -30°/oo, to as high as +20°/oo) (Scott et al., 1994; Rice, 1993). Analysis of Drunkards Wash coalbed gasses show $\delta^{13}C$ values ranging from -45.59^o/oo to -50.07^o/oo for methane and +15°/oo to +20°/oo for carbon dioxide. These data imply mixing of biogenic and thermogenic gases in Ferron coals (Lamarre and Burns, 1997; Collett, 1999).

Such information, in conjunction with results from a multiyear sampling program of gases produced from Ferron Sandstone and coalbed reservoirs, suggests that at least three major sources of gas exist: (1) thermogenic methane generated from Ferron coals; (2) thermogenic methane and heavier hydrocarbon gases (mainly ethane) that have migrated into the Ferron from more mature sources; and (3) biogenic methane (Collett, 1999).

Hydrological Conditions and Water Composition

Because most coalbed reservoirs are also aquifers and require dewatering for optimal gas production, analysis of hydrological conditions in a CBM play is essential. In many instances, a complex interaction exists between a regional groundwater flow system and local influences related to faults, folds, and cleat development. In central Utah, mapping has helped establish preliminary flow gradients along the length of the CBM fairway (Tabet, 1998). This mapping, revised here in Figure 12, combines measured Ferron water levels from the south with a few estimated levels in the north that were derived from oil and gas production tests that isolated the Ferron interval. A few water level estimates that were particularly low fell below the top of the Ferron and probably resulted from tests of tight intervals for an inadequate amount of time for the pressure to equilibrate. These estimates were discarded for mapping purposes. The hydrodynamic gradient mapped shows the general groundwater flow direction is to the east and south along the length of the Ferron trend. Recharge on a regional basis is believed to occur along the grabens that cut the Wasatch Plateau to the west; however, local patterns are almost certainly affected by variations in structure and lithology (Applied Hydrology Associates, 1998). Lines and Morrissey (1983) calculated that the prominent north-south Joes Valley fault system supplies subsurface inflow to the Ferron at a rate of 2.4 ft^3/sec (0.07 m^3/sec) to the area around the town of Emery.

It has also been proposed that southward flow of groundwater from the Uinta Basin into the northern fairway area may have contributed an added component of thermogenic gas to this area (Tabet, 1998). Such introduction might help explain the large difference in observed gas content between northern and southern coals. In addition, more immediate proximity to recharge from the Joes Valley fault system to the west might have resulted in some degree of flushing to the south. Overall, more study is needed in this area.

The composition of waters derived from Ferron coal provides another clue to differences within the overall fairway. As discussed by Rice (1999), produced water shows considerable differences among the three existing fields (Drunkards Wash, Helper, and Buzzard Bench). The most recent available data indicate an average of 8900 mg/L total dissolved solids (TDS) for Drunkards Wash field, increasing to 26,000 mg/L TDS only 6 mi (9.6 km) to the north, at Helper field (Rice and Nuccio, 2000). To the south, in Buzzard Bench field, values average 11,000 mg/L TDS, but range from higher than at Drunkards Wash to less than 1000 mg/L TDS. Analysis of isotope values and chloride concentrations also indicate that recharge waters are distinct from those at Drunkards Wash. The implication is that Ferron coalbed reservoirs are quite heterogeneous and possibly compartmentalized hydrologically (Rice, 1999).

THE CURRENT CBM PLAY: BRIEF DRILLING HISTORY

Early exploration in central Utah discovered gas in conventional Ferron sandstone reservoirs. These sandstone reservoirs produce gas that averages about 95% methane from structural-stratigraphic traps located on low-amplitude, faulted anticlines, such as those at Clear Creek field (discovered 1951, cumulative production 115 bcf) and Ferron field (discovered 1957, cumulative production 9 bcf). The existence of pipelines to these fields has proven very supportive to the development of the new CBM play.

Until recently, gas from sandstone reservoirs at Clear Creek and Ferron fields was thought to have come from coals in the Ferron Sandstone. Analysis, however, have shown distinct differences in the composition of gases produced from sandstone and coal (Tabet et al., 1995). In particular, coalbed gas is significantly drier (C_2<0.3%, no C_2-C_5, versus, for sandstones, C_2 >1.2% and C_2-C_5 0.1–4.3%) and richer in nitrogen and carbon dioxide (average 4% and 1.9%, respectively, compared to 1.5% and 0.2% for sandstones). Differences in the mixing

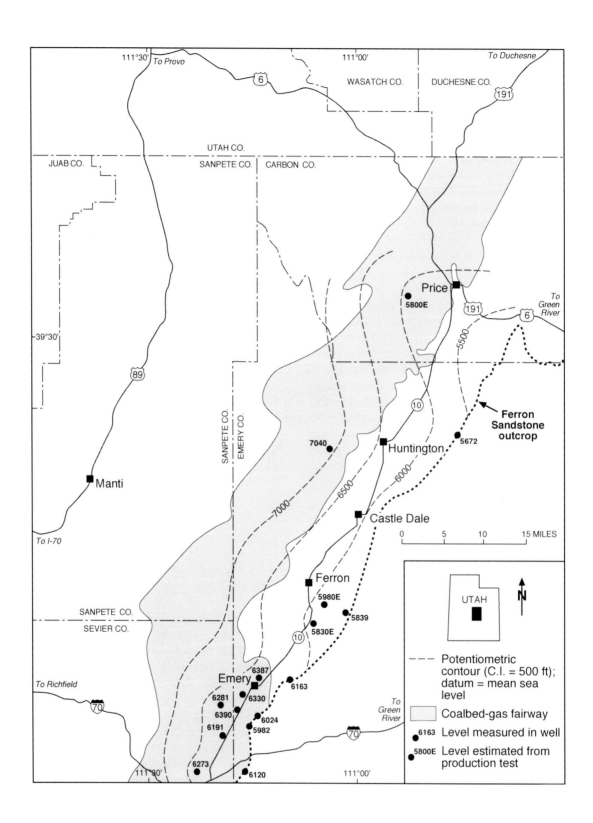

Figure 12. Map showing the elevation of the potentiometric surface for the Ferron Sandstone in the vicinity of the CBM fairway, central Utah. Map modified from Tabet (1998).

of gas from two or more sources may account for these variations.

Exploration for CBM began in 1988 with two wells drilled by Texaco Exploration and Production in the northern part of the present fairway, close to the Drunkards Wash area. These wells were tested over a several-month period at daily rates between 90 and 230 mcf, and with up to 500 bbl water. One of the wells was cored, and analysis indicated a very high coalbed gas content of about 470 scf/ton (Lyle, 1991).

The current play was opened in 1991, when River Gas Corporation took a 92,000-acre (37,232 ha) farmout from Texaco and drilled a core hole in what later became Drunkards Wash field. Desorption tests on 37 ft (11 m) of net coal from this well, the River Gas unit 1 (Sec. 36, T14S, R9E), confirmed the high gas content found by Texaco and led to the drilling of several producers in the ensuing months. The first of these wells recovered gas at a rate of more than 2 mmcf/day before reaching target depth (Lyle, 1991). By mid-1992, three producers had been connected to an existing 6-in (15-cm) pipeline. Three years later, a total of 89 producing wells existed in Drunkards Wash field and yearly production (for 1995) stood at 11.1 bcf and 5.7 million bbl water. Continued dewatering of these wells had raised the average daily gas flow per well from 310 mcf to 428 mcf during that same year, and to 491 mcf by 1997 (Lamarre and Burns, 1997). River Gas sold its interest in the Drunkards Wash field to Phillips Petroleum in 2000; Phillips merged with Conoco in early 2003. As of the end of 2003, Anadarko Petroleum and Marathon Oil had independently drilled their leases within the Drunkards Wash field, and 62 of these wells were in production.

Between 1994 and 1997, drilling to the north and south of Drunkards Wash opened two new CBM fields in the Ferron. Exploratory drilling by Texaco along the central fairway between Huntington and Ferron opened the Buzzard Bench field and established production at several other wildcats within this general area. These wells are all located in the central portion of the fairway and were tested at rates of 75–300 mcf per day from depths of less than 2000 ft (606 m). By year-end 2003, a total of 90 producers had been completed in the Buzzard Bench field, with monthly production at 650.2 mmcf and cumulative production was 18.8 bcf. North of Price, meanwhile, Anadarko Petroleum Corporation established Helper field in T13S, R10E. By December 2003, this field had 102 producers and was yielding 1,082.7 mmcf per month with a cumulative production of 47.1 bcf. Depth to production in this field averages about 1000 ft (303 m) deeper than at Drunkards Wash. Gas contents in Helper field are in the range of 400–600 scf/ton (12.5 to 18.8 cm³/g). During 2000, Marathon Oil Company developed its acreage within the Drunkards Wash field. At the end of 2003, the company had 17 producing

wells with combined monthly production of 173.4 mmcf, and cumulative production of 6.7 bcf.

In 1997 and again in 1999, the play entered new phases of development with release by the U.S. Bureau of Land Management (BLM) of its records of decision on two separate CBM environmental impact statements (EIS). The earlier EIS, approved by the BLM in 1997, concerned the area immediately within and surrounding Drunkards Wash field and allowed for development of up to 550 wells, along with associated infrastructure. Work related to the first EIS had been in progress since 1994 and had postponed full-scale exploration and development within the unit.

Following the approval from the BLM, development within Drunkards Wash increased from around 95 wells in mid-1997 to 494 at the end of December 2001. Completion of the second EIS in mid-1999 added a total of 200 mi² (520 km²) to the area approved for future development, covering much of the northern and central CBM fairway. This second EIS sanctioned an additional 335 natural gas wells beyond the 550 approved in 1997, as well as new infrastructure including a 27 mi (43 km) gas transmission line. By the end of 2003, a total of 754 producing CBM wells existed in the combined Drunkards Wash, Helper, and Buzzard Bench fields. We predict that the number of wells will increase by at least one hundred during the next several years in the portion of the fairway approved for development by the EIS studies (see Figure 1). This prediction is based on the continued use of 160-ac (64 ha) spacing; if this is found insufficient for proper drainage in certain areas, a larger number of wells may result.

NORTHERN FAIRWAY EXAMPLE: DRUNKARDS WASH FIELD

General Field Data

Discovered in 1991, Drunkards Wash field now encompasses 127,000 acres (51,396 ha) and includes 549 producing wells on 160-ac (64 ha) spacing as of December 2003. Depth to production ranges from 1100–3500 ft (330–1060 m). Dip of the beds is to the west-northwest, away from the axis of the San Rafael Swell. Reservoirs include between three and six coalbeds that vary from 4-48 ft (1.2–14.4 m) in total net thickness, averaging 24 ft (7.2 m). The rank of these coals is high-volatile B bituminous, with an average gas content of 400 scf/ton (12.5 cm³/g). Analysis of gas from productive wells indicate dry gas compositions, including 94–98% methane, 3–5% nitrogen, 1–2.5% carbon dioxide, and minor amounts of ethane (0.1–0.3%). Heat content is 987–1000 Btu/cf (7.04–7.14 kcal/m³).

Most wells in the unit have exhibited initial negative declines over time, that is, increasing gas production as

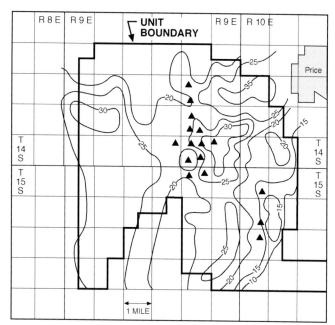

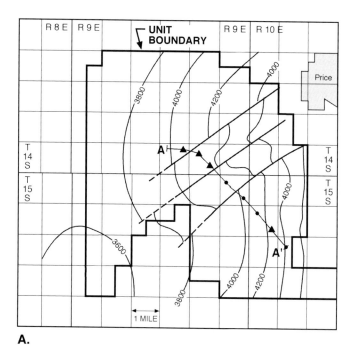

Figure 13. Total net coal isopach map for the Ferron Sandstone interval, Drunkards Wash field. Contour interval is 5 ft. Black triangles indicate wells that have produced at rates of 1 mmcf/day or greater. Modified from Burns and Lamarre (1997).

A.

a result of progressive dewatering of the reservoir coals. Wells are cased, perforated and hydraulically stimulated (Conway et al., 1997). Pumping units are commonly installed; however, according to Lamarre and Burns (1997), as many as 27 producers have flowed gas and water unassisted to the surface, suggesting greater than average pressures, permeabilities, or both.

Reservoir Character

Both interval and net coal isopach data show maximum thickness patterns that trend north-northeast, parallel with interpreted paleoshorelines. Figure 13 reveals that local thickness trends can be complex and that highest well productivity does not correspond to areas of thicker coals (Lamarre and Burns, 1997). Individual coalbeds are observed to split or merge over short distances, with some thicker beds exhibiting good continuity between wells (Bunnell and Holberg, 1991).

Structurally, Ferron coals in the Drunkards Wash field are deformed by a southwest-plunging anticline cut by several northeast-striking reverse faults that have up to 150 ft (45 m) of vertical displacement (Figure 14). As indicated by the structure map and cross section of Figure 14, the better wells within the field do not display any obvious relationship to structure. Detailed information on cleating and bottom hole pressure conditions in the Drunkards Wash unit have not been released.

Pressure transient and other well tests indicate coal permeability ranges from 4–20 md (Conway et al., 1997; Lamarre and Burns, 1997). This is relatively high for

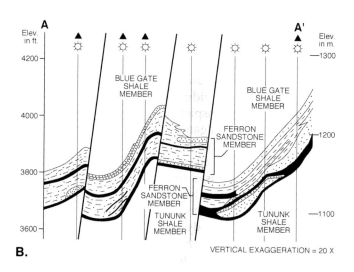

B.

Figure 14. (A) Structure contour map, and (B) northwest-southeast cross section for the Drunkards Wash field. Contour interval is 200 ft; black triangles indicate wells that have produced at rates of 1 mmcf/day or greater. Map modified from Burns and Lamarre (1997).

coals, even at shallow depths. The precise reasons for such permeabilities are not understood, but may be related to a combination of regional and local stress patterns, changes in cleat patterns, and the presence of over-pressured conditions. Pressure gradients measured in Ferron coals at Drunkards Wash range from 0.45–0.53 psi/ft (10.3–12.3 kPa/m). This gradient is somewhat elevated compared with the average hydrostatic gradient of 0.46 psi/ft (10.5 kPa/m) estimated for most petroleum basins (Hunt, 1996). Such pressures in the Ferron constitute a significant artesian head that although inadequate to counteract the lithostatic gradient entirely, may nonetheless help increase the susceptibility of coals to respond to local stress patterns. Moreover, changes in the detailed composition of reservoir coals appears to influence their relative strength: fracture gradients measured for individual coal horizons have varied from 0.6–1.4 psi/ft (13.7–31.9 kPa/m) even within single wells. The precise determinants of reservoir quality in Ferron coals therefore remain a topic for further investigation.

With regard to log characteristics, coals have unique lithologic features that make them readily identifiable. On mud logs, Ferron coals exhibit pronounced gas kicks, usually stronger than those exhibited by sandstones within the Ferron. Typical open-hole log signatures of Ferron coals include low density, low gamma ray, and high resistivity values. As discussed by Burns and Lamarre (1997), however, meteoric deposition of uranium in some of the coals, as well as the presence of thin (< 6 in [14 cm]) shales, can result in gamma ray readings above 200 API units. Such shales are too thin to be resolved by conventional gamma ray tools, but can be delineated by high-resolution density tools. Open-hole logging is thus performed with two passes, the second conducted at a reduced speed over the coal-bearing intervals to yield high-resolution data. This also allows for the identification of tonstein layers, which are thin yet highly continuous, and thus potential barriers to vertical flow within individual coalbeds. Fortunately, these layers are rather conspicuous on high-resolution logs because of their high gamma ray signature and density values above 2.2 g/cm³. For coals having significant reservoir potential, an upper density cutoff of 1.75 g/cm³ was initially used (Burns and Lamarre, 1997). An example log is given in Figure 15 for a well that has multiple perforations and excellent production (>1 mmcf/day). Recent study indicates other lithologies surrounding the coalbeds are also gas bearing, particularly carbonaceous ones, and should be included in the reservoir package; an upper density cutoff for reservoir lithologies of 2.55 g/cm³ is now suggested (Lamarre, 2001). Lamarre (2001) notes that the coals only contain 43% of the gas within the Ferron interval, and thus including all the gas-bearing lithologies can more than double the in-place gas resources per well.

Reservoir Distribution

Drilling outside of Drunkards Wash field suggests that productive coals within the Ferron thin and pinch out to the east, west, and south, but continue a significant distance to the north. This northward continuation is confirmed by the successful drilling in Helper field. Good potential likely exists to the west of this field, but at increasingly greater depths, that is below 4000 ft (1210 m). North of Helper, the Ferron coals dip into the Uinta Basin and become more deeply buried as they pass under the striking Book Cliffs, that rise approximately 1200–1500 ft (360–450 m) above the valley floor. It is unclear whether the deeply buried Ferron coals, which exist at depths of 6000 ft (1820 m) or more, will be developed in this area.

To the south of Drunkards Wash, where drill hole data are more sparse, some coalbeds appear to pinch out along what may be an important facies boundary that begins near Castle Dale and continues to the west. This boundary seems marked by the presence of a major east-west distributary channel, within which the regional potentiometric surface reaches its minimum (see Figure 12). The percentage of coal in the Ferron decreases considerably along this channel trend, whereas the proportion of sandstone increases commensurately. In areas where drill hole data are absent, Lyons (2001) has indicated that 2-D seismic data can be used to delineate the areal extent of the Ferron coals, and identify structures that enhance reservoir permeability.

Drilling and Completion

Wells in Drunkards Wash field are drilled through the Ferron, logged, cased and perforated (Lamarre and Burns, 1997). Air drilling is used to minimize damage to the coalbed reservoirs and to reduce overall costs. Because the overlying Blue Gate Shale is stable, drilling can proceeded at high rates, for example, up to 130 ft (39 m) per hour, allowing wells to be drilled in less than two days after surface casing has been set (Lamarre and Burns, 1997). The typical procedure is to drill at least 200 ft (60 m) below the base of the lowest coal in order to facilitate logging, provide a "trap" for coal fines during production, and allow installation of pumping equipment below the productive zones.

Following initial log evaluation, wells are cased, perforated, and hydraulically stimulated. Where several reservoir zones are identified, coals are artificially fractured in two or more stages; this has proved much more effective in optimizing gas production and minimizing water cut than single treatments (Conway et al., 1997). According to Lamarre and Burns (1997), an average stimulation employs 83,000 lbs (37,649 kg) of sand proppant in 30 lb/gal (3.6 kg/dm³) cross-linked gel. Most wells are completed on pump. Approximately half of the wells produce water through tubing and gas up the annulus between tubing and casing; the other half, in order to avoid slugging problems, produce both water and gas through tubing and separate them at the surface. However, several wells have been completed flowing or have increased sufficiently in production during the first year to be taken off pump altogether, thus improving economics still further.

Conway et al. (1997) have published a performance evaluation on a series of wells in the unit, and the data presented in that article, as well as the information discussed by Lyle (1991), indicate that both coals and adjacent sandstones contribute to gas production. This is shown by the gamma ray and bulk density log data in Figure 15, which reveals perforations that extend above and below individual coal beds. Early estimates supplied by River Gas Corporation suggest that gas reserves in sandstones commonly contribute between 10 and 15% of the total reserves (Lyle, 1991). However, Conway et al. (1997) conclude that significant improvements in well completion can be made by better aligning perforations with stimulated fractures.

Production

Gas production from Drunkards Wash field has been consistently impressive during the first nine years for which data are available. Pipeline sales began in January 1993 and have risen abruptly and continuously, as new wells have been added (Figure 16). In April of 1999, daily flow rates for a total of 193 wells averaged 625 mcf of gas, and 263 bbl water per well. For producers that had been on-stream more than five years (a total of 33), this average was significantly higher, at 974 mcf of gas and 85 bbl water per day. Cumulative production through December 2003 was 438 bcf for 549 wells.

A normalized production curve for the first 33 wells drilled in the unit is given in Figure 17. Only the first 37 months of production are shown. The graph indicates several important features: (1) increasing gas and decreasing water production over time; (2) particularly rapid increase in gas production after about 20 months; (3) steady decrease in the proportion of water versus gas production over time; (4) an apparent flattening in gas production after about three years. These features are

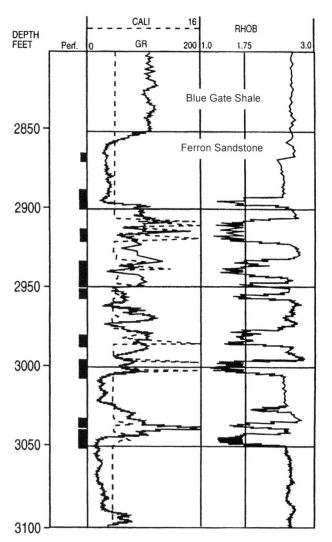

Figure 15. Gamma ray and bulk density log data and perforated intervals for Phillips Petroleum 19-151 Telonis well, a good producer (>1 mmcf/day) in Drunkards Wash field. Modified from Conway et al. (1997).

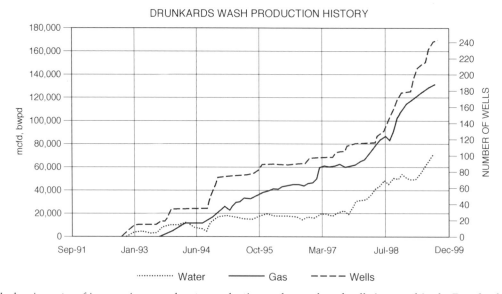

Figure 16. Graph showing rates of increase in gas and water production as the number of wells increased in the Drunkards Wash field for the period January 1993 to August 1999. Graph modified from Burns and Lamarre (1997).

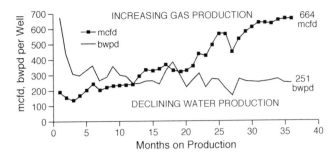

Figure 17. Normalized production curve for the initial, 33 coalbed gas wells completed in the Drunkards Wash field. Graph modified from Burns and Lamarre (1997).

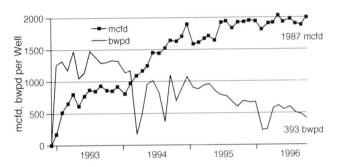

Figure 18. Production data for Phillips Petroleum 35-1-9 Utah Federal (Sec. 35, T14S, R9E), showing negative decline and decreasing water production typical for many good producers in Drunkards Wash field. Data modified from Burns and Lamarre (1997).

also demonstrated by data from the Phillips Petroleum 35-1-9 Utah Federal (Sec. 35, T14S, R9E), a typical good producer in Drunkards Wash (Figure 18). Note that, for this well, an abrupt increase in gas production occurred after about 14 months. This rapid increase is interpreted to be a result of continuous dewatering, which lowers reservoir pressure to a point where gas desorption takes place throughout the contacted fracture/cleat system. The Phillips Petroleum 35-1-9 Utah Federal has 28 ft (8.5 m) of net coal and was the ninth completion in the unit. It has produced in excess of 5.8 bcf in approximately nine years (as of December 2003).

Productivity maps reveal that the best wells are generally concentrated in the central and western parts of the unit and may comprise one or more intra-field sweet spots. The reasons for this concentration of better productivity are not well understood, but may be related to increased permeability. Lamarre and Burns (1997) state that a pressure monitoring well in section 26 of T14S, R9E., one of the high-productivity areas, indicates an average decline of about 9.6 psi (65.9 kPa) per month, suggesting that wells in this section are in pressure communication. Despite this conclusion, these wells are estimated to have ultimate recoveries in the range of 1.5 to 4 bcf, implying that 160-ac (64-ha) spacing is sufficient to adequately drain the reservoir in this area.

CONCLUSIONS

The summary of information presented in this paper on the Ferron CBM play in central Utah highlights several major conclusions.

1. The Ferron play is one of the most successful and profitable examples of CBM development in North America. Shallow target depths, low drilling and pumping costs, and existing infrastructure, coupled with high gas flow rates, have made for favorable economics.

2. The play as a whole is divided into northern and southern parts, with the boundary roughly marked by a significant thinning of coals west of Castle Dale. To date, northern coals have been the dominant target of exploration and development. This reflects both high overall potential in these coals and access to existing infrastructure, which continues only as far south as the town of Ferron.

3. Northern coals are unexposed, have very high gas contents, and appear to be overpressured due to artesian conditions. Southern coals exposed along an extensive outcrop belt exhibit somewhat lower rank and have significantly lower gas contents. Southern coals, however, are slightly thicker, cover a much wider area, and are relatively unexplored. Thus, their ultimate potential is unknown.

4. The high gas contents measured in, and large volumes of gas produced from, northern coals suggest a unique set of circumstances. In particular, there may have existed multiple sources for the gas now found in these coals, namely, in-situ thermogenic gas, migrated thermogenic gas, and late-stage biogenic gas.

5. High flow rates (500 mcf-1.5 mmcf/day) and large reserves (1–4 bcf/well) characterize many wells completed in northern coals. Such rates and reserves may be helped by a combination of artesian conditions and other factors, which combine to enhance cleat permeability and thus allow for excellent long-term drainage.

6. Although there appear to be "sweet spots" in areas of development, such as Drunkards Wash field, the nature of these is not yet fully understood. A combination of factors, such as conventional trapping, groundwater flow, coal composition, and local structure may well be involved.

Although highly successful, the Ferron CBM play continues to have important, unanswered questions. Continued development in Buzzard Bench, Drunkards Wash, and Helper fields, as well as drilling in adjacent areas, should provide a much expanded database to clarify controls on CBM occurrence and producibility in the existing play, and to advance exploration and development in other basins where similar conditions exist.

ACKNOWLEDGMENTS

The authors appreciate the thoughtful comments of William F. Haslebacher (U.S. Department of Energy) and Michael D. Hylland (Utah Geological Survey). James W. Parker (Utah Geological Survey) redrafted the figures for this paper.

REFERENCES CITED

Anderson, P. B., T. C. Chidsey, Jr., and T. A. Ryer, 1997, Fluvial-deltaic sedimentation and stratigraphy of the Ferron Sandstone, in P. K. Link and B. J. Kowallis, eds., Mesozoic to Recent geology of Utah: Brigham Young University Geology Studies, v. 42, part 2, p. 135–142.

Applied Hydrology Associates, 1998, Aquifer modeling report: U. S. Bureau of Land Management Final Environment Impact Statement — Ferron Natural Gas Project, supplemental report, 5 sections, variously paginated.

Armstrong, R. L., 1968, Sevier orogenic belt in Nevada and Utah: Geological Society of America Bulletin, v. 79, p. 429–458.

Barker, C. E., and T. A. Dallegge, 1998, Approaches to coalbed methane gas-in-place analysis using sorption isotherms and burial history reconstruction — an example from the Ferron Sandstone, Utah: 15th Annual International Pittsburgh Coal Conference, Technical Papers, unpaginated.

Barker, C. E., and B. S. Pierce, 1999, Deposition and burial history models of coalbed gas generation and retention as an exploration guide in low rank coals — Ferron production fairway, Utah: Rocky Mountain Association of Geologists Symposium on the Future of Coal Bed Methane in the Rocky Mountains, paper 5, 3 p.

Barker, C. E., J. R. Garrison, Jr., and T. C. V. van den Bergh, 1999, The stratigraphy, lithology, and compositions of coal zones as a function of stratigraphic position within the Turonian-Coniacian upper Ferron Sandstone Last Chance Delta, east-central Utah (abs.): Geological Society of America Annual Meeting, Abstracts with Program, v. 31, abstract 14980.

Buck, K. F., W. K. Camp, and J. Allison, 2001, Helper field — a multidisciplinary approach to coalbed methane development, Uinta Basin, Utah (abs.): AAPG Annual Meeting, Abstracts with Program, v. 10, p. A28.

Bunnell, M. D., and R. J. Holberg, 1991, Coal beds of the Ferron Sandstone Member in northern Castle Valley, east-central Utah, in T. C. Chidsey, Jr., Geology of east-central Utah: Utah Geological Association Publication 19, p. 157–172.

Burns, T. D., and R. A. Lamarre, 1997, Drunkards Wash project — coalbed methane production from Ferron coals in east-central Utah: Proceedings of the International Coalbed Methane Symposium, University of Alabama, Tuscaloosa, paper 9709, p. 507–520.

Collett, T. S., 1999, Composition and source of the gas associated with coalbed gas production from the Ferron coals in eastern Utah (abs.): Geological Society of America Annual Meeting, Abstracts with Program, v. 31, abstract 51255.

Condon, S. M., 1999, Fracture patterns in the Ferron Sandstone Member of the Mancos Shale, Western San Rafael Swell and eastern Wasatch Plateau, east-central Utah (abs.): Geological Society of America Annual Meeting, Abstracts with Program, v. 31, abstract 50207.

—2003, Fracture network of the Ferron Sandstone Member of the Mancos Shale, east-central Utah: International Journal of Coal Geology, v. 56, p. 111–139.

Conway, M. W., R. D. Barree, J. Hollingshead, C. Willis, and M. Farrens, 1997, Characterization and performance of coalbed methane wells in Drunkards Wash, Carbon County, Utah: Proceedings of the International Coalbed Methane Symposium, University of Alabama, Tuscaloosa, paper 9736, p. 195–212.

Cotter, E., 1975, Deltaic deposits in the Upper Cretaceous Ferron Sandstone, Utah, in M. L. S. Broussard, ed., Deltas, models for exploration: Houston Geological Society, p. 471–484.

Dallegge, T. A., and C. E. Barker, 2000, Coal-bed methane gas-in-place resource estimates using sorption isotherms and burial history reconstruction — an example from the Ferron Sandstone Member of the Mancos Shale, Utah, in M. A. Kirschbaum, L. N. R. Roberts, and L. R. H. Biewick, eds., Chapter L of Geologic assessment of coal in the Colorado Plateau — Arizona, Colorado, New Mexico, and Utah: U.S. Geological Survey Professional Paper 1625-B, p. L1-24.

Doelling, H. H., 1972, Central Utah coal fields — Sevier-Sanpete, Wasatch Plateau, Book Cliffs, and Emery: Utah Geological and Mineralogical Survey, Monograph Series 3, 571 p.

Fisher, R. S., M. D. Barton, and N. Tyler, 1993, Quantifying reservoir heterogeneity through outcrop Characterization — 2. Architecture, lithology, and permeability distribution of a seaward-stepping fluvial-deltaic sequence, Ferron Sandstone (Cretaceous), central Utah: Topical Report GRI-93-0023 for Gas Research Institute, contract no. 5089-260-1902, 83 p.

Gardner, M. H., and T. A. Cross, 1994, Middle Cretaceous paleogeography of Utah, in M. V. Caputo, J. A.

Peterson, and K. J. Franczyk, eds., Mesozoic Systems of the Rocky Mountain Region, U.S.A.: Rocky Mountain Section, Society of Economic Paleontologists and Mineralogists, p. 471–503.

Gardner, M. H., M. D. Barton, N. Tyler, and R. S. Fisher, 1992, Architecture and permeability Structure of fluvial-deltaic sandstones, Ferron Sandstone, east-central Utah, in R. M. Flores, ed., Mesozoic of the Western Interior: Society of Economic Paleontologists and Mineralogists Guidebook, p. 5–21.

Garrison, J. R., Jr., and T. C. V. van den Bergh, 1999, The evolution and depositional sequence stratigraphy of the Middle Turonian-Middle Coniacian foreland basin, south-central and east-central Utah — the Notom, Last Chance, and "A" deltas (abs.): Geological Society of America Annual Meeting, Abstracts with Program, v. 31, abstract 14981.

Garrison, J. R., Jr., T. C. V. van den Bergh, C. E. Barker, and D. E. Tabet, 1997, Depositional Sequence stratigraphy and architecture of the Cretaceous Ferron Sandstone: implications for coal and coalbed methane resources — a field excursion, in P. K. Link and B. J. Kowallis, eds., Mesozoic to Recent geology of Utah: Brigham Young University Geology Studies, v. 42, part 2, p. 155–202.

Hale, L. A., 1972, Depositional history of the Ferron Formation, central Utah, in J. L. Baer and E. Callaghan, eds., Plateau-Basin and Range transition zone: Utah Geological Association Publication 2, p. 115–138.

Hale, L. A., and F. R. Van DeGraaff, 1964, Cretaceous stratigraphy and facies patterns-northeastern Utah and adjacent areas: Intermountain Association of Petroleum Geologists, 13th Annual Field Conference Guidebook, p. 115–138.

Hintze, L. F., 1988, Geologic History of Utah: Brigham Young University Geology Studies Special Publication 7, 202 p.

Hucka, B. P., 1991, Analysis and regional implication of cleat and joint systems in selected coal Seams, Carbon, Emery, Sanpete, Sevier, and Summit Counties, Utah: Utah Geological Survey Special Study 74, 48 p.

Hucka, B. P., S. N. Sommer, and D. E. Tabet, 1997, Petrographic and physical characteristics of Utah coals: Utah Geological Survey Circular 94, 80 p., 1 diskette.

Hunt, J. M., 1996, Petroleum geochemistry and geology, 2nd edition: San Francisco, W. H. Freeman, 743 p.

Lamarre, R. A., 2001, The Ferron play — a giant coalbed methane field in east-central Utah: Independent Petroleum Association of Mountain States, 2001 Coalbed Methane Symposium, Denver, Colorado, October 16, 2001, 35 slides from http;www.ipams.org/cbm16.htm.

Lamarre, R. A., and T. D. Burns, 1997, Drunkard's Wash project — coalbed methane production from Ferron coals in east-central Utah, in E. B. Coalson, J. C. Osmond, and E. T. Williams, eds., Innovative applications of petroleum technology in the Rocky Mountain area: Rocky Mountain Association of Geologists, p. 47–559.

Lamarre, R. A., T. Pratt, and T. D. Burns, 2001, Reservoir characterization study significantly increases coalbed methane reserves at Drunkards Wash unit, Carbon County, Utah (abs.): AAPG Annual Meeting, Abstracts with Program, v. 10, p. A111.

Lines, G. C., and D. J. Morrissey, 1983, Hydrology of the Ferron Sandstone aquifer and the effects of proposed surface-coal mining in Castle Valley, Utah: U.S. Geological Survey Water-Supply Paper 2195, 40 p.

Lupton, C. T., 1916, Geology and coal resources of Castle Valley in Carbon, Emery, and Sevier Counties, Utah: U.S. Geological Survey Bulletin 628, 88 p.

Lyle, D., 1991, First well in Utah coal-gas program strikes pay on 92,000-acre Texaco farmout to River Gas: Oil and Gas Journal, v. 47, no. 8, p. 8–9.

Lyons, W. S., 2001, Seismic maps Ferron coalbed sweetspots: AAPG Explorer, December 2001, p. 32–37.

Mattson, A., 1997, Characterization, facies relationships, and architectural framework in a fluvial-deltaic sandstone, Cretaceous Ferron Sandstone, central Utah: Master's thesis, University of Utah, Salt Lake City, 174 p.

Neuhauser, K. R., 1988, Sevier-age ramp-style thrust faults at Cedar Mountain, northwestern San Rafael Swell (Colorado Plateau), Emery County, Utah: Geology, v. 16, p. 299–302.

Quick, J. C., and D. E. Tabet, 1999, Suppressed vitrinite reflectance in the Ferron coalbed gas fairway, central Utah (abs.): Geological Society of America Annual Meeting, Abstracts with Program, v. 31, abstract 51006.

—2003, Suppressed vitrinite reflectance in the Ferron coalbed gas fairway, central Utah: International Journal of Coal Geology, v. 56, p. 49–67.

Rice, C. A., 1999, Waters co-produced with coalbed methane from the Ferron Sandstone in east-central Utah — chemical and isotopic composition, volumes, and impacts of disposal (abs.): Geological Society of America Annual Meeting, Abstracts with Program, v. 31, abstract 6245.

Rice, C. A., and Nuccio, V., 2000, Waters produced with coalbed methane: U.S. Geological Survey Fact Sheet FS-156-00, 2 p.

Rice, D. D., 1993, Composition and origins of coalbed gas, *in* B. E. Law and D. D. Rice, eds., Hydrocarbons from coal: AAPG Studies in Geology 38, p. 159–184.

Ryer, T. A., 1981, Deltaic coals of Ferron Sandstone Member of Mancos Shale — predictive model for Cretaceous coal-bearing strata of Western Interior: AAPG Bulletin, v. 65, p. 2323–2340.

—1991, Stratigraphy, facies, and depositional history of the Ferron Sandstone near Emery, Utah, *in* T. C. Chidsey, Jr., ed., Geology of east-central Utah: Utah Geological Association Publication 19, p. 45–54.

Ryer, T. A., and J. R. Lovekin, 1986, The Upper Cretaceous Vernal Delta of Utah — depositional or paleotectonic feature?, *in* J. A. Peterson, ed., Paleotectonics and sedimentation: AAPG Memoir 41, p. 497–510.

Ryer, T. A., and M. McPhillips, 1983, Early Late Cretaceous paleogeography of east-central Utah, *in* M. W. Reynolds and E. D. Dolly, eds., Mesozoic paleogeography of west-central United States: Rocky Mountain Section, Society of Economic Paleontologists and Mineralogists, Paleogeography Symposium No. 2, p. 253–271.

Schwans, P., and K. M. Campion, 1997, Sequence architecture and stacking patterns in the Cretaceous foreland basin, Utah — tectonism versus eustasy, *in* P. K. Link and B. J. Kowallis, eds., Mesozoic to Recent geology of Utah: Brigham Young University Geology Studies, v. 42, part 2, p. 105–125.

Scott, A. R., 1993, Composition and origin of coalbed gases from selected basins in the United States, *in* D. A. Thompson, ed., Proceedings from the 1993 International Coalbed Methane Symposium, Birmingham, Alabama: v. 1, paper 9370, p. 207–222.

Scott, A. R., Kaiser, W. R., and W. B. Ayers, Jr., 1994, Thermogenic and secondary biogenic gases, San Juan Basin, Colorado and New Mexico — implications for coalbed gas producibility: AAPG Bulletin, v. 78, no. 8, p. 1186–1209.

Smith, A. D., 1986, Utah coal core methane desorption project, final report: Utah Geological and Mineral Survey Open-File Report 88, 59 p.

Sprinkel, D. A., 1993, Wasatch Plateau [WP] play — overview, *in* C. A. Hjellming, project ed., Atlas of major Rocky Mountain gas reservoirs: New Mexico Bureau of Mines and Mineral Resources, p. 89.

Tabet, D. E., 1998, Migration as a process to create abnormally high gas contents in the Ferron Sandstone coal beds, central Utah (abs.): AAPG Annual Convention, Abstracts with Program, v. 7, p. A646.

Tabet, D. E., B. P. Hucka, and S. N. Sommer, 1995, Depth, vitrinite reflectance, and coal thickness maps, Ferron Sandstone, central Utah: Utah Geological Survey, Open-File Report 329, 3 plates.

Tingey, D. G., 1989, Late Oligocene and Miocene minette and olivine nephelinite dikes, Wasatch Plateau, Utah: Master's thesis, Brigham Young University, Provo, 60 p., 1 plate (1:100,000).

Tingey, D. G., E. H. Christiansen, M. R. Best, J. Ruiz, and D. R. Lux, 1991, Tertiary minette and melanephelinite dikes, Wasatch Plateau, Utah — records of mantle heterogeneities and changing tectonics: Journal of Geophysical Research, v. 96, no. B8, p. 13,529–13,544.

Tripp, C. N., 1989, A hydrocarbon exploration model for the Cretaceous Ferron Sandstone Member of the Mancos Shale, and the Dakota Group in the Wasatch Plateau and Castle Valley of east-central Utah, with emphasis on post-1980 subsurface data: Utah Geological Survey Open-File Report 160, 81 p.

Tyler, R., W. A. Ambrose, A. R. Scott, and W. R. Kaiser, 1995, Geologic and hydrologic assessment of natural gas from coal — Greater Green River, Piceance, Powder River and Raton Basins: University of Texas at Austin, Bureau of Economic Geology, Report of Investigations 228, 219 p.

White, C. D., and M. D. Barton, 1998, Comparison of the recovery behavior of contrasting reservoir analogs in the Ferron Sandstone using outcrop studies and numerical simulation: University of Texas at Austin, Bureau of Economic Geology, Report of Investigations No. 249, 46 p.

Analog for Fluvial-Deltaic Reservoir Modeling: Ferron Sandstone of Utah
AAPG Studies in Geology 50
T. C. Chidsey, Jr., R. D. Adams, and T. H. Morris, editors

Hydrodynamic and Stratigraphic Controls for a Large Coalbed Methane Accumulation in Ferron Coals of East-Central Utah

Robert A. Lamarre[1]

ABSTRACT

Upper Cretaceous coals within the Ferron Sandstone Member of the Mancos Shale contain large volumes of coalbed methane. Coals at the northern end of the 80-mi (129 km)-long Ferron trend have produced 469 bcf of coalbed methane, as of the end of July 2003. At that time, 735 wells were producing 275 mmcfd. Analysis of cores from exploratory wells indicate that the gas content of the coals decreases dramatically from north to south, even though the rank and maturity of the coals decreases only slightly. The area with lowest gas content correlates with the outcrop of the Ferron coals in the southern portion of the trend. Hydrodynamic studies have shown that the Ferron coals are aquifers that are recharged from the Wasatch Plateau to the west. Regional mapping indicates that the productive fields are large stratigraphic traps in the central and northern part of the trend where the coals pinch-out updip into tight marine shales to the east. Much of the produced gas is secondary biogenic gas and migrated thermogenic gas that has moved from the Wasatch Plateau and south margin of the Uinta Basin, respectively. In the southern part of the trend, near the town of Emery, coals are present at the surface, but most of the gas has been flushed out of the coals due to reduction of reservoir pressure and active water flow from the west. Therefore, the entire Ferron trend probably contained tremendous volumes of stratigraphically trapped coalbed methane before uplift and erosion exposed the southern coals to the atmosphere.

[1]Lamarre Geological Enterprises, LLC, Denver, Colorado

INTRODUCTION

The Ferron coalbed methane play occupies an 80-mi (129 km)-long trend from north of the city of Price, Utah to south of Interstate 70 (Figure 1). The northeast-southwest oriented play is bounded on the west by the Wasatch Plateau and on the east by the 70 mi-long (113 km) San Rafael uplift. The coalbed methane potential of the area was first evaluated in the early 1980s (Mroz et al., 1983; Adams and Kirr, 1984). The play was discovered in 1988 by Texaco Exploration and Production, Inc. at the northern end of the trend near Price. Subsequent exploratory drilling tested the Ferron coals along trend to the south. The Ferron coals had produced 469 bcf by the end of July 2003. At that time, daily production was 275 mmcfd from 735 wells.

Productive coals are found in the Ferron Sandstone Member of the Upper Cretaceous Mancos Shale (Figure 2). The Ferron Sandstone was deposited in a river-dominated and wave-modified deltaic system during middle to late Turonian time (88.8 to 91.3 Ma) (Ryer and McPhillips, 1983; Ryer, 1991). Rivers carried sediment eastward into the Cretaceous Western Interior Seaway from the thrust-faulted Sevier orogenic belt located to the west in western Utah and eastern Nevada (Figure 3). The Ferron eastward-thinning clastic wedge was deposited in response to a widespread regression that has been mapped throughout the Western Interior of the U.S. and Canada (Ryer, 1994). This regression was initiated by lowering of sea level and enhanced by increased sediment supply from tectonic activity in the orogenic belt to the west. A series of stacked clastic wedges was deposited near the western shoreline as a result of repeated transgressions and regressions of the seaway (Figure 4). The thickest and best quality coals were formed in aggradational coastal plain settings in back-barrier/lagoonal environments (Barker et al., 1999). Hale and Van DeGraff (1964) identified two Ferron deltaic systems: a northern ("Vernal") delta, and a southern ("Last Chance") delta. All coalbed methane production comes from coals associated with the older river- and wave-dominated Vernal delta.

The Ferron coals do not crop out in the northern two-thirds of the trend, but do occur in excellent exposures east of the town of Emery, Utah. These exposures have been extensively studied (Cotter, 1975; Ryer, 1981a; Ryer and McPhillips, 1983). More recent work (Gardner et al., 1992; Gardner, 1993; Gardner and Cross, 1994; Garrison and van den Bergh, 1996) interpreted the Ferron in terms of high-resolution sequence stratigraphy. These studies provide a comprehensive understanding of the

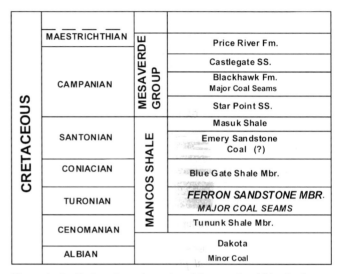

Figure 2. Coalbed methane is produced from coals within the Ferron Sandstone Member of the Mancos Shale.

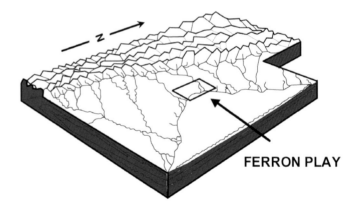

Figure 3. Paleogeographic reconstruction of Utah during Turonian time. The box represents the position of the Ferron trend. The Sevier orogenic belt in the western part of the state was the source for clastic debris that was carried eastward to the Western Interior Seaway. Peat swamps formed landward of (to the west of) the shoreline sandstones. Modified from Ryer and McPhillips (1983).

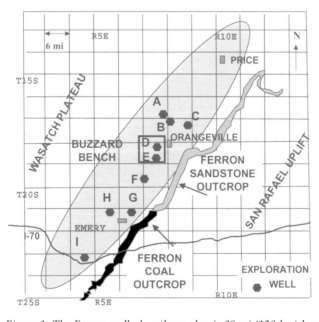

Figure 1. The Ferron coalbed methane play is 80 mi (128 km) long from north of Price to south of Interstate 70, Utah. Letters identify exploratory wells listed in Table 1.

depositional setting and facies architecture of the Ferron Sandstone. Van den Bergh (1995) and Garrison and van den Bergh (1997) examined outcrops of coal zones in the Last Chance delta near Emery. According to their terminology, coal zones consist of laterally extensive intervals of coal and correlative carbonaceous shales, root zones, and organic-rich mudstones and siltstones.

Altered volcanic ash layers (tonsteins) are commonly found within coal seams in the Ferron Member (Ryer et al., 1980). These tonsteins represent volcanic events of short duration and therefore are nearly isochronous horizons that are used to correlate coal zones.

Detailed mapping by Gardner (1993), in terms of sequence stratigraphy, shows that the coals are slightly younger to the south. Coals associated with the southern Last Chance delta near Emery are within the upper part of the Ferron Sandstone. These coals were deposited between 90.5 and 88.8 Ma (Gardner and Cross, 1994). The northern Vernal delta prograded to the east and southeast between 91.3 and 90.5 Ma. In both deltas the thickest coals are found in zones parallel and adjacent to the landward pinch-outs of the delta-front sandstones (Ryer, 1981b; Tabet et al., 1995). Current coalbed methane

production is from coals within the lower part of the Ferron Sandstone; these coals are in the Vernal delta in the northern part of the trend (Figure 5).

CHARACTERISTICS OF FERRON COALS

The thickest Ferron coals in the northern Vernal delta trend northeast-southwest and were formed in back-barrier lagoons, parallel to the paleoshoreline. In the Last Chance delta near Emery, the coals trend northwest-southeast, reflecting the northeast progradation of the shoreline. The coals become thinner and pinch out into marine shales and shoreline sandstones to the east, and into fluvial sandstones and non-marine shales to the west (Figure 4). The coals are usually found in three to six seams over a stratigraphic interval of 150 to 200 feet (46–61 m). Figure 6 is the wireline log of a representative well in the Buzzard Bench field with 30 ft (9 m) of coal. During September 2001, the well produced 3657 mcfd and 589 bwpd, bringing its cumulative production to 2.3 bcf. Coals are easily recognized on the standard geophysical wireline log suite of borehole-compensated density and neutron, induction, caliper, and gamma ray logs. Log responses of coals include low bulk density, high resistivity, and low gamma ray values, with washouts on the caliper log.

In many cases the coals have higher than expected gamma ray readings (>200 API units). Gamma ray spectroscopy logs often indicate high uranium concentrations in other coals (Scholes and Johnston, 1993). Any uranium may have been deposited in the coals from

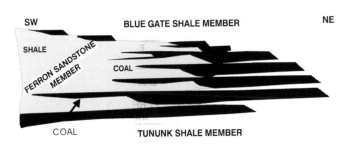

Figure 4. Diagrammatic cross section through the Ferron Sandstone. Coals pinchout into alluvial plain deposits to the west and into marine shale and delta-front sandstones to the east. From Ryer, 1981a.

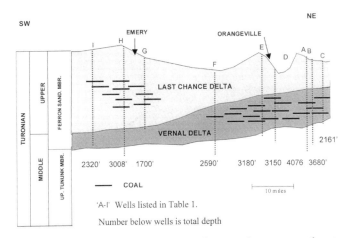

Figure 5. Diagrammatic cross section from southwest to northeast, showing Last Chance and Vernal deltas. Wells shown are listed in Table 1. The Vernal delta produces coalbed methane from the Ferron Sandstone Member of the Mancos Shale. Diagram modified from Gardner and Cross (1994).

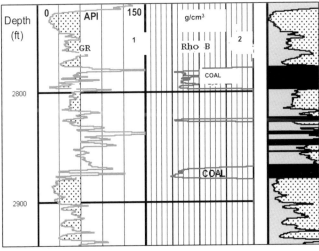

Figure 6. Geophysical wireline logs of Utah Power and Light 14-55 well in Sec. 14, T18S, R7E, in Buzzard Bench field. Curves shown are gamma ray and bulk density, with an interpreted lithology curve in the far right track. Coal is shaded black, sandstone is represented by coarse stippling and shale is depicted by solid gray. The well contains 30 ft (9 m) of coal, (using a bulk density cutoff of 1.75 g/cm³), which is shown in black. The well has produced 5.0 bcf of gas and produced 5577 mcfd and 79 bwpd during September 2003.

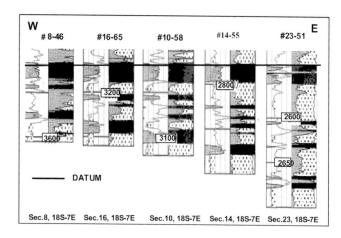

Figure 7. Cross section showing continuity of Ferron coals across the Buzzard Bench field in T18S, R7E, near Orangeville, Utah. Coals are shown in black and sandstones are stippled. The datum is a tonstein at the top of the upper coal seam.

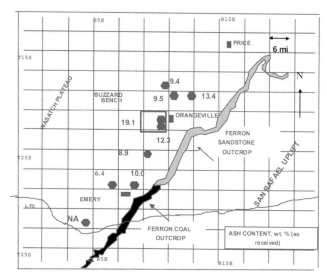

Figure 8. Map showing the weighted average ash content of coals in exploratory wells in the Ferron trend. Values are on an as-received basis.

migrating ground water after the coalification process was complete. Alternatively, the high gamma ray spikes may be caused by thin ash layers within the coals.

STRUCTURE AND STRATIGRAPHY

Ferron coals dip to the west away from the San Rafael uplift at approximately 200 ft per mi (40 m/km). Drill depths range from 1700 ft (518 m) on the east, to 4100 ft (1250 m) in the west due to structural dip and higher surface elevations on the west side of the trend. In Buzzard Bench field, large north-trending normal faults are parallel to the major fault zones in the Wasatch Plateau (Lamarre, this volume). Smaller faults, with throws of less than 100 ft (30 m), trend northeast-southwest. Figure 7 is a stratigraphic cross section showing the continuity of the coals in Buzzard Bench field; the datum is an areally extensive tonstein. On outcrop, such tonsteins are typically about 6 in (15 cm) thick. Figure 7

shows how the coal-bearing interval thickens to the east as the shoreline sandstones become thicker and better developed.

COALBED METHANE EXPLORATION PROGRAM

From 1994 to 1996, nine exploratory wells were drilled in the central and southern portions of the Ferron trend, from just north of the town of Orangeville to Interstate 70 (Figure 8). Location data for these wells are shown in Table 1. This exploratory program was designed to assess the coalbed methane potential of the southern two-thirds of the Ferron trend. Among the issues addressed were: does the gas content, coal quality, and thermal maturity of the coals vary between the Last Chance delta and the Vernal delta, and is one delta more prospective than the other for coalbed methane production?

Table 1. Data for all exploratory wells used in this study. Locations are shown in Figure 1. KB = elevation of kelly bushing, TD = total depth at bottom of hole.

LETTER	OPERATOR	WELL NAME	LOCATION	DATE LOGGED	KB ELEV. (Feet)	API NO.
A	TEXACO	State of Utah 'T' # 36-10	Sec. 36, T16S-R7E	Dec. 15, 1995	6515	43-015-30268
B	TEXACO	Utah Federal 'M' # 6-25	Sec. 6, T17S-R8E	Aug. 27, 1996	6325	43-015-30292
C	TEXACO	L.M. Lemmon # 10-1	Sec. 10, T18S-R7E	Jan. 9, 1995	5929	43-015-30293
D	TEXACO	Utah Federal 'C' # 23-8	Sec. 23, T18S-R7E	Aug. 30, 1995	6192	43-015-30245
E	TEXACO	Utah Federal 'A' # 26-2	Sec. 26, T18S-R7E	Dec. 4, 1994	6578	43-015-30244
F	TEXACO	Utah Federal 'H' # 6-21	Sec. 6, T20S-R7E	Dec. 2, 1996	6234	43-015-30294
G	TEXACO	A.L. Jensen # 27-9	Sec. 27, T21S-R6E	Oct. 3, 1995	6344	43-015-30259
H	ANADARKO	Ferron Federal # A-1	Sec. 36, T21S-R5E	Mar. 13, 1994	6934	43-041-30027
I	TEXACO	State of Utah 'V' # 36-16	Sec. 36, T23S-R4E	Aug. 11, 1996	6880	43-041-30028

Table 2. Coal analysis results for wells in this study. (ar) = as received, (dmmf) = dry, mineral matter-free, (mmmf) = moist, mineral matter free, VM = weight percent volatile matter, Btu/lb = British Thermal units per pound, scf/t = standard cubic feet per short ton, R$_o$ = mean maximum reflectance of vitrinite.

Letter	Ash (ar)	VM (dmmf)	Btu/lb (mmmf)	R$_o$ %	Gas Content (scf/t ash-free)
A	9.4	56.9	15,668	NA	429
B	9.5	47.7	15,573	NA	457
C	13.4	44.6	15,243	NA	407
D	19.1	43.1	15,027	0.74	409
E	12.3	45.1	15,315	0.70	370
F	8.9	45.8	NA	NA	288
G	10	43.6	14,639	0.65	132
H	6.4	NA	NA	NA	72
I	NA	NA	NA	0.53	12

Most of the exploratory wells penetrated the entire Ferron section and the coals were sampled in all wells. Samples consisted of whole core or rotary sidewall cores. Proximate analysis were performed to determine ash, moisture, volatile matter, and fixed carbon. The heating value (Btu/lb) was also measured. Vitrinite reflectance values were determined for many samples. A summary of the coal chemistry data is shown in Table 2. To measure the gas content of the coals, all whole and sidewall core samples were placed in desorption canisters immediately upon reaching the surface. The core samples were desorbed at reservoir temperatures of approximately 80 to 95°F (27–35°C). The volume of gas lost between the time the bit penetrated the coal and when the coal was sealed in a canister is called lost gas. The amount of lost gas was calculated by the U.S. Bureau of Mines Direct Method (McLennan, et al., 1995). The average lost gas is 20.1 scf/t (0.63 cm^3/g). Measurements of residual or undesorbed gas were made on samples after desorption was complete. The average residual gas content is 25.3 scf/t (0.79 cm^3/g), or approximately 7.4% of the total gas. Residual gas values for coals in other basins range from 12.6 to 15.8% of the total gas content, although residual gas in the San Juan Basin is negligible (Mavor et al., 1993). All gas content values have been normalized to an ash-free basis for comparative purposes.

RESULTS OF THE CORING PROGRAM

Figure 8 is a map of the Ferron trend. The squares are townships, 6 mi (9.6 km) on a side. The Wasatch Plateau is northwest of the play and the San Rafael uplift is located to the southeast. The outcrop of the Ferron Sandstone member is shaded. In the gray area, only shales and sandstones are present. However, in the areas shaded black, the coals are present at the surface and have been mined locally from the outcrop. The nine core

holes are shown by hexagons. The weighted average ash content (yield) of the coals is shown for most wells. The values are in weight percent, on an as-received basis. Notice that the data show no apparent trend, and vary from 6.4 to 19.1%.

Volatile matter shows only minor variations along trend, as shown on Figure 9. These weighted average values are reported on a dry, mineral matter-free (dmmf) basis. The volatile matter data indicate no appreciable change along trend, and range from 43.1 to 56.9% (dmmf).

Ferron coals are all high-volatile A bituminous apparent rank, based on heating (calorific) values in Btu/lb (Table 3), expressed on a moist, mineral matter-free (mmmf) basis (American Society for Testing Materials standard, D388, 1983). Figure 10 shows the heating values for the exploratory wells. These data suggest that the coals are slightly less mature to the south.

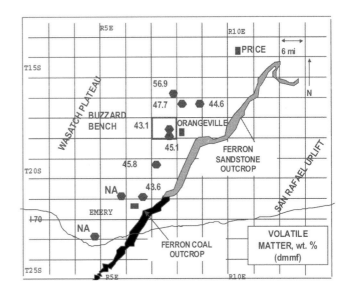

Figure 9. Map showing volatile matter percentage in coals in the Ferron trend. Values are reported on dry, mineral matter-free basis.

Table 3. Apparent coal rank classification (ASTM standard, D388, 1983). Ferron coals are high volatile A bituminous, as shown by bold italics. (dmmf) = dry, mineral matter-free, (mmmf) = moist, mineral matter-free.

Class	Group		Fixed Carbon Limits, % (dmmf FC)		Volatile Matter Limits, % (dmmf VM)		Calorific Value Limits, Btu/lb (mmmf BTU)	
			=>	<	>	<=	>=	<
Anthracite	1. Meta-anthracite		98	---	---	2	---	---
	2. Anthracite	(an)	92	---	2	8	---	---
	3. Semi-anthracite	(sa)	86	92	8	14	---	---
Bituminous	1. Low volatile bituminous	(lvb)	78	86	14	22	---	---
	2. Medium volatile bituminous	(mvb)	69	78	22	31	---	---
	3. High volatile A bituminous	***(hvAb)***	---	69	31	---	14,000	---
	4. High volatile B bituminous	(hvBb)	---	---	---	---	13,000	14,000
	5. High volatile C bituminous	(hvCb)	---	---	---	---	11,500	13,000
Sub-bituminous	1. Sub-bituminous A	(subA)	---	---	---	---	10,500	11,500
	2. Sub-bituminous B	(subB)	---	---	---	---	9,500	10,500
	3. Sub-bituminous C	(subC)	---	---	---	---	8,300	9,500
Lignite	1. Lignite A	(ligA)	---	---	---	---	6,300	8,300
	2. Lignite B	(ligB)	---	---	---	---	---	6,300

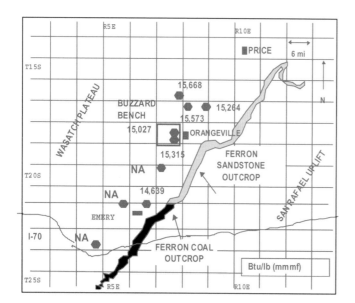

Figure 10. Map of heating value of coals in Btu/lb on a moist, mineral matter-free (mmmf) basis.

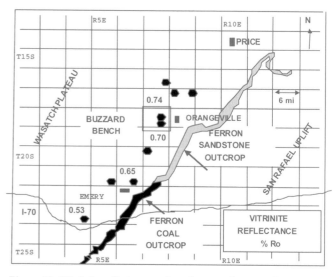

Figure 11. Vitrinite reflectance values decrease from north to south.

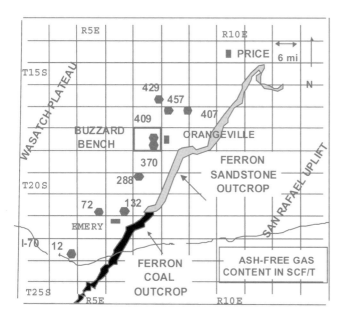

Figure 12. Map showing dramatic decrease in gas content along trend to the south. Values are scf/t, normalized to an ash-free basis.

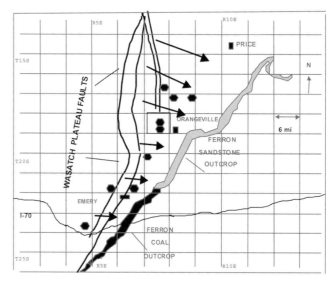

Figure 13. The Ferron potentiometric head is highest on the Wasatch Plateau. Arrows show the direction of subsurface fluid flow through the Ferron Sandstone.

The Ferron coals are high-volatile A to high-volatile B bituminous based on vitrinite reflectance (R_o%) (Meissner, 1984) as shown on Figure 11. Previous studies (Quick and Tabet, 1999) suggest that the vitrinite reflectance values may be somewhat suppressed in the northern part of the trend, so the absolute values may be too low by as much as 0.2% R_o. However, the values show a trend toward lower rank to the south, which substantiates data from Figure 10 on heating values.

The mapped data indicate no significant variation in chemical composition of the coals from north to south along the Ferron trend, except for a slight decrease of coal rank. This also suggests that there are no major coal quality differences between the Last Chance and Vernal deltas. However, a map of gas contents (scf/t) in the coal (Figure 12) shows major changes from north to south. Note that the values shown are normalized to an ash-free (af) basis. The lowest value is 12 scf/t (0.4 cm^3/g) (af) and occurs in the south, near Interstate 70. The highest value is 457 scf/t (14.3 cm^3/g) (af) and is found near the northern end of the trend. There is a very strong trend of decreasing gas content to the south.

INTERPRETATION

The gas content values in the northern two-thirds of the Ferron trend (Figure 12) are much higher than anticipated for these relatively low rank coals (Lamarre and Burns, 1997). Meissner (1984) reported that the threshold for significant generation of thermogenic gas begins at thermal maturity levels of approximately 0.74% R_o. Laboratory pyrolysis of lignite suggests that significant generation of gas (300 scf/t, [9.3 cm^3/g]) occurs at thermal maturity levels of 1.0% R_o (Tang et al., 1991). Previous studies have shown that abnormally high gas contents in low rank coals indicate generation of *in-situ* biogenic gas as well as gas migration and conventional trapping of thermogenic gas (Tang et al., 1991; Scott and Ambrose, 1992).

The highest gas contents in the Ferron coals are found in the northern two-thirds of the trend where the coals do not outcrop (Figure 12). Near the southern end of the trend, where the Ferron coals are exposed at the surface, gas contents are dramatically lower. There may be two explanations for this gas content trend. The first is shown on Figure 13. Studies near Emery by the United States Geological Survey (Lines and Morrissey, 1983) indicate that hydrodynamic flow through the Ferron Sandstone aquifer is from the Wasatch Plateau on the west to the San Rafael uplift on the east. Tests indicate that the transmissivity of the aquifer ranges from 200 to 700 ft3/d/ft (19–65 m^2/d/m). They calculate the subsurface recharge to the Ferron aquifer in that area to be about 2.4 ft^3/s (6.8 x 10^{-2} m^3/s) during 1979 or about 1700 acre-feet (2.1 ha^3) per year. The potentiometric surface is highest in the west due to precipitation of up to 40 in (102 cm) per year on the Wasatch Plateau. Meteoric water flows downward along major fault zones and migrates eastward, downgradient, in highly fractured sandstones, shales, and coals of the Ferron Sandstone in the subsurface. The Joes Valley-Paradise Valley fault system is 2 to 4 mi (3.2–6.4 km) wide, over 75 mi (121 km) long and has vertical displacements reaching 2500 ft (762 m) (Doelling, 1972). The fault zone is an area with very high hydraulic conductivity and enhanced permeability. Two other fault zones, the Pleasant Valley and Gordon Creek faults, are located farther east and therefore closer to the Ferron coalbed methane trend.

535

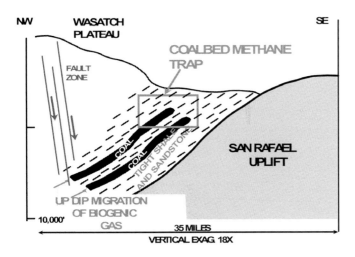

Figure 14. Diagrammatic section showing movement of meteoric water and updip migration of biogenic gas.

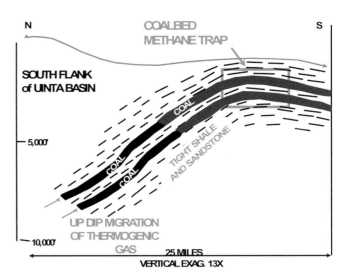

Figure 15. North to south diagrammatic section showing updip movement of thermogenic coalbed methane out of the Uinta Basin into traps on the southern margin of the basin (and northern portion of the Ferron trend).

Carbon-14 dating of two Ferron aquifer water samples from a well near Emery indicated geologically young formation water with ages of 28,000 and 31,000 years bp (Lines and Morrissey, 1983). A similar study of Fruitland Formation water in the San Juan Basin indicated an age of 33,176 years bp (Mavor et al., 1991).

This eastward-flowing hydrodynamic system affects the gas content in two ways. First, as water from meteoric recharge flows through the permeable coals in the southern area, the advancing groundwater flushes gas to the surface (Scott et al., 1991; Scott, 1993a, b; Scott et al., 1994). Since the coals are exposed at the surface, the reservoir pressure is reduced and natural gas is desorbed from the coals and lost to the atmosphere. This mechanism may explain the very low gas contents in the southern portion of the trend. But hydrodynamics also impacts the gas content at the northern end of the trend.

Figure 14 is a diagrammatic cross section running northwest to southeast near the town of Orangeville, in T18S, R7E. The figure shows the Ferron coals encased in shale, siltstone, and tight sandstones. Notice that the coals pinchout updip and do not reach the present day land surface. Meteoric water migrates down the faults and carries bacteria with it. When bacteria enter the coaly zones they metabolize the organic compounds generated during coalification and produce secondary biogenic methane and carbon dioxide. This reaction typically occurs at relatively low temperatures (less than 150°F [65°C]) (Scott et al., 1994). Isotopic studies of the gases produced from the coals at Buzzard Bench indicate that some of the gas has an isotopic signature of biogenic gas (Lamarre and Burns, 1997). The microbial activity in the Ferron aquifer altered the composition of the existing thermogenic gas and generated late-stage, isotopically light methane and isotopically heavy CO_2 (Rice et al., 1988; Rice, 1993a, b).

This methane migrated updip until it became trapped at the subsurface pinchout of the coals. Therefore, the northern portion of the Ferron trend is a large stratigraphic trap. This movement of water through the coals is also responsible for the artesian overpressure encountered in portions of the trend. Near Emery, where the coals are exposed at the surface, secondary biogenic gas has also probably been generated in the coals, but there is no trapping mechanism to enrich the gas content of the coals.

Isotopic studies have demonstrated that Fruitland coalbed gases in the San Juan Basin of southwestern Colorado and northwestern New Mexico are composed of 25–58% thermogenic gas, 15–30% secondary biogenic gas and 12–60% migrated thermogenic gases (Scott et al., 1994; Scott and Kaiser, 1995). Isotopic data from Ferron coalbed gases also suggest that Ferron coalbed gases are a mixture of thermogenic gas, migrated thermogenic gas, and secondary biogenic gas (Burns and Lamarre, 1997).

This hypothesis helps to explain the higher than expected gas contents in the northern portion of the trend. However, there is an additional mechanism to explain these higher gas contents. Figure 15 is a cross section from north to south, with the location of producing coalbed methane fields shown within the box. The vertical exaggeration is 13 times. On the south flank of the Uinta Basin the Ferron coals and organic-rich shales are buried to depths greater than 8000 ft (2438 m). These coals have generated much more thermogenic gas due to their higher thermal maturity (Tabet, 1998). At greater depths in the basin the coals are probably generating gas today. Some of this gas has migrated updip within the coals and is structurally trapped as shown in Figure 15.

Farther to the south, the Ferron Sandstone is at a lower structural position and further from the source of migrated gas so it contains no (or very little) migrated thermogenic gas. This theory may explain the higher gas contents in coals in the northern part of the Ferron trend.

CONCLUSIONS

Exploratory drilling along the 80-mi (129 km)-long Ferron coalbed methane trend indicates that the coal chemistry, quality, and maturity change slightly, but do not vary enough to explain the marked change in gas content. There are no appreciable bulk chemical differences between the coals of the Last Chance delta near Emery and those deposited in the Vernal delta near Price. However, the gas content of the coals decreases dramatically from north to south. Higher gas contents at the northern end of the trend result from fortuitous hydrodynamic conditions as well as migration of thermogenic gas from deeper portions of the Uinta Basin.

The Ferron trend forms a very large coalbed methane accumulation where the coals do not outcrop, resulting in a large stratigraphic trap near their updip pinchout into marine shale and near-shore sandstones. Gas within this large trap comes from three different sources: (1) thermogenic gas generated within the coals, (2) migrated thermogenic gas generated in deeper coals within the Uinta Basin, and (3) biogenic gas generated downdip within the coal zones and transported updip by meteoric water from the Wasatch Plateau.

This study indicates that rank of coal and the amount of *in-situ* generated coalbed methane are not necessarily the prime factors to consider when exploring for coalbed methane. Biogenic and migrated thermogenic gases can significantly increase the gas content of relatively low rank coals. Hydrodynamics plays a large role by increasing the gas content in areas of permeability barriers, or by reducing the gas content by flushing gas to the atmosphere in places where the coals are exposed at the surface.

ACKNOWLEDGMENTS

I thank Steve Ruhl (Anadarko Petroleum Corp.), David E. Tabet and Jeffery C. Quick (Utah Geological Survey), and Charles W. Bryer and William F. Haslebacher (U.S. Department of Energy) for their thoughtful reviews and suggestions for improvement. The manuscript benefited greatly from their suggestions. I also thank ChevronTexaco (formerly Texaco Exploration and Production, Inc.) for permission to publish this data.

REFERENCES CITED

Adams, M. A., and J. N. Kirr, 1984, Geologic overview, coal deposits, and potential for methane recovery from coalbeds of the Uinta Basin-Utah and Colorado, *in* C. T. Rightmire, G. E. Eddy, and J. N. Kirr eds., Coalbed methane resources of the United States: AAPG Studies in Geology, no. 17 p. 253–270.

American Society for Testing Materials (ASTM), 1983, Standard classification of coals by rank; ASTM designation D388-82, in gaseous fuels, coal and coke: 1983 Book of Standards, v. 0505 ASTM, Philadelphia, 5 p.

Barker, C. E., J. R. Garrison, Jr., and T. C. V. van den Bergh, 1999, The stratigraphy, lithology and composition of coal zones as a function of stratigraphic position within the Turonian-Coniacian Upper Ferron Sandstone, Last Chance delta, east-central Utah (abs.): Geological Society of America Annual Meeting, Abstracts with Programs, v. 31, abstract 14980.

Burns T. D., and R. A. Lamarre, 1997, Drunkard's Wash project-coalbed methane production from Ferron coals in east-central Utah: Proceedings of the 1997 International Coalbed Methane Symposium, paper 9709, p. 507–520.

Cotter, E., 1975, Deltaic deposits in the Upper Cretaceous Ferron Sandstone, Utah, *in* M. L. S. Broussard, ed., Deltas, models for exploration: Houston Geological Society, p. 471–484.

Doelling, H. H., 1972, Central Utah coal fields-Sevier-Sanpete, Wasatch Plateau, Book Cliffs and Emery: Utah Geological and Mineralogical Survey, Monograph Series No. 3, p. 245–417.

Gardner, M. H., 1993, Sequence stratigraphy and facies architecture of the Upper Cretaceous Ferron Sandstone Member of the Mancos Shale, east-central Utah: Ph.D. dissertation T3975, Colorado School of Mines, Golden, 528 p.

Gardner, M. H., M. D. Barton, N. Tyler, and R. S. Fisher, 1992, Architecture and permeability structure of fluvial-deltaic sandstones, Ferron Sandstone, east-central Utah, *in* R. M. Flores, ed., Mesozoic of the Western Interior: SEPM Theme Meeting Field Guidebook, p. 5–21.

Gardner, M. H., and T. A. Cross, 1994, Middle Cretaceous paleogeography of Utah, *in* M. V. Caputo, J. A. Peterson, and K. J. Franczyk, eds., Mesozoic systems of the Rocky Mountain region, U.S.A.: Rocky Mountain Section SEPM, p. 471–503.

Garrison, J. R., Jr., and T. C. V. van den Bergh, 1996, Coal zone stratigraphy — a new tool for high-resolution depositional sequence stratigraphy in near-marine fluvial-deltaic facies associations-a case study from the Ferron Sandstone, east-central Utah: AAPG Rocky Mountain Section meeting, Billings, Montana, expanded abstracts volume, Montana Geological Society, p. 31–36.

—1997, Coal zone and high-resolution depositional sequence stratigraphy of the Upper Ferron Sandstone, *in* P. K. Link and B. J. Kowallis, eds., Mesozoic to Recent Geology of Utah: Brigham Young University Geology Studies, v. 42, pt. 2, p. 160–178.

Hale, L. A., and F. R. Van De Graaff, 1964, Cretaceous stratigraphy and facies patterns — northeastern Utah and adjacent areas: Intermountain Association of Petroleum Geologists, 13th Annual Field Conference Guidebook, p. 115–138.

Lamarre, R. A., and T. D. Burns, 1997, Drunkard's Wash Unit-coalbed methane production from Ferron coals in east-central Utah, *in* E. B. Coalson, J. C. Osmond, and E. T. Williams, eds., Innovative applications of petroleum technology in the Rocky Mountain area: Rocky Mountain Association of Geologists Guidebook, p. 47–59.

Lines, G. C., and D. J. Morrissey, 1983, Hydrology of the Ferron Sandstone aquifer and effects of proposed surface-coal mining in Castle Valley, Utah: U.S. Geological Survey Water-Supply Paper 2195, 40 p.

McLennan, J. D., P. S. Schafer, and T. J. Pratt, 1995, A guide to determining coalbed gas content: Gas Research Institute, GRI-94/0396, p. 1.1–11.32.

Mavor, M. J., T. J. Pratt, A. Crandlemire, and G. Ellerbrok, 1993, Assessment of coalbed methane resources at the Donkin Mine site, Cape Breton, Nova Scotia, Canada: Proceedings of the 1993 International Coalbed Symposium, paper 9638, p. 471–482.

Mavor, M. J., J. C. Close, and T. J. Pratt, 1991, Western Cretaceous coal seam project summary of the Completion Optimization and Assessment Laboratory (COAL) site: Gas Research Institute Topical Report GRI-91/0377, 120 p.

Meissner, F. F., 1984, Cretaceous and lower Tertiary coals as sources for gas accumulations in the Rocky Mountain area, *in* J. Woodward, F. F. Meissner, and J. L. Clayton, eds., Hydrocarbon source rocks of the greater Rocky Mountain region: Rocky Mountain Association of Geologists Guidebook, p. 401–432.

Mroz, T. H., J. G. Ryan, and C. W. Byrer, 1983, Methane recovery from coalbeds-a potential energy source: U.S. Department of Energy, DOE/METC/83-76, p. 291–324.

Quick, J. C., and D. E. Tabet, 1999, Suppressed vitrinite reflectance in the Ferron coalbed gas fairway, central Utah (abs.): Geological Society of America Annual Meeting, Abstract with Program, v. 31, abstract 51006.

Rice, D. D., 1993a, Controls of coalbed gas composition: 1993 International Coalbed Methane Symposium Proceedings, p. 577–589.

—1993b, Composition and origins of coalbed gas, *in* B. E. Law and D. D. Rice, eds., Hydrocarbons from coal: AAPG Studies in Geology 38, p. 159–185.

Rice, D. D., C. N. Threlkeld, A. K. Vuletich, and M. J. Pawlewicz, 1988, Identification and significance of coal-bed gas, San Juan Basin, northwestern New Mexico and southwestern Colorado, *in* J. E. Fassett, ed., Geology and coal-bed methane resources of the northern San Juan Basin, Colorado and New Mexico: Rocky Mountain Association of Geologists Guidebook, p. 51–60.

Ryer, T. A., 1981a, Deltaic coals of Ferron Sandstone Member of Mancos Shale- predictive model for Cretaceous coal-bearing strata of Western Interior: AAPG Bulletin, v. 65, p. 2323–2340.

—1981b, The Muddy and Quitchupah projects-a progress report with descriptions of cores of the I, J and C coal beds from the Emery coal field, central Utah: U.S. Geological Survey Open-File Report 81-460, 34 p.

—1991, Origin of mid-Cretaceous clastic wedges, central Rocky Mountain region (extended abstract), *in* J. Dolson, (compiler), Unconformity related hydrocarbon exploitation and accumulation in clastic and carbonate settings: Rocky Mountain Association of Geologists Course Notes, p.16–20.

—1994, Interplay of tectonics, eustatics and sedimentation in the formation of Mid-Cretaceous clastic wedges, Central and Northern Rocky Mountain regions, *in* J. C. Dolson, M. L. Hendricks, and W. A. Wescott, eds., Unconformity related hydrocarbons in sedimentary sequences: Rocky Mountain Association of Geologists Guidebook, p. 35–44.

Ryer, T. A., and M. McPhillips, 1983, Early Late Cretaceous paleogeography of east-central Utah, *in* M. W. Reynolds and E. D. Dolly, eds., Mesozoic paleogeography of west-central United States: Rocky Mountain Section SEPM, p. 253–272.

Ryer, T. A., R. E. Phillips, B. F. Bohor, and R. M. Pollastro, 1980, Use of altered volcanic ash falls in stratigraphic studies of coal-bearing sequences-an example from the Upper Cretaceous Ferron Sandstone Member of the Mancos Shale in central Utah: Geological Society of America Bulletin, v. 91, p. 579–586.

Scholes, P. L., and D. Johnston, 1993, Coalbed methane applications of wireline logs, *in* B. E. Law and D. D. Rice, eds., Hydrocarbons from coal: AAPG Studies in Geology 38, p. 287–302.

Scott, A. R., 1993a, Coal rank, gas content, and composition and origin of coal gases, Mesaverde Group, Sand Wash Basin, *in* W. R. Kaiser, A. R. Scott, D. S. Hamilton, R. Tyler, R. G. McMurray, N. Zhou, and C. M. Tremain, eds., Geologic and hydrologic controls on coalbed methane: Sand Wash Basin, Colorado and Wyoming: Gas Research Institute Topical Report GRI-92/0420, p. 51–62.

—1993b, Composition and origin of coalbed gases from selected basins in the United States: 1993 International Coalbed Methane Symposium Proceedings, paper 9370, p. 207–222.

Scott, A. R., and W. A. Ambrose, 1992, Thermal maturity and coalbed methane potential of the Greater Green River, Piceance, Powder River and Raton Basins (abs.): AAPG Annual Convention, Official Program with Abstracts, Calgary, Alberta, Canada, p. 116.

Scott, A. R., and W. R. Kaiser, 1995, Hydrogeologic factors affecting dynamic open-hole cavity completions in the San Juan Basin, U.S.A.: INTERGAS '95, International Unconventional Gas Symposium Proceedings, paper 9510, p. 247–257.

Scott, A. R., W. R. Kaiser, and W. B. Ayers, Jr., 1991, Composition, distribution and origin of Fruitland Formation and Pictured Cliffs Sandstone gases, San Juan Basin, Colorado and New Mexico, in S. D. Schwochow, D. K. Murray, and M. F. Fahy, eds., Coalbed methane of western North America: Rocky Mountain Association of Geologists Guidebook, p. 93–108.

Scott, A. R., W. R. Kaiser, and W. B. Ayers, Jr., 1994,

Thermogenic and secondary biogenic gases, San Juan Basin, Colorado and New Mexico-implications for coalbed gas producibility: AAPG Bulletin, v.78, p. 1186–1209.

Tabet, D. E., 1998, Migration as a process to create abnormally high gas contents in the Ferron Sandstone coal beds, central Utah: AAPG Annual Convention (extended abstracts), v. 2, paper A646.

Tabet, D. E., B. P. Hucka, and S. N. Sommer, 1995, Maps of total Ferron coal, depth to the top, and vitrinite reflectance for the Ferron Sandstone Member of the Mancos Shale, central Utah: Utah Geological Survey Open-File Report 329, 3 plates, 1:250,000.

Tang, Y., P. D. Jenden, and S. C. Teerman, 1991, Thermogenic methane formation in low-rank coals-published models and results from pyrolysis of lignite, in D. A. C. Manning, ed., Organic geochemistry-advances and applications in the natural environment; Manchester, England, Manchester University Press, p. 329–331.

van den Bergh, T. C. V., 1995, Facies architecture and sedimentology of the Ferron Sandstone Member of the Mancos Shale, Willow Springs Wash, east-central Utah: M.S. thesis, University of Wisconsin, Madison, 225 p.

Main Street, Price, Utah, circa 1924. Photograph used by permission, Utah State Historical Society, all rights reserved.

Analog for Fluvial-Deltaic Reservoir Modeling: Ferron Sandstone of Utah
AAPG Studies in Geology 50
T. C. Chidsey, Jr., R. D. Adams, and T. H. Morris, editors

Helper Field: An Integrated Approach to Coalbed Methane Development, Uinta Basin, Utah

Andre Klein, Keith Buck, and Steve Ruhl[1]

ABSTRACT

Helper field, located on the southwestern flank of the Uinta Basin in central Utah, produces gas from multiple coal seams and interbedded sandstone within the Upper Cretaceous Ferron Sandstone Member of the Mancos Shale. Subsurface mapping, along with coal and sandstone petrophysical properties obtained from two exploratory wells drilled in 1993, led to the drilling of a five well exploratory pilot program that established gas sales in 1994. Drilling from 1995 through 2003 resulted in the completion of 111 additional coalbed methane wells. In mid-2003, Helper field was producing 35 mmcfg/day and production was increasing as pressure draw down continued. Full field development is anticipated at 125 wells.

Drilling success and production optimization at Helper field has been accomplished through a multidisciplinary approach that integrates detailed geological mapping, reservoir simulation modeling, and completion techniques. Detailed decompacted stratigraphic well-log cross sections, and mapping of eight coal seams and interbedded sandstone identified two sea-level transgressions within the Ferron Sandstone. A multi-layer reservoir simulation model was developed to account for total fluid production, improving production forecasts. Hydraulic fracture stimulation designs and completion practices have evolved over time based on new drilling, production, and 3-D fracture simulation modeling.

INTRODUCTION

Of the estimated 530 tcf of coal-derived methane trapped in Rocky Mountain basins (Montgomery et al., 2001), as much as 10 tcf is thought to be contained in the Uinta Basin of northeastern Utah. Productive intervals include the upper Cretaceous Blackhawk Formation, and the Ferron Sandstone Member of the Mancos Shale Formation. Of these, the Ferron Sandstone has attracted a great deal of attention, owing to excellent continuous exposures and analogies of the Ferron with Gulf Coast deltaic reservoirs (cf., Tyler et al., 1991). In recent years, the Ferron has also caught the attention of petroleum geologists, with the discovery of major coalbed gas fields throughout the greater Castle Valley region (e.g., Drunkards Wash field, estimated 1–4 tcf [Montgomery et al., 2001]) (Figure 1).

Anadarko Petroleum Corporation became interested in Ferron coals in the early 1990s as part of a program aimed at targeting unconventional reservoirs. At that time, coalbed methane (CBM) was emerging as an important economic component in many exploration

[1]Unconventional Reservoirs Group, Anadarko Petroleum Co., The Woodlands, Texas

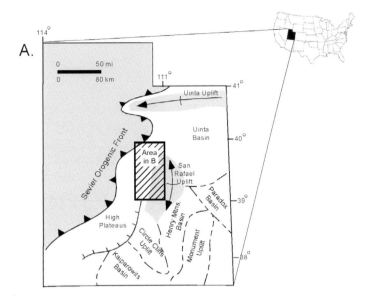

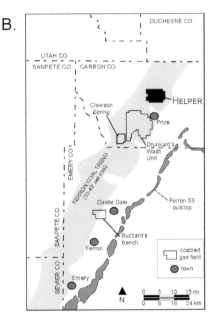

Figure 1. (A) Generalized tectonic map of Utah illustrating major basins and uplifts. Castle Valley region (seen in B) is represented by diagonal hatch pattern. (B) Map of greater Castle Valley region showing Ferron coal trend (light shading) and Ferron outcrop belt (dark gray). Helper coalbed methane (CBM) field is shown north of the town of Price, Utah.

portfolios. Numerous prospects were identified, primarily in Wyoming, Colorado, Utah, and New Mexico. These prospects were leased and subsequently drilled and evaluated. To date, Anadarko has interest in over 300 CBM wells throughout out the Rocky Mountain region.

Helper field was part of an Anadarko exploration program that was designed to evaluate the potential of the Castle Valley Ferron CBM trend (Figure 1). During 1993–1994, four exploratory wells were drilled at strategically located sites over a 50 mi (80 km) transect parallel to the depositional strike of the trend. Based on the

results from these wells, an exploratory CBM pilot project was drilled in the northern portion of Helper field in late 1994. Between 1994 and 2003, Anadarko drilled and completed between 12 and 33 wells each year as environmental assessments were submitted and approved. As of mid-2003, the Helper field had 116 CBM wells producing in excess of 36 mmcf/day and has yet to reach peak production. The combination of shallow (1500–3600 ft [460–1095 m]) coal horizons with relatively high gas contents (approximately 300–400 scf/ton [9.5–12.5 cm^3/g]) makes the Ferron coals in Helper field especially attractive drilling targets. This field continues to be a benchmark that Anadarko uses when evaluating the geologic, drilling, production, and reservoir parameters of other CBM prospects.

The ability to properly characterize and produce CBM reservoirs is an important task. Characterization challenges at Helper include accurate delineation of reservoir lithologies within the entire lower delta-plain package. Recent studies (Mavor and Nelson, 1997; Lamarre et al., 2001) indicate that non-coal lithologies (bulk densities >1.75 g/cm^3) can contain significant volumes of methane. This possibility may also exist in Helper field. Early well production results have exceeded model expectations, suggesting that either: (a) methane gas-storage content of the coals has been underestimated, or (b) methane gas is present in non-coal lithologies. Work is underway to better constrain the gas content of the coals, and to examine the gas content of interbedded carbonaceous shales. In addition, a multidisciplinary approach to reservoir characterization, which integrates detailed geological mapping with finite element modeling, will further reduce discrepancies between predicted and observed production rates and allow for better reserves estimates. Technological advances in fracture stimulation and well completion methods have reduced costs and increased well performance. The combination of a better understanding of CBM systems along with evolving technologies tailored toward CBM production allows the experience gained at Helper field to provide lessons that can be applied toward future CBM exploration and production efforts.

GEOLOGIC SETTING OF FERRON COALS IN THE HELPER FIELD AREA

Structural Setting

The most prominent structural features in the Helper area are the San Rafael uplift and the Uinta Basin (Figure 1A). Both structures are related to Laramide crustal warping. Situated on the southwest flank of the Uinta Basin, Helper field dips monoclinally north at 7–8°, away from the northward-plunging nose of the San Rafael uplift. The Ferron Sandstone is encountered at depths of about 1500–3600 ft (460–1095 m). No significant faults or folds are apparent from well log correlation studies through the field. Farther south in Drunkard's Wash field, however, reverse faults and normal faults display up to 150 ft (45 m) and 450 ft (136 m) of vertical offset, respectively (Doelling, 1972; Burns and Lamarre, 1997; Montgomery et al., 2001). In Helper field, traps are inferred to be predominantly stratigraphic (i.e., intraformational fluid pressure keeps the gas adsorbed within the coal).

Stratigraphy and Lithology

The Ferron Sandstone comprises an approximately 250–300 ft (60–75 m) gross section of lower alluvial-plain, delta-plain, and delta-front strata that interfinger with prodelta and marine deposits of the Mancos Shale (Ryer, 1981). The Ferron constitutes the lowest of a series of regressive coarser clastic units that prograded from west to east in response to a complex interplay of eustasy, tectonics, and sediment supply during the Late Cretaceous. These coarser clastic units are separated by tongues of marine shale that accumulated during relative transgressions of the Mancos sea (Ryer, 1981). Along strike, depositional variability within the Ferron has been suggested by Stalkup and Eubanks (1986), who document a barrier island-tidal channel-tidal delta complex northeast of the town of Moore, Utah. Thus, the Ferron coastline probably somewhat resembled the modern-day Texas coast, with a series of deltaic promontories separated by interdeltaic embayments (cf., Ryer, 1991).

For practical purposes, the Ferron Sandstone at Helper field can be subdivided into several lithologic units defined by the predominance of a particular rock type (Figure 2). The lower Ferron consists predominantly of sandstone, arranged in two seaward-stepping intervals known as the Clawson and Washboard sandstone beds. These sands were not differentiated in the present study. Together, these units are as much as 100 ft (30 m) thick and consist of shelf-margin sandstone that are concentrated primarily along the eastern side of the northeast-trending Ferron clastic wedge (Montgomery et al., 2001). Above the shelf-margin sands are a series of seaward-stepping, river-dominated delta packages capped by relatively thick (5–15 ft [2–5 m]) coals. A middle unit that probably represents a delta-plain environment consists predominantly of channel, crevasse splay, and overbank deposits interbedded with thin coals of varying thickness. The paludal environment returns near the top of the Ferron section, although coals here are rarely as thick as they are in the lower, coal-dominated unit. The top of the Ferron is usually but not always capped by sandstone.

These deposits are illustrated by a decompacted stratigraphic cross section (Figure 3). The decompacted

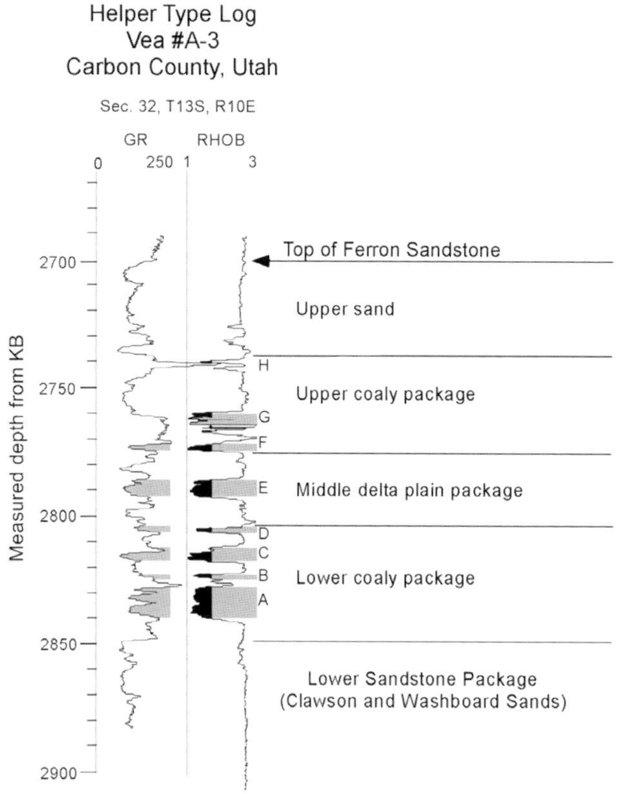

Figure 2. Type log from Helper field illustrating generalized lithologic units in the Ferron Sandstone. Shaded intervals represent coals. Letters A-H are coalbed labels. GR = gamma ray in API units, RHOB = bulk density in g/cc.

cross section assumes a peat:coal thickness ratio of 5:1, based on porosity differences between Ferron coals and average peats. Digital well-log data were exported into a data storage and analysis software package. The thicknesses of sand, shale, and coal units were picked from well logs based on predetermined gamma and density cutoffs. The thickness column derived from the digital logs was multiplied by one of three values, with one value representing either sandstone (1.4), shale (3), or coal (5). The adjusted, or decompacted, thickness column and curves were then reloaded into Geoplus Petra as decompacted log data. The decompacted curves better illustrate continuity of coals, as well as local absence of coal owing to erosion by distributary or fluvial channels.

Detailed mapping of the coal-bearing intervals in Helper field indicate up to eight sequences of paludal deposits (Figure 4). Based on available age constraints, these sequences likely represent higher (4th or 5th) order climatically driven cyclicity (cf., Moiola, et al., 2000) superimposed onto a longer-term tectonic or eustatic signal. The basal Ferron unconformity (below the Clawson sandstone bed) is interpreted as a regressive surface of marine erosion (RSME of Plint and Nummedal, 2000) of Turonian age (Montgomery et al., 2001, and references therein). The contact is sharp and juxtaposes lower shoreface sandstones onto marine shales of the Tununk

543

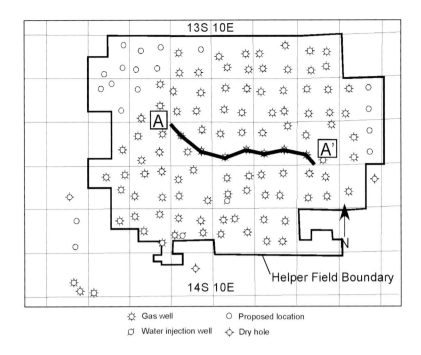

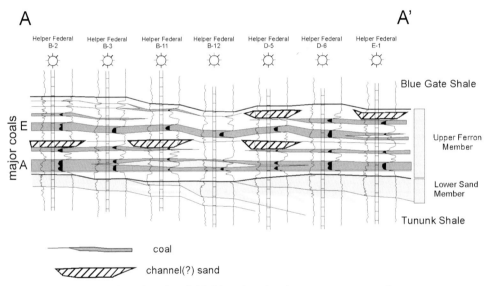

Figure 3. Decompacted well-log cross section through Helper field. Line of section shown on map at top. Squares on map represent congressional sections (1 mi^2 [1.6 km^2]). Cross section roughly parallels depositional dip. Datum on base of lowest coal.

Shale Member of the Mancos Shale. A minor transgression of the sea occurred during deposition of the Washboard sandstone bed (cf., White and Barton, 1997). Paleocurrent data indicate that both of the lower sandstones are derived from a northerly source (Ryer, 1981). A sequence boundary is interpreted above the Washboard sandstone, and commonly juxtaposes fluvial/deltaic sediments of the upper Ferron over shelf-margin sandstones and shales (White and Barton, 1997; Montgomery et al., 2001).

Within the coal-bearing upper Ferron Sandstone, transgressions and regressions are interpreted to be largely autocyclic. Using facies relationships and land-ward pinchout stacking patterns, Dewey (1997) identified parasequences (*sensu* Van Wagoner et al., 1990) within the Ferron that represent progradational packages separated by flooding surfaces (parasequence boundaries). Moiola et al. (2000) used biostratigraphy to identify a brackish water signature for the flooding surfaces in the upper Ferron, suggesting that autocyclic lobe switching is the dominant control on Ferron architecture. Detailed coal mapping in Helper field generally agrees with these interpretations. Isopach maps of Ferron coals (Figure 5) illustrate two distinct episodes of progradation, aggradation, and transgression that may be due to autocyclic processes.

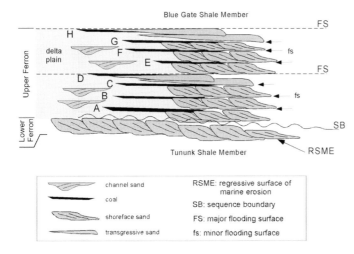

Figure 4. Schematic cross section, showing interpreted stratigraphic and facies relationships, and depositional cycles of the Ferron Sandstone. Letters A-H represent coals identified by Lupton (1916). Modified from Montgomery et al. (2001).

Thickness Trends

In general, Ferron coals in Helper field show thickness patterns that are similar to patterns mapped throughout the Ferron trend in central Utah (Montgomery et al., 2001). Helper wells commonly penetrate three to eight coalbeds over a stratigraphic interval of approx. 200–250 ft (60–75 m). A net coal isopach map for all coals in Helper field (Figure 6), coupled with the individual coal isopach maps (Figure 5), illustrates the following points.

1. The net coal interval ranges in thickness from 9–35 ft (3–10.5 m), with an average of 19.3 ft (6 m).
2. The thickest coals are in the north and west part of the field, thinning to the south and east. The southward thinning, coupled with increased sandstone thickness in that area, suggests the presence of a distributary channel system separating Helper field from Drunkards Wash field (Figure 1). At present, it is not known how far west the coals might extend; a drilling program is underway (2002–2003) to target this region covered by Anadarko acreage.
3. Local thinner coal intervals within the central part of Helper field are likely due to the presence of smaller distributary channels in those areas.

Helper Field General Field Data

Discovered in 1993, Helper field now encompasses 15,520 ac (6280 ha) and includes more than 116 producing wells on 160-ac (64-ha) spacing. Depth to production ranges from 1500–3600 ft (460–1095 m). Dip is to the north, away from the axis of the San Rafael uplift. Reservoirs include between four and eight coalbeds that range from 9–35 ft (3.0–10.5 m) in total net thickness and aver-

age 19.3 ft (6 m). The rank of these coals is high-volatile B bituminous, having an average gas content of 372 scf/ton (11.6 cm^3/g). Analysis of gas from productive wells indicate dry gas compositions, including 85–90% methane, 3–5% nitrogen, 3.5–4.5% carbon dioxide, ethane (1–6%), and minor amounts of propane. Heat content is 930 to 1115 Btu/cf.

GEOLOGIC EVALUATION AND STRUCTURAL IMPLICATIONS

For the more than 116 wells producing methane in Helper field as of 2003, there appears to be little correlation between well productivity and coal thickness. Montgomery et al. (2001) recognized a similar phenomenon in Drunkards Wash field (Figure 1). The more productive wells within Helper field occur in clusters (Figure 7), which contain wells of different vintages (i.e., slightly variable completion methods). The orientation of the clusters, which generally parallels the orientations of good producers in Drunkards Wash field (see Montgomery et al, 2001; Lamarre and Burns, 1997), suggests the influence of a structural grain or fabric on permeability. The general monoclinal northward dip of the Ferron interval at Helper field is well-constrained, and suggests that post-depositional folding and faulting are minimal. However, farther south in Drunkards Wash field, reverse faults that display up to 150 ft (45 m) of vertical displacement are common (Montgomery et al., 2001; Lamarre and Burns, 1997). Clawson Spring field, southwest of Drunkards Wash (Figure 1), is developed on the northern end of a plunging anticlinal structure. In both Drunkards Wash and Clawson Spring fields, well performance appears to be somewhat related to structure and/or local and regional stresses. Farther south in Buzzard's Bench field (Figure 1), Lyons (2001) recognized a similar relationship between faulting (observed with high-resolution seismic data) and increased production. Structural modeling of the reservoir interval should help determine if the stresses responsible for generating faults in the Ferron Sandstone might also generate fracture swarms that would increase permeability without offsetting the stratigraphy.

RESERVOIR MODELING

Reservoir simulation models were used in the pilot evaluation and have been modified throughout development of the field. Early models were simple, but became more complex with continued analysis of production data over time. The initial single layer model (net coal thickness), designed in 1996, estimated an initial water rate of 175 bbl of water per day per well. As more wells were drilled and performance analyzed, it became obvious that water production was exceeding predicted rates (Figure 8). In order to match performance, the single-

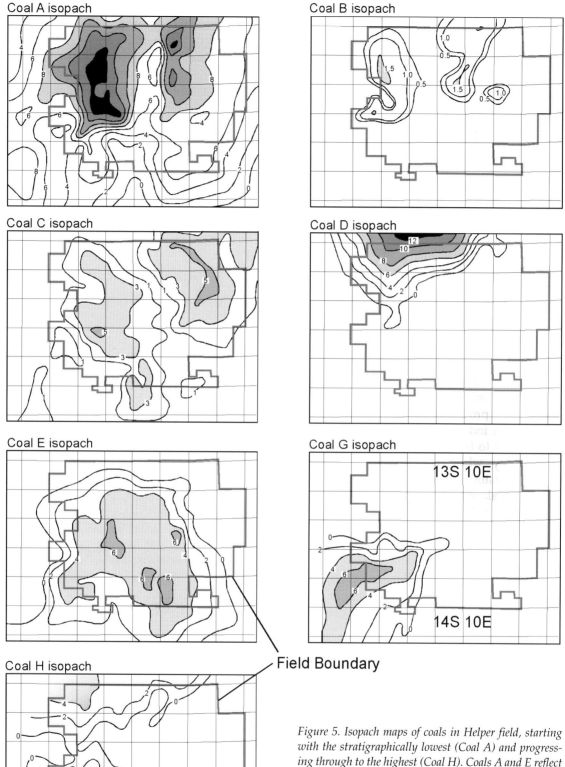

Field Boundary

Figure 5. Isopach maps of coals in Helper field, starting with the stratigraphically lowest (Coal A) and progressing through to the highest (Coal H). Coals A and E reflect progradation. Coals B, C, and G were deposited during aggradation. Coals D and H reflect transgression. Thicknesses shown are in feet. Contour interval is 2 ft, except for Coal B map where the interval is 0.5 ft. Shading indicates areas of thickest coal.

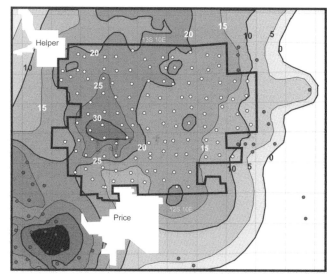

Figure 6. Net coal isopach map of Helper field. Thicknesses shown are in feet. Contour interval is 5 ft.

layer model was modified to include a second, water-bearing layer. The thickness and permeability values of the second layer were adjusted to more closely match actual water rates, within the constraints on sandstone thickness and permeability provided by wireline logs and conventional core analysis. The updated, two-layer model matches field production quite well, yet commonly fails to match individual well production. One variable that has yet to be modeled in Helper field is fracture permeability, which may help explain some anomalously high water rates. The current simulation model predicts a peak gas production rate for Helper field of 40 mmcf/day in late 2003 (Figure 9).

WELL COMPLETION PLAN

Drilling

Average drill depth to the top of the Ferron Sandstone in Helper field is 2800 ft (850 m), with TD's that range from 1800–4200 ft (550–1290 m). Typically, surface casing is 8.625 in. (22 cm), set to 300 ft (90 m) below surface. The interval from surface casing to the top of the Ferron is primarily Mancos Shale, which enables drilling operations to be conducted with air or air mist. A production string of 5.5-in. (14-cm) casing is set to total depth. Figure 10 illustrates a typical wellbore diagram.

Cementing the Ferron is designed to reduce slurry leak off into coal cleats and to avoid formation breakdown. Occasionally the holes were blown dry with air after logging but before running and cementing casing, in order to remove any hydrostatic head. This practice was dropped in 2002. Spud to rig release averages 58 hours, with 36 hours of rotating time.

Due to U.S. government restrictions regarding wildlife, oil and gas activity on federal lands in this field

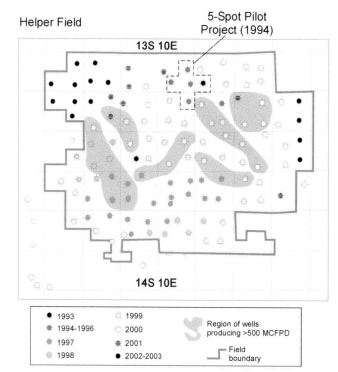

Figure 7. Map of Helper field illustrating well drilling and completion history. Gray regions represent clusters of wells that individually produce more than 500 mcfg/day.

A.

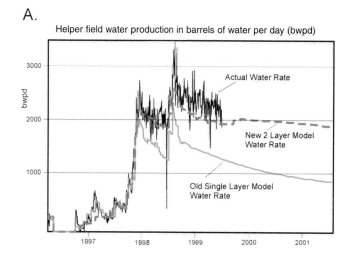

B.

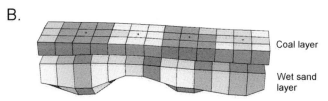

Figure 8. (A) Example of reservoir model results during early two-layer model run (gray dashed line). Plot is water production in bbl of water per day (bwpd). Significant improvement is seen over old single layer model (gray solid line). (B) Schematic example of reservoir model cell design. Cells represent about 50 ac (20 ha). Dots are well locations in model. Intensity of cell shading diagrammatically depicts a model parameter, such as permeability. More recent models incorporate exact location of wells relative to one another, rather than equal spacing pattern of older models.

A.

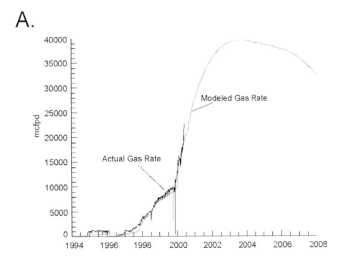

B.

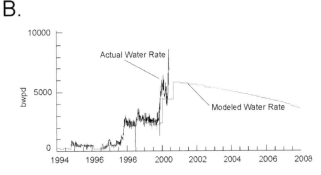

Figure 9. Reservoir model predictions for gas (A) and water (B) production at Helper field. Early predictions underestimated both the gas and water production.

is allowed only from July 30 to November 16. This allows for a 109-day window to perform any construction, drilling, and completion activity on federal lands.

Development

Development efforts at Helper field have attempted to maximize gas flow while minimizing cost and formation damage. Large perforations of 0.6 in. (1.5 cm) or larger are generally used to optimize stimulation. The majority of fracture stimulation jobs have been multi-staged and range in size from 163,000–365,000 lb (74,000–165,500 kg) of sand. A sand mesh of 16/30 is the standard used for proppant. Fracture stimulation fluid is pumped at a rate of 50 bbl per minute with fluid volumes averaging 250,000 gals (946,250 L). There was a wide variety in the quantity of material pumped between 1997 and 1999 over a large sampling of wells. The relative performance of each program seemed to be related to the average job size of the respective program, with higher-volume fracture stimulation programs generally leading to higher and faster production. Proppant volume plays a key role in the deliverability potential for these coalbed wells.

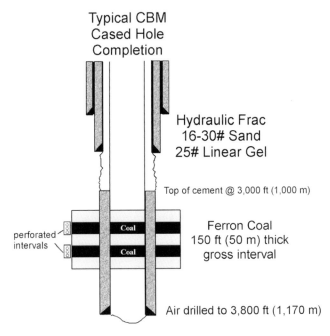

Figure 10. Typical well design for a CBM cased-hole operation.

Definitive results from the various fracture designs are very slow to achieve in CBM fields due to the lengthy dewatering phase. As a result, it takes time before we can fully evaluate whether a fracture stimulation design was effective. During this time lag, wells were fracture stimulated using as many as three stages. However, three-dimensional fracture models indicate that a single, large stage should be adequate to cover the entire Ferron interval. Of the wells that were recently fracture stimulated using the single, large stage method, most of them appear to be performing at the same levels as their offset neighbors. These results are encouraging, given that a single large fracture stimulation is more cost effective than two or more small ones.

Production

A critical part of any successful CBM project is the efficiency of operations to maximize the coal desorption process through the reduction of formation pressure. Key to overall pressure reduction of the field generally is water production, with a few exceptions. Wells in Helper field have a typical initial production of 200 bbl of water per day during first year of production, and decline to less than 100 bbl of water per day after 18 months. Concurrently, gas production increases (Figure 11) as effective dewatering continues.

Beam pumps are utilized throughout the field with occasional use of progressive cavity pumps. Fluid levels are generally kept at the perforated interval in order to reduce the hydrostatic pressure effects on the desorption process. Early in field development, produced water was trucked to existing commercial water disposal sites. As economic viability became apparent, injector wells

Helper Field Production

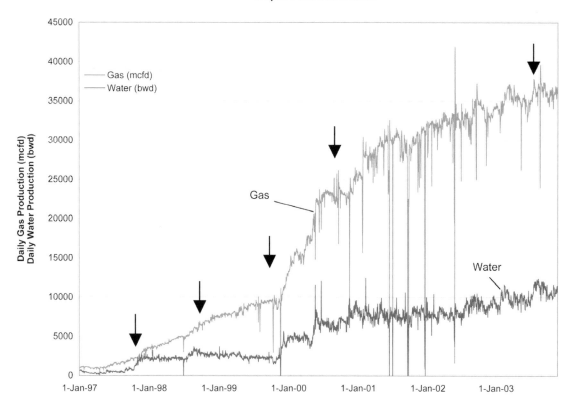

Figure 11. Historical production plot from Helper field showing gas and water rates from 1997 to 2001. Major increases in the rates are due to annual drilling programs (arrows).

were drilled to 7000 ft (2130 m) for injection into the Jurassic-age Navajo Sandstone. Helper field has two water disposal wells in the Navajo with a combined injection capacity of 15,000 bbl of water per day at 1250 psi (8619 kPa).

The gathering system consists of separate water and gas lines. Conventional gas separators are utilized at each well site. Gas is routed to a central production facility where it is compressed. Carbon dioxide (CO_2) is removed by amine, and moisture is removed by dehydration before sales. Water is sent to respective saltwater disposal sites for disposal. Due to topographic issues, tank and pump stations are situated throughout the field to assist in water disposal (Figure 7).

SUMMARY

Helper field produces coalbed gas from the upper Cretaceous Ferron Sandstone Member of the Mancos Shale. Production was initiated in 1994 following a pilot program that defined the play north of the existing Drunkards Wash field. Gas production rates continue to increase at Helper field, owing to a typical CBM negative decline curve during initial dewatering and continued field development.

Detailed geologic mapping of the field defined a thick (up to 35-ft [10-m]) net section of high-volatile B

bituminous coal distributed over as many as eight seams. The coals at Helper field have high gas contents (average 372 scf/ton [11.6 cm^3/g]) and are encountered at depths of less than 4700 ft (1430 m), making them economical. Sequence stratigraphic analysis suggests that the Ferron Sandstone at Helper field comprises two parasequences capped by flooding surfaces. Each parasequence contains progradational and aggradational components, creating an ideal environment for the deposition and preservation of coal-forming sediments.

Reservoir modeling methods have evolved over time to better match observed early production rates, especially on a field-wide basis. The current simulation model predicts a peak gas production rate for Helper field of 40 mmcf/day in late 2003, and production is expected to continue for at least 15 or 20 years.

Advances in completion (and recompletion) techniques continue to increase production at Helper field. The introduction of progressively higher fracture stimulation fluid volumes through single, large-staged jobs has reduced costs, and increased deliverability to the wellbore.

ACKNOWLEDGMENTS

The authors wish to thank Lloyd Stutz, Rob Dunleavy, and Mark Beiriger (APC) for their contributions to the manuscript. Reviews by Roger Bon, James Fouts, and David Tabet significantly improved the text.

REFERENCES CITED

Burns, T. D., and R. A. Lamarre, 1997, Drunkard's Wash project — coalbed methane production from Ferron coals in east-central Utah: Proceedings of the International Coalbed Methane Symposium, University of Alabama, Tuscaloosa, Paper 9709, p. 507–520.

Dewey, J. A., 1997, Facies analysis, sequence stratigraphy, and depositional history of a portion of the Ferron Sandstone, Indian Canyon, east-central Utah: M.S. thesis, Brigham Young University, Provo, 137 p.

Doelling, H. H., 1972, Central Utah coal fields — Sevier-Sanpete, Wasatch Plateau, Book Cliffs, and Emery: Utah Geological and Mineralogical Survey Monograph Series 3, p. 245–417.

Lamarre, R. A., and T. D. Burns, 1997, Drunkard's Wash Unit — coalbed methane production from Ferron coals in east-central Utah, in E. B. Coalson, J. C. Osmond, and E. T. Williams, eds., Innovative applications of petroleum technology in the Rocky Mountain area: Rocky Mountain Association of Geologists, p. 47–59.

Lamarre, R. A., T. Pratt, and T. D. Burns, 2001, Reservoir characterization study significantly increases coalbed methane reserves at Drunkard's Wash Unit, Carbon County, Utah (abs.): AAPG Annual Convention Official Program, v. 10, p. A11.

Lupton, C. T., 1916, Geology and coal resources of Castle Valley in Carbon, Emery, and Sevier Counties, Utah: U.S. Geological Survey Bulletin 628, 88 p.

Lyons, W. S., 2001, Seismic maps Ferron coalbed sweetspots: AAPG Explorer, v. 22, p. 32–37.

Mavor, M., and C. R. Nelson, 1997, Coalbed reservoir gas-in-place analysis: Gas Research Institute Report No. GRI-97/0263.

Moiola, R. J., J. E. Welton, J. B. Wagner, L. B. Fearn, M. E. Farrell, R. J. Enrico, and R. J. Echols, 2000, Integrated analysis of the Ferron deltaic complex, Utah (abs.): AAPG Annual Convention Official Program, v. 9, p. A100.

Montgomery, S. L., D. E. Tabet, and C. E. Barker, 2001, Upper Cretaceous Ferron Sandstone — major coalbed methane play in central Utah: AAPG Bulletin, v. 86, no. 2, p. 199–219.

Plint, A. G., and D. Nummedal, 2000, The falling stage systems tract: recognition and importance in sequence stratigraphic analysis, in D. Hunt and R. L. Gawthorpe, eds., Sedimentary responses to forced regressions: Geological Society London Special Publication 172, p. 1–17.

Ryer, T. A., 1981, Deltaic coals of Ferron Sandstone Member of Mancos Shale — predictive model for Cretaceous coal-bearing strata: AAPG Bulletin, v. 65, p. 2323–2340.

—1991, Stratigraphy, facies, and depositional history of the Ferron Sandstone near Emery, Utah, in T. C. Chidsey, Jr., ed., Geology of east-central Utah: Utah Geological Association Publication 19, p. 45–54.

Stalkup, F. I., and W. J. Eubanks, Jr., 1986, Permeability variation in a sandstone barrier island-tidal channel-tidal delta complex, Ferron Sandstone (Lower Cretaceous), central Utah: Society of Petroleum Engineers, SPE paper 15532, p. 1–13.

Tyler, N., M. D. Barton, and R. J. Finley, 1991, Outcrop characterization of flow unit and seal properties and geometries, Ferron Sandstone, Utah: Society of Petroleum Engineers, SPE paper 22670, p. 127–134.

Van Wagoner, J. C., R. M. Mitchum, K. M. Campion, and V. D. Rahmani, 1990, Siliciclastic sequence stratigraphy in well logs, cores, and outcrops: techniques for high-resolution correlation of time and facies: AAPG Bulletin Methods in Exploration Series, No. 7, 55 p.

White, C. D., and M. D. Barton, 1997, Translating outcrop data to flow models, with applications to the Ferron Sandstone: Society of Petroleum Engineers, SPE paper 38741, p. 233–248.

Analog for Fluvial-Deltaic Reservoir Modeling: Ferron Sandstone of Utah
AAPG Studies in Geology 50
T. C. Chidsey, Jr., R. D. Adams, and T. H. Morris, editors

Coalbed Methane Production from Ferron Coals at Buzzard Bench Field, Utah

Robert A. Lamarre[1]

ABSTRACT

Buzzard Bench field is one of three fields that produce coalbed methane from high volatile A bituminous coal within the Upper Cretaceous Ferron Sandstone Member of the Mancos Shale. The field is 1 mi (1.6 km) west of Orangeville, Emery County, Utah. It was discovered in 1994 and produced 10,384 mcfd and 6277 bwpd from 49 wells during July 2003.

The average depth to the coal is 3250 ft (991 m) and the average total coal thickness is 28 ft (8.5 m). Vitrinite reflectance and coal chemistry data indicate that the coals are high volatile A bituminous in rank with a weighted average ash content of 15.2%, on an as-received basis. The in-situ gas content is 282 scf/t (8.8 cm^3/g). The best wells are within a narrow fairway adjacent to north-south-trending normal faults. Apparent tectonic fracturing of the coals significantly increases the permeability of the coals and deliverability to the wellbore. Production plots for many of the wells show a typical negative decline curve with the gas rate increasing as the water rate decreases. The wells have not been producing long enough to reach their peak gas rate. All wells are produced through casing after perforating and hydraulic fracture stimulation with up to 120,000 lbs (54,420 kg) of 16/30 sand carried in borate gel.

INTRODUCTION

Buzzard Bench field is located 1 mi (1.6 km) west of the town of Orangeville, Emery County, Utah (Figure 1). The field is on the northwestern flank of the Laramide-age (Late Cretaceous-Eocene) San Rafael uplift, at the southwestern edge of the Uinta Basin. The coalbed methane discovery well at Buzzard Bench field, the Federal 'A' 26-2, (Figure 2) was drilled in Sec. 26, T18S, R7E by Texaco Exploration and Production, Inc. in late 1994.

REGIONAL GEOLOGIC SETTING

The productive coals are in the Ferron Sandstone Member of the Upper Cretaceous Mancos Shale. Coals were formed in peat swamps behind (to the northwest of) delta-front shoreline sandstones of the Vernal delta (Hale and Van DeGraff, 1964) that prograded eastward into the Cretaceous Interior Seaway during middle to late Turonian time (88.8 to 91.3 Ma) (Ryer, 1991).

The Ferron Sandstone is exposed in a narrow outcrop belt east of Buzzard Bench field on the northwest-

[1]Lamarre Geological Enterprises, LLC, Denver, Colorado

dipping flank of the San Rafael uplift. Coals are found in outcrop east of the town of Emery, approximately 22 mi (35.4 km) south of the field. East of Buzzard Bench field, the Ferron outcrop contains delta-front sandstones and pro-delta siltstones and shales, but no coals (Figure 1).

The Ferron coalbed methane trend is approximately 80 mi (129 km) long and 10 mi (16 km) wide (Figure 1). The trend is limited on the north and west by deep drilling depths (greater than 6000 ft [1829 m]) beneath the Book Cliffs and Wasatch Plateau, respectively. The trend is limited on the east by the exposure of the coals at the surface or the pinch-out of coals into marine shales. To the south, the gas content of the coals decreases (Lamarre, this volume) and the coals are overlain by volcanics of the Fish Lake Plateau.

The first coalbed methane production in the Ferron trend was established at Drunkard's Wash field, 10 mi (16 km) to the north of Buzzard Bench field, by River Gas Corp. in early 1993 (Figure 1). Drunkard's Wash field is the largest coalbed methane producer in Utah, with 477 wells producing 186 mmcfd from Ferron coals during July 2003. Additional exploratory drilling along trend to the south from 1994 to 1996 tested the coals in nine core holes (Lamarre, this volume). Coalbed methane production at Buzzard Bench field resulted from that exploratory program.

STRUCTURE

In the Buzzard Bench field area, the Ferron Sandstone Member dips to the west at approximately 100

ft/mi (20 m/km) (Figure 2). A major north-south trending normal fault system has been mapped near the center of the field. This fault system is parallel to the Wasatch Plateau to the west. It was identified by interpretation of two-dimensional seismic data and mapping of subsurface well log data. This fault system has been mapped on the surface (Doelling, 1972) to the north of Buzzard Bench field, but is not visible in the easily weathered shales of the Blue Gate Shale Member that are present on the surface in the field area. Interpretation of dipole sonic logs from two wells indicates that the present day maximum horizontal stress direction is almost north-south, parallel to the normal faults. This fault system is probably related to the north-south grabens formed by the Joe's Valley-Pleasant Valley faults, located in the Wasatch Plateau to the west. These formed as a result of early Tertiary extensional tectonism caused by Basin and Range deformation (Hunt, 1982). Unlike the Drunkard's Wash Unit (Burns and Lamarre, 1997) and Helper field (Ruhl and Buck, this volume), structure appears to have a significant impact on coalbed methane production at Buzzard Bench field. Most of the highly productive wells are located adjacent to the major north-south faults.

COAL RESERVOIR CHARACTERISTICS

The average depth to the Ferron coals within Buzzard Bench field is 3250 ft (991 m). The shallowest coals

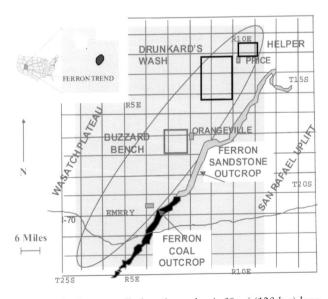

Figure 1. The Ferron coalbed methane play is 80 mi (128 km) long, extending from north of Price to south of Interstate 70. The three coalbed methane fields (Helper, Drunkard's Wash and Buzzard Bench) are located in the northern half of the trend. The black shaded portion of the Ferron outcrop represents the area where the coals are exposed. In the gray shaded portion, coals are present only in the subsurface.

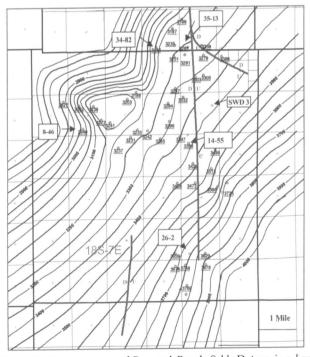

Figure 2. Structure map of Buzzard Bench field. Datum is a bentonite marker in marine shales approximately 100 ft (300 m) above the top of the Ferron Sandstone. All values shown are above sea level. Contour interval is 50 ft (15 m). Named wells are mentioned in the text.

are found at depths of 2845 ft (867 m) on the east side of the field near the San Rafael uplift. Coals plunge to the west under the Wasatch Plateau, with the deepest coals found at a depth of 4100 ft (1250 m).

The initial reservoir pressure at the time of discovery was 1258 psi at a depth of 3128 ft (953 m). The reservoir is underpressured with a simple pressure gradient of 0.40 psi/ft. Reservoir temperature is 93°F (34°C) at 3150 ft (960 m).

The average total coal thickness in the Ferron Sandstone is 28 ft (8.5 m), based on a bulk density cutoff of 1.75 g/cm^3. This density cutoff includes all intervals with more than 50% by weight carbonaceous material, which is the commonly accepted definition of coal (Close, 1993). Coal thickness is measured on high-resolution bulk-density logs with a vertical resolution of 6 in (15.2 cm). A density of 1.75 g/cm^3 is used as a cutoff for identifying coals for mapping purposes. This density cutoff value results in an average coal density of 1.31 g/cm^3.

The total coal thickness ranges from 5 to 36 ft (1.5–11.0 m) as shown on Figure 3. The thickest coals trend northeast-southwest, parallel to the paleo-shoreline. Coals pinch-out into marine shales and shoreline sandstones to the east. Coals are found in three to five seams over a stratigraphic interval of 150 to 200 ft (46–61 m). Figure 4 is a west-east-oriented stratigraphic cross section across Buzzard Bench field. This cross section shows the continuity of the coals and thickening of delta-front sandstones to the east. The datum is a tonstein (altered volcanic ash layer) within the upper coal in the Ferron Sandstone Member. Figure 5 shows the wireline logs of the Utah Power and Light (UP&L) 14-55 well with a 226-ft (69 m)-thick Ferron interval containing 30 ft (9 m) of coal.

Although we have no measured permeability values for the coals, visual examination of cores suggests that reservoir permeability is low due to a poorly developed cleat system. Some coal cores contained very few cleats while others showed well-developed face cleats with a few butt cleats. However, highly productive wells located adjacent to faults provide evidence of fracture-enhanced permeability. The UP&L 14-55 well in Sec. 14, T18S, R7E (Figure 2) is 0.5 mi (0.8 km) west of a fault and it produced 5577 mcfd and 79 bwpd during September 2003. These rates are unusually high for Ferron coals (Lamarre and Burns, 1997). Additional data suggesting fracture-enhanced permeability resulted from shutting-

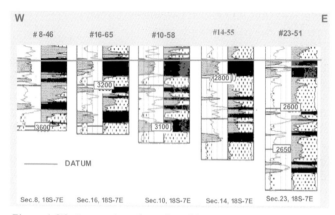

Figure 4. West-east oriented stratigraphic cross section showing continuity of Ferron coals across Buzzard Bench field. Coals are shown in black and sandstones are stippled. Datum is a tonstein (altered volcanic ash layer) near the top of the upper coal seam. The Ferron interval thickens to the east as delta-front sandstones become better developed. Line of section is shown on Figure 3.

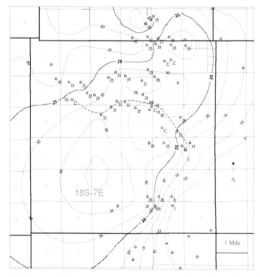

Figure 3. Total coal isopach map showing northeast-southwest trend and thinning of the Ferron coals to the east. Contour interval is 5 ft (1.5 m). Coal thickness values are underlined and were determined using a bulk density cutoff of 1.75 g/cm^3. Cross section (west-east) is shown on Figure 4.

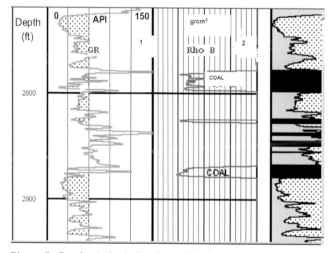

Figure 5. Geophysical wireline logs of Utah Power and Light 14-55 well in Sec. 14, T18S, R7E, in Buzzard Bench field (location shown on Figure 2). Curves shown are gamma ray and bulk density, with an interpreted lithology curve in the track on the far right. Coal is shaded black, sandstone is represented by coarse stippling, and shale is depicted by solid gray. The well contains 30 ft (9 m) of coal, (using a bulk density cutoff of 1.75 g/cm^3) which is shown in black. The well has produced 5.0 bcf of gas and it produced 5577 mcfd and 79 bwpd during September 2003.

in the Utah Federal 'D' 35-13 well, in Sec. 13, T18S, R7E, (Figure 2), 2 mi (3.2 km) north of the 14-55 well. When the 35-13 well is shut-in, gas production decreases in the 14-55 well. This suggests that the 35-13 well is helping to de-water the coals 2 mi (3.2 km) to the south. The gas rate in the 14-55 well declines immediately after shutting-in the 35-13 well, suggesting open fracture connectivity between the two wells.

Whole core from the Utah Federal 'S' 8-46 well, in Sec. 8, T18 S, R7E (Figure 2), showed very few natural cleats. Producing rates from 15 wells that are not close to the faults are very low (<100 bwpd and <50 mcfd per well). Typically, wells that produce less than 100 bwpd cannot effectively remove sufficient water to lower the reservoir pressure and allow large volumes of gas to desorb from the coals at high rates.

Coal chemistry and rank data were determined from proximate analysis and vitrinite reflectance measurements on whole core samples of coal from four wells and sidewall core samples from two wells. Representative coal chemistry data for the Utah Federal 34-82 well (Figure 2) are shown in Table 1. The coal rank is high volatile A bituminous based on a mean maximum reflectance value of 0.72% Ro. Proximate analysis indicate a field-wide weighted average volatile matter content of 44.8%,

on a dry, ash-free basis. The weighted average ash content of the coals is 15.2% (dry) and the average fixed carbon content is 55.2% (dry, ash-free). The gross calorific content or heating value is 15,058 Btu/ft^3 (moist, mineral matter-free). The proximate analysis characterize the coals as high volatile A bituminous (see Table 3 in Lamarre, this volume).

Analysis of the maceral composition of the Ferron coals indicates that they are humic coals with vitrinite being the most common maceral group. Table 2 compares the coal petrography for Ferron coals in two wells (locations shown on Figure 2). Within Buzzard Bench field, the average maceral content is 68% vitrinite, 13% semifusinite, and 12% fusinite, on a mineral-matter-free basis. The average field-wide mineral matter composition consists of 84% clay, 9% quartz, 3% carbonate, and 3% sulfide, on a maceral-free basis. Regression analysis of bulk density data indicates that the clays have an apparent dry density of 2.68 g/cm^3.

GAS AND WATER ANALYSIS

The methane content of produced gas ranges from 71.85% to 96.68% (88.0% average). The gas also contains 0.40% to 2.34% nitrogen (0.89% average) and 0.86% to

Table 1. Chemistry of coals in Utah Federal 34-82 well in Sec. 34, T18S, R7E (location shown on Figure 2). Data are from a composite sample of 14 1-ft intervals between 3825 and 3839 ft (1148–1152 m). Based on ASTM rank classification, the coals are high volatile A bituminous.

Utah Federal # 34-82	As-Received	Dry	Dry Ash-Free	Moist	Moist Mineral-Matter-Free	Dry, Mineral-Matter-Free
Gross Calorific Value (Btu/lb.)	11,826	11,972	14,932	11,864	15,027	
Moisture Holding Capacity (90°F)^2				0.90		
	weight %	weight %	weight %	weight %	weight %	weight %
Proximate:						
Moisture	1.22					
Ash	19.58	19.82		19.64		
Volatile Matter	31.67	32.06	39.99	31.77		38.49
Fixed Carbon	47.53	48.12	60.01	47.68		61.51
Total	100.00	100.00	100.00	100.00		100.00
Ultimate:						
Moisture	1.22					
Hydrogen	4.68	4.74	5.91	4.70		
Carbon	67.64	68.48	85.40	67.86		
Nitrogen	0.81	0.82	1.02	0.81		
Sulfur	0.67	0.68	0.85	0.67		
Oxygen	5.40	5.47	6.82	5.42		
Ash	19.58	19.82		19.64		
Total	100.00	100.00	100.00	100.00		
Apparent ASTM Rank Classification	(dry, mineral-matter-free)		**High Volatile A Bituminous**			

Table 2. Petrography of Ferron coals in Utah Federal 34-82 (Sec. 34, T18S, R7E) and SWD 3 (Sec. 11, T18S, R7E). Locations are shown on Figure 2.

Well Name & Number	Utah Federal #34-82	SWD #3
Drill Depth, ft	3825-3839 (14 of 30 feet).	3064-3080 (7 of 16 ft).
Sample Type	Upper Ferron Composite	Basal Ferron Composite
Maceral Composition, Volume % (mineral-matter-free)		
Vitrinite	72.1	79.8
Pseudovitrinite	1.8	3.9
Fusinite	8.8	3.7
Semifusinite	11.3	8.1
Micrinite	0.6	0.2
Macrinite	0.3	0.2
Exinite	2.5	3.1
Resinite	2.6	1.0
Total	100.0	100.0
Mineral Composition, Volume % (maceral-free)		
Clay	90.2	92.2
Quartz	1.6	3.5
Carbonate	7.4	1.4
Sulfide	0.8	2.9
Total	100.0	100.0
Vitrinite Reflectance (% in oil)		
Mean Maximum Vitrinite Reflectance	0.77	0.71
Standard Deviation	0.03	0.06
Mean Random Vitrinite Reflectance	0.74	0.68
Standard Deviation	0.03	0.06

40.17% carbon dioxide (10.16% average). The heating value ranges from less than 800 Btu/ft^3 to 980 Btu/ft^3. In order to meet pipeline gas quality specifications, all gas is processed in the field by an amine unit to remove CO_2. The resulting gas that is sold contains 96.68% methane, 1.15% ethane, 0.77% nitrogen and 1.004% carbon dioxide, with a heating value of 1008 Btu /ft^3.

The average total dissolved solids content of the produced water is 13,986 ppm. The water is sodium bicarbonate-rich, with 7549 ppm HCO_3 and 2316 ppm Cl.

Most of the methane produced from coals is thermogenic in origin, having been produced as a by-product of the coalification process (Rice, 1993a, b). Some basins, such as the Powder River Basin in northeastern Wyoming, produce secondary biogenic methane resulting from bacterial activity in the coals (Rice, 1993b). Carbon and oxygen isotope data from Ferron gases at Buzzard Bench suggest that these gases may be a combination of thermogenic and biogenic origin (Lamarre and Burns, 1977; Collett, 1999). Geologically young (28,000 to

31,000 years bp) meteoric water from the Wasatch Plateau may have migrated downward along faults into the Ferron coals, and then updip into the field area (Lines and Morrissey, 1983). Bacteria that are carried in the water metabolize the organic compounds in the coal, producing secondary biogenic gas that is trapped in the coals (Lamarre, this volume).

GAS CONTENT

Total gas content of coals consists of desorbed, lost, and residual gas. The gas content in Ferron coals was measured from whole core from four wells and sidewall cores from three wells. All cores were placed in desorption canisters and desorbed at reservoir temperatures of approximately 85 to 95°F (30–35°C). Lost gas was calculated by the U.S. Bureau of Mines Direct Method (Diamond et al., 1986). The average residual gas content is 27.7 scf/t (0.86 cm^3/g), which is 9.8% of the total gas content. Lost gas was calculated to be 24 scf/t (0.75

cm^3/g), and the total average *in-situ* gas content is 282 scf/t (8.8 cm^3/g). The average total gas content, on an ash-free basis, is 474 scf/t (14.8 cm^3/g).

ADSORPTION ISOTHERM

An adsorption isotherm indicates the maximum amount of gas that a coal can contain at a given temperature, pressure, and equilibrium moisture content. Figure 6 is the dry, ash-free adsorption isotherm for a composite sample from 3571 to 3577.5 ft (1088–1090 m) in the Utah Federal 'S' 8-46 well in Sec. 8, T18S, R7E. Two adsorption isotherms, one for 100% methane and one for 100% carbon dioxide, were measured on the sample. Testing was done at a temperature of 92°F (34°C) and equilibrium moisture content of 1.8%. The average ash content of the sample was 12.4 wt% and the average grain density was 1.345 g/cm3. The curve shown is for a gas composition of 90% CH$_4$ and 10% CO$_2$, which is a good approximation of the average produced gas composition. This curve was modeled from the pure methane and carbon dioxide curves.

The measured total gas content, on an ash-free basis, is 276 scf/t (8.6 cm^3/g) for this well and is also plotted on the chart. Since this desorption point plots below the isotherm for the reservoir pressure, the coals in this well are undersaturated and contain less gas than they are capable of storing. These samples are 59% saturated compared to a fully saturated value of 465 scf/t (14.5 cm^3/g). This implies that the reservoir pressure needs to be reduced to 338 psi before these coals will begin desorbing gas.

The average depth to the top of the coal in the field is 3250 ft (991 m), with an average pressure of 1300 psi, assuming a pressure gradient of 0.4 psi/ft. The average ash-free gas content in the field of 474 scf/t (14.8 cm^3/g) is plotted on the chart with a triangle. This value plots on the adsorption isotherm curve, indicating that the average coal in the field is fully saturated. The apparent undersaturation in the 'S' 8-46 well may be due to incorrect desorption measurements or local de-gassing of the coals at that location.

DRILLING AND COMPLETION

Twelve and one-quarter in (31 cm) holes are drilled to approximately 300 ft (91 m) and 8 ⅞ in (21.9 cm) surface casing is cemented in place. To minimize damage to the coal reservoir, a 7 ⅞ in (20 cm) hole is drilled with air or an air-mist mixture to total depth. Penetration rates typically exceed 110 ft/hr (34 m/hr) using 2100 cfm of air. Wells are drilled at least 200 ft (61 m) deeper than the bottom coal to provide sufficient rathole for logging and collection of coal fines during production. The Ferron coals are evaluated with an open-hole log suite consisting of gamma ray, neutron/density porosity, sponta-

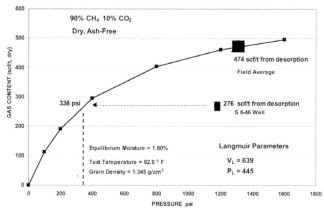

Figure 6. Adsorption isotherm plot from a composite sample in Utah Federal 'S' 8-46 well in Sec. 8, T18S, R7E (location shown on Figure 2). Curve shown is modeled for 90% CH$_4$ and 10% CO$_2$, on an ash-free basis. The oval represents the average measured desorption value for this well of 276 scf/t (8.6 cm^3/g). The triangle represents the average field desorption value of 474 scf/t (14.8 cm^3/g).

neous potential, and resistivity tools. Five and one-half in (14 cm) casing is cemented in place with 175 sacks (50 bbls) of thixotropic cement.

Open-hole cavity completions were attempted in three wells. In all cases, the cavitation process was unsuccessful with very little coal returning to the surface. A sonic tool was run in one well and it measured a cavity of only approximately 20 in (51 cm) in diameter. Production from these wells was no better than in the cased wells. Two of these wells have subsequently been fracture stimulated in the open hole interval. Approximately six months after the fracture treatment, these wells began to show an increase in gas and water production.

Coals are perforated with 0.88-in (2.2 cm) diameter holes, six shots per foot, with 60-degree phasing. The formation is broken down with 300 gals of 7½% HCL. Coals are hydraulically fractured with 80,000 to 120,000 lbs (36,280–54,420 kg) of 16/30 sand carried in 1200 to 1600 bbls of YF125 gel. Each well requires two or three frac stages. Two and seven-eighths in (7.3 cm) tubing is placed in each well and the wells are pumped with 2 ⅞ in (7.3 cm) rods and a 2 in (5 cm) pump. Wells with high permeability that produce more than 500 bwpd are pumped with electric submersible pumps capable of handling more than 5000 bwpd.

Gas is treated with an amine unit to lower the CO$_2$ content to pipeline specifications of less than 3% total inert gases. Produced water is injected into high- porosity (15–20%) eolian sandstones of the Jurassic Navajo Sandstone in two wells at depths of 6500 to 7200 ft (1981–2195 m).

PRODUCTION

Buzzard Bench field is being developed with one well per 160-acre (64 ha) spacing unit. Figure 7 is a field production plot of daily gas and water since sales began. Buzzard Bench field is a typical coalbed methane field with the gas rate increasing as water is removed from the reservoir. The gas rate increased significantly after November of 2000 when water-bearing sandstones were cement-squeezed. The decrease in gas production from August through November 2001 was due to high gas pipeline pressures. The field produced 10,384 mcfd and 6277 bwpd from 49 wells in July 2003. Cumulative production at that time was 13.1 bcf.

Figure 8 is a production plot for the UP&L 14-55 well. This plot shows a typical "negative decline curve" with the gas rate increasing as water is produced from the reservoir. Water is produced through the tubing with a rod pump. In June 1998, the well started flowing and the water rate increased to approximately 3500 bwpd. The water rate reached a peak of 7431 bwpd in August 1999 and has been declining since then. Water production has been choked back significantly during August and September 2001 due to water disposal problems. Note that the reduction in the water production rate results in a reduction of the gas rate as well. This is the most prolific well in the field and it produced 5577 mcfd and 79 bwpd during September 2003. By that time, it had produced 5.0 bcf since completion on August 27, 1997.

CONCLUSIONS

Coalbed methane production at Buzzard Bench field is increasing as the Ferron coals are de-watered. Unlike the Ferron coals at Drunkard's Wash, the coals at Buzzard Bench appear to be poorly cleated. Coalbed methane producibility is controlled by fracture permeability associated with normal faults that are parallel to the Wasatch Plateau. Gas content and coal thickness are sufficient to provide adequate reserves to justify additional drilling. However, locations must be carefully selected because natural fractures are required for commercially viable daily producing rates.

ACKNOWLEDGMENTS

I thank ChevronTexaco (formerly Texaco Exploration and Production, Inc.) for their support and permission to publish this data. I also acknowledge my teammates at Texaco who have worked with me on this project for many years. I thank Charles W. Byrer and William F. Haslebacher (U.S. Department of Energy), Paul Baclawski (Devon Energy), and Thomas C. Chidsey, Jr. (Utah Geological Survey) for their review and recommendations for improvement of the manuscript. I give special thanks to Bill Nagel of Pason Systems USA

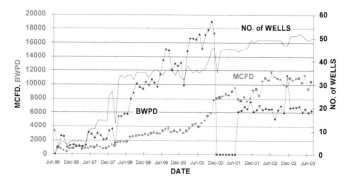

Figure 7. Daily production chart for Buzzard Bench field. Gas production increases as water is produced from the coals. The field produced 10,384 mcfd and 6277 bwpd from 49 wells during July 2003. Note the large decrease in water production and increase in the gas rate after November 2000 when water-bearing sandstones were cement-squeezed.

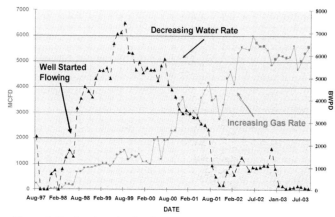

Figure 8. Production plot for the UP&L 14-55 well in Sec. 14, T18S, R7E (location shown on Figure 2). This well shows a typical negative decline curve with the gas rate increasing as gas desorbs from the coal matrix. Desorption results from the reduction of reservoir pressure as water is produced from the coals. Note the increase in gas and water rates in July 1998 when the well started flowing.

Corp. and Tim Pratt and MaryAnna Hatch of TICORA Geosciences, Inc. for their field and laboratory work that provided the data for this study. TerraTek, Inc. provided the isotherm data and Langmuir modeling.

REFERENCES CITED

American Society for Testing Materials (ASTM), 1983, Standard classification of coals by rank; ASTM designation D388-82, in Gaseous fuels, coal and coke: 1983 Book of Standards, v. 0505 ASTM, Philadelphia, 5 p.

Burns T. D. and R. A. Lamarre, 1997, Drunkard's Wash project-coalbed methane production from Ferron coals in east-central Utah: Proceedings of the 1997 International Coalbed Methane Symposium, paper 9709, p. 507–520.

Close, J. C., 1993, Natural fractures in coal, *in* B. E. Law and D. D. Rice, eds., Hydrocarbons From Coal: AAPG Studies in Geology 38, p. 119–132.

Collett, T. S., 1999, Composition and source of the gas associated with coalbed gas production from the Ferron coals in eastern Utah (abs.): Geological Society of America Annual Meeting, Denver, Colorado.

Diamond, W. P., J. C. LaScola, and D. M. Hyman, 1986, Results of direct-method determination of the gas content of U.S. coalbeds: U.S. Bureau of Mines Information Circular 9067, 95 p.

Doelling, H. H., 1972, Central Utah coal fields-Sevier-Sanpete, Wasatch Plateau, Book Cliffs and Emery: Utah Geological and Mineralogical Survey, Monograph Series no. 3, p. 245–417.

Hale, L. A., and F. R. Van De Graaff, 1964, Cretaceous stratigraphy and facies patterns-northeastern Utah and adjacent areas: Intermountain Association of Petroleum Geologists, 13th Annual Field Conference Guidebook, p. 115–138.

Hunt, C. B., 1982, The anomalous transverse canyons of the Wasatch Range, *in* D. L. Nielson, ed., Overthrust belt of Utah, Utah Geological Association Publication 10, p. 81–89.

Lamarre, R. A., and T. D. Burns, 1997, Drunkard's Wash Unit-coalbed methane production from Ferron coals in east-central Utah, *in* E. B. Coalson, J. C. Osmond, and E. T. Williams, eds., Innovative applications of petroleum technology in the Rocky Mountain area: Rocky Mountain Association of Geologists guidebook, p. 47–59.

Lines, G. C., and D. J. Morrissey, 1983, Hydrology of the Ferron Sandstone aquifer and effects of proposed surface-coal mining in Castle Valley, Utah: U.S. Geological Survey Water-Supply Paper 2195, 40 p.

Rice, D. D., 1993a, Controls of coalbed gas composition: 1993 International Coalbed Methane Symposium Proceedings, p. 577–589.

—1993b, Composition and origins of coalbed gas, *in* B. E. Law and D. D. Rice, eds., Hydrocarbons from coal: AAPG Studies in Geology, no. 38, p. 159–185.

Ryer, T. A., 1991, Origin of mid-Cretaceous clastic wedges, central Rocky Mountain region, (extended abstract), *in* J. Dolson, compiler, Unconformity related hydrocarbon exploitation and accumulation in clastic and carbonate settings: Rocky Mountain Association of Geologists Course Notes, p. 16–20.

Index

A

accommodation: 76, 83–84, 89, 98–100, 105, 114, 120, 194–196, 203, 206, 222, 301, 379, 399, 496, 510
  accommodation space: 60, 80, 83, 96, 98, 128, 181–183, 217, 465, 470, 471, 479, 485, 488, 489, 496, 510
  accommodation-to-sediment supply: 96
accretion: 111, 155, 435, 445, 483–484, 488, 497
  downstream: 103, 290, 310, 312, 385
  lateral: 47, 73– 76, 88, 143, 310, 312, 313, 314, 385, 477, 478, 482, 484, 485, 497
  upstream: 290, 441
  vertical: 102, 485, 488
adsorption:
  capacity: 516, 518
  isotherm: 518, 556
aggradational: 43, 51, 76, 100, 111, 126, 128, 131, 142, 152, 158, 165, 172–173, 175, 182–183, 201, 203–204, 206, 209, 384, 401, 429, 455, 456, 463, 470, 471, 473, 474, 485, 489, 493, 496, 507, 510, 530, 549
  parasequence set: 133, 157, 162, 221, 223, 452, 465–471, 475, 479, 489–493, 496
  unit: 100, 103–105, 109, 114
  vertical: 470, 479, 485, 488
allocyclic, allocyclic processes: 22, 79–80, 83, 306, 315–316, 326–328
alluvial plain: 12, 26, 74–75, 186, 188, 495
amalgamated braided fluvial deposits: 161, 165–166
ammonite zonation, Inoceramid: 97, 132, 171, 175–176
analytical petrography: 409–410
anisotropy: 230, 244, 246, 362, 368–369, 373–374, 376–379, 416–417, 423
architecture:
  internal: 79–80, 89, 182–183, 189, 194, 307, 328, 384, 388, 396, 398, 400, 452, 458, 473, 475–476, 481–482, 485–486, 490–491, 497, 506
  stratal: 105, 223, 360
arithmetic: 254, 257, 361, 365, 369–372, 406, 414–415, 417–418, 420–423
aspect ratios: 452, 465, 473–474, 479, 489, 496
autocyclic, autocyclic processes: 22, 79–80, 82–83, 85, 89, 128, 217, 306, 315–317, 328, 544

channel avulsion: 82, 89, 494
  lobe switching: 83, 85, 89, 306, 315–316, 544
avulsion: 43, 82, 282, 315, 328, 341, 399, 453, 469, 479, 485, 489–490, 494, 497
  channel: 82, 89, 494
  local: 470, 479
  regional: 177, 179, 183, 489–490, 507

B

baffle: 355, 380
bankfull:
  channel: 45–46, 441
  depth: 46–47, 443
  width: 443, 445
barrier: 13, 43, 48, 55, 69, 101, 164, 185, 206, 320, 355, 530–531, 543
barrier shorelines: 101
base-level rise and fall: 23, 99
bay:
  bay-fill: 102, 109, 120, 280, 311–313, 327, 340, 342, 346, 353
  bay-head delta: 45, 152, 176, 183, 353, 491
  brackish-bay deposits: 151, 153
  brackish-water bay: 146, 345–346, 478
bedform-phase diagram: 40
bedforms: 40–41, 47, 71, 96, 105–106, 111, 114, 116, 120, 263, 281, 290, 312–313, 325, 336, 476–478, 481, 483, 486
bedset: 51, 158–159, 213, 215, 217, 219–220, 222–223, 230, 252, 263–267, 270, 273, 312, 332–337, 340, 344, 346, 348–349, 351–352, 362, 372, 374–375, 430, 435–438, 440, 443–445, 460–461, 483–484, 487, 491, 496
bifurcation: 43–44, 48, 55, 322, 473–474, 476, 479, 495–496
biostratigraphy analysis:
  foraminiferal: 85, 87
  palynofacies: 85–88
  palynological: 86–87
biozones: 97, 132, 175–176
Blue Gate Shale: 4–7, 27, 62, 76, 80, 85–87, 103, 127, 175, 179, 194, 197, 216, 252, 456, 505, 523, 552
bounding layer: 265, 335–340, 360–362, 367–371

bounding surface: 117, 142, 150–151, 170, 217, 263, 336–337, 355, 412, 436–437, 444
braided stream: 151–152, 159, 176, 183, 490–491
breakthrough: 367–368, 371, 374–376, 379, 423
burial: 74, 84, 228, 241, 245, 294, 502, 505, 513–514, 516
burial history: 228, 505, 513, 516

C

Calico bed: 20
capillary pressure: 367, 369, 371, 373–374, 378, 416
carbonate: 84, 233–236, 238–239, 241–242, 244–245, 554
Castle Valley: 4–14, 16, 18, 20–21, 23, 26, 28–29, 31–32, 60, 62–63, 79–80, 91, 104, 126–128, 132, 172–173, 175–177, 209, 456, 542
Castle Valley bar: 10–12, 16, 20
cement: 84, 230, 233–236, 242, 245, 302
cementation: 228, 235, 238–239, 242, 246, 254
channel:
    branching patterns: 43–44
    Caprock Channel: 485–489
    County Line Channel: 476–479, 482, 489, 494–495, 496
    depth: 45–47, 82, 441, 493, 495
    geometry: 55, 485
    Kokopelli Channel: 479, 481–482, 484–485, 489, 496
    multi-storied: 143, 147, 158, 452, 475, 482, 484, 491, 496
    multilateral: 322, 324, 327
    single-storied: 24, 76
channelbelt: 24–26, 29, 46, 74–76, 82, 88–89, 96, 101–102, 105, 111, 113–115, 117–118, 120, 215, 452–453, 472, 474–479, 482–497
channelform sandstone bodies: 101–102
characterization: 26, 40, 48, 252, 333, 365, 406
    outcrop: 189, 223, 246
    reservoir: 33, 55, 246, 273, 355, 380, 542
clastic wedge: 40–41, 60, 100, 119, 127–129, 133, 141, 157, 181–183, 188, 194, 197, 203, 280, 282, 384, 429, 444, 457, 468–469, 471, 473, 475, 496, 506, 530, 543
Clawson sandstone unit, Clawson unit: 12, 63, 128
clay content: 235, 239, 244, 413, 434, 445
clays injected: 300
clinoform: 53, 55, 108, 231, 234, 252, 263–268, 270, 273, 282, 335–341, 343–346, 348, 350–351, 355, 360–374, 377–380
clinoform facies: 263–268, 270, 273, 337–338, 343–344, 355, 361, 365–369, 371–373
    clinoform cap: 263, 336, 344–346, 351, 372, 374
    clinoform distal: 231, 263, 266, 336, 338–339, 341, 344–345
    clinoform medial: 231, 267, 336, 338–339, 344–345
    clinoform proximal: 231, 263, 265–267, 336, 338–339, 344–346, 348, 351
    model: 252, 270–273
coal: 4–5, 7, 9, 12–14, 17, 25, 28–33, 60, 62, 67–68, 72–75, 82, 84, 88, 100–103, 109, 111, 116, 121, 126, 129–130, 132–133, 142–144, 146, 148, 150–151, 153

ash content: 532–533, 554, 556
Btu values: 522, 533, 545, 554–555
burial history: 505, 513, 516
cleating: 513, 522
compaction: 507
field: 28–29, 33, 130
quality: 29, 516, 532, 535
gas content: 33, 502, 510, 516, 518–519, 521, 532–533, 535–537, 542, 545, 552, 555–556, 557
maceral: 516, 554
mine: 5, 28, 113, 146, 152–154, 167, 171, 176, 513, 515
rank: 510, 513, 516–518, 521, 525, 533–535, 537, 545, 554
vitrinite reflectance: 84, 228, 510, 512, 513, 516, 518, 533, 534, 535, 554
zone: 28, 68, 72, 74–75, 82, 126, 129–130, 142–144, 146, 148, 150–151, 153–158, 161–162, 164–167, 169–170, 173–176, 179, 185–188, 216, 220, 223, 306, 322, 335, 340–342, 346, 429–430, 476, 478, 484, 487, 537
coal zone:
    A-C coal: 186–187
    C coal: 130, 142, 153–156, 161, 185–187
    G coal: 130, 158, 161–162, 165, 484
    I coal: 72, 162, 164–166, 216
    J coal 72, 167, 169, 216
    M coal: 167, 170
    Sub-A coal: 68, 142–144, 146, 187, 220, 322, 335, 340–341, 346
coalbed gas, coalbed methane: 29, 32–33, 502–503, 512, 518–519, 521, 525, 530, 532–533, 536–537, 542, 549, 552, 557
    adsorption: 516, 518, 556
    biogenic gas: 510, 525, 535–537, 555
    coring program, core: 521, 533, 547, 552, 554–555
    development: 29, 32, 543, 545, 547, 549
    development: 32, 507, 510, 513, 516, 519, 521, 525, 543, 545, 547, 548, 549
    dewatering: 502, 519, 521–522, 525, 548–549
    drilling: 502, 519, 521, 523, 525, 530, 537, 542, 545, 547–549, 552, 556–557
    fracture stimulation: 542, 548–549
    gas isotope data: 519
    hydrological conditions: 519
    production: 502, 519, 522–525, 530, 533, 547–549, 552, 554, 557
    reservoir modeling: 510, 545, 549
    thermogenic gas: 518–519, 525, 535–537
coalbed methane and gas fields:
    Buzzard Bench: 506, 510, 516, 518–519, 521, 525, 531–532, 536, 552–555, 557
    Clear Creek: 8–9, 12, 31–32, 505, 519
    Drunkards Wash: 502, 505–506, 510, 513, 516, 518–519, 521–525, 545, 549
    Ferron: 31, 519
    Flat Canyon: 31
    Gordon Creek: 31, 536
    Helper: 506, 513, 518–519, 521, 523, 525, 542–549, 552

Coalville conglomerate: 20

coastal plain: 16, 24–25, 43, 62, 73, 80, 105–106, 109, 129, 196, 345–346, 355, 399, 452–453, 530

coastal-plain strata: 101–105, 109, 111, 215

compaction: 67, 82–83, 111, 219, 233, 238, 242, 244, 246, 288, 290, 292, 294, 300, 336, 342, 351, 401

composite sequence: 126, 129, 132–133, 152, 157, 172, 175, 177, 179, 184, 488

compressibility: 373, 416

concretions: 9, 13, 173, 176

Coon Spring Sandstone: 21

Corbula Gulch: 143, 170, 176, 421, 428–431, 433–434, 441, 443–445

core: 17–18, 22, 40, 49, 51, 55, 80, 84–85, 87, 100, 103, 126, 132, 189, 229–230, 234–235, 238–246, 252, 254, 273, 280, 301–302, 311, 327–328, 332–333, 338, 340, 343–346, 348, 355, 360, 364, 367, 370, 373, 380, 406–412, 423, 431, 456, 506–507, 521, 533, 547, 552, 554–555

core plug: 360, 408, 431

Coyote Basin: 27, 143, 148, 165, 186, 216, 234, 406–409, 413, 421, 423, 461, 475–476

crevasse: 43, 76, 82, 101–102, 109, 111, 117, 129, 133, 143, 146, 155, 158, 160, 163, 167, 185, 187–188, 280, 282, 311, 313, 322, 324, 327–328, 407, 453, 478, 543

cross-bedding, cross-stratification: 288, 343, 346, 441

   hummocky: 40, 65, 70–72, 83, 105–106, 108, 111, 143, 146, 155, 158, 163, 167, 196, 201, 232–233, 309, 344, 347, 351, 386, 391, 394, 397, 440, 460–461, 464–465, 468

   low-angle: 14, 82, 108–109, 111, 163, 167, 231, 263, 266, 273, 309, 336, 338, 344–345, 351, 407, 435–436, 460, 483–484

   swaley: 65, 70, 72, 83, 106, 108, 111, 146, 155, 309, 347, 391, 397

   tabular: 68–69, 73, 82, 106, 111, 147–149, 155, 159, 167–168, 228–229, 254, 273, 309–310, 312, 314, 325, 360, 373, 413, 435, 444, 478, 487, 506

   trough: 67–71, 73, 75–76, 81–82, 106, 108, 111, 114, 116–117, 143, 146–149, 155, 158–159, 163, 167–168, 197, 203, 206, 220, 231, 254, 266, 309–310, 312–314, 325–326, 340, 342–346, 386, 389, 392, 395, 398, 407–408, 435–437, 444, 460–461, 464, 468, 478, 483–485, 487, 490–491, 493

cycle(s): 14–15, 23, 25, 40, 51, 60, 62, 79–85, 87–89, 96–97, 99–109, 111–117, 118–121, 126–128, 132, 164, 179, 185, 198, 201, 203–204, 206, 209, 213, 228, 280, 282, 290, 294, 333, 394, 397, 428–429, 510

   landward-stepping: 43, 76, 79–84, 88–89, 96, 100–116, 119–120, 195–197, 209, 221–222, 290, 385, 399–401, 496, 507

   seaward-stepping: 43, 76, 79–84, 89, 96, 100–117, 119–120, 195, 197–198, 209, 219, 221–223, 246, 290, 300, 326, 385–386, 399, 401, 496, 506–507, 543

   stacking pattern: 51, 79–80, 83–84, 89, 100, 105, 109, 111, 119–120, 128, 130–131, 133, 142, 152, 154, 167, 173, 194, 201, 206, 212, 401, 453, 455–456, 458, 460–461, 463, 465–466, 470, 474, 483, 490, 493, 496

   subcycle: 335–336, 338, 345, 348, 351

   transgressive-regressive: 14, 60, 62, 355, 506

   vertically stacked: 46, 79–80, 83, 89, 100, 102–103, 105, 111, 119–120, 487–488, 507

D

delta: 9, 12–17, 19–20, 22–26, 40, 42–45, 47–49, 51, 53, 55, 62–65, 69–72, 74, 79–89, 100–103, 105–106, 109

   "A" Delta: 172, 175, 179, 183

   Atchafalaya: 43, 49

   Brazos: 48, 55, 488

   Colville: 49

   crevasse deltas: 43

   Cubits Gap: 49

   Danube: 45, 48–49, 55, 220

   delta front: 17, 43, 53, 55, 71–72, 81–82, 84–85, 87, 103, 108–109, 111, 143, 146, 158, 163, 167, 196, 263, 267, 281–282, 285–287, 290, 300–302, 335, 337–338, 348, 351, 360, 458–460, 461, 484, 506, 510

   delta lobe switching: 315–316

   Ebro: 42, 44, 46–49, 55

   flood-tidal: 148, 153

   fluvial-dominated, river-dominated: 15, 17, 20, 22–24, 33, 43–45, 48, 53, 55, 63–64, 69–73, 79–81, 83–84, 88, 102, 107, 109, 120–121, 127, 129, 142–143, 146, 150–151, 154, 172, 195–196, 201, 203–204, 206, 209, 213–214, 217, 219–220, 222–223, 228–230, 232–234, 246, 252, 254–256, 259–261, 263–265, 267, 270, 273, 280–281, 287, 290, 300–302, 306, 315, 317, 320, 326–327, 332–333, 335, 343–344, 355, 360, 399, 452, 456, 458, 460–461, 465–473, 475–476, 478–479, 496, 507, 530, 543

   Ganges-Brahmaputra: 42

   Gilbert: 4–5, 7, 14, 335

   Last Chance: 12–14, 16–17, 20, 42–43, 48, 53, 62, 126–133, 136–138, 141, 155, 165, 171–173, 175–177, 179–188, 428, 452–460, 463, 465–476, 488–490, 496, 507–508, 510, 531–532, 537

   Lena: 44

   Magdalena: 49

   Mississippi: 40, 43, 45, 49, 51, 53, 83, 91, 478–479

   Notom: 42, 127, 129, 172–173, 177, 184

   Paraibo do Sul: 44

   Rhône: 48–49, 55

   São Francisco: 49

   Vernal: 9, 12–13, 16–17, 19–20, 42, 62–63, 126–127, 209, 507–508, 510, 528, 530–532, 535, 537

   wave-dominated: 22–25, 44, 62–64, 68–73, 80–81, 83–84, 102, 107, 109, 120–121, 129, 157–158, 162–163, 166–167, 195–196, 201, 203–204, 206, 209, 213–217, 219–220, 222–223, 228–229, 246, 252, 254, 314–317, 319–321, 326–327, 353, 400, 453, 456,

463–464, 468–472, 475, 484–485, 488, 491–492, 507, 530

    wave-influenced: 43–44, 48, 55, 220, 223, 334

    wave-modified: 15, 22, 43, 48, 55, 64–65, 69, 71–73, 127, 151, 154, 157, 162, 166, 183, 213–214, 217, 219, 230, 232, 234, 252, 254–261, 263, 267, 273, 332, 335, 342–343, 347, 352, 355, 456, 460, 467, 491–492, 507, 530

    wave-reworked deltaic: 142, 153, 161, 165, 167, 475

    Wax Lake: 49

    West Delta: 49

delta front:

    distal: 158, 163, 287, 458–460

    distal bar: 82, 254, 353, 461

    distributary-mouth bar : 82, 84, 322, 327

    facies: 108

    proximal: 143, 290, 300, 302, 458

    turbidites: 287, 301

delta plain: 26, 43, 81, 186, 219, 327, 430, 441, 445, 452–453, 473, 507, 510

    interdistributary bays: 81–82, 102, 106

    marshes: 81–82,

    splays: 40, 81–82, 129, 133, 185, 311

depositional history: 8–9, 12, 30, 60, 177, 209, 306, 317, 348, 502, 507

dewatering: 116, 238, 288–289

diagenesis: 228, 230, 233–236, 238, 245, 267, 273, 360, 408

diapiric: 293

discharge: 40, 46–48, 53, 109, 238, 351, 441, 444–445, 453, 488

dissolution: 228, 233–236, 238, 242–245

distributary:

    channelbelt deposits: 24–25, 29, 75, 96, 114, 497

    channel: 14, 40, 43–45, 47–49, 51, 53, 55, 65, 70–73, 81–85, 88–89, 100–102, 106, 109, 111, 113–118, 120, 143, 148, 152, 158, 161–162, 166–167, 189, 195–196, 203, 214, 223, 230, 232, 234, 246, 254, 267, 273, 281–284, 288, 290, 292, 300, 302, 310–311, 313–314, 320, 322–324, 327, 341–344, 348, 351–353, 355, 360, 423, 435, 441, 444, 452–453, 455, 458, 472–479, 485, 489, 491–497, 507, 510, 523, 545

    complex: 72, 343–344, 355, 476

    mouth bars: 49, 51, 81, 146, 283–284, 290, 300, 302, 313, 322, 325–326, 355, 468

    mouth-bar sandstones: 461

drainage networks: 42

drilling: 29, 228, 373, 375, 377, 380, 407, 452

    drill hole: 18, 232, 241–242, 244, 254, 340, 342, 346, 352, 361, 406, 410–411, 523

    history: 519

    well data: 49, 51, 236, 238, 384, 494, 496, 506, 510

Dry Wash: 25, 29, 63, 73, 75, 85, 101–102, 108, 111–112, 142, 148, 152–157, 161, 173, 175–176, 188, 198, 200, 214, 216

E

Emery Sandstone: 5, 7, 505

estuary: 26, 100,

eustatic sea level: 20, 26

exhumation: 228, 236, 238–239, 241–246, 505

external geometry: 183, 473, 477, 485

extrabasinal pebbles: 143, 158–160, 167–168, 170, 175–176, 478, 483–484, 487, 491

F

facies:

    alluvial plain: 12, 26, 74–75, 186, 188, 495

    barrier shorelines: 101

    beach ridge: 327

    brackish-bay: 151–153, 159–160, 162, 166, 176, 183

    braided: 55, 119, 148, 151–152, 159, 161, 165–166, 172, 174–176, 183, 441, 452, 455, 472, 490–491, 493, 497

    channel belt: 24–26, 29, 46, 74–76, 82, 88–89, 96, 101–102, 105, 111, 113–115, 117–118, 120, 215, 452–453, 472, 474–479, 482–497

    clinoform: 263–268, 270, 273, 337–338, 343–344, 355, 361, 365–369, 371–373

    coastal plain: 16, 24–25, 43, 62, 73, 80, 105–106, 109, 129, 196, 345–346, 355, 399, 452–453, 530

    crevasse splay: 82, 109, 111, 313, 322, 324, 327, 543

    delta front : 17, 55, 71–72, 81–82, 84, 108–109, 111, 143, 146, 267, 281–282, 285–286, 290, 300–302, 337–338, 484, 510

    delta plain: 26, 43, 81, 186, 219, 327, 430, 441, 445, 452–453, 473, 507, 510

    distal bar: 82, 254, 353, 461

    distributary-mouth bar: 82, 84–85, 281, 322, 327

    estuarine: 131, 152, 162, 166–167, 172, 174, 176–177, 179, 183, 196, 206, 455, 491–492, 507

    flood-tidal delta: 148, 153

    foreshore: 64, 67–68, 196, 199, 220, 254, 267, 273, 309–310, 312, 317, 320, 325, 327, 341–342, 348, 353, 355

    interdistributary bay: 83, 85, 100–102, 106, 109, 129, 143, 311–313, 316, 320, 322–324, 327–328, 453, 491

    lagoonal: 53, 73, 129, 148, 152–153, 157, 162, 164, 166, 185, 453, 530

    lower shoreface: 65–66, 129, 155, 158, 163, 167, 189, 196, 199, 217, 254, 309, 311–312, 317, 327, 341, 344, 347, 353, 452–453, 461, 463–468, 470–471, 543

    marine: 24, 53, 55, 86, 103, 126, 129–131, 133, 143–144, 146–147, 149–153, 155, 157–158, 160–161, 162, 163, 164, 165, 167, 169, 170, 173, 176, 185, 186, 187, 188, 196, 219, 221, 223, 254, 342, 391, 453, 456, 458, 460, 463, 465, 469, 472

    marsh: 74, 82, 85, 87–88,

    meanderbelt: 75, 222, 354

    middle shoreface: 16, 65–68, 129, 153, 156, 158, 163, 166–167, 196, 213, 232, 254, 267, 273, 309–310, 313,

316–317, 320, 325, 327, 341–342, 344, 347, 353, 452–453, 463–464, 466–467, 471

offshore: 81–82, 87, 129, 345, 453

overbank: 60, 74, 76, 82, 129, 143, 146, 155, 163, 453, 478, 543

paludal: 12, 109, 543

point-bar: 5, 47, 421, 428, 439, 441–442, 445

prodelta/offshore: 47, 53, 81–83, 85–87, 129, 143, 146, 150, 155, 158, 162–163, 167, 196, 217, 254, 280, 282, 286–288, 290, 293, 300–302, 308–309, 312, 317, 327, 341, 345, 347, 453, 460–461, 463, 468, 543

shallow marine: 55, 80–81, 83, 109, 196, 203, 348, 353, 355, 384, 507

shelf: 87, 100, 102, 286

shoreface: 96, 100–109, 111, 114, 119–121, 129, 148, 153, 155, 167, 185, 267, 327, 342, 344, 347, 453, 463–464, 470–471

strand plain: 214, 314, 320, 322–323, 327

tidal inlet: 68, 73, 112, 310, 314, 320–321, 325, 327

transgressive: 345–346

turbidite: 287, 301

upper shoreface: 16, 64, 67, 106, 108, 129, 148, 153, 155–156, 158, 163, 167, 189, 196, 199, 220, 254, 267, 273, 309–310, 317, 322, 325, 327, 342, 347–348, 353, 453, 461, 463–464, 466–467, 471

washover fan: 101, 148–149, 153, 164, 166, 185

facies architecture: 46, 194, 209, 252, 291, 361–364, 374, 378, 384, 386, 387, 392, 395, 397, 399, 406, 411, 531

facies associations: 49, 51, 53, 55, 79–81, 96, 105, 107, 109, 119–121, 126, 133, 143, 150, 158, 185–188, 194, 203, 328, 456, 472, 478–479

fault: 42, 288, 294

antithetic: 291, 293, 296, 298–299

conjugate: 291–293, 295, 298–300

fault growth: 299–300

growth: 83, 111, 114, 143, 146, 150, 197, 280, 282–285, 291–294, 300–302

listric: 111, 282, 290, 292–293, 296

normal: 513, 552

synsedimentary: 280, 290, 302

synthetic: 87–89, 291, 296, 298

thrust: 181, 185

Ferron Composite Sequence: 126, 129, 132–133, 152, 157, 172, 175, 177, 179, 184, 488

Ferron Creek: 4, 26, 60, 132, 142, 153, 157, 161, 173–176, 186, 456

Ferron trend: 518–519, 530, 532–533, 535–537, 545, 552

Ferronensis sequence: 23, 27, 80, 104, 128, 212, 282

flood-plain strata: 102

flooding surfaces: 51, 53, 60, 63, 79, 82, 85–86, 88, 126, 130, 143–144, 146, 153, 165, 185, 194, 213, 215, 217, 219, 222, 316, 326, 328, 380, 386, 455, 472, 544, 549

abandonment: 83, 87

marine: 55, 80, 83, 87, 196, 198–199, 209, 212

flow:

depth: 40–41, 47

velocity: 40–41, 46–47, 53

fluid flow: 228, 235–236, 238, 244–245, 257, 263, 314–315, 327, 355, 371, 406, 414, 416, 423, 535

fluid flux: 238, 244

foreland basin: 42, 62, 80, 100, 184–185, 189–194, 219, 252–253, 280, 341, 352, 354–355, 384, 489, 505

Frontier Formation: 20, 61–62, 234, 245

G

genetic sequence: 24, 100, 126, 128–129, 131

geologic modeling: 452, 471–472

geometric: 100, 132, 385, 456, 494

geostatistics: 413

arithmetic: 254, 257, 361, 365, 369–372, 406, 414–415, 417–418, 420–423

average grain size : 257, 260, 266–267, 273, 337, 484

cumulative probability: 257, 266–267, 390–391

geometric: 254, 257, 265–267, 273, 361, 369–371, 414–415, 417–418, 420–423

harmonic: 257, 361, 369–371, 373, 406, 414–415, 417–418, 420–423

kriging: 273

Lilliefors test: 266

log-normal distribution: 266, 273

net/gross ratio: 493

scaling: 273, 369, 495

semivariogram: 267, 270

variogram: 270, 393, 411–413, 439–440

grain size: 40–41, 47, 84, 159, 230–232, 234, 244, 246, 252, 254, 256–257, 260–261, 263, 265–267, 269, 273, 307, 310, 315–316, 325–327, 337–338, 342–345, 355, 360, 365, 385, 388–389, 392, 395–396, 398, 400–402, 408, 428, 433–434, 437, 444, 460–461, 471, 478, 483–484, 487–488

grain-rolling: 294

ground water: 532

ground-penetrating radar, GPR: 89, 378–380, 406–415, 417, 419, 421. 423, 428, 431, 433–435, 437–445

growth sands: 282, 288, 299

H

heavy mineral content: 9

Henry Mountains: 4–5, 7, 9, 62, 127, 129, 132, 172–173, 177, 184

Henry Mountains Basin: 4–5, 7, 9, 62

heterogeneity: 55, 96, 105, 120, 236–237, 244, 267, 273, 308–310, 312–314, 328, 373–375, 401–402, 452, 471, 493, 496

highstand systems tract, HST: 131, 152, 157, 162, 172–173, 175–176, 178, 183, 491

Horn Silver Gulch: 5

Huntington anticline: 29, 36

Hyatti Composite Sequence: 126, 129, 132, 172, 177, 184

Hyatti sequence: 23, 128–129, 171–172, 212
hydrodynamic: 360, 369, 519, 535–537

I

ichnofacies: 158, 169, 461, 465, 468
ichnofossil: 146, 155, 168, 464, 487
   Arenicolites: 70, 83, 106, 158, 167, 286, 436, 441, 445, 460–461
   Chondrites: 83, 106, 158
   Diplocraterion: 106, 158, 460–461, 465
   Ophiomorpha: 66–67, 72–73, 82, 106, 148–149, 155, 158–164, 166–167, 196, 286, 290, 309–310, 344, 347–348, 430, 460–461, 464–465
   Planolites: 70, 82–83, 106, 158, 160, 168–169, 286, 435–436, 441, 460–461, 465, 483–484, 487
   Rosselia: 82–83, 106, 158, 286
   Scoyenia: 148, 158, 484
   Skolithos: 106, 146, 148, 155, 158, 163, 286–287, 290, 435–436, 441, 445, 460–461, 465, 468, 483–484, 490
   Teichichnus: 82–83, 148, 155, 158, 166, 169, 460–461, 465
   Teredolites: 148, 163, 286–287, 429–430, 436, 441, 445, 488
   Thalassinoides: 66–67, 72, 82–83, 106, 148–149, 155, 158, 160–164, 166–167, 169–170, 196, 347, 435–436, 441, 445, 460–461, 464–465
incised valley: 44, 53, 82, 151, 161–162, 166, 176, 183, 189, 399, 474–475, 491–493, 497
incised-valley fill deposits: 161–162, 165–167, 172, 183, 455, 458, 475, 491–492
incised-valley fluvial systems: 472, 474, 490
inclined bedsets: 355, 435–438, 444–445
Indian Canyon: 22, 66, 68, 70, 73, 143, 154, 161, 165, 167, 213–214, 217, 306–307, 309, 312–313, 316–317, 320, 322, 325–327, 476
interdistributary bay: 109, 311–313, 327
   deposits: 83, 85, 106, 143
   strata: 100–102, 106
interwell connectivity: 379, 471, 493
Ivie Creek, Ivie Creek Canyon: 4, 14, 18, 25, 28, 49, 51, 53, 70–72, 116, 142–144, 146, 150, 155, 170, 176, 189, 214–215, 217, 222, 229–231, 234–236, 238–239, 241–244, 246, 252–254, 264–271, 332–343, 345–355, 360–361, 363–364, 369–371, 374, 376–380
Ivie Creek amphitheater: 146, 264, 267–269, 271, 332, 334, 336–338, 342, 345–346, 348–351, 352, 360, 361, 376

J

John Henry Member: 179
Juana Lopez Member: 21

K

Kaiparowits Plateau: 20, 27, 129, 172, 175–179, 183

L

lagoon: 26, 45, 51, 72–73, 100, 216
landward pinchout: 22, 72–73, 131, 133, 142–143, 149–150, 154–155, 158, 160–161, 163–165, 167, 169, 186, 213–217, 219–220, 222–223, 306, 310, 316, 320, 335, 342, 353, 355, 455, 544
landward-stepping cycles: 79–81, 82–84, 88, 96, 105–106, 109–111, 113, 115, 119–120, 209
Last Chance Creek: 4, 12, 23, 26, 28, 60, 132–133, 142–143, 161, 212, 220, 456
lithofacies: 33, 40, 143, 146, 155, 158, 164, 167, 194, 230, 234, 238, 246, 267, 269–270, 306–315, 317, 320, 322, 325–328, 332–337, 340, 355, 435
lithology: 103–104, 219, 236, 242, 256, 263, 266–267, 273, 334, 337, 365, 408, 413, 431, 439, 505, 519, 531, 543, 553
littoral drift: 473
Lower Ferron Sandstone: 16, 127–128, 201–202, 206, 456
lowstand systems tract, LST: 131, 157, 162, 166, 172–173, 175, 177–179, 474–475

M

Mancos Shale: 4, 6–7, 33, 40, 60–63, 80, 126–128, 194, 209, 212–213, 222, 228, 252, 282, 333, 384, 428, 445, 502, 505, 528–531, 543–544, 547, 549–550
marine-flooding surface: 22, 65, 73, 131, 142–143, 150–151, 155, 165, 167, 170, 213–215, 217, 219, 222, 316, 341, 460
meandering: 46–47, 119, 310, 314, 322, 399, 402, 441, 444, 453, 477, 493–494, 497
microcrack: 241, 244
Miller Canyon: 25, 65–68, 73, 85, 111, 142, 148, 150–151, 153, 155, 157–158, 161–163, 165, 167, 173, 175–176, 186, 200, 215, 220, 463–464, 475, 491
mini-permeameter: 230, 232, 242–244, 254, 266
mobile prodelta muds: 280, 300
model: 412–419, 421, 423, 430–431, 434, 439, 444–445, 471, 494–496, 528, 542, 545, 547–549
modeling: 79–80, 87, 89, 244, 252, 265–267, 273, 332–333, 337, 355, 360–364, 366–374, 378, 380, 401, 412, 415, 439, 452, 471–472, 490, 493–495, 510, 542, 545, 549
   acoustic impedance: 89
   anisotropic: 362, 372–377, 416
   geologic: 270, 273, 336, 494
   homogenization: 369, 378–380
   isotropic: 361–362, 366, 371–373, 375, 377–378, 411, 415–416, 418, 423
   object: 471, 494–495
   perturbation: 361, 369–371
   reservoir: 362, 371, 377, 431, 547–548
   scaling: 273, 369, 495
   seismic: 79–80, 87–89
   three-dimensional, 3-D: 362, 368–369, 371, 376, 378–37
   two-dimensional, 2-D: 361–362, 369, 373, 376, 378–379

upscaling: 361, 364, 369–371, 378, 406, 410, 414–415, 418, 420–421, 423

modern analogs: 40, 44, 49–52

molluscan faunal: 126, 132, 173, 179, 189

Muddy Creek: 22, 25, 28, 49, 51, 53, 65–68, 72–73, 75, 80–82, 84, 102, 104, 108, 111, 142, 147, 150–153, 155, 161–165, 167, 169, 173–175, 177, 215–216, 220, 236, 238, 280–284, 286, 341, 385–386, 389–391, 393, 395–397, 399–401, 458, 461, 463, 474–475, 490

Muddy Creek Canyon: 22, 25, 65–68, 72–73, 75, 102, 104, 108, 111, 142, 147, 150–152, 162–163, 165, 167, 169, 173–174, 177, 215–216, 220, 236, 238, 280, 282, 391, 463, 474

mudstone drapes: 106, 116, 386, 397, 399, 402, 431, 434–437, 439, 444–445

mudstone-intraclast: 431, 434

N

near-marine facies association: 129–131, 186, 453

non-marine facies association: 126

O

offshore marine: 87, 126, 196, 199, 201, 213, 215, 217

overbank deposits: 60, 76, 478, 543

P

paleoclimate: 41, 56

paleohydraulics: 40, 46

palynofacies types: 87

parasequence: 22–23, 25–27, 33, 40, 43, 49, 51, 53, 65–70, 72, 75, 80, 82, 99–100, 126, 128–133, 142–144, 146–151

boundary: 215, 217, 219, 341

intervals: 252, 254

stacking patterns: 201, 203, 455, 465

parasequence set: 53, 126, 130–133, 142–144, 146–155, 157–177, 179, 181–183, 188–189, 201, 206, 209, 213–215, 221–222, 228–230, 232, 244, 246, 252, 254, 263, 265–267, 270, 272–273, 306–307, 316, 327, 332–335, 341, 347, 352, 355, 384, 455–456, 458, 460–461, 463–470, 473–476, 479, 482–485, 488–490, 492–493, 496

aggradational: 133, 157, 162, 221, 223, 452, 465–471, 475, 479, 489–493, 496

boundary: 188, 265, 341, 347, 352

progradational: 142, 182, 188, 201, 206, 209, 384, 466–467, 474, 479, 493

retrogradational: 151, 167, 170, 182, 206, 209, 485, 489, 493

peak-flood discharge: 46, 48

peninsula: 7–9, 12, 20

peripheral bulge: 62, 352

permeability: 22, 25, 55, 84–85, 87, 89, 222–223, 228, 230–234, 238, 242–245, 246, 252, 254, 257, 260, 261, 262, 265–266, 267, 269, 270, 273, 294

barrier: 537

data: 22, 25, 242, 244, 254, 265, 374, 391, 393, 409–411, 416

distribution: 246, 413

structure: 25, 223, 246, 371–378, 380, 384, 386, 406, 411–415, 418, 423

transects: 254, 267, 269–270, 336, 361

relative: 242, 366–367, 369, 371, 373–374, 376, 378, 416

permeameter data: 230, 242, 244, 408

petrography: 80, 84, 189, 230, 234, 409–410, 507, 554–555

petrophysical: 33, 76, 96, 99, 105, 120, 223, 228–230, 236, 244–245, 294, 333, 355, 360–362, 364, 366–368, 370, 375–377, 379–380, 385, 397, 406, 409–410, 423

photomosaic: 71, 112, 117, 132, 189, 202, 205, 220, 252, 254, 264–265, 267, 273, 280, 282–284, 307, 309, 312–313, 316–318, 335–337, 341–343, 360, 362–363, 407, 456

point bars: 75, 428, 441, 444–445

porosity: 49, 84, 87, 222, 228–230, 232–236, 238–239, 241–246

intergranular: 235–236, 242, 245

microporosity: 84, 234, 239, 409

secondary: 233–236, 238–239, 242, 244–246, 267, 273

prodelta: 47, 53, 81–83, 85–87, 129, 143, 146, 150, 155, 158, 162–163, 167, 196, 217, 254, 280, 282, 286–288, 290, 293, 300–302, 308–309, 312, 317, 327, 341, 345, 347, 453, 460–461, 463, 468, 543

progradational: 51, 53, 81, 100–101, 103–105, 109, 114, 126–127, 131, 133, 142, 150, 152, 154, 157, 162, 171–173, 175, 177, 182–183, 185, 188, 194, 201, 203–204, 206, 209, 220, 252, 254, 270, 317, 352, 384, 401, 429, 452, 455–456, 458, 460–461, 465–471, 473–476, 479, 489–491, 493, 496, 507, 544, 549

progradation rate: 43, 454, 468, 478, 485, 488, 490

Q

Quitchupah Canyon: 73–74, 264–265, 267–268, 270–271, 335, 342, 348, 352–353, 371

Quitchupah Creek: 28, 142–143, 146–147, 149–155, 161, 167, 170, 173–174, 187–188, 252, 332, 348, 353, 460, 475

R

radar sequence: 434, 438, 440

recurved: 101, 112

regional structure: 504

relay ramps: 296

reservoir: 367–369, 371–374, 379–380, 384, 401, 407, 409, 411, 413, 415–417, 419, 421, 423, 431, 434, 445, 471–472, 494, 496, 502, 510, 518, 522–523, 525–528, 533, 536, 542, 545, 549

characterization: 33, 55, 246, 273, 355, 380, 542

continuity: 306, 471

facies: 25, 84–85, 88–89, 252, 254, 261, 265, 267, 273, 327, 406
    heterogeneity: 246, 314
    modeling: 362, 371, 377, 431, 547–548
    potential: 234, 246, 326, 471, 496, 523
    quality: 22, 79–80, 84–85, 194, 327, 471–472, 523
    simulation: 252, 265–266, 270–273, 362, 379, 406, 545
rivers:
    Amazon: 42, 48
    Colville: 43
    Ebro: 42–46
    Fly: 48
    Ganges-Brahmaputra: 42
    Lena: 43–44
    Mississippi: 40, 42, 46, 48, 53, 83
    Nile: 48
    Niger: 48
    Po: 42, 48
    Red: 42
    Rhône: 42, 48
    Saganavirtok: 43
    trunk river: 42, 53
    Yellow: 42
Rochester Creek: 150–153, 155, 161, 164–165, 168, 173–175, 187, 475
Rock Canyon: 75, 149–150, 155, 170, 176, 186, 188, 475–476
rock physics : 87, 89
rotated block: 300, 340
rotated-slump blocks: 106

S

Sanpete Valley embayment: 10, 12, 14, 16, 280, 286
saturation: 239, 362, 367–369, 372, 374–377, 416, 418, 434
Scabby Canyon: 150, 173, 175, 460, 475
sea level:
    relative change: 182, 184–185, 219–220, 452, 468–469, 471, 480, 489, 497
    relative fall: 181, 492–493, 497
    relative rise: 129, 131, 181–184, 453–454, 468, 478–479, 484–485, 488, 491
    relative: 12, 16, 20, 22, 24–26, 74, 181, 206–207, 213, 316, 328, 384, 452, 468, 470, 479, 488–490, 492, 496
seaward pinchout: 155, 186–187, 219, 335, 352
seaward-stepping cycles: 80–82, 83–84, 89, 96, 102, 105, 106, 109, 111, 114, 115, 119, 120, 209
sediment volume partitioning: 96, 100, 103, 105–106, 111, 119–121
sediment volumes: 83, 96, 99–100, 103–105, 111, 114, 119
sedimentary structure: 75, 168, 231–232, 252, 256–257, 261, 263, 266–267, 269, 273, 336–337, 344, 365, 483–484, 487
sedimentation rate: 128–129, 150, 171, 173, 181–185, 344, 452–454, 456, 468–471, 478–479, 480, 484, 485, 488, 489, 490, 491, 492, 496, 497

seismic: 79–80, 82, 87–89, 230, 244, 280, 301–302, 327–328, 353, 378–379, 408, 434, 445, 523, 545, 552
    acoustic impedance model: 89
    modeling: 89
    seismic response: 79–80, 82, 88–89, 230
    synthetic seismograms: 87
sequence stratigraphy: 21, 26–27, 80–81, 96, 126–133, 135–137, 139, 141, 143, 145, 147, 149, 151, 153, 155, 157, 159, 161, 163, 165, 167, 169, 171, 173, 175, 177, 179–181, 183–185, 187, 189, 212, 219, 306, 315, 326, 378, 455, 472, 497, 502, 530, 531
    bedset: 51, 158–159, 213, 215, 217, 219–220, 222–223, 230, 252, 263–267, 270, 273, 312, 332–337, 340, 344, 346, 348–349, 351–352, 362, 372, 374–375, 430, 435–438, 440, 443–445, 460–461, 483–484, 487, 491, 496
    composite sequence: 126, 129, 132–133, 152, 157, 172, 175, 177, 179, 184, 488
    depositional sequence: 64, 69, 126, 128–133, 152, 158, 162, 165–166, 171–177, 179, 181–182, 184–185, 187–189, 429, 455–456, 472, 474, 497
    depositional sequence boundary: 131, 142, 152, 162, 165–166, 172–177, 179, 185, 472
    flooding surfaces: 51, 53, 60, 63, 79, 82, 85–86, 88, 126, 130, 143–144, 146, 153, 165, 185, 194, 213, 215, 217, 219, 222, 316, 326, 328, 380, 386, 455, 472, 544, 549
    genetic sequence: 24, 100, 126, 128–129, 131
    parasequence: 22–23, 25–27, 33, 40, 43, 49, 51, 53, 65–70, 72, 75, 80, 82, 99–100, 126, 128–133, 142–144, 146–167
    parasequence boundary: 215, 217, 219, 341
    parasequence intervals: 252, 254
    parasequence set: 53, 126, 130–133, 142–144, 146–155, 157–177, 179, 181–183, 188–189, 201, 206, 209, 213–215, 221–222, 228–230, 232, 244, 246, 252, 254, 263, 265–267, 270, 272–273, 306–307, 316, 327, 332–335, 341, 347, 352, 355, 384, 455–456, 458, 460–461, 463–470, 473–476, 479, 482–485, 488–490, 492–493, 496
    ravinement: 83, 126, 131, 133, 150–153, 156–157, 161, 165, 172–173, 175–177, 179, 182–189, 196, 213, 456, 491
    stacking patterns: 201, 203, 455, 465
    systems tract: 131, 158, 162, 171–172, 175, 456, 474–475, 488, 491–492
shelf sand: 19–20, 63, 91
shelf sand plume: 19–20
shoreface: 5, 16, 51, 53, 64–69, 71–73, 75, 96, 99–109, 111, 114, 119–121, 129, 146, 148, 151, 153, 155–158, 161–163, 165–167, 176–177, 183, 185, 189, 196, 199, 201, 209, 213–214, 217, 220, 222, 232–234, 254, 267, 273, 306, 309–310
shoreface facies: 129, 153, 167, 327, 344, 347, 453, 463–464, 470–471
    foreshore: 64, 67–68, 196, 199, 220, 254, 267, 273,

309–310, 312, 317, 320, 325, 327, 341–342, 348, 353, 355

lower shoreface: 65–66, 129, 155, 158, 163, 167, 189, 196, 199, 217, 254, 309, 311–312, 317, 327, 341, 344, 347, 353, 452–453, 461, 463–468, 470–471, 543

middle shoreface: 16, 65–68, 129, 153, 156, 158, 163, 166–167, 196, 213, 232, 254, 267, 273, 309–310, 313, 316–317, 320, 325, 327, 341–342, 344, 347, 353, 452–453, 463–464, 466–467, 471

prodelta: 47, 53, 81–83, 85–87, 129, 143, 146, 150, 155, 158, 162–163, 167, 196, 217, 254, 280, 282, 286–288, 290, 293, 300–302, 308–309, 312, 317, 327, 341, 345, 347, 453, 460–461, 463, 468, 543

upper shoreface: 6, 64, 67, 106, 108, 129, 148, 153, 155–156, 158, 163, 167, 189, 196, 199, 220, 254, 267, 273, 309–310, 317, 322, 325, 327, 342, 347–348, 353, 453, 461, 463–464, 466–467, 471

Short Canyon: 5

simulation: 33, 223, 246, 252, 263, 265–266, 270, 273, 333, 336–337, 355–356, 360–362, 364, 366–367, 369–372, 374–380, 406, 410–411, 413, 415–419, 421, 423, 431, 445, 528, 545, 547, 549

fluid-flow: 406, 412, 415–416, 423

numerical: 360–362, 371, 528

reservoir: 252, 265–266, 270–273, 362, 379, 406, 545

single-phase: 242, 415–417, 423

two-phase: 361, 371, 374, 406, 415–416, 418, 423

waterflood: 361–362, 367, 371–372, 374, 376, 415–416, 418, 423

slumps: 114, 301

smear: 294–295

Smoky Hollow Member: 178

soft-sediment deformation: 83, 106, 111, 114, 116, 167, 197, 287, 289–290, 298, 353

sorting: 84, 230–231, 234, 254, 266–267, 273, 343, 401, 434, 471

stacking patterns: 96–97, 99–101, 103, 105, 107, 109, 111, 113, 115, 117, 119, 121, 131, 189, 201, 203, 455–456, 465, 474, 489, 496, 507, 528, 544

statistics: 49, 223, 379, 409

Straight Cliffs Formation: 27, 129, 172, 177, 179

stratal architecture: 105, 223, 360

stratigraphic rise: 15–16, 131, 144, 146, 149–153, 161, 165, 169–170, 179, 185, 213

stratigraphic sections: 252, 254–255, 267, 273, 307, 333, 407, 431, 433

stream-mouth bar: 129, 143, 146, 148, 153, 155, 158, 163, 167, 188, 458, 460–461, 466–467, 471–472

structural restorations: 295

subsidence: 17, 20, 25, 41, 61–62, 74, 83, 100, 114, 126, 179, 181, 183–185, 206, 219, 246, 252, 315, 327, 341, 351, 373, 489–490

curve: 126, 181

rate: 185, 489–490

successions: 40, 64, 67, 70–72, 80–83, 85, 87, 96, 100, 102, 105–107, 109, 119–121, 181, 188, 194, 198, 219, 289–290, 300, 326, 384, 386, 398, 440, 444, 474

systems tract: 131, 158, 162, 171–172, 175, 456, 474–475, 488, 491–492

highstand systems tract, HST: 131, 152, 157, 162, 172–173, 175–176, 178, 183, 491

lowstand systems tract, LST: 131, 157, 162, 166, 172–173, 175, 177–179, 474–475

transgressive systems tract, TST: 131, 157, 162, 166, 170, 172–173, 175–179, 182–183, 488, 491–492

T

Tibbet Canyon Member: 177

tidal: 43, 51, 63–64, 68–70, 73, 100–102, 109, 112, 120, 129, 148, 152–153, 155–157, 159, 194, 203, 206, 216, 302, 309–310, 312, 314, 320–322, 325–327, 342, 399, 401, 428, 441, 444, 453, 461, 488, 507, 543

inlet: 12, 51, 68–69, 73, 112, 155–157, 310, 312, 314, 320–321, 325–327

deposits: 63, 109, 120, 325–326

strata: 100–101

tonstein: 102, 155, 157, 186, 523, 532, 553

trace fossils: 82–83, 85, 169, 196, 287, 430, 436

transfer zone: 291, 296

transgressive lag: 131, 133, 150–151, 153, 156–158, 161, 165, 167, 172–173, 175–176, 183, 188–189, 216, 290, 430

transgressive ravinement: 126, 131, 133, 150–151, 153, 156–157, 161, 165, 172–173, 176–177, 179, 182, 184–189, 196, 456

transgressive systems tract, TST: 131, 157, 162, 166, 170, 172–173, 175–179, 182–183, 488, 491–492

transgressive-regressive cycle: 60

transition zone: 64–66, 209, 505

trunk rivers: 43, 53

Tununk Shale Member: 68, 127–128, 194, 282, 333, 544

type locality: 4

U

unconformity: 5, 26–28, 62, 82, 102, 120, 133, 142, 147–152, 154, 158–167, 173–174, 175, 176, 177, 188, 203, 209, 384, 492, 543

uplift: 4–5, 9, 17, 20–21, 26, 62, 184–185, 228, 238, 242, 244, 322, 502, 505, 513, 519, 530, 532–533, 535, 542–543, 545, 553

Upper Ferron Sandstone: 96, 100–103, 119, 126–128, 130–133, 136–138, 141–142, 144, 146–149, 156–160, 162–164, 168–173, 175, 177, 179–182, 185, 187–188, 197, 200–201, 452–453, 455–458, 460, 463, 465–467, 469, 472–475, 488–490, 492, 544

upstream: 281, 283–284, 288, 290, 300, 360, 401, 435, 440–441, 444–445, 470, 479, 485, 488, 493, 497

accretion: 290, 441

migration: 441, 445

V

valleys: 26, 43–44, 46, 51, 53–55, 79–80, 82, 89, 100, 151, 161, 176, 203, 206, 209, 327, 384, 399, 452, 455, 475

velocity: 40–41, 46–47, 53, 87–88, 228–229, 236–242, 244–245, 431, 434, 440, 444

velocity/porosity relationship: 240–241, 245

volcanic ash layer: 126, 143, 153–154, 157, 171, 187, 429, 430, 553

volumetric partitioning: 24, 99–100, 121,

W

Washboard sandstone unit, Washboard unit: 5, 12–13, 16, 20, 27, 62–63, 70, 171, 173, 175, 212–213, 334, 384, 506, 543–544

washover fans, deposits: 101, 148–149, 153, 164, 166, 185

wave reworking: 64, 72, 129, 150, 320, 453, 469

weathering: 76, 216, 233–234, 338, 362, 380, 409–410, 423

well logs: 9, 100, 103, 229, 237, 240, 245, 302, 316, 379, 493–494, 543

    density: 87, 236, 524

    sonic: 87, 229, 242, 245, 333, 552, 556

Western Interior Basin: 172, 181, 184–185, 189

Western Interior Seaway: 80, 105, 126–128, 194, 252, 384, 505, 530

Willow Springs Wash: 22, 27, 29, 70, 76, 81, 84, 117, 132, 142, 146, 149–150, 155, 167, 169–171, 173–174, 176, 187, 213, 306, 306–307, 315, 316, 320, 322, 327, 458, 460, 461, 473, 474, 475, 476, 477, 478, 479, 482, 484, 485, 487, 488, 489, 493, 494

Woodside unit: 20–21